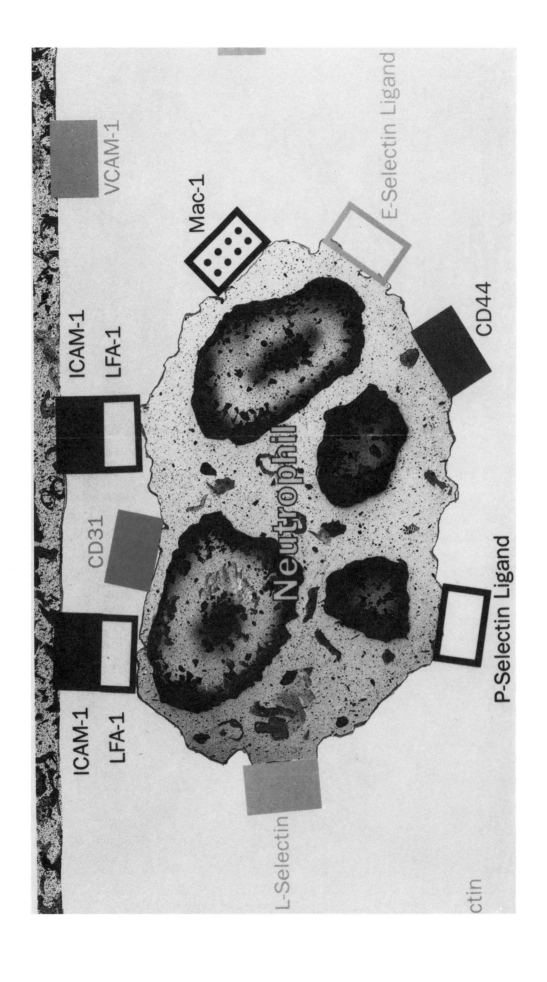

ILLUSTRATED DICTIONARY of

IMMUNOLOGY

Julius M. Cruse, B.A., B.S., D.Med.Sc., M.D., Ph.D.
Professor of Pathology
Director of Immunopathology and Transplantation Immunology
Director of Graduate Studies in Pathology
Department of Pathology, Associate Professor of Medicine
    and Associate Professor of Microbiology
University of Mississippi Medical Center
Jackson, Mississippi

Robert E. Lewis, B.A., M.S., Ph.D.
Professor of Pathology
Co-Director of Immunopathology and Transplantation Immunology
Department of Pathology
University of Mississippi Medical Center
Jackson, Mississippi

CRC Press
Boca Raton   New York   London   Tokyo

DISCLAIMER

The authors and the publisher have exerted every effort to ensure that drug selection and dosage set forth in this text are in accord with current recommendations and practice at the time of publication. However, in view of ongoing research, changes in government regulations, and the constant flow of information relating to drug therapy and drug reactions, the reader is urged to check the package insert for each drug for any change in indications and dosage and for added warnings and precautions. This is particularly important when the recommended agent is a new and/or infrequently employed drug.

Library of Congress Cataloging-in-Publication Data

Cruse, Julius M., 1937–
  Illustrated dictionary of immunology / Julius M. Cruse, Robert E. Lewis.
      p.   cm.
  Includes bibliographical references and index.
  ISBN 0-8493-4557-X
  1. Immunology—Dictionaries.   I. Lewis, R. E. (Robert Edwin), 1947–   .   II. Title.
  [DNLM: 1. Allergy and Immunology—dictionaries.   QW 513 C957i 1994]
  QR180.4.C78 1994
  574.2′9—dc20
  DNLM/DLC
  for Library of Congress   94-5345
                                                                                                                    CIP

This book contains information obtained from authentic and highly regarded sources. Reprinted material is quoted with permission, and sources are indicated. A wide variety of references are listed. Reasonable efforts have been made to publish reliable data and information, but the author and the publisher cannot assume responsibility for the validity of all materials or for the consequences of their use.

Neither this book nor any part may be reproduced or transmitted in any form or by any means, electronic or mechanical, including photocopying, microfilming, and recording, or by any information storage or retrieval system, without prior permission in writing from the publisher.

CRC Press, Inc.'s consent does not extend to copying for general distribution, for promotion, for creating new works, or for resale. Specific permission must be obtained in writing from CRC Press for such copying.

Direct all inquiries to CRC Press, Inc., 2000 Corporate Blvd. N.W., Boca Raton, FL 33431.

Dedicated to

Professor Marcus Eugene Morrison, M.D.,
former Chairman, Department of Bacteriology,
University of Mississippi School of Medicine.

Master teacher and visionary who correctly foretold, in the 1940s,
the future renaissance in
immunological research and challenged several generations of
students to unravel some of
Nature's most jealously guarded secrets through
scientific investigation.

EDITORIAL STAFF FOR THE AUTHORS

AUTHORS

Julius M. Cruse, B.A., B.S., D.Med.Sc., M.D., Ph.D., is Professor of Pathology, Director of Immunopathology and Transplantation Immunology, and Director of Graduate Studies in Pathology at the University of Mississsippi Medical Center in Jackson. He also serves as Associate Professor of Medicine and of Microbiology. Formerly, Dr. Cruse was Professor of Immunology and of Biology at the University of Mississippi Graduate School.

Dr. Cruse graduated in 1958 with his B.A. and B.S. degrees in chemistry from the University of Mississippi. He was a Fulbright Fellow at the University of Graz (Austria) Medical Faculty, where he wrote a thesis on Russian Tickborne Encephalitis Virus and received his D.Med.Sc. degree summa cum laude in 1960. On his return to the U.S. he entered the M.D. and Ph.D. program at the University of Tennessee College of Medicine, Memphis completing his M.D. degree in 1964 and his Ph.D. in pathology (immunopathology) in 1966. Dr. Cruse also trained in pathology at the University of Tennessee Center for the Health Sciences in Memphis.

Dr. Cruse is a member of numerous professional societies which include the American Association of Immunologists (Historian), the American Society for Investigative Pathology, the American Society for Histocompatibility and Immunogenetics (Chairman, Publications Committee), and the Societe Francaise d'Immunologie, among many others. He is a Fellow of the American Academy of Microbiology and is a Fellow of the Royal Society of Health (U.K.).

Dr. Cruse's research has centered on transplantation and tumor immunology, autoimmunity, MHC genetics in the pathogenesis of AIDS, and neuroendocrine immune interactions. He has received many research grants during his career and is presently funded by the Earl R. and Martha Lyles Wilson Foundation for neuroendocrine-immune system interactions in spinal cord injury and stroke patients. He is the author of more than 200 publications in scholarly journals and 35 books, and has directed dissertation and thesis research for more than 40 graduate students during his career.

Robert E. Lewis, B.A., M.S., Ph.D., is Professor of Pathology and Co-Director of Immunopathology and Transplantation Immunology in the Department of Pathology at the University of Mississippi Medical Center in Jackson.

Dr. Lewis received his B.A. and M.S. in microbiology degrees from the University of Mississippi and earned his Ph.D. in pathology (immunopathology) degree from the University of Mississippi Medical Center. Following specialty postdoctoral training at several medical institutions, Dr. Lewis has risen through the academic ranks from instructor to professor at the University of Mississippi Medical Center.

Dr. Lewis is a member of numerous professional societies which include the American Association of Immunologists, the American Society for Investigative Pathology, the Society for Experimental Biology and Medicine, the American Society for Microbiology, the Canadian Society for Immunology, and the American Society for Histocompatibility and Immunogenetics, among numerous other professional scientific organizations. He is a Fellow of the Royal Society of Health of Great Britain. Dr. Lewis has been the recipient of a number of research grants in his career and is currently funded by the Earl R. and Martha Lyles Wilson Foundation for his research on neuroendocrine immune system interactions in spinal cord injury and stroke patients.

Dr. Lewis has authored or co-authored more than 75 papers and 100 abstracts and has made numerous scientific presentations at both the national and international levels. His current research also includes immunogenetic aspects of AIDS progression.

PREFACE

Immunology is an eclectic science with foundations in molecular biology, biophysical chemistry, microbiology, and clinical medicine. From the beginning, immunologists have maintained a unique nomenclature that has often mystified and even baffled their colleagues in other fields, causing them to liken immunology to a "black box". This book is designed to offer immunologists and nonimmunologists alike a resource for many of the basic terms encountered in contemporary immunological literature. Simple illustrations clarify the explanations and enhance the terms or concepts described. This book is addressed to students, teachers, physicians, and basic science researchers in all fields of biomedicine. The subject matter of immunlogy intersects essentially all basic biomedical and clinical sciences. If immunology becomes meaningful and demystified to the reader, the effort in preparing this book will have been well spent.

ACKNOWLEDGMENTS

Although many individuals have offered help or suggestions in the preparation of this book, several deserve special mention. We are very grateful to Dr. Sherman Bloom, Chairman, Department of Pathology, University of Mississippi Medical Center, Jackson, for his genuine interest and generous support of our academic endeavors at this institution. Dr. Fredrick H. Shipkey, Professor of Pathology and Chief of Surgical Pathology at the University of Mississippi Medical Center, provided valuable assistance in selecting and photographing appropriate surgical pathology specimens to illustrate immunological lesions. We thank Professor Albert Wahba for offering constructive criticism related to a number of the chemical definitions and express genuine appreciation to Dr. Virginia Lockard for providing the electron micrographs that appear in the book. We also thank Dr. Robert Peace for a case of Job's syndrome, Dr. Ray Shenefelt for a photomicrograph of cytomegalovirus, and Dr. David Debauche for providing an illustration of the Philadelphia chromosome. We thank Dr. Julian L. Ambrus, Jr. of the Department of Rheumatology at Washington University School of Medicine for his contribution concerning interleukin-14. We also thank Dr. Andrew G. Farr of the Department of Biological Structure at the University of Washington, Seattle and Dr. Albert Zlotnik of the DNAX Research Institute for providing the latest information on lymphokines.

We express gratitude to doctoral candidates Martha Brackin and Andrew Achord for suggesting recent molecular terms to be included in the book. We appreciate the constructive suggestions of Patsy Foley, M.T., C.H.T.; Paula Hymel, M.T., C.H.T.; Elisa Smith, M.T., C.H.T.; and Kathy Vanlandingham, M.T., C.H.T. The DAKO Corporation, Carpinteria, CA has generously consented for us to use their summaries of the leukocyte differentiation antigens in the definitions of these terms. We are also grateful for permission granted by Serotec Corporation, Oxford, England to use their summary of the latest CD (cluster of differentiation) nomenclature. We would also like to commend the individuals at CRC Press—Harvey Kane, Acquiring Editor; Becky McEldowney, Marketing Manager; Jim Labeots, our Project Editor for this book; and Carol Messing, Editorial Assistant—for their professionalism and unstinting support in bringing this book to publication. To these individuals we offer our grateful appreciation.

The authors' cellular immunology research is funded through the generous support of the Earl R. and Martha Lyles Wilson Foundation, Mississippi Methodist Rehabilitation Center, Jackson, Mississippi 39216.

ILLUSTRATION CREDITS

Figure for abzyme. Redrawn from Haber, E. et al., Innovative approaches to plasminogen activator therapy, *Science,* 243:51–56, 1989.

Figure for actin. Redrawn from Kabsch, W., Mannherz, H.G., Suck, D., Pai, E.F., and Homes, K.C., Atomic structure of the actin: DNase I complex, *Nature,* 347:37, 1990.

Figure for allotype. Courtesy of Dr. Leon Carayannopoulos, Department of Microbiology, University of Texas, Southwestern Medical School, Dallas.

Figure for autobody. Adapted from Kang, C. and Kohler, H., *Immunoregulation and Autoimmunity,* Vol. 3, Cruse, J.M. and Lewis, R.E., Eds., S. Karger, Basel, Switzerland, 1986.

Figure for Bence-Jones protein. Adapted from Edmundson, A.B., Ely, K.R., Abola, E.E., Schiffer, M., and Panagotopoulos, N., Rotational allomerism and divergent evolution of domains in immunoglobulin light chains, *Biochemistry,* 14:3953–3961, 1975.

Figure for C1. Redrawn from Arlaud, G.J., Colomb, M.G., and Gagnon, J., A functional model of the human C1 complex, *Immunol. Today,* 8:107–109, 1987.

Figures for C3b, Fc receptors, FcγR, FcεRI, immunoglobulin genes, immunoglobulin superfamily, interleukin 2 receptor, and selectins. Redrawn from Lachmann, P.J., *Clinical Aspects of Immunology*, Blackwell Scientific Publications, Cambridge, MA, 1993.

Figures for CALLA, CD1, CD2, CD4, CD5, CD8, CD10, CD11/CD18, CD13, CD19, CD21, CD22, CD33, CD42, and CD45. Redrawn from Barclay, A.N., Birkeland, M.L., Brown, M.H., Beyers, A.D., Davis, S.J., Somoza, C., and Williams, A.F., *The Leucocyte Antigen Facts Book*, Academic Press, Orlando, FL, 1993.

Figures for CD3 and systemic lupus erythematosus. Redrawn from Davis, M.M., T cell receptor gene diversity and selections, *Annu. Rev. Biochem.,* 59:477, 1990.

Figure for CD9. Redrawn from Boucheix, C. et al., Molecular cloning of the CD9 antigen, *J. Biol. Chem.,* 226(1):121, 1991.

Figure for CD16. Redrawn from Ravetch, J.V. and Kinet, J.P., Fc receptors, *Annu. Rev. Immunol.,* 9:462, 1991.

Figure for CD20. Redrawn from Tedder, T.F., Structure of the gene encoding the human B lymphocyte differentiation antigen CD20 (B1)[1], *J. Immunol.,* 142(7):2567, 1989.

Figure for CD59. Redrawn from Rooney, I.A., Oglesby, T.J., and Atkinson, J.P., Complement in human reproduction: activation and control, *Immunol. Res.,* 7:282, 1993.

Figure for chromatography. Redrawn from Hudson, L. and Hay, F.C., *Practical Immunology*, Blackwell Scientific Publications, Cambridge, MA, 1989.

Figure for class I MHC molecule. Adapted form Nikolic-Zugic, J. and Carbone, F.R., Peptide presentation by class I major histocompatibility complex molecules, *Highlights in Immunology*: 55, 1991.

Figure for complement receptor 1 and 2. Redrawn from Kinoshita, T., *Complement Today*, Cruse, J.M. and Lewis, R.E., Eds., S. Karger, Basel, Switzerland, 1993.

Figures for complement receptors, membrane immunoglobulin, and Western blot. Redrawn from Paul, W.E., *Fundamental Immunology*, Raven Press, New York, 1993.

Figure for Elek plate. Redrawn from Elek, S.D., *Staphylococcus Pyogenes and its Relation to Disease*, E & S Livingstone, Edinburgh and London, 1959.

Figure for FcεRII. Redrawn from Conrad, D.H., Keegan, A.D., Kalli, K.R., Van Dusen, R., Rao, M., and Levine, A.D., Superinduction of low affinity IgE receptors on murine B lymphocytes by LPS and interleukin-4, *J. Immunol.,* 141:1091–1097, 1988.

Figures for Heymann nephritis and Philadelphia chromosome. Redrawn from Cotran, R.S., Kumar, V., and Robbins, S.L., *Robbins Pathologic Basis of Disease*, W. B. Saunders, Philadelphia, PA, 1989.

Figure for HLA-A2. Redrawn from Bjorkman et al., Structure of the human class I histocompatibility antigen, HLA-A2, *Nature,* 329:508, 1987. (Figure redrawn from Michael Silver.)

Figure for immunoglobulins. Redrawn from Oppenheim, J.J., Rosenstreich, D.L., and Peter, M., *Cellular Functions in Immunity and Inflammation*, Elsevier Science Publishing, New York, 1984.

Figures for IgM, lymphoid tissues, secretory IgA, type III complex-mediated hypersensitivity, and type IV cell-mediated hypersensitivity. Redrawn from Murray, P.R., *Medical Microbiology*, Mosby-Yearbook, St. Louis, MO, 1994.

Figure for immunoglobulin superfamily. Redrawn from Hunkapiller, T. and Hood, L., Diversity of the immunoglobulin gene superfamily, *Adv. Immunol.,* 44:1–63, 1989.

Figure for integrin. Redrawn from Argraves, W.S. et al., Amino acid sequence of the human fibronectin receptor, *J. Cell Biol.*, 105:1189, 1987; Larson, R.S. et al., Primary structure of the leukocyte function-associated molecule-1α subunit: an integrin with an embedded domain defining a protein superfamily, *J. Cell Biol.*, 108:711, 1989; Kishimoto, T.K. et al., Cloning of the β subunit of the leukocyte adhesion proteins: homology to an extracellular matrix receptor defines a novel supergene family, *Cell,* 48:685, 1987.

Figure for interferon γ receptors. Redrawn from Ealick, S.E., Cook, W.J., and Vijay-Kumar, S., Three-dimensional structure of recombinant human interferon-γ, *Science,* 252:698–702, 1991.

Figure for *Listeria*. Adapted from Tilney, L.G. and Portnoy, D.A., Actin filaments and the growth, movement, and spread of the intracellular bacterial parasite, Listeria monocytogenes, *J. Cell Biol.*, 109:1597–1608, 1989.

Figure for membrane attack complex. Redrawn from Podack, E.R., Molecular mechanisms of cytolysis by complement and cytolytic lymphocytes, *J. Cell. Biochem.*, 30:133–170, 1986.

Figures for monocytes and front and back inside cover. Adapted from a poster prepared by KMP Associates Ltd., R&D Systems, Minneapolis, MN, 1992.

Figure for oligoclonal bands. Redrawn from Stites, D.P., *Basic and Clinical Immunology,* Appleton & Lange, East Norwalk, CT, 1991.

Figures for oncofetal antigen, orthopic graft, patching and synthetic antigen. Redrawn from Bellanti, J.A., *Immunology II,* W. B. Saunders, Philadelphia, PA, 1978.

Figure for precipitation curve. Redrawn from Eisen, H., *Immunology*, Harper and Row, Hagerstown, MD, 1980.

Figure for structure of cellular prion protein. Courtesy of Dr. Ziwei Huang, Thomas Jefferson University, Philadelphia, PA, A plausible model for the three-dimensional structure of the cellular prion protein proposed by Z. Huang et al., *Proc. Natl. Acad. Sci. U.S.A.*, 91, 7139–7143, 1994. The ribbon drawing was generated using the molscript program (P. J. Kraulis, *J. Appl. Crystallogr.*, 24, 946, 1991).

Figures for psoriasis vulgaris and relapsing polychondritis. Redrawn from Valenzuela, R., *Interpretation of Immunofluorescent Patterns in Skin Disease,* American Society of Clinical Pathologists Press, Chicago, 1984.

Figure for rheumatoid nodule. Redrawn from Dieppe, P.A., Bacon, P.A., Bamji, A.N., and Watt, I., *Atlas of Clinical Rheumatology,* Lea & Febiger, Philadelphia, PA, 1986.

Figure for side chain theory. Adapted from Hemmelweit, F., *Collected Papers of Paul Ehrlich,* Pergamon Press, Tarrytown, NY, 1956–1960.

Figure for single radial immunodiffusion. Redrawn from Miller, L.E., *Manual of Laboratory Immunology*, Lea & Febiger, Malvern, PA, 1991.

Figure for splits. Adapted from *Splits, Associated Antigens and Inclusions,* PEL-FREEZE Clinical Systems, Brown Deer, WI, 1992.

Figure for Wu-Kabat plot. Redrawn from Capra, J.D. and Edmundson, A.B., The antibody combining site, *Sci. Am.,* 236:50–54, 1977.

Figure for Appendix II reproduced with permission from Janeway, C.A., Jr. and Travers, P., *Immunology: The Immune System in Health and Disease*, Current Biology, Ltd., London and Garland Publishing, Inc., New York, 1994.

α-1 antichymotrypsin

A histiocytic marker. By immunoperoxidase staining, it is demonstrable in tumors derived from histiocytes. It may also be seen in various carcinomas.

α-1 antitrypsin (A1AT)

A glycoprotein in circulating blood that blocks trypsin, chymotrypsin, and elastase, among other enzymes. The gene on chromosome 14 encodes 25 separate allelic forms which differ according to electrophoretic mobility. The PiMM phenotype is physiologic. The PiZZ phenotype is the most frequent form of the deficiency which is associated with emphysema, cirrhosis, hepatic failure, and cholelithiasis, with an increased incidence of hepatocellular carcinoma. It is treated with prolastin. Adenoviruses may be employed to transfer the A1AT gene to lung epithelial cells, after which A1AT mRNA and functioning A1AT become demonstrable.

A blood group

Refer to ABO blood group system.

α chain

The immunoglobulin (Ig) class-determining heavy chain found in IgA molecules.

α-fetoprotein

A principal plasma protein in the α globulin fraction present in the fetus. It bears considerable homology with human serum albumin. It is produced by the embryonic yolk sac and fetal liver and consists of a 590-amino acid residue polypeptide chain structure. It may be elevated in pregnant women bearing fetuses with open neural tube defects, central nervous system defects, gastrointestinal abnormalities, immunodeficiency syndromes, and various other abnormalities. After parturition, the high levels in fetal serum diminish to levels that cannot be detected. α-Fetoprotein induces immunosuppression, which may facilitate neonatal tolerance. Based on *in vitro* studies, it is believed to facilitate suppressor T lymphocyte function and diminish helper T lymphocyte action. Liver cancer patients reveal significantly elevated serum levels of α-fetoprotein. In immunology, however, it is used as a marker of selected tumors such as hepatocellular carcinoma. It is detected by the avidin-biotin-peroxidase complex (ABC) immunoperoxidase technique using monoclonal antibodies.

α heavy chain disease

A rare condition in individuals of Mediterranean extraction who may develop gastrointestinal lymphoma and malabsorption with loss of weight and diarrhea. The aberrant plasma cells infiltrating the lamina propria of the intestinal mucosa and the mesenteric lymph nodes synthesize α chains alone, usually α-1, with no production of light chains. Even though the end-terminal sequences are intact, a sequence stretching from the V region through much of the $C_{\alpha-1}$ domain is deleted. Thus, there is no cysteine residue to crosslink light chains. α Heavy chain disease is more frequent than the γ type. It has been described in North Africa, the Near East and the Mediterranean area, and in some regions of southern Europe. Rare cases have been reported in the U.S. The condition may prove fatal, even though remissions may follow antibiotic therapy.

α helix

A spiral or coiled structure present in many proteins and polypeptides. It is defined by intrachain hydrogen bonds between –CO and –NH groups that hold the polypeptide chain together in a manner which results in 3.6 amino acid residues per helical turn. There is a 1.5-Å rise for each residue. The helix has a pitch of 5.4 Å. The helical backbone is formed by a peptide group and the α carbon. Hydrogen bonds link each –CO group to the –NH group of the fourth residue forward in the chain. The α helix may be left or right handed. Right-handed α helices are the ones found in proteins.

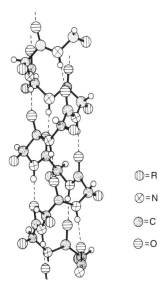

= R
= N
= C
= O

Right-handed α-helix

α₂ macroglobulin (α₂ M)

A 725-kD plasma glycoprotein that plays a major role in inhibition of proteolytic activity generated during various extracellular processes. α_2M is synthesized in the liver and reticuloendothelial system. α_2M is produced by lymphocytes and is found associated with the surface membrane of a subpopulation of B cells. It has the unique property of binding all active endopeptidases. Other enzymes or even the inactive forms of proteinases are not bound. Complexes of α_2M with proteinase are rapidly cleared from circulation (minutes), in contrast to the turnover of α_2M which requires several days. Some of the roles of α_2M include (1) regulation of the extracellular proteolytic activity resulting from clotting, fibrinolysis, and proteinases of inflammation; and (2) specific activity against some proteinases of fungal or bacterial origin. It is elevated significantly in nephrotic syndrome. Increased levels have been reported also in atopic dermatitis and ataxia telangiectasia.

α₁-microglobulin

A 30 kD protein that belongs to the lipocalin family and possesses hydrophobic prosthetic groups. It is synthesized in the liver and is present in the urine and serum. α_1M may be complexed with monomeric IgA. It may be increased in IgA nephropathy. Elevated serum α_1-M in patients with AIDS may signify renal pathology. α_1-M blocks antigen stimulation and migration of granulocytes. It has a role in immunoregulation and functions as a mitogen.

α₂-plasmin inhibitor-plasmin complexes (α₂PIPC)

Complexes formed by the combination of α_2-PI or α_2-macroglobulin with plasmin, the active principle in fibrinolysis. These complexes are found in elevated quantities in plasma of systemic lupus erythematosus (SLE) patients with vasculitis compared to plasma of SLE patients without vasculitis.

AA amyloid

A nonimmunoglobulin amyloid fibril of the type seen following chronic inflammatory diseases such as tuberculosis and osteomyelitis or, more recently, chronic noninfectious inflammatory disorders. Kidneys, liver, and spleen are the most significant areas of AA (amyloid-associated) deposition. The precursor for AA protein is

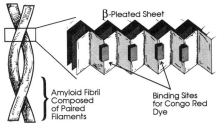

AA Amyloid

apo-SAA (serum amyloid associated), with a monomer mol wt of 12.5 kD which is found in the circulation as a 220- to 235-kD molecular complex because it is linked to high-density lipoproteins. Interleukin-6 stimulates its synthesis. AA deposition is either associated with an amyloidogenic isotypical form of SAA or results from the inability to completely degrade SAA.

Amyloid consists of nonbranching fibrils 7.5 to 10 nm in width and of indefinite length. Chemically, amyloid occurs in two classes. The AL (amyloid light) chain type consists of immunoglobulin light chains or parts of them. The AA-type is derived from the SAA protein in the serum. SAA acts like an acute-phase reactant increasing greatly during inflammation. Thus, AA protein is the principal type of amyloid deposited in chronic inflammatory diseases. AL amyloid consists of either whole immunoglobulin light chains or their N-terminal fragments or combination of the two. The λ light chain especially gives rise to AL. AL amyloid protein is often deposited following or during B cell disorders. Other biochemical forms of amyloid include transthyretin, β_2 microglobulin, β_2 amyloid protein or additional forms. Amyloid filaments stained with Congo red exhibit green birefringence with polarized light.

AB blood group
Refer to ABO blood group system.

Front Typing		Back Typing			
Reaction of Cells Tested with		Reaction of Serum Tested Against			Interpretation
		A	B	O	ABO
Anti-A	Anti-B	Cells	Cells	Cells	Group
O	O	+	+	O	O
+	O	O	+	O	A
O	+	+	O	O	B
+	+	O	O	O	AB

+ = agglutination
O = no agglutination

ABO Blood Grouping

αβ T cells
T lymphocytes that express αβ chain heterodimers on their surface. The vast majority of T cells are of the αβ variety.

Abelson murine leukemia virus (A-MuLV)
A B cell murine leukemia-inducing retrovirus that bears the v-abl oncogene. The virus has been used to immortalize immature B lymphocytes to produce pre-B cell or less differentiated B cell lines in culture. These have been useful in unraveling the nature of immunoglobulin differentiation, such as H and L chain immunoglobulin gene assembly, as well as class switching of immunoglobulin.

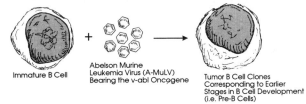

Immature B Cell   +   Abelson Murine Leukemia Virus (A-MuLV) Bearing the v-abl Oncogene   →   Tumor B Cell Clones Corresponding to Earlier Stages in B Cell Development (i.e. Pre-B Cells)

Abelson Murine Leukemia Virus

ablastin
An antibody with the exclusive property of preventing reproduction of such agents as the rat parasite *Trypanosoma lewisi*. It does not demonstrate other antibody functions.

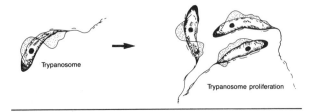

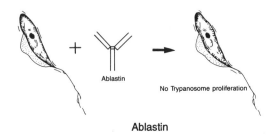

Ablastin

ABO blood group substances
Glycopeptides with oligosaccharide side chains manifesting ABO epitopes of the same specificity as those present on red blood cells of the individual in whom they are detected. Soluble ABO blood group substances may be found in mucous secretions of man such as saliva, gastric juice, ovarian cyst fluid, etc. Such persons are termed secretors, whereas those without the blood group substances in their secretions are nonsecretors.

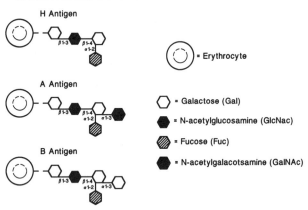

ABO Blood Group Substances

ABO blood group system
The first described of the human blood groups based upon carbohydrate alloantigens present on red cell membranes. Anti-A or anti-B isoagglutinins (alloantibodies) are present only in the blood sera of individuals not possessing that specificity, i.e., anti-A is found in the serum of group B individuals and anti-B is found in the serum of group A individuals. This serves as the basis for

Blood Type	RBC surface antigen	Antibody in Serum
A	A antigen	Anti-B
B	B antigen	Anti-A
AB	AB antigens	no antibody
O	no A or B antigens	Both Anti-A and Anti-B

ABO Blood Group System

grouping humans into phenotypes designated A, B, AB, and O. Type AB subjects possess neither anti-A nor anti-B antibodies, whereas group O persons have both anti-A and anti-B antibodies in their serum. Blood group methodology to determine the ABO blood type makes use of the agglutination reaction. The ABO system remains the most important in the transfusion of blood and is also critical in organ transplantation. Epitopes of the ABO system are found on oligosaccharide terminal sugars. The genes designated as *A/B, Se, H,* and *Le* govern the formation of these epitopes and of the Lewis (Le) antigens. The two precursor substances type I and type II differ only in that the terminal galactose is joined to the penultimate *N*-acetylglucosamine in the β 1-3 linkage in type I chains, but in the β 1-4 linkage in type II chains.

aboriginal mouse
An animal that has lived apart from man.

abrin
A powerful toxin and lectin used in immunological research by Paul Ehrlich (*circa* 1900). It is extracted from the seeds of the jequirity plant and causes agglutination of erythrocytes.

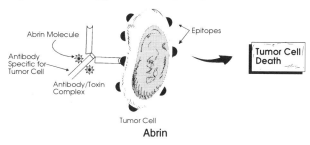

Abrin

absorption
The elimination of antibodies from a mixture by adding soluble antigens or the elimination of soluble antigens from a mixture by adding antibodies.

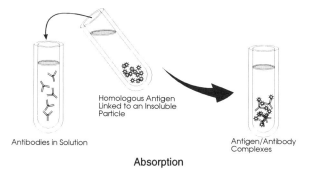

Absorption

absorption elution test
The identification of the ABO type in stains of semen or blood on clothing. ABO blood group antibodies are applied to the stain after it has been exposed to boiling water. Following washing to remove unfixed antibody, the preparation is heated to 56°C in physiological saline. An antibody that may be eluted from the stain is tested with erythrocytes of known ABO specificity to determine the ABO type. This is being replaced by DNA analysis of such specimens by forensics experts.

abzymes (see figure, top right)
The union of antibody and enzyme molecules to form a hybrid catalytic molecule. Specificity for a target antigen is provided through the antibody portion and for a catalytic function through the enzyme portion. Thus, these molecules have numerous potential uses. These molecules are capable of catalyzing various chemical reactions and show great promise as protein-clearing antibodies, as in the dissolution of fibrin clots from occluded coronary arteries in myocardial infarction.

acanthosis nigricans
A condition in which the afflicted subject develops insulin receptor autoantibodies associated with insulin-resistant diabetes mellitus, as well as thickened and pigmented skin.

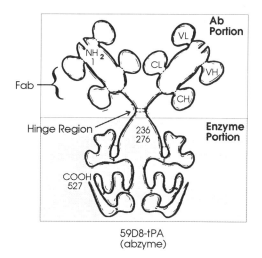

59D8-tPA
(abzyme)

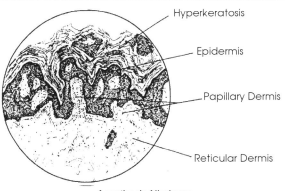

Acanthosis Nigricans

accessory cell
A cell such as a monocyte, macrophage, dendritic cell, or Langerhans cell that facilitates the generation of an immune response through antigen presentation to helper T cells. B cells may also act as antigen-presenting cells, thereby serving an accessory cell function.

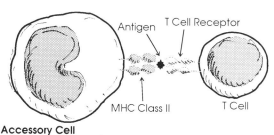

Accessory Cell
(Monocyte, Macrophage,
Dendritic Cell or Langerhan's
Cell)

Accessory Cell

accessory molecules
Molecules other than the antigen receptor and major histocompatability complex (MHC) that participate in cognitive, activation, and effector functions of T lymphocyte responsiveness. Adhesion molecules facilitating the interaction of T lymphocytes with other cells that signal transducing molecules which participate in T cell activation are classified as accessory molecules.

acetylcholine receptor (AChR) antibodies
IgG autoantibodies that cause loss of function of AChRs that are critical to chemical transmission of the nerve impulse at the neuromuscular junction. This represents a type II mechanism of hypersensitivity, according to the Coombs and Gell classification. AChR antibodies are heterogeneous with some showing specificity for antigenic determinants other than those that serve as acetylcholine

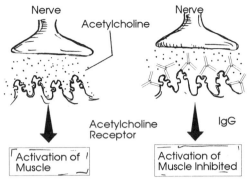

Acetylcholine Receptor Antibodies

or α-bungarotoxin binding sites. As many as 85 to 95% of myasthenia gravis patients may manifest AChR antibodies.

acquired agammaglobulinemia
Refer to common variable immunodeficiency.

acquired C1 inhibitor deficiency
A condition in which the C1 inhibitor is inactivated, resulting in elevated C4 and C2 cleavage as C1 is activated. Patients experience repeated laryngeal, intestinal, and subcutaneous tissue swelling. A kinin-like peptide derived from C2b promotes vascular permeability and produces symptoms. Subjects with B lymphocyte or plasma cell monoclonal proliferation may develop acquired C1 inhibitor deficiency. These syndromes include multiple myeloma, Waldenström's macroglobulinemia, and B cell lymphoma. Subjects may develop antiidiotypic antibodies against membrane immunoglobulins or myeloma proteins. Idiotype-antiidiotype interactions at the cell membrane may result in C1 fixation and may increase utilization of C4 and C2 as well as of C1 inhibitor.

acquired immunity
Protective resistance against an infectious agent generated as a consequence of infection with a specific microorganism or as a result of deliberate immunization.

acquired tolerance
Immunologic tolerance induced by the inoculation of a neonate or fetus *in utero* with allogeneic cells prior to maturation of the recipient's immune response. The inoculated antigens are accepted as self. Immunologic tolerance may be induced to some soluble antigens by low-dose injections of neonates with the antigen or to older animals by larger doses, the so-called low-dose and high-dose tolerance, respectively. Refer also to immunologic tolerance.

acridine orange
A fluorescent substance that binds nonspecifically to RNA with red fluorescence and to DNA with green fluorescence. It also interacts with polysaccharides, proteins, and glycosaminoglycans. It is a nonspecific tissue stain which identifies increased mitoses and shows greater sensitivity, but less specificity than the Gram stain. It is carcinogenic and of limited use in routine histology.

ACT-2
A human homolog of murine MIP-1β that chemoattracts monocytes but prefers activated CD4+ cells to CD8+ cells. T cells and monocytes are sources of ACT-2.

actin (see figure, top right)
In immunology, the principal muscle protein, which together with myosin causes muscle contraction, is used in surgical pathology as a marker for the identification of tumors of muscle origin. Actin is identified through immunoperoxidase staining of surgical pathology tissue specimens.

activated lymphocytes
A lymphocyte whose cell surface receptors have interacted with a specific antigen or with a mitogen such as phytohemagglutinin, concanavalin A, or staphylococcal protein A. The morphologic appearance of activated (or stimulated) lymphocytes is characteristic and in this form the cells are called immunoblasts. The cells increase in size from 15 to 30 μm in diameter; show increased cytoplasmic basophilia; and develop vacuoles, lysosomes, and ribosomal aggregates. Pinocytotic vesicles are present on the cell membrane. The nucleus contains little chromatin, which is limited to a thin marginal layer, and the nucleolus becomes conspicuous. The array of changes that follows stimulation is called transformation. Such cells are called transformed cells. An activated B lym-

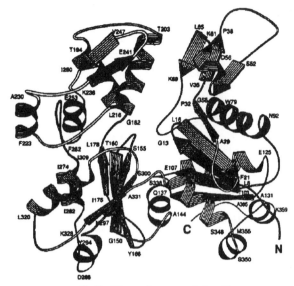

3-D Structure of Actin

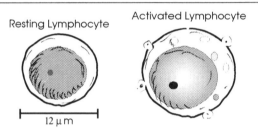

Resting Lymphocyte

Activated Lymphocyte

12 μm

phocyte may synthesize antibody molecules, whereas an activated T cell may mediate a cellular immune reaction.

activated macrophage
A macrophage that has been stimulated in some manner or by some substance to increase its functional efficiency with respect to phagocytosis, intracellular bactericidal activity, or lymphokine, i.e., IL-1, production. A lymphokine-activated mononuclear phagocyte is double the size of resting macrophages. MHC class II antigen surface expression is elevated, and lysosomes increase. The latter changes facilitate antimicrobial defense.

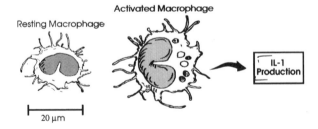

Activated Macrophage

Resting Macrophage

IL-1 Production

20 μm

activation
(1) The stimulation of lymphocytes or macrophages to increase their functional activity, or (2) the initiation of the multicomponent complement cascade in serum consisting of a series of enzyme-substrate reactions leading to the generation of functionally active effector molecules.

activation unit
Interaction of C3b with C4b2a bound to the cell membrane.

active anaphylaxis
The anaphylactic state induced by natural or experimental sensitization in atopic subjects or experimental animals. See also anaphylaxis.

active immunity
Protection attained as a consequence of clinical or subclinical infection or deliberate immunization with an infectious agent or its products.

active immunization

The induction of an immune response either through exposure to an infectious agent or by deliberate immunization with products of the microorganism inducing the disease to develop protective immunity. A clinical disease or subclinical infection or vaccination may be used to induce the desired protective effect. Booster immunization injections given at intervals after primary exposure may lead to long-lasting immunity through the activation of immunological memory cells.

active kinins

The active kinin compounds are characterized by a nonapeptide amino acid sequence whose prototype is bradykinin, the active kinin generated from plasma kininogen. Generation of the other forms depends on the enzyme and substrate used, but they differ in length and the additional residues. Tissue kallikreins are best activated by enzymes like trypsin and hydrolyze both low mol wt kininogen (LMK) and high mol wt kininogen (HMK) to give kallidin, a tissue form of kinin. Bradykinin is formed by the action of plasma kallikrein on HMK or by the action of trypsin on both LMK and HMK. Met-lys-bradykinin, another kinin, results from hydrolysis by plasma kallikrein activated by acidification. Other active kinins have also been described. Besides being the precursors for the generation of kinins, kininogens also affect the coagulation system, with the activation of the Hageman factor (HF) being the link between the two systems.

Primary Structure of
Serum Bradykinin

Active Kinins

active site

A crevice formed by the V_L and V_H regions of an immunoglobulin's Fv region. It may differ in size or shape from one antibody molecule to another. Its activity is governed by the amino acid sequence in this variable region and differences in the manner in which V_H and V_L regions relate to one another. Antibody molecule specificity is dependent on the complementary relationship between epitopes on antigen molecules and amino acid residues in the recess comprising the antibody active site. The V_L and V_H regions contain hypervariable areas that permit great diversity in the antigen-binding capacity of antibody molecules.

acumentin

A neutrophil and macrophage motility protein that links to the actin molecule to control actin filament length.

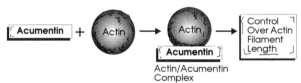

Actin/Acumentin
Complex

Acumentin

acute AIDS syndrome

Within the first to sixth week following HIV-1 infection, some subjects develop the flu-like symptoms of sore throat, anorexia, nausea and vomiting, lymphadenopathy, maculopapular rash, wasting, and pain in the abdomen, among other symptoms. The total leukocyte count is slightly depressed with possible CD4 to CD8 ratio inversion. Detectable antibodies with specificity for HIV constituents gp120, gp160, p24, and p41 are not detectable until at least 6 months following infection. Approximately 33% of the infected subjects manifest the acute AIDS syndrome.

acute disseminated encephalomyelitis

Brain inflammation that may be a sequela of certain acute viral infections such as measles in children or following vaccination. It was reported in some subjects following smallpox vaccination and in early recipients of rabies vaccine containing nervous system tissues. Symptoms include neck stiffness, headache, disorientation, and coma. Elevated quantities of protein and lymphocytes appear in the cerebrospinal fluid. Histopathologically, lymphocytes, plasma cells, and polymorphonuclear leukocytes may form perivascular infiltrates. The pathological changes are probably attributable to immune reactivity against the central nervous system constituent myelin basic protein and may represent the human equivalent of experimental allergic encephalomyelitis produced in animals. Refer also to the animal model of this condition termed experimental allergic encephalomyelitis (EAE).

acute graft-vs.-host reaction

The immunopathogenesis of acute graft-vs.-host disease (GVHD) consists of recognition, recruitment, and effector phases. Epithelia of the skin, gastrointestinal tract, small intrahepatic biliary ducts, and the lymphoid system constitute primary targets of acute GVHD. GVHD development may differ in severity based on relative antigenic differences between donor and host and the reactivity of donor lymphocytes against non-HLA antigens of recipient tissues. The incidence and severity of GVHD has been ascribed also to HLA-B alleles, i.e., an increased GVHD incidence associated with HLA-B8 and HLA-B35. Epithelial tissues serving as targets of GVHD include keratinocytes, erythrocytes, and bile ducts, which may express Ia antigens following exposure to endogeneous interferon produced by T lymphocytes. When Ia antigens are expressed on nonlymphoid cells, they may become antigen-presenting cells for autologous antigens and aid perpetuation of autoimmunity.

Cytotoxic T lymphocytes mediate acute GVHD. While most immunohistological investigations have implicated CD8+ (cytotoxic/suppressor) lymphocytes, others have identified CD4+ (T helper lymphocytes) in human GVHD, whereas natural killer (NK) cells have been revealed as effectors of murine, but not human GVHD. Following interaction between effector and target cells, cytotoxic granules from cytotoxic T or NK cells are distributed over the target cell membrane, leading to perforin-induced large pores across the membrane and nuclear lysis by deoxyribonuclease. Infection, rather than failure of the primary target organ (other than gastrointestinal bleeding), is the major cause of mortality in acute GVHD. Within the first few months posttransplant, all recipients demonstrate diminished immunoglobulin synthesis, decreased T helper lymphocytes, and increased T suppressor cells.

Acute GVHD patients manifest an impaired ability to combat viral infections. They demonstrate an increased risk of cytomegalovirus (CMV) infection, especially CMV interstitial pneumonia. GVHD may also reactivate other viral diseases such as herpes simplex. Immunodeficiency in the form of acquired B cell lymphoproliferative disorder (BCLD) represents another serious complication of postbone marrow transplantation. Bone marrow transplants treated with pan-T cell monoclonal antibody or those in which T lymphocytes have been depleted account for most cases of BCLD, which is associated with severe GVHD. All transformed B cells in cases of BCLD have manifested the Epstein Barr viral genome.

acute lymphocytic leukemia (ALL)

Patients may experience a profound reduction in the concentration of the serum immunoglobulins, possibly due to malignant expansion of suppressor T lymphocytes.

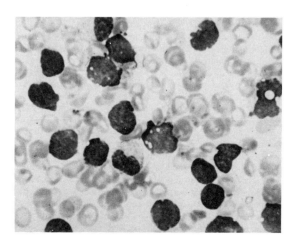

acute-phase reactants

Serum proteins that increase during acute inflammation. These proteins, which migrate in the α-1 and α-2 electrophoretic regions, include α-1 antitrypsin, α-1 glycoprotein, amyloid A&P, antithrombin III, C-reactive protein, C1 esterase inhibitor, C3 complement, ceruloplasmin, fibrinogen, haptoglobin, orosomucoid, plasminogen, and transferrin.

acute-phase response (APR)

A nonspecific response by an individual stimulated by interleukin-1, interleukin-6, interferons and tumor necrosis factor. C-reactive protein may show a striking rise within a few hours. Infection, inflammation, tissue injury, and, very infrequently, neoplasm may be associated with APR. The liver produces acute-phase proteins at an accelerated rate, and the endocrine system is affected with elevated gluconeogenesis, impaired thyroid function, and other changes. Immunologic and hematopoietic system changes include hypergammaglobulinemia and leukocytosis with a shift to the left. There is diminished formation of albumin, elevated ceruloplasmin, and diminished zinc and iron.

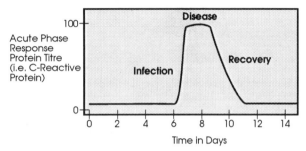

Acute Phase Response (APR)

acute-phase serum

A serum sample drawn from a patient with an infectious disease during the acute phase.

acute poststreptococcal glomerulonephritis

A disease of the kidney with a good prognosis for most patients that follows streptococcal infection of the skin or streptococcal pharyngitis by 10 days to 3 weeks in children. The subject presents with hematuria, fever, general malaise, facial swelling, and smokey urine. There is mild proteinuria and red blood cells in the urine. These children often have headaches, are hypertensive, and have an elevated blood pressure. Serum C3 levels are decreased. The erythrocyte sedimentation rate (ESR) is elevated, and an abdominal X-ray may reveal enlarged kidneys. Renal biopsy reveals infiltration of glomeruli by polymorphonuclear leukocytes and monocytes and a diffuse proliferative process. Immunofluorescence shows granular deposits of IgG and C3 on the epithelial side of peripheral capillary loops. Electron microscopy shows subepithelial "humps". The process usually resolves spontaneously within 1 week after onset of renal signs and symptoms, the patient becomes afebrile, and the malaise disappears. It is attributable to nephritogenic streptococci of types 1, 4, 12, and 49.

acute rheumatic fever

Following a group A β hemolytic streptococcal infection of the throat, inflammation of the heart, joints, and connective tissue may follow. Arthritis may affect several joints in a migratory pattern. Carditis, inflammation of the heart, may be associated with the development of high-titer heart-reactive antibodies (HRA), which have been implicated in the pathogenesis of rheumatic fever. A patient may develop HRA, rheumatic fever, or both. HRA development in rheumatic fever represents molecular mimicry. Selected antistreptococcal cell wall M protein antibodies are cross-reactive through molecular mimicry with myocardial epitopes of the human heart. These serve as sites of attachment for IgG and complement molecules that are detectable by immunofluorescence examination. The antibodies on the cardiac muscle are found at sarcolemmal and subsarcolemmal sites and in the pericardium. The crossreactivity also involves heart valve glycoproteins and the myocardial conduction system. There may also be a cell-mediated immune attack, as revealed by the accumulation of CD4 T lymphocytes in valvular tissues. There appears to be a positive correlation between the development of rheumatic fever and HRA titers.

acyclovir 9(2-hydroxyethoxy-methylguanine)

An antiviral nucleoside analog that blocks herpes simplex virus-2 (HSV-2), the causative agent of genital herpes. HSV thymidine kinase activates acyclovir through monophosphorylation, followed by triple phosphorylation with host enzymes to yield a powerful blocking action of the DNA polymerase of HSV-2. Acyclovir is prescribed for the treatment of HSV-2 genital infection.

adaptive differentiation

Acquisition of the ability to identify MHC class II antigens by thymocytes undergoing differentiation and maturation to CD4+ T helper/inducer cells in the thymus.

adaptive immunity

Protection from an infectious disease agent as a consequence of clinical or subclinical infection with that agent or by deliberate immunization against that agent with products from it. This type of immunity is in contrast to natural immunity.

Addison's disease

Destruction of the adrenal cortex by tuberculosis or another infection or by autoimmune mechanisms. Lymphocytes infiltrate the adrenal, and autoantibodies against cells of the adrenal cortex appear in the blood serum in many of the cases. The disease results from the loss of functioning adrenal cortical tissue and associated diminished adrenal cortical-stimulating hormones. The disease may occur in association with autoimmune thyroiditis, insulin-dependent type I diabetes melitus, hypoparathyroidism, and pernicious anemia. It may be reproduced experimentally in animals by immunizing them with tissue incorporated into Freund's complete adjuvant to induce experimental allergic adrenalitis.

addressin

A molecule such as a peptide or protein that serves as a homing device to direct a molecule to a specific location (an example is ELAM-1). Lymphocytes from Peyer's patches home to mucosal endothelial cells bearing ligands for the lymphocyte homing receptor.

adenosine

Adenosine is normally present in the plasma in a concentration of 0.03 μM in man and 0.04 μM in the dog. In various clinical states associated with hypoxia, the adenosine level increases five- to tenfold, suggesting that it may play a role in the release of mediators. Experimentally, adenosine is a powerful potentiator of mast cell function. The incubation of mast cells with adenosine does not induce the release of mediators. However, by preincubation with adenosine and subsequent challenge with a mediator-releasing agent, the response is markedly enhanced.

adenosine deaminase (ADA)

A 38-kD deaminating enzyme that prevents increased levels of adenosine, adenosine trisphosphate (ATP), deoxyadenosine, deoxy-ATP, and S-adenosyl homocysteine. It is encoded by the chromosome 20q13-ter gene. Elevated adenosine levels block DNA methylation within cells, leading to their death. Increased levels of deoxy-ATP block ribonucleoside-diphosphate reductase, which participates in the synthesis of purines.

ADA Deficiency

Adenosine   Adenosine   Inosine
deaminase

adenosine deaminase (ADA) deficiency

A form of severe combined immunodeficiency (SCID) in which affected individuals lack an enzyme, adenosine deaminase (ADA), which catalyzes the deamination of adenosine as well as deoxyadenosine to produce inosine and deoxyinosine, respectively.

Cells of the thymus, spleen, and lymph node, as well as red blood cells, contain free ADA enzyme. In contrast to the other forms of SCID, children with ADA deficiency possess Hassall's corpuscles in the thymus. The accumulation of deoxyribonucleotides in various tissues, especially thymic cells, is toxic and is believed to be a cause of immunodeficiency. As deoxyadenosine and deoxy-ATP accumulate, the latter substance inhibits ribonucleotide reductase activity, which inhibits formation of the substrate needed for synthesis of DNA. These toxic substances are especially injurious to T lymphocytes. The autosomal recessive ADA deficiency leads to death. Two fifths of severe combined immunodeficiency cases are of this type. The patient's signs and symptoms reflect defective cellular immunity with oral candidiasis, persistent diarrhea, failure to thrive, and other disorders, with death occurring prior to 2 years of age. T lymphocytes are significantly diminished. There is eosinophilia and elevated serum and urine adenosine and deoxyadenosine levels. As bone marrow transplantation is relatively ineffective, gene therapy is the treatment of choice.

adherent cell

A cell such as a macrophage (mononuclear phagocyte) that attaches to the wall of a culture flask, thereby facilitating the separation of such cells from B and T lymphocytes which are not adherent.

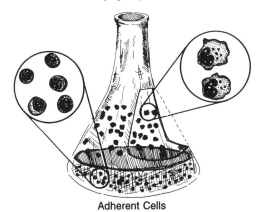

Adherent Cells

adhesins

Bacterial products that split proteins. They combine with human epithelial cell glycoprotein or glycolipid receptors, which could account for the increased incidence of pulmonary involvement attributable to *Pseudomonas aeruginosa* in patients who are intubated.

adhesion proteins

Extracellular matrix proteins that attract leukocytes from the circulation. For example, T and B lymphocytes possess lymph node homing receptors on their membranes which facilitate their passage through high endothelial venules. Neutrophils migrate to areas of inflammation in response to endothelial leukocyte adhesion molecule 1 (ELAM-1) stimulated by TNF and IL-1 on the endothelium of vessels. B and T lymphocytes that pass through high endothelial venules have lymph node homing receptors.

adhesion receptors

Proteins in cell membranes that facilitate the interaction of cells with matrix. They play a significant role in adherence and chemoattraction in cell migration. They are divided into three groups that include the immunoglobulin superfamily, which contains the T cell receptor/CD3, CD4, CD8, MHC class I, MHC class II, sCD2/LFA-2, LFA-3/CD58, ICAM-1, ICAM-2, and V-CAM, 2. The second group of adhesion receptors is made up of the integrin family, which contains LFA-1, Mac-1, p150,95, VLA-5, VLA-4/LPAM-1, LPAM-2, and LPAM-3. The third family of adhesion receptors consists of selectin molecules that include Mel-14/LAM-1, ELAM-1, and CD62.

adjuvant

Substance that facilitates or enhances the immune response to an antigen with which it is combined. Various types of adjuvants have been described, including Freund's complete and incomplete adjuvants, aluminum compounds, and muramyl dipeptide. Some of these act by forming a depot in tissues from which an antigen is slowly released. In addition, Freund's adjuvant attracts a large number of cells to the area of antigen deposition to provide in-

creased immune responsiveness to it. Modern adjuvants include such agents as muramyl dipeptide. The ideal adjuvant is one that is biodegradable with elimination from the tissues once its immunoenhancing activity has been completed. An adjuvant nonspecifically facilitates an immune response to antigen. An adjuvant usually combines with the immunogen, but is sometimes given prior to or following antigen administration. Adjuvants represent a heterogenous class of compounds capable of augmenting the humoral or cell-mediated immune response to a given antigen. They are widely used in experimental work and for therapeutic purposes in vaccines. Adjuvants comprise compounds of mineral nature, products of microbial origin, and synthetic compounds. The primary effect of some adjuvants is postulated to be the retention of antigen at the inoculation site so that the immunogenic stimulus persists for a longer period of time. However, the mechanism by which adjuvants augment the immune response is poorly understood. The macrophage may be the target and mediator of action of some adjuvants, whereas others may require T cells for their response augmenting effect. Adjuvants such as lipopolysaccharide (LPS) may act directly on B lymphocytes.

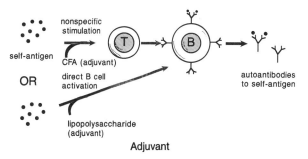

Adjuvant

adjuvant disease

The injection of rats with Freund's complete adjuvant, a water-in-oil emulsion containing killed, dried mycobacteria, e.g., *Mycobacterium tuberculosis,* leads to the production of aseptic synovitis, which closely resembles rheumatoid arthritis (RA) in man. Sterile inflammation occurs in the joints and lesions of the skin. In addition to swollen joints, inflammatory lesions of the tail may also result in animals developing adjuvant arthritis, which represents an animal model for RA.

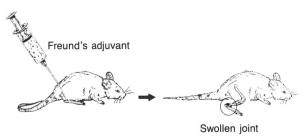

Adjuvant Disease

adjuvant granuloma

A tissue reaction that occurs at a local site following the injection of such adjuvant materials as Freund's complete adjuvant or alum, both of which have been used extensively in immunologic research in past years.

adoptive immunity

The term assigned by Billingham, Brent, and Medawar (1955) to transplantation immunity induced by the passive transfer of specifically immune lymph node cells from an actively immunized animal to a normal (previously nonimmune) syngeneic recipient host.

adoptive immunization

The passive transfer of immunity by the injection of lymphoid cells from a specifically immune individual to a previously nonimmune recipient host. The resulting recipient is said to have adoptive immunity.

adoptive immunotherapy

The experimental treatment of terminal cancer patients with metastatic tumors unresponsive to other modes of therapy by the inoculation of lymphokine-activated killer (LAK) cells or tumor-infiltrating lymphocytes (TIL) together with IL-2. This mode of therapy has shown some success in approximately one-tenth of treated individuals with melanoma or renal cell carcinoma.

adoptive tolerance
The passive transfer of immunologic tolerance with lymphoid cells from an animal tolerant to that antigen to a previously nontolerant and irradiated recipient host animal.

adoptive transfer
A synonym for adoptive immunization.

adrenergic receptors
Structures on the surfaces of various types of cells that are designated α or β and interact with adrenergic drugs.

adsorption
The elimination of antibodies from a mixture by adding particulate antigen or the elimination of particulate antigen from a mixture by adding antibodies. The incubation of serum-containing antibodies such as agglutinins with red blood cells or other particles may remove them through sticking to the particle surface.

adsorption chromatography
A method to separate molecules based on their adsorptive characteristics. Fluid is passed over a fixed-solid stationary phase.

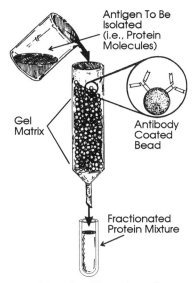

Adsorption Chromatography

adult respiratory distress syndrome (ARDS)
Embarrassed respiratory function as a consequence of pulmonary edema caused by increased vascular permeability.

adult T cell leukemia-lymphoma (ATLL)
A lymphoproliferative neoplasm of mature lymphocytes that progresses rapidly. This has been linked to the HTLV-1 retrovirus infection, which has been observed in Japan, Africa, the Caribbean, and the southeastern U. S. Patients develop hypercalcemia, progressive skin changes, enlarged hilar, and retroperitoneal and peripheral lymph nodes without mediastinal node enlargement. There may be involvement of the lungs, gastrointestinal tract, and central nervous system, as well as opportunistic infections. The condition occurs in five clinical forms.

AET rosette test (historical)
A technique used previously to enumerate human T cells based upon the formation of sheep red cell rosettes surrounding them. The

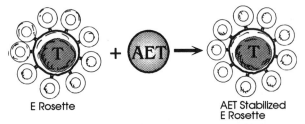

E Rosette   AET Stabilized E Rosette

AET Rosette Test

use of sheep red cells treated with aminoethylthiouridium bromide renders the rosettes more stable than when using sheep red cells untreated by this technique. This technique was later replaced by the use of anti-CD2 monoclonal antibodies and flow cytometry.

affinity
The strength of binding between antigen and antibody molecules. It increases with linkage stability. The affinity constant reflects the strength of binding. The paratope of an antibody molecule views the epitope as a three-dimensional structure.

affinity chromatography
A method to isolate antigen or antibody based upon antigen-antibody binding. Antibody molecules fixed to plastic beads in a column, constituting the solid phase, may capture antigen molecules in solution passed over the column. Subsequent elution of the antigen is then accomplished with acetate buffer at pH 3.0 or diethylamine at pH 11.5.

affinity constant
Determination of the equilibrium constant though application of the Law of Mass Action to interaction of an epitope or hapten with its homologous antibody. The lack of covalent bonds between the interacting antigen and antibody permits a reversible reaction.

affinity maturation
The sustained increase in affinity of antibodies for an antigen with time following immunization. The genes encoding the antibody variable regions undergo somatic hypermutation with the selection of B lymphocytes whose receptors express high affinity for the antigen. The IgG antibodies that form following the early, heterogeneous IgM response manifest greater specificity and less heterogeneity than do the IgM molecules.

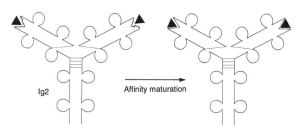

Ig2   Affinity maturation →

agammaglobulinemia
Refer to hypogammaglobulinemia. Agammaglobulinemia was used in earlier years before the development of methods sufficiently sensitive to detect relatively small quantities of gamma globulin in the blood. Primary agammaglobulinemia was attributed to defective immunoglobulin formation, whereas secondary agammaglobulinemia referred to immunoglobulin depletion as in loss through inflammatory bowel disease or through the skin in burn cases.

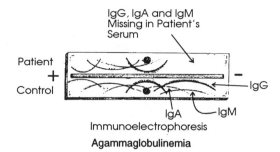

IgG, IgA and IgM Missing in Patient's Serum

Patient + Control

IgG
IgA   IgM
Immunoelectrophoresis
Agammaglobulinemia

agarose
A neutral polygalactoside consisting of alternating D-galactose and 3,6-anhydrogalactose linear polymer, the principal constituent of agar. Gels made from agarose are used for the hemolytic plaque assay and for leukocyte chemotaxis assays, as well as for immunodiffusion and nucleic acid/protein electrophoresis.

agglutination
The combination of soluble antibody with particulate antigens in an aqueous medium containing electrolyte, such as erythrocytes, latex particles bearing antigen, or bacterial cells to form an aggregate which may be viewed either microscopically or macroscopically. If antibody

is linked to insoluble beads or particles, they may be agglutinated by soluble antigen by reverse agglutination. Agglutination is the basis for multiple serological reactions including blood grouping, diagnosis of infectious diseases, rheumatoid arthritis (RA) test, etc. To carry out an agglutination reaction, serial dilutions of antibody are prepared, and a constant quantity of particulate antigen is added to each antibody dilution. Red blood cells may serve as carriers for adsorbed antigen, e.g., tanned red cell or bis-diazotized red cell technique. Like precipitation, agglutination is a secondary manifestation of antigen-antibody interaction. As specific antibody crosslinks particulate antigens, aggregates form that become macroscopically visible and settle out of suspension. Thus, the agglutination reaction has a sensitivity 10 to 500 times greater than that of the precipitin test with respect to antibody detection.

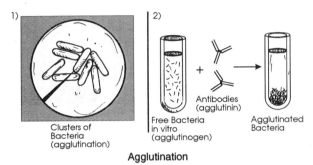

Agglutination

agglutinin

An antibody that interacts with antigen on the surface of particles such as erythrocytes, bacteria, or latex cubes to cause their aggregation or agglutination in an aqueous environment containing electrolyte. Substances other than agglutinin antibody that cause agglutination or aggregation of certain specificities of red blood cells include hemagglutinating viruses and lectins.

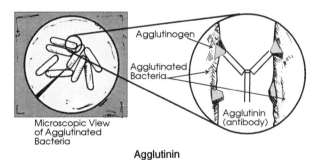

Agglutinin

agglutinogen

Antigens on the surfaces of particles such as red blood cells that react with the antibody known as agglutinin to produce aggregation or agglutination. The most widely known agglutinogens are those of the ABO and related blood group systems.

aggregate anaphylaxis

Form of anaphylaxis caused by aggregates of antigen and antibody in the fluid phase. The aggregates bind complement-liberating complement fragments C3a, C5a, and C4a, also called anaphylatoxins, which induce the release of mediators. Preformed aggregates of antigen-antibody complexes in the fluid phase fix complement. Fragments of complement components, the anaphylatoxins, may induce the release of mediators from mast cells. There is no evidence that these components play a role in anaphylactic reactions *in vivo*. Aggregates of antigen-IgG antibody, however, may induce anaphylaxis, whose manifestations are different in the various species.

agranulocytosis

A striking decrease in circulating granulocytes including neutrophils, eosinophils, and basophils as a consequence of suppressed myelopoiesis. The deficiency of polymorphonuclear leukocytes leads to decreased resistance and increased susceptibility to microbial infection. Patients may present with pharyngitis. The etiology may be either unknown or follow exposure to cytotoxic drugs such as nitrogen mustard or following administration of the antibiotic chloramphenicol.

agretope

The region of a protein antigen that combines with an MHC class II molecule during antigen presentation. This is then recognized by the T cell receptor MHC class II complex. Amino acid sequences differ in their reactivity with MHC class II molecules.

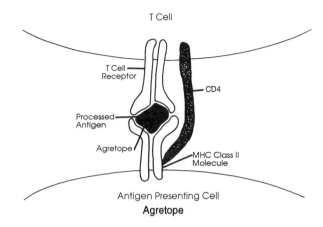

Agretope

AH$_{50}$

A technique to assay alternate complement pathway activity expressed as complement-mediated hemolysis. The AH$_{50}$ can be used to reveal C3, factor H, and factor I homozygous deficiencies.

AIDS (acquired immune deficiency syndrome)

A disease induced by the human immunodeficiency retrovirus designated HIV-1. Although first observed in homosexual men, the disease affects both males and females equally in central Africa and is beginning to affect an increasing number of heterosexuals with cases in both males and females in the Western countries including North America and Europe. Following exposure to the AIDS virus, the incubation period is variable and may extend to 11 years before clinical AIDS occurs in HIV-positive males in high-risk groups. It is transmitted by blood and body fluids, but is not transmitted through casual contact or through air, food, or other means. Besides homosexual and bisexual males, others at high risk include intravenous drug abusers, hemophiliacs, the offspring of HIV-infected mothers, and sexual partners of any HIV-infected individuals in the above groups.

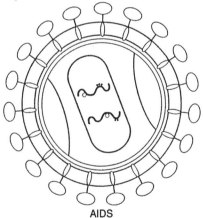

AIDS

AIDS belt (see figure, page 10)

A geographic area across central Africa that describes a region where multiple cases of heterosexual AIDS, related to sexual promiscuity, was reported. Nations in this belt include Burundi, Kenya, Central African Republic, Rwanda, the Congo, Malawi, Zambia, Tanzania, and Uganda.

AIDS dementia complex

Up to two thirds of AIDS patients may develop CNS signs and symptoms such as sustained cognitive behavior and motor impairment believed to be associated with infection of microglial cells

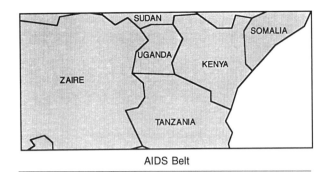

AIDS Belt

with the HIV-1 virus. This could be due to the structural similarity of gp120 of HIV-1 to neuroleukin. Patients have memory loss, are unable to concentrate, have poor coordination of gait, and have altered psychomotor function, among other symptoms. The subcortical white matter and deep gray matter degenerate; lateral and posterior spinal cord columns show white matter vacuolization; and the gp120 of HIV serves as a calcium channel inhibitor, causing toxic levels of calcium within neurons.

AIDS embryopathy
A condition in children born to HIV-infected mothers who are intravenous drug abusers. Affected children have craniofacial region defects that include microcephaly, hypertelorism, a cube-shaped head, a saddle nose, widened palpebral fissures with bluish sclera, triangular philtrum, and widely spreading lips.

AIDS encephalopathy
AIDS dementia. Refer to AIDS dementia complex.

AIDS enteropathy
A condition that may be seen in AIDS-related complex patients marked by diarrhea, especially nocturnal; wasting; possibly fever; and defective D-xylose absorption, leading to malnutrition. The small intestine may demonstrate atrophy of villi and hyperplasia of crypts. Both small and large intestines may reveal diminished plasma cells, elevated intraepithelial lymphocytes, and viral inclusions.

AIDS-related complex (ARC)
A preamble to AIDS that consists of a constellation of symptoms and signs which include a temperature of greater than 38°C, a greater than 10% loss of body weight, lymphadenopathy, diarrhea, night sweats of greater than a 3-month duration, and fatigue. Laboratory findings include CD4+ T lymphocyte levels of less than 0.4×10^9, a CD4 to CD8 T lymphocyte ratio of less than 1.0, leukopenia, anemia, and thrombocytopenia. There may be a decreased response to PHA, principally a T cell mitogen, and anergy, manifested as failure to respond to skin tests. In contrast, there may be a polyclonal gammopathy. A diagnosis of ARC requires at least two of the clinical manifestations and two of the laboratory findings listed above.

AIDS serology
Three to six weeks after infection with HIV-1 there are high levels of HIV p24 antigen in the plasma. One week to three months following infection there is an HIV-specific immune response resulting in the formation of antibodies against HIV envelope protein gp-120 and HIV core protein p24. HIV-specific cytotoxic T lymphocytes are also formed. The result of this adaptive immune

AIDS Serology

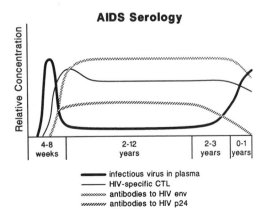

infectious virus in plasma
HIV-specific CTL
antibodies to HIV env
antibodies to HIV p24

response is a dramatic decline in viremia and a clinically asymptomatic phase lasting from two to twelve years. As CD4+ T cell numbers decrease the patient becomes clinically symptomatic. HIV-specific antibodies and cytotoxic T lymphocytes decline, and p24 antigen increases.

AIDS treatment
Although no drug is curative, zidovudine (azidothymidine-AZT), ddC (dideoxycytidine), and ddI (dideoxyinosine) are effective in delaying progression of the disease. Many experimental preparations are under investigation, such as DAB/486 IL-2, which is cytotoxic for high-affinity IL-2 receptors expressed on HIV-infected T lymphocytes.

AIDS vaccine
Several experimental AIDS vaccines are under investigation. HIV-2 inoculation into cynomologus monkeys apparently prevented them from developing simian AIDS following injection of the SIV virus.

AIDS virus
See human immunodeficiency virus (HIV).

AILA
Abbreviation for angioimmunoblastic lymphadenopathy.

Alanyl-tRNA synthetase antibodies
Rare antibodies most often found in patients with arthritis, myositis and interstitial lung disease as observed in the Jo-1 syndrome.

albumin agglutinating antibody (see facing page)
An antibody that does not agglutinate erythrocytes in physiological saline solution, but does cause their aggregation in 30% bovine serum albumin (BSA). Antibodies with this property have long been known as "incomplete antibodies" and are of interest in red blood cell typing.

albumin, serum
The principal protein of human blood serum that is soluble in water and in 50% saturated sodium sulfate. At pH 7.0, it is negatively charged and migrates toward the anode during electrophoresis. It is important in regulating osmotic pressure and binding of anions. Serum albumin (e.g., bovine serum albumin, BSA) is commonly used as an immunogen in experimental immunology.

aleutian mink disease
A parvovirus-induced disease of mink and ferrets that is associated with a polyclonal proliferation of B cells and a significant elevation of γ globulin in the blood. It bears similarity both clinically and pathologically to human systemic lupus erythematosus and polyarteritis nodosa. The viscera reveal lymphocytes and plasma cells, and renal glomeruli trap immune complexes, which are also present in small- and medium-sized arteries of the kidney, brain, and heart. The immune deposits lead to inflammation, resulting in glomerulonephritis, hepatitis, and arteritis, in addition to the hypergammaglobulinemia.

alexine (or alexin)
Historical synonym for complement, Hans Buchner (1850–1902) found that the bactericidal property of cell-free serum was destroyed by heat. He named the active principle alexine (which in Greek means to ward off or protect). Jules Bordet (1870–1961) studied this thermolabile and nonspecific principle in blood serum that induces the lysis of cells, e.g., bacterial cells, sensitized with specific antibody. Bordet and Gengou went on to discover the complement fixation reaction. (Bordet received the Nobel Prize in Medicine in 1919.)

ALG (see facing page)
Abbreviation for antilymphocyte globulin.

alkaline phosphatase (AP)
A Zn^{2+} metalloenzyme present in many living organisms. In man, it consists of three tissue-specific isozymes encoded by three separate genes. The separate forms of AP are detectable in intestine, placenta, kidney, bone, and liver. They are glycoproteins comprised of a single polypeptide chain containing 500 amino acid residues and are membrane bound. Calf intestine alkaline phosphatase is used in immunology in ELISA assays, immunoblotting, and molecular cloning. AP deletes 5′-phosphate from linear DNA terminals. It is also used in DNA recombinants to inhibit recircularization, which diminishes background for DNA. It is also used to treat genomic DNA fragments to prevent two or more fragments from ligating to each other. It may be employed to prepare 5′-^{32}P-N-label DNA.

allele
One of the alternative forms of a gene at a single locus on a chromosome that encodes the phenotypic features of a certain inherited characteristic.

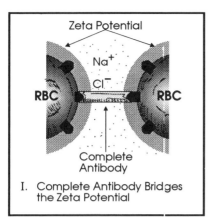

I. Complete Antibody Bridges
the Zeta Potential

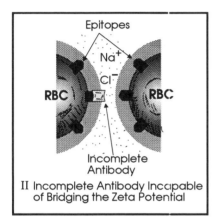

II Incomplete Antibody Incapable
of Bridging the Zeta Potential

III. Albumin Reduces the
Zeta Potential and
Thus Allows the
Cells to Agglutinate

Albumin Agglutinating Antibody

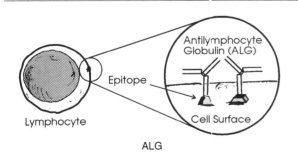

ALG

allelic dropout

In the amplification of a DNA segment by the polymerase chain reaction, one of the alleles may not be amplified, leading to the false impression that the allele is absent. The phenomenon takes place at 82 to 90°C in the thermocycler.

allelic exclusion

Only one of two genes for which the animal is heterozygous is expressed, whereas the remaining gene is not. Immunoglobulin genes manifest this phenomenon. Allelic exclusion accounts for a B cell's ability to express only one immunoglobulin or a T cell's capacity to express a T cell receptor of a single specificity. Investigations of allotypes in rabbits established that individual immunoglobulin molecules have identical heavy chains and light chains. Immunoglobulin-synthesizing cells produce only a single class of H chain and one type of light chain at a time. Thus, by allelic exclusion, a cell that is synthesizing antibody expresses just one of two alleles encoding an immunoglobulin chain at a particular locus.

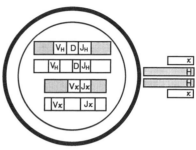

Allelic Exclusion

allergen

An antigen that induces an allergic or hypersensitivity response in contrast to a classic immune response produced by the recipient host in response to most immunogens. Allergens include such environmental substances as pollens, i.e., their globular proteins, from trees, grasses, and ragweed, as well as certain food substances, animal danders, and insect venom. Selected subjects are predisposed to synthesizing IgE antibodies in response to allergens and are said to be atopic. The crosslinking of IgE molecules anchored to the surfaces of mast cells or basophils through their Fc regions results in the release of histamine and other pharmacological mediators of immediate hypersensitivity from mast cells/basophils.

allergen immunotherapy

Desensitization treatment. Refer to desensitization.

allergic alveolitis

Refer to farmer's lung.

allergic contact dermatitis

Delayed-type hypersensitivity mediated by specifically sensitized T lymphocytes (type IV hypersensitivity) in response to the covalent linkage of low mol wt chemicals, often of less than 1000 M_r to proteins in the skin. The inflammation induced by these agents is manifested as erythema and swelling at approximately 12 h after contact and is maximal at 24 to 48 h. Blisters form that are filled with serum, neutrophils, and mononuclear cells. There is perivascular cuffing with lymphocytes, vesiculation, and necrosis of epidermal cells. Basophils, eosinophils, and fibrin deposition appear together with edema of the epidermis and dermis. Langerhans cells in the skin serve as antigen-processing cells where the allergen has penetrated. Sensitization lasts for many years and becomes generalized in the skin. Chemicals become conjugated to skin proteins and serve as haptens. Therefore, the hapten alone can elicit the hypersensitivity once sensitization is established. After blistering, there is crust formation and weeping of the lesion. It is intensely pruritic and painful. Metal dermatitis, such as that caused by nickel, occurs as a patch which corresponds to the area of contact with the metal or jewelry. Dyes in clothing may produce skin lesions at points of contact with the skin. The patch test is used to detect sensitivity to contact allergens. Rhus dermatitis represents a reaction to urushiols in poison oak or ivy which elicit vesicles and bullae on affected areas. Treatment is with systemic corticosteroids or the application of topical steroid cream to localized areas. Dinitrochlorobenzene (DNCB) and dinitrofluorobenzene (DNFB) are chemicals that have been used to induce allergic contact dermatitis in both experimental animals and in man.

allergic granulomatosis

A type of pulmonary necrotizing vasculitis with granulomas in the lung and pulmonary vessel walls. There may be infiltrates of eosinophils in the tissues and asthma. Also called Churg-Strauss syndrome.

allergic orchitis

The immunization of guinea pigs with autologous extracts of the testes incorporated into Freund's complete adjuvant leads to lymphocytic infiltrate in the testis and antisperm cytotoxic antibodies in the serum 2 to 8 weeks after inoculation. Human males who have been vasectomized may also develop allergic orchitis.

allergic response

A response to antigen (allergen) that leads to a state of increased reactivity or hypersensitivity rather than a protective immune response.

allergoids

Allergens that have been chemically altered to favor the induction of IgG rather than IgE antibodies to diminish allergic manifestations in the hypersensitive individual. These formaldehyde-modified allergens are analogous to toxoids prepared from bacterial exotoxins. Some of the physical and chemical characteristics of allergens are similar to those of other antigens. However, the mol wt of allergens is lower.

allergy

A term coined by Clemens von Pirquet in 1906 to describe the altered reactivity of the animal body to antigen. Presently, the term allergy refers to altered immune reactivity to a spectrum of environmental antigens which includes pollen, insect venom, and food. Allergy is also referred to as hypersensitivity and usually describes type I immediate hypersensitivity of the atopic/anaphylactic type.

allergy, infection

Hypersensitivity, especially of the delayed T cell type, that develops in subjects infected with certain microorganisms such as *Mycobacterium tuberculosis* or certain pathogenic fungi.

alloantibody

An antibody that interacts with an alloantigen, such as the antibodies generated in the recipient of an organ allotransplant (such as kidney or heart) which then may react with the homologous alloantigen of the allograft.

alloantigen

An antigen present in some members or strains of a species, but not in others. Alloantigens include blood group substances on erythrocytes and histocompatibility antigens present in grafted tissues that stimulate an alloimmune response in the recipient not possessing them, as well as various proteins and enzymes. Two animals of a given species are said to be allogeneic with respect to each other.

alloantiserum

An antiserum generated in one member or strain of a species not possessing the histocompatibility antigen with which they have been challenged that is derived from another member or strain of the same species.

allogeneic (or allogenic)

An adjective that describes genetic variations or differences among members or strains of the same species. The term refers to organ or tissue grafts between genetically dissimilar humans or unrelated members of other species.

allogeneic disease

Pathologic consequences of immune reactivity of bone marrow allotransplants in immunosuppressed recipient patients as a result of graft-vs.-host reactivity in genetically dissimilar members of the same species.

allogeneic effect

The synthesis of antibody by B cells against a hapten in the absence of carrier-specific T cells, provided allogeneic T lymphocytes are present. Interaction of allogeneic T cells with the MHC class II molecules of B cells causes the activated T lymphocytes to produce factors that facilitate B cell differentiation into plasma cells without the requirement for helper T lymphocytes. There is allogeneic activation of T cells in the graft-vs.-host reaction.

Allogeneic Effect Factor

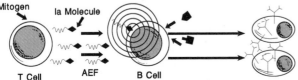

allogeneic inhibition

The better growth of homozygous tumors when they are transplanted to homozygous syngeneic hosts of the strain of origin than when they are transplanted to F_1 hybrids between the syngeneic (tumor) strain and an allogeneic strain. This is manifested as a higher frequency of tumor and shorter latency period in syngeneic hosts. The better growth of tumor in syngeneic than in heterozygous F_1 hybrid hosts was initially termed syngeneic preference. When it became apparent that selective pressure against the cells in a mismatching environment produced the growth difference, the phenomenon was termed allogeneic inhibition.

allograft

An organ, tissue, or cell transplant from one individual or strain to a genetically different individual or strain within the same species. Also called homograft.

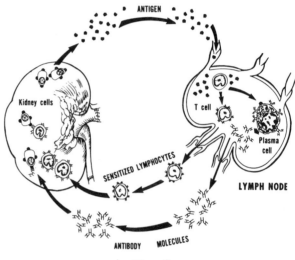

Allograft

allogroup

Several allotypes representing various immunoglobulin classes and subclasses inherited as a unit. Alleles that are closely linked encode the immunoglobulin heavy chains in an allogroup. An allogroup is a form of a haplotype.

alloimmunization

An immune response provoked in one member or strain of a species with an alloantigen derived from a different member or strain of the same species. Examples include the immune response in man following transplantation of a solid organ graft such as a kidney or heart from one individual to another. Alloimmunization with red blood cell antigens in humans may lead to pathologic sequelae such as hemolytic disease of the newborn (erythroblastosis fetalis) in a third Rh(D)+ baby born to an Rh(D)− mother.

allophenic mouse

A tetraparental, chimeric mouse whose genetic makeup is derived from four separate parents. It is produced by the association of two early eight-cell embryos that differ genetically. A single blastocyst forms, is placed in a pseudopregnant female uterus, and is permitted to develop to term. Tetraparental mice are widely used in immunological research.

allotope

An allotype's antigenic determinant. Also called allotypic determinant.

allotype

A distinct antigenic form of a serum protein that results from allelic variations present on the immunoglobulin heavy chain constant region. Allotypes were originally defined by antisera which differentiated allelic variants of Ig subclasses. The allotype is due to the existence of different alleles at the genetic locus which determines the expression of a given determinant. Immunoglobulin allotypes have been extensively investigated in inbred rabbits. Currently, allotypes are usually defined by DNA techniques. To be designated as an official allotype, the polymorphism must be present in a reasonable subset of the population (approximately 1%) and follow

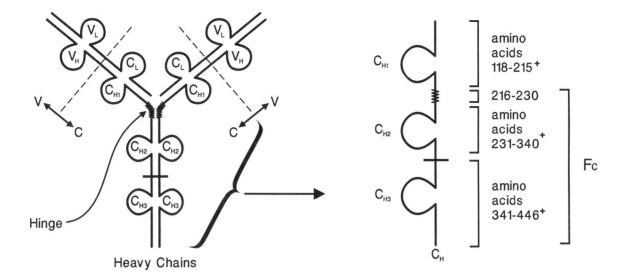

Allotype Examples:

IgG3m^b | IgG3m^g

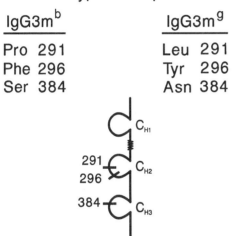

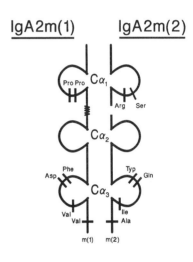

Allotype

Mendelian genetics. Allotype examples include the IgG3 Caucasian allotypes G3m^b and G3m^g. These two alleles vary at positions 291, 296, and 384. Another example is the allotype at the IgA2 locus. The IgA2m(1) allele is European/Near Eastern, while IgA2m(2) is African/East Asian. The allotypic differences are in $C\alpha_1$ and $C\alpha_3$, and the IgA2m(2) allele has a shorter hinge than the IgA2m(1) allele.

allotype suppression
The failure to produce antibodies of a given allotype by offspring of mothers who have been immunized against the paternal Ig allotype before mating. It may also occur after administration of antibodies specific for the paternal allotype to heterozygous newborn rabbits. Animals conditioned in this way remain deficient for the suppressed allotype for months and possibly years thereafter. The animal switches to a compensatory increase in the production of the nonsuppressed allotype. In homozygotes, suppression involves heavy chain switching from one Ig class to another.

allotypic determinant
Epitope on serum immunoglobulin molecules present in selected members of a species. Allotypic determinants are present in addition to other markers characteristic of the molecule as a whole for the respective species. The allotypic determinants or allotopes are characteristic of a given class and subclass of immunoglobulin and have been demonstrated in several species. The inheritance of genes controlling these determinants is strictly Mendelian and is not sex

linked. Both H and κ L chains contain such determinants and the encoding genes are not linked and are codominant in an individual. This means that markers present both in the father and mother are expressed phenotypically in a heterozygote. However, an individual immunoglobulin-producing cell expresses only one of a pair of allelic genes, since a single cell can utilize only one parental chromosome. In an individual, some cells use the information encoded on the chromosome derived paternally and other cells use that derived from the maternal contribution. So the individual is heterozygous, but a single cell only secretes products of one allele because of allelic exclusion. The allotypic markers of human IgG are designated Gm determinants. The Km(Inv) markers are characteristic for the C region of κ light chains.

allotypic marker
Refer to allotypic determinant or allotope.

allotypic specificities
Genetically different antibody classes and subclasses produced within individuals of the same species. They are detected as changes in the amino acid residues present in specific positions in various polypeptide chains. The allotypic specificities are also called genetic markers. Gm and Km(Inv) markers are examples of allotypes of human IgG H chains and κ light chains, respectively.

allotypy
A term that describes the various allelic types or allotypes of immunoglobulin molecules.

alopecia areata

Hair loss in subjects who demonstrate autoantibodies against hair follicle capillaries.

ALS (antilymphocyte serum) or ALG (antilymphocyte globulin)

Refer to antilymphocyte serum.

alternative complement pathway

A nonantibody-dependent pathway for complement activation in which the early components C1, C2, and C4 are not required. It involves the protein properdin factor D, properdin factor B, and C3b leading to C3 activation and continu-

Alopecia Areata

ing to C9 in a manner identical to that which takes place in the activation of complement by the classical pathway. Substances such as endotoxin, human IgA, microbial polysaccharides, and other agents may activate complement by the alternative pathway. The C3bB complex forms as C3b combines with factor B. Factor D splits factor B in the complex to yield the Bb active fragment that remains linked to C3b and Ba, which is inactive and is split off. C3bBb, the alternative pathway C3 convertase, splits C3 into C3b and C3a, thereby producing more C3bBb, which represents a positive feedback loop. Factor I, when accompanied by factor H, splits C3b's heavy chain to yield C3bi, which is unable to anchor Bb, thereby inhibiting the alternative pathway. Properdin and C3 nephritic factor stabilize C3bBb. C3 convertase stabilized by properdin activates complement's late components, resulting in opsonization, chemotaxis of leukocytes, enhanced permeability, and cytolysis. Properdin, IgA, IgG, lipopolysaccharide and snake venom can initiate the alternate pathway. Trypsin-related enzymes can activate both pathways.

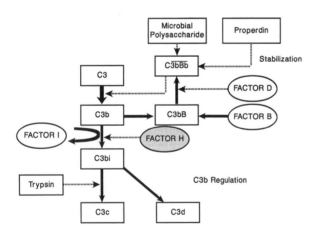

C3b Regulation

Alternative Complement Pathway

alternative pathway C3 convertase

Alternate unstable C3bB complex that splits C3 into C3a and C3b. Factor P, also known as properdin, stabilizes C3bB to yield C3bBbP. C3 nephritic factor can also stabilize C3bB.

alum granuloma

A tissue reaction in the form of a granuloma produced at the local site of intramuscular or subcutaenous inoculation of a protein antigen precipitated from solution by an aluminum salt acting as adjuvant. Slow release of antigen from the granuloma has been considered to facilitate enhanced antibody synthesis to the antigen.

Alternative Pathway C3 Convertase

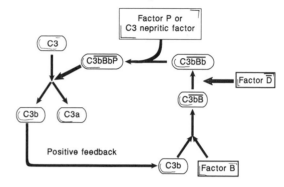

$\overline{Cn}$ - activated complement component
➡ - enzymatic action

alum-precipitated antigen

A soluble protein antigen such as a toxoid adsorbed to aluminum salts during precipitation from solution. The aluminum salt acts as an adjuvant that facilitates an immune response to such antigens as diphtheria and tetanus toxoids. Soluble protein antigen is combined with 1% potassium aluminum sulfate, and sodium hydroxide is added until floccules are produced.

alum-precipitated toxoid

Refer to alum-precipitated antigen.

aluminum adjuvant

Aluminum-containing substances that have a powerful capacity to adsorb and precipitate protein antigens from solution. The use of these preparations as immunogen causes depot formation in the tissues at the site of inoculation from which the antigen is slowly released, thereby facilitating greater antibody production than if the antigen is dissipated and rapidly lost from the body. Substances used extensively in the past for this purpose include aluminum hydroxide gel, aluminum sulfate, and ammonium alum, as well as potassium alum.

aluminum hydroxide gel

Aluminum hydroxide [Al(OH)$_3$] was widely used in the past as an immunologic adjuvant by reacting antigen with 2% hydrated AL(OH)$_3$ to adsorb and precipitate the protein antigen from solution. See also aluminum adjuvant.

alums

Aluminum salts employed to adsorb and precipitate protein antigens from solution, followed by the use of the precipitated antigen as an immunogen which forms a depot in animal tissues. See also aluminum adjuvant.

alveolar macrophage

A macrophage in the lung alveoli that may remove inhaled particulate matter.

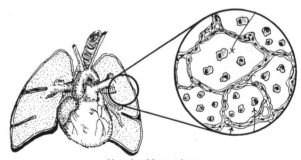

Alveolar Macrophage

ALZ-50

A monoclonal antibody that serves as an early indicator of Alzheimer's disease by reacting with Alzheimer's brain tissue, specifically protein A-68.

Am allotypic marker

An allotypic antigenic determinant located on the a heavy chain of the IgA molecule in man. Of the two IgA subclasses, the IgA1 subclass has no known allotypic determinant. The IgA2 subclass has two allotypic determinants designated A2m(1) and A2m(2) based on differences in α-2 heavy chain primary structures. Allelic genes at the A2m locus encode these allotypes which are expressed on the α-2 heavy chain constant regions.

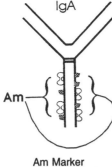

Am Marker
(Allotypic Marker)

amboceptor (historical)

Paul Ehrlich (*circa* 1900) considered anti-sheep red blood cell antibodies known as amboceptors to have one receptor for sheep erythrocytes and another receptor for complement. The term gained worldwide acceptance with the popularity of complement fixation tests for syphilis, such as the Wasserman reaction. The term is still used by some when discussing complement fixation.

aminoethylcarbazole (AEC)

3-Amino-9-ethyl carbazole is used in the ABC immunoperoxidase technique to produce a visible reaction product detectable by light microscopy when combined with hydrogen peroxide. AEC is oxydized to produce a reddish-brown pigment that is not water soluble. Peroxidase catalyzes the reaction. Because peroxidase is localized only at sites where the PAP is bound via linking antibody and primary antibody to antigen molecules, the antigen is identified by the reddish-brown pigment.

aminophylline

Refer to theophylline.

ammonium sulfate precipitation

The ammonium sulfate method is a means of measuring the primary antigen-binding capacity of antisera and detects both precipitating and nonprecipitating antibodies. If offers an advantage over equilibrium dialysis in that large, nondializable protein antigens may be used. This assay is based on the principle that certain proteins are soluble in 50% saturated ammonium sulfate, whereas antigen-antibody complexes are not. Thus, complexes may be separated from unbound antigen. Spontaneous precipitation will occur if a precipitating-type antibody is used, until a point of antigen excess is reached where complex aggregation no longer occurs and soluble complexes are formed. Upon the addition of an equal volume of saturated ammonium sulfate solution (SAS), these complexes become insoluble, leaving radiolabeled antigen in solution. SAS fractionation does not significantly alter the stoichiometry of the antibody-antigen reaction and inhibits the release or exchange of bound antigen. Thus, radioactivity of this "induced" precipitate is a measure of the antigen-binding capacity of the antisera as opposed to a measure of the amount of antigen or antibody spontaneously precipitated.

amyloid

Amyloid is an extracellular, homogenous eosinophilic material deposited in various tissues in disease states designated as primary and secondary amyloidosis. It is composed chiefly of protein and shows a green birefringence when stained with Congo red and observed by polarizing light microscopy. By electron microscopy, the fibrillar appearance is characteristic. By X-ray crystallography, it shows a β-pleated sheet structure arranged in an antiparallel fashion. The amino termini of the individual chains face opposite directions, and the chains are bound by hydroxyl bonds.

Amyloid consists of two principal and several minor biochemical varieties. Pathogenetic mechanisms for its deposition differ, although the deposited protein appears similar from one form to another. Amyloid consists of nonbranching fibrils 7.5 to 10 nm wide and are of indefinite length. X-ray crystallography reveals a β-pleated sheet configuration which gives the protein its optical and staining properties. It also has a P component that is nonfibrillary, is pentagonal in structure, and constitutes a minor component of amyloid. Chemically, amyloid falls into two principal classes, i.e., AL, consisting of amyloid light chains, and AA (amyloid associated), comprising a nonimmunoglobulin protein called AA. These molecules are antigenically different and have dissimilar deposition patterns based on the clinical situation. AL amyloid is comprised of whole immunoglobulin light chains, their N-terminal fragments, or a combination of the two. λ Light chains rather than κ are the ones usually found in AL. Proliferating immunoglobulin-producing B cells, as in B cell dyscrasias, produce AL amyloid protein. AA amyloid fibroprotein is not an immunoglobulin and has

a molecular weight of 8.5 kD. Serum amyloid-associated protein (SAA) is the serum precursor of AA amyloid. It constitutes the protein constituent of a high-density lipoprotein and acts as an acute-phase reactant. Thus, its level rises remarkably within hours of an acute inflammatory response. AA protein is the principal type of amyloid deposited in the tissues during chronic inflammatory diseases. Several other distinct amyloid proteins exist also.

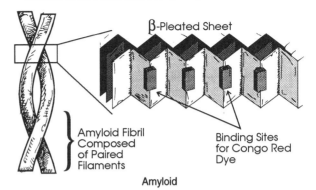

β-Pleated Sheet

Amyloid Fibril Composed of Paired Filaments

Binding Sites for Congo Red Dye

Amyloid

amyloid β fibrillosis

All amyloids have a β-pleated sheet structure, which accounts for the ability of Congo red to stain them and the ability of proteolytic enzymes to digest them. Refer also to amyloid and amyloidosis.

amyloid P component

The P component has a mol wt of 180 kD. It migrates in electrophoresis with the α globulin fraction, and by electron microscopy it has a pentagonal shape, suggesting that it consists of subunits linked by hydrogen bonds. It is a minor component of all amyloid deposits and is nonfibrillar. It is a normal α_1 glycoprotein and has close structural homology with the C-reactive protein. It has an affinity for amyloid fibrils and accounts for their PAS positive staining quality.

amyloidosis

A constellation of diseases characterized by the extracellular deposition of fibrillar material that has a homogeneous and eosinophilic appearance in conventional staining methods. It may compromise the function of vital organs. Diseases with which it is associated may be inflammatory, hereditary, or neoplastic. All types of amyloid link to Congo red and manifest an apple-green birefrigence when viewed by polarizing light microscopy after first staining with Congo red. By electron microscopy, amyloid has a major fibrillar component and a minor rod-like structure which is shaped like a pentagon with a hollow core when observed on end, i.e., the P component. All forms of amyloid share the P component in common. It is found as a soluble serum protein in the circulation (SAP). Amyloid has a β-pleated sheet structure; it is insoluble in physiologic saline, but is soluble in distilled water.

The classification of amyloidosis depends upon the clinical presentation, anatomic distribution, and chemical content of the amyloid. In the U.S., AL amyloid is the most common type of amyloidosis which occurs in association with multiple myeloma and Waldenström's macroglobulinemia. These patients have free light chain production in association with the development of Bence-Jones proteins in myeloma. The light chain quality and degradation mechanism are critical in determining whether or not Bence-Jones proteins will be deposited as amyloid. Chronic inflammation leads to increased levels of SAA, which are produced by the liver following IL-6 and IL-1 stimulation. Normally, SSA is degraded by the enzymes of monocytes. Thus, individuals with a defect in the degradation process could generate insoluble AA molecules. Likewise, there could be a defect in the degradation of immunoglobulin light chains in subjects who develop AL amyloidosis.

Amyloidosis secondary to chronic inflammation is severe with kidney, liver, spleen, lymph node, adrenal, and thyroid involvement. These secondary amyloidosis deposits consist of amyloid A protein (AA), which makes up 85 to 90% of the deposits, and serum amyloid P component, which accounts for the remainder of the deposit. The AL type of amyloidosis more often involves the heart, gastrointestinal tract, respiratory tract, peripheral nerves, and tongue. Amyloidosis may also be heredofamilial or associated with aging.

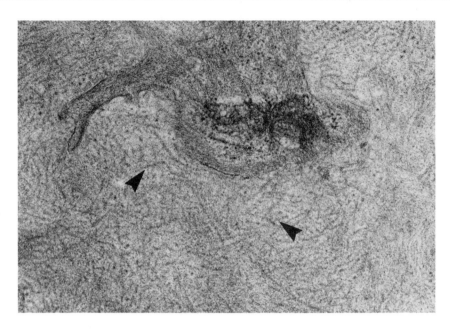

amyloid fibrils

Amyloidosis

ANA
Abbreviation for antinuclear antibodies.

ANAE (α-naphthyl acetate esterase)
Refer to nonspecific esterase.

AnaINH
Anaphylatoxin inhibitor.

anamnesis
Immunologic memory. Refers to the elevated immune response following secondary or tertiary administration of immunogen to a recipient previously primed or sensitized to the immunogen, i.e., the secondary response.

anamnestic response
Accentuated immune response that occurs following exposure of immunocompetent cells to an immunogen to which they have been exposed before. Commonly called the secondary or anamnestic response. It occurs rapidly, i.e., within hours, following secondary immunogen inoculation and does not have the lag period observed with primary immunization. Immunologic memory is involved in the production of this response which generally consists of IgG antibodies of high titer and high affinity. There may also be heightened T cell (cell-mediated) immune reactivity. Also called memory or booster response or secondary immune response.

anaphylactoid reaction
A response resembling anaphylaxis, except that it is not attributable to an allergic reaction mediated by IgE antibody. It is due to the nonimmunologic degranulation of mast cells such as that caused by drugs or chemical compounds like aspirin, radiocontrast media, chymopapain, bee or snake venom, and gum acacia which cause release of the pharmacological mediators of immediate hypersensitivity including histamine and other vasoactive molecules.

anaphylatoxin inactivator
A 300-kD α globulin carboxy peptidase in serum that destroys the anaphylatoxin activity of C5a, C3a, and C4a by cleaving their carboxy terminal arginine residues.

anaphylatoxin inhibitor (AnaINH)
A 300-kD α globulin carboxy peptidase that cleaves anaphylatoxin's carboxy terminal arginine. The enzyme acts on all three forms including C3a, C4a, and C5a, inactivating rather than inhibiting them.

anaphylatoxins
Substances generated by the activation of complement that lead to increased vascular permeability as a consequence of the degranulation of mast cells with the release of pharmacologically active mediators of immediate hypersensitivity. These biologically active peptides of low mol wt are derived from C3, C4, and C5. They are generated in serum during fixation of complement by Ag-Ab

Anaphylatoxins

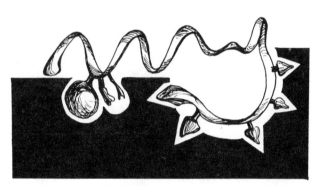

C3a/C3a57-77 --
Receptor interactions

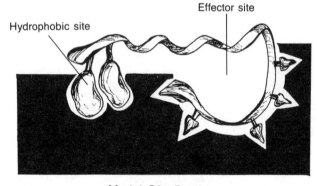

Model C3a Peptide --
Receptor interactions

complexes, immunoglobulin aggregates, etc. Small blood vessels, mast cells, smooth muscle, and leukocytes in peripheral blood are targets of their action. Much is known about their primary structures. These complement fragments are designated C3a, C4a, and

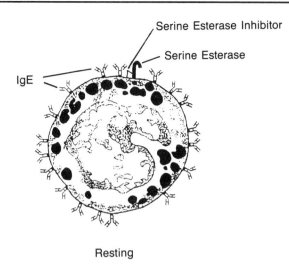

Resting

Antigen Bound by IgE

Release of Pharmacological
Mediators of Type I
Hypersensitivity

Anaphylaxis

C5a. They cause smooth muscle contraction, mast cell degranulation with histamine release, increased vascular permeability, and the triple response in skin. They induce anaphylactic-like symptoms upon parenteral inoculation.

anaphylaxis

A shock reaction that occurs within seconds following the injection of an antigen or drug or after a bee sting to which the susceptible subject has IgE-specific antibodies. There is embarrassed respiration due to laryngeal and bronchial constriction and shock associated with decreased blood pressure. Signs and symptoms differ among species based on the primary target organs or tissues. Whereas IgE is the anaphylactic antibody in man, IgG1 may mediate anaphylaxis in selected other species. Type I hypersensitivity occurs following the crosslinking of IgE antibodies by a specific antigen or allergen on the surfaces of basophils in the blood or mast cells in the tissues. This causes the release of the pharmacological mediators of immediate hypersensitivity with a reaction occurring within seconds of contact with antigen or allergen. Eosinophils, chemotactic factor, heparin, histamine, and serotonin, together with selected other substances, are released during the primary response. Acute-phase reactants are formed and released in the secondary response. Secondary mediators include slow-reacting substance of anaphylaxis (SRS-A), bradykinin, and platelet-activating factor. In addition to systemic anaphylaxis described above, local anaphylaxis may occur in the skin, gut, or nasal mucosa following contact with the antigen. The skin reaction, called urticaria, consists of a raised wheal surrounded by an area of erythema. Cytotoxic anaphylaxis follows the interaction of antibodies with cell surface antigens. See also aggregate anaphylaxis.

anatoxin

Antibody specific for exotoxins produced by certain microorganisms such as the causative agents of diphtheria and tetanus. Prior to the antibiotic era, antitoxins were the treatment of choice for diseases produced by the soluble toxic products of microorganisms, such as those from *Corynebacterium diphtheriae* and *Clostridium tetani.*

anavenom

A toxoid consisting of formalin-treated snake venom which destroys the toxicity, but preserves immunogenicity of the preparation.

anergy

Diminished or absent delayed-type hypersensitivity, i.e., type IV hypersensitivity, as revealed by lack of responsiveness to commonly used skin test antigens including PPD, histoplasmin, candidin, etc. Decreased skin test reactivity may be associated with uncontrolled infection, tumor, Hodgkin's disease, sarcoidosis, etc. There is decreased capacity of T lymphocytes to secrete lymphokines when their T cell receptors interact with a specific antigen.

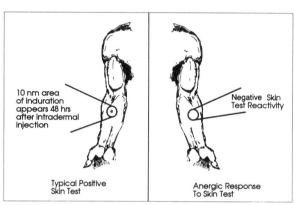

Anergy

angioedema

Significant localized swelling of tissues as a consequence of complement activation which takes place when C1 esterase inhibitor is lacking.

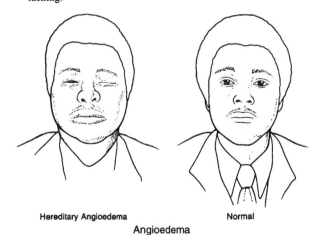

Hereditary Angioedema   Normal

Angioedema

angiogenesis factor

A macrophage-derived protein that facilitates neovascularization through stimulation of vascular endothelial cell growth. Among the five angiogenesis factors known, basic fibroblast growth factor may facilitate neovascularization in type IV delayed-hypersensitivity responses.

angioimmunoblastic lymphadenopathy (AILA)

Proliferation of hyperimmune B lymphocytes. Immunoblasts, both large and small, form a pleomorphic infiltrate together with plasma cells in lymph nodes revealing architectural effacement. There is arborization of newly formed vessels and proliferating vessels with hyperplasia of endothelial cells. In the interstitium, amorphous eosinophilic PAS positive deposits, possibly representing debris from cells, are found. Fever, night sweats, hepatosplenomegaly, generalized lymphadenopathy, weight loss, hemolytic anemia, polyclonal gammopathy, and skin rashes may characterize the disease in middle-aged to older subjects. Patients live approximately 15 months, with some developing monoclonal gammopathy or immunoblastic lymphomas. AILA must be differentiated from AIDS, Hodgkin's disease, immunoblastic lymphoma, histocytosis X, and a variety of other conditions affecting the lymphoid tissues.

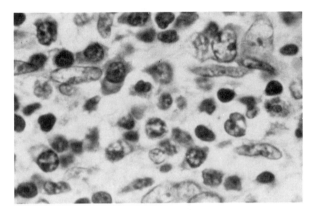

Angioimmunoblastic Lymphadenopathy

angry macrophage

A term sometimes used to refer to activated macrophages.

ankylosing spondylitis

A chronic inflammatory disease affecting the spine, sacroiliac joints, and large peripheral joints. There is a strong male predominance with onset in early adult life. The erythrocyte sedimentation rate is elevated, but subjects are negative for rheumatoid factor and antinuclear antibodies. Pathologically, there is chronic proliferative synovitis which resembles that seen in rheumatoid arthritis. The sacroiliac joints and interspinous and capsular ligaments ossify when the disease advances. There is a major genetic predisposition, as revealed by increased incidence in selected families. Ninety percent of ankylosing spondylitis patients are positive for HLA-B27, compared to 8% among Caucasians in the U.S. The HLA-B27 genes may be linked to genes that govern pathogenic autoimmunity. There may be increased susceptibility to infectious agents or molecular mimicry between HLA-B27 and an infectious agent such as *Klebsiella pneumoniae,* leading to the synthesis of a cross-reacting antibody. Treatment is aimed at diminishing inflammation and pain and providing physical therapy.

ankylosing spondylitis

ANNA

Abbreviation for antineutrophil nuclear antibodies.

anti-Clq antibody

Present in the majority of patients with hypocomplementemic urticarial vasculitis syndrome (HUVS) and in 30 to 60% of systemic lupus erythematosus patients. Clq is strikingly decreased in the blood sera of HUVS patients, even though their Clr and Cls levels are within normal limits and C5-C9 are slightly activated.

anti-D

Antibody against the Rh blood group D antigen. This antibody is stimulated in RhD– mothers by fetal RhD+ red blood cells that enter her circulation at parturition. Anti-D antibodies become a problem usually with the third pregnancy, resulting from the booster immune response against the D antigen to which the mother was previously exposed. IgG antibodies pass across the placenta, leading to hemolytic disease of the newborn (erythroblastosis fetalis). Anti-D antibody (Rhogam®) administered up to 72 h following parturition may combine with the RhD+ red blood cells in the mother's circulation, thereby facilitating their removal by the reticuloendothelial system. This prevents maternal immunization against the RhD antigen.

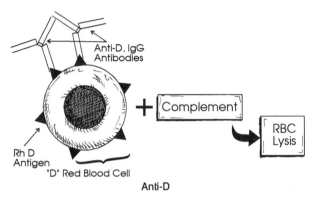

Anti-D

anti-DEX antibodies

Murine α1-3 dextran specific antibodies

anti-I

Antibodies against the I blood group antigen, which is present on the majority of adult red blood cells in man. The Ii antigens are present in the subterminal portions of the oligosaccharides which are ultimately converted to H and A or B antigens. I and i configurations are present on membrane-associated glycoproteins and glycosphingolipids. The heterogeneity observed with different anti-I antisera may reflect the recognition of different parts of the branched oligosaccharide chain. Fetal erythrocytes contain abundant i antigen, but few branched oligosaccharides and little I antigen. The I antigen develops during the first 2 years of life with simultaneous loss of i. Anti-I is a common autoantibody that is frequently present as a cold-reacting agglutinin. Anti-I is of pathologic significance in many cases of CHD when it acts as a complement-binding monoclonal antibody. Autoanti-i is of less significance in cold hemagglutinin disease than is anti-I. Thus, anti-I acting as a cold agglutinin may be detected as an autoantibody in a number of cases of cold antibody type hemolytic anemia and in patients with *Mycoplasma pneumoniae* infection.

anti-phospholipid antibodies

See lupus anticoagulant.

anti-Purkinje cell antibody

An antibody that has been detected in the circulation of subacute cerebellar degeneration patients and in those with ovarian neoplasms and other gynecologic malignancies.

antianaphylaxis

Inhibition of anaphylaxis through desensitization. This is accomplished by repeated injections of the sensitizing agent too minute to produce an anaphylactic reaction.

antiantibody

In addition to their antibody function, immunoglobulin molecules serve as excellent protein immunogens when inoculated into another species or they may become autoantigenic even in their own host. The Gm antigenic determinants in the Fc region of an IgG molecule may elicit autoantibodies, principally of the IgM class, known as rheumatoid factor in individuals with rheumatoid arthritis. Antiidiotypic antibodies, directed against the antigen-binding N-terminal variable regions of antibody molecules, represent another type of antiantibody. Rabbit anti-human IgG (the Coombs' test reagent) is an antiantibody used extensively in clinical immunology to reveal autoantibodies on erythrocytes.

antiagglutinin

A specific antibody that interferes with the action of an agglutinin.

antibodies

Antibodies are glycoprotein substances produced by B lymphoid cells in response to stimulation with an immunogen. They possess the ability to react *in vitro* and *in vivo* specifically and selectively with the antigenic determinants or epitopes eliciting their production or with an antigenic determinant closely related to the homologous antigen. Antibody molecules are immunoglobulins found

Antibody

in the blood and body fluids. Thus, all antibodies are immunoglobulins formed in response to immunogens. Antibodies may be produced by hybridoma technology in which antibody secreting cells are fused by polyethylene glycol (PEG) treatment with a mutant myeloma cell line. Monoclonal antibodies are widely used in research and diagnostic medicine and have potential in therapy. Antibodies in the blood serum of any given animal species may be grouped according to their physicochemical properties and antigenic characteristics. Immunoglobulins are not restricted to the plasma, but may be found in other body fluids or tissues, such as urine, spinal fluid, lymph nodes, spleen, etc. Immunoglobulins do not include the components of the complement system. Immunoglobulins (antibodies) constitute approximately 1 to 2% of the total serum proteins in health. γ Globulins comprise 11.2 to 20.1% of the total serum content in man. Antibodies are in the γ globulin fraction of serum. Electrophoretically they are the slowest migrating fraction.

antibody absorption test

A serological assay based upon the ability of a cross-reactive antigen to diminish a serum sample's titer of antibodies against its homologous antigen, i.e., the antigen that stimulated its production. Cross-reactive antibodies, as well as cross-reactive antigens, may be detected in this way.

antibody affinity

The force of binding of one antibody molecule's paratope with its homologous epitope on the antigen molecule. It is a consequence of positive and negative portions affecting these molecular interactions.

Antibody Affinity

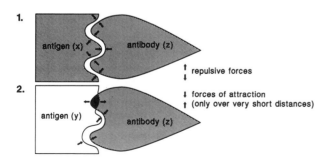

antibody-binding site

The antigen-binding site of an antibody molecule, known as a paratope, that is comprised of heavy chain and light chain variable

regions. The paratope represents the site of attachment of an epitope to the antibody molecule. The complementarity-determining hypervariable regions play a significant role in dictating the combining site structure together with the participation of framework region residues. The T cell receptor also has an antigen-binding site in the variable regions of its α and β (or γ and δ) chains.

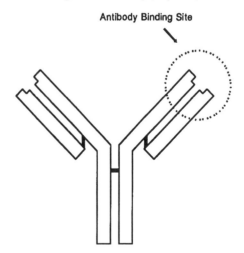

antibody deficiency syndrome

A few patients have been observed in which normal immunoglobulin levels are present, but the ability to mount an immune response to immunogenic challenge is impaired. This condition is associated with several separate disease states and might more properly be considered a syndrome. Some may present clinically as severe combined immunodeficiency with diminished cell-mediated immunity, lymphopenia, and infection by microorganisms of low pathogenicity. There are normal or even elevated numbers of plasma cells, and there may be no demonstrable T cell deficiency, both of which are in contrast to the usual clinical picture of severe combined immunodeficiency. These individuals may develop autoimmune reactions and show reduced numbers of lymphoid cells with surface immunoglobulin in the circulating blood. One possible explanation for normal immunoglobulin levels and an inadequate humoral immune response to antigenic challenge could be accounted for by a defect in clonal diversity, resulting in an antibody response to only a limited number of antigens. Some investigators have associated the defect with T cell clonal diversity. This combined B and T cell system disorder resembles, but is less pronounced than severe combined immunodeficiency (SCID). Some patients with this defect develop paraproteins with subsequent agammaglobulinemia and clinical manifestations closely resembling severe combined immunodeficiency.

antibody-dependent cell-mediated cyotoxicity (ADCC)

A reaction in which T lymphocytes; NK cells, including large granular lymphocytes; neutrophils; and macrophages may lyse tumor cells, infectious agents, and allogeneic cells by combining through their Fc receptors with the Fc region of IgG antibodies bound through their Fab regions to target cell surface antigens. Following linkage of Fc receptors with Fc regions, destruction of the target is accomplished through released cytokines. It represents an example of participation between antibody molecules and immune system cells to produce an effector function.

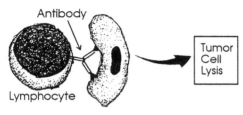

Tumor Cell
Antibody-Dependent Cell-Mediated Cytotoxicity

antibody excess immune complexes (ABIC)

May also result from the alteration of the immunoglobulin molecules, such as that seen in rheumatoid arthritis, or may be produced locally, such as in type B hepatitis. ABIC have a short intravascular life and, in contrast to antigen-excess immune complexes, adhere to platelets and cause platelet aggregation. The aggregation of platelets is not due to crosslinking of ABIC to platelets, but to changes in the adhesive properties of the latter. ABIC also bind to neutrophils and induce the release of lysosomal enzymes without prior phagocytic activity.

antibody feedback

The negative feedback system whereby antigen-specific antibodies downregulate further immune responses to that antigen. Several mechanisms may be responsible for this including:
1. Removal of the initiating stimulus by the antibody.
2. Binding of antigen/IgG antibody immune complexes to the Fcγ receptor of B cells.
3. Inhibition of T cell responses by antigen/antibody complexes.

The use of Rh immune globulin to prevent erythroblastosis fetalis in the infants of Rh negative mothers is an example of antibody feedback.

Antibody Feedback

Ag-Ab complex
(binds to
antigen-specific
B cells)

Anti-Ig antibody
(binds to all
B cells)

Fc receptor

Fc receptor

Membrane Ig

Membrane Ig

antibody fragment

A product of enzymatic treatment of an antibody immunoglobulin molecule with an enzyme such as papain or pepsin. For example, papain treatment leads to the production of two Fab and one Fc fragments, whereas the use of pepsin yields the $F(ab')_2$ fragment. Refer to the individual fragments for further information.

antibody half-life

The mean survival time of any particular antibody molecule after its formation. It refers to the time required to rid the animal body of one half of a known amount of antibody. Thus, antibody half-life differs according to the immunoglobulin class to which the antibody belongs.

antibody-mediated suppression

The feedback inhibition that antibody molecules exert on their own further synthesis.

antibody screening

Candidates for organ transplants, especially renal allografts, are monitored with relative frequency for changes in their percent reactive antibody (PRA) levels. Obviously, those with relatively high PRA values are considered to be less favorable candidates for renal allotransplants than are those in whom the PRA values are low. PRA determinations may vary according to the composition of the cell panel. If the size of the panel is inadequate, it may affect the relative frequency of common histocompatibility antigens found in the population.

antibody titer

The amount or level of circulating antibody in a patient with an infectious disease. For example, the reciprocal of the highest dilution of serum (containing antibodies) that reacts with antigen, e.g., agglutination, is the titer. Two separate titer determinations are required to reflect an individual's exposure to an infectious agent.

antibody units

Refer to titer.

anticardiolipin antibody syndrome

Circulating lupus anticoagulant syndrome (CLAS). A clinical situation in which circulating anticardiolipin antibodies may occur in patients with lupus erythematosus in conjunction with thromboembolic events linked to repeated abortions caused by placenta vasculothrombosis, repeated myocardial infarction, pulmonary hypertension, and possibly renal and cerebral infarction. There is neurologic dysfunction, including a variety of manifestations such as myelopathy, transient ischemic attacks, chorea, epilepsy, etc. There may be hemolytic anemia, thrombocytopenia, and Coombs' positive reactivity. IgG anticardiolipin antibodies manifest 80% specificity for the anticardiolipin antibody syndrome. Anticardiolipin antibodies and DNA show crossreactivity.

anticentriole antibodies

Antibodies that may occur in sera specific for the mitotic spindle apparatus (MSA). They are found rarely in the sera of subjects developing connective tissue diseases such as scleroderma.

anticentromere antibody

Twenty-two percent of systemic sclerosis patients, most of whom have limited scleroderma, i.e., CREST syndrome, have anticentrome/antikinetophore antibodies. Twelve percent of primary biliary cirrhosis patients, half of whom also have manifestations of systemic sclerosis, also have anticentromere antibodies. Anticentromere antibodies do not affect survival and pulmonary hypertension in patients with limited scleroderma. However, survival is much longer in anticentromere positive patients with limited scleroderma than in anti-Scl-70 positive diffuse scleroderma patients.

anticomplementary

The action of any agent or treatment that interferes with complement fixation through removal or inactivation of complement components or cascade reactants. Multiple substances may exhibit anticomplementary activity. These are especially significant in complement fixation serology, since anticomplementary agents may impair the evaluation of test results.

anticytoplasmic antibody

Antineutrophil cytoplasmic antibody occurs in 84 to 100% of active generalized Wegener's granulomatosis patients. This antibody is assayed by flow cytometry and indirect fluorescence microscopy. HIV-1 infected patients may be biologically false-positive for neutrophil cytoplasmic antibody.

antidouble-stranded DNA (anti-dsDNA)

Antibodies present in the blood sera of systemic lupus erythematosus (SLE) patients. Among the detection methods is an immunofluorescence technique (IFT) using *Crithidia luciliae* as the substrate. In this method, fluorescence of the kinetoplast, which contains mitochondrial DNA, signals the presence of anti-dsDNA antibodies. This technique is useful for assaying SLE serum, which is usually positive in patients with active disease. A rim or peripheral pattern of nuclear staining of cells interacting with antinuclear antibody represents morphologic expression of antidouble-stranded DNA antibody.

antifibrillarin antibodies

Antibodies to this 34-kD protein constituent of U3 ribonucleoproteins (RNP) are detectable in the serum of about 8% of patients with diffuse and limited scleroderma. Also called anti-U3 RNP antibodies.

antigen

A substance that reacts with the products of an immune response stimulated by a specific immunogen, including both antibodies and/or T lymphocyte receptors. Presently considered to be one of many kinds of substances with which an antibody molecule or T cell receptor may bind. These include sugars, lipids, intermediary metabolites, autocoids, hormones, complex carbohydrates, phospholipids, nucleic acids, and proteins. By contrast, the "traditional" definition of antigen is a substance that may stimulate B and/or T cell limbs of the immune response and react with the products of that response, including immunoglobulin antibodies and/or specific receptors on T cells. See immunogen definition. The "traditional" definition of antigen more correctly refers to an immunogen. A complete antigen is one that both induces an immune response and reacts with the products of it, whereas an incomplete antigen or hapten is unable to induce an immune response alone, but is able to react with the products of it, e.g., antibodies. Haptens could be rendered immunogenic by covalently linking them to a carrier molecule.

Following the administration of an antigen (immunogen) to a host animal, antibody synthesis and/or cell-mediated immunity or immunologic tolerance may result. To be immunogenic, a substance usually needs to be foreign, although some autoantigens represent an exception. They should usually have a mol wt of at least 1000 and be either proteins or polysaccharides. Nevertheless, immunogenicity depends also upon the genetic capacity of the host to respond to rather than merely upon the antigenic properties of an injected immunogen.

antigen-antibody complex

The union of antibody with soluble antigen in solution containing electrolyte. When the interaction takes place *in vitro,* it is called the precipitin reaction, but it may take place also *in vivo.* The relative proportion in which antigen and antibody combine varies their molar ratio. Excess antigen may lead to soluble complexes, whereas excess antibody may lead to insoluble complexes. *In vivo,* soluble complexes are more likely to produce tissue injury, whereas larger

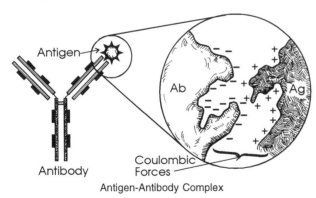

Antigen-Antibody Complex

insoluble complexes are often removed by reticuloendothelial system cells. Also called immune complex.

antigen-binding capacity

Assay of the total capacity of antibody of all immunoglobulin classes to bind antigen. This refers to primary as opposed to secondary or tertiary manifestations of the antigen-antibody interaction. Equilibrium dialysis measures the antigen-binding capacity of antibodies with the homologous hapten, and the Farr test measures primary binding of protein antigens with the homologous antibody.

antigen-binding cell (ABC) assays

The principle of this assay is the binding of cells bearing receptors for antigen to a gelatin dish in which antigen is incorporated. After incubation for a specified time and temperature, the unbound cells are washed out, and the bound cells are collected following melting of the gelatin layer at 37°C. The harvested cells are washed, counted, and used for various other assays.

antigen-binding site

The location on an antibody molecule where an antigenic determinant or epitope combines with it. The antigen-binding site is located in a cleft bordered by the N-terminal variable regions of heavy and light chain parts of the Fab region. Also called paratope.

Antigen-Binding Site

antigen capture assay

A method to identify minute quantities of antigen in blood sera or supernatants. Antibodies of high titer are linked to an insoluble solid support, and the specimen containing the antigen to be evaluated is passed over the solid phase. This will bind or capture the antigen, making it available for reaction with a separate enzyme-labeled antibody which reacts with and reveals the captured antigen.

antigen clearance

The liver has important immunologic functions by virtue of its mass of Kupffer cells, which represent the major part (90%) of the body's phagocytic capacity. Antigens escaping the intestinal barrier by passage through the liver are removed. The liver's anatomical position at the border between the splanchnic and systemic circulations substantiates its function as a filter for noxious substances, whether antigen or otherwise. The same removal mechanism is operative during liver passage in situations in which the antigen circulates in the blood.

antigen excess

The interaction of soluble antigen and antibody in the precipitin reaction leads to occupation of all the antibody molecules' antigen-binding sites and leaves additional antigenic determinants free to combine with more antibody molecules if excess antigen is added to the mixture. This leads to the formation of soluble antigen-antibody complexes *in vitro,* i.e., the postzone in the precipitin reaction. A similar phenomenon may take place *in vivo* when immune complexes form in the presence of excess antigen. These are of clinical significance in that soluble immune complexes may induce tissue injury, leading to immunopathologic sequelae.

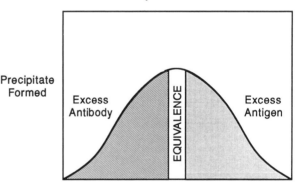

Precipitation Curve

Antigen Used

antigen masking

The ability of some parasites (ex. *S. mansoni*) to become coated with host proteins, theoretically rendering them "invisible" to the host's immune system.

antigen presentation

The expression of antigen molecules on the surface of a macrophage or other antigen-presenting cell in association with MHC class II molecules when the antigen is being presented to a CD4+ T helper lymphocyte or in association with cell surface MHC class I or cell surface MHC class II molecules when presentation is to CD8+ cytotoxic T lymphocytes. Antigen-presenting cells, known also as accessory cells, include macrophages, dendritic cells, and Langerhans cells of the skin, as well as B lymphocytes. Target cells such as fibroblasts present antigen to CD8+ cytotoxic T lymphocytes.

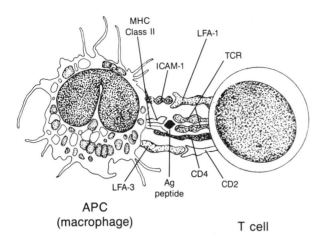

APC
(macrophage)

T cell

Antigen Presentation

Mononuclear phagocytes ingest proteins and split them into peptides in endosomes. These 8- to 10-amino acid residue peptides link to cell surface MHC class II molecules. For appropriate presentation, it is essential that peptides bind securely to the MHC class II molecules, since those that do not bind or are bound only weakly are not presented and fail to elicit an immune response. Following interaction of the presented antigen and MHC class II molecules with the CD4[+] T helper T cell receptor, the CD4[+] lymphocyte is activated, IL-2 is released, and IL-2 receptors are expressed on the CD4[+] lymphocyte surface. The IL-2 produced by the activated cell stimulates its own receptors, as well as those of mononuclear phagocytes, increasing their microbicidal activity. IL-2 also stimulates B cells to synthesize antibody. Whereas B cells may recognize a protein antigen in its native state, T lymphocytes recognize the peptides that result from antigen processing.

antigen-presenting cell (APC)

A cell that can process a protein antigen, break it into peptides, and present it in conjunction with class II MHC molecules on the cell surface where it may interact with appropriate T cell receptors. Macrophages, Langerhans cells, B cells, and dendritic reticulum cells process and present antigen to immunoreactive lymphocytes such as CD4[+], helper/inducer T cells. An MHC transporter gene-encoded peptide supply factor may mediate peptide antigen presentation. Other antigen-presenting cells that serve mainly as passive antigen transporters include B cells, endothelial cells, keratinocytes, and Kupffer cells. APCs include cells that present exogenous antigen processed in their endosomal compartment and presented together with MHC class II molecules. Other APCs present antigen that has been endogenously produced by the body's own cells with processing in an intracellular compartment and presentation together with class I MHC molecules. A third group of APCs present exogenous antigen that is taken into the cell and processed, followed by presentation together with MHC class I molecules.

T Cell

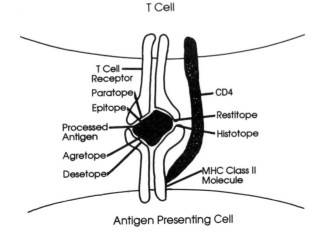

Antigen Presenting Cell

antigen recognition activation motif

A conserved sequence of 17 amino acid residues which contains two tyrosine-X-X-leucine regions. This motif is found in the cytoplasmic tails of the FcεRI-β and -α chains, the ζ and η chains of the TCR complex, the Igβ and Igα proteins of membrane IgD and IgM, and the α, δ, and ε chains of CD3. The antigen recognition activation motif is thought to be involved in signal transduction.

antigen-specific suppressor cells

Antigen-specific Ts cells can be demonstrated both in humoral and cell-mediated immunity. The Ts cells active in the humoral response can be generated after priming with the carrier to be used in subsequent experiments with hapten-carrier conjugates. These Ts cells can suppress the hapten-specific IgM and IgG antibody response if recipient animals are immunized with the hapten coupled to the homologous carrier. This type of suppression may have a differential effect on IgM and IgG antibody responses according to the time frame in which Ts cells are administered to the recipient animal. The early IgG response is relatively independent of T cell function and, accordingly, less susceptible to Ts cell effects. The late IgM and IgG responses are more T cell dependent and, accordingly, more susceptible to Ts cell inhibition.

antigen, supertypic

An inclusive term to describe an antigenic mosaic that can be separated into smaller, but related parts, called inclusions, splits, and subtypic antigens. Bw4 and Bw6 are classic examples of supertypic antigens. This implies that an antibody that detects Bw4 will also react with all antigens associated with Bw4 and an antibody that detects Bw6 will also react with all antigens associated with Bw6.

antigenic

An adjective that refers to the ability of a substance to induce an immune response and to react with its products, which include antibodies and T lymphocyte receptors. The term "antigenic" has been largely replaced by "immunogenic".

antigenic competition

The simultaneous injection of two closely related antigens may lead to suppression or a decrease of the immune response to one of them compared to the antigen's ability to elicit an immune response if injected alone. Proteins that are thymus-dependent antigens are the ones with which antigenic competition occurs. The phenomenon has been claimed to be due in part to the competition by antigenic peptides for one binding site on class II MHC molecules. Antigenic competition was observed in the early days of vaccination when it was found that the immune response of a host to the individual components of a vaccine might be less than if they had been injected individually.

antigenic deletion

Antigenic deletion describes antigenic determinants that have been lost or masked in the progeny of cells that usually contain them. Antigenic deletion may take place as a consequence of neoplastic transformation or mutation of parent cells, resulting in the disappearance or repression of the parent cell genes.

antigenic determinant (see facing page)

The site on an antigen molecule that is termed an epitope and interacts with the specific antigen-binding site in the variable region of an antibody molecule known as a paratope. The excellent fit between epitope and paratope is based on their three-dimensional interaction and noncovalent union. An antigenic determinant or epitope may also react with a T cell receptor for which it is specific. A lone antigen molecule may have several different epitopes available for reaction with antibody or T cell receptors.

antigenic diversion

The replacement of a cell's antigenic profile by the antigens of a different normal tissue cell. Used in tumor immunology.

antigenic drift

Spontaneous variation, as in influenza virus, expressed as relatively minor differences exemplified by slow antigenic changes from one year to the next. Antigenic drift is believed to be due to mutation of the genes encoding the hemagglutinin or the neuraminidase components. Antigenic variants represent those viruses that have survived exposure to the host's neutralizing antibodies. Minor alterations in a viral genome might occur every few years, especially in influenza A subtypes that are made up of H1, H2, and H3 hemagglutinins and N1 and N2 neuraminidases. Antigenic shifts follow point mutations of DNA encoding these hemagglutinins and neuraminidases.

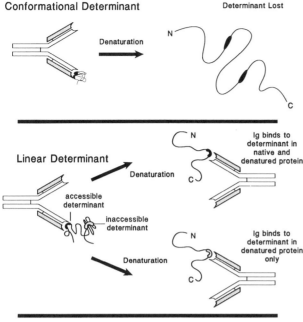

Conformational Determinant   Determinant Lost

Denaturation

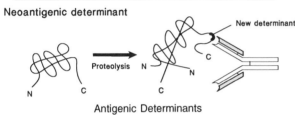

Linear Determinant

Denaturation

accessible determinant

inaccessible determinant

Denaturation

Ig binds to determinant in native and denatured protein

Ig binds to determinant in denatured protein only

Neoantigenic determinant

New determinant

Proteolysis

Antigenic Determinants

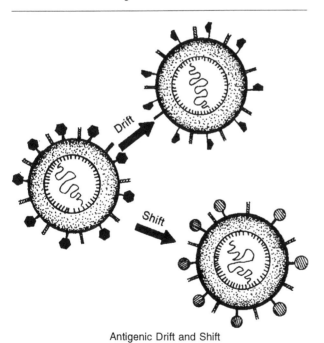

Drift

Shift

Antigenic Drift and Shift

antigenic modulation

The loss of epitopes or antigenic determinants from a cell surface following combination with an antibody. The antibodies either cause the epitope to disappear or become camouflaged by covering it.

antigenic mosaicism

Antigenic variation first discovered in pathogenic *Neisseria*. It is the result of genetic transformation between gonococcal strains. This is also observed in penicillin resistance of several bacterial species where the resistant organism contains DNA from a host commensal organism.

antigenic profile

The total antigenic content, structure, or distribution of epitopes of a cell or tissue.

antigenic reversion

The change in antigenic profile characteristic of an adult cell to an antigenic mosaic that previously existed in the immature or fetal cell stage of the species. Antigenic reversion may accompany neoplastic transformation.

antigenic shift

A major antigenic change in which a strain with distinctive new antigens may appear, such as Asian or A2 influenza in 1957. Antigenic variants of type A influenza virus are known as subtypes. Influenza virus antigenic shift is attributable mainly to alterations in the hemagglutinin antigens with less frequent alterations in the neuraminidase antigens. The appearance of a new type A influenza virus signals the addition of a new epitope, even though several original antigenic determinants are still present.

In contrast to antigenic drift, antigenic shift involves a principal alteration in a genome attributable to gene rearrangement between two related microorganisms. Since antigenic shift involves the acquisition of totally new antigens against which the host population is not immune, this alteration may lead to an epidemic of significant proportions.

antigenic sin, doctrine of original

When the immune response against a virus, such as a parental strain, to which an individual was previously exposed is greater than it is against the immunizing agent, such as type A influenza virus variant, the concept is referred to as the "doctrine of original antigenic sin".

antigenic transformation

Antigenic transformation refers to changes in a cell's antigenic profile as a consequence of antigenic gain, deletion, reversion, or other process.

antigenic variation

Antigenic variation represents a mechanism whereby selected viruses, bacteria, and animal parasites may evade the host immune response, thereby permitting antigenically altered etiologic agents of disease to produce a renewed infection. The variability among infectious disease agents is of critical significance in the development of effective vaccines. Antigenic variation affects the surface antigens of the viruses, bacteria, or animal parasite in which it occurs. By the time the host has developed a protective immune response against the antigens originally present, the latter have been replaced in a few surviving microorganisms by new antigens to which the host is not immune, thereby permitting survival of the microorganism or animal parasite and its evasion of the host immune response. Thus, from these few surviving viruses, bacteria, or animal parasites, a new population of infectious agents is produced. This cycle may be repeated, thereby obfuscating the protective effects of the immune response.

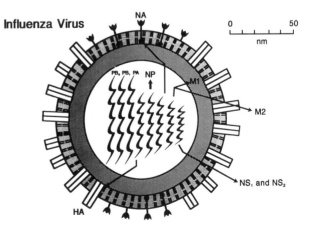

Influenza Virus

Antigen Variation

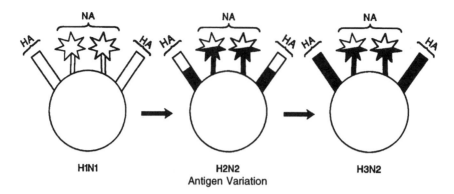

Antigen Variation

antigenicity

A property of a substance that renders it immunogenic or capable of stimulating an immune response. Antigenicity was more commonly used in the past to refer to what is now known as immunogenicity, although the two are still used interchangably by various investigators. An antigen is considered by many to be a substance that reacts with the products of immunogenic stimulation. It is a substance that combines specifically with antibodies formed or receptors of T cells stimulated during an immune response.

antigliadin antibodies (AGA)

Antibodies specific for gliadin, a protein present in wheat and rye grain gluten. α-gliadins are requisite for the development of celiac disease with associated jejunal mucosal flattening in genetically prone subjects. Thus, antibodies against gliadin may be used for population screening for gluten-sensitive enteropathies such as celiac disease and dermatitis herpetiformis. Environmental agents, such as an adenovirus in the intestine, may induce an aberrant immune response to gluten in genetically susceptible subjects such as in HLA-DR3-DQw2; HLA-B8 individuals.

antiglobulin

An antibody raised by immunization of one species, such as a rabbit, with immunoglobulin from another species, such as man. Rabbit anti-human globulin has been used for many years in an antiglobulin test to detect incomplete antibodies coating red blood cells, as in erythroblastosis fetalis or autoimmune hemolytic anemia. Antiglobulin antibodies are specific for epitopes in the Fc region of immunoglobulin molecules used as immunogen, rendering them capable of agglutinating cells whose surface antigens are combined with the Fab regions of IgG molecules whose Fc regions are exposed.

antiglobulin consumption test

An assay to test for the presence of an antibody in serum which is incubated with antigen-containing cells or antigen-containing particles. After washing, the cells or particles are treated with antiglobulin reagents and incubated further. If any antibody has complexed with the cells or particles, antiglobulin will be taken up. Antiglobulin depletion from the mixture is evaluated by assaying the free antiglobulin in the supernatant through combination with incomplete antibody-coated erythrocytes. No hemagglutination reveals that the antiglobulin reagent was consumed in the first step of the reaction and shows that the original patient's serum contained the antibody in question.

antiglobulin inhibition test

An assay based upon interference with the antiglobulin test through reaction of the antiglobulin reagent with antibody against it prior to combination with incomplete antibody-coated erythrocytes. This is the basis for the so-called antiglobulin consumption test.

antiglobulin test

When red blood cells are coated with antibodies that are not agglutinable in saline, such as those from an infant with erythroblastosis fetalis, a special anti-human immunoglobulin prepared by immunizing rabbits with human IgG may be employed to crosslink the antibody-coated red cells to produce agglutination. Although previously considered to be incomplete antibodies, they are known to be bivalent, but may be of a smaller size than saline agglutinable type antibodies. R.R.A. Coombs developed this test in England in the 1940s. In addition to its usefulness in hemolytic disease of the newborn, the Coombs' test detects incomplete antibody-coated erythrocytes from patients with autoimmune hemolytic anemia. In the direct Coombs' test, red blood cells linked to saline nonaggglutinable antibody are first washed, combined with rabbit anti-human immunoglobulin serum, and then observed for agglutination. In the indirect Coombs' test, serum containing the saline nonagglutinable antibodies is combined with red blood cells which are coated, but not agglutinated. The rabbit anti-human immunoglobulin is then added to these antibody-coated red cells, and agglutination is observed as in the direct Coombs' reaction. A third assay termed the "non-gamma" test requires the incubation of erythrocytes with anti-C3 or anti-C4 antibodies. Agglutination reflects the presence of these complement components on the red blood cell surface. This is an indirect technique to identify IgM antibodies that have fixed complement, such as those that are specific for Rh blood groups.

antiglutinin

Mammalian seminal plasma substance that prevents washed spermatozoa from spontaneously agglutinating, i.e., autoagglutinating.

Antigranulocyte antibodies

Antigranulocyte antibodies play a significant role in the pathogenesis of febrile transfusion reactions; drug-induced neutropenia; isoimmune neonatal neutropenia; autoimmune neutropenia, including Felty's syndrome; Graves' disease; Evans syndrome; SLE; and primary autoimmune neutropenia of children. Antigranulocyte antibodies are best detected and quantitated by flow cytometry.

antiheat shock protein antibodies

Heat shock proteins (hsp) have a broad phylogenetic distribution and share sequence similarities in molecules derived from bacteria, humans, or other animals. They play a significant role in inflammation. Heat shock proteins of mycobacteria are important in the induction of adjuvant arthritis by these microorganisms. 40% of SLE and 10 to 20% of RA patients have antibodies of IgM, IgG, and IgA classes to a 73-kD protein of the hsp70 group. RA synovial fluid contains T lymphocytes that react with a 65-kD mycobacterial heat shock protein. The significance of these observations of immune reactivity to heat shock proteins remains to be determined.

antihistamine

A substance that links to histamine receptors, thereby inhibiting histamine action. Antihistamine drugs derived from ethylamine block H_1 histamine receptors, whereas those derived from thiourea block the H_2 variety.

antiidiotypic antibodies

An antibody that interacts with antigenic determinants (idiotopes) at the variable N-terminus of the heavy and light chains comprising the paratope region of an antibody molecule where the antigen-binding site is located. The idiotope antigenic determinants may be situated either within the cleft of the antigen-binding region or on the periphery or outer edge of the variable region of heavy and light chain components.

antiidiotypic vaccine

An immunizing preparation of antiidiotypic antibodies that are internal images of certain exogenous antigens. To develop an effective antiidiotypic vaccine, epitopes of an infectious agent that induce protective immunity must be identified. Antibodies must be identified which confer passive immunity to this agent. An antiidiotypic antibody prepared using these protective antibodies as the immunogen, in some instances, can be used as an effective vaccine. Antiidiotypic vaccines have effectively induced protective immunity against such viruses as rabies, coronavirus, cytomegalovirus, and hepatitis B; such bacteria as *Listeria monocytogenes, Escherichia coli,* and *Streptococcus pneumoniae;* and such parasites as *Schistosoma mansoni* infections. Antiidiotypic vaccination is especially desirable when a recombinant vaccine is not feasible. Monoclonal antiidiotypic vaccines represent a uniform and reproducible source for an immunizing preparation.

antiimmunoglobulin antibodies

Antibodies produced by immunizing one species with immunoglobulin antibodies derived from another.

antilymphocyte serum (ALS) or antilymphocyte globulin (ALG)

An antiserum prepared by immunizing one species, such as a rabbit or horse, with lymphocytes or thymocytes from a different species, such as a human. Antibodies present in this antiserum combine with T cells and other lymphocytes in the circulation to induce immunosuppression. ALS is used in organ transplant recipients to suppress graft rejection. The globulin fraction known as ALG rather than whole antiserum produces the same immunosuppressive effect.

antimalignin antibodies

Specific for the 10-kD protein malignin comprised of 89 amino acids. These antibodies are claimed to be increased in cancer patients without respect to tumor cell type. It has been further claimed that antibody levels are related to survival. These claims will require additional confirmation and proof to be accepted as fact.

antimetabolite

A drug that interrupts the normal intracellular processes of metabolism, such as those essential to mitosis. Antimetabolite drugs such as azathioprine, mercaptopurine, and methotrexate induce immunosuppression in organ transplant recipients and diminish autoimmune reactivity in patients with selected autoimmune diseases.

antimyocardial antibodies (AMyA)

Occur in elevated titers in two thirds of coronary artery bypass patients and do not have to be related to postcardiotomy syndrome. These antibodies are also found in a majority of acute rheumatic fever patients.

antineutrophil cytoplasmic antibodies (ANCA)

Antibodies detected in 84 to 100% of generalized active Wegener's granulomatosis patients. These antibodies react with the cytoplasm of fixed neutrophils. They may also be detected in patients with microscopic polyarteritis. ANCA may be quantified by flow cytometry in conjunction with indirect immunofluorescence microscopy, which permits observation of antibody reactivity with the cytoplasm. There is a positive correlation between antibody levels and disease activity, with a decrease following therapy. The staining pattern is to be distinguished from that produced by antimyeloperoxidase antibodies, which display perinuclear staining. One of the antibodies producing diffuse cytoplasmic fluorescence is against proteinase 3. Sera from some HIV-positive subjects may prove false-positive for ANCA.

antinuclear antibodies (ANA)

Antibodies found in the circulation of patients with various connective tissue disorders. They may show specificity for various nuclear antigens, including single- and double-stranded DNA, histones, and ribonucleoprotein. To detect antinuclear antibodies, the patient's

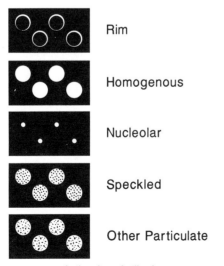

Rim

Homogenous

Nucleolar

Speckled

Other Particulate

Antinuclear Antibody

serum is incubated with Hep-2 cells, and the pattern of nuclear staining is determined by fluorescence microscopy. The homogeneous pattern of staining represents the morphologic expression of antinuclear antibodies specific for ribonucleoprotein which is positive in systemic lupus erythematosus, progressive systemic sclerosis, rheumatoid arthritis, and other connective tissue disorders. Peripheral nuclear staining represents the morphologic expression of DNA antibodies associated with systemic lupus erythematosus. Nucleolar fluorescence signifies anti-RNA antibodies of the type that occurs in progressive systemic sclerosis (scleroderma). The speckled pattern of staining is seen in several connective tissue diseases.

antiphospholipid antibodies

Refer to anticardiolipin antibody syndrome and to lupus anticoagulant.

antiseptic paint

A colloquial designation for the coating effect of secretory IgA, such as that produced locally in the gut, on mucosal surfaces, thereby barring antigen access.

antiserum

A preparation of serum containing antibodies specific for a particular antigen, i.e., immunogen. A therapeutic antiserum may contain antitoxin, antilymphocyte antibodies, etc.

antisperm antibody

An antibody specific for any one of several sperm constituents. Antisperm agglutinating antibodies are detected in blood serum by the Kibrick sperm agglutination test, which uses donor sperm. Sperm-immobilizing antibodies are detected by the Isojima test. The subject's serum is incubated with donor sperm, and motility is examined. Testing for antibodies is of interest to couples with infertility problems. Treatment with relatively small doses of prednisone is sometimes useful in improving the situation by diminishing antisperm antibody titers. One half of infertile females manifest IgG or IgA sperm-immobilizing antibodies which affect the tail of the spermatozoa. By contrast, IgM antisperm head agglutinating antibodies may occur in homosexual males.

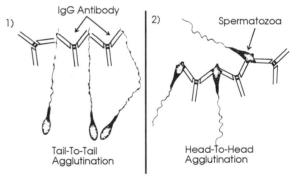

Antisperm Antibody

antithymocyte serum (ATS)

Antibody raised by immunizing one species, such as a rabbit or horse, with thymocytes derived from another, such as a human. The resulting antiserum has been used to induce immunosuppression in organ transplant recipients. It acts by combining with the surface antigens of T lymphocytes and suppressing their action.

antitoxin

Antibody specific for exotoxins produced by certain microorganisms such as the causative agents of diphtheria and tetanus. Prior to the antibiotic era, antitoxins were the treatment of choice for diseases produced by the soluble toxic products of microorganisms, such as those from *Corynebacterium diphtheriae* and *Clostridium tetani*.

antitoxin assay (historical)

Antitoxins are assayed biologically by their capacity to neutralize homologous toxins as demonstrated by production of no toxic manifestations following inoculation of the mixture into experimental animals, e.g., guinea pigs. They may be tested serologically by their ability to flocculate (precipitate) toxin *in vitro*.

antitoxin unit

A unit of antitoxin is that amount of antitoxin present in 1/6000 g of a certain dried unconcentrated horse serum antitoxin which has been maintained since 1905 at the National Institutes of Health in Bethesda, MD. The standard antitoxin unit contained sufficient antitoxin to neutralize 100 MLD of the special toxin prepared by Ehrlich and used by him in titration of standard antitoxin. Both the American and the international unit of antitoxin are the same.

antivenom

Antitoxin prepared specifically for the treatment of bite or sting victims of poisonous snakes or arthropods. Antibodies in this immune serum preparation neutralize the snake or arthropod venom. Also called antivenin or antivenene.

antrypol

Alternative name for suramin.

AP-1

A transcription factor that binds the IL-2 promoter therefore regulating the induction of the IL-2 gene. Immediately following T cell stimulation, *c-fos* mRNA is increased, and the *c-fos* gene product combines with the *c-jun* gene product to form AP-1. A similar series of events occurs following B cell stimulation, however the genes regulated by B cell AP-1 are not known.

APC

Abbreviation for antigen-presenting cell.

APECED (autoimmune polyendocrinopathy-candidiasis-ectodermal dystrophy)

An autosomal recessive disorder characterized by hypoparathyroidism, adrenal cortical failure, gonadal failure, candidiasis, and malabsorption. Antithymocyte globulin (ATG) IgG isolated from the blood serum of rabbits or horses hyperimmunized with human thymocytes is used in the treatment of aplastic anemia patients and to combat rejection in organ transplant recipients. The equine ATG contains 50mg/ml of immunoglobulin and has yielded 50% recovery of bone marrow and treated aplastic anemia patients.

apheresis

The technique whereby blood is removed from the body, its components are separated and some are retained for therapeutic or other use, and the remaining elements are recombined and returned to the donor. Also called hemapheresis.

aplasia

The disappearance of a particular population of cells because of their failure to develop.

APO-1

Synonym for *fas* gene. FAS membrane protein ligation has been shown to initiate apoptosis. This is the reverse action of bcl-2 protein, which blocks apoptosis.

apolipoprotein E

A 33-kD protein produced by nonactivated macrophages, but not monocytes. It binds low-density lipids as well as high-density cholesterol esters.

apoptosis

Programmed cell death in which the chromatin becomes condensed and the DNA is degraded. The immune system employs apoptosis

for clonal deletion of cortical thymocytes by antigen in immunologic tolerance.

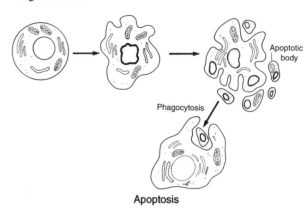

Apoptosis

appendix, vermiform (see below)

A lymphoid organ situated at the ileocecal junction of the gastrointestinal tract.

APT (alum-precipitated toxoid)

Refer to alum-precipitated antigen.

Aquaphor®

An emulsifying preparation of lanolin used extensively in the past to prepare the water-in-oil emulsion immunologic adjuvants of the Freund type.

aqueous adjuvants

Freund's adjuvants are not used in man because of the ease with which hypersensitivity is induced and the unpredictability of the local reaction. A number of water-soluble synthetic components comprising active moieties of the mycobacterial cell wall have been synthesized in search of more adequate adjuvants. One of them is muramyl dipeptide (MDP), which is active when administered by the oral route. MDP in water is extremely active as an adjuvant and is not generally toxic when administered at high doses. It is neither mitogenic, immunogenic, nor antigenic and is rapidly eliminated from the animal body. The simplicity of its chemical structure allows the study of the targets of its action in the immune system. Other synthetic adjuvant compounds are polynucleotides such as poly-inosine-poly-cytidine (poly I:C), whose structure is similar to that of native nucleotides. Their mechanism of action involves signals which are rapidly received by the immune system, since such compounds are destroyed in 5 to 10 min by the nucleases of the serum. The prevalent concept is that adjuvants have a number of other regulatory activities on the immune response, and the term "adjuvant" may be replaced by "immunoregulatory molecule".

Appendix

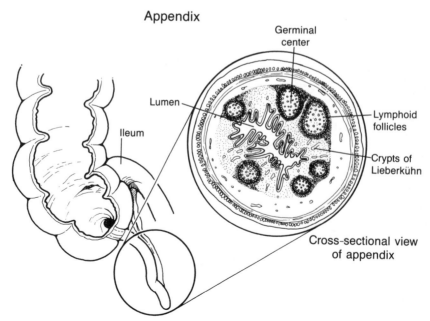

Cross-sectional view of appendix

Arlacel® A

A mannide monooleate used as an emulsifier to stabilize water-in-oil emulsions employed as adjuvants, e.g., Freund's adjuvant, in experimental immunology.

armed macrophages

Macrophages bearing surface IgG or IgM cytophilic antibodies or T cell lymphokines that render them capable of inducing antigen-specific cytotoxicity.

Arthus, Nicolas Maurice (1862–1945)

Paris physician. He studied venoms and their physiological effects; first to describe local anaphylaxis or the Arthus reaction 1903. *De l'Anaphylaxie a l'Immunite,* 1921.

Arthus reaction

Induced by repeated intradermal injections of antigen into the same skin site. It is dependent upon the development of humoral antibodies of the precipitin type which react *in vivo* with specific antigen at a local site. It may also be induced by the inoculation of antigen into a local skin site of an animal possessing performed IgG antibodies specific for the antigen. Immune complexes are comprised of antigen, antibody, and complement formed in vessels. The chemotactic complement fragment C5a and other chemotactic peptides produced attract neutrophils to antigen-antibody-complement complexes. This is followed by lysosomal enzyme release, which induces injury to vessel walls with the development of thrombi, hemorrhage, edema, and necrosis. Events leading to vascular necrosis include blood stasis, thrombosis, capillary compression in vascular injury which causes extravasation, venule rupture, hemorrhage, and local ischemia. There is extensive infiltration of polymorphonuclear cells, especially neutrophils, into the connective tissue. Grossly, edema, erythema, central blanching, induration, and petechiae appear. Petechiae develop within 2 h, reach a maximum between 4 and 6 h, and then may diminish or persist for 24 h or longer with associated central necrosis, depending on the severity of the reaction. If the reaction is more prolonged, macrophages replace neutrophils; histiocytes and plasma cells may also be demonstrated. The Arthus reaction is considered a form of immediate-type hypersensitivity, but it does not occur as rapidly as does anaphylaxis. It takes place during a 4-h period and diminishes after 12 h. Thereafter, the area is cleared by mononuclear phagocytes. The passive cutaneous Arthus reaction consists of the inoculation of antibodies intravenously into a nonimmune host, followed by local cutaneous injection of antigen. The reverse passive cutaneous Arthus reaction requires the intracutaneous injection of antibodies, followed by the intravenous or incutaneous (at the same site) administration of antigen. The Arthus reaction is a form of type III hypersensitivity since it is based upon the formation of immune complexes with complement fixation. Clinical situations for which it serves as an animal model include serum sickness, glomerulonephritis, and farmer's lung.

artificial antigen

An antigen prepared by chemical modification of a natural antigen. Compare with synthetic antigen.

artificial passive immunity

The transfer of immunoglobulins from an immune individual to a nonimmune, susceptible recipient.

artificially acquired immunity

The use of deliberate active or passive immunization or vaccination to elicit protective immunity as opposed to immunity which results from unplanned and coincidental exposure to antigenic materials, including microorganisms in the environment.

Aschoff bodies

Areas of fibrinoid necrosis encircled first by lymphocytes and macrophages with a rare plasma cell. The mature Aschoff body reveals prominent modified histiocytes termed Anitschkow cells or Aschoff cells in the inflammatory infiltrate. These cells have round to oval nuclei with wavy ribbon-like chromatin and amphophilic cytoplasm. Aschoff bodies are pathognomonic of rheumatic fever. They may be found in any of the heart's three layers, i.e., pericardium, myocardium, or endocardium.

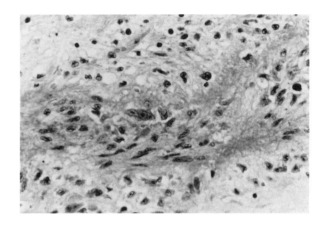

Aschoff Body

Ascoli's test

A ring precipitin assay used in the past to identify anthrax antigen in tissues, skins, and hides of animals infected with *Bacillus anthracis.* The simple test was considered useful in that it could identify anthrax antigen in decaying material from which anthrax bacilli could no longer be cultured.

ASLT

Abbreviation for the antistreptolysin O test.

ASO (antistreptolysin O)

A laboratory technique that serves as an indicator of infection by group A β hemolytic streptococci. IgM antibody titers, expressed in Todd units (TU), increase fourfold within 3 weeks after infection in untreated subjects. Penicillin treatment decreases the ASO titer. Less than 166 TU is normal, whereas greater than 333 TU in children and greater than 250 TU in adults suggests recent infection. The ASO assay depends upon hemolysis inhibition. The greatest dilution of a patient's blood combined with 1 U of streptolysin O that prevents the lysis of erythrocytes determines the Todd units, the reciprocal of endpoint dilution.

aspirin (ASA) acetyl salicylic acid

An antiinflammatory, analgesic, and antipyretic drug that blocks the synthesis of prostaglandin. It may induce atopic reactions such as asthma and rhinitis due to intolerance and idiosyncratic reactivity against the drug.

association constant (K$_A$)

A mathematical measurement of the reversible interaction between two molecular forms at equilibrium. The AB complex, free A and B concentrations at

Aspirin

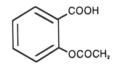

acetylsalicylic acid

Arthus Reaction

equilibrium, are expressed in K_A liters per mole by [AB], [A], and [B]. s (molecules of substance A) interact reversibly with t (molecules of substance B), i.e., $sA + tB \rightleftharpoons A_sB_t$, the association constant is $[A_sB_t]/[A]^s[B]^t$. Molar concentrations at equilibrium are indicated by the symbols in brackets.

asthma

A disease of the lungs characterized by reversible airway obstruction (in most cases), inflammation of the airway with prominent eosinophil participation, and increased responsiveness by the airway to various stimuli. There is bronchospasm associated with recurrent paroxysmal dyspnea and wheezing. Some cases of asthma are allergic, i.e., bronchial allergy, mediated by IgE antibody to environmental allergens. Other cases are provoked by nonallergic factors that are not discussed here.

Asthma

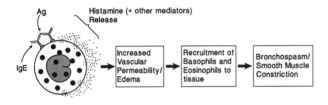

Mast Cell

ataxia telangiectasia

A disorder characterized by cerebellar ataxia, oculocutaneous telangiectasis, variable immunodeficiency which affects both T and B cell limbs of the immune response, the development of lymphoid malignancies, and recurrent sinopulmonary infections. Clinical features may appear by 2 years of age. Forty percent of patients have selective IgA deficiency. The disease has an autosomal recessive mode of inheritance. There may be lymphopenia, normal or decreased T lymphocyte numbers, and a normal or diminished lymphocyte response to PHA and allogeneic cells. The delayed-type hypersensitivity skin test may not stimulate any response. Some individuals may have an IgG2, IgG4, or IgA2 subclass deficiency. Other patients may reveal no IgE antibody level. There is diminished antibody responsiveness to selected antigens. B cell numbers are usually normal, and NK cell function is within physiologic limits. The level of T cell deficiency varies. Defects in DNA repair mechanisms lead to multiple breaks, inversions, and translocations within chromosomes, rendering them highly susceptible to the injurious action of ionizing radiation and radiomimetic chemicals. The chromosomal breaks are especially apparent on chromosome 7 and 14 in the regions that encode immunoglobulin genes and T cell receptor genes. The multiple chromosomal breaks are believed to be linked to the high incidence of lymphomas in these patients. α-Fetoprotein is also elevated. Endocrine abnormalities associated with the disease include glucose intolerance associated with antiinsulin receptor antibodies and hypogonadism in males. Patients may experience retarded growth and hepatic dysfunction. Death may occur in many of the patients related to recurrent respiratory tract infections or lymphoid malignancies.

ATG

Abbreviation for antithymocyte globulin.

athymic nude mice

A mouse strain with no thymus and no hair. T lymphocytes are absent. Therefore, no manifestations of T cell immunity are present, i.e., they do not produce antibodies against thymus-dependent

Athymic Mouse (Nude Mouse)

antigens and fail to reject allografts. They possess a normal complement of B and NK cells. These nude or *nu nu* mice are homozygous for a mutation, *v* on chromosome 11, which is inherited as an autosomal recessive trait. These features make the strain useful in studies evaluating thymic-independent immune responses.

atopic dermatitis

Chronic eczematous skin reaction marked by hyperkeratosis and spongiosis especially in children with a genetic predisposition to allergy. These are often accompanied by elevated serum IgE levels, which are not proved to produce the skin lesions.

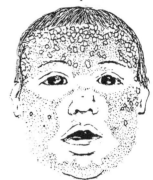

Infantile Atopic Dermatitis

atopic hypersensitivity

Refer to atopy.

atopy (see facing page)

A type of immediate (type I) hypersensitivity to common environmental allergens in man mediated by humoral antibodies of the IgE class formerly termed reagins, which are able to passively transfer the effect. Atopic hypersensitivity states include hay fever, asthma, eczema, urticaria, and certain gastrointestinal disorders. There is an hereditary predisposition to atopic hypersensitivities, which affect more than 10% of the human population. Antigens that sensitize atopic individuals are termed allergens. They include (1) grass and tree pollens; (2) dander, feathers, and hair; (3) eggs, milk, and chocolate; and (4) house dust, bacteria, and fungi. IgE antibody is a skin-sensitizing homocytotropic antibody which occurs spontaneously in the sera of human subjects with atopic hypersensitivity. IgE antibodies are nonprecipitating (*in vitro)*, are heat sensitive (destroyed by heating to 60°C for 30 to 60 min), are not able to pass across the placenta, remain attached to local skin sites for weeks after injection, and fail to induce passive cutaneous anaphylaxis (PCA) in guinea pigs.

attenuated

An adjective that denotes diminished virulence of a microorganism.

attenuation

Decrease of a particular effect, such as exposing a pathogenic microorganism to conditions which destroy its virulence, but leave its antigenicity or immunogenicity intact.

AtxBm

Abbreviation for a so-called B cell mouse, which refers to a thymectomized irradiated adult mouse that has received a bone marrow transplant.

Auer's colitis

An Arthus reaction in the intestine produced by the inoculation of albumin serving as antigen into the colon of rabbits that have developed antialbumin antibodies. An inflammatory lesion is produced in the colon and is marked by hemorrhage and necrosis.

Australia antigen (AA)

Hepatitis B viral antigen. The name is derived from detection in an Australian aborigine. Australia antigen is demonstrable in the cytoplasm of an infected hepatocyte. In early hepatitis B, there is sublobular cell involvement, but later in the disease, only some hepatocytes are antigen positive. There is a positive correlation between the presence of hepatitis B antigen in the liver of a group of people and that group's incidence of hepatocellular carcinoma.

autoagglutination

The spontaneous aggregation of erythrocytes, microorganisms, or other particulate antigens in a saline suspension, thereby confusing interpretation of bacterial agglutination assays. The term refers also to the aggregation of an individual's cells by his own antibody.

autoallergy

Tissue injury or disease induced by immune reactivity against self antigens.

autoantibodies

An antibody that recognizes and interacts with an antigen present as a natural component of the individual synthesizing the autoantibody. The ability of these autoantibodies to "crossreact" with corresponding antigens from other members of the same species provides a method for *in vitro* detection of such autoantibodies.

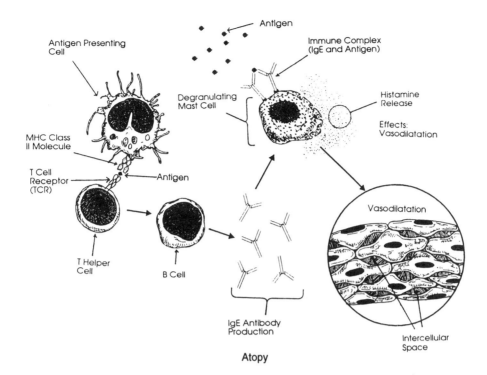

Atopy

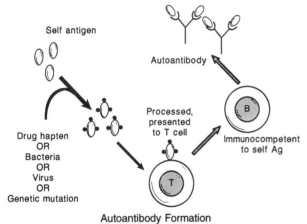

Autoantibody Formation

autoantibodies, virus infection associated

Viral infections may stimulate the production of autoantibodies in three ways:

1. By complexing with cell surface histocompatibility antigens to form new immunogenic units
2. By nonspecifically stimulating the proliferation of lymphocytes, e.g., after infection with Epstein-Barr virus; the nonspecific response includes clones of cells specific for autoantigens
3. By inducing the expression of antigens normally repressed in the host cells

autoantigens

Normal body constituents recognized by autoantibodies specific for them. T cell receptors may also identify autoantigens (self antigen) when the immune reactivity has induced a cell-mediated T lymphocyte response.

autobody (see figure, page 30)

An antibody that exhibits the internal image of an antigen as well as a binding site for an antigen. It manifests dual binding to both idiotope and epitope. It bears an idiotope that is complementary to its own antigen-binding site or paratope. Thus, it has self-binding potential. This type of antiidiotypic antibody has features of Ab1 and Ab2 on the same molecule, causing it to be designated Ab1-2 or "autobody". The name points to the potential for self-aggregation of the molecules and the potential participation of autobodies in autoimmune phenomena. Antibodies to phosphorylcholine (PC)

epitope raised in Balb/c mice expressing the T15 idiotype self aggregate, i.e., bind to one another.

autochthonous

Occuring in the same subject. Also called autologous.

autocrine

The action of a hormone on the same cell that synthesized it.

autogenous vaccine

The isolation and culture of microorganisms from an infected subject. The microorganisms in culture are killed and used as an immunogen, i.e., a vaccine, to induce protective immunity in the same subject from which they were derived. In earlier years, this was a popular method to treat *Staphylococcus aureus*-induced skin infections.

autograft

A graft of tissue taken from one area of the body and placed in a different site on the body of the same individual, e.g., grafts of skin from unaffected areas to burned areas in the same individual.

autoimmune complement fixation reaction

The ability of human blood serum from patients with certain autoimmune diseases such as systemic lupus erythematosus, chronic active hepatitis, etc. to fix complement when combined with kidney, liver, or other tissue suspensions in saline.

autoimmune disease

Pathogenic consequences, including tissue injury, produced by autoantibodies or autoreactive T lymphocytes interacting with self epitopes, i.e., autoantigens. The mere presence of autoantibodies or autoreactive T lymphocytes does not prove that there is any cause and effect relationship between these components and a patient's disease. To show that autoimmune phenomena are involved in the etiology and pathogenesis of human disease, Witebsky suggested that certain criteria be fulfilled. (See Witebsky's criteria). In addition to autoimmune reactivity against self constituents, tissue injury in the presence of immunocompetent cells corresponding to the tissue distribution of the autoantigen, duplication of the disease features in experimental animals injected with the appropriate autoantigen, and passive transfer with either autoantibody or autoreactive T lymphocytes to normal animals offer evidence in support of an autoimmune pathogenesis of a disease. Individual autoimmune diseases are discussed under their own headings, such as systemic lupus erythematosus, autoimmune thyroiditis, etc.

autoimmune hemolytic anemia

Although both warm-antibody and cold-antibody types are known, the warm-antibody type is the most common and is characterized by a positive direct antiglobulin (Coombs' test) associated with

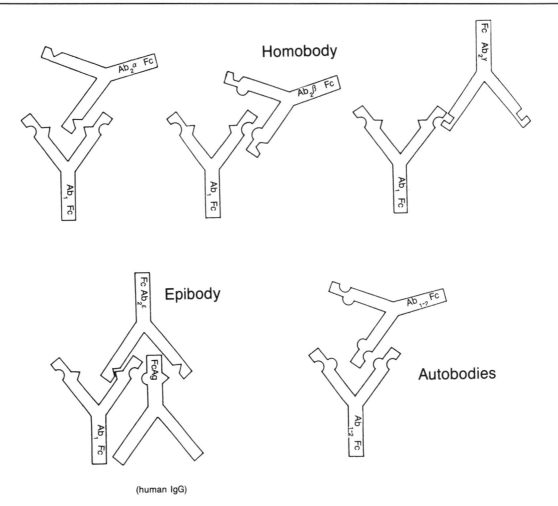

Homobody

Epibody

Autobodies

(human IgG)

lymphoreticular cancer or autoimmune disease and splenomegaly. Patients may have anemia, hemolysis, lymphadenopathy, hepatosplenomegaly, or features of autoimmune disease. They commonly have a normochromic, normocytic anemia with spherocytosis and nucleated red blood cells in the peripheral blood. Leukocytosis and thrombocytosis may also occur. There is a significant reticulocytosis and an elevated serum indirect (unconjugated) bilirubin. IgG and complement adhere to red blood cells. Antibodies are directed principally against Rh antigens. There is a positive indirect antiglobulin test in 50% of cases and agglutination of enzyme-treated red blood cells in 90% of cases. In the cold agglutinin syndrome, IgM antibodies with an anti-I specificity are involved. Warm-autoantibody autoimmune hemolytic anemia has a fairly good prognosis.

autoimmune neutropenia
This can be either an isolated condition or secondary to autoimmune disease. Patients may have either recurrent infections or remain asymptomatic. Antigranulocyte antibodies may be demonstrated. There is normal bone marrow function with myeloid hyperplasia and a shift to the left as a result of increased granulocyte destruction. The autoantibody may suppress myeloid cell growth. The condition is treated by immunosuppressive drugs, corticosteroids, or splenectomy. Patients with systemic lupus erythematosus and Felty's syndrome (rheumatoid arthritis, splenomegaly, and severe neutropenia), as well as other autoimmune diseases, may manifest autoimmune neutropenia.

autoimmunity
Immune reactivity involving either antibody-mediated (humoral) or cell-mediated limbs of the immune response against the body's own (self) constituents, i.e., autoantigens. When autoantibodies or autoreactive T lymphocytes interact with self epitopes, tissue injury may occur (e.g., in rheumatic fever the autoimmune reactivity against heart muscle sacrolemmal membranes occurs as a result of crossreactivity with antibodies against streptococcal antigens [molecular mimicry]. Thus, the immune response can be a two-edged sword, producing both beneficial (protective) effects, while also leading to severe injury to host tissues. Reactions of this deleterious

nature are referred to as hypersensitivity reactions, which are subgrouped into four types.

autologous
An adjective that refers to derivation from self. The term describes grafts or antigens taken from an individual and returned to the same subject from which they were derived.

autologous bone marrow transplantation (ABMT)
Leukemia patients in relapse may donate marrow which can be readministered to them following a relapse. Leukemic cells are removed from the bone marrow which is cryopreserved until needed. Prior to reinfusion of the bone marrow, the patient receives supralethal chemoradiotherapy. This mode of therapy has improved considerably the survival rate of some leukemia patients.

autolymphocyte therapy (ALT)
An unconfirmed immunotherapeutic treatment for metastatic renal carcinoma. Leukocytes from the patient are isolated and activated with monoclonal antibodies to induce the leukocytes to synthesize and secrete cytokines. Cytokines produced in the supernatant are combined with a sample of the patient's own lymphocytes and reinjected. Preliminary reports claim success, but these are not confirmed.

autoradiography
A method employed to localize radioisotopes in tissues or cells from experimental animals injected with radiolabeled substances. The radioisotopes serve as probes bound to specific DNA or RNA segments. Radioactivity is detected by placing the X-ray or photographic emulsion into contact with the tissue sections or nylon/nitrocellulose membranes in which they are localized to record sites of radioactivity. The technique permits the detection of radioactive substances by analytical methods involving electrophoresis, Southern blotting, and Northern blot hybridization.

autosensitization
Developing reactivity against one's own antigens, i.e., autoantigens, that occurs in autoimmunity or in autoimmune disease.

autosome
The nonsex (non-X and non-Y) chromosomes.

avidin

A 68-kD tetrameric egg white glycoprotein that has a very high affinity for and binds biotin, a water-soluble vitamin. Four identical 128-amino acid residue subunits, which bind a single biotin molecule each, comprise the avidin molecule. It also has an N-linked oligosaccharide and one disulfide bridge. Avidin's strong affinity for biotin has made it useful as an indicator molecule in a number of experimental methods. An enzyme can be linked to avidin and the complex can be bound to an antibody linked to biotin. The avidin-biotin-peroxidase complex (ABC) method is an immunoperoxidase reaction used extensively in antigen identification in histopathological specimens, especially in surgical pathological diagnosis. See also streptavidin.

avidin-biotin-peroxidase complex (ABC) technique

A method useful for the localization of peptide hormones or other antigens in formalin-fixed tissues. After incubating the tissue section with primary antibody specific for the antigen being sought, biotin-labeled secondary antibody is applied. This is followed by avidin-biotinylated horseradish peroxidase complex, which then binds to the biotinylated secondary antibody. The specific antigenic markers may be visualized in tissue sections by conventional light microscopy following incubation in a solution of peroxidase substrate. The technique makes use of the very high affinity which the 68-kD glycoprotein avidin, from egg white, has for biotin. The ease with which biotin may be covalently linked to antibody makes the ABC staining system feasible. In addition to the widespread use of the ABC technique in surgical pathologic diagnosis, the principle can be applied to *in situ* hybridization, gene mapping, double labeling, immunoelectron microscopy, southern blotting, radioimmunoassay, solid phase ELISA, hybridoma screening, etc.

avidity

The strength of binding between an antibody and its specific antigen. The stability of this union is a reflection of the number of binding sites which they share. Avidity is the binding force or intensity between multivalent antigen and a multivalent antibody. Multiple binding sites on both the antigen and the antibody, e.g., IgM or multiple antibodies interacting with various epitopes on the antigen, and reactions of high affinity between each of the antigens and its homologous antibody all increase the avidity. Such nonspecific factors such as ionic and hydrophobic interactions also increase avidity. Whereas affinity is described in thermodynamic terms, avidity is not, since it is described according to the assay procedure employed. The sum of the forces contributing to the avidity of an antigen and antibody interaction may be greater than the strength of binding of the individual antibody-antigen combinations contributing to the overall avidity of a particular interaction. K_a, the association constant for Ab + Ag = AbAg interaction, is frequently used to indicate avidity.

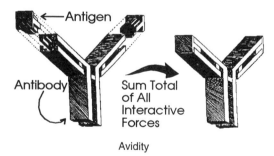

Avidity

axenic

An adjective implying freedom from association with or contamination by other organisms. For example, germ-free mice are raised in isolation. Axenic refers also to pure cultures of microorganisms. Called also gnotobiotic and germ free.

azathioprine

A nitroimidazole derivative of 6-mercaptopurine, a purine antagonist. Following administration, it is converted to 6-mercaptopurine *in vivo*. Its principal action is to interfere with DNA synthesis. Of less significance is its ability to impair RNA synthesis. Azathioprine has a greater inhibitory effect on T cell than on B cell responses, even though it suppresses both cell-mediated and humoral immunity. It diminishes circulating NK and killer cell numbers. It has been used to treat various autoimmune disorders, including rheumatoid arthritis, other connective tissue diseases, autoimmune blood diseases, and immunologically mediated neurological disorders. It is active chiefly against reproducing cells. The drug has little effect on immunoglobulin levels or antibody titers, but it does diminish neutrophil and monocyte numbers in the circulation.

Azathioprine

(6-[(1-methyl-4-nitro-1H-imidazol-5-yl)thio]-1H-purine)

azidothymidine

Synonym for zidovudine.

azoprotein

The joining of a substance to a protein through a diazo linkage –N=N–. Karl Landsteiner (in early 1900s) made extensive use of diazotization to prepare hapten-protein conjugates to define immunochemical specificity. See also diazo reaction.

AZT

3′-azido-3′-deoxythymidine. Refer to zidovudine.

Azidothymidine (AZT)

(3′-Azido-3′-deoxythymidine)

B

b allotype
A rabbit immunoglobulin κ light chain allotype encoded by alleles at the *K1* locus.

B antigen, acquired
The alteration of A1 erythrocyte membrane through the action of such bacteria as *Escherichia coli, Clostridium tertium,* and *Bacteroides flagilis* to make it react as if it were a group B antigen. The named microorganisms can be associated with gastrointestinal infection or carcinoma.

B blood group
Refer to ABO blood group system.

B cell antigen receptor
An antibody expressed on antigen reactive B cells that is similar to secreted antibody but is membrane-bound due to an extra domain at the Fc portion of the molecule. Upon antigen recognition by the membrane-bound immunoglobulin, noncovalently associated accessory molecules mediate transmembrane signaling to the B cell nucleus. The immunoglobulin and accessory molecule complex is similar in structure to the antigen receptor–CD3 complex of T lymphocytes.

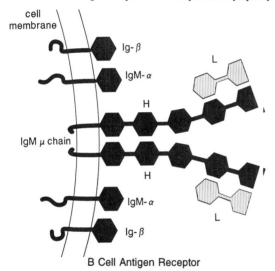

B Cell Antigen Receptor

B cell co-receptor
A three-protein complex that consists of CR2, TAPA-1, and CD19. CR2 unites not only with an activated component of complement, but also with CD23. TAPA-1 is a serpentine membrane protein.

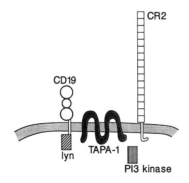

B-cell co-receptor complex

CD19's cytoplasmic tail is the mechanism through which the complex interacts with lyn, a tyrosine kinase. Activation of the co-receptor by ligand binding leads to union of phosphatidyl inositol-3′ kinase with CD19 resulting in activation. This produces intracellular signals that facilitate B cell receptor signal transduction.

B cell differentiation and growth factors
T lymphocyte-derived substances that promote differentiation of B lymphocytes into antibody producing cells. They can facilitate the growth and differentiation of B cells *in vitro*. Interleukins 4, 5, and 6 belong in this category of factors.

B cell growth factor (BCGF)
See interleukins 4, 5, and 6.

B cell growth factor I (BCGF-1)
An earlier term for interleukin-4.

B cell growth factor II (BCGF-2)
An earlier term for interleukin-5.

B cell lymphoproliferative syndrome (BLS)
A rare complication of immunosuppression in bone marrow or organ transplant recipients. Epstein-Barr virus appears to be the etiologic agent. It occurs in less than 1% of HLA-identical bone marrow recipients and is more likely in those where anti-CD3 monoclonal antibodies were used to treat graft-vs.-host disease. Clinically, it may be either a relatively mild infectious mononucleosis or a proliferating and relentless lymphoma that produces high mortality. Monoclonal antibodies to the B cell antigens, CD21 and CD24, have proven effective in controlling the B cell proliferation, but further studies are needed.

B cell-stimulating factor 1 (BSF-1)
An earlier term for interleukin-4.

B cell-stimulating factor 2 (BSF-2)
An earlier term for interleukin-6.

B cell tolerance
B cell tolerance is manifested as a decreased number of antibody-secreting cells following antigenic stimulation, compared to a normal response. Hapten-specific tolerance can be induced by inoculation of deaggregated haptenated gamma globulins (Ig). Induction of tolerance requires membrane Ig crosslinking. Tolerance may have a duration of 2 months in B cells of the bone marrow and 6 to 8 months in T cells. Whereas prostaglandin E enhances tolerance induction, IL-1, LPS, or 8-bromoguanosine block tolerance instead of an immunogenic signal. Tolerant mice carry a normal complement of hapten-specific B cells. Tolerance is not attributable to a diminished number or isotype of antigen receptors. It has also been shown that the six normal activation events related to membrane Ig turnover and expression do not occur in tolerant B cells. Whereas tolerant B cells possess a limited capacity to proliferate, they fail to do so in response to antigen. Antigenic challenge of tolerant B cells induces them to enlarge and increase expression, yet they are apparently deficient in a physiologic signal requisite for progression into a proliferative stage.

B cells
The B lymphocytes that derive from the fetal liver in the early embryonal stages of development and from the bone marrow thereafter. In birds, maturation takes place in the bursa of Fabricius, a lymphoid structure derived from an outpouching of the hindgut near the cloaca. In mammals, maturation is in the bone marrow. Plasma cells that synthesize antibody develop from precursor B cells.

β cells
Insulin secreting cells in the islet of Langerhans of the pancreas.

9–12μm

B Cell

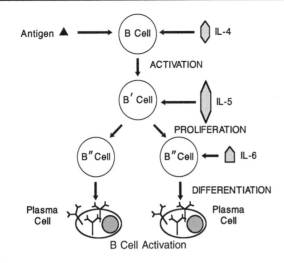

B Cell Activation

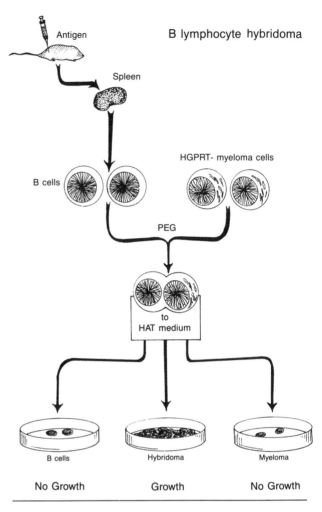

B lymphocyte hybridoma

No Growth   Growth   No Growth

B complex
The major histocompatibility complex (MHC) in chickens. Genes at these loci determine MHC class I and II antigens and erythrocyte antigens.

B locus
The major histocompatibility locus in the chicken.

B lymphocyte
Lymphocytes of the B cell lineage that mature under the influence of the bursa of Fabricus in birds and the bursa equivalent (bone marrow) in mammals. B cells occupy follicular areas in lymphoid tissues and account for 30% of the lymphocytes in the circulating blood. They synthesize antibodies and provide defense against microorganisms including bacteria and viruses. Surface and cytoplasmic markers reveal the stage of development and function of lymphocytes in the B cell lineage. Pre-B cells contain cytoplasmic immunoglobulins, whereas mature B lymphocytes express surface immunoglobulins and complement receptors. B lymphocyte markers include CD9, CD19, CD20, CD24, Fc receptors, B1, BA-1, B4, and Ia.

B lymphocyte hybridoma
A clone formed by the fusion of a B lymphocyte with a myeloma cell. Activated splenic B lymphocytes from a specifically immune mouse are fused with myeloma cells by polyethylene glycol. Thereafter, the cells are plated in HAT medium in tissue culture plates containing multiple wells. The only surviving cells are the hybrids, since the myeloma cells employed are deficient in HAT medium and fail to grow in HAT medium. Wells with hybridomas are screened for antibody synthesis. This is followed by cloning which is carried out by limiting the dilution or in soft agar. The hybridomas are maintained either in tissue culture or through inoculation into the peritoneal cavity of a mouse that corresponds genetically to the cell strain. The antibody-producing B lymphocyte confers specificity and the myeloma cell confers immortality upon the hybridoma. B lymphocyte hybridomas produce monoclonal antibodies.

B lymphocyte receptor
Immunoglobulin anchored to the B lymphocyte surface. Its combination with antigen leads to B lymphocyte division and differentiation into memory cells, lymphoblasts, and plasma cells. The original antigen specificity of the immunoglobulin is maintained in the antibody molecules subsequently produced. B lymphocyte receptor immunoglobulins are to be distinguished from those in the surrounding medium that adhere to the B cell surface through Fc receptors. Refer to membrane immunoglobulin.

B lymphocyte stimulatory factors
See interleukins 4, 5, and 6.

B lymphocyte tolerance
Immunologic nonreactivity of B lymphocytes induced by relatively large doses of antigen. It is a relatively short duration. By contrast, T cell tolerance requires less antigen and is a longer duration. Exclusive B cell tolerance leaves T cells immunoreactive and unaffected.

β lysin
A thrombocyte-derived antibacterial protein that is effective mainly against Gram-positive bacteria. It is released when blood platelets are disrupted, as occurs during clotting. β lysin acts as a nonantibody humoral substance that contributes to nonspecific immunity.

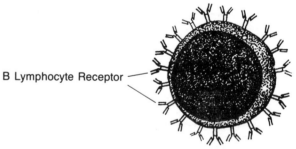

B Lymphocyte Receptor

β-pleated sheet
A protein configuration in which the β sheet polypeptide chains are extended and have a 35-nm axial distance. Hydrogen bonding between NH and CO groups of separate polypeptide chains stabilize the molecules. Adjacent molecules may be either parallel or antiparallel. The β-pleated sheet configuration is characteristic of amyloidosis and is revealed by Congo red staining followed by

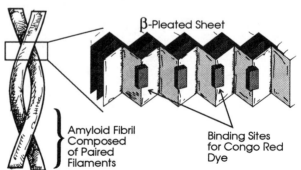

β-Pleated Sheet

Amyloid Fibril Composed of Paired Filaments

Binding Sites for Congo Red Dye

polarizing light microscopy which yields an apple-green birefringence and ultrastructurally consists of nonbranching fibrils.

β propiolactone

A substance employed to inactivate the nucleic acid core of pathogenic viruses without injuring the capsids. This permits the development of an inactivated vaccine, as the immunizing antigens that induce protective immunity are left intact.

B-type virus (Aspergillus macaques)

An Old World monkey virus that resembles herpes simplex. Clinical features include intermittent shedding and reactivation in the presence of stress and immunosuppression. Humans who tend these monkeys may become infected with fatal consequences. B-type viruses possess an eccentric nuclear core.

β₁A globulin

A breakdown product of β_1C globulin. It has a mol wt less than that of β_1C globulin, and its electrophoretic mobility is more rapid than that of the β_1C globulin. β_1A degradation is linked to the disappearance of C3 activity.

β₁E globulin

The globulin fraction of serum that contains complement component C4 activity.

β₁F globulin

The globulin fraction of serum that contains complement component C5 activity.

β₁H

Refer to factor H.

β₂ microglobulin (β₂M)

A thymic epithelium derived polypeptide that is 11.8 kD and makes up part of the MHC class I molecule which appears on the surfaces of nucleated cells. It is noncovalently linked to the MHC class I polypeptide chain. It promotes maturation of T lymphocytes and serves as a chemotactic factor. β_2M makes up part of the peptide-antigen class I-β_2 microglobulin complex involved in antigen presentation to cytotoxic T lymphocytes. Nascent β_2M facilitates the formation of antigenic complexes that can stimulate T lymphocytes. This monomorphic polypeptide accumulates in the serum in renal dialysis patients and may lead to β_{-2} microglobulin-induced amyloidosis.

b4, b5, b6, and b9

The four alleles whose κ chains vary in multiple constant region amino acid residues.

B7

A homodimeric immunoglobulin superfamily protein whose expression is restricted to the surface of cells that stimulate growth of T lymphocytes. The ligand for B7 is CD28. B7 is expressed by accessory cells and is important in costimulatory mechanisms.

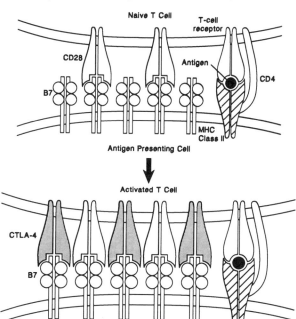

Some APCs may upregulate expression of B7 following activation by various stimuli including IFN-α, endotoxin, and MHC class II binding. B7 is also termed BB1, B7.1, or CD80.

B7.2

A costimulatory molecule whose sequence resembles that of B7. Dendritic cells, monocytes, activated T cells, and activated B lymphocytes may express B7.2.

back typing

The interaction of antibodies in an individual's serum with known antigens of an erythrocyte panel. To ascertain whether or not the person's serum contains antierythrocyte antibodies. Also called reversed typing.

backcross

Breeding an F₁ hybrid with either one of the strains that produced it.

bacterial agglutination

Antibody-mediated aggregation of bacteria. This technique has been used for a century in the diagnosis of bacterial diseases through the detection of an antibody specific for a particular microorganism or for the identification of a microorganism isolated from a patient.

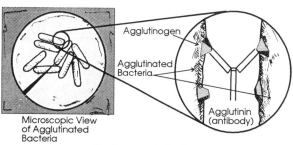

Bacterial Agglutination

bacterial allergy

Delayed-type hypersensitivity of infection such as in tuberculosis.

bacterial hypersensitivity

Refer to delayed-type hypersensitivity (type IV).

bactericidin

An agent such as an antibody or nonantibody substance in blood plasma which destroys bacteria.

bacterin

A vaccine comprised of killed bacterial cells in suspension. Inactivation is by either chemical or physical treatment.

bacteriolysin

An agent such as an antibody or other substance that lyses bacteria.

bacteriolysis

The disruption of bacterial cells by such agents as antibody and complement or lysozyme, causing the cells to release their contents.

bacteriophage λ

λ bacteriophage.

bacteriophage neutralization test

Refer to phage neutralization assay.

bagassosis

Hypersensitivity among sugar cane workers to a fungus, *Thermoactinomyces saccharic,* that thrives in the pressings from sugar cane. The condition is expressed as a hypersensitivity pneumonitis. Subjects develop type III (Arthus reaction) hypersensitivity following inhalation of dust from molding hot sugar cane bagasse.

Bakᵃ

A normal human platelet (thrombocyte) antigen. Anti-Bakᵃ IgG antibody synthesized by a Bakᵃ negative pregnant female may be passively transferred across the placenta to induce immune thrombocytopenia in the neonate.

BALB/c mice

An inbred white mouse strain that responds to an intraperitoneal inoculation of mineral oil and Freund's complete adjuvant with a myeloproliferative reaction.

BALT

See bronchial-associated lymphoid tissue.

band test

Antigen-antibody deposits at the dermal-epidermal junction in patients with lupus erythematosus. They may consist of IgG, IgM,

IgA, and C3. Deposits are not found in uninvolved areas of the dermal-epidermal junction in discoid lupus erythematosus (DLE), but are present in both involved and uninvolved areas of the dermal-epidermal junction in systemic lupus erythematosus (SLE). Immune complexes at the dermal-epidermal junction appear in 90 to 95% of SLE patients. Ninety percent reveal them in skin exposed to sunlight, and 50% have deposits in skin that is not exposed to the sun. Also called lupus band test. Immunofluorescence bands also occur in some cases of anaphylactoid purpura, atopic dermatitis, contact dermatitis, autoimmune thyroiditis, bullous pemphigoid, cold agglutinin syndrome, dermatomyositis, hypocomplementemic vasculitis, polymorphous light eruption, rheumatoid arthritis, scleroderma, and a number of other conditions.

bare lymphocyte syndrome

Failure to express class I HLA-A, -B, or -C major histocompatibility antigens due to defective ß$_2$ microglobulin expression on the cell surface. This immune deficiency is inherited as an autosomal recessive trait. In some individuals, the class II HLA-DR molecules are likewise not expressed. Patients may be asymptomatic or manifest respiratory tract infections, mucocutaneous candidiasis, opportunistic infections, chronic diarrhea and malabsorption, aplastic anemia, inadequate responsiveness to antigen, decreased T lymphocytes, leukopenia, and normal or elevated B lymphocytes. The mechanism appears to be related to either defective gene activation or inaccessibility of promoter protein. DNA techniques are required for tissue typing.

barrier filter

A device in the eyepiece of an ultraviolet light source microscope employed for fluorescent antibody techniques to protect the viewer's eyes from injury by ultraviolet radiation. It also facilitates observation of the fluorescence by blocking light of the wavelength that produces fluorochrome excitation.

bas

Mutation in an mRNA splicing acceptor site in rabbits that leads to diminished expression of the principal type of immunoglobulin κ light chain (MO-κ-1). Rabbits with the bas mutation have λ light chains, although a few have the κ-2 isotype of light chains.

basement membrane antibody;

Antibodies specific for the basement membrane of various tissues such as the lung basement membranes, the glomerular basement membrane, etc. This antibody is usually observed by immunofluorescence and less often by immunoperoxidase technology.

basophil

A polymorphonuclear leukocyte of the myeloid lineage with distinctive basophilic secondary granules in the cytoplasm that frequently overlie the nucleus. These granules are storage depots for heparin, histamine, platelet-activating factor, and other pharmacological mediators of immediate hypersensitivity. Degranulation of the cells with release of these pharmacological mediators takes place following crosslinking by allergen or antigen of Fab regions of IgE receptor molecules bound through Fc receptors to the cell surface. They comprise less than 0.5% of peripheral blood leukocytes. Following crosslinking of surface-bound IgE molecules by specific allergen or antigen, granules are released by exocytosis. Substances liberated from the granules are pharmacological mediators of immediate (type I) anaphylactic hypersensitivity.

Basophil

basophil-derived kallikrein (BK-A)

BK-A represents the only known instance where an activator of the kinin system is generated directly from a primary immune reaction. The molecule is a high mol wt enzyme with arginine esterase activity. It is stored in the producing cells in a preformed state. Its release depends on basophil-IgE interactions with antigen and parallels the release of histamine.

basophilic

Adjective that refers to an affinity of cells or tissues for basic stains leading to a bluish tint.

β$_1$C globulin

The globulin fraction of serum that contains complement component C3. On storage of serum, β$_1$C dissociates into β$_1$A globulin, which is inactive.

BCDF

B cell differentiation factors.

BCG (bacille Calmette-Guerin)

A *Mycobacterium bovis* strain maintained for more than 75 years on potato, bile glycerine agar, which preserves the immunogenicity, but dissipates the virulence of the microorganism. It has long been used in Europe as a vaccine against tuberculosis, although it never gained popularity in the U.S. It has also been used in tumor immunotherapy to nonspecifically activate the immune response in selected tumor-bearing patients, such as those with melanoma. It has been suggested as a possible vector for genes that determine HIV proteins such as *gag, pol, env,* gp20, gp40, reverse transcriptase, and tetanus toxin.

BCGF (B cell growth factors)

See interleukins 4, 5, and 6.

BDB

Refer to bis-diazotized benzidine.

Behcet's disease

Oral and genital ulcers, vasculitis, and arthritis that recur as a chronic disease in young men. It is postulated to have an immunologic basis and possibly be immune complex mediated. There are perivascular infiltrates of lymphocytes. The serum contains immune complexes, and immunofluorescence may reveal autoantibodies against the oral mucous membrane. It is associated with HLA-B5 in subjects from the Middle East or Japan, but not Caucasians.

Behring, Emil Adolph von (1854–1917)

German bacteriologist. With Kitasato, he demonstrated that circulating antitoxins against diphtheria and tetanus conferred immunity. Received the first Nobel Prize in medicine in 1901 for this work. *Die Blutserumtherapie,* 1902; *Gesammelte Abhandlungen,* 1915; *Behring, Gestalt und Werk,* 1940; *Emil von Behring zum Gedachtnis,* 1942.

beige mice

A mutant strain of mice that develops abnormalities in pigment, defects in natural killer cell function, and heightened tumor incidence. This serves as a model for the Chediak-Higashi disease in man.

Benacerraf, Baruj (1920–)

American immunologist born in Caracas, Venezuela. His multiple contributions include the carrier effect in delayed hypersensitivity, lymphocyte subsets, MHC, and Ir immunogenetics, for which he received the Nobel Prize in 1980. *Textbook of Immunology (with E. Unanue),* 1979.

Bence-Jones (B-J) proteins

Represent the light chains of either the κ or λ variant excreted in the urine of patients with a paraproteinemia as a result of excess synthesis of such chains or from mutant cells which make only such chains. Both mechanisms appear operative, and over 50% of patients with multiple myeloma, a plasma cell neoplasm, have B-J proteinuria. The highest frequency of B-J excretion is seen in IgD myeloma; the lowest is seen in IgG myeloma. The daily amount excreted parallels the severity of the disease. The B-J proteins are

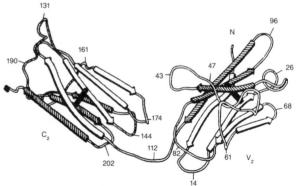

Bence-Jones Protein

secreted mostly as dimers and show unusual heat solubility properties. They precipitate at temperatures between 40 to 60°C and redissolve again near 100°C. With proper pH control and salt concentration, precipitation may detect as low as 30 mg/100 ml of urine. Better identification is by protein electrophoresis.

benign lymphadenopathy

Lymph node enlargement that is not associated with malignant neoplasms. Histologic types of benign lymphadenopathy include nodular, granulomatous, sinusoidal, paracortical, diffuse or obliterative, mixed, and depleted, whereas clinical states are associated with each histologic pattern.

benign lymphoepithelial lesion

Autoimmune lesion in lacrimal and salivary glands associated with Sjögren's syndrome. There are myoepithelial cell aggregates together with extensive lymphocyte infiltration.

benign monoclonal gammopathy

A paraproteinemia that occurs in normal healthy subjects who develop the serum changes characteristic of myeloma, i.e., a myeloma protein-type immunoglobulin spike on electrophoresis. They have none of the clinical signs and symptoms of multiple myeloma and have an excellent prognosis.

bentonite ($Al_2O_3 \cdot 4SiO_2 \cdot H_2O$)

Aluminum silicate that is hydrated and colloidal. This insoluble particulate substance has been used to adsorb proteins, including antigens. It was used in the past in the bentonite flocculation test.

bentonite flocculation test

An assay in which bentonite particles were used as carriers to adsorb antigens. These antigen-coated bentonite particles were then agglutinated by the addition of a specific antibody.

berylliosis

A disease induced by the inhalation of beryllium in dust. Subjects may develop a delayed-type hypersensitivity (type IV) reaction to beryllium-macromolecular complexes. Either an acute chemical pneumonia or chronic pulmonary granulomatous disease that resembles sarcoidosis may develop. Granulomas may form and lead to pulmonary fibrosis. The skin, lymph nodes, or other anatomical sites may be affected. Subjects who develop the chronic progressive granulomatous pulmonary disease appear to be sensitized by beryllium.

Besredka, Alexandre (1870–1940)

Parisian immunologist who worked with Metchnikoff at the Pasteur Institute. He was born in Odessa. He contributed to studies of local immunity, anaphylaxis, and antianaphylaxis. *Anaphylaxie et Antianaphylaxie*, 1918; *Histoire d'une Idee: L'Oeuvre de Metchnikoff*, 1921; *Etudes sur l'Immunite dans les Maladies Infectieuses*, 1928.

beta-gamma bridge

Patients with chronic liver disease such as that caused by alcohol, chronic infection, or connective tissue disease may synthesize sufficient polyclonal proteins whose electrophoretic mobilities are in the beta-gamma range to cause obliteration of the beta and gamma peaks, forming a "bridge" from one to the other.

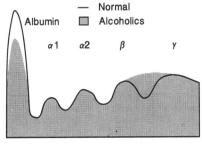

Beta-Gamma Bridge

BFPR

Abbreviation for biological false-positive reaction.

BI-RG-587

A powerful inhibitor of reverse transcriptase in man. This dipyridodiazepinone can prevent the replication of HIV-1 *in vitro*. It can be used in conjunction with such nucleoside analogs as zidovudine, ddI, and ddC, as well as in subjects whose HIV-1 infection no longer responds to these drugs.

biclonality

In contrast to uncontrolled proliferation of a single clone of neoplastic cells which is usually associated with tumors, rarely two neoplastic cell clones may proliferate simultaneously, leading to a biclonality. For example, either neoplastic B or T lymphocytes could demonstrate this effect.

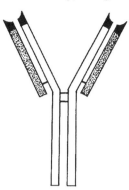

binding constant

Refer to association constant.

binding protein

Also called immunoglobulin heavy chain-binding protein.

binding site

In immunology, the paratope area of an antibody molecule that binds antigen or that part of the T cell receptor which is antigen binding.

biochemical sequestration

Antigenic determinants that are hidden in a molecule may be unable to act as immunogens or to react with antibody. Structural alterations in the molecule may render them identifiable and capable of serving as immunogens.

biological false-positive reaction

A positive serological test for syphilis, such as in the VDRL (Venereal Disease Research Laboratory) serological test for syphilis produced by the serum of an individual who is not infected with *Treponema palladium*. It is attributable to antibodies reactive with antigens of tissues such as the heart from which cardiolipin antigen used in the test is derived. The blood sera of patients with selected autoimmune diseases, including systemic lupus erythematosus, may contain antibodies which give a biological false-positive test for syphilis.

biological response modifiers (BRM)

A wide spectrum of molecules that alter the immune response. They include substances such as interleukins, interferons, hematopoietic colony-stimulating factors, tumor necrosis factor, B lymphocyte growth and differentiating factors, lymphotoxins, and macrophage-activating and chemotactic factors, as well as macrophage inhibitory factor, eosinophil chemotactic factor, osteoclast activating factor, etc. BRM may modulate the immune system of the host to augment antitumor defense mechanisms. Some have been produced by recombinant DNA technology and are available commercially. An example is α interferon used in the therapy of hairy cell leukemia.

biologicals

Substances used for therapy that include antitoxins, vaccines, products prepared from pooled blood plasma, and biological response modifiers (BRM). BRMs are prepared by recombinant DNA technology and include lymphokines such as interferons, interleukins, tumor necrosis factor, etc. Monoclonal antibodies for therapeutic purposes also belong in this category. Biologicals have always presented problems related to chemical and physical standardization to which drugs are subjected. Biologicals are regulated by the Food and Drug Administration within the U.S.

biovin antigens

Salmonella O antigens. These carbohydrate-lipid-protein complexes withstand trichloroacetic acid treatment.

BiP

A chaperonin that binds unassembled heavy and light chains after they are synthesized in the endoplasmic reticulum. Chains that are malformed are not allowed to leave the ER and are thus not used in immunoglobulin assembly.

Birbeck granules

A 10- to 30-nm diameter round cytoplasmic vesicle present in the cytoplasm of Langerhans cells in the epidermis.

bird fancier's lung

Respiratory distress in subjects who are hypersensitive to plasma protein antigens of birds following exposure of the subject to bird feces or skin and feather dust. Hypersensitive subjects have an Arthus type of reactivity or type III hypersensitivity to the plasma albumin and globulin components. Precipitates may be demonstrated in the blood sera of hypersensitivity subjects.

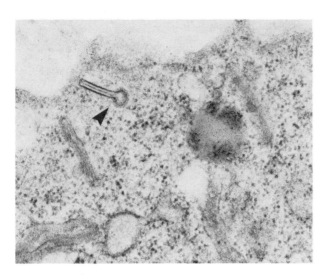

Birbeck Granules

bis-diazotized benzidine

A chemical substance that serves as a bivalent coupling agent which can link to protein molecules. This method was used in the past to conjugate erythrocytes with antigens for use in the passive agglutination test.

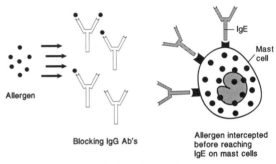

bis-diazotized benzidine

bispecific antibodies

Molecules that have two separate antigen-binding specificities. They may be produced by either cell fusion or chemical techniques. An immunoglobulin molecule in which one of two antigen-binding sites is specific for one antigen-binding specificity, whereas the other antigen-binding site is specific for a different antigen specificity. This never occurs in nature, but it can be produced *in vitro* by treating two separate antibody specificities with mild reducing agents converting the central disulfide bonds of both antibody molecules to sulfhydryl groups, mixing the two specificities of half molecules together, and allowing them to reoxidize to form whole molecules, some of which will be bispecific.

BLA-36

An antigen demonstrable by immunoperoxidase staining in Reed-Sternberg cells of all types of Hodgkin's disease and in activated B lymphocytes and B cell lymphomas.

blast cell

A relatively large cell that is greater than 8 μm in diameter with abundant RNA in the cytoplasm, a nucleus with loosely arranged chromatin and a prominent nucleolus. Blast cells are active in synthesizing DNA and contain numerous polyribosomes in the cytoplasm.

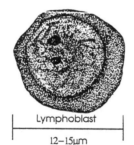

Lymphoblast

12–15μm

Blast Cell

blast transformation

The activation of small lymphocytes to form blast cells.

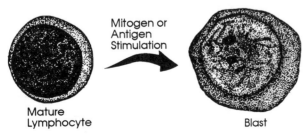

Blastogenesis

Mature Lymphocyte

Mitogen or Antigen Stimulation

Blast

blocking

Prevention of nonspecific interaction of an antibody with a certain antigenic determinant, whose identification is sought, by washing with mammalian serum other than that being used in the test system. For example, enzyme-linked immunosorbent assays (ELISA) employ blocking.

blocking antibody

(1) An incomplete IgG antibody that, when diluted, may combine with red blood cell surface antigens and inhibit agglutination reactions used for erythrocyte antigen identification. This can lead to errors in blood grouping for Rh, K, and k blood types. Pretreatment of red cells with enzymes may correct the problem. (2) An IgG antibody specifically induced by exposure of allergic subjects to specific allergens, to which they are sensitive, in a form that favors IgG rather than IgE production. The IgG, specific for the allergens to which they are sensitized, competes within IgE molecules bound to mast cell surfaces, thereby preventing their degranulation and inhibiting a type I hypersensitivity response. (3) A specific immunoglobulin molecule that may inhibit the combination of a competing antibody molecule with a particular epitope. Blocking antibodies may also interfere with the union of T cell receptors with an epitope for which they are specific, as occurs in some tumor-bearing patients with blocking antibodies which may inhibit the tumoricidal action of cytotoxic T lymphocytes.

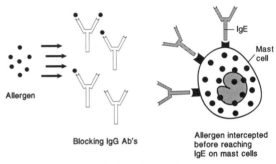

Allergen

Blocking IgG Ab's

IgE

Mast cell

Allergen intercepted before reaching IgE on mast cells

Blocking Antibodies

blocking factor

Agents such as immune complexes in the serum of tumor-bearing hosts that interfere with the capacity of immune lymphoid cells to mediate cytotoxicity of tumor target cells.

blocking test

An assay in which the interaction between an antigen and its homologous antibody is inhibited by the previous exposure of the antigen to a different antibody which has the same specificity as the first one, but does not have the same biological function. In a different situation, a hapten may be used to prevent the reaction of an antibody with its intended antigen. This is referred to as the hapten inhibition test. An example would be blood group substance soluble molecules equivalent to erythrocyte surface isoantigen epitopes found in the body fluids. Refer to ABO blood group substances.

blood group

The classification of erythrocytes based on their surface isoantigens. Among the well-known human blood groups are the ABO, Rh, and MNS systems.

blot

The transfer of DNA, RNA, or protein molecules from an electrophoretic gel to a nitrocellulose or nylon membrane by osmosis or vacuum, followed by immersing the membrane in a solution containing a complementary, i.e., mirror-image molecule corresponding to the one on the membrane. This is known as a hybridization blot.

Bombay phenotype

The O_h phenotype is an ABO blood group antigen variant on human erythrocytes in rare subjects. These red blood cells do not possess A, B, or H antigens on their surfaces, even though the subject does have anti-A, anti-B, and anti-H antibodies in the serum. The Bombay phenotype may cause difficulties in crossmatching for transfusion.

bombesin

A neuropeptide of 14 residues that is analogous to a gastrin-releasing peptide that is synthesized in the gastrointestinal tract and induces GI smooth muscle contraction and the release of stomach acid and the majority of GI hormones with the exception of secretin. Bombesin injection into the brain may induce hyperglucagonemia, hyperglycemia, analgesia, and hypothermia. Bombesin facilitates bronchial epithelial cell proliferation and pancreas and small cell carcinoma. Antibombesin antibodies might prove useful in the future for the treatment of lung small cell carcinoma whose cells bear bombesin receptors.

bone marrow
Soft tissue within bone cavities that contains hematopoietic precursor cells and hematopoietic cells that are maturing into erythrocytes, the five types of leukocytes, and thrombocytes. Whereas red marrow is hemopoietic and is present in developing bone, ribs, vertebrae, and long bones, some of the red marrow may be replaced by fat and become yellow marrow.

bone marrow cell
Stem cells from which the formed elements of the blood, including erythrocytes, leukocytes, and platelets, are derived. B lymphocyte and T lymphocyte precursors are abundant. The B lymphocytes and pluripotent stem cells in bone marrow are important for reconstitution of an irradiated host. Bone marrow transplants are useful in the treatment of aplastic anemia, leukemias, and immunodeficiencies. Patients may donate their own marrow for subsequent bone marrow autotransplantation if they are to receive intensive doses of irradiation.

bone marrow transplantation
A procedure used to treat both nonneoplastic and neoplastic conditions not amenable to other forms of therapy. It has been especially used in cases of aplastic anemia, acute lymphocytic leukemia, and acute nonlymphocytic leukemia. Using an HLA-matched donor, 750 ml of bone marrow are removed from the iliac crest. Following appropriate treatment of the marrow to remove bone spicules, the cell suspension is infused intravenously into an appropriately immunosuppressed recipient who has received whole body irradiation and immunosuppressive drug therapy. Graft-vs.-host episodes, acute graft-vs.-host disease, or chronic graft-vs.-host disease may follow bone marrow transplantation in selected subjects. See graft-vs.-host disease. The immunosuppressed patients are highly susceptible to opportunistic infections.

booster
A second administration of immunogen to an individual primed months or years previously by a primary injection of the same immunogen. The purpose is to deliberately induce a secondary or anamnestic immune response to facilitate protection against an infectious disease agent.

booster injection
The administration of a second inoculation of an immunizing preparation, such as a vaccine, to which the individual has been previously exposed. The booster inoculation elicits a recall or anamnestic response through stimulation of memory cells that have encountered the same antigen previously. Booster injections are given after the passage of time sufficient for a primary immune response specific for the immunogen to have developed. Booster injections are frequently given to render the subject immune prior to the onset of a particular disease or to protect the individual when exposed to subjects infected with the infectious disease agent against which immunity is desired.

booster phenomenon
An expansion in the diameter of a tuberculin reaction following the administration of a subsequent PPD skin test for tuberculosis. This is usually greater than 6 mm and shows an increase in size from below 10 mm to greater than 10 mm in diameter following the secondary challenge. A positive test suggests an increased immunologic recall as a consequence of either previous infection with *Mycobacterium tuberculosis* or other mycobacteria. It is seen in older subjects with previous *M. tuberculosis* infections who fail to convert to active disease.

booster response
The secondary antibody response produced during immunization of subjects primed by earlier exposure to the same antigen. Also called an anamnestic response and secondary response.

Bordet, Jules Jean Baptiste Vincent (1870–1961)
Belgian bacteriologist and immunologist. He received the Nobel Prize in 1919 "for his studies in regard to immunity." His contributions include work on complement-mediated bacteriolysis and specific hemolysis. With Octave Gengou he discovered complement fixation and described its potential in infectious disease diagnosis. *Trait's de l'Immunite dans les Maladies Infectieuses,* 1920.

Bordetella pertussis
The etiologic agent of whooping cough in children. Killed *B. pertussis* microorganisms are administered in a vaccine together with diphtheria toxoid and tetanus toxoid as DPT. The endotoxin of *B. pertussis* has an adjuvant effect that can facilitate antibody synthesis.

botulinum toxin
A toxin formed by *Clostridium botulinum.* The 150-kD type A toxin is available in purified form and is employed to treat a neuromuscular junction disease such as dystonias. It acts by combining with the presynaptic cholinergic nerve terminals where it is internalized and prevents exocytosis of acetylcholine. Subsequently, sprouting takes place, and new terminals are formed which reinnervate the muscle.

Bovet, Daniel (1907–)
Primarily a pharmacologist and physiologist, Bovet received the Nobel Prize in 1957 for his contributions to the understanding of the role histamine plays in allergic reactions and the development of antihistamines. *Structure chimique et Activite Pharmacodynamique des Medicaments du Systems Nerveux Vegetatif,* 1948; *Curare and Curare-Like Agents,* 1959.

bovine serum albumin (BSA)
Albumin in the serum of cows that has been used extensively as an antigen in experimental immunologic research.

Boyden chamber
A two-compartment structure used in the laboratory to assay chemotaxis. The two chambers in the apparatus are separated by a micropore filter. The cells to be tested are placed in the upper chamber and a chemotactic agent such as F-met-leu-phe is placed in the lower chamber. As cells in the upper chamber settle to the filter surface, they migrate through the pores if the agent below chemoattracts them. On staining of the filter, cell migration can be evaluated.

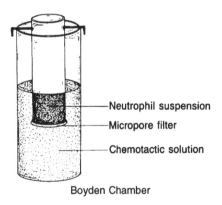

—Neutrophil suspension

—Micropore filter

—Chemotactic solution

Boyden Chamber

bradykinin
A 9-amino acid peptide split by plasma kallikrein from plasma kininogens. It produces slow, sustained, smooth muscle contraction. Its action is slower than is that of histamine. It is produced in experimental anaphylaxis in animal tissues. Its sequence is Arg-Pro-Pro-Gly-Phe-Ser-Pro-Phe-Arg. Besides anaphylaxis, bradykinin is also increased in endotoxin shock. Lysyl-bradykinin (kallidin), which is split from kininogens by tissue kallikreins, also has a lysine residue at the amino terminus.

Brequinar sodium (BQR)
A recently developed antineoplastic and immunosuppressive agent. Its major activity is inhibition of the de novo biosynthesis of pyrimidine nucleosidases, resulting in inhibition of both DNA and RNA synthesis. BQR has also been shown to interfere with IgM production by IL-6 stimulated SKW6.4 cells, although in a manner independent of DNA synthesis. In transplantation studies, BQR has been shown to inhibit both the humoral and the cellular immune responses of the host, thereby significantly suppressing acute and antibody-mediated graft rejection.

CO$_2$Na

Brequinar Sodium

BRMs

Abbreviation for biological response modifiers.

bromelin

An enzyme that has been used to render erythrocyte surfaces capable of being agglutinated by incomplete antibody.

bronchial-associated lymphoid tissue (BALT)

BALT is present in both mammals, including man, and birds. In many areas it appears as a collar containing nodules located deep around the bronchus and connected with the epithelium by patches of loosely arranged lymphoid cells. Germinal centers are absent (except in the chicken), although cells in the center of nodules stain lighter than do those at the periphery. Plasma cells are present occasionally beneath the epithelium. The cells in BALT have a high turnover rate and apparently do not produce IgG. BALT development is independent of that of the peripheral lymphoid tissues or antigen exposure. The cells of BALT apparently migrate there from other lymphoid areas.

***Brucella* vaccine**

A preparation used for the prophylactic immunization of cattle. It contains live, attenuated *Brucella abortus* microorganisms. A second vaccine comprised of McEwen strain 45/20 killed microorganisms in a water-in-oil emulsion (adjuvant) has also been used.

brucellin

A substance similar to tuberculin, but derived from a culture filtrate of *Brucella abortus* that is used to test for the presence of delayed-type hypersensitivity to brucella antigens. The test is of questionable value in diagnosis.

Bruton's agammaglobulinemia

Synonym for X-linked agammaglobulinemia.

BSA

Abbreviation for bovine serum albumin.

BSF (B lymphocyte stimulatory factors)

Refer to interleukins 4, 5, and 6.

btk

A protein tyrosine kinase coded for by the defective gene in X-linked agammaglobulinemia (XLA). B lymphocytes and polymorphonuclear neutrophils express the btk protein. In XLA (Bruton's disease) patients, only the B lymphocytes manifest the defect, and the maturation of B lymphocytes stops at the pre-B cell stage. There is rearrangement of heavy chain genes but not of the light chain genes. The btk protein might have a role in linking the pre-B-cell receptor to nuclear changes that result in growth and differentiation of pre-B cells.

bubble boy

A 12-year-old male child maintained in a germ-free (gnotobiotic) environment in a plastic bubble from birth because of his severe combined immunodeficiency. A bone marrow transplant from a histocompatible sister was treated with monoclonal antibodies and complement to diminish alloreactive T lymphocytes. He died of a B cell lymphoma as a consequence of Epstein-Barr virus-induced polyclonal gammopathy that transformed into monoclonal proliferation, leading to lymphoma.

Buchner, Hans (1850–1902)

German bacteriologist who was professor of hygiene in Munich in 1894. He discovered complement. Through his studies of normal serum and its bactericidal effects, he became an advocate of the humoral theory of immunity.

buffy coat

The white cell layer that forms between the red cells and plasma when anticoagulated blood is centrifuged.

bullous pemphigoid

A blistering skin disease with fluid filled bullae developing at flexor surfaces of extremities, groin, axillae, and inferior abdomen. IgG is deposited in a linear pattern at the lamina lucida of the dermal-epidermal junction in most (50 to 90%) patients and at linear C3 in

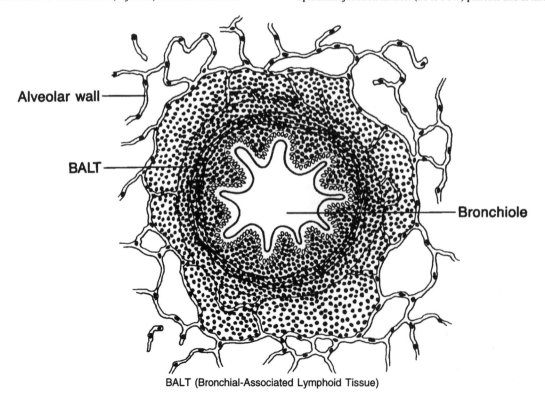

BALT (Bronchial-Associated Lymphoid Tissue)

nearly all cases. The blisters are subepidermal bullae filled with fluid containing fibrin, neutrophils, eosinophils, and lymphocytes. Antigen-antibody-complement interaction and mast cell degranulation release mediators that attract inflammatory cells and facilitate dermal-epidermal separation.

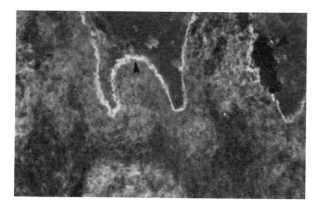

▲ bulla formation

▲ linear IgG and C3
at dermal-epidermal junction

Bullous Pemphigoid

bullous pemphigoid antigen
The principal antigen is a 230-kD basic glycoprotein produced by keratinocytes in the epidermis. Autoantibody and complement react with this antigen to produce bullous pemphigoid skin lesions.

bungarotoxin
An anticholinergic neurotoxin extracted from the venom of Australian snakes belonging to the genus *Bungarus*. It binds to acetylcholine receptors of the nicotinic type and inhibits depolarization of the neuromuscular junction's postsynaptic membrane. This leads to muscular weakness.

Burkitt's lymphoma
An Epstein-Barr virus-induced neoplasm of B lymphocytes. It affects the jaws and abdominal viscera. It is seen especially in African children. The Epstein-Barr virus is present in tumor cells which may reveal rearrangement between the c-*myc*-bearing chromosome and the immunoglobulin heavy chain gene-bearing chromosome. Burkitt's lymphoma patients have antibodies to the Epstein-Barr virus in their blood sera. The disease occurs in geographic regions that are hot and humid and where malaria is endemic. It occurs in subjects with acquired immunodeficiency and in other immunosuppressed individuals. There is an effective immune response against the lymphoma that may lead to remission.

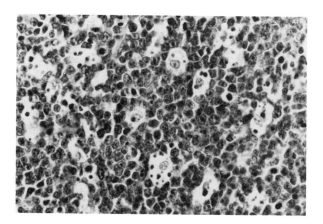

Burkitt's Lymphoma

Burnet, Frank Macfarlane (1899–1985)
Australian virologist and immunologist who shared the Nobel Prize with Peter B. Medawar in 1960 for the discovery of acquired immunological tolerance. Burnet was a theoretician who made major contributions to the developing theories of self tolerance and clonal selection in antibody formation. *Production of Antibodies (with Fenner)*, 1949; *Natural History of Infectious Diseases*, 1953; *Clonal Selection Theory of Antibody Formation*, 1959; *Autoimmune Diseases (with Mackay)*, 1962; *Cellular Immunology*, 1969; *Changing Patterns (autobiography)*, 1969.

bursa of Fabricius
An outpouching of the hindgut located near the cloaca in avian species that governs B cell ontogeny. This specific lymphoid organ is the site of migration and maturation of B lymphocytes. The bursa is located near the terminal portion of the cloaca and, like the thymus, is a lymphoepithelial organ. The bursa begins to develop after the 5th day of incubation and becomes functional around the 10th to 12th day. It has an asymmetric sac-like shape and a star-like lumen, which is continuous with the cloacal cavity. The epithelium of the intestine covers the bursal lumen, but lacks mucous cells. The bursa contains abundant lymphoid tissue, forming nodules beneath the epithelium. The nodules show a central medullary region containing epithelial cells and project into the epithelial coating. The center of the medullary region is less structured and also contains macrophages, large lymphocytes, plasma cells, and granulocytes. A basement membrane separates the medulla from the cortex; the latter comprises mostly small lymphocytes and plasma cells. The bursa is well developed at birth, but begins to involute around the fourth month; it is vestigial at the end of the first year. There is a direct relationship between the hormonal status of the bird and involution of the bursa. Injections of testosterone may lead to premature regression or even lack of development, depending on the time of hormone administration. The lymphocytes in the bursa originate from the yolk sac and migrate there via the blood stream. They comprise B cells which undergo maturation to immunocompetent cells capable of antibody synthesis. Bursectomy at the 17th day of incubation induces agammaglobulinemia, with the absence of germinal centers and plasma cells in peripheral lymphoid organs.

Bursa of Fabricius

bursacyte
A lymphocyte that undergoes maturation and differentiation under the influence of the bursa of Fabricius in avian species. This cell synthesizes the antibody which provides humoral immunity in this species. A bursacyte is a B lymphocyte.

bursal equivalent
The anatomical site in mammals and other nonavian species that resembles the bursa of Fabricius in controlling B cell ontogeny. Mammals do not have a specialized lymphoid organ for maturation of B lymphocytes. Although lymphoid nodules are present along the gut, forming distinct structures called Peyer's patches, their role in B cell maturation is no different from that of lymphoid structures in other organs. After commitment to B cell lineage, the B cells of mammals leave the bone marrow in a relatively immature stage; likewise, after education in the thymus, T cells migrate from the thymus also in a relatively immature stage; both populations continue their maturation process away from the site of origin and are subject to influences originating in the environment in which they reside.

bursectomy

The surgical removal or ablation of the bursa of Fabricius, an outpouching of the hindgut near the cloaca in birds. Surgical removal of the bursa prior to hatching or shortly thereafter followed by treatment with testosterone *in vivo* leads to failure of the B cell limb of the immune response responsible for antibody production.

busulfan (1,4-butanediol dimethanesulfonate)

An alkylating drug that is toxic to bone marrow cells and is used to condition bone marrow transplant recipients.

Busulfan

$$H_3C-\overset{\overset{O}{\|}}{\underset{\underset{O}{\|}}{S}}-O-CH_2-CH_2-CH_2-CH_2-O-\overset{\overset{O}{\|}}{\underset{\underset{O}{\|}}{S}}-CH_3$$

1,4-butanediol dimethanesulfonate

butterfly rash

A facial rash in the form of a butterfly across the bridge of the nose. Seen especially in patients with lupus erythematosus. These areas are photosensitive and consist of erythematous and scaly patches that may become bulbous or secondarily infected. The rash is not specific for lupus erythematosus, since butterfly-type rashes may also occur in various other conditions including AIDS, dermatomyositis, ataxia-telangiectasia, erysipelas, pemphigus erythematosus, pemphigus foliaceous, etc.

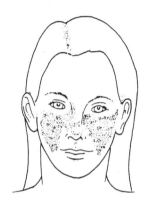

Butterfly Rash

BXSB mice

A mouse strain genetically prone to developing lupus erythematosus-like disease spontaneously. The BXSB strain manifests serologic aberrations and immune complex glomerulonephritis, but demonstrates a distinct and significant acceleration of the disease in males. Among other features, BXSB strains develop moderate lymphadenopathy, which reaches 10 to 20 times greater than normal. The B cell content of these proliferating male lymph nodes may reach 70%. B cell content also develop significant levels of antinuclear antibodies, including anti-DNA, diminished complement, and immune complex-mediated renal injury. Acceleration of this autoimmune disease in the male rather than in the female has been shown not to be hormone mediated.

byssinosis

A disease of cotton, flax, jute, and hemp employees, probably attributable to hypersensitivity to vegetable fiber dust. They develop tightness in the chest upon returning to work after several days absence.

bystander B cells

Non-antigen-specific B cells in the area of B cells specific for antigen. Released cytokines activate bystander B cells that synthesize nonspecific antibody following immunogenic challenge.

bystander lysis

Tissue cell lysis that is nonspecific. The tissue cells are not the specific targets during an immune response, but are killed as innocent bystanders because of their close proximity to the site where nonspecific factors are released near the actual target of the immune response.

C

c allotype

A rabbit immunoglobulin λ light chain allotype 100 designated as c7 and c21.

C gene

DNA encoding the constant region of immunoglobulin heavy and light polypeptide chains. The heavy chain *C* gene is comprised of exons that encode the heavy chain's different homology regions.

C gene segment

DNA coding for a T cell receptor or an immunoglobulin polypeptide chain constant region. One or more exons may be involved.

c-kit ligand

A cytokine, also termed "stem cell factor", that interacts with a tyrosine-kinase membrane receptor of pluripotent stem cells. The receptor, which contains a five Ig domain extracellular structure, is encoded by the cellular oncogene, *c-kit*. Bone marrow stromal cells such as fibroblasts, endothelial cells, and adipocytes, produce a 27-kD transmembrane form and a 24-kD secreted form of a c-kit ligand. This cytokine alone apparently does not induce colony formation but is postulated to render stem cells reactive with other colony stimulating factors.

c-*myb* gene

A gene that encodes formation of a DNA-binding protein that acts during early growth and differentiation stages of normal cells. A c-*myb* gene is expressed mainly in hematopoietic cells, especially bone marrow hematopoietic precursor cells, but it is greatest in the normal murine thymus. The highest c-*myb* expression is in the double-negative thymocyte subpopulation.

c-*myc*

Refer to *myc*.

C-reactive protein (CRP)

One of the pentraxin proteins in serum that is 115 kD and is comprised of five 206-amino acid polypeptide subunits that are all the same and arranged in a disk conformation without covalent bonds. Although present in normal individuals in only trace amounts in the plasma, with a median level of <8 µg/ml, inflammation induced by bacterial infection, necrosis of tissue, trauma, or malignant tumors may cause a striking increase in the serum concentration to levels of up to 2000 times the reference range within 48 h of the inducing condition. CRP is encoded by a gene on chromosome 1. IL-6 regulates its production. Only hepatocytes express it. Once the condition that induced its elevation has resolved, the CRP concentration returns to normal within a short time. CRP levels signify the extent of the disease activity. CRP has the greatest sensitivity of any nonspecific test used in screening for organic disease. It is much more reliable than is the erythrocyte sedimentation rate (ESR). CRP reacts with phosphoryl choline in the C polysaccharide of *Streptococcus pneumoniae* (pneumococcus) in the presence of Ca^{2+} ions. CRP binds firmly with platelet-activating factor (PAF). Following CRP binding to its ligands, C-reactive protein may produce effects similar to those of antibodies, such as activation of complement through the classic pathway by binding to C1q, agglutination of particulate ligands, and participation of insoluble ones, as well as neutralization of biological activity.

C region (constant region)

Abbreviation for the constant region carboxy terminal portion of immunoglobulin heavy or light polypeptide chain that is identical in a particular class or subclass of immunoglobulin molecules. C_H designates the constant region of the heavy chain of immunoglobulin, and C_L designates the constant region of the light chain of immunoglobulin.

C-terminus

The carboxy terminal end of a polypeptide chain containing a free –COOH group.

C1

A 750-kD multimeric molecule comprised of one C1q, two C1r, and two C1s subcomponents. The classical pathway of complement activation begins with the binding of C1q to IgM or IgG molecules. C1q, C1r, and C1s form a macromolecular complex in a Ca^{2+}-dependent manner. The 400-kD C1q molecule possesses three separate polypeptide chains that unite into a heterotrimeric structure resembling stems which contain an amino terminal in triple helix and a globular structure at the carboxy terminus resembling a tulip. Six of these tulip-like structures with globular heads and stems form a circular and symmetric molecular complex in the C1q molecule. There is a central core. The serine esterase molecules designated C1r and C1s are needed for the complement cascade to progress. These 85-kD proteins that are single chains unite, in the presence of calcium, to produce a tetramer comprised of two C1r and two C1s subcomponents to form a structure that is flexible and has a C1s-C1r-C1r-C1s sequence. When at least two C1q globular regions bind to IgM or IgG molecules, the C1r in a tetramer associated with the C1q molecule becomes activated, leading to splitting of the C1r molecules in the tetramer with the formation of a 57- and a 28-kD chain. The latter, termed C1r, functions as a serine esterase splitting the C1s molecules into 57- and 28-kD chains. The 28-kD chain derived from the cleavage of C1s molecules, designated C1s, also functions as a serine esterase, cleaving C4 and C2 and causing progression of the classical complement pathway cascade.

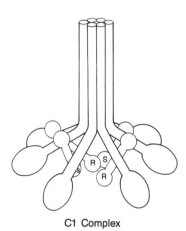

C1 Complex

C1 deficiencies

Only a few cases of C1q, C1r, or C1r and C1s deficiencies have been reported. These have an autosomal recessive mode of inheritance. Patients with these defects may manifest systemic lupus erythematosus, glomerulonephritis or pyogenic infections. They have an increased incidence of type III (immune complex) hypersensitivity diseases. Half of C1q-deficient persons may contain physiologic levels of mutant C1q that are not functional.

C1 esterase inhibitor

A serum protein that counteracts activated C1. This diminishes the generation of C2b which facilitates development of edema.

C1 inhibitor (C1 INH)

A 478-amino acid residue single polypeptide chain protein in the serum. It blocks C1r activation, prevents C1r cleavage of C1s, and inhibits C1s splitting of C4 and C2. The molecule is highly glycosylated with carbohydrates making up approximately one half of its content. It contains seven O-linked oligosaccharides linked to serine and six N-linked oligosaccharides tethered to an asparagine residue. Besides its effects on the complement system, C1 INH blocks factors in the blood clotting system that include kallikrein, plasmin, factor XIa, and factor XIIa. C1 INH is an α_2 globulin and is a normal serum constituent which inhibits serine protease. The 104-kD C1 INH interacts with activated C1r or C1s to produce a stable complex. This prevents these serine protease molecules from splitting their usual substrates. Either C1s or C1r can split C1 INH to uncover an active site in the inhibitor which becomes bound to the proteases through a covalent ester bond. By binding to most of the C1 in the blood, C1 INH blocks the spontaneous activation of C1. C1 INH binding blocks conformational alterations that would lead to spontaneous activation of C1. When an antigen-antibody complex binds C1, C1 INH's inhibitory influence on C1 is relinquished. Genes on chromosome 11 in man encode C1 INH. C1r and C1s subcomponents disengage from C1q following their interaction with C1 INH. In hereditary angioneurotic edema, C1 INH formation is defective. In acquired C1 INH deficiency, there is elevated catabolism of C1 INH.

C1 inhibitor (C1 INH) deficiency

The absence of C1 INH is the most frequently found deficiency of the classic complement pathway and may be seen in patients with hereditary angioneurotic edema. This syndrome may be expressed as either a lack of the inhibitor substance or a functionally inactive C1 INH. The patient develops edema of the face; respiratory tract, including the glottis and bronchi; and the extremities. Severe abdominal pain may occur with intestinal involvement. Since C1 INH can block Hagemann factor (factor XII) in the blood clotting mechanism, its absence can lead to the liberation of kinin and fibrinolysis which results from the activation of plasmin. The disease is inherited as an autosomal dominant trait. When edema of the larynx occurs, the patient may die of asphyxiation. When abdominal attacks occur, there may be watery diarrhea and vomiting. These bouts usually span 48 h and are followed by a rapid recovery. During an attack of angioedema, C1r is activated to produce C1s, which depletes its substrates C4 and C2. The action of activated C1s on C4 and C2 leads to the production of a substance that increases vascular permeability, especially that of postcapillary venules. C1 and C4 cooperate with plasmin to split this active peptide from C2. Of the families of patients with hereditary angioneurotic edema, 85% do not contain C1 INH. Treatment is by preventive maintenance. Patients are given inhibitors of plasmin such as ε-aminocaproic acid and tranexamic acid. Methyl testosterone, which causes synthesis of normal C1 INH in angioneurotic edema patients, is effective by an unknown mechanism.

C1q

An 18-polypeptide chain subcomponent of C1, the first component of complement. It commences the classical complement pathway. The three types of polypeptide chain are designated A chain, B chain, and C chain. Disulfide bonds link these chains. The C1q molecule's triple helix structures are parallel and resemble the stems of six tulips in the amino terminal half of their structure. They then separate into six globular regions, which resemble the heads of a tulip. The molecule is arranged in a heterotrimeric rod-like configuration, bearing a collagen-like triple helix at its amino terminus and a tulip-like globular region at its carboxyl terminus. The combination of six of the rod-shaped structures leads to a symmetric molecular arrangement comprised of three helices at

one terminus and the globular (tulip-like) heads at the other terminus. The binding of antibody to C1q initiates the classic complement pathway. It is the globular C-terminal region of the molecule that binds to either IgM or IgG molecules. A tetramer comprised of two molecules of C1r and two molecules of C1s bind by Ca^{2+} to the collagen-like part of the stem. C1q A chain and C1q B chain are coded for by genes on chromosome 1p in man. The interaction of C1q with antigen-antibody complexes represents the basis for assays for immune complexes in patients' serum. IgM, IgG1, IgG2, and IgG3 bind C1q, whereas IgG4, IgE, IgA, and IgD do not.

C1q deficiency

Deficiency of C1q may be found in association with lupus-like syndromes. C1r deficiency, which is inherited as an autosomal recessive trait, may be associated with respiratory tract infections, glomerulonephritis, and skin manifestations which resemble a systemic lupus erythematosus (SLE)-like disease. C1s deficiency is transmitted as an autosomal dominant trait and patients may again show SLE-like signs and symptoms. Their antigen-antibody complexes can persist without resolution.

C1q receptors (C1q-R)

Receptors that bind the collagen segment of C1q fixed to antigen-antibody complexes. The C1q globular head is the site of binding of immunoglobulin's Fc region. Thus, the C1q-R can facilitate the attachment of antigen-antibody complexes to cells expressing C1q-R and Fc receptors. Neutrophils, B cells, monocytes, macrophages, NK cells, endothelial cells, platelets, and fibroblasts all express C1q-R. C1q-R stimulation on neutrophils may lead to a respiratory burst.

C1r

A subcomponent of C1, the first component of complement in the classical activation pathway. It is a serine esterase. Ca^{2+} binds C1r molecules to the stem of a C1q molecule. Following binding of at least two globular regions of C1q with IgM or IgG, C1r is split into a 463-amino acid residue α chain, the N-terminal fragment, and a 243-amino acid residue carboxy-terminal β chain fragment where the active site is situated. C1s becomes activated when C1r splits its arginine-isoleucine bond. Refer also to C1.

C1s

A serine esterase that is a subcomponent of C1, the first component of complement in the classical activation pathway. Ca^{2+} binds two C1s molecules to the C1q stalk. Following activation, C1r splits the single chain 85-kD C1s molecule into a 431-amino acid residue A chain and a 243-amino acid residue B chain where the active site is located. C1s splits a C4 arginine-alanine bond and a C2 arginine-lysine bond.

C2 (complement component 2)

The third complement protein to participate in the classical complement pathway activation. C2 is a 110-kD single polypeptide chain that unites with C4b molecules on the cell surface in the presence of Mg^{2+}. C1s splits C2 following its combination with C4b at the cell surface. This yields a 35-kD C2b molecule and a 75-kD C2a fragment. Whereas C2b may leave the cell surface, C2a continues to be associated with surface C4b. The complex of C4b2a constitutes classical pathway C3 convertase. This enzyme is able to bind and split C3. C4b facilitates combination with C3. C2b catalyzes the enzymatic cleavage. C2a contains the active site of classical pathway C3 convertase (C4b2a). C2 is encoded by genes on the short arm of chromosome 6 in man. C2A, C2B, and C2C alleles encode human C2. Murine C2 is encoded by genes at the S region of chromosome 17.

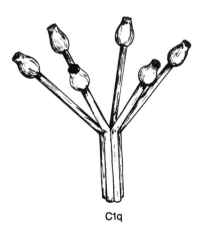

C1q

C2 Gene

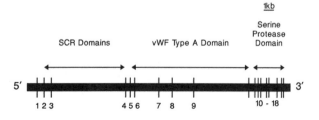

numbered vertical bars are exons

C2a

The principal substance produced by $\overline{\text{C1s}}$ cleavage of C2. N-linked oligosaccharides may combine with C2a at six sites. The 509 carboxy terminal amino acid residues of C2 constitute C2a. The catalytic site for C3 and C5 cleavage is located in the 287 residue carboxy terminal sequence. The association of C2a with C4b yields the C3 convertase (C4b2a) of the classical pathway.

C2b

A 223-amino acid terminal residue of C2 that represents a lesser product of C1s cleavage of C2. There are three abbreviated 68-amino acid residue homologous repeats in C2b that are present in C3- or C4-binding proteins. N-linked oligosaccharides combine with C2b at two sites. A peptide split from the carboxy terminus of C2b by plasmin has been implicated in the formation of edema in hereditary angioneurotic edema patients.

C2 and B

The C2 and B genes are situated within the MHC locus on the short arm of chromosome 6. They are termed MHC class III genes. TNF-α and TNF-β genes are situated between the C2 and HLA-B genes. Another gene designated FD lies between the Bf and C4a genes. C2 and B complete primary structures have been deduced from cDNA and protein sequences. C2 is comprised of 732 residues and is an 81-kD molecule, whereas B contains 739 residues and is an 83-kD molecule. Both proteins have a three-domain globular structure. During C3 convertase formation, the amino terminal domains, C2b or Ba, are split off. They contain consensus repeats that are present in CR1, CR2, H, DAF, and C4bp, which all combine with C3 and/or C4 fragments and regulate C3 convertases. The amino acid sequences of the C2 and B consensus repeats are known. C2b contains site(s) significant for C2 binding to C4b. Ba, resembling C2b, manifests binding site(s) significant in C3 convertase assembly. Available evidence indicates that C2b possesses a C4b-binding site and that Ba contains a corresponding C3b-binding site.

In considering assembly and decay of C3 convertases, initial binding of the three-domain structures C2 or B to activator-bound C4b or C3b, respectively, requires one affinity site on the C2b/Ba domain and another on one of the remaining two domains. A transient change in C2a and Bb conformation results from C2 or B cleavage by C1s or D. This leads to greater binding affinity, Mg^{++} sequestration, and acquisition of proteolytic activity for C3. C2a or Bb dissociation leads to C3 convertase decay. Numerous serum-soluble and membrane-associated regulatory proteins control the rate of formation and association of C3 convertases.

C2 deficiency

Rare individuals may demonstrate a failure to express C2. Although no symptoms are normally associated with this trait, which has an autosomal recessive mode of inheritance, autoimmune-like manifestations that resemble features of certain collagen-vascular diseases, such as systemic lupus erythematosus, may appear. Thus, many genetically determined complement deficiencies are not associated with signs and symptoms of disease. When they do occur, it is usually manifested as an increased incidence of infectious diseases which affect the kidneys, respiratory tract, skin, and joints.

C3 (complement component 3)

A 195-kD glycoprotein heterodimer that is linked by disulfide bonds. It is the fourth complement component to react in the classical pathway, and it is also a reactant in the alternative complement pathway. C3 contains α and β polypeptide chains and has an internal thioester bond which permits it to link covalently with surfaces of cells and proteins. Much of the C3 gene structure has now been elucidated. It is believed to contain approximately 41 exons. Eighteen of 36 introns have now been sequenced. The C3 gene of man is located on chromosome 19. Hepatocytes, monocytes, fibroblasts, and endothelial cells can synthesize C3. More than 90% of serum C3 is synthesized in the liver. The concentration of C3 in serum exceeds that of any other complement component. Human C3 is generated as a single chain precursor which is cleaved into the two-chain mature state. C3 molecules are identical antigenically, structurally, and functionally, regardless of cell source. Hepatocytes and monocytes synthesize greater quantities of C3 than do epithelial and endothelial cells. C3 convertases split a 9-kD C3a fragment from C3's α chain. The other product of the reaction is C3b, which is referred to as metastable C3b and has an exposed thioester bond. Approximately 90% of the metastable C3b thioester bonds interact with

H_2O to form inactive C3b byproducts that have no role in the complement sequence. Ten percent of C3b molecules may bind to cell substances through covalent bonds or with the immunoglobulin bound to C4b2a. This interaction leads to the formation of C4b2a3b, which is classical pathway C5 convertase, and serves as a catalyst in the enzymatic splitting of C5 which initiates membrane attack complex (MAC) formation. When C3b, in the classical complement pathway, interacts with E (erythocyte), A (antibody), C1 (complement 1), and 4b2a, EAC14b2a3b is produced. As many as 500 C3b molecules may be deposited at a single EAC14b2a complex on an erythrocyte surface. C3S (slow electrophoretic mobility) and C3F (fast electrophoretic mobility) alleles on chromosome 19 in man encode 99% of C3 in man, with rare alleles accounting for the remainder. C3 has the highest concentration in serum of any complement system protein with a range of 0.552 to 1.2 mg/ml. Following splitting of the internal thioester bond, it can form a covalent link to amino or hydroxyl groups on erythrocytes, microorganisms or other substances. C3 is an excellent opsonin. C3 was known in the past as $\beta_1 C$ globulin.

C3a

A low molecular weight (9-kD) peptide fragment of complement component C3. It is comprised of the 77 N-terminal end residues of C3 α chain. This biologically active anaphylatoxin, which induces histamine release from mast cells and causes smooth muscle contraction, is produced by the cleavage of C3 by either classical pathway C3 convertase, i.e., C4b2a, or alternative complement pathway C3 convertase, i.e., C3bBb. Anaphylatoxin inactivator, a carboxy peptidase N, can inactivate C3a by digesting the C-terminal arginine of C3a.

C3a receptor (C3a-R)

A protein on the surface membrane of mast cells and basophils. It serves as a C3a anaphylatoxin receptor.

C3a/C4a receptor (C3a/C4a-R)

C3a and C4a share a common receptor on mast cells. When a C-terminal arginine is removed from C3a and C4a by serum carboxy peptidase N (SCPN), these anaphylatoxins lose their ability to activate cellular responses. Thus, $C3a_{des\ Arg}$ and $C4a_{des\ Arg}$ lose their ability to induce spasmogenic responses. C3a-R has been demonstrated on guinea pig platelets. Eosinophils have been found to bind C3a.

C3b

The principal fragment produced when complement component C3 is split by either classical or alternative pathway convertases, i.e., C4b2a or C3bBb, respectively. It results from C3 convertase digestion of C3's α chain. It is an active fragment as revealed by its combination with factor B to produce C3bBb, which is the alternative pathway C3 convertase. Classical complement pathway C5 convertase is produced when C3b combines with C4b2a to yield C4b2a3b. Factor I splits the arginine-serine bonds in C3b, if factor H is present, to yield C3bi. This produces

Native C3 (fluid phase)

C3b fragment (bound)

iC3b fragment (bound)

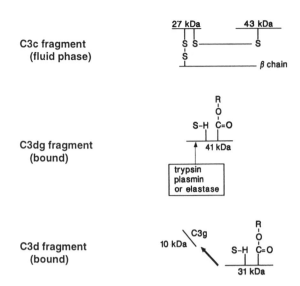

C3c fragment (fluid phase)

C3dg fragment (bound)

trypsin plasmin or elastase

C3d fragment (bound)

the C3f peptide. Particle-bound C3b interacts with complement receptor 1. C3b interacts with C3b receptors on macrophages, B lymphocytes, polymorphonuclear neutrophils (PMNs), and possibly T cells. It promotes phagocytosis and immune adherence and may function as an opsonin.

C3bi (iC3b)

The principal molecular product when factor I cleaves C3b. If complement receptor 1 or factor H is present, factor I can split C3bi's arginine-glutamic acid bond at position 954-955 to yield C3c and C3dg. C3bi attached to particles promotes phagocytosis when combined with complement receptor 3 on the surface of polymorphonuclear neutrophils (PMNs) and monocytes. It also promotes phagocytosis by binding to conglutinin in the serum of cows.

C3b inactivator

Refer to factor I.

C3b receptors

Refer to complement receptor 1 (CR1).

C3c

The principal molecule that results from factor I cleavage of C3bi when factor H or complement receptor 1 is present. C3c is comprised of 27- and 43-kD α chain fragments linked through disulfide bonds to a whole β chain.

C3 convertase

An enzyme that splits C3 into C3b and C3a. There are two types: one in the classical pathway designated C4b2a and one in alternative pathway of complement activation termed C3bBb. An amplification loop with a positive feedback is stimulated by alternative pathway C3 convertase. Each of the two types of C3 convertase lacks stability, leading to the ready disassociation of their constituents. However, C3 nephritic factor can stabilize both classical and alternative pathway C3 convertases. Properdin may stabilize alternative pathway C3 convertase. C2a and Bb contain the catalytic sites.

C3d

A 33-kD B cell growth factor that is formed by proteolytic enzyme splitting of a lysine-histidine bond in C3dg at position 1001-1002. C3d is comprised of the carboxy terminal 301-amino acid residues of C3dg. It interacts with complement receptor 2 on the surface of B cells. C3d contains the C3 α chain's thioester.

C3 deficiency

An extremely uncommon genetic disorder that may be associated with repeated serious pyogenic bacterial infections and may lead to death. The C3 deficient individuals are deprived of appropriate opsonization, prompt phagocytosis, and the ability to kill infecting microorganisms. There is defective classical and alternative pathway activation. Besides infections, these individuals may also develop an immune complex disease such as glomerulonephritis. C3 levels that are one half normal in heterozygotes are apparently sufficient to avoid the clinical consequences induced by a lack of C3 in the serum.

C3dg

A 41-kD, 349-amino acid residue molecule formed by the cleavage of C3bi with factor H or complement receptor 1 present. Polymor-

phonuclear neutrophil leukocytes express complement receptor 4, which is reactive with C3dg. Complement receptor 2 on B cells is also a C3dg receptor.

C3e

A C3c α chain nonapeptide that causes leukocytosis. The peptide is comprised of Thr-Leu-Asp-Pro-Glu-Arg-Leu-Gly-Arg.

C3f

A 17-amino acid residue peptide split from the α chain of C3b by factor I with factor H or complement receptor 1 present.

C3g

An 8-kD molecule comprised of C3dg's amino terminal 47-amino acid residues. Trypsin digestion of C3dg yields C3g, whose function is unknown.

C3H/HeJ mouse

A mutant substrain of C3H mice that manifests a suppressed response by macrophages and B cells to challenge with lipopolysaccharide. Their macrophages do not produce interleukin-1 and tumor necrosis factor following the lipopolysaccharide challenge. This mutation has an autosomal dominant mode of inheritance and is encoded by genes on chromosome 4. This immunosuppression leads to an increased incidence of microbial infections in these mice.

C3 nephritic factor (C3NeF)

An IgG autoantibody to the alternate complement pathway C3 convertase that mimics the action of properdin. C3NeF is present in the serum of patients with membranoproliferative glomerulonephritis type II (dense deposit disease). It stabilizes the alternate pathway C3 convertase, thereby enhancing the breakdown of C3, and produces hypocomplementemia. Rarely, C3NeF may be IgG autoantibodies to C3 convertase (C4b2a) of the classical pathway. Systemic lupus erythematosus patients may contain antibodies against C4b2a which stabilize the classical pathway C3 convertase, leading to increased *in vivo* cleavage of C3.

C3 PA (C3 proactivator)

An earlier designation for factor B.

C3 tickover

Alternative pathway C3 convertase perpetually generates C3b. C3 internal thioester bond hydrolysis is the initiating event.

Tick-Over Mechanism

C3 Tickover

C4 (complement component 4)

A 210-kD molecule comprised of α, β and γ chains. The α chain has an internal thioester bond linking a cysteine residue and adjacent glutamate residue. C4 reacts immediately following C1 in the classical pathway of complement activation. C1s splits the α chain of C4 at position 76-77, where an arginine-alanine bond is located. This yields a 8.6-kD C4a fragment, an anaphylatoxin, and C4b, which is a larger molecule. C4b remains linked to C1. Many C4b molecules can be formed through the action of a single C1s molecule. Enzymatic cleavage renders the α chain thioester bond of the C4b fragment very unstable. The molecule's chemically active form is termed metastable C4b. C4bi intermediates form when C4b thioester bonds and water molecules react. C4b molecules may become bound covalently to cell surfaces when selected C4b thioester bonds undergo transesterification, producing covalent amide or ester bonds with proteins or carbohydrates on the cell surface. This enables complement activation to take place on the surfaces of cells where antibodies bind. C4b may also link covalently with antibody. C4 is first formed as a 1700-amino acid residue chain which contains β, α, and γ chain components joined through connecting peptides. *C4A* and

C4B genes located at the major histocompatibility complex on the short arm of chromosome 6 in man encode C4. *Slp* and *Ss* genes located on chromosome 17 in the mouse encode murine C4.

C4a

A 76-amino acid terminal residue peptide produced by C1s cleavage of C4. Together with C3a and C5a, C4a is an anaphylatoxin which induces degranulation of mast cells and basophils associated with histamine release and the features of anaphylaxis. However, the anaphylatoxin activity of C4a is 100 times weaker than is that of the other two anaphylatoxin molecules.

C4A

A very polymorphic molecule expressing the Rodgers epitope that is encoded by the *C4A* gene. The equivalent murine gene encodes a sex-limited protein (SLP). It has less hemolytic activity than does C4B. C4A and C4B differ in only four amino acid residues in the α chain's C4d region. C4A is Pro-Cys-Pro-Bal-Leu-Asp, whereas C4B is Leu-Ser-Pro-Bal-Ile-His.

C4b

The principal molecule produced when C1s splits C4. C4b is that part of the C4 molecule that remains after C4a has been split off by enzymatic digestion. C4b unites with C2a to produce C4b2a, an enzyme which is known as the classical pathway C3 convertase. Factor I splits the arginine-asparagine bond of C4b at position 1318-1319 to yield C4bi, if C4b binding protein is present. C4b linked to particulate substances reacts with complement receptor 1.

C4B

A polymorphic molecule that usually expresses the Chido epitope and is encoded by the *C4B* gene. The murine equivalent gene encodes an Ss protein. It shows greater hemolytic activity than does C4A.

C4b-binding protein (C4bp)

A 600-kD protein in serum capable of binding six C4b molecules at once by means of seven spokes extending from a core at the center. C4b halts progression of complement activation. Factor I splits C4b molecules captured by C4bp. C4bp belongs to the regulators of complement activity molecules. C4bp interferes with C2a association with C4b. It also promotes C4b2a dissociation into C4b and C2a. It is also needed for the action of factor I in splitting C4b to C4bi and of C4bi into C4c and C4d. The C4bp gene is located on chromosome 1q3.2.

C4bi (iC4b)

When factor I splits C4b, this is the principal product of the reaction. When C4b-binding protein is present, C4bi splits an α chain arginine-threonine bond to yield C4c and C4d.

C4b inactivator

Refer to factor I.

C4c

The principal product of factor I cleavage of C4bi when C4b-binding protein is present. This 145-kD molecule, of unknown function, is comprised of β and γ chains of C4 and two α chain fragments.

C4d

A 45-kD molecule produced by factor I cleavage of C4bi when C4b-binding protein is present. C4d is the molecule where Chido and Rogers epitopes are located. It is also the location of the C4 α chain's internal thioester bond.

C4 deficiency

An uncommon genetic defect with an autosomal recessive mode of inheritance. Affected individuals have defective classical complement pathway activation. Those who manifest clinical consequences of the defect may develop systemic lupus erythematosus (SLE) or glomeruloneophritis. Half of the patients with C4 and C2 deficiencies develop SLE, but deficiencies in these two complement components are not usually linked to increased infections.

C5 (complement component 5)

A component comprised of α and β polypeptide chains linked by disulfide bonds that react in the complement cascade following C1, C4b, C2a, and C3b fixation to complexes of antibody and antigen. The 190-kD dimeric C5 molecule shares homology with C3 and C4, but does not possess an internal thioester bond. C5 combines with C3b of C5 convertase of either the classical or the alternative pathway. C5 convertases split the α chain at an arginine-leucine bond at position 74-75, producing an 11-kD C5a fragment, which has both chemotactic action for neutrophils and anaphylatoxin activity. It also produces a 180-kD C5b fragment that remains anchored to the cell surface. C5b maintains a structure that is able to bind with C6. C5 is a β_1F globulin in man. C5b complexes with C6, C7, C8, and C9 to form the membrane attack complex (MAC) which mediates immune lysis of cells. Murine C5 is encoded by genes on chromosome 2.

C5a

A peptide split from C5 through the action of C5 convertases, C4b2a3b or C3bBb3b. It is comprised of the C5 α chain's 74 amino terminal residues. It is a powerful chemotactic factor and is an anaphylatoxin, inducing mast cells and basophils to release histamine. It also causes smooth muscle contraction. It promotes the production of superoxide in polymorphonuclear neutrophils (PMNs) and accentuates CR3 and Tp150,95 expression in their membranes. In addition to chemotaxis, it may facilitate PMN degranulation. Human serum contains anaphylatoxin inactivator that has carboxy peptidase N properties. It deletes C5a's C-terminal arginine which yields $C5a_{des\ Arg}$. Although deprived of anaphylatoxin properties, $C5a_{des\ Arg}$ demonstrates limited chemotactic properties.

C5a$_{74des\ Arg}$

That part of C5a that remains following deletion of the carboxy terminal arginine through the action of anaphylatoxin inactivator. Although deprived of C5a's anaphylatoxin function, $C5a_{74des\ Arg}$ demonstrates limited chemotactic properties. This very uncommon deficiency of C2 protein in the serum has an autosomal recessive mode of inheritance. Affected persons have an increased likelihood of developing type III hypersensitivity disorders mediated by immune complexes, such as systemic lupus erythematosus. Whereas affected individuals possess the C2 gene, mRNA for C2 is apparently absent. Individuals who are heterozygous possess 50% of the normal serum levels of C2 and manifest no associated clinical illness.

C5a receptor (C5a-R)

A receptor found on phagocytes and mast cells that binds the anaphylatoxin C5a, which plays an important role in inflammation. Serum carboxy peptidase N (SCPN) controls C5a function by eliminating the C-terminal arginine. This produces $C5a_{des\ Arg}$. Neutrophils are sites of C5a catabolism. C5a-R is a 150- to 200-kD oligomer comprised of multiple 40- to 47-kD C5a-binding components. C5a-R mediates chemotaxis and other leukocyte reactions.

C5b

The principal molecular product that remains after C5a has been split off by the action of C5 convertase on C5. It has a binding site for C6 and complexes with it to begin generation of the membrane attack complex (MAC) of complement, which leads to cell membrane injury and lysis.

C5 convertase

A molecular complex that splits C5 into C5a and C5b in both the classical and the alternative pathway of complement activation. Classical pathways C5 convertase is comprised of C4b2a3b, whereas alternative pathway C5 convertase is comprised of C3bBb3b. C2a and Bb contain the catalytic sites.

C5 deficiency

A very uncommon genetic disorder that has an autosomal recessive mode of inheritance. Affected individuals have only trace amounts of C5 in their plasma. They have a defective ability to form the membrane attack complex (MAC), which is necessary for the efficient lysis of invading microorganisms. They have an increased susceptibility to disseminated infections by *Neisseria* microorganisms such as *N. meningitidis* and *N. gonorrhoeae*. Heterozygotes may manifest 13 to 65% of C5 activity in their plasma and usually show no clinical effects of their partial deficiency. C5 deficient mice also have been described.

C6 (complement component 6)

A 128-kD single polypeptide chain that participates in the membrane attack complex (MAC). It is encoded by *C6A* and *C6B* alleles. It is a β_2 globulin.

C6 deficiency

A highly uncommon genetic defect with an autosomal recessive mode of inheritance in which affected individuals have only trace amounts of C6 in their plasma. They are defective in the ability to form a membrane attack complex (MAC) and have increased susceptibility to disseminated infections by *Neisseria* microorganisms which include gonococci and meningococci. C6-deficient rabbits have been described.

C7 (complement component 7)

An 843-amino acid residue polypeptide chain that is a β_2 globulin. C5b67 is formed when C7 binds to C5b and C6. The complex has the appearance of a stalk with a leaf type of structure. C5b constitutes the leaf, and the stalk consists of C6 and C7. The stalk facilitates introduction of the C5b67 complex into the cell membrane, although no transmembrane perforation is produced. C5b67 anchored to the cell membrane provides a binding site for C8 and C9 in formation of the membrane attack complex (MAC). N-linked oligosaccharides bind to asparagine at positions 180 and 732 in C7.

C7 deficiency

A highly uncommon genetic disorder with an autosomal recessive mode of inheritance in which the serum of affected persons contains only trace amounts of C7 in the plasma. They have a defective ability to form a membrane attack complex (MAC) and show an increased incidence of disseminated infections caused by *Neisseria* microorganisms. Some may manifest an increased propensity to develop immune complex (type III hypersensitivity) diseases such as glomerulonephritis or systemic lupus erythematosus.

C8 (complement component 8)

A 155-kD molecule comprised of a 64-kD α chain, a 64-kD β chain and a 22-kD γ chain. Disulfide bonds join the α and γ chains. Noncovalent bonds link α and γ chains to the β chain. The C5b678 complex becomes anchored to the cell surface when the γ chain inserts into the membrane's lipid bilayer. When the C8 β chain combines with C5b in C5b67 complexes, the α chain regions change in conformation from β-pleated sheets to α helixes. The C5b678 complex has a limited capacity to lyse the cell to which it is anchored, since the complex can produce a transmembrane channel. The α chain of C8 combines with a single molecule of C9, thereby inducing C9 polymerization in the membrane attack complex (MAC). Genes at three different loci encode C8 α, γ, and β chains. One third of the amino acid sequences are identical between C8 α and β chains. These chains share the identity of one fourth of their amino acid sequences with C7 and C9. C8 is a β_1 globulin. In humans, the C8 concentration is 10 to 20 μ/ml.

C8 deficiency

A highly uncommon genetic disorder with an autosomal recessive mode of inheritance in which affected individuals are missing C8 α, γ, or β chains. This is associated with a defective ability to form a membrane attack complex (MAC). Individuals may have an increased propensity to develop disseminated infections caused by *Neisseria* microorganisms such as meningococci.

C9 (complement component 9)

A 535-amino acid residue single chain protein that binds to the C5b678 complex on the cell surface. It links to this complex through the α chain of C8, changes in conformation, significantly increases its length, and reveals hydrophobic regions that can react with the cell membrane lipid bilayer. With Zn^{2+} present, a dozen C9 molecules polymerize to produce 100-nm diameter hollow tubes that are positioned in the cell membrane to produce transmembrane channels. The interaction of 12 to 15 C9 molecules with one C5b678 complex produces the MAC. When viewed by an electron microscope, the pores in the plasma membrane produced by the poly-C9 have a 110-Å internal diameter, a 115-Å stalk anchored in the membrane's lipid bilayer, and a 100-Å structure above the membrane that gives an appearance of a doughnut when viewed from above. Similar pores are produced by proteins released from cytotoxic T lymphocytes and natural killer cells called perforin or cytolysin. Sodium and water quickly enter the cells through these pores, leading to cell swelling and lysis. C9 shares one fourth the amino acid sequence identity with C7 and C8's α and β chains. It resembles perforin structurally. No polymorphism is found in C9, which is encoded by genes on chromosome 5 in man.

C9 deficiency

A highly uncommon genetic disorder with an autosomal recessive mode of inheritance in which only trace amounts of C9 are present in the plasma of affected persons. There is a defective ability to form the membrane attack complex (MAC). The serum of C9-deficient subjects retains its lytic and bactericidal activity, even though the rate of lysis is decreased compared to that induced in the presence of C9. There are usually no clinical consequences associated with this condition. The disorder is more common in the Japanese than in most other populations.

CA-15-3

An antibody specific for an antigen frequently present in the blood serum of metastatic breast carcinoma patients.

CA-19-9

A tumor-associated antigen found on the Lewis A blood group antigen that is sialylated or in mucin-containing tissues. In individuals whose serum levels exceed 37 U/ml, 72% have carcinoma of the pancreas. In individuals whose levels exceed 1000 U/ml, 95% have pancreatic cancer. Anti-CA-19-9 monoclonal antibody is useful to detect the recurrence of pancreatic cancer following surgery and to distinguish between neoplastic and benign conditions of the pancreas. However, it is not useful for pancreatic cancer screening.

CA-125

A mucinous ovarian carcinoma cell surface glycoprotein detectable in the patient's blood serum. Increasing serum concentrations portend a grave prognosis. It may also be found in the blood sera of patients with other adenocarcinomas, such as breast, gastrointestinal tract, uterine cervix, and endometrium.

cachectin

An earlier name for tumor necrosis factor α found in the blood serum and associated with wasting in these individuals. See tumor necrosis factor α (TNF-α).

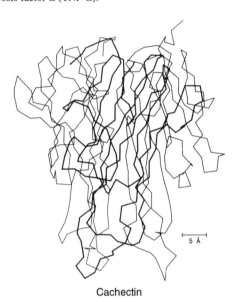

5 Å

Cachectin

cadherins

One of four specific families of cell adhesion molecules that enable cells to interact with their environment. Cadherins help cells to communicate with other cells in immune surveillance, extravasation, trafficking, tumor metastasis, wound healing, and tissue localization. Cadherins are calcium dependent. The five different cadherins include N-cadherin, P-cadherin, T-cadherin, V-cadherin, and E-cadherin. Cytoplasmic domains of cadherins may interact with proteins of the cytoskeleton. They may bind to other receptors based on homophilic specificity, but they still depend on intracellular interactions linked to the cytoskeleton.

caecal tonsils

Lymphoid aggregates containing germinal centers found in the gut wall in birds, specifically in the wall of the caecum.

calcineurin

A protein phosphatase that is serine/threonine-specific. Activation of T cells apparently requires deletion of phosphates from serine or threonine residues. Its action is inhibited by the immunosuppressive drugs, cyclosporin-A and FK506. Cyclosporin-A and FK506 combine with immunophilin intracellular molecules to form a complex that combines with calcineurin and inhibits its activity.

calcitonin

A hormone that influences calcium ion transport. Immunoperoxidase staining demonstrates calcitonin in thyroid parafollicular or C cells. It serves as a marker characteristic of medullary thyroid carcinoma and APUD neoplasms. Lung and gastrointestinal tumors may also form calcitonin.

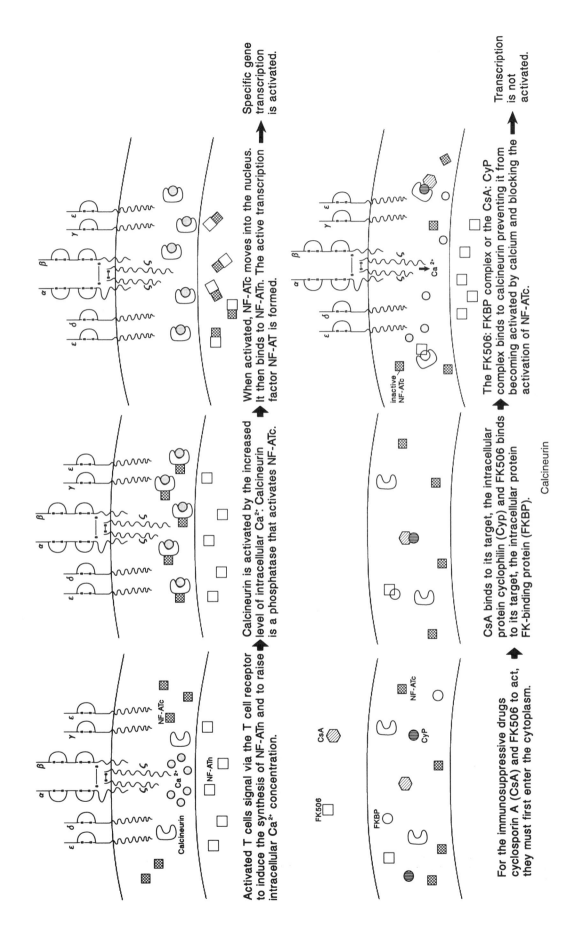

Activated T cells signal via the T cell receptor to induce the synthesis of NF-ATn and to raise intracellular Ca²⁺ concentration.

Calcineurin is activated by the increased level of intracellular Ca²⁺. Calcineurin is a phosphatase that activates NF-ATc.

When activated, NF-ATc moves into the nucleus. It then binds to NF-ATn. The active transcription factor NF-AT is formed.

Specific gene transcription is activated.

For the immunosuppressive drugs cyclosporin A (CsA) and FK506 to act, they must first enter the cytoplasm.

CsA binds to its target, the intracellular protein cyclophilin (Cyp) and FK506 binds to its target, the intracellular protein FK-binding protein (FKBP).

The FK506: FKBP complex or the CsA: CyP complex binds to calcineurin preventing it from becoming activated by calcium and blocking the activation of NF-ATc.

Transcription is not activated.

Calcineurin

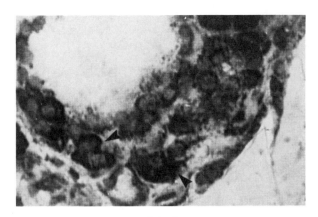

calcitonin
staining in
thyroid

CALLA

Common acute lymphoblastic leukemia antigen. Also known as CD10.

calnexin

An 88-kD membrane molecule that combines with newly formed α chains that also interact with nascent β_2-microglobulin. Calnexin maintains partial folding of the MHC-class I molecule in the endo-plasmic reticulum. It also interacts with MHC-class II molecules, T cell receptors, and immunoglobulins that are partially folded.

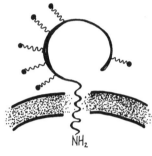

NH₂

CALLA (CD10)

CAM (cell adhesion molecule)

Cell-selective proteins that promote adhesion of cells to one another and are calcium independent. They are believed to help direct migration of cells during embryogenesis.

Campath-1 (CD52)

The majority of lymphocytes and monocytes express this antigen, which is not found on other cells. The "humanized" antibodies specific for this epitope are termed Campath-1H. Refer to CD52.

canine distemper virus

A virus that induces disease in dogs and is associated with demyelination, probably induced by myelin-sensitive lymphocytes.

canine parvovirus vaccine

Initially, a feline enteritis vaccine that was live and attenuated was used based on its crossreactivity with canine parvovirus. Canine parvovirus may have originated from the feline enteritis organism by mutation. This vaccine was later replaced with attenuated canine parvovirus vaccine.

capillary leak syndrome

The therapeutic administration of GM-CSF (granulocyte-macrophage colony-stimulating factor) may lead to progressive dyspnea and pericarditis in the treatment of patients bearing metastatic solid tumors. Interleukin-2 (IL-2) may likewise induce the effect when it is used to treat tumor patients. IL-2 treatment may lead to the accumulation of 10 to 20 l of fluid in peripheral tissues with resultant disorientation, confusion, and pronounced fever.

Caplin's syndrome

Rheumatoid nodules in the lungs of subjects such as hard coal miners in contact with silica.

capping

The migration of antigens on the cell surface to a cell pole following crosslinking of antigens by specific antibody. These antigen-antibody complexes coalesce or aggregate into a "cap" produced by interaction of antigen with cell surface IgM and IgD molecules at sites distant from each other, as revealed by immunofluorescence. Capping is followed by interiorization of the antigen. Following internalization, the cell surface is left bereft of immunoglobulin receptors until they are reexpressed.

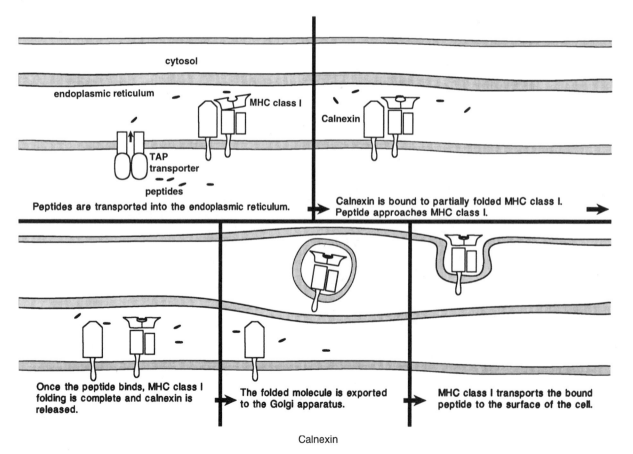

Calnexin

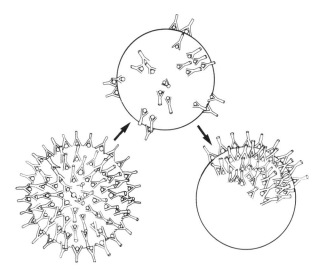

Capping

capping phenomenon

The migration of surface membrane proteins toward one pole of a cell following crosslinking by a specific antibody, antigen, or mitogen. Bivalent or polyvalent ligands cause the surface molecules to aggregate into patches. This passive process is referred to as patching. The ligand-surface molecule aggregates in patches move to a pole of the cell where they form a cap. If a cell with patches becomes motile, the patches move to the rear, forming a cluster of surface molecule-ligand aggregates that constitutes a cap. The process of capping requires energy and may involve interaction with microfilaments of the cytoskeleton. In addition to capping in lymphocytes, the process occurs in numerous other cells.

caprinized vaccine

A preparation used for therapeutic immunization which contains microorganisms attenuated by passage through goats.

capsid

A virus protein envelope comprised of subunits that are called capsomers.

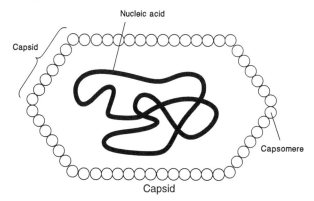

capsular polysaccharide

A constituent of the protective coating around a number of bacteria such as the pneumococcus (*Streptococcus pneumoniae*) which is a polysaccharide, chemically, and stimulates the production of antibodies specific for its epitopes. In addition to the pneumococcus, other microorganisms such as *Streptococci* and certain *Bacillus* species have polysaccharide capsules.

capsule swelling reaction

Pneumococcus swelling reaction. Refer to Quellung reaction.

carbon clearance test

An assay to judge the mononuclear phagocyte system's activity *in vivo*. Blood samples are collected at designated intervals following the intravenous inoculation of a colloidal carbon particle suspension. Following dissolution of the erythrocytes in blood samples, their carbon particle content is determined colorimetrically. This permits determination of the rate at which blood phagocytes remove the carbon. The logarithms of the readings in the colorimetry are plotted against time to yield the desired slope.

carcinoembryonic antigen (CEA)

A 200-kD membrane glycoprotein epitope that is present in the fetal gastrointestinal tract in normal conditions. However, tumor cells, such as those in colon carcinoma, may reexpress it. CEA was first described as a screen for identifying carcinoma by detecting nanogram quantities of the antigen in serum. It was later shown to be present in certain other conditions as well. CEA levels are elevated in almost one third of patients with colorectal, liver, pancreatic, lung, breast, head and neck, cervical, bladder, medullarythyroid, and prostatic carcinoma. However, the level may be elevated also in malignant melanoma, lymphoproliferative disease, and smokers. Regrettably, CEA levels also increase in a variety of non-neoplastic disorders, including inflammatory bowel disease, pancreatitis, and cirrhosis of the liver. Nevertheless, determination of CEA levels in the serum is valuable for monitoring the recurrence of tumors in patients whose primary neoplasm has been removed. If the patient's CEA level reveals a 35% elevation compared to the level immediately following surgery, this may signify metastases. This oncofetal antigen is comprised of one polypeptide chain with one variable region at the amino terminus and six constant region domains. CEA belongs to the immunoglobulin superfamily. It lacks specificity for cancer, thereby limiting its diagnostic usefulness.

carcinoma-associated antigens

Self antigens whose epitopes have been changed due to effects produced by certain tumors. Self antigens are transformed into a molecular structure for which the host is immunologically intolerant. Examples include the T antigen, which is an MN blood group precursor molecule exposed by the action of bacterial enzymes, and Tn antigen, which is a consequence of somatic mutation in hematopoietic stem cells caused by inhibition of galactose transfer to *N*-acetyl-D-galactosamine.

carcinomatous neuropathy

Neurological findings in tumor-bearing patients who have no nervous system metastasis. Sensory carcinomatous neuropathy, in which patients develop autoantibodies specific for neurone cytoplasm RNA-protein, is an example of a carcinomatous neuropathy.

cardiolipin

Diphosphatidyl glycerol, a phospholipid, extracted from beef heart as the principal antigen in the Wasserman complement fixation test for syphilis used earlier in the century.

cardiolipin antibodies

Anticardiolipin antibodies (ACA) may be linked to thrombocytopenia, thrombotic events and repeated fetal loss in systemic lupus erythematosus patients. Caucasian but not Chinese patients with SLE may have a relatively high incidence of valve defects associated with the presence of these antibodies. Other conditions associated with ACA include adrenal hemorrhage and Addison's disease, livedo reticularis (LR), livedo reticularis with cardiovascular disease (Sneddon syndrome), possibly polymyalgia rheumatica/ giant cell arteritis, and possibly focal CNS lupus.

carrier (see figure, page 52)

An immunogenic macromolecular protein such as ovalbumin to which an incomplete antigen termed a hapten may be conjugated either *in vitro* or *in vivo*. Whereas the hapten is not immunogenic by itself, conjugation to the carrier molecule renders it immunogenic.

carrier specificity

That part of an immune response, either humoral antibody or cell-mediated immunity, that is specific for the carrier portion of a hapten-carrier complex that has been used as an immunogen. The carrier-specific part of the immune response does not react with the hapten either by itself or conjugated to a different carrier.

cartwheel nucleus

A descriptor for the arrangement of chromatin in the nucleus of a typical plasma cell based on the more and less electron dense areas observed by electron microscopy. Euchromatin makes up the less electron-dense spokes of the wheel, whereas heterochromatin makes up the more electron-dense areas.

cascade reaction

A sequential event in such enzymatic reactions as complement fixation and blood coagulation in which each stage of the reaction triggers the next appropriate step.

caseous necrosis

Tissue destruction, as occurs in tuberculosis, that has the appearance of cottage cheese.

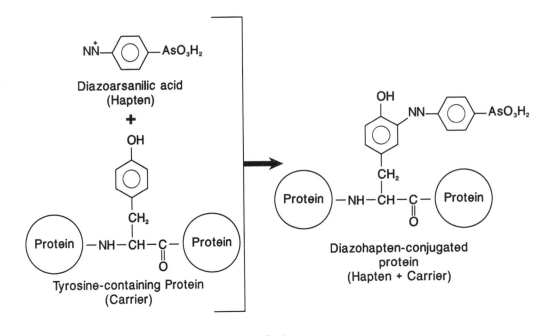

Carrier

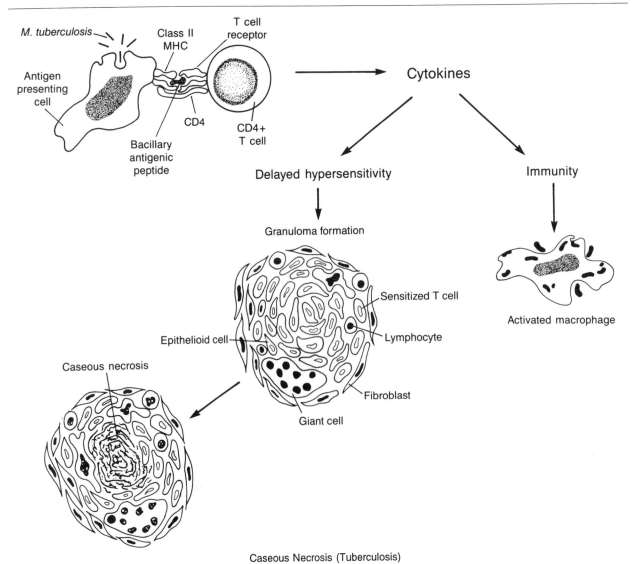

Caseous Necrosis (Tuberculosis)

Casoni test

A diagnostic skin test for hydatid disease in humans induced by *Echinococcus granulosus* infection. In sensitive individuals, a wheal and flare response develops within 30 min following intradermal inoculation of a tapeworm or hydatid cyst fluid extract. This is followed within 24 h by an area of induration produced by this cell-mediated delayed-hypersensitivity reaction.

Castleman's disease

A polyclonal hyperplasia of B lymphocytes that is benign. It is marked by a significant vascular component. Also called giant lymph node hyperplasia.

cat scratch disease

Regional lymphadenitis, common usually in children following a cat scratch. The condition is induced by a small Gram-negative bacillus. Erythematous papules may appear on the hands or forearms at the site of the injury. Patients may develop fever, malaise, swelling of the parotid gland, lymphadenapathy that is regional or generalized, maculopapular rash, anorexia, splenic enlargement, and encephalopathy. There may be hyperplasia of lymphoid tissues, formation of granulomas, and abscesses. A positive skin test together with the history establishes the diagnosis. Gentamycin and ciprofloxacin have been used in treatment.

catalase

An enzyme present in activated phagocytes that causes degradation of hydrogen peroxide and superoxide dismutase.

catalytic antibody

A monoclonal antibody into whose antigen-binding site the catalytic activity of a specific biological enzyme has been introduced. This permits enzymatic catalysis of previously arranged specificity to take place. Site-directed mutagenesis, in which a catalytic residue is added to a combining site by amino acid substitution, is used to attain the specificity. Specific catalysts can be generated by other mechanisms, such as alternation of enzyme sites genetically or chemical alternation of receptors with catalytic properties.

cationic proteins

Phagocytic cell granule constituents that have antimicrobial properties.

CBA mouse

A strain of inbred mice from which many substrains have been developed, such as CBA/H-T6.

CBA/N mouse

A CBA murine mutant that is incapable of responding immunologically to linear polysaccharides and selected other thymus-independent immunogens. The mutant strain's B cells are either defective or immature. The Lyb3, Lyb5, and Lyb7 B lymphocyte subset is not present in these mice. This mutation is designated *xid,* which has an X-linked recessive mode of inheritance. Their serum IgG concentrations are diminished. They mount only weak immune responses to thymus-dependent immunogens.

CD (cluster of differentiation)

Refer to cluster of differentiation. Molecular weights for the CD designations that follow are given for reduced conditions.

CD1

An antigen that is a cortical thymocyte marker, which disappears at later stages of T cell maturation. The antigen is also found on interdigitating cells, fetal B cells, and Langerhans cells. These chains are associated with β_2-microglobulin and the antigen is thus analogous to classical histocompatibility antigens, but coded for by

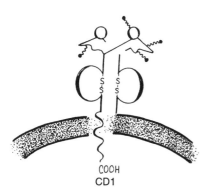

COOH
CD1

a different chromosome. More recent studies have shown that the molecule is coded for by at least five genes on chromosome 1, of which three produce recognized polypeptide products. CD1 may participate in antigen presentation.

CD1a

An antigen that is a membrane glycoprotein with a mol wt of 49,12 kD. It is expressed strongly on cortical thymocytes.

CD1b

An antigen that is a membrane glycoprotein with a mol wt of 45,12 kD. It is expressed moderately on thymocytes.

CD1c

An antigen that is a membrane glycoprotein with a mol wt of 43,12 kD. It is expressed weakly on cortical thymocytes.

CD2

A T cell antigen that is the receptor molecule for sheep red cells and is also referred to as the T11 antigen, the leukocyte function associated antigen-3 (LFA-3), or CD58. The molecule has a mol wt of 50 kD. The antigen also seems to be involved in cell adherence, probably binding LFA-3 as its ligand. CD2 can activate T lymphocytes.

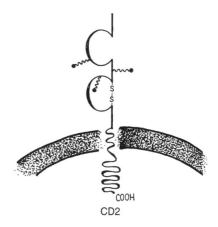

COOH
CD2

CD2R

A molecule, restricted to activated T cells and some NK cells, that has a mol wt of 50 kD. The CD2R epitope is unrelated to LFA-3 sites. The antigen is of importance in T cell maturation, since certain CD2 antibodies in combination with CD2R antibodies or LFA-3 may induce T cell proliferation. CD2R is a conformational form of CD2 that is activation dependent.

CD3

A molecule, also referred to as the T3 antigen, that consists of five different polypeptide chains with mol wt ranging from 16 to 28 kD. The five chains are designated γ, δ, ε, ζ, and η, with most CD3 antibodies being against the 20-kD ε-chain. They are closely associated with each other physically, and also with the T cell antigen receptor, in the T cell membrane. Incubation of T cells with CD3 antibodies induces calcium flux and proliferation. This group of

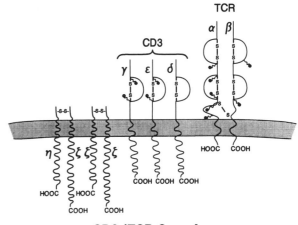

CD3/TCR Complex

molecules may therefore transmit a signal to the cell interior following binding of the antigen to the antigen receptor.

CD4

A single chain glycoprotein, also referred to as the T4 antigen, that has a mol wt of 56 kD and is present on approximately two-thirds of circulating human T cells, including most T cells of helper/inducer type. The antigen is also found on human monocytes and macrophages. The molecule appears to be a receptor for gp120 of HIV-1 and HIV-2 (AIDS viruses). This antigen binds to class II MHC molecules on antigen-presenting cells (APC), and may stabilize APC/T cell interactions. It is physically associated with the intracellular tyrosine protein kinase, known as p56^{lck}, which phosphorylates nearby proteins. This antigen is thereby relaying a signal to the cells. Cross-linking of CD4 may induce activation of this enzyme and phosphorylation of CD3.

CD4, it appears to be physically associated with the p56^{lck} protein tyrosine kinase which phosphorylates nearby proteins. This antigen is thereby relaying a signal to the cell. It is widely used as a marker for the subpopulation of T cells which includes suppressor/cytotoxic cells. The antigen is also present on splenic sinusoidal lining cells.

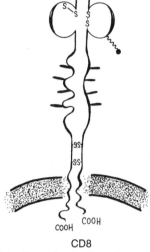

CD8

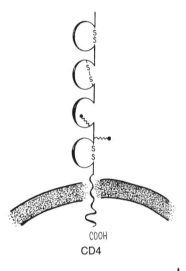

CD4

CD5

A molecule, also referred to as the T1 antigen, that is a single chain glycopolypeptide with a mol wt of 67 kD, and is present on the majority of human T cells. Its density increases with maturation of T cells. It is also found on a subpopulation of B cells. Sensitive immunohistochemical techniques have demonstrated its presence on immature B cells in the fetus and at low levels on mantle zone B cells in adult human lymphoid tissue. The majority of cases of B cell chronic lymphocytic leukemia expresses easily detectable levels of CD5. CD5 binds to CD72.

CD6

A molecule, sometimes referred to as the T12 antigen, that is a single chain glycopolypeptide

COOH

CD5

with a mol wt of 105 kD, and is present on the majority of human T cells (similar in distribution to CD3). It stains some B cells weakly.

CD7

An antigen, with a mol wt of 40 kD, that is present on the majority of T cells, and is useful as a marker for T cell neoplasms when other T cell antigens are absent. The CD7 antigen is probably an Fc receptor for IgM.

CD8

An antigen, also referred to as the T8 antigen, that has a mol wt of 32 to 34 kD. The CD8 antigen consists of two polypeptide chains, α and β, which may exist in the combination α/α homodimer or α/β heterodimer. Most antibodies are against the α chain. This antigen binds to class I MHC molecules on antigen presenting cells (APC), and may stabilize APC/class I cell interactions. Like

CD9

A single chain protein, with a mol wt of 24 kD, that is present on pre-B cells, monocytes, granulocytes, and platelets. Antibodies against the molecule can cause platelet aggregation. The CD9 antigen has protein kinase activity. It may be significant in aggregation and activation of platelets.

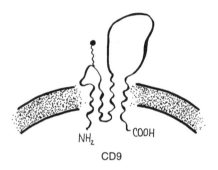

CD9

CD10

An antigen, also referred to as common acute lymphoblastic leukemia antigen (CALLA), that has a mol wt of 100 kD. CD10 is now known to be a neutral endopeptidase (enkephalinase). It is present on many cell types, including stem cells, lymphoid progenitors of B and T cells, renal epithelium, fibroblasts, and bile canaliculi.

CD11

A "family" of three leukocyte-associated single chain molecules, that has been identified in recent years [sometimes referred to as the leukocyte function associated antigen (LFA)/Mac-1 family]. They all consist of two polypeptide chains; the larger of these chains (α) is different for each member of the family; the smaller chain (β) is common to all three molecules (see CD11a, CD11b, CD11c).

CD11a

α chain of the leukocyte function associated-1 antigen (LFA-1) molecule with a mol wt of 180 kD. It is

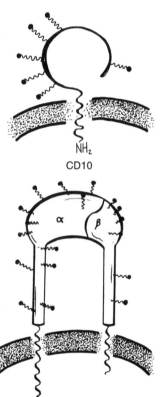

CD10

CD11/CD18

present on leukocytes, monocytes, macrophages, and granulocytes but negative on platelets. LFA-1 binds the intercellular adhesion molecules ICAM-1 (CD54), ICAM-2, and ICAM-3.

CD11b

α chain of Mac-1 (C3bi receptor). It has a mol wt of 170 kD and is present on granulocytes, monocytes, and NK cells.

CD11c

α chain of the p150,95 molecule. It has a mol wt of 150 kD and is present on granulocytes, monocytes, and NK cells. Furthermore, it reacts very strongly with hairy cell leukemia. The antigen is weakly expressed on B and T cell subsets.

CD12

Little is known about this antigen, which has a mol wt of 90 to 120 kD. It is present on some monocytes, granulocytes, and platelets.

CD13

An antigen that is a single chain membrane glycoprotein with a mol wt of 130 kD. It is present on monocytes, granulocytes, some macrophages, and connective tissue. CD13 has recently been shown to be aminopeptidase-N. It functions as a zinc metalloproteinase.

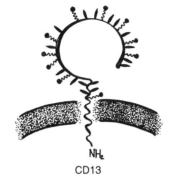

CD13

CD14

An antigen that is a single chain membrane glycoprotein with a mol wt of 55 kD. It is found principally on monocytes, but also on granulocytes, dendritic reticulum cells, and some tissue macrophages. It serves as a receptor for lipopolysaccharide (LPS) and for lipopolysaccharide binding protein (LBP).

CD15

A carbohydrate antigen that is often referred to as hapten X and consists of galactose, fucose, and *N*-acetyl-glucosamine linked in a specific sequence. The antigen appears to be particularly immunogenic in the mouse, and numerous monoclonal antibodies of this oligosaccharide specificity have been produced. CD15 is present in neutrophil secondary granules of granulocytic cells which express the antigen strongly late in their maturation. It is also present on eosinophils, monocytes, Reed-Sternberg and Hodgkin's cells, but can also be found on non-Hodgkin's cells. CD15 is also termed Lewis-x (Le^x).

CD16

An antigen that is also known as the low-affinity Fc receptor for complexed IgG-FcγRIII. It is expressed on NK cells, granulocytes, (neutrophils), and macrophages. Structural differences in the CD16 antigen from granulocytes and NK cells have been reported. This apparent polymorphism suggests two different genes for the FcγRIII molecule in polymorphonuclear leukocytes (PMN) and in NK cells. The CD16 molecule

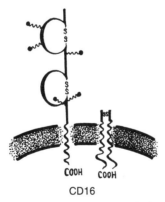

CD16

in NK cells has a transmembrane form, whereas it is phosphatidylinositol (PI)-linked in granulocytes. CD16 mediates phagocytosis. It is the functional receptor structure for performing antibody-dependent-cell-mediated cytotoxicity (ADCC). CD16 is also termed FcγRIII.

CDw17

An antigen that is a cell surface glycosphingolipid moiety, lactosylceramide, and is present principally on granulocytes, but also on monocytes and platelets.

CD18

An antigen that is an integral membrane glycoprotein with a mol wt of 95 kD. This integrin β2 subunit is noncovalently linked to CD11a, CD11b, and CD11c molecules and is expressed on leukocytes. It is important for cell adhesion. The immunodeficiency termed leukocyte adhesion deficiency (LAD) has been shown to be caused by a genetic defect of the CD18 gene.

CD19

An antigen, with a mol wt of 90 kD, that has been shown to be a transmembrane polypeptide with at least two immunoglobulin-like domains. The CD19 antigen is the most broadly expressed surface marker for B cells, appearing at the earliest stages of B cell differentiation. The CD19 antigen is expressed at all stages of B cell maturation, from the pro-B cell stage until just before the terminal differentiation to plasma cells. CD19 complexes with CD21 (CR2) and CD81 (TAPA-1). It is a co-receptor for B lymphocytes.

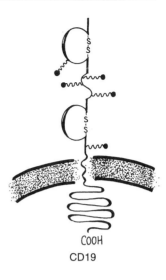

CD19

CD20

A B cell marker, with a mol wt of 33, 35, 37 kD, that appears relatively late in the B cell maturation (after the pro-B cell stage), and then persists for some time before the plasma cell stage. Its molecular structure resembles that of a transmembrane ion channel. The gene is on chromosome 11 at band q12-q13. It may be involved in regulating B cell activation.

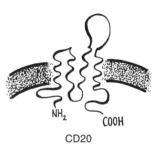

CD20

CD21

An antigen, with a mol wt of 145 kD, that is expressed on B cells and, even more strongly, on follicular dendritic cells. It appears when surface Ig is expressed after the pre-B cell stage and is lost during early stages of terminal B cell differentiation to the final plasma cell stage. CD21 is coded for by a gene found on chromosome 1 at band q32. The antigen functions as a receptor for the C3d component and also for Epstein-Barr virus. CD21, together with CD19 and CD81, constitutes the co-receptor for B cells. It is also termed CR2.

CD22 (see figure page 56)

A molecule, with a mol wt of α130 and β140 kD, that is expressed in the cytoplasm of B cells of the pro-B and pre-B cell stage and on the cell surface on mature B cells with surface Ig. The antigen is lost shortly before the terminal plasma cell phase. The molecule has five extracellular immunoglobulin domains and shows homology with myelin adhesion glycoprotein and with N-CAM (CD56). It participates in B cell adhesion to monocytes and T cells. Also called BL-CAM.

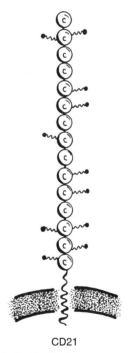

CD21

CD23

An antigen that is an integral membrane glycoprotein, with a mol wt of 50 to 45 kD. The CD23 antigen has been identified as the low-affinity Fc-IgE receptor (FcεRII). Two species of FcεRII/CD23 have been found. FcεRIIa and FcεRIIb differ only in the N-terminal cytoplasmic region, but share the same C-terminal extracellular region. FcεRIIa is strongly expressed on IL-4-activated B cells and weakly on mature B cells. FcεIIa is not found on circulating B cells and bone marrow B cells positive for surface IgM, IgD. FcεRIIb is

expressed weakly on mono-
cytes, eosinophils, and T cell
lines. However, IL-4-treated
monocytes show a stronger
staining. The CD23 molecule
may also be a receptor for B
cell growth factor (BCGF).

CD24

An antigen that is a glycoprotein found on B cells, granulocytes, interdigitating cells, and
on epithelial cells. The CD24
antigen, with a mol wt of 41/38
kD, is expressed on B cells
from the late pro-B cell stage
and is lost before the plasma
cell stage. It may be a homologue in man of murine heat
stable antigen (HSA) or J11d.

CD25

A single chain glycoprotein,
often referred to as the α-chain
of the interleukin-2-receptor
(IL-2R) or the Tac antigen, that
has a mol wt of 55 kD, and is
present on activated T and B
cells and activated macrophages. It functions as a receptor

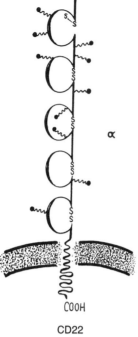

α

COOH

CD22

for IL-2. Together with the β-chain of the IL-2R (p75, mol wt 75
kD), the CD25 antigen forms a high-affinity receptor complex for
IL-2. The gene for IL-2R has been located as a single gene on
chromosome 10. It associates with CD122 and complexes with the
IL-2Rβγ high-affinity receptor. It facilitates T cell growth.

CD26

An antigen that is a single chain glycoprotein with a mol wt of 110
kD. It is present on activated T and B cells, but also on macrophages,
bile canaliculi, and splenic sinusoidal lining cells. It has recently
been shown to be a dipeptidyl-peptidase IV (DPP-IV), which is a
serine-type protease, and may play a role in cell proliferation
together with the CD25 antigen. Although the CD4 molecule is
requisite for binding HIV particles, it is not sufficient for efficient
entry of the virus and infection. The cofactor is dipeptidyl peptidase
IV (DPP-IV), also termed CD26.

CD27

A T cell antigen (mol wt: 110 kD) that is a disulfide-linked
homodimer of two polypeptide chains, each with a mol wt of 55 kD.
It is found on a subpopulation of thymocytes, on the majority of
mature T cells, and on transformed B lymphocytes. CD27 antigen
expression is unregulated on activated cells.

CD28

A T cell antigen that is a disulfide-linked homodimer of two chains,
each with a mol wt of 44 kD. It is found on a subpopulation of T cells
and activated B cells. Antibodies against CD28 can maintain the
growth of phorbol myristate acetate (PMA)-activated T cells. Lymphocytes bearing both CD8 and CD28 are cytotoxic T lymphocyte
precursors. It has a role in naive T cell activation. CD28 is the
receptor for co-stimulatory (signal 2), binds CD80 (B7.1) and B7.2.
Also called Tp44.

CD29

An antigen, with a mol wt of 130 kD, that is very widely distributed
on human tissues (e.g., nerve, connective tissue, endothelium), as
well as on many white cell populations. The structure recognized is
now known to be the platelet gpIIa, the integrin β-1 chain, the
common β-subunit of very late antigens (VLA) 1 to 6, and the
fibronectin receptor. The CD29 antigen is termed gpIIa in contrast
to CD31 which is called gpIIa'. Together with α¹ chain (CDw49),
it forms a heterodimeric complex. CD29 associates with CD49a in
VLA-1 integrin. Antibodies against CD29 appear to define a subset
of cells among CD4 positive T cells which provide help for antibody
production. Like the antibody belonging to CD45RO (UCHL1),
CD29 antibodies appear to be reciprocal (when T4 positive cells are
analyzed) to those of CD45RA antibodies against the B cell restricted form of LCA. This form recognizes a T cell subset of CD4-
positive cells suppressing antibody production. The molecule is
involved in the mediation of cell adhesion to cells or matrix,
especially in conjunction with CD49.

CD30

A molecule that is a single chain glycoprotein, also referred to as the
Ki-1 antigen. It has a mol wt of 105 kD, and is present on activated
T and B cells, embryonal carcinoma cells, and Hodgkin's and Reed-
Sternberg cells. It is also found on a minority of non-Hodgkin's
lymphomas (referred to as anaplastic large cell lymphomas) with a
characteristic anaplastic morphology.

CD31

An antigen that is a single chain membrane glycoprotein with a mol
wt of 140 kD. It is found on granulocytes, monocytes, macrophages,
B cells, platelets, and endothelial cells. Although it is termed gpIIa',
it is different from the CD29 antigen. At present, the function of
CD31 is unknown. It may be an adhesion molecule. Also called
PECAM-1 (platelet endothelial cell adhesion molecule-1).

CD32

An antigen, also known as FcγRII receptor, which is a low-affinity Fc
receptor for IgG, aggregated immunoglobulin/immune complexes,
that has a mol wt of 40 kD. Several different isoforms exist, which are
coded for by closely related genes. It is found on granulocytes, B
cells, monocytes, platelets, and endothelial cells. FcγRII functions as
the receptor for aggregated IgG and is involved in cell activation.

CD33

An antigen that is a single chain
transmembrane glycoprotein with
a mol wt of 67 kD. It is restricted
to myeloid cells and is found on
early progenitor cells, monocytes,
myeloid leukemias, and weakly
on some granulocytes.

CD34

A molecule (mol wt: 105 to 120
kD) that is a single chain transmembrane glycoprotein, present
on immature haematopoietic cells
and endothelial cells. Its gene is
on chromosome 1. It is the ligand
for L-selectin (CD62L).

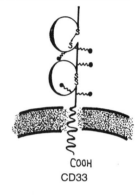

COOH

CD33

CD35

An antigen that is known as CR1 that binds C3b and C4b. The CD35
antigen includes four different allotypes termed C, A, B, and D, with
mol wts of 160 (C), 190 (A), 220 (B), and 250 kD (D), respectively.
The antigen is widely distributed on various cell types, including
erythrocytes, B cells, monocytes, granulocytes (negative on basophils), some NK cells, and follicular dendritic cells. The functions
of CR1 are processing of immune complexes and promotion of
binding and phagocytosis of C3b-coated particles/cells.

CD36

An antigen (mol wt: 88 kD) that is a single chain membrane glycoprotein, found on monocytes, macrophages, platelets, some endothelial
cells, and weakly on B cells. The CD36 antigen is also referred to as
gpIV or gpIIIb. The antigen acts as a receptor for thrombospondin and
collagen. Furthermore, it is the endothelial cell receptor for erythrocytes infected with *Plasmodium falciparum*.

CD37

A B cell antigen that is an integral glycoprotein with a mol wt of 40
to 52 kD which appears at the late pre-B cell stage and then persists
throughout B cell maturation until shortly before the terminal plasma
cell stage. The antigen is also weakly expressed on macrophages,
neutrophils and monocytes, and on resting and activated T cells.
The function of CD37 antigen is unknown.

CD38

A molecule that is an integral glycoprotein and is also referred to as
the T10 antigen. It has a mol wt of 45 kD and is found on plasma
cells, pre-B cells, immature T cells, and activated T cells.

CD39

An antigen, with a mol wt of 100 to 70 kD, that is present on a
number of cell types, including B cells (mantle zone), most mature
B cells with surface-bound IgG, monocytes, some macrophages,
and vascular endothelial cells. Its function is yet to be determined,
but CD39 might mediate B cell adhesion. Also called gp80.

CD40

An integral membrane glycoprotein that has a mol wt of 48/44 kD
and is also referred to as gp50. The antigen shares similarities with
many nerve growth factor receptors. The CD40 antigen is expressed
on peripheral blood and tonsillar B cells from the pre-B cell stage
until the plasma cell stage, where it is lost. It is also expressed on B

cell leukemias and lymphomas, some carcinomas, weakly on monocytes, and interdigitating cells. It has been shown that the CD40
antigen is active in B cell proliferation. The CD40 ligand is gp39.
CD40 binds CD40-L, the CD40 ligand. It is the receptor for the costimulatory signal for B cells.

CD40-L

A 39-kD antigen present on activated CD4⁺ T cells. It is the ligand
for CD40. It is also called T-BAM or gp39.

CD41

An antigen, equivalent to glycoprotein IIb/IIIa, that is found on
platelets and megakaryocytes. This structure is a Ca^{2+}-dependent
complex between the 110-kD gpIIIa (CD61) and the 135-kD gpIIb
molecules. The gpIIb molecule consists of two chains: an α-chain
of 120 kD and a β-chain of 23 kD. The gpIIIa (CD61) is a single
chain protein, but both gpIIIa (CD61) and gpIIb contain transmembrane domains. The molecule is a receptor for fibrinogen, but it also
binds von Willebrand factor, thrombospondin, and fibronectin. It is
absent or reduced in Glanzmann's thrombasthenia (hereditary gpIIb/
IIIa deficiency). It has a role in platelet aggregation and activation.

CD42a

An antigen, equivalent to glycoprotein IX, that is a single
chain membrane glycoprotein
with a mol wt of 23 kD. It is
found on megakaryocytes and
platelets. CD42a forms a noncovalent complex with CD4b
(gpIb) which acts as a receptor
for von Willebrand factor. It is
absent or reduced in the Bernard-Soulier syndrome.

CD42b

An antigen, equivalent to glycoprotein Ib, that is a two chain
membrane glycoprotein with a
mol wt of 170 kD. CD42b has
an α-chain of 135 kD and a β-
chain of 23 kD. It is found on
platelets and megakaryocytes.
CD42b forms a noncovalent complex with CD42a (gpIX) which
acts as a receptor for von
Willebrand factor. The antigen is absent or reduced in the
Bernard-Soulier syndrome.

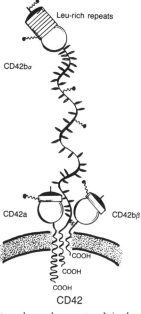

CD42

CD42c

A 22-kD antigen found on platelets and megakaryocytes. It is also
referred to as GPIB-β.

CD42d

An 85-kD antigen present mainly on platelets and megakaryocytes.
It is also referred to as GPV.

CD43

A 90- to 100-kD molecule that is a sialylated glycoprotein. It is
found on leukocytes, including T cells, granulocytes, erythrocytes,
epithelial cell lines, and brain cells. The CD43 antigen is not present
on peripheral B cells. The molecule is involved in activation of T
cells, B cells, NK cells, and monocytes. It binds CD54 (ICAM-1).

CD44

A molecule that is a transmembrane molecule with a mol wt of 80
to 90 kD. It is found on some white and red cells. It is weakly
expressed on platelets. It functions probably as a homing receptor.
CD44 is a receptor on cells for hyaluronic acid. It binds to hyaluron-
ate. CD44 mediates leukocyte adhesion.

CD45

An antigen that is a single chain glycoprotein referred to as the
leukocyte common antigen (or "T200"). It consists of at least five
high molecular weight glycoproteins present on the surface of the
majority of human leukocytes (mol wts: 180, 190, 205, and 220 kD).
The different isoforms arise from a single gene via alternative mRNA
splicing. The variation between the isoforms is all in the extracellular
region. The larger (700-amino acid) intracellular portion is identical
in all isoforms and has protein tyrosine phosphatase activity. It can
potentially interact with intracellular protein kinases such as p56ˡᶜᵏ,
which may be involved in triggering cell activation. By dephospho-
rylating proteins, CD45 would act in an opposing fashion to a protein
kinase. It facilitates signaling through B and T cell antigen receptors.

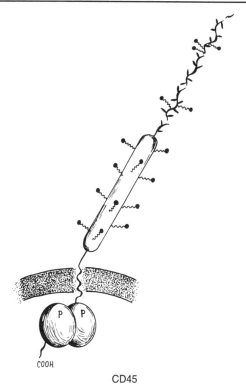

CD45

CD45R

A subfamily that is now divided into three isoforms: CD45RO,
CD45RA, and CD45RB. The designation CD45R has been maintained for those antibodies which have not been tested on appropriate transfectants.

CD45RA

A 220,205 kD isoform of CD45 (sequence encoded by exon A) that
is found on B cells, monocytes, and a subtype of T cells. T cells
expressing this isotype are naive or virgin T cells and nonprimed
CD4⁺ and CD8⁺ cells.

CD45RB

A molecule, which consists of four isoforms of CD45 (sequence
encoded by exon B) with mol wts of 220, 205, and 190 kD, that is
found on B cells, subset of T cells, monocytes, macrophages, and
granulocytes.

CD45RO

A 180-kD isoform of CD45 (sequence not encoded by either exons
A, B, or C), that is found on T cells, subset of B cells, monocytes,
and macrophages. T cells expressing this antigen are T memory
cells or primed T cells.

CD46

An antigen that is a dimeric protein with a mol wt of 66/56 kD. It has
a broad tissue distribution. It is present on hematopoietic and
nonhematopoietic nucleated cells. The CD46 antigen shows sequence homology to complement-associated molecules such as decay-accelerating factor (DAF, CD55) and the complement receptors
CR1 (CD35) and CR2 (CD21). It is a membrane cofactor protein that
binds C3b and C4b, allowing their degradation by Factor I.

CD47

A single chain glycoprotein, with a mol wt of 47 to 52 kD, that is widely
distributed and may be associated with the Rhesus blood group.

CD48

An antigen that is a phosphatidylinositol (PI)-linked glycoprotein
with a mol wt of 43 kD, and is widely distributed on white blood cells.

CD49a

A 210-kD antigen present on activated T cells and monocytes. It
associates with CD29, binds collagen and laminin. It is also referred
to as VLA-1 and α¹ integrin.

CD49b

A 165-kD antigen present on B cells, platelets, and monocytes. It
associates with CD29, binds collagen and laminin. It is also referred
to as VLA-2 and α² integrin.

CD49c

A 125-kD antigen present on B cells. It associates with CD29, binds
laminin and fibronectin. It is also referred to as VLA-3 and α³ integrin.

CD49d

A 150-kD antigen present on B cells and thymocytes. It associates with CD29, binds fibronectin, Peyer's patch HEV, and VCAM-1. It is also referred to as VLA-4 and as α^4 integrin.

CD49e

A 135,25 dimer antigen present on memory T cells, platelets, and monocytes. It associates with CD29 and binds fibronectin. It is also referred to as VLA-5 and as α^5 integrin.

CD49f

A 120,25 dimer antigen present on memory T cells, monocytes, and thymocytes. It associates with CD29 and binds laminin. It is also referred to as VLA-6 and as α^6 integrin.

CD50

A molecule, with a mol wt of 124 kD, that is broadly distributed on leukocytes including thymocytes, T cells, B cells (not plasma cells), monocytes, and granulocytes. However, it is not present on platelets and erythrocytes. CD50 is also called ICAM3.

CD51

A molecule that is the α-chain of the vitronectin receptor (VNR) with a mol wt of 140 kD. The CD51 antigen consists of a disulfide-linked large subunit of 125 kD and a smaller one of 25 kD. The VNR α-chain is noncovalently associated with the VNR β3-chain, also termed gpIIIa or CD61. The α-chain is weakly expressed on platelets. The VNR mediates cell adhesion to arg-gly-arp containing sequences in vitronectin, von Willebrand factor, fibrinogen, and thrombospondin.

CD51/61 complex

An antigen found on platelets and on endothelial cells. It is also referred to as vitronectin receptor or as integrin αVβ3.

CD52

A cluster that corresponds to the leukocyte antigen (mol wt: 21 to 28 kD) recognized by antibody Campath-1. Anti-CD52 antibodies have been used to deplete T lymphocytes therapeutically.

CD53

A single chain glycoprotein, with a mol wt of 32 to 42 kD, that is widely distributed, e.g., in monocytes, T cells, B cells, and NK cells.

CD54

An intercellular adhesion molecule (ICAM-1) that is found on phytohemagglutinin (PHA) blasts, endothelial cells, and follicular dendritic cells. The mol wt of the antigen is 90 kD. ICAM-1 is a lymphokine inducible molecule and has been shown to be a ligand for LFA-1 or the CD11/CD18 complex. It is also the receptor for rhinovirus. CD54 binds CD11a/CD18 integrin (LFA-1) and CD11b/CD18 integrin (Mac-1).

CD55

A 75-kD molecule that is a PI (phosphatidylinositol)-linked single chain glycoprotein, also known as the decay accelerating factor (DAF). The antigen is involved in complement degradation on the cell surface. It is broadly distributed on hematopoietic and many nonhematopoietic cells. CD55 binds C3b and disassembles C3/C5 convertase.

CD56

A 220/135 kD molecule that is an isoform of the neural adhesion molecule (N-CAM). It is used as a marker of NK cells, but it is also present on neuroectodermal cells.

CD57

A 110-kD myeloid-associated glycoprotein that is recognized by the antibody HNK1. It is a marker for NK cells, but it is also found on some T and B cells. It is an oligosaccharide present on multiple cell surface glycoproteins.

CD58

A 65- to 70-kD molecule that is a single chain glycoprotein, also known as LFA-3 (leukocyte

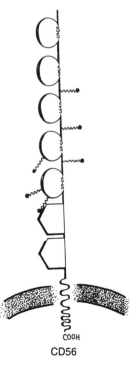

CD56

function associated-3 antigen), and is the ligand to which CD2 binds. The CD58 antigen is expressed on many hematopoietic and nonhematopoietic cells.

CD59

An 18- to 20-kD glycoprotein that is present on many hematopoietic and nonhematopoietic cells. It may be involved in the inhibition of the complement membrane attack complex. It binds C8 and C9 complement components.

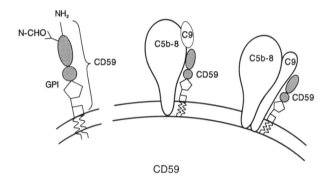

CD59

CDw60

An oligosaccharide found on gangliosides. It is present on a subpopulation of lymphocytes, platelets, and monocytes.

CD61

An antigen that is also termed gpIIIa, the integrin β3-chain, or the β-chain of the vitronectin receptor (VNR β3-chain). GpIIIa can associate with either gpIIb to form the gpIIb/IIIa complex (CD41) or with the VNR α-chain (CD51). The CD61 molecule has a mol wt of 110 kD and is expressed on platelets and megakaryocytes.

CD62E

A 140-kD antigen present on endothelium. CD62E is endothelium leuckocyte adhesion molecule (ELAM). It mediates neutrophil rolling on the endothelium. It also binds sialyl-Lewis x.

CD62L

A 150-kD antigen present on B and T cells, monocytes, and NK cells. CD62L is leukocyte adhesion molecule (LAM). It mediates cell rolling on the endothelium. It also binds CD34, GlyCAM. It is also referred to as L-selectin, LECAM-1, or LAM-1.

CD62P

A 75- to 80-kD antigen present on endothelial cells, platelets, and megakaryocytes. CD62P is an adhesion molecule that binds sialyl-Lewis x. It is a mediator of platelet interaction with monocytes and neutrophils. It also mediates neutrophil rolling on the endothelium. It is also referred to as P-selectin, GMP-140, PADGEM, or LECAM-3.

CD63

A 53-kD antigen that is a platelet activation antigen associated with lysosomes. The CD63 molecule is also present on monocytes, macrophages, and weakly on granulocytes, T cells, and B cells. Following activation, the CD63 lysosomal membrane protein is translocated to the cell surface.

CD64

A 75-kD receptor that is a single chain glycoprotein. It is a high affinity receptor IgG, also known as FcγR1, present on monocytes and some tissue macrophages.

CDw65

A 180- to 200-kD fucoganglioside (NeuAc-Gal-GlcNac-Gal-GlcNac(Fuc)-Gal-GlcNac-Gal-GlcNac-Gal-Glc-Cer) that is present on granulocytes and monocytes.

CD66a

A 160- to 180-kD antigen present on cells of neutrophil lineage. It is also referred to as BGP (biliary glycoprotein).

CD66b

A 95- to 100-kD antigen present on granulocytes. It is also referred to as CGM6 (CEA gene member 6), p100. It was previously referred to as CD67.

CD66c

A 90-kD antigen present on neutrophils and colon carcinoma cells. It is also referred to as NCA (nonspecific cross-reacting antigen).

CD66d

A 30-kD antigen present on neutrophils. It was formerly called CGM1 (CEA gene member 1).

	Cluster designation	Main cellular expression of antigen	Other names	Antigen molecular weight (kD)
B	CD79a	B-cell specific	mb-1, Igα	33, 40
	CD79b	B-cell specific	B29, Igβ	33, 40
	CD80	B-cell subset *in vivo*, most activated B-cells *in vitro*	B7, BB1	60
	CD81	B-cells, (Broad Expression including lymphocytes)	TAPA-1 (Target of an antiproliferative antibody)	22
	CD82	Broad expression on leucocytes (weak) not erythrocytes	R2, IA4, 4F9	50-53
	CD83	Specific marker for Circulating Dendritic cells, activated B and T-cells, Germinal centre cells	HB15	43
	CDw84	Platelets and Monocytes (strong), circulating B-cells (weak)	GR6	73
	CD85	Circulating B-cells (weak), Monocytes (strong)	VMP-55, GH1/75,GR4	120, 83
	CD86	Circulating monocytes, Germinal centre cells (histol.), Activated B-cells	FUN-1, GR65, BU63	80
M	CD87	Granulocytes, Monocytes, Macrophage, Activated T-cells	UPA-R (Urokinase Plasminogen Activator Receptor)	50-65
	CD88	Polymorphonuclear leucocytes, mast cells, macrophage, smooth muscle	C5a Receptor, GR10	42
	CD89	Neutrophils, monocytes, macrophage, T and B-cell subpopulation	FcαR, IgA receptor	55-70
	CDw90	CD34+ve, subset on bone marrow, cord blood, fetal liver	Thy-1	25-35
	CD91	Monocytes and some non haemopoietic cell lines	α2M-R (α2 macroglobulin receptor)	600
	CDw92	Neutrophils, monocytes, endothelial cells, platelets	GR9	70
	CD93	Neutrophils, monocytes, endothelial cells	GR11	120
NK	CD94	NK-cells, α/β,γ/δ, T-cell subsets	KP43	43
Ac	CD95	Variety of cell lines including myeloid and T lymphoblastoid	APO-1, FAS	42
	CD96	Activated T-cells	TACTILE-(T-cell activation increased late expression)	160
	CD97	Activated cells	GR1, BL-KDD/F12	74, 80, 89
T	CD98	T-cells and B-cells (weak), monocytes (strong), most human cell lines	4F2,2F3	80,40
	CD99	Peripheral blood lymphocytes, thymocytes	MIC2, E2	32
	CD99R	B and T lymphocytes, some leukaemias	CD99 mAb restricted	32
	CD100	Broad expression on haemopoietic cells	BB18, A8, GR3	150
	CDw101	Granulocytes, macrophage	BB27, BA27, GR14	140
Ad	CD102	Resting lymphocytes, monocytes. Vascular endothelial cells (strongest)	ICAM-2	60
	CD103	Intraepithelial lymphocytes, 2-6%, PBL	HML-1, αE integrin, α6	150, 25
	CD104	Epithelia, Schwann cells, some tumor cells	β4 integrin chain, beta4	220
E	CD105	Endothelial cells, bone marrow cell subset, *in vitro* activated macrophage	Endoglin, TGF B1 and β3 receptor, GR7	95
	CD106	Endothelial cells	VCAM-1, INCAM-110	100, 110
P	CD107a	Activated platelets	LAMP 1 (Lysosomal associated membrane protein)	110
	CD107b	Activated platelets	LAMP 2	120
Ad	CDw108	Activated T-cells in spleen, some stromal cells	GR2	80
E	CDw109	Activated T-cells, platelets, endothelial cells	Platelet activation factor, 8A3, 7D1, GR56	170/150
BP	CD110-CD114	Nothing yet assigned to these numbers		
M	CD115	Monocytes, macrophage, placenta	M-CSFR (Macrophage colony stimulating factor receptor), CSF-1R	150
	CDw116	Monocytes, neutrophils, eosinophils, fibroblasts endothelial cells	GM-CSF R (Granulocyte, macrophage colony stimulating factor receptor), HGM-CSFR	75-85
	CD117	Bone marrow progenitor cells	Stem cell factor receptor (SCF R), cKIT	145
	CD118	Broad cellular expression	IFNα,β receptor	
	CD119	Macrophage, monocyte, B-cells, epithelial cells	IFNγR (Interferon γ receptor)	90
	CD120a	Most cell types, higher levels on epithelial cell lines	TNFR; 55kD (Tumor necrosis factor receptor)	55
	CD120b	Most cell types, higher levels on myeloid cell lines	TNF R; 75kD	75
C	CDw121a	T-cells, thymocytes, fibroblasts, endothelial cells	IL-1R (Interleukin-1 receptor), Type 1	80
	CDw121b	B-cells, macrophages, monocytes	IL-1R, Type II	68
	CD122	NK-cells, resting T-cell sub population, some B-cell lines	IL-2Rβ, IL2R; 75kD	75
	CD123	Bone marrow stem cells, granulocytes, monocytes, megakaryocytes	IL-3R	
	CDw124	Mature B and T-cells, haemopoietic precursor cells	IL-4R	140
	CD125	Eosinophils and basophils	IL-5R	
	CD126	Activated B-cells and plasma cells (strong), most leucocytes (weak)	IL-6R	80 (αsubunit)
	CDw127	Bone marrow lymphoid precursors, Pro-B-cells, mature T-cells, monocytes	IL-7R	75
BP	CDw128	Neutrophils, basophils, T-cell subset	IL-8R	58-67
	CD129	Nothing yet assigned to this number		
C	CDw130	Activated B-cells and plasma cells (strong), most leucocytes (weak), endothelial cells	IL-6R-gp 130SIG	130

T = T Cells   B = B Cells   M = Myeloid   P = Platelets   Ac = Activation
Ad = Adhesion   C = Cytokine   NK = NK Cells   E = Endothelial   BP = Blind Panel (Multi-lineage)

CD66e

A 180- to 200-kD antigen present on adult colon epithelial cells and colon carcinoma cells. It was previously referred to as CEA (carcinoembryonic antigen).

CD67

Refer to CD66b.

CD68

A 110-kD macrophage/myeloid marker that is principally intracellular, but has weak expression under some conditions. Macrophages, monocytes, neutrophils, basophils, and large lymphocytes express it. Also called macrosialin.

CD69

A 60-kD phosphorylated glycoprotein that is a homodimer of 34/28 kD. CD69 is also known as AIM (activation inducer molecule), and is found on activated B and T cells, activated macrophages, and NK cells.

CD70

A 75-, 95-, 170-kD antigen present on activated T and B cells, Reed-Sternberg cells, and weakly expressed on macrophages. It has also been referred to as Ki-24 antigen or as the CD27 ligand.

CD71

A molecule that is otherwise known as transferrin receptor (mol wt: 95 kD). The antigen is a homodimeric glycoprotein present on activated T and B cells, macrophages, and in proliferating cells. The main function of this receptor is the binding of transferrin. By internalization of the receptor and its ligand within cells, iron is delivered for cellular metabolism. Also called T9.

CD72

An antigen that is a heterodimeric glycoprotein (mol wt: 43 and 39 kD). The CD72 antigen is a pan-B cell marker which occurs like the CD19 antigen at the earliest stages of B cell differentiation. The antigen is expressed from the pro-B cell stage until the plasma cell stage. It is the ligand for CD5.

CD73

A 69-kD molecule that is an ecto-5′ nucleotidase found on B and T cell subsets, dendritic reticulium cells, epithelial cells, and endothelial cells. It is expressed weakly on pre-B cells until the final plasma cell stage, where it is lost. CD73 dephosphorylates nucleotides to permit uptake of nucleoside.

CD74

An antigen, also known as Ii, that is invariant chain of HLA-class II. It is found on B cells, macrophages, and some epithelial cells. The expression of CD74 begins during the pre-B cell stage and is lost before the plasma cell stage. Several forms (mol wt: 41, 35, and 33 kD) are recognized arising via alternative splicing.

CDw75

A marker of mature B cells expressing surface immunoglobulin and of a T cell subset, but is also detectable on epithelial cells. The antigen is lost before the plasma cell stage. it may be an oligosaccharide.

CDw76

An antigen (mol wt: 85/67 kD) that is a marker of mature B cells expressing surface immunoglobulin and a T cell subset. The CD76 antigen is lost before the plasma cell stage. Mantle zone B cells are positive, and germinal centers are negative. it may be an oligosaccharide.

CD77

A molecule that corresponds to a sugar moiety of globotriaosylceramide (Gb3) with the formula Galα1-4Galβ1-4Glc1-1 ceramide and is also recognized as the p^k blood group antigen. The CD77 antigen is only expressed on activated B cells. It is present on follicular center B cells, follicular dendritic cells, endothelium, and a variety of epithelial cell types. Also called BLA.

CDw78

A pan-B cell and macrophage subset marker that increases its expression on peripheral B cells after activation. The CDw78 antigen is expressed on pre-B cells and is lost before the plasma cell stage. Also called Ba.

CD79a

A 33,40 kD antigen present on mature B cells. It is part of the B cell antigen receptor. It is a analogous to CD3 in T cells and is needed for cell-surface expression and signal transduction. Also called MB1 or Igα.

CD79b

A 33,40 kD antigen present on mature B cells. It is part of the B cell antigen receptor and is analogous to CD3 in T cells. It is needed for cell-surface expression and signal transduction. Also called B29 or Igβ.

CD80

A 60-kD antigen present on a B cell subset in vivo and most activated B cells in vitro. It may also be found on dendritic cells and macrophages. It serves as a co-stimulator for T cell activation. It is a ligand for CTLA-4 and for CD28. Also called B7 (B7.1) or BB1.

CD81

A 22-kD antigen present on B cells. It has broad expression including lymphocytes. It associates with CD19 and CD21 to produce the B cell coreceptor. Also called TAPA-1.

CD82

A 50- to 53-kD antigen. It has broad expression on leukocytes (weak) but not on erythrocytes. Also called R2, IA4, or 4F9. Its function is unknown.

CD83

A 43-kD antigen that serves as a specific marker for circulating dendritic cells, activated B and T cells and germinal center cells. Its function is unkown. Also called HB15.

CDw84

A 73-kD antigen present on platelets and monocytes (strong), and circulating B cells (weak). Its function is unkown. Also called GR6.

CD85

A 120,83-kD antigen present on circulating B cells (weak) and monocytes (strong). Its function is unknown. Also called VMP-55, GH1/75, or GR4.

CD86

An 80-kD antigen present on ciruclation monocytes, germinal center cells, and activated B cells. Its function is unknown. Also called FUN-1, GR65, or BU63.

CD87

A 50- to 65-kD antigen present on monocytes, macrophages, granulocytes, and activated T cells. It is a urokinase plasminogen activator receptor. Also called UPA-R (urokinase plasminogen activator receptor).

CD88

A 42-kD antigen present on polymorphonuclear leukocytes, mast cells, macrophages, and smooth muscle cells. It is a receptor for complement component C5a. Also called C5a receptor or GR10.

CD89

A 55- to 70-kD antigen present on neutrophils, monocytes, macrophages, and T and B cell subpopulations. It is the receptor for IgA. Also called FcαR, IgA receptor.

CDw90

A 25- to 35-kD molecule present on human CD34+ prothymocytes and on murine thymocytes and T cells. It is also expressed on bone marrow, cord blood, and fetal liver. Also called Thy-1.

CD91

A 600-kD antigen present on monocytes and some nonhematopoietic cell lines. It is the α$_2$ macroglobulin receptor. Also called α2M-R (α$_2$ macroglobulin receptor).

CDw92

A 70-kD antigen present on neutrophils, monocytes, endothelial cells, and platelets. Its function is unknown. Also called GR9.

CD93

A 120-kD antigen present on neutrophils, monocytes, and endothelial cells. Its function is unknown. Also called GR11.

CD94

A 43-kD antigen present on NK cells, α/β, γ/δ, T cell subsets, and on natural killer cells. Its function is unknown. Also called KP43.

CD95

A 42-kD antigen present on myeloid and T lymphoblastoid cell lines as well as various other cell lines. It binds the TNF-like ligand. It causes apoptosis. Also called APO-1 or FAS.

CD96

A 160-kD antigen present on activated T cells. Its function is unknown. Also called TACTILE-(T cell activation incresed late expression).

CD97

A 74,80,89-kD antigen present on activated cells. Its function is unknown. Also called GR1, GL-KDD/F12.

CD98

An 80,94-kD heterodimeric antigen present on T cells and B cells (weak), monocytes (strong), natural killer cells, granulocytes, and most human cell lines. Its function is unknown. Also called 4F2 or 2F3.

CD99

A 32-kD antigen present on peripheral blood lymphocytes and thymocytes. Also called MIC2 or E2.

CD99R

A 32-kD antigen present on B and T lymphocytes, some leukemias. Also called CD99 mAb restricted.

CD100

A 150-kD antigen that has broad expression on hematopoietic cells. Its function is unknown. Also called BB18, A8, GR3.

CDw101

A 140-kD antigen present on granulocytes and macrophages. Its function is unknown. Also called BB27, BA27, GR14.

CD102

A 60-kD antigen present on resting lymphocytes, monocytes, and vascular endothelial cells (strongest). It binds CD11α/CD18 (LFA-1) but not CD11β/CD18 (Mac-1). Also called ICAM-2.

CD103

A 150/25-kD antigen present on intraepithelial lymphocytes, 2 to 6%, and on peripheral blood lymphocytes. It is the αE integrin. Also called HML-1, αE integrin, or α6.

CD104

A 220-kD antigen present on epithelia, Schwann cells, and some tumor cells. It functions as a β4 integrin. Also called β4 integrin chain or β4.

CD105

A 95-kD homodimeric antigen present on endothelial cells, bone marrow cell subset, and in vitro-activated macrophages. Its function is unknown except it might be a ligand for an integrin. Also called Endoglin, TGF B1 and β3 receptor, GR7.

CD106

A 100,110-kD antigen present on endothelial cells that functions as an adhesion molecule. It is a ligand for VLA-4. Also called VCAM-1 or INCAM-110.

CD107a

A 110-kD antigen present on activated platelets. Its function is unknown, but it is a lysosomal membrane protein that is translocated to the cell surface following activation. Also called LAMP-1 (lysosomal associated membrane protein-1).

CD107b

A 120-kD antigen present on activated platelets. Its function is unknown, but it is a lysosomal membrane protein that is translocated to the cell surface following activation. Also called LAMP-2.

CDw108

An 80-kD antigen present on activated T cells in spleen, some stromal cells. Its function is unknown. Also called GR2.

CDw109

A 170/150-kD antigen present on activated T cells, platelets, and endothelial cells. Its function is unknown. Also called platelet activation factor, 8A3, 7D1, GR56.

CD110

Nothing yet assigned to this number.

CD111

Nothing yet assigned to this number.

CD112

Nothing yet assigned to this number.

CD113

Nothing yet assigned to this number.

CD114

Nothing yet assigned to this number.

CD115

A 150-kD antigen present on monocytes, macrophages, and placenta. It is a macrophage colony stimulating factor (M-CSF) receptor. Also called M-CSFR (macrophage colony stimulating factor receptor), CSF-1R, or cFMS.

CDw116

A 75- to 85-kD antigen present on monocytes, neutrophils, eosinophils, fibroblasts, and endothelial cells. It is a granulocyte macrophage colony stimulating factor (GM-CSF) receptor α chain. Also called GM-CSF R (granulocyte, macrophage colony stimulating factor receptor) or HGM-CSFR.

CD117

A 145-kD antigen present on bone marrow progenitor cells. It is a stem cell factor (SCF) receptor. Also called c-kit or stem cell factor receptor (SCF-R).

CD118

An antigen with broad cellular expression that serves as the interferon α, β receptor. Also called IFN-α,βR.

CD119

A 90-kD antigen present on macrophages, monocytes, B cells, and epithelial cells. It is the interferon-γ receptor. Also called IFN-γR.

CD120a

A 55-kD antigen present on most cell types, higher levels on epithelial cell lines. It is the TNF receptor. It binds both TNF-α and TNF-β. Also called TNFR-1.

CD120b

A 75-kD antigen present on most cell types, higher levels on myeloid cell lines. It is the TNFβ receptor, binding both TNFα and TNFβ. Also called TNFR-II.

CDw121a

An 80-kD antigen present on T cells, thymocytes, fibroblasts, and endothelial cells. It is the type I interleukin-1 receptor. It binds IL-1α and IL-1β. Also called IL-1R (interleukin-1 receptor), type I.

CD121b

A 68-kD antigen present on B cells, macrophages, and monocytes. It is type II interleukin-1 receptor, binding IL-1α and IL-1β. Also called IL-1R type II.

CD122

A 75-kD antigen present on NK cells, resting T cell subpopulation, and some B cell lines. It is the IL-2 receptor β chain. Also called IL-2Rβ.

CD123

A 70-kD antigen present on bone marrow stem cells, granulocytes, monocytes, and megakaryocytes. It is the IL-3 receptor α chain. Also called IL-3Rα.

CDw124

A 140-kD antigen present on mature B and T cells, and hematopoietic precursor cells. It is the IL-4 receptor. Also called IL-4R.

CD125

A 55- to 60-kD antigen present on eosinophils and basophils. It is the IL-5 receptor α chain. Also called IL-5 receptor α chain.

CD126

An 80 (a subunit)-kD antigen present on activated B cells and plasma cells (strong) and most leukocytes (weak). It is the IL-6 receptor α subunit. Also called IL-6Rα.

CDw127

A 75-kD antigen present on bone marrow lymphoid presursors, pro-B cells, mature T cells, and monocytes. It serves as the IL-7 receptor. Also called IL-7R.

CDw128

A 58- to 67-kD antigen present on neutrophils, basophils, and a T cell subset. It is the IL-8 receptor. Also called IL-8R.

CD129

Nothing has yet been assigned to this number.

CDw130

A 130-kD antigen present on activated B cells and plasma cells (strong), most leukocytes (weak), and endothelial cells. It is a common subunit of IL-6, IL-11, oncostatin M (OSM), and leukemia inhibitory factor (LIF) receptors. Also called IL-6Rβ, IL-11Rβ, OSMRβ, LIFRβ, or IL-6R-gp 130SIG.

CEA

Abbreviation for carcinoembryonic antigen.

cecropin

An antibacterial protein derived from immunized cecropia moth pupae. It is also found in butterflies. Cecropin is a basic protein that induces prompt lysis of selected Gram-negative and Gram-positive bacteria.

celiac disease

Refer to gluten-sensitive enteropathy.

celiac sprue (gluten-sensitive enteropathy)

Gluten-sensitive enteropathy results from hypersensitivity to cereal grain storage proteins, including gluten, or its product gliadin present in oats, wheat, and barley. It is characterized by villous atrophy and malabsorption in the small intestine. It occurs mostly in Caucausians and occasionally in African-Americans, but not in Asians. Individual patients may have the disease limited to the intestines or associated with dermatitis herpetiformis, a vesicular eruption of the skin. The muscosa of the small intestine shows the greatest reactivity in areas in contact with gluten-containing food. Antigliadin antibodies are formed, and lymphocytes and plasma cells appear in the lamina propria in association with villous atrophy. Gluten-sensitive enteropathy is associated with HLA-DR3, -DR7, and -DQ2, as well as with HLA-B8. Diagnosis is made by showing villous atrophy in a biopsy of the small intestine. Administering a gluten-free diet leads to resolution of the disease. Antigliadin antibodies are of the IgA class. Both T and B cell limbs of the immune response participate in the pathogenesis of this disease. Patients may also develop IgA or IgG reticulin antibodies. An α gliadin amino acid sequence shares homology with adenovirus 12Elb early protein. Celiac disease patients develop immune reactivity for both gliadin and this viral peptide sequence. The disease occurs more frequently in individuals exposed earlier to adenovirus 12. Patients develop weight loss, diarrhea, anemia, petechiae, edema, and dermatitis herpetiformis, among other signs and symptoms. Lymphomas such as immunoblastic lymphoma may develop in 10 to 15% of untreated patients.

cell adhesion molecules

Refer to CAM.

cell-bound antibody (cell-fixed antibody)

An antibody anchored to the cell surface either through its paratopes binding to cell epitopes or attachment of its Fc region to Fc receptors. An example is cytophilic antibody or IgE which may then react with antigen as their Fab regions are available.

cell cooperation

Refers to T lymphocyte and B lymphocyte cooperation.

cell-mediated hypersensitivity

Refer to delayed-type hypersensitivity and to type IV cell-mediated hypersensitivity.

cell-mediated immunity (CMI)

The limb of the immune response that is mediated by specifically sensitized T lymphocytes that produce their effect through direct reaction in contrast to the indirect effect mediated by antibodies of the humoral limb produced by B lymphocytes. The development of cell-mediated immunity to an exogenous antigen first involves processing of the antigen by an antigen-presenting cell such as a macrophage. Processed antigen is presented in the context of MHC class II molecules to a CD4+ T lymphocyte. IL-1 β is also released from the macrophage to induce IL-2 synthesis in CD4+ lymphocytes. The IL-2 has an autocrine effect on the cells producing it, causing their proliferation and also causing proliferation of other lymphocyte subsets including CD8+ suppressor/cytotoxic T cells, B lymphocytes that form antibody, and natural killer (NK) cells. Cell-mediated immunity is of critical importance in the defense against mycobacterial and fungal infections, resistance to tumors, and its significance as a role in allograft rejection.

cell-mediated immunodeficiency syndrome

Conditions in which cell-mediated immunity is defective. This may be manifested as negative skin tests following the application of tuberculin, histoplasmin, or other common skin test antigens; failure to develop contact hypersensitivity following application of sensitizing substances such as DNCB to the skin; or failure to reject an allograft, such as a skin graft. Severe combined immunodeficiency is characterized by defective T lymphocyte as well as B lymphocyte limbs of the immune response. DiGeorge syndrome is characterized by failure of development of the T cell-mediated limb of the immune response.

cell-mediated lympholysis (CML) test

Responder (effector) lymphocytes are cytotoxic for donor (target) lymphocytes after the two are combined in culture. Target cells are labeled by incubation with ^{51}Cr at 37°C for 60 min. Following combination of effector and target cells in tissue culture, the release of ^{51}Cr from target cells injured by cytotoxicity represents a measure of cell-mediated lympholysis (CML). The CML assay gives uniform results, is relatively simple to perform, and is rather easily controlled. The effector cells can result from either *in vivo* sensitization following organ grafting or can be induced *in vitro*. Variations in effector to target cell ratios can be employed for quantification.

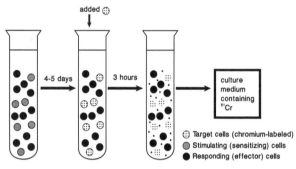

Cell-Mediated Lympholysis

cellular allergy

Refer to delayed-type hypersensitivity, type IV cell-mediated hypersensitivity, and cell-mediated immunity.

cellular hypersensitivity

Refer to delayed-type hypersensitivity, type IV cell-mediated hypersensitivity, and cell-mediated immunity.

cellular immunity

Refer to cell-mediated immunity.

cellular interstitial pneumonia

Inflammation of the lung in which the alveolar walls are infiltrated by mononuclear cells.

cellular oncogene

Refer to protooncogene.

centiMorgan (cM)

A chromosomal unit of physical distance that corresponds to a 1% recombination frequency between two genes that are closely linked. Also termed a map unit.

central lymphoid organs

Lymphoid organs that are requisite for the development of the lymphoid and therefore of the immune system. These include the thymus, bone marrow, and bursa of Fabricus. Also termed primary lymphoid organs.

central tolerance

The mechanism involved in the functional inactivation of cells requisite for the initiation of an immune response. Central tolerance affects the afferent limb of the immune response, which is concerned with sensitization and cell proliferation.

centriole antibodies

Antibodies sometimes detected in blood sera also containing antibodies against mitotic spindle apparatus (MSA). They are only rarely found in patients developing connective tissue disease of the scleroderma category. Selected centriole antibodies may be directed against the glycolytic enzyme enolase.

centromere antibodies

Antibodies specific for centromeres/kinetochores are detectable in 22% of systemic sclerosis patients, most of whom have CREST syndrome rather than diffuse scleroderma. Approximately 12% of primary biliary cirrhosis patients, half of whom also manifest scleroderma, have centromere antibodies.

cerebrospinal fluid (CSF) immunoglobulins

In normal individuals, CSF immunoglobulins are derived from plasma by diffusion across the blood-brain barrier. The amount present is dependent on the immunoglobulin concentration in the serum, the molecular size of the immunoglobulin, and the permeability of the blood-brain barrier. IgM is normally excluded by virtue of its relatively large molecular size and low plasma concentration. However, in certain disease states, such as demyelinating diseases and infections of the central nervous system, immunoglobulins may be produced locally. The permeability of the blood-brain barrier is accurately reflected by the CSF total protein or albumin levels relative to those in the serum. By comparing these data, it is possible to derive information about deviation from normal. The comparative method is called Ig quotient and is calculated in various ways:

1. CSF - IgG/albumin (normal 13.9 + 14%)
2. CSF - IgG/total protein
3. CSF - IgA/albumin
4. CSF - κ/λ (ratio)

To correct for variations in the blood-brain barrier, the calculation can be modified to give a more sensitive quotient, which is represented as,

$$\frac{CSF\ IgG/serum\ IgG}{CSF\ albumin/serum\ albumin}$$

The ratio of κ to λ light chains in CSF in comparison with that of these light chains in serum is significant in that some patients with local immunoglobulin production show a change in the ratio. An increase in the IgA present in CSF appears in some viral infections of the CNS in which antiviral antibodies are also detectable.

CFA

Abbreviation for complete Freund's adjuvant.

CFT

Abbreviation for complement fixation test.

CFU

Abbreviation for colony-forming unit.

CFU-GEMM

A colony-stimulating factor that acts on multiple cell lines which include erythroid cells, granulocytes, megakaryocytes, and macrophages. The pancytopenia observed in myelodysplasia and Fanconi's disease has been attributable to the total lack of CFU-GEMM.

CFU-S (colony-forming units, spleen)

A mixed-cell population considered to contain the ideal stem cell that is pluripotent and capable of proliferating and of renewing itself.

C$_\gamma$

Immunoglobulin γ chain constant region that is further subdivided into four isotypes in man that are indicated as C$_{\gamma1}$, C$_{\gamma2}$, C$_{\gamma3}$, and C$_{\gamma4}$. The corresponding exons are shown by the same designations in italics.

CGD

Abbreviation for chronic granulomatous disease.

C$_H$

Immunoglobulin heavy chain constant region encoded by the *C$_H$* gene.

C$_H$1

An immunoglobulin heavy chain's first constant domain encoded by the *C$_H$1* exon.

C$_H$2

An immunoglobulin heavy chain's second constant domain encoded by the *C$_H$2* exon.

C$_H$3

An immunoglobulin heavy chain's third constant domain encoded by the *C$_H$3* exon.

C$_H$4

An immunoglobulin heavy chain's fourth constant domain encoded by the *C$_H$4* exon. Of the five immunoglobulin classes in man, only the μ heavy chain of IgM and the ε heavy chain of IgE possess a fourth domain.

C$_H$50 unit

The amount of complement (serum dilution) that induces lysis of 50% of erythrocytes sensitized (coated) with specific antibody. More specifically, the 50% lysis should be of 5×10^8 sheep erythrocytes sensitized with specific antibody during 60 min of incubation at 37°C. To obtain the complement titer, i.e., the number CH_{50} present in 1 ml of serum that has not been diluted, the log $y/1 - y$ (y =% lysis) is plotted against the log of the quantity of serum. At 50% lysis, the plot approaches linearity near $y/1 - y$.

CHAD
Abbreviation for cold hemagglutinin disease.

challenge
Antigen deliberately administered to induce an immune reaction in an individual previously exposed to that antigen to determine the state of immunity.

challenge stock
An antigen dose that has been precisely measured and administered to an individual following earlier exposure to an infectious microorganism.

chancre immunity
Resistance to reinfection with *Treponema pallidum* that develops 3 months following a syphilis infection that is untreated.

chaperones
A group of proteins that includes BiP, a protein that binds the immunoglobulin heavy chain. Chaperones aid the proper folding of oligomeric protein complexes. They prevent incorrect conformations or enhance correct ones. Chaperones are believed to combine with the surfaces of proteins exposed during intermediate folding and to restrict further folding to the correct conformations. They take part in transmembrane targeting of selected proteins. Chaperones hold some proteins that are to be inserted into membranes in intermediate conformation in the cytoplasm until they interact with the target membrane. Besides BiP, they include heat shock protein 70 and 90 and nucleoplasmins.

Charcot-Leyden crystals
Crystals present in asthmatic patients' sputum that are hexagonal and bipyramidal. They contain a 13-kD lysophospholipase derived from the eosinophil cell membrane.

Chase, Merrill (1905–)
American immunologist who worked with Karl Landsteiner at the Rockefeller Institute for Medical Research, New York. He investigated hypersensitivity, including delayed type hypersensitivity and contact dermatitis. He was the first to demonstrate the passive transfer of tuberculin and contact hypersensitivity and also made contributions in the fields of adjuvants and quantitative methods.

Chase-Sulzberger phenomenon
Refer to Sulzberger-Chase phenomenon.

Chediak-Higashi syndrome
A childhood disorder with an autosomal recessive mode of inheritance that is identified by the presence of large lysosomal granules in leukocytes that are very stable and undergo slow degranulation. Multiple systems may be involved. Repeated bacterial infections with various microorganisms, partial albinism, central nervous system disorders, hepatosplenomegaly, and an inordinate incidence of malignancies of the lymphoreticular tissues may occur. The large cytoplasmic granular inclusions that appear in white blood cells may also be observed in blood platelets and can be seen by regular light microscopy in peripheral blood smears. There is defective neutrophil chemotaxis and an altered ability of the cells to kill ingested microorganisms. There is a delay in the killing time, even though hydrogen peroxide formation, oxygen consumption, and hexose monophosphate shunt are all within normal limits. There is also defective microtubule function, leading to defective phagolysosome formation. Cyclic AMP levels may increase. This causes decreased neutrophil degranulation and mobility. High doses

of ascorbic acid have been shown to restore normal chemotaxis, bactericidal activity, and degranulation. Natural killer cell numbers and function are decreased. There is an increased incidence of lymphomas in Chediak-Higashi patients. There is no effective therapy other than the administration of antibiotics for the infecting microorganisms. The disease carries a poor prognosis because of the infections and the neurological complications. The majority of affected individuals die during childhood, although occasional subjects may live longer.

chemical "splenectomy"
Deliberate suppression of the immune system by the administration of high-dose corticosteroids (1 mg/kg/day) or intravenous immunoglobulin (0.4 g/kg/day). This prevents endocytosis of cells or microorganisms opsonized by a coating of immunoglobulin or complement which blocks Fc receptors. Although the opsonized particles are bound, they are not endocytosed. This procedure has been used in the management of hypersplenism associated with certain immune disorders such as autoimmune hemolytic anemia, Felty's syndrome, or autoimmune neutropenia.

chemiluminescence
The conversion of chemical energy into light by an oxidation reaction. A high-energy peroxide intermediate, such as luminol, is produced by the reaction of a precursor substance exposed to peroxide and alkali. The emission of light energy by a chemical reaction may occur during reduction of an unstable intermediate to a stable form. Chemiluminescence measures the oxidative formation of free radicals such as superoxide anion by polymorphonuclear neutrophils and mononuclear phagocytes. Light is released from these cells after they have taken up luminol (5-amino-2,3-dihydro-1,4-phthalazinedione). This is a mechanism to measure the respiratory burst in phagocytes. The oxidation of luminol increases intracellular luminescence. Chronic granulomatous disease may be diagnosed by this technique.

chemokines
A family of 8- to 10-kD chemotactic cytokines that share structural homology. They are chemokinetic and chemotactic, stimulating leukocyte movement and directed movement. Two internal disulfide loops are present in chemokine molecules, that may be subdivided according to the position of the two amino terminal cysteine residues, i.e., adjacent (cys-cys) or separated by a single amino acid (cys-X-cys). Activated mononuclear phagocytes as well as fibroblasts, endothelium, and megakaryocytes synthesize cys-X-cys chemokines, including interleukin-8, that act mainly on polymorphonuclear neutrophils as acute inflammatory mediators. Activated T lymphocytes synthesize cys-cys chemokines that act principally on mononuclear inflammatory cell subpopulations. Cys-X-cys and cys-cys chemokines combine with heparan sulfate proteoglycans on endothelial cell surfaces. They may activate chemokinesis of leukocytes that adhere to endothelium via adhesion molecules. Chemokine receptors are being characterized, and selected ones interact with more than one chemokine.

chemokinesis
Determination of the rate of movement or random motion of cells by chemical substances in the environment. The direction of cellular migration is determined by chemotaxis, not chemokinesis.

chemotactic assays
The chemotactic properties of various substances can be determined by various methods. The most popular is the Boyden technique. This consists of a chamber separated into two compartments by a Millipore® filter of appropriate porosity, through which cells can migrate actively, but not drop passively. The cell preparation is placed in the upper compartment of the chamber, and the assay solution is placed in the lower compartment. The chamber is incubated in air at 37°C for 3 h, after which the filter is removed and the number of cells migrating to the opposite surface of the filter are counted.

chemotactic deactivation
Represents the reduced chemotactic responsiveness to a chemotactic agent caused by prior incubation of leukocytes with the same agent, but in the absence of a concentration gradient. It can be tested by adding first the chemotactic factor to the upper chamber, washing, and then testing the response to the chemotactic factor placed in the lower chamber (no gradient being present). The mechanism of deactivation has been postulated as obstruction of the membrane channels involved in cation fluxes. Deactivation phenomena are used to discriminate between chemokinetic factors which enhance random

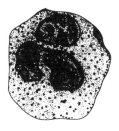
Normal
PMN

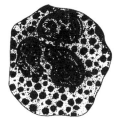
Chediak-Higashi Syndrome
PMN

migration and true chemotactic factors which cause directed migration. Only true chemotactic factors are able to induce deactivation.

chemotactic factors

The list of chemotactic factors is very long and includes substances of both endogenous and exogenous origin. Among them are bacterial extracts, products of tissue injury, chemical substances, various proteins, and secretory products of cells. The most important among them are those generated from complement and described as anaphylatoxins. This name is related to their concurrent ability of stimulating the release of mediators from mast cells. Some chemotactic factors act specifically in directing the migration of certain cell types. Others have a broader spectrum of activity. Many of them have additional activities besides acting as chemotactic factors. Such effects of aggregation and adhesion of cells, discharge of lysosomal enzymes, and phagocytosis by phagocytic cells may be concurrently stimulated. Participation in various immunologic phenomena such as cell triggering of cell-cell interactions is known for certain chemotactic factors. The structure of chemotactic factors and even the active region in their molecules have been determined in many instances. However, advances in the clarification of their mechanism of action have been facilitated by the use of synthetic oligopeptides with chemotactic activities. The specificity of such compounds depends both on the nature of the amino acid sequence and the position of amino acids in the peptide chain. Methionine at the NH_2-terminal is essential for chemotactic activity. Formylation of Met leads to a 3000- to 30,000-fold increase in activity. The second position from the NH_2-terminal is also essential, and Leu, Phe, and Met in this position are essentially equivalent. Positively charged His and negatively charged Glu in this position are significantly less active, substantiating the role of a neutral amino acid in the second position at the N-terminal.

Chemotactic Factor

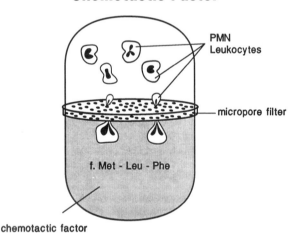

PMN
Leukocytes

micropore filter

f. Met - Leu - Phe

chemotactic factor

chemotactic peptide

A peptide that attracts cell migration, such as formyl-methionyl-leucyl-phenylalanine.

chemotactic receptor

Cellular receptors for chemotactic factors. In bacteria such receptors are designated sensors and signalers and are associated with various transport mechanisms. The cellular receptors for chemotactic factors have not been isolated and characterized. In leukocytes, the chemotactic receptor appears to activate a serine proesterase enzyme, which sets in motion the sequence of events related to cell locomotion. The receptors appear specific for the chemotactic factors under consideration, and apparently the same receptors mediate all types of cellular responses inducible by a given chemotactic factor. However, these responses can be dissociated from each other, suggesting that binding to the putative receptor initiates a series of parallel, interdependent, and coordinated biochemical events leading to one or another type of response. Using a synthetic peptide, *N*-formyl-methionyl-leucyl-phenylalanine, about 2000 binding sites have been demonstrated per PMN leukocyte. The binding sites are specific, have a high affinity for the ligand, and are saturable. Competition for the binding sites is shown only by the parent or related compounds; the potency of the latter varies. Positional isomers may inhibit binding. Full occupancy

of the receptors is not required for a maximal response, and occupancy of only 10 to 20% of them is sufficient. The presence of spare receptors may enhance the sensitivity in the presence of small concentrations of chemotactic factors and may contribute to the detection of a gradient. There also remains the possibility that some substances with chemotactic activity do not require specific binding sites on cell membranes.

chemotaxis

The process whereby chemical substances direct cell movement and orientation. The orientation and movement of cells in the direction of a chemical's concentration gradient is positive chemotaxis, whereas movement away from the concentration gradient is termed negative chemotaxis. Substances that induce chemotaxis are referred to as chemotaxins and are often small molecules, such as C5a, formyl peptides, lymphokines, bacterial products, leukotriene B_4, etc., that induce positive chemotaxis of polymorphonuclear neutrophils, eosinophils, and monocytes. These cells move into inflammatory agents by chemotaxis. A dual chamber device called a Boyden chamber is used to measure chemotaxis in which phagocytic cells in culture are separated from a chemotactic substance by a membrane. The number of cells on the filter separating the cell chamber from the chemotaxis chamber reflect the chemotactic influence of the chemical substance for the cells.

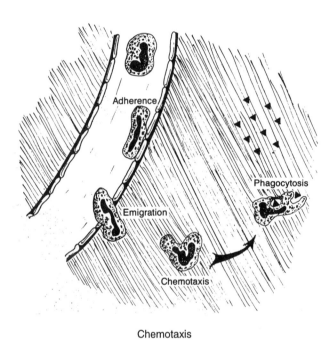

Adherence

Phagocytosis

Emigration

Chemotaxis

Chemotaxis

chickenpox (varicella)

Human herpes virus type 3 (HHV-3) induced in acute infection that occurs usually in children less than 10 years of age. There is anorexia, malaise, low fever, and a prodromal rash following a 2-week incubation period. Erythematous papules appear in crops and intensify for 3 to 4 days. They are very pruritic. Complications include viral pneumonia, secondary bacterial infection, thrombocytopenia, glomerulonephritis, myocarditis, and other conditions. HHV-3 may become latent when chickenpox resolves. Its DNA may become integrated into the dorsal route ganglion cells. This may be associated with the development of Herpes zoster or shingles later in life.

Chido (Ch) and Rodgers (Rg) antigens

Epitopes of C4d fragments of human complement component C4. They are not intrinsic to the erythrocyte membrane. The Chido epitope is found on C4d from C4B, whereas the Rodgers epitope is found on C4A derived from C4d. The Rodgers epitope is Val-Asp-Leu-Leu, and the Chido epitope is Ala-Asp-Leu-Arg. They are situated at residue positions 1188 and 1191 in the C4 α chain's C4d region. Antibodies against Ch and Rg antigenic determinants agglutinate saline suspensions of red blood cells coated with C4d. Since C4 is found in human serum, anti-Ch and anti-Rg are neutralized by sera of most individuals which contain the relevant antigens. Ficin and papain destroy these antigens.

chimera

The presence in an individual of cells of more than one genotype. This can occur rarely under natural circumstances in dizygotic twins, as in cattle, who share a placenta in which the blood circulation has become fused, causing the blood cells of each twin to circulate in the other. More commonly, it refers to humans or other animals who have received a bone marrow transplant that provides a cell population consisting of donor and self cells. Tetraparental chimeras can be produced by experimental manipulation. The name chimera derives from a monster of Greek mythology that had the body of a goat, the head of a lion, and the tail of a serpent.

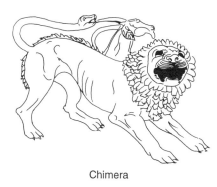

Chimera

chimeric antibodies

Antibodies that have, e.g., mouse Fv fragments for the Ag-binding portion of the molecule, but Fc regions of human Ig which convey effector functions.

chimerism

The presence of two genetically different cell populations within an animal at the same time.

chlorambucil (4-[*bis*(2-chloroethyl)amino-phenylbutyric acid)

An alkylating and cytotoxic drug. Chlorambucil is not as toxic as is cyclosphosphamide and has served as an effective therapy for selected immunological diseases such as rheumatoid arthritis, SLE, Wegener's granulomatosis, essential cryoglobulinemia, and cold agglutinin hemolytic anemia. Although it produces bone marrow suppression, it has not produced hemorrhagic cystitis and is less irritating to the GI tract than is cyclosphosphamide. Chlorambucil increases the likelihood of opportunistic infections and the incidence of some tumors.

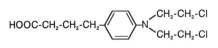

Chlorambucil

chlorodinitrobenzene (1-chlor-2,4-dinitrobenzene)

More often termed dinitrochlorobenzene (DNCB). A chemical substance used to test for a patient's ability to develop the type of delayed-type hypersensitivity referred to as contact hypersensitivity. This is a type IV hypersensitivity reaction. The chemical is applied to a patient's forearm. Following sufficient time for sensitization to develop, the patient's other forearm is exposed to a second (test) dose of the same chemical. In an individual with an intact cell-mediated limb of the immune response, a positive reaction develops at the second challenge site within 48 to 72 h. Individuals with cell-mediated immune deficiency disorders fail to develop a positive delayed-type hypersensitivity reaction.

cholera toxin

A *Vibro cholerae* enterotoxin comprised of five B subunits that are cell-binding 11.6-kD structures that encircle a 27-kD catalase that conveys ADP-ribose to G protein, leading to continual adenyl cyclase activation. Other toxins that resemble cholera toxin in function include diptheria toxin, exotoxin A, and pertussis toxin.

cholera vaccine

An immunizing preparation comprised of *Vibrio cholerae* smooth strains Inaba and Ogawa in addition to El Tor vibrio, which have been killed by heat or formalin treatment. It is designed to induce protective active immunity against cholera in regions where it is endemic as well as in travelers to those locations. The immunity induced is effective for only about 12 weeks.

cholinergic urticaria

Skin edema induced by an aberrant response to acetylcholine that occurs following diminished cholinesterase activity.

chorea

Involuntary muscle twitching that occurs in cases of acute rheumatic fever and is commonly known as St. Vitus' dance.

choriocarcinoma

An unusual malignant neoplasm of the placenta trophoblast cells in which the fetal neoplastic cells are allogeneic in the host. On rare occasions, these neoplasms have been "rejected" spontaneously by the host. Antimetabolites have been used in the treatment of choriocarcinoma.

chromium release assay

The release of chromium (^{51}Cr) from labeled target cells following their interaction with cytotoxic T lymphocytes or antibody and K cells (ADCC) or NK cells. The test measures cell death, which is reflected by the amount of radiolabel released according to the number of cells killed.

chromogranins

Acidic glycoproteins of neurosecretory granules in multiple tissue sites, with a mol wt of 20 to 100 kD, that are used as a general endocrine indicator of neuroendocrine tumors using the immunoperoxidase reaction. The chromogranins are designated as A, B, and C. B and C are also termed secretogranin I and II.

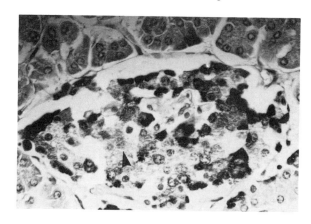

chromogranin

chromosomal translocations

DNA sequence rearrangement between chromosomes, which is frequently associated with neoplasia.

chromatography

A group of methods employed for the separation of proteins.

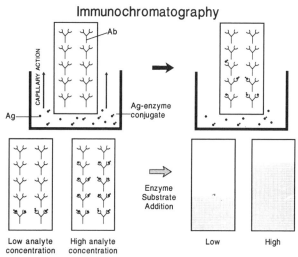

Chromatography

Column Chromatography
Arrows indicate direction of flow

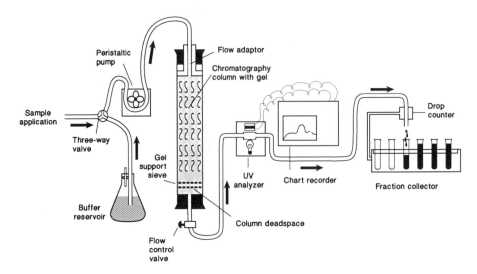

chronic active hepatitis, autoimmune

A disease that occurs in young females who may develop fever, arthralgias, and skin rashes. They may be of the HLA-B8 and DR3 haplotype and suffer other autoimmune disorders. Most develop antibodies to smooth muscle, principally against actin, and autoantibodies to liver membranes. They also have other organ- and nonorgan-specific autoantibodies. A polyclonal hypergammaglobulinemia may be present. Lymphocytes infiltrating portal areas destroy hepatocytes. Injury to liver cells produced by these infiltrating lymphocytes produces piecemeal necrosis. The inflammation and necrosis are followed by fibrosis and cirrhosis. The T cells infiltrating the liver are CD4+. Plasma cells are also present, and immunoglobulins may be deposited on hepatocytes. The autoantibodies against liver cells do not play a pathogenetic role in liver injury. There are no serologic findings that are diagnostic. Corticosteroids are useful in treatment. The immunopathogenesis of autoimmune chronic active hepatitis involves antibody, K cell cytotoxicity, and T cell reactivity against liver membrane antigens. Antibodies and specific T suppressor cells reactive with LSP are found in chronic active hepatitis patients, all of whom develop T cell sensitization against asialo-glycoprotein (AGR) antigen. Chronic active hepatitis has a familial predisposition.

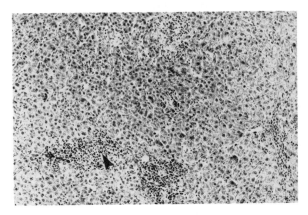

Chronic Active Hepatitis, Autoimmune

chronic and cyclic neutropenia

A syndrome characterized by recurrent fever, mouth ulcers, headache, sore throat, and furunculosis occurring every 3 weeks in affected individuals. This chronic agranulocytosis leads to premature death from infection by pyogenic microorganisms in affected children who may have associated pancreatic insufficiency, dysostosis, and dwarfism. Antibodies can be transmitted from the maternal to the fetal circulation to induce an isoimmune neutropenia. This may consist of either a transitory type in which the antibodies are against neutrophil antigens determined by the father or a type produced by autoantibodies against granulocytes.

chronic fatigue syndrome (CFS)

A disabling fatigue that persists for at least 6 months. Although the etiology is idiopathic, laboratory studies on patients reveal a consistent observation of immune system dysfunction primarily affecting the cellular immune response. CD4+ T helper cells and CD8+ suppressor/cytotoxic cells may be normal, increased, or decreased but the CD4/DC8 ratio is usually elevated. This has been attributed to a diminished number of suppressor cells with a concomitant increase in cytotoxic T cell (CD8+, CD28+, CD11b−) numbers. The increased cytotoxic T cells express HLA-DR and/or CD38 activation markers. Manifestations of altered T cell functions also include decreased delayed-type hypersensitivity, diminished responsiveness in mitogen-stimulation assays *in vitro*, increased suppression of immunoglobulin synthesis by T cells, and elevated spontaneous suppressor activity. NK cells may be normal, increased or decreased but there may be qualitative alterations in NK cell function. Elevated IgG antibody titers to Epstein-Barr virus (EBV) early antigen and capsid antigen are demonstrable in many CFS patients. Occasionally, increased antibodies against cytomegalovirus (CMV), herpes simplex, HHV-6, coxsackie B or measles may be observed. Some CFS patients have abnormal levels of IgG, IgM, IgA or IgD. Approximately one-third of CFS patients have antibodies against smooth muscle or thyroid. Laboratory test results from CFS patients should be interpreted as one battery and not as individual tests.

chronic graft-vs.-host disease (GVHD)

Chronic GVHD may occur in as many as 45% of long-term bone marrow transplant recipients. Chronic GVHD differs both clinically and histologically from acute GVHD and resembles autoimmune connective tissue diseases. For example, chronic GVHD patients may manifest skin lesions resembling scleroderma; sicca syndrome in the eyes and mouth; inflammation of the oral, esophageal, and vaginal mucosa; bronchiolitis obliterans; occasionally myasthenia gravis; polymyositis; and autoantibody synthesis. Histopathologic alterations in chronic GVHD, such as chronic inflammation and fibrotic changes in involved organs, resemble changes associated with naturally occurring autoimmune disease. The skin may reveal early inflammation with subsequent fibrotic changes. Infiltration of lacrimal, salivary, and submucosal glands by lymphoplasmacytic cells leads ultimately to fibrosis. The resulting sicca syndrome, which resembles Sjögren's syndrome, occurs in 80% of chronic GVHD patients. Drying of mucous membranes in the sicca syndrome affects the mouth, esophagus, conjunctiva, urethra, and vagina. The pathogenesis of chronic GVHD involves the interaction of alloimmunity, immune dysregulation, and resulting immunodeficiency and autoimmunity. The increased incidence of infection among chronic GVHD patients suggests immunodeficiency. The dermal fibrosis is associated with increased numbers of activated fibroblasts in the papillary dermis. T lymphocyte or mast cell cytokines may activate this fibroplasia, which leads to dermal fibrosis in chronic GVHD.

chronic granulomatous disease (CGD)
A disorder that is inherited as an X-linked trait in two thirds of the cases and as an autosomal recessive trait in the remaining one third. Clinical features are usually apparent before the end of the second year of life. There is an enzyme defect associated with NADPH oxidase. This enzyme deficiency causes neutrophils and monocytes to have decreased consumption of oxygen and diminished glucose utilization by the hexose monophosphate shunt. Although neutrophils phagocytize microorganisms, they do not form superoxide and other oxygen intermediates that usually constitute the respiratory burst. Neutrophils and monocytes also form a smaller amount of hydrogen peroxide, have decreased iodination of bacteria, and have diminished production of superoxide anions. All of this leads to decreased intracellular killing of bacteria and fungi. Thus, these individuals have an increased susceptibility to infection with microorganisms that normally are of relatively low virulence. These include *Aspergillus* sp., *Serratia marcescens,* and *Staphylococcus epidermidis.* Patients may have hepatosplenomegaly, pneumonia, osteomyelitis, abscesses, and draining lymph nodes. The quantitative nitroblue tetrazolium (NBT) test and the quantitative killing curve are both employed to confirm the diagnosis. Most microorganisms that cause difficulty in CGD individuals are catalase positive. Therapy includes interferon γ, antibiotics, and surgical drainage of abscesses.

chronic lymphocytic leukemia (CLL)
A B cell leukemia in which long-lived small lymphocytes continually collect in the spleen, lymph nodes, bone marrow, and blood. The CLL cells are B lymphocytes bearing monoclonal immunoglobulin on their surface. However, their function is defective, leading to development of hypogammaglobulinemia. This lack of antibody globulin facilitates the development of recurrent infection by pyogenic microorganisms. The course of the disease may be relatively uneventful, and patients may live for 30 years.

chronic lymphocytic thyroiditis
Profound infiltration of the thyroid by lymphocytes, leading to the extensive injury of thyroid follicular structure. Even though the gland becomes enlarged, its function diminishes, leading to hypothyroidism. Women are affected much more commonly than are men. Antibodies detectable in the serum are specific for the 107-kD thyroid microsomal peroxidase, the thyrotropin receptor, and thyroglobulin. Also called Hashimoto's thyroiditis. Thyroid hormone replacement therapy is the usual approach to treatment.

chronic mucocutaneous candidiasis
Infection of the skin, mucous membranes, and nails by *Candida albicans* associated with defective T cell-mediated immunity that is specific to *Candida.* Skin tests for delayed-hypersensitivity to the *Candida* antigen are negative. There may also be an associated endocrinopathy. The selective deficiency in T lymphocyte immunity leads to increased susceptibility to chronic *Candida* infection. T cell immunity to non-*Candida* antigens is intact. B cell immunity is normal, which leads to an intact antibody response to *Candida* antigens. T lymphocytes form migration inhibitor factor (MIF) to most of the antigens, except for those of *Candida* microorganisms. The most common endocrinopathy that develops in these patients is hypoparathyroidism. Clinical forms of the disease may be either granulomatous or nongranulomatous. *Candida* infection of the skin may be associated with the production of granulomatous lesions. The second most frequent endocrinopathy associated with this condition is Addison's disease. The disease is difficult to treat. The antifungal drug ketoconazole has proven effective. Intravenous amphotericin B has led to improvement. Transfer factor has been administered with variable success in selected cases.

chronic myeloid leukemia
Leukemia characterized by cell types in the circulation that are in the late stages of granulocyte maturation. These include mature granulocytes, myelocytes, and metamyelocytes.

chronic progressive vaccinia (vaccinia gangrenosa) (historical)
An unusual sequela of smallpox vaccination in which the lesions produced by vaccinia on the skin became gangrenous and spread from the vaccination site to other areas of the skin. This occurred in children with cell-mediated immunodeficiency.

chrysotherapy
Refer to gold therapy.

Churg-Strauss syndrome (allergic granulomatosis)
A combination of asthma associated with necrotizing vasculitis, eosinophilic tissue infiltrates, and extravascular granulomas.

CIC
Abbreviation for circulating immune complexes.

CID
Abbreviation for cytomegalic (CMV) inclusion disease.

CIE
Abbreviation for counterimmunoelectrophoresis or crossed immunoelectrophoresis.

ciliary neurotrophic factor (CNTF)
A protein hormone related to the IL-6 family. It has several functions in the nervous system which are similar to those of leukemia inhibitory factor.

cimetidine
A well-known treatment for peptic ulcers. This histamine H_2-receptor antagonist is of interest to immunologists because of its efficacy in treating common variable immunodeficiency, possibly through suppressor T lymphocyte inhibition.

Cimetidine

circulating anticoagulant
Antibodies specific for one of the blood coagulation factors. They may be detected in the blood serum of patients treated with penicillin, streptomycin, or isoniazid; in systemic lupus erythematosus patients; or following treatment of hemophilia A or B patients with factor VIII or factor IX. These are often IgG4 antibodies.

circulating lupus anticoagulant syndrome (CLAS)
The occurrence in lupus patients, who are often ANA negative of thromboses that are recurrent, kidney disease, and repeated spontaneous abortions. There is an IgM gammapathy and fetal wastage that occurs repeatedly.

cis-pairing
Association of two genes on a single chromosome encoding the protein.

cisterna chyli
Refer to thoracic duct.

cisternal space
The endoplasmic reticulum lumen of plasma cells that contains immunoglobulin molecules prior to secretion.

Cκ
An immunoglobulin κ light chain constant region. The corresponding exon is designated as $C\kappa$.

C_L
An immunoglobulin light chain constant domain. The corresponding exon is designated C_L.

Cλ
An immunoglobulin λ light chain constant region. The corresponding exon is designated $C\lambda$. There is more than one isotype in mouse and man.

C_μ
An immunoglobulin μ chain constant region. The corresponding exon is designated as $C\mu$.

class I antigen
A major histocompatibility complex (MHC) antigen found on nucleated cells on multiple tissues. In man, class I antigens are encoded by genes at A, B, and C loci and in mice by genes at D and K loci.

class IB genes
Genes linked to the MHC class I region that code for class I-like α chains. These genes that encode molecules on the cell surface that associate with β2 microglobulin vary in their cell surface expression and tissue distribution from one species to another. An individual animal may have multiple class IB molecules. One such molecule has a role in presentation of peptides bearing N-formylated

amino termini. Other class IB molecules may also be active in antigen presentation.

class I MHC molecules

Glycoproteins that play an important role in the interactions among cells of the immune system. Class I molecules occur on essentially all nucleated cells of the animal body, but are absent from trophoblast cells and sperm. The cell membrane of T lymphocytes is rich in class I molecules that are comprised of two distinct polypeptide chains, i.e., a 44-kD α (heavy) chain and a 12-kD β chain (β_2 microglobulin). There is a 40-kD core polypeptide in the human α chain that has one N-linked oligosaccharide. Approximately 75% of the α chain is extracellular, including the amino terminus and the oligosaccharide group. The membrane portion is an abbreviated hydrophobic segment. The cytoplasm contains the 30-amino acid residue that comprises the carboxy terminus. The β_2 microglobulin component is neither linked to the cell surface nor to the α chain by covalent bonds. Its association with the α chain is noncovalent. Class I molecules consist of four parts that include an extracellular amino terminal peptide-binding site, an immunoglobulin (Ig)-like

region, a transmembrane segment, and a cytoplasmic portion. The main function of MHC molecules is to bind foreign peptides to form a complex that T cells can recognize. The class I molecular site that binds protein antigens is a 180-amino acid residue segment at the class I α chain's amino terminus. The heavy chain's α-3 segment contains approximately 90 amino acid residues between the α-2 segment's carboxy terminal end and the point of entrance into the plasma membrane. The α-3 segment joins the plasma membrane through a short connecting region and spans the membrane as a segment of 25 hydrophobic amino acid residues. This stabilizes the α chain of MHC class I in the membrane. The carboxy terminal region emerges as a 30-amino acid stretch that is present in the cytoplasm. Class I histocompatibility antigens are products of the major histocompatibility complex (MHC) locus. HLA-A, -B, and -C genes located in the MHC region on the short arm of chromosome 6 in man encode these molecules. K, D, and L genes located on chromosome 17 in the H-2 complex in mice encode murine class I MHC antigens. The Tla complex situated near H-2 encodes additional class I molecules in mice. In T cell-mediated cytotoxicity, CD8+ T lymphocytes kill antigen-bearing target cells. The cytotoxic T lymphocytes play a significant role in resistance to viral infection. Class I MHC molecules present viral antigens to CD8+ T lymphocytes as a viral peptide class I molecular complex, which is transported to the infected cell surface. Cytotoxic CD8+ T cells recognize this and lyse the target before the virus can replicate, thereby stopping the infection.

Class II antigens

MHC histocompatibility antigen with limited distribution on such cells as B lymphocytes and macrophages. In man, these antigens are encoded by genes at the DR, DP, and DQ loci.

class II MHC molecules

Glycoprotein histocompatibility antigens that play a critical role in immune system cellular interactions. Each class II MHC molecule is comprised of a 32- to 34-kD α chain and a 29- to 32-kD β chain, each of which possess N-linked oligosaccharide groups, amino termini that are extracellular, and carboxy termini that are intracellular. Approximately 70% of both α and β chains are extracellular. Separate MHC genes encode the class II molecule α and β chains, which are polymorphic. Class II molecules resemble class I molecules structurally as revealed by class II molecule nucleotide and amino acid sequences. Class II MHC molecules consist of a peptide-binding region, a transmembrane segment, and an intracytoplasmic portion. The extracellular portion of α and β chains consist of α-1 and α-2 and β-1 and β-2 segments, respectively. The α-1 and α-2 segments constitute the peptide-binding region and consist of approximately 90 amino acid residues each. The immunoglobulin-like region is comprised of α-2 and β-2 segments that are folded into immunoglobulin domains in the class

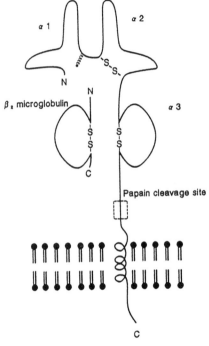

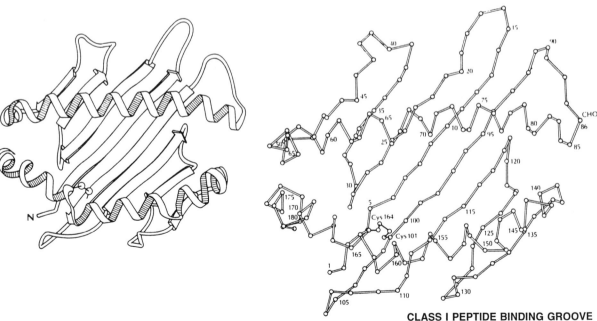

CLASS I PEPTIDE BINDING GROOVE

Class I MHC Molecule

II molecule. The transmembrane region consists of approximately 25 hydrophobic amino acid residues. The transmembrane portion ends with a group of basic amino acid residues immediately followed by hydrophilic tails that extend into the cytoplasm and constitute the carboxy terminal ends of the chains. The α chain is more heavily glycosylated than is the β chain. Of the five exons in the α genes, one encodes the signal sequence and two code for the extracellular domains. The transmembrane portion and a portion of the 3′ untranslated segment are encoded by a fourth exon. The remaining part of the 3′ untranslated region is coded for by a fifth exon. Six exons are present in the β genes. They resemble α gene exons 1 through 3. The transmembrane domain and a portion of the cytoplasmic domain are encoded by a fourth exon, the cytoplasmic domain is coded for by the fifth exon, and the sixth exon encodes the 3′ region that is untranslated.

B lymphocytes, macrophages, or other accessory cells express MHC class II antigens. γ Interferon or other agents may induce an aberrant expression of class II antigen by other types of cells. Antigen-presenting cells (APC), such as macrophages, present antigen at the cell surface to immunoreactive CD4+ helper/inducer T cells in the context of MHC class II antigens. For appropriate presentation, the peptide must bind securely to the MHC class II molecules. Those that do not fail to illicit an immune response. Following interaction of the peptide and the CD4+ helper T lymphocyte receptor, the CD4 cell is activated, interleukin-2 (IL-2) is released, and the immune response is initiated. In man, the class II antigens, DR, DP, and DQ, are encoded by HLA-D region genes. In the mouse, class II antigens, designated as Ia antigens, are encoded by I region genes. The I invariant chain (Ii) represents an essentially nonpolymorphic polypeptide chain that is associated with MHC class II molecules of man and mouse.

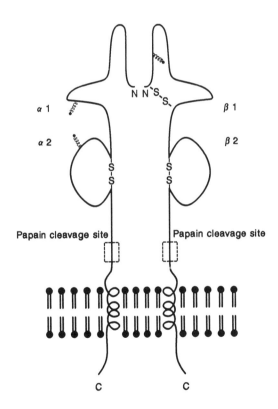

class III molecules

Substances that include factors B, C2, and C4 that are encoded by genes in the major histocompatibility complex (MHC) region. Although adjacent to class I and class II molecules that are important in histocompatibility, C3 genes are not important in this regard. The 100-kB region is located between HLA-B and HLA-D loci on the short arm of chromosome 6 in man and between the I and H-2D regions on chromosome 17 in mice. The genes encoding C4 and P-450 21-hydroxylase are closely linked.

class switching (isotype switching)

A change in the isotype or class of an immunoglobulin synthesized by a B lymphocyte undergoing differentiation. IgM is the main antibody produced first in a primary humoral response to thymus-dependent antigens, with IgG being produced later in the response. A secondary antibody response to the same antigen results in the production of only small amounts of IgM, but much larger quantities of IgG, IgA, or IgE antibodies. T_H cell lymphokines have a significant role in controlling class switching. Only heavy chain constant regions are involved in switching, with the light chain type and heavy chain variable region remaining the same. The specificity of the antigen-binding region is not altered. Mechanisms of class switching during B cell differentiation include the generation of transcripts processed to separate mRNAs and the rearrangements of immunoglobulin genes that lead to constant region gene segment transposition. Membrane IgM appears first on immature B cells, followed by membrane IgD as cell maturation proceeds. A primary transcript bearing heavy chain variable region, μ chain constant region, and δ chain constant region may be spliced to form mRNA that codes for each heavy chain. Following stimulation of B cells by antigen and T lymphocytes, class switching is probably attributable to immunoglobulin gene rearrangements. During switching, B cells may temporarily express more than one class of immunoglobulin. Class switching in B cells mediated by IL-4 is sequential, proceeding from C_μ to $C_{\gamma-1}$ to C_ϵ. IgG1 expression replaces IgM expression as a consequence of the first switch. IgE expression replaces IgG1 expression as a result of the second switch. TGF-β and IL-5 have been linked to the secretion of IgA.

classic pathway of complement

A mechanism to activate C3 through participation by the serum proteins C1, C4, and C2. Either IgM or a doublet of IgG may bind the C1 subcomponent C1q. Following subsequent activation of C1r and C1s, the two C1s substrates, C4 and C2, are cleaved. This yields C4b and C2a fragments that produce C4b2a, known as C3 convertase. It activates opsonization, chemotaxis of leukocytes, increased permeability of vessels, and cell lysis. Activators of the classic pathway include IgM, IgG, staphylococcal protein A, C-reactive protein, and DNA. C1 inhibitor blocks the classical pathway by separating C1r and C1s from C1q. The C4-binding protein also blocks the classical pathway by linking to C4b, separating it from C2a, and permitting factor I to split the C4b heavy chain to yield C4bi, which is unable to unite with C2a, thereby inhibiting the classical pathway.

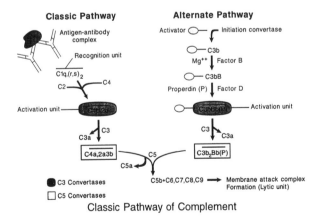

Classic Pathway of Complement

clathrin

The principal protein enclosing numerous coated vesicles. The molecular structure consists of three 180-kD heavy chains and three 30- to 35-kD light chains arranged into typical lattice structures comprised of pentagons or hexagons. These structures encircle the vesicles.

Cleveland procedure

A form of peptide map in which protease-digested protein products, with sodium dodecyl sulfate (SDS) present, are subjected to SDS-PAGE. This produces a characteristic peptide fragment pattern that is typical of the protein substrate and enzyme used.

clonal anergy

The interaction of immune system cells with an antigen, without a second antigen signal, that is usually needed for a response to an immunogen. This leads to functional inactivation of the immune system cells in contrast to the development of antibody formation or cell-mediated immunity.

clonal balance

In explaining autoimmunity in terms of clonal balance, it is convenient to describe it as an alteration in the helper to suppressor ratio with a slight predominance of helper activity. Factors that influence the balance of helper to suppressor cells include aging, steroid hormones, viruses, and chemicals. The genetic constitution of the host and the mechanism of antigen presentation are the two most significant factors that govern clonal balance. Immune response genes associated with MHC determine class II MHC antigen expression on cells presenting antigen to helper CD4+ lymphocytes. Thus, the MHC class II genotype may affect susceptibility to autoimmune disease. Other genes may be active as well. Antigen presentation exerts a major influence on the generation of an autoimmune response. Whereas a soluble antigen administered intravenously with an appropriate immunologic adjuvant may induce an autoimmune response leading to immunopathologic injury, the same antigen injected intravenously without the adjuvant may induce no detectable response. Animals rendered tolerant to foreign antigens possess suppressor T lymphocytes associated with the induced unresponsiveness. Thus, self-tolerance could be due, in part, to the induction of suppressor T cells. This concept is called clonal balance rather than clonal deletion. Self antigens are considered to normally induce mostly suppressor rather than helper T cells, leading to a negative suppressor balance in the animal body. Three factors with the potential to suppress immune reactivity against self include nonantigen-specific suppressor T cells, antigen-specific suppressor T cells, and antiidiotypic antibodies. Suppressor T lymphocytes may leave the thymus slightly before the corresponding helper T cells. Suppressor T cells specific for self antigens are postulated to be continuously stimulated and usually in greater numbers than the corresponding helper T cells.

clonal deletion (negative selection)

The elimination of self-reactive T lymphocyte in the thymus during the development of natural self-tolerance. T cells recognize self antigens only in the context of major histocompatibility complex (MHC) molecules. Autoreactive thymocytes are eliminated following contact with self antigens expressed in the thymus before maturation is completed. The majority of CD4+ T lymphocytes in the blood circulation that survived clonal deletion in the thymus failed to respond to any stimulus. This reveals that clonal anergy participates in suppression of autoimmunity. Clonal deletion represents a critical mechanism to rid the body of autoreactive T lymphocytes. This is brought about by minor lymphocyte stimulation (Mls) antigens that interact with the T cell receptor's V β region of the T lymphocyte receptor, thereby mimicking the action of bacterial super antigen. Intrathymic and peripheral tolerance in T lymphocyte can be accounted for by clonal deletion and functional inactivation of T cells reactive against self.

clonal ignorance

Lymphocytes that survive the principal mechanisms of self tolerance and remain functionally competent but are unresponsive to self antigens and do not cause autoimmune reactions.

clonal restriction

An immune response that is limited to the expression of a few lymphoid cell clones.

clonal selection theory

A selective theory of antibody formation proposed by F.M. Burnet who postulated the presence of numerous antibody-forming cells, each capable of synthesizing its own predetermined antibody. One of the cells, after having been selected by the best-fitting antigen, multiplies and forms a clone of cells which continue to synthesize the same antibody. Considering the existence of many different cells, each capable of synthesizing an antibody of a different specificity, all known facts of antibody formation are easily accounted for. An important element of the clonal selection theory was the hypothesis that many cells with different antibody specificities arise through random somatic mutations during a period of hypermutability early in the animal's life. Also early in life, the "forbidden" clones of antibody-forming cells (i.e., the cells that make antibody to the animal's own antigen) are still destroyed after encountering these autoantigens. This process accounted for an animal's tolerance of its own antigens. Antigen would have no effect on most lymphoid cells, but it would selectively stimulate those cells already synthesizing the corresponding antibody at a low rate. The cell surface antibody would serve as receptor for antigen and proliferate into a clone of cells, producing antibody of that specificity. Burnet introduced the "forbidden clone" concept to

Selection Theory of Antibody Production

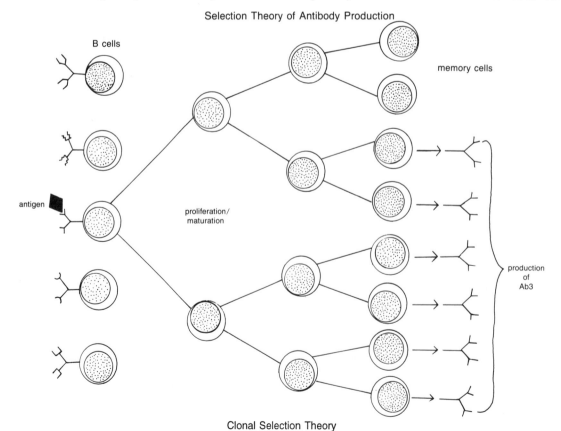

Clonal Selection Theory

explain autoimmunity. Cells capable of forming antibody against a normal self antigen were "forbidden" and eliminated during embryonic life. During fetal development, clones that react with self antigens are destroyed or suppressed. The subsequent activation of suppressed clones reactive with self antigens in later life may induce autoimmune disease. D.W. Talmage proposed a cell selection theory of antibody formation, which was the basis for Burnet's clonal selection theory.

clone

A cell or organism that develops from one parent and has exactly the same genotype and phenotype of that parent. Malignant proliferation of a clone of plasma cells in multiple myeloma represents a type of monoclonal gammopathy. The fusion of an antibody-producing B cell with a mutant myeloma cell *in vitro* by the action of polyethylene glycol to form a hybridoma that is immortal and produces monoclonal antibody is an example of the *in vitro* production of a clone.

cloned DNA

A DNA fragment or gene introduced into a vector and replicated in eukaryotic cells or bacteria.

clonotypic

An adjective that defines the features of a specific B cell population's receptors for antigen that are products of a single B lymphocyte clone. Following release from the B cells, these antibodies should be very specific for antigen, have a restricted spectrotype, and should possess at least one unique private idiotypic determinant. Clonotypic may also be used to describe the features of a particular clone of T lymphocytes' specific receptor for antigen with respect to idiotypic determinants, specificity for antigen, and receptor similarity from one daughter cell of the clone to another.

cluster of differentiation (CD)

Cell surface molecules comprising epitopes, identifiable by monoclonal antibodies, on the surfaces of hematopoetic (blood) cells in man, as well as in mice and other animals. CD markers are given number designations in man, but separate designations equivalent to the human CD numbers are given to animal determinants. Some individuals use the CD designation to refer to the antibodies which identify a particular antigen.

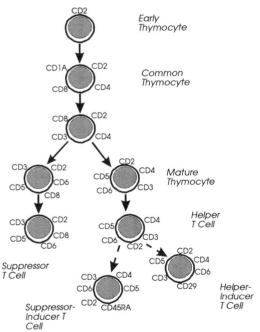

Cluster of Differentiation

clusterin — (serum protein SP-40,40)

A complement regulatory protein that inhibits membrane attack complex (MAC) formation by blocking the fluid-phase of the MAC. Its substrate is C5b67.

CMI

Abbreviation for cell-mediated immunity.

coagglutination

The interaction of IgG antibodies with the surface of protein A-containing *Staphylococcus aureus* microorganisms through their Fc regions, followed by interaction of the Fab regions of these same antibody molecules with surface antigens of bacteria for which they are specific. Thus, when the appropriate reagents are all present, coagglutination will take place in which the Y-shaped antibody molecule will serve as the bridge between staphylococci and the coagglutinated microorganism for which it is specific.

coagulation system

A cascade of interaction among 12 proteins in blood serum that culminates in the generation of fibrin, which prevents bleeding from blood vessels whose integrity has been interrupted.

coated pit

A depression in the cell membrane coated with clathrin. Hormones, such as insulin and epidermal growth factor may bind to their receptors in the coated pit or migrate toward the pit following binding of the ligand at another site. After the aggregation of complexes of receptor and ligand in the coated pits, they invaginate and bud off as coated vesicles containing the receptor-ligand complexes. These structures, called receptosomes migrate into the cell by endocytosis. Following association with GERL structures, they fuse with lysosomes where receptors and ligands are degraded.

coated vesicle

Vesicles in the cytoplasm usually encircled by a coat of protein-containing clathrin molecules. Coated vesicles originate from coated pits and are important for protein secretion and receptor-mediated endocytosis. Coated vesicles convey receptor-macromolecule complexes from an extracellular to an intracellular location. Clathrin-coated vesicles convey proteins from one intracellular organelle to another. Refer also to coated pit.

cobra venom factor (CVF)

A protein in the venom of the Indian cobra, *Naja naja*. It is the equivalent of mammalian C3b, which means that it can activate the alternative pathway of complement. Mammalian factor I does not inactivate cobra venom, which leads to the production of a stable alternative pathway C3 convertase if CVF is injected intravenously into a mammal. Thus, the injection of cobra venom factor into mammals has been used to destroy complement activity for experimental purposes.

Coca, Arthur Fernandez (1875–1959)

American allergist and immunologist. He was a major force in allergy and immunology. He named atopic antibodies and was a pioneer in the isolation of allergens. Together with Robert A. Cooke, Coca classified allergies in humans.

co-capping

If two molecules are associated in a membrane, capping of one induced by its ligand may lead also to capping of the associated molecule. Antibodies to membrane molecule x may induce capping of membrane molecule y, as well as of x, if x and y are associated in the membrane. In this example, the capping of the associated y molecule is termed co-capping.

coccidioidin

A *Coccidioides immitis* culture extract that is used in a skin test for cell-mediated immunity against the microorganism in a manner analogous to the tuberculin skin test.

codominant

The expression of both alleles of a pair in the heterozygote. The traits which they determine are codominant as in the expression of blood group A and B epitopes in type AB persons.

codon

A three-adjacent nucleotide sequence mRNA that acts as a coding unit for a specific amino acid during protein synthesis. The codon controls which amino acid is incorporated into the protein molecule at a certain position in the polypeptide chain. Out of 64 codons, 61 encode amino acids and 3 act as termination codons.

coelomocyte

A circulating or fixed-ameboid phagocytic leukocyte that participates in the defense of invertebrate animals that have a coelom by phagocytosis and encapsulation.

Cogan's syndrome

Corneal inflammation (interstitial keratitis) and inflammation of the ear, leading to nausea, vomiting, vertigo, and ringing in the ears. This may be associated with connective tissue disease or occur following an infection.

cognate interaction

Processed antigen on a B cell surface interacting with a T cell receptor for antigen resulting in B cell differentiation into an antibody-producing cell.

cognate recognition

Refer to cognate interaction.

Cohn fraction II

Principally, gamma globulin isolated by ethanol fractionation of serum by the method of Cohn.

Cohnheim, Julius (1839–1884)

German experimental pathologist who was the first proponent of inflammation as a vascular phenomenon. *Lectures on General Pathology,* 1889.

coisogenic

Refer to congenic strains.

coisogenic strains

Inbred mouse strains that have an identical genotype except for a difference at one genetic locus. A point mutation in an inbred strain provides the opportunity to develop a coisogenic strain by inbreeding the mouse in which the mutation occurred. The line carrying the mutation is coisogenic with the line not expressing the mutation. Considering the problems associated with developing coisogenic lines, congenic mouse strains were developed as an alternative. Refer to congenic strains.

cold agglutinin

An antibody that agglutinates particulate antigen, such as bacteria or red cells, optimally at temperatures less than 37°C. In clinical medicine, the term usually refers to antibodies against red blood cell antigens as in the cold agglutinin syndrome.

cold agglutinin syndrome

An immune condition in which IgM autoantibodies agglutinate erythrocytes most effectively at 4°C. Normal individuals may have cold agglutinins in low titer (<1:32). Certain infections, such as cytomegalovirus, trypanosomias, mycoplasma, malaria, and Epstein-Barr virus infection, are followed by the development of polyclonal cold agglutinins. These antibodies are of concern only if they are hemolytic. Acquired hemolytic anemia patients with a positive direct Coombs' test should be tested for cold agglutinins. For example, they might have anti-Pr, anti-I, anti-i, anti-Sd^a, or anti-Gd. Aged individuals suffering from monoclonal κ proliferation or simultaneous large cell lyphoma may develop cold agglutinin syndrome. It also occurs in the younger age group in whom anti-I antibodies have been synthesized following an infection with *Mycoplasma pneumoniae* or in whom anti-i antibodies associated with infectious mononucleosis have formed. C3d coats the cells. Agglutination and complement fixation may take place intravascularly in parts of the body exposed to the cold. When the red blood cells with attached complement are warmed to 37°C (normal body temperature), mild hemolysis occurs.

cold antibodies

Antibodies that occur at higher titers at 4°C rather than at 37°C.

cold ethanol fractionation

A technique used to fractionate serum proteins by precipitation with cold ethanol. One of the fractions obtained is Cohn fraction II, which contains the immunoglobulins. This method has been largely replaced by more modern and sophisticated techniques.

cold hemoglobinuria

Refer to paroxysmal cold hemoglobinuria.

cold hypersensitivity

A localized wheal and flare reaction and bradycardia that follow exposure to low temperatures that induce overstimulation of the autonomic nervous system.

cold-reacting autoantibodies

Cold-reacting autoantibodies represent a special group of both naturally occurring and pathologic antibodies characterized by the unusual property of reacting with the corresponding antigen at low temperature. Those reacting with red blood cells are also called cold agglutinins, although they may also react with other cells. Those with a more restricted range of targets are called cytotoxins.

cold target inhibition

The introduction of unlabeled target cells to inhibit radioisotope release from labeled target cells through the action of antibody or cell-mediated immune mechanisms.

cold urticaria

Urticaria that occurs soon after exposure to cold. The lesions are usually confined to the exposed areas. The condition has been observed in patients with underlying conditions that include cryoglobulinemia, cryofibrinogenemia, cold agglutinin disease, and paroxysmal cold hemoglobinuria. Although the mechanism is unknown, cold exposure has been shown to cause the release of histamine and other mediators. Cold sensitivity has been passively transferred in individuals with abnormal proteins. Cryoprecipitates may fix complement and lead to the generation of anaphylatoxin. The condition can be diagnosed by placing ice cubes on the forearm for 4 min and observing for the following 10 min for the appearance of urticaria when the area is rewarmed. Treatment consists of limiting exposure to cold and the administration of antihistamines such as oral cyproheptadine. Individuals with an abnormal protein should have the underlying disease treated.

collagen disease and arthritis panel

A cost-effective battery of tests to diagnose rheumatic disease that includes the erythrocyte sedimentation rate and assays for rheumatoid factor (RA test), antinuclear antibody, uric acid levels, and C-reactive protein.

collagen disease/lupus erythematosus diagnostic panel

A battery of serum tests for the diagnosis of collagen vascular disease that yields the most information for the least cost.

collagen vascular diseases

A category of connective tissue diseases in which type III hypersensitivity mechanisms with immune complex of deposition play a major role. These diseases are characterized by inflammation and fibrinoid necrosis in tissues. Patients may manifest involvement of multiple systems, including the vasculature, joints, skin, kidneys, and other tissues. These are classic, systemic autoimmune diseases in most cases. The prototype of this category is systemic lupus erythematosus. Of the multiple disorders included in this category are dermatomyositis, polyarteritis nodosa, progressive systemic sclerosis (scleroderma), rheumatoid arthritis, and mixed connective tissue disease. They are treated with immunosuppressive drugs, especially corticosteroids.

collectin receptor

The receptor of C1q, a subcomponent of the complement component C1.

colon antibodies

IgG antibodies in the blood sera of 71% of ulcerative colitis patients may be shown by flow cytometry to react with rat colon epithelial cells. Antibodies reactive with a 40 kD constituent of normal colon extracts have been found in the blood sera of 79% of ulcerative colitis patients. Anti-neutrophil cytoplasmic antibodies, distinct from the P-ANCA of systemic vasculitis and the C-ANCA of Wegener's granulomatosis are detectable in 70% of ulcerative colitis patients.

colon-ovary tumor antigen (COTA)

A type of mucin demonstrable by immunoperoxidase staining in all colon neoplasms and in some ovarian tumors. COTA occurs infrequently in other neoplasms. Normal tissues express limited quantities of COTA.

colony-forming unit (CFU)

The hematopoietic stem cell and the progeny cells that derive from it. Mature (end-stage) hematopoietic cells in the blood are considered to develop from one CFU. Some progenitor cells are precursors of erythrocytes, others are precursors of polymorphonuclear leukocytes and monocytes, and still others are megakaryocyte and platelet precursors.

colony-forming units, spleen (CFU-S)

A hematopoietic precursor cell that can produce a tiny nodule in the spleen of mice that have been lethally irradiated. These small nodular areas are sites of cellular proliferation. Each arises from a single cell or colony-forming unit. The CFU-S form colonies of pluripotent stem cells.

colony stimulating factors (CSF)

Glycoproteins that govern the formation, differentiation and function of granulocytes and monomacrophage system cells. CSF promotes the growth, maturation, and differentiation of stem cells to produce progenitor cell colonies. They facilitate the development of functional end-stage cells. They act on cells through specific receptors on the target cell surface. T cells, fibroblasts, and endothelial cells produce CSF factors. Different colony-stimulating factors act on cell line progenitors that include CFU-E (red blood cell precursors), GM-CFC (granulocyte-macrophage colony forming cells), MEG-CFC (megakaryocyte-colony forming cells), EO-CFC (eosinophil-leukocyte colony forming cells), T cells, and B cells. Colony-stimulating factors promote the clonal growth of cells. Colony-stimulating factors include granulocyte CSF that is synthesized by endothelial cells, macrophages, and fibroblasts. It activates the formation of granulocytes and is synergistic with IL-3 in the generation of megakaryocytes and granulocytes-macrophages. Endothelial cells, T lymphocytes, and fibroblasts form granulocyte-macrophage CSF, which stimulates granulocyte and macrophage colony formation. It also stimulates megakaryocyte blast cells. Colony-stimulating factor 1 is produced by endothelial cells, macrophages, and fibroblasts and induces the generation of macrophage colonies. Multi-CSF (interleukin-3) is produced by T lymphocytes and activates the generation of granulocytes, macrophages, eosinophils, and mast cell colonies. It is synergistic with other factors in activating hematopoietic precursor cells. Renal interstitial cells synthesize erythropoietin that activates erythroid colony formation.

colostrum
Immunoglobulin-rich first breast milk formed in mammals after parturition. The principal immunoglobulin is IgA with lesser amounts of IgG. It provides passive immune protection of the newborn prior to maturation of its own immune competence.

combinatorial joining
A mechanism for one exon to unite alternatively with several other gene regions, increasing the diversity of products encoded by the gene.

combined immunodeficiency
A genetically determined or primary immunodeficiency that may affect T cell-mediated immunity and B cell (humoral antibody)-mediated immunity. The term is usually reserved for immunodeficiency that is less profound than severe combined immunodeficiency. Combined immunodeficiency may occur in both children and adults.

combined prophylactic
Refer to mixed vaccine.

combining site
Refer to antigen binding site.

commensal mice
Mice that associate closely with man.

common acute lymphocytic leukemia antigen (CALLA)
A 100-kD surface membrane glycoprotein present on human leukemia cells and, to a lesser degree, on other cells that include granulocytes and kidney cells. It is a zinc metalloproteinase and is classified as CD10/neutral endopeptidase 24.11. Four fifths of non-T cell leukemias express CALLA. Under physiologic conditions, 1% of cells in the bone marrow express CALLA. The presence of CALLA is revealed by monoclonal antibodies and flow cytometry using bone marrow or other cells. Bone marrow to be used for autologous transplants may be purged of CALLA positive lymphocytic leukemia cells by using anti-CALLA monoclonal antibodies

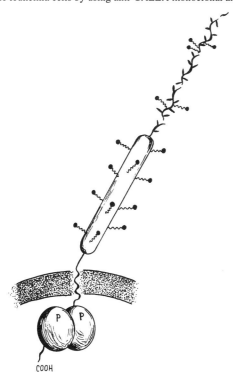

COOH
Common Leukocyte Antigen

and complement. It is a pre-B lymphoblast marker that is the most frequent type of cell in childhood acute lymphocytic leukemia (ALL). If Ia antigen is present together with CALLA, this portends a favorable prognosis. CALLA may also be positive in Burkitt's lymphoma, B cell lymphomas, and 40% of T cell lymphoblastic lymphomas. All blasts are usually positive, not only for CALLA and the Ia antigen, but also for TdT.

common leukocyte antigen (LCA) (CD45)
An antigen shared in common by both T and B lymphocytes and expressed, to a lesser degree, by histiocytes and plasma cells. By immunoperoxidase staining, it can be demonstrated in sections of paraffin-embedded tissues containing these cell types. Thus, it is a valuable marker to distinguish lymphoreticular neoplasms from carcinomas and sarcomas.

common variable antibody deficiency
Refer to common variable immunodeficiency (CVID).

common variable immunodeficiency (CVID)
Common variable immunodeficiency is a relatively common congenital or acquired immunodeficiency that may be either familial or sporadic. The familial form may have a variable mode of inheritance. Hypogammaglobulinemia is common to all of these patients and usually affects all classes of immunoglobulin, but in some cases only IgG is affected. The World Health Organization (WHO) classifies three forms of the disorder: (1) an intrinsic B lymphocyte defect, (2) a disorder of T lymphocyte regulation that includes deficient T helper lymphocytes or activated T suppressor lymphocytes, and (3) autoantibodies against T and B lymphocytes. The majority of patients have an intrinsic B cell defect with normal numbers of B cells in the circulation that can identify antigens and proliferate, but cannot differentiate into plasma cells. The ability of B cells to proliferate when stimulated by antigen is evidenced by hyperplasia of B cell regions of lymph nodes, spleen, and other lymphoid tissues. Yet, differentiation of B cells into plasma cells is blocked. The deficiency of antibody that results leads to recurrent bacterial infections, as well as intestinal infestation by *Giardia lamblia,* which produces a syndrome that resembles sprue. Noncaseating granulomas occur in many organs. There is an increased incidence of autoimmune diseases, such as pernicious anemia, rheumatoid arthritis, and hemolytic anemia. Lymphomas also occur in these immunologically deficient individuals.

competency, immunologic
The capacity of an animal's immune system to generate a response to an immunogen.

competitive inhibition
Prevention by one group of cells or molecules of the interaction of antibody or cells with a separate population of cells or molecules.

complement (C) (see figure, page 332)
A system of 20 soluble plasma and other body fluid proteins together with cellular receptors for many of them and regulatory proteins found on blood and other tissue cells. These proteins play a critical role in aiding phagocytosis of immune complexes, which activate the complement system. These molecules and their fragments, resulting from the activation process, are significant in the regulation of cellular immune responsiveness. Once complement proteins identify and combine with target substance, serine proteases are activated. This leads ultimately to the assembly of C3 convertase, a protease on the surface of the target substance. The enzyme cleaves C3, yielding a C3b fragment that is bound to the target through a covalent linkage. C3b or C3bi bound to phagocytic cell surfaces become ligands for C3 receptors, as well as binding sites for C5. The union of C5b with C6, C7, C8, and C9 generates the membrane attack complex (MAC) which may associate with the cell's lipid bilayer membrane to produce lysis, which is critical in resistance against certain species of bacteria. The complement proteins are significant, nonspecific mediators of humoral immunity. Multiple substances may trigger the complement system. There are two pathways of complement activation designated the classical pathway in which an antigen, e.g., red blood cell, and antibody combine and fix the first subcomponent designated C1q. This is followed in sequence: C1qrs,4,2,3,5,6,7,8,9 to produce lysis. The alternative pathway does not utilize C1, 4, and 2 components. Bacterial products such as endotoxin and other agents may activate this pathway through C3. There are numerous biological activities associated with complement besides immune lysis. These include the formation of anaphylatoxin, chemotaxis, opsonization,

phagocytosis, bacteriolysis, hemolysis, and other amplification mechanisms.

complement deficiency conditions

Inherited complement deficiencies are rare. In healthy Japanese blood donors, only 1 in 100,000 persons had no C5, C6, C7, and C8. No C9 was contained in 3 of 1000 individuals. Most individuals with missing complement components do not manifest clinical symptoms. Additional pathways provide complement-dependent functions that are necessary to preserve life. If C3, factor I, or any segment of the alternative pathway is missing, the condition may be life threatening with markedly decreased opsonization in phagocytosis. C3 is depleted when factor I is absent. C5, C6, C7, or C8 deficiencies are linked with infections, mainly meningococcal or gonococcal, which usually succumb to complement's bactericidal action. Deficiencies in classical complement pathway activation are often associated with connective tissue or immune complex diseases. Systemic lupus erythematosus may be associated with C1qrs, C2, or C4 deficiencies. Hereditary angioedema (HAE) patients have a deficiency of C1 inactivator. A number of experimental animals with specific complement deficiencies has been described, such as C6 deficiency in rabbits and C5 deficiency in mice. Acquired complement deficiencies may be caused by either accelerated complement consumption in immune complex diseases with a type III mechanism or by diminished formation of complement proteins as in acute necrosis of the liver.

complement deviation (Neisser-Wechsberg phenomenon)

Blocking of complement fixation or of complement-induced lysis when excess antibody is present. There is deviation of the complement by the antibody.

complement fixation

The primary union of an antigen with an antibody in the complement fixation reaction takes place almost instantaneously and is invisible. A measured amount of complement present in the reaction mixture is taken up by complexes of antigen and antibody. The consumption or binding of complement by antigen-antibody complexes. This serves as the basis for a serologic assay in which antigen is combined with a serum specimen suspected of containing the homologous antibody. Following the addition of a measured amount of complement, which is fixed or consumed only if antibody was present in the serum and has formed a complex with the antigen, sheep red blood cells sensitized (coated) with specific antibody are added to determine whether or not complement has been fixed in the first phase of the reaction. Failure of the sensitized sheep red blood cells to lyse constitutes a positive test since no complement is available. However, sheep red blood cell lysis indicates that complement was not consumed during the first phase of the reaction, implying that homologous antibody was not present in the serum, and complement remains free to lyse the sheep red blood cells sensitized with antibody. Hemolysis constitutes a negative reaction. The sensitivity of the complement fixation test falls between that of agglutination and precipitation. Complement fixation tests may be carried out in microtiter plates, which are designed for the use of relatively small volumes of reagents. The lysis of sheep red blood cells sensitized with rabbit antibody is measured either in a spectrophotometer at 413 nm or by the release of ^{51}Cr from red cells that have been previously labeled with the isotope. Complement fixation can detect either soluble or insoluble antigen. Its ability to detect virus antigens in impure tissue preparations makes the test still useful in diagnosis of virus infections.

complement fixation assay

A serologic test based on the fixation of complement by antigen-antibody complexes. It has been applied to many antigen-antibody systems and was widely used earlier in the century as a serologic test for syphilis.

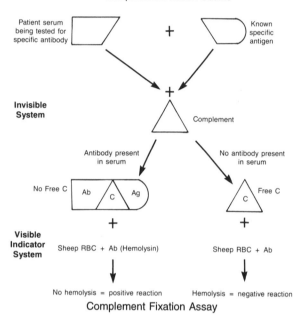

Complement Fixation Assay

complement fixation inhibition test

An assay in which a known substance prevents interaction of antibody and antigen, thereby preventing complement uptake and fixation.

complement fixing antibody

An antibody of the IgG or IgM class that binds complement after reacting with its homologous antigen. This represents complement fixation by the classic pathway. Nonantibody mechanisms as well as IgA fix complement by the alternative pathway.

complement inhibitors

Protein inhibitors that occur naturally and block the action of complement components include factor H, factor I, C1 inhibitor, and C4-binding protein (C4BP). Also included among complement inhibitors are heating to 56°C to inactivate C1 and C2; combining with hydrazine and ammonia to block the action of C3 and C4; and the adding of zymosan or cobra venom factor to induce alternate pathway activation of C3, which consumes C3 in the plasma.

complement multimer

A doughnut-shaped configuration as a part of the complement reaction sequence.

complement receptor 1 (CR1)

CR1 is a membrane glycoprotein found on human erythrocytes, monocytes, polymorphonuclear leukocytes, B cells, a T cell subset, mast cells, and glomerular podocytes. On red cells, CR1 binds C3b or C4b components of immune complexes, facilitating their transport to the mononuclear phagocyte system. CR1 facilitates attachment, endocytosis, and phagocytosis of C3b/C4b-containing complexes to macrophages or neutrophils and may serve as a cofactor for factor I-mediated C3 cleavage. The identification of CR1 cDNA has made possible the molecular analysis of CR1 biological properties.

complement receptor 2 (CR2)

A receptor for C3 fragments that also serves as a binding site for Epstein-Barr virus (EBV). It is a receptor for C3bi, C3dg, and C3d based on its specificity for their C3d structure. B cells, follicular dendritic cells of lymph nodes, thymocytes, and pharyngeal epithelial cells, but not T cells express CR2. EBV enters B lymphocytes by way of CR2. The gene encoding CR2 is linked closely with that of CR1. A 140-kD single polypeptide chain makes up CR2, which has a short consensus repeat (SCR) structure similar to that of CR1.

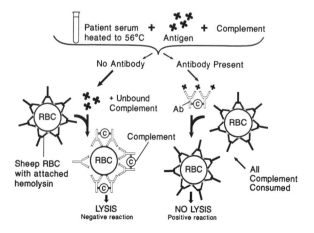

CR2 may be active in B cell activation. Its expression appears restricted to late pre-B and to mature B cells. CR2 function is associated with membrane IgM. Analysis of cDNA clones has provided CR2's primary structure.

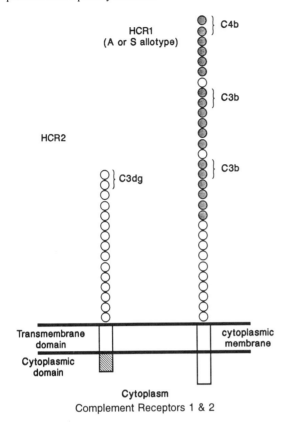

Complement Receptors 1 & 2

complement receptor 3 (CR3)

A principal opsonin receptor expressed by monocytes, macrophages, and neutrophils. It plays an important role in the removal of bacteria. CR3 binds fixed C3bi in the presence of divalent cations. It also binds bacterial lipopolysaccharides and β glucans of yeast cell walls. The latter are significant in the ability of granulocytes to identify bacteria and yeast cells. CR3 is an integrin type of adhesion molecule that facilitates the binding of neutrophils to endothelial cells in inflammation. CR3 enables phagocytic cells to attach to bacteria or yeast cells with fixed C3bi, β glucans, or lipopolysaccharide on their surface. This facilitates phagocytosis and the respiratory burst. CR3 is comprised of a 165-kD α glycoprotein chain and a 95-kD β glycoprotein chain. CR3 appears related to LFA-1 and P150,95 molecules and shares a β chain with them. All three of these molecules are of critical significance in antigen-independent cellular adhesion, which confines leukocytes to inflammatory areas, among other functions. Deficient surface expression of these molecules occurs in leukocyte adhesion deficiency (LAD) in which patients experience repeated bacterial infections. The primary defect appears associated with the common β chain. Besides C3bi binding, CR3 is of critical significance in IgG- and CR1-facilitated phagocytosis by neutrophils and monocytes. CR3 has a more diverse function than does either CR1 or CR2.

complement receptor 4 (CR4)

Glycoprotein membrane receptor for C3dg on polymorphonuclear neutrophils (PMNs), monocytes, and platelets. CR4 facilitates Fc receptor-mediated phagocytosis and mediates Fc-independent phagocytosis. It consists of a 150-kD α chain and a 95-kD β chain. Chromosome 16 is the site of genes that encode the α chain, whereas chromosome 21 is the site of genes that encode the β chain. Tissue macrophages express CR4. It is an integrin with a β chain in common with CR3 and LFA-1.

complement receptor 5 (CR5)

A receptor that binds C3bi, C3dg, and C3d fragments based on its specificity for their C3d component. Reactivity is only in the fluid phase and not when the fragments are fixed. CR5 is the C3dg-dimer receptor. Neutrophils and platelets manifest CR5 activity.

complement receptors

Receptors for products of complement reactions. Proteolytic cleavage of human complement component C3 takes place following activation of either the classical or the alternative complement pathway. Following the generation of C3a and C3b, the C3b covalently binds to bacteria, immune complexes, or some other target and then unites with a high-affinity receptor termed the C3b/C4b receptor currently known as CR1. Subsequent proteolytic cleavage of the bound C3b is attributable to factor I and a cofactor. This action yields C3bi, C3d, g, and C3c, which interact with specific receptors. CR2 is the C3dg receptor, and CR3 is the C3bi receptor.

Complement Receptor Complexes
on the Surface of B Cells

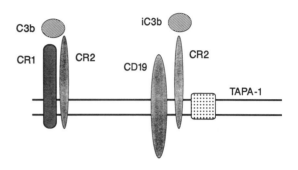

complement receptors, membrane

Receptors expressed on blood cells and tissue macrophages of man. These include C1q-R (C1q receptor; CR1 (C3b/C4b receptor; CD35), CR2 (C3d/Epstein-Barr virus (EBV) receptor; CD21), CR3 (iC3b receptor; CD11b/CD18), CR4 (C3bi receptor; CD11c/CD18), CR5 (C3dg-dimer receptor), fH-R (factor H receptor), C5a-R (C5a receptor), and C3a/C4a-R (C3a/C4a receptor). Ligands for C receptors generated by either the classic or alternate pathways include fluid-phase activation peptides of C3, C4, and C5 designated C3a, C4a, and C5a, which are anaphylatoxins that interact with either C3a/C4a-R or C5a-R and participate in inflammation. Other ligands for C receptors include complement proteins deposited on immune complexes that are either soluble or particulate. Fixed C4 and C3 fragments (C4b, C3b, C3bi, C3dg, and C3d), C1q, and factor H constitute these ligands. These receptors play a major role in facilitating improved recognition of pathogenic substances. They aid in the elimination of bacteria and soluble immune complexes. Neutrophils, monocytes, and macrophages express C3 receptors on their surface. Neutrophils and erythrocytes express immune adherence receptors, termed CR1, on their surface. Four other receptors for CR3 are designated CR2, CR3, CR4, and CR5. Additional receptors for complement components other than C3 and a receptor for C3a are termed C1q-R (C1q receptor), C5a-R (C5a receptor), C3a-R (C3a receptor), and fH-R (factor H receptor).

complementarity

A genetic term that indicates the requirement for more than one gene to express a trait.

complementarity-determining region (CDR)

The hypervariable regions in an immunoglobulin molecule that form the three-dimensional cavity where an epitope binds to the antibody molecule. The heavy and light polypeptide chains each contribute three hypervariable regions to the antigen binding region of the antibody molecule. Together, they form the site for antigen binding. Likewise, the T cell receptor α and β chains each have three regions with great diversity that are analogous to the immunoglobulin's CDRs. These hypervariable areas are sites of binding for foreign antigen and self MHC molecular complexes.

complete Freund's adjuvant (CFA)

A suspension of killed and dried mycobacteria suspended in light weight mineral oil, which when combined with an aqueous-phase antigen, as a water-in-oil emulsion, enhances immunogenicity. CFA facilitates stimulation of both humoral (B cell) and cell-mediated (T cell) limb's of the immune response.

complex allotype

An allotype with multiple amino acid residue positions that are not the same as those of a different allotype at that same locus.

complex release activity

Binding of injected preformed complexes to the endothelial cell membranes immediately after their injection into experimental animals. The amount of such binding decreases with age, favoring deposition of such complexes within tissues.

complotype

An MHC class III haplotype. It is a precise arrangement of linked alleles of MHC class III genes that encode C2, factor B, C4A, and C4B MHC class III molecules in man. Caucasians have 12 ordinary complotypes that may be in positive linkage disequilibrium.

Concanavalin A (con A)

A jack bean (*Canavalia ensiformis*) lectin that induces erythrocyte agglutination and stimulates T lymphocytes to undergo mitosis and proliferate. Con A interacts with carbohydrate residues rich in mannose. Macrophages must be present for T lymphocytes to proliferate in response to con A stimulation. There are four 237-amino acid residue subunits in con A. There is one binding site for saccharide, one for Ca^{2+}, and one for a metallic ion such as Mn^{2+} in each con A subunit. T lymphocytes stimulated by con A release interleukin-2. Cytotoxic T lymphocytes stimulated by con A induce lysis of target cells without regard to the antigen specificity of either the effector or target cell. This could be induced by crosslinking of the effector and target cells by con A, which is capable of linking to high-mannose oligosaccharides on target cell surfaces as well as to high-mannose sugars on the T cell receptor. Con A binds readily to ordinary cell membrane glycoproteins such as glucopyranosides, fructofuranosides, and mannopyranosides.

concomitant immunity

(1) In tumor immunology, resistance to a tumor that has been transplanted into a host already bearing that tumor. Immunity to the reinoculated neoplasm does not inhibit growth of the primary tumor. (2) Resistance to reinfection of a host currently infected with that parasite.

conformational determinant

An epitope composed of amino acid residues that are not contiguous and represent separated parts of the linear sequence of amino acids that are brought into proximity to one another by folding of the molecule. A conformational determinant is dependent on three-dimensional structure. Conformational determinants, are therefore, usually associated with native rather than denatured proteins. Antibodies specific for linear determinants and others specific for conformational determinants may give clues as to whether a protein is denatured or native, respectively.

Conformational Determinant   Determinant Lost

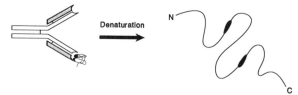

Denaturation

congenic

Signifies a mouse line that is identical or almost identical to other inbred stains with the exception of the substitution of an alien allele at a single histocompatibility locus that crosses with a second inbred strain permitting introduction of the foreign allele.

congenic strains

Inbred mouse strains that are believed to be genetically identical except for a single genetic locus difference. Congenic strains are produced by crossing a donor strain and a background strain. Repeated backcrossing is made to the background strain, selecting in each generation for heterozygosity at a certain locus. Following 12 to 14 backcrosses, the progeny are inbred through brother-sister matings to yield a homozygous inbred strain. Mutation and genetic linkage may lead to random differences at a few other loci in the congenic strain. Designations for congenic strains consist of the symbol for the background strain followed by a period and then the symbol for the donor strain.

congenital agammaglobulinemia

Refer to X-linked agammaglobulinemia.

conglutinating complement absorption test

An assay based on the removal of complement from the reaction medium if an antigen-antibody complex develops. This is a test for antibody. As in the complement fixation test, a visible or indicator combination must be added to determine whether any unbound complement is present. This is accomplished by adding sensitized erythrocytes and conglutinin, which is prepared by combining sheep erythrocytes with bovine serum that contains natural antibody against sheep erythrocytes as well as conglutinin. Horse serum may be used as a source of nonhemolytic complement for the reaction. Aggregation of the erythrocytes constitutes a negative test.

conglutination

The strong agglutination of antigen-antibody-complement complexes by conglutinin, a factor present in normal sera of cows and other ruminants. The complexes are similar to EAC1423 and are aggregated by conglutinin in the presence of Ca^{2+}, which is a required cation. Conglutination is a sensitive technique for detecting complement-fixing antibodies.

conglutinin

A bovine serum protein that reacts with fixed C3. It causes the tight aggregation, i.e., conglutination, of red blood cells coated with complement. Conglutinin, which is confined to sera of Bovidae, is not to be confused with immunoconglutinin, which has a similar activity, but is produced in other species by immunization with complement-coated substances or may develop spontaneously following activation of complement *in vivo*. Conglutinin reacts with antigen-antibody-complement complexes in a medium containing Ca^{2+}. The N-linked oligosaccharide of the C3bi α chain is its ligand. The phagocytosis of C3bi-containing immune complexes is increased by interaction with conglutinin. Conglutinin contains 12 33-kD polypeptide chains that are indistinguishable and are grouped into four subunits. Following a 25-residue amino terminal sequence, there is a 13-kD sequence that resembles collagen in each chain. The 20-kD carboxy terminal segments form globular structures that contain disulfide-linked chains.

conglutinin solid phase assay

A test that quantifies C3bi-containing complexes that may activate complement by either the classical or the alternate pathways.

conjugate

Immunologically, the term usually refers to the covalent bonding of a protein carrier with a hapten or it may refer to the labeling of a molecule such as an immunoglobulin with fluorescein isothiocyanate, ferritin, or an enzyme used in the enzyme-linked immunoabsorbent assay.

conjugate vaccine

An immunogen comprised of polysaccharide bound covalently to proteins. The conjugation of weakly immunogenic, bacterial polysaccharide antigens with protein carrier molecules considerably enhances their immunogenicity. Conjugate vaccines have reduced morbidity and mortality for a number of bacterial diseases in vulnerable populations such as the very young or adults with immunodeficiencies. An example of a conjugate vaccine is *Haemophilus influenzae* 6 polysaccharide polyribosyl-ribitol-phosphate vaccine.

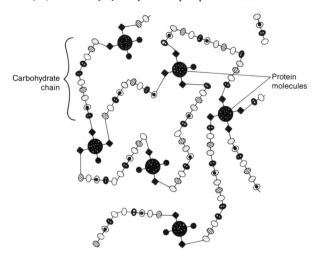

Carbohydrate chain   Protein molecules

Conjugate Vaccine

conjugated antigen

Refer to conjugate.

connective tissue disease

One of a group of diseases, formerly known as collagen vascular diseases, that affect blood vessels producing fibrinoid necrosis in connective tissues. The prototype of a systemic connective tissue disease is systemic lupus erythematosus. Also included in this classification are systemic sclerosis (scleroderma), rheumatoid arthritis, dermatomyositis, polymyositis, Sjögren's syndrome, polyarteritis nodosa, and a number of other disorders that are believed to have an immunological etiology and pathogenesis. They are often accompanied by the development of autoantibodies, such as antinuclear antibodies, or antiimmunoglobulin antibodies, such as rheumatoid factors.

consensus sequence

The typical nucleic acid or protein sequence where a nucleotide or amino acid residue present at each position is that found most often during comparison of numerous similar sequences in a specific molecular region.

constant domain

The immunoglobulin C_H and C_L regions. A globular compact structure that consists of two antiparallel twisted β sheets. There are differences in the number and the irregularity of the β strands and bilayers in variable (V) and constant (C) subunits of immunoglobulins. C domains have a tertiary structure that closely resembles that of the domains, which are comprised of a five-strand β sheet and a four-strand β sheet packed facing one another. However, the C domain does not have a hairpin loop at the edge of one of the sheets. Thus, the C domain has seven or eight β strands rather than the nine that are found in the V domains.

constant region

That part of an immunoglobulin polypeptide chain that has an invariant amino acid sequence among immunoglobulin chains belonging to the same isotype and allotype. There is a minimum of two and often three to four domains in the constant region of immunoglobulin heavy polypeptide chains. The hinge region "tail end piece" (a carboxy terminal region) constitutes part of the constant region in selected classes of immunoglobulin. A few exons encode the constant region of an immunoglobulin heavy chain, and one exon encodes the constant region of an immunoglobulin light chain. The constant region is the location for the majority of isotypic and allotypic determinants. It is associated with a number of antibody functions. T cell receptor α, β, γ, and δ chains have constant regions coded for by three to four exons. MHC class I and II molecules also have segments that are constant regions in that they show little sequence variation from one allele to another.

consumption test

An assay in which antigen or antibody disappears from the reaction mixture as a result of its interaction with the homologous antibody or antigen. By quantifying the amount of unreacted antigen or antibody remaining in the reaction system and comparing it with the quantity of that reagent that was originally present, the result can be ascertained. The antiglobulin consumption test is an example.

contact dermatitis

A type IV, T lymphocyte-mediated hypersensitivity reaction of the delayed type that develops in response to an allergen applied to the skin.

Contact Dermatitis due to poison ivy

contact hypersensitivity

A type IV delayed-type hypersensitivity reaction in the skin characterized by a delayed-type hypersensitivity (cell-mediated) immune reaction produced by cytotoxic T lymphocytes invading the epidermis. It is often induced by applying a skin-sensitizing simple chemical such as dinitrochlorobenzene (DNCB) that acts as a hapten uniting with proteins in the skin leading to the delayed-type hypersensitivity response mediated by T cells. Although substances such as DNCB alone are not antigenic, they may combine with epidermal proteins which serve as carriers for these simple chemicals acting as haptens. There is spongiosis, vesiculation, and pleuritis.

contact sensitivity (CS) (or allergic contact dermatitis)

A form of DTH reaction limited to the skin and consisting of eczematous changes. It follows sensitization by topical drugs, cosmetics, or other types of contact chemicals. The causative agents, usually simple, low molecular weight compounds (mostly aromatic molecules), behave as haptens. The development of sensitization depends on the penetrability of the agent and its ability to form covalent bonds with protein. Part of the sensitizing antigen molecule is thus represented by protein, usually the fibrous protein of the skin. Local skin conditions that alter local proteins, such as inflammation, stasis, and others, facilitate the development of CS, but some chemicals such as penicillin, picric acid, or sulfonamides are unable to conjugate to proteins. It is believed that in this case the degradation products of such chemicals have this property. CS may also be induced by hapten conjugates given by other routes in adjuvant. The actual immunogen in CS remains unidentified. CS may also have a toxic, nonimmunologic component, and frequently both toxic and sensitizing effects can be produced by the same compound. With exposure to industrial compounds, an initial period of increased sensitivity is followed by a gradual decrease in reactivity. This phenomenon is called hardening and could represent a process of spontaneous desensitization. The histologic changes in CS are characteristic.

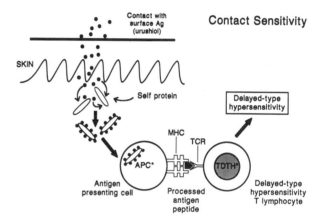

contact system

A system of proteins in the plasma that engage in sequential interactions following contact with surfaces of particles that bear a negative charge, such as glass, or with substances such as lipopolysaccharides, collagen, etc. Bradykinin is produced through their sequential interaction. C1 inhibitor blocks the contact system. Anaphylactic shock, endotoxin shock, and inflammation are processes in which the contact system has a significant role.

contrasuppression

A part of the immunoregulatory circuit that prevents suppressor effects in a feedback loop. This is a postulated mechanism to counteract the function of suppressor cells in a feedback-type mechanism. Proof of contrasuppressor and suppressor cell circuits awaits confirmation by molecular biologic techniques.

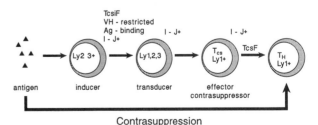

Contrasuppression

contrasuppressor cell

A T cell that opposes the action of a suppressor T lymphocyte.

control

A specimen of known content used together with an unknown specimen during an analysis in order that the two may be compared. A positive control known to contain the substance under analysis and a negative control known not to contain the substance under analysis are required.

control tolerance

The mechanism that involves the absence or functional inactivation of cells requisite for the initiation of an immune response. These cells are defective or inactivated. Control tolerance affects the afferent limb of the immune response, which is concerned with sensitization and cell proliferation.

convalescent serum

A patient's blood serum sample obtained 2 to 3 weeks following the beginning of a disease. The finding of an antibody titer to a pathogenic microorganism that is higher than the titer of a serum sample taken earlier in the disease is considered to signify infection produced by that particular microorganism. For example, a fourfold or greater elevation in antibody titer would represent presumptive evidence that a particular virus, for example, had induced the infection in question.

conventional (holoxenic) animals

Experimental animals exclusive of those that have been raised in a gnotobiotic or germ-free environment.

conventional mouse

A mouse maintained under ordinary living conditions and provided water and food on a regular basis.

Cooke, Robert Anderson (1880–1960)

American immunologist and allergist who was instrumental in the founding of several allergy societies. With Coca he classified allergies in humans. Cooke also pioneered skin test methods and desensitization techniques.

Coombs, Robin (1921–)

British pathologist and immunologist who is best known for the Coombs' test as a means for detecting autoimmune hemolytic anemia. He has also contributed much to serology, immunohematology, and immunopathology. *The Serology of Conglutination and Its Relation to Disease*, 1961; *Clinical Aspects of Immunology (with Gell)*, 1963.

Coombs' test

An antiglobulin assay that detects immunoglobulin on the surface of a patient's red blood cells. The test was developed in the 1940s by Robin Coombs to demonstrate autoantibodies on the surface of red blood cells that fail to cause agglutination of these red blood cells. In the direct Coombs' test, rabbit anti-human immunoglobulin is added to a suspension of patient's red cells, and if they are coated with autoantibody, agglutination results. In the indirect Coombs' test, the patient's serum can be used to coat erythroyctes, which are then washed and the antiimmunoglobulin reagent added to produce agglutination if the antibodies in question had been present in the serum sample. The Coombs' test has long been a part of an autoimmune disease evaluation of patients. Refer to antiglobulin test.

Coombs' Test

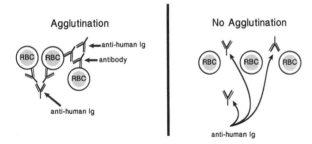

Coons, Albert Hewett (1912–1978)

American immunologist and bacteriologist who was an early leader in immunohistochemistry with the development of fluorescent antibodies. Coons received the Lasker medal in 1959, the Ehrlich prize in 1961, and the Behring prize in 1966.

cooperation

Refer to T lymphocyte-B lymphocyte cooperation.

cooperative determinant

Carrier determinant.

copolymer

A polymer such as a polypeptide comprised of at least two separate chemical specificities such as two different amino acids.

copper deficiency

Trace amounts of copper are requisite for the ontogeny and proper functioning of the immune system. Neonates or malnourished children with copper deficiency may have associated neutropenia and an increased incidence of infection. Antioxidant enzyme levels are diminished in copper deficiency. This may render immunologically competent cells unprotected against elevated oxygen metabolism associated with immune activation.

coprecipitation

The addition of an antibody specific for either the antigen portion or the antibody portion of immune complexes to effect their precipitation. Protein A may be added instead to precipitate soluble immune complexes. The procedure may be employed to quantify low concentrations of radiolabeled antigen that are combined with excess antibody. After soluble complexes have formed, antiimmunoglobulin or protein A is added to induce coprecipitation.

coproantibody

A gastrointestinal tract antibody, commonly of the IgA class, which is present in the intestinal lumen or feces.

cords of Billroth

Splenic medullary cords.

corneal response

In an animal that has been previously sensitized to an antigen, the cornea of the eye may become clouded (or develop opacities) after injection of the same antigen into it. There is edema and lymphocytic and macrophage infiltration into the area. The response has been suggested to represent cell-mediated immunity.

corneal test

Refer to corneal response.

corticosteroids

Lympholytic steroid hormones such as cortisone derived from the adrenal cortex. Glucocorticoids such as prednisone or dexamethasone can diminish the size and lymphocyte content of lymph nodes and spleen, while sparing proliferating myeloid or erythroid stem cells of the bone marrow. Glucocortoids may interfere with the activated lymphocyte's cell cycle. Glucocorticoids are cytotoxic for selected T lymphocyte subpopulations and are also able to suppress cell-mediated immunity and antibody synthesis, as well as the formation

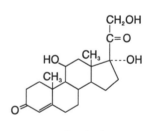

Cortisone   Corticosterone

Cortisol
(hydrocortisone)

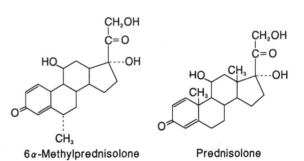

6α-Methylprednisolone   Prednisolone

Corticosteroid

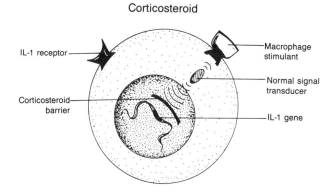

of prostaglandin and leukotrienes. Corticosteroids may lyse either suppressor or helper T lymphocytes, but plasma cells may be more resistant to their effects. However, precursor lymphoid cells are sensitive to the drug which may lead in this way to decreased antibody responsiveness. The repeated administration of prednisone diminishes the concentration of specific antibodies in the IgG class whose fractional catabolic rate is increased by prednisone. Corticosteroids interfere with the phagocytosis of antibody-coated cells by macrophages. Glucocorticoids have been widely administered for their immunosuppressive properties in autoimmune diseases such as autoimmune hemolytic anemia, systemic lupus erythematosus, Hashimoto's thyroiditis, idiopathic thrombocytopenic purpura, and inflammatory bowel disease. They are also used in the treatment of various allergic reactions and for bronchial asthma. They have been widely used in organ transplantation, especially prior to the introduction of cyclosporine and related drugs. They have been used to manage rejection crises without producing bone marrow toxicity. Long-term usage has adverse effects that include adrenal suppression.

corticotrophin receptor antibodies

Adrenal antibodies that have a role in the pathogenesis of Addison's disease. Corticotrophin receptor antibodies (CRA) may block ACTH binding to specific receptors on cells of the adrenal cortex. Corticotrophin receptor antibodies of the stimulatory type may be found in Cushing's syndrome attributable to primary pigmented nodular adrenocortical disease.

costimulatory molecules

Membrane bound or secreted products of accessory cells that activate signal transduction events in addition to those induced by MHC/TCR interactions. They are required for full activation of T

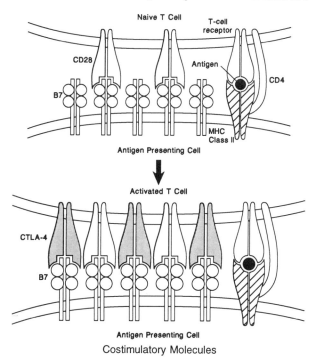

Costimulatory Molecules

cells, and it is thought that adjuvants may work by enhancing the expression of costimulator molecules by accessory cells. The interaction of CD28/CTLA-4 with B7 to induce full transcription of IL-2 mRNA is an example of costimulator mechanisms.

counter current electrophoresis

Refer to counter immunoelectrophoresis.

counter electrophoresis

Refer to counter immunoelectrophoresis.

counter immunoelectrophoresis (CIE)

An immunoassay in which antigen and antibody are placed into wells in agar gel and followed by electrophoresis in which the antigen, that carries a negative charge, migrates toward the antibody which moves toward the antigen by electroendosomis. Interaction of antigen and antibody molecules in the gel leads to the formation of a precipitin line. The method has been used to identify serotypes of *Streptococcus pneumoniae, Neisseria meningitidis* groups, and *Haemophilus influenzae* type b.

counter migration electrophoresis

Refer to counter immunoelectrophoresis.

cowpox

A bovine virus disease that induces vesicular lesions on the teats. It is of great historical significance in immunology because Edward Jenner observed that milkmaids who had cowpox lesions on their hands failed to develop smallpox. He used this principle in vaccinating humans with the cowpox preparation to produce harmless vesicular lesions at the site of inoculation (vaccination). This stimulated protective immunity against smallpox (variola) because of shared antigens between the vaccinia virus and the variola virus.

Coxsackie

A picornavirus family from Enteroviridiae. Coxsackie A viruses have 23 virotypes and Coxsackie B virus have 6 types. Clinical conditions produced by Coxsackie viruses include herpangina, epidemic pleurodynia, aseptic meningitis, summer grippe, and acute nonspecific pericarditis and myocarditis.

CR3 deficiency syndrome

Refer to leukocyte adhesion deficiency.

CREST complex

Calcinosis, Raynaud's phenomenon, esophageal dysmotility, sclerodactyly, and telangiectasia associated with mixed connective tissue disease. The prognosis of CREST is slightly better than that of other connective tissue diseases, but biliary cirrhosis and pulmonary hypertension are complications. CREST patients may develop anticentromere antibodies, which may also occur in progressive systemic sclerosis patients, in aged females, or in individuals with HLA-DR1. CREST represents a mild form of systemic sclerosis.

CREST syndrome

A relatively mild clinical form of scleroderma (progressive systemic sclerosis). CREST is an acronym for calcinosis, Raynaud's phenomenon, esophageal dysmotility, sclerodactyly, and telangiectasis. Skin lesions are usually limited to the face and fingers, with only later visceral manifestations. Most (80 to 90%) of CREST patients have anticentromere antibodies.

Creutzfeldt-Jakob syndrome

A slow virus infection of brain cells that reveal membrane accumulations.

Crithidia assay

The use of a hemoflagellate termed *Crithidia luciliae* to measure anti-dsDNA antibodies in the serum of systemic lupus erythematosus patients by immunofluorescence methods. The kinetoplast of this organism is an altered mitochondrion that is rich in double-stranded DNA.

Crithidia luciliae

A hemoflagellate possessing a large mitochondrion that contains concentrated mitochondrial DNA in a single large network called the kinetoplast. Used in immunofluorescence assays for the presence of anti-dsDNA antibodies in the blood sera of systemic lupus erythematosus patients.

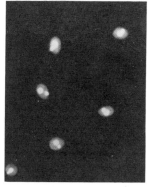

Crithidia luciliae

CRM 197

A carrier protein used in vaccines. It is a nontoxic mutant protein related to diphtheria toxin.

Crohn's disease

A condition usually expressed as ileocolitis, but it can affect any segment of the gastrointestinal tract. Crohn's disease is associated with transmural granulomatous inflammation of the bowel wall characterized by lymphocyte, plasma cell, and eosinophil infiltration. Goblet cells and gland architecture are not usually affected. Granuloma formation is classically seen in Crohn's disease, appearing in 70% of patients. The etiology is unknown. *Mycobacterium paratuberculosis* has been found in a few patients with Crohn's disease, although no causal relationship has been established. An immune effector mechanism is believed to be responsible for maintaining chronic disease in these patients. Their serum immunoglobulins and peripheral blood lymphocyte counts are usually normal except for a few diminished T cell counts in selected Crohn's disease patients. Helper/suppressor ratios are also normal. Active disease has been associated with reduced suppressor T cell activity, which returns to normal during remission. Patients have complexes in their blood that are relatively small and contain IgG, although no antigen has been identified. The complexes may be merely aggregates of IgG. Complexes in Crohn's disease patients are associated with involvement of the colon and are seen less often in those with the disease confined to the ileum. During active disease, serum concentrations of C3, factor B, C1 inhibitor, and C3b inactivator are elevated, but return to normal during remission. Patients with long-standing disease often develop high titers of immunoconglutinins which are antibodies to activated C3. High-titer antibodies against bacterial antigens such as those of *Escherichia coli* and *Bacteroides* crossreact with colonic goblet cell lipopolysaccharides. Patients' peripheral blood lymphocytes can kill colonic epithelial cells *in vitro*. Colonic mucosa lymphocytes in these patients are also cytotoxic for colonic epithelial cells.

cromolyn (1,3-*bis*[2-carboxychromon-5-yloxy-2-hydroxypropane])

A therapeutic agent that prevents mast cell degranulation. It has proven effective in the therapy of selected allergies that include allergic rhinitis and asthma.

cromolyn sodium

A drug that blocks the release of pharmacological mediators from mast cells and diminishes the symptoms and tissue reactions of type I hypersensitivity (i.e., anaphylaxis) mediated by IgE. Although cromolyn sodium's mechanism of action remains to be determined, it apparently inhibits the passage of calcium through the cell membrane. The drug inhibits mast cell degranulation, but has no adverse effect on the linkage of IgE to the mast cell surface or to its interaction with antigen. Cromolyn sodium is inhaled as a powder or applied topically to mucous membranes. This is usually the treatment for asthma, allergic rhinitis, and allergic conjunctivitis and has low cytotoxicity.

Cromolyn

HOOC OCH₂CHCH₂O COOH OH

crossabsorption

The use of cross-reacting antigens or cross-reacting antibodies to absorb antibodies or antigens, respectively.

crossed immunoelectrophoresis

A gel diffusion method employing two-dimensional immunoelectrophoresis. Protein antigens are separated by gel electrophoresis. This is followed by the insertion of a segment of the gel into a separate gel into which specific antibodies have been incorporated. The gel is then electrophoresed at right angles to the first electroporesis, forcing the antigen into the gel containing antibody. This results in the formation of precipitin arcs in the shape of a rocket that resembles bands formed in the Laurell rocket technique.

cross-matching procedures

The conventional cross-matching procedure for organ transplants involves the combination of donor lymphocytes with recipient serum. There are three major variables in the standard cross-match procedure that predominantly affect the reactivity of the cell/sera sensitization. These include (1) incubation time and temperature; (2) wash steps after cell/sera sensitization; and (3) the use of additional reagents, such as antiglobulin in the test. Variations in these steps can cause wide variations in results. Lymphocytes can be separated into T and B cell categories for cross-match procedures that are conducted at cold (4°C), room (25°C), and warm (37°C) temperatures. These permit the identification of warm anti-T cell antibodies that are almost always associated with graft rejection.

Crossmatch Procedure

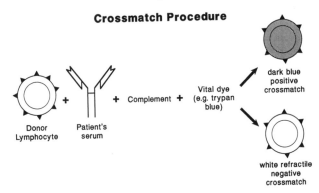

Donor Lymphocyte + Patient's serum + Complement + Vital dye (e.g. trypan blue) → dark blue positive crossmatch / white refractile negative crossmatch

cross-reacting antibody

An antibody that reacts with epitopes on an antigen molecule different from the one that stimulated its synthesis. The effect is attributable to shared epitopes on the two antigen molecules.

cross-reacting antigen

An antigen that interacts with an antibody synthesized following immunogenic challenge with a different antigen. Epitopes shared between these two antigens or epitopes with a similar stereochemical configuration may account for this type of crossreactivity. The presence of the same or of a related epitope between bacterial cells, red blood cells, or other types of cells may crossreact with an antibody produced against either of them.

cross-reaction

The reaction of an antigenic determinant with an antibody formed against another antigen. A laboratory technique used for matching blood for transfusion and organs for transplantation. In blood transfusion, donor erythrocytes are combined with the recipient's serum. If there is antibody in the recipient's serum that is specific for donor red cells, agglutination occurs. This represents the major part of the crossmatch. In a negative test, the presence of incomplete antibodies may be detected by washing the red blood cells and adding rabbit anti-human globulin. Agglutination signifies the presence of incomplete antibodies. The minor part of the crossmatch test consists of mixing recipient's red blood cells with donor serum. It is of less importance than the major crossmatch because of the limited amount of serum in a unit of blood compared to the red cell volume of the recipient.

cross-sensitivity

Induction of hypersensitivity to a substance by exposure to another substance containing cross-reacting antigens.

cross-tolerance

The induction of immunologic tolerance to an antigen by exposure of the host to a separate antigen containing cross-reacting epitopes under conditions that favor tolerance induction.

cryofibrinogenemia

Cryofibrinogen in the blood that is either primary or secondary to lymphoproliferative and autoimmune disorders, tumors, acute or chronic inflammation. Cryofibrinogenemia is often associated with IgA nephropathy.

cryoglobulin

A serum protein that precipitates or gels in the cold. It is an immunoglobulin that gels or precipitates when the temperature falls below 37°C. Cryoglobulins undergo reversible precipitation at cold temperatures. Most of them are complexes of immunoglobulin molecules, but nonimmunologic cryoprecipitate proteins such as cryofibrinogen or C-reactive protein-albumin complexes may also

Activated T Cell

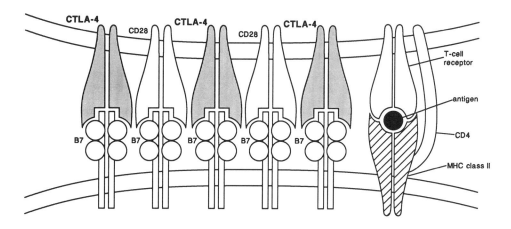

Antigen Presenting Cell

CTLA-4

occur. When the concentration of cryoglobulins is relatively low, precipitation occurs near 4°C, but if the concentration is greater, precipitation occurs at a higher temperature. Cryoglobulins are usually associated with infectious, inflammatory, and neoplastic processes. They are found in different body fluids and also appear in the urine. Cryoglobulins are divisible into three groups: type I are monoclonal immunoglobulins, usually IgM, associated with malignant B cell neoplasms; type II consists of mixed cryoglobulins with a monoclonal constituent specific for polyclonal IgG; and type III are mixed cryoglobulins comprised of polyclonal immunoglobulins as immunoglobulin-antiimmunoglobulin complexes. Cryoglobulin is not present in normal serum. Refer to cryoglobulinemia.

cryoglobulinemia
Cryoglobulins in the blood that are usually monoclonal immunoglobulins, i.e., IgG or IgM. Polymeric IgG3 may be associated with cryoglobulinemia in which the protein precipitates in those parts of the body exposed to cooling. Cryoglobulinemia patients develop embarrassed circulation following precipitation of the protein in peripheral blood vessels. This may lead to ulcers on the skin and to gangrene. More commonly, patients may manifest Raynaud's phenomenon following exposure of the hands or other parts of the anatomy to cold. Cryoglobulins may be detected in patients with Waldenström's macroglobulinemia, multiple myeloma, or systemic lupus erythematosus. Cryoglobulinemias are divisible into three types: type I monoclonal cryoglobulinemia, which is often associated with a malignant condition, i.e., IgG-multiple myeloma, IgM-macroglobulinemia, lymphoma or chronic lymphocytic leukemia, and benign monoclonal gammopathy; type II polymonoclonal cryoglobulinemia with mixed immunoglobulin complexes such as IgM-IgG, IgG-IgG, and IgA-IgG that may be linked to connective tissue disease such as rheumatoid arthritis or Sjögren's syndrome or with lymphoreticular disease; and type III mixed polyclonal-polyclonal cryoglobulinemia with IgG and IgM mixtures, rarely including IgA, associated with infections, lupus erythematosus, rheumatoid arthritis, Epstein-Barr and cytomegalovirus inclusion virus, Sjögren's syndrome, crescentic and membranoproliferative glomerulonephritis, subacute bacterial infections, biliary cirrhosis, diabetes mellitus, and chronic active hepatitis.

cryostat®
A microtome in a refrigerated cabinet used by pathologists to prepare frozen tissue sections for surgical pathologic diagnosis. Immunologists use this method of quick frozen thin sections for immunofluorescence staining by fluorochrome-labeled antibody to identify antigens, antibodies, or immune complexes in tissue sections such as renal biopsies.

cryptantigens
Surface antigens of red cells not normally detectable, but demonstrable by microbial enzyme action that leads to the modification of cell surface carbohydrates. Naturally occurring IgM antibodies in normal serum may agglutinate these exposed antigens.

cryptodeterminant
Refer to hidden determinant.

CSF
Abbreviation for (1) colony-stimulating factor, or (2) cerebrospinal fluid.

CSIF
Abbreviation for cytokine synthesis inhibitory factor. Refer to interleukin-10.

CTL
Abbreviation for cytotoxic T lymphocyte.

CTLA-4 (see above)
Molecule that is homologous to CD28 and expressed on activated T cells. The genes for CD28 and CTLA-4 are closely linked on chromosome 2. The binding of CTLA-4 to its ligand B7 is an important costimulatory mechanism (refer to CD28 and costimulatory molecules).

CTLA4-Ig
A soluble protein composed of the CD28 homolog CTLA and the constant region of an IgG1 molecule. It is used experimentally to inhibit the immune response by blocking CD28-B7 interaction.

Cu-18
A glycoprotein of breast epithelium. Immunoperoxidase staining identifies this marker in most breast tumors and a few tumors of the ovary and lung. Stomach, pancreas, and colon tumors do not express this antigen.

Cunningham plaque technique
A modification of the hemolytic plaque assay in which an erythrocyte monolayer between a glass slide and cover slip is used without agar for the procedure.

cutaneous anaphylaxis
A local reaction specifically elicited in the skin of an actively or passively sensitized animal. Causes of cutaneous anaphylaxis include immediate wheal and flare response following prick tests with drugs or other substances; insect stings or bites; and contact urticaria in response to food substances such as nuts, fish, or eggs or other substances such as rubber, dander, or other environmental agents. The signs and symptoms of anaphylaxis are associated with the release of chemical mediators that include histamine and other substances from mast cell or basophil granules following crosslinking of surface IgE by antigen or nonimmunological degranulation of these cells. The pharmacological mediators act principally on the blood vessels and smooth muscle. The skin may be the site where an anaphylytic reaction is induced or it can be the target of a systemic anaphylactic reaction resulting in itching (pruritus), urticaria, and angiodema.

cutaneous basophil hypersensitivity (Jones-Mote hypersensitivity)
A type of delayed (type IV) hypersensitivity in which there is prominent basophil infiltration of the skin immediately beneath the epidermis. It can be induced by the intradermal injection of a

soluble antigen such as ovalbumin incorporated into Freund's incomplete adjuvant. Swelling of the skin reaches a maximum within 24 h. The hypersensitivity reaction is maximal between 7 and 10 days following induction and vanishes when antibody is formed. Histologically, basophils predominate, but lymphocytes and mononuclear cells are also present. Jones-Mote hypersensitivity is greatly influenced by lymphocytes that are sensitive to cyclophosphamide (suppressor lymphocytes).

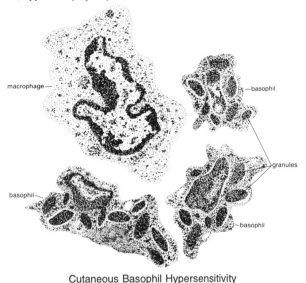

Cutaneous Basophil Hypersensitivity

cutaneous lymphocyte antigen
HECA-452 epitope expressed on a skin associated subset of memory T cells that are active in recirculation and homing to skin sites.

cutaneous sensitization
Application of antigen to the skin to induce hypersensitivity.

cyanogen bromide
A chemical that specifically breaks methionyl bonds. Approximately one half of the methionine residues in an IgG molecule, e.g., those in the Fc region, are cleaved by treatment with cyanogen bromide.   **Br-CN**

cycle-specific drugs
Immunosuppressive and cytotoxic drugs that lead to death of mitotic and resting cells.

cyclic adenosine monophosphate (cAMP)
Adenosine 3',5'-(hydrogen phosphate). A critical regulator within cells. It is produced through the action of adenylate cyclase on adenosine triphosphate. It activates protein kinase C. It serves as a "second messenger" when hormones activate cells. Elevated cAMP concentrations in mast cells diminish their response to degranulation signals.

cyclic guanosine monophosphate (cGMP)
Guanosine cyclic 3',5'-(hydrogen phosphate). A cAMP antagonist produced by the action of guanylatecyclase on guanosine triphosphate. Elevated cGMP concentrations in mast cells accentuate their response to degranulation signals.

cyclooxygenase pathway
The method whereby prostaglandins are produced by enzymatic metabolism of arachidonic acid derived from the cell membrane, as in type I (anaphylactic) hypersensitivity reactions.

cyclophilin
An 18-kD protein in the cytoplasm that has peptidyl-prolyl isomerase functions. It has a unique and conserved amino acid sequence that has a broad phylogenetic distribution. It represents a protein kinase with a postulated critical role in cellular activation. It serves as a catalyst in *cis-trans-rotamer* interconversion. It catalyzes phosphorylation of a substrate which then serves as a cytoplasmic messenger associated with gene activation. Genes coding for the synthesis of lymphokines would be activated in helper T lymphocyte responsiveness. Cyclophilin has a high affinity for cyclosporine (CSA), which accounts for the drug's immunosuppressive action. Inhibition of cyclophilin-mediated activities as a consequence of CSA-cyclophilin interaction could lead to inhibition of the synthesis and release of lymphokines. CSA not only inhibits primary immunization, but it may halt an ongoing immune response. This has been postulated to occur through inhibition of continued lymphokine release and by suppression of continued effector cell activation and recruitment.

cyclophosphamide
(*N,N-bis*-[2-choroethyl]-tetrahydro-2H-1,3,2-oxazaphosphorine-2-amine-2-oxide). A powerful immunosuppressive drug that is more toxic for B than for T lymphocytes. Therefore, it is a more effective suppressor of humoral antibody synthesis than of cell-mediated immune reactions. It is administered orally or intravenously and mediates its cytotoxic activity by crosslinking DNA strands. This alkylating action mediates target cell death. It produces dose-related lymphopenia and inhibits lymphocyte proliferation *in vitro*. The greater effect on B than on T cells is apparently related to B cells' lower rate of recovery. Cyclosphosphamide is beneficial for therapy of various immune disorders including rheumatoid arthritis, systemic lupus erythematosus-associated renal disease, Wegener's granulomatosis and other vasculitides, autoimmune hematologic disorders including idiopathic thrombocytopenic purpura, pure red cell aplasia, and autoimmune hemolytic anemia. It is used also to treat Goodpasture's syndrome and various glomerulonephritides. Its beneficial use as an immunosuppressive agent is tempered by the finding of its significant toxicity, such as its association with hemorrhagic cystitis, suppression of hematopoiesis, gastrointestinal symptoms, etc. It also may increase the chance of opportunistic infections and be associated with an increased incidence of malignancies such as non-Hodgkin's lymphoma, bladder carcinoma, and acute myelogenous leukemia.

Cyclophosphamide

cyclosporine (cyclosporin A) (ciclosporin)
A cyclic endecapeptide of 11 amino acid residues isolated from soil fungi which has revolutionized organ transplantation. Rather than acting as a cytotoxic agent, which defines the activity of a number of currently available immmunosuppressive drugs, cyclosporine (CSA) produces an immunomodulatory effect principally on the helper/inducer (CD4) lymphocytes which orchestrate the generation of an immune response. A cyclic polypeptide, CSA blocks T cell help for both humoral and cellular immunity. A primary mechanism of action is its ability to suppress interleukin-2 (IL-2) synthesis. It fails to block activation of antigen-specific suppressor T cells, thereby assisting development of antigen-specific tolerance. Side effects include nephrotoxicity and hepatotoxicity with a possible increase in B cell lymphomas. Some individuals may also develop hypertension. CSA's mechanism of action appears to include inhibition of the synthesis and release of lymphokines and alteration of expression of MHC gene products on the cell surface. CSA inhibits IL-2 mRNA formation. This does not affect IL-2 receptor expression on the cell surface. Although CSA may diminish the number of low-affinity binding sites, it does not appear to alter high-affinity binding sites on the cell surface. CSA inhibits the early increase in cytosolic-free calcium, which occurs in beginning activation of normal T lymphocytes. It appears to produce its effect in the cytoplasm rather than on the cell surface of a lymphocyte. It could reach the cytoplasmic location because of its ability to dissolve in the plasma membrane lipid bilayer. CSA's cytosolic site of action may involve calmodulin and/or cyclophilin, a protein kinase. Although immunosuppressive action cannot be explained based upon CSA-calmodulin interaction, this association closely parallels the immunosuppressive effect. CSA produces a greater suppressive effect upon class II than upon class I antigen expression in at least some experiments. While decreasing T helper lymphocytes, the T suppressor cells appear to be spared following CSA therapy. Not only sparing, but amplification of T lymphocyte suppression has been reported during CSA therapy. This is a powerful immunosuppressant that selectively affects CD4⁺ helper T cells without altering the activity of suppressor T cells, B cells, granulocytes, and macrophages. It alters lymphocyte function,

Cyclosporine

but it does not destroy the cells. Its principal immunosuppressive action is to inhibit IL-2 production and secretion. Thus, the suppression of IL-2 impairs the development of suppressor and cytotoxic T lymphocytes that are antigen specific. It has a synergistic immunosuppressive action with corticosteroids. Corticosteroids interfere with IL-2 synthesis by inhibiting IL-1 release from monocytes and macrophages. Cyclosporine, although water insoluble, has been successfully employed as a clinical immunosuppressive agent principally in preventing rejection of organ and tissue allotransplants including kidney, heart, lung, pancreas, and bone marrow. It has also been succcessful in preventing graft-vs.-host reactions. The drug has some nephrotoxic properties, which may be kept to a minimum by dose reduction. As with other long-term immunosuppressive agents, there may be an increased risk of lymphoma development such as Epstein-Barr (EBV) associated B cell lymphomas.

cytochalasins

Metabolites of various species of fungi that affect microfilaments. They bind to one end of actin filaments and block their polymerization. Thus, they paralyze locomotion, phagocytosis, capping, cytokinesis, etc.

cytochrome b deficiency

Refer to chronic granulomatous disease.

cytokine receptor families

A classification system of cytokine receptors according to conserved sequences or folding motifs. Type I receptors share a tryptophan-serine-X-tryptophan-serine, or WSXWS sequence, on the proximal extracellular domains. Type I receptors recognize cytokines with a structure of four α-helical strands, including IL-2 and G-CSF. Type II receptors are defined by the sequence pattern of Type I and Type II interferon receptors. Type III receptors are those found as receptors for TNF (p55 and p75). CD40, nerve growth factor receptor, and Fas protein have sequences homologous to those of Type III receptors. A fourth family of receptors has extracellular domains of the Ig superfamily. IL-1 receptors as well as some growth factors and colony stimulating factors have Ig domains. The fifth family of receptors displays a seven-transmembrane α-helical structure. This motif is shared by many of the receptors linked to GTP-binding proteins.

cytokine synthesis inhibitory factor

Refer to interleukin-10.

cytokines

Immune system proteins that are biological response modifiers. They coordinate antibody and T cell immune system interactions and amplify immune reactivity. Cytokines include monokines synthesized by macrophages and lymphokines produced by activated T lymphocytes and natural killer cells. Monokines include interleukin-1, tumor necrosis factor, α and β interferons, and colony-stimulating factors. Lymphokines include interleukins, gamma interferon, granulocyte-macrophage colony-stimulating factor, and lymphotoxin. Endothelial cells and fibroblasts and selected other cell types may also synthesize cytokines.

cytolysin

A substance such as perforin that lyses cells.

cytolytic

An adjective describing the property of disrupting a cell.

cytolytic reaction

Cell destruction produced by antibody and complement or perforin released from cytotoxic T lymphocytes.

cytomegalovirus (CMV)

A herpes (DNA) virus group that is distributed worldwide and is not often a problem, except in individuals who are immunocompromised, such as the recipients of organ or bone marrow transplants or individuals with acquired immunodeficiency syndrome (AIDS). Histopathologically, typical inclusion bodies that resemble an owl's eye are found in multiple tissues. CMV is transmitted in the blood.

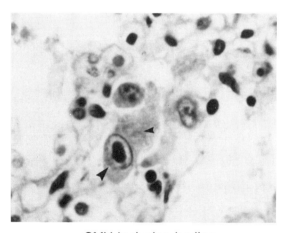

CMV inclusion bodies

▲ intranuclear and
▲ intracytoplasmic

Cytomegalovirus

cytophilic antibody

(1) An antibody that attaches to a cell surface through its Fc region. It binds to Fc receptors on the cell surface. For example, IgE molecules bind to the surface of mast cells and basophils in this manner. Murine IgG1, IgG2a, and IgG3 bind to mononuclear phagocytic cell surface Fc receptors through their Fc regions. IgG1 and IgG3 may also attach through their Fc regions to mononuclear phagocytic cell Fc receptors in humans. Immunoglobulin molecules that bind to macrophage surfaces through their Fc regions represent a type of cytophilic antibody. (2) Described in the 1960s

as a globulin fraction of serum which is adsorbed to certain cells *in vitro* in a manner that allows them to specifically adsorb antigen. Sorkin, in 1963, suggested the possible significance of cytophilic antibody in anaphylaxis and other immunologic and/or hypersensitivity reactions.

cytosine arabinoside

An antitumor substance that is inactive by itself, but following intracellular conversion to the nucleoside triphosphate acts as a competitive inhibitor with regard to dCPP of DNA polymerase. It has an immunosuppressive effect on antibody formation in both the primary and secondary immune responses and also depresses the generation of cell-mediated immunity.

cytoskeleton

A framework of cytoskeletal filaments present in the cell cytoplasm. They maintain the cell's internal arrangement, shape, and motility. This framework interacts with the membrane of the cell and with organelles in the cytoplasm. Microtubules, microfilaments, and intermediate filaments constitute the varieties of cytoskeletal filaments. Microtubules help to determine cell shape by polymerizing and depolymerizing. They are 24-nm diameter hollow tubes whose walls are comprised of protofilaments that contain α and β tubulin dimers. The 7.5-nm diameter microfilaments are actin polymers. In addition to their interaction with myosin filaments in muscle contraction, actin filaments may affect movement or cell shape through polymerization and depolymerization. Microfilaments participate in cytoplasmic streaming, ruffling of membranes, and phagocytosis. They may be responsible for limiting protein mobility in the cell membrane. The proteins of the 10-nm intermediate filaments differ according to the cells in which they occur. Vimentin intermediate filaments occur in macrophages, lymphocytes, and endothelial cells, whereas desmin occurs in muscle and epithelial cells containing keratin.

cytoskeletal antibodies

Antibodies that are specific for cytoskeletal proteins that include cytokeratins, desmin, actin, titin, vimentin and tropomyosin. They have been demonstrated in some patients with various diseases that include autoimmune diseases, chronic active hepatitis and other liver disease, infection, myasthenia gravis and Crohn's disease. These antibodies are not helpful in diagnosis.

cytotoxic

An adjective describing an injurious effect on the cell.

cytotoxic antibody

Antibody that combines with cell surface epitopes followed by complement fixation that leads to cell lysis or cell membrane injury without lysis.

cytotoxic drugs

Agents that kill self-replicating cells such as immunocompetent lymphocytes. Cytotoxic drugs have been used for anticancer therapy as well as for immunosuppression in the treatment of transplant rejection and aberrant immune responses. The four cytotoxic drugs commonly used for immunosuppression include cyclosphosphamide, chlorambucil, azathioprine, and methotrexate.

cytotoxic T lymphocyte (CTL)

Specifically sensitized T lymphocytes that are usually CD8[+] and recognize antigens, through the T cell receptor, on cells of the host infected by viruses or that have become neoplastic. CD8[+] cell recognition of the target is in the context of MHC class I histocompatibility molecules. Following recognition and binding, death of the target cell occurs a few hours later. CTLs secrete lymphokines that attract other lymphocytes to the area, releasing serine proteases and perforins that produce ion channels in the membrane of the target leading to cell lysis. Interleukin-2, produced by CD4[+] T cells, activates cytotoxic T cell precursors. Interferon γ generated from CTLs activates macrophages. CTLs have a significant role in the rejection of allografts and in tumor immunity. A minor population of CD4[+] lymphocytes may also be cytotoxic, but they recognize target cell antigens in the context of MHC class II molecules.

CTL-mediated Target Cell Lysis

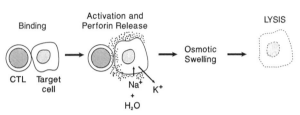

Cytotoxic T Lymphocyte (CTL)

cytotoxic T lymphocyte precursor (CTLp)

A progenitor that develops into a cytotoxic T lymphocyte after it has reacted with antigen and inducer T cells.

cytotoxicity

The fatal injury of target cells by either specific antibody and complement or specifically sensitized cytotoxic T cells, activated macrophages, or natural killer (NK) cells. Dye exclusion tests are used to assay cytotoxicity produced by specific antibody and complement. Measurement of the release of radiolabel or other cellular constituents in the supernatant of the reacting medium is used to determine effector cell-mediated cytotoxicity.

cytotoxicity tests

(1) Assays for the ability of specific antibody and complement to interrupt the integrity of a cell membrane, which permits a dye to enter and stain the cell. The relative proportion of cells stained, representing dead cells, is the basis for dye exclusion tests. Refer to microlymphocytotoxicity. (2) The ability of specifically sensitized T lymphocytes to kill target cells whose surface epitopes are the targets of their receptors. Loss of the structural integrity of the cell membrane is signified by the release of a radioisotope such as ^{51}Cr, which was taken up by the target cells prior to the test. The amount of isotope released into the supernatant reflects the extent of cellular injury mediated by the effector T lymphocytes.

cytotrophic antibodies

IgE and IgG antibodies that sensitize cells by binding to Fc receptors on their surface, thereby sensitizing them for anaphylaxis. When the appropriate allergen crosslinks the Fab regions of the molecules, it leads to the degranulation of mast cells and basophils bearing IgE on their surface.

cytotropic anaphylaxis

Form of anaphylaxis caused by antigen binding to reaginic IgE antibodies. The latter are cytotropic, that is, they bind to cells. Cytotropic antibodies (IgE) bind to specific receptors on the mast cell surface. The receptors are in close proximity to a serine esterase enzyme causing the release of mast cell granules and to a natural inhibitor of this enzyme. As long as the surface IgE has not bound antigen, the status quo of the cell is maintained. It is believed that binding of the antigen induces a conformational alteration in the IgE with displacement of the inhibitor from its steric relationship to the enzyme. The inhibition-free enzyme mediates the release. The process requires energy and is Ca^{2+} dependent.

D

D-amino acid polymers

Synthetic peptides (and polypeptides) that are antigenic. They are found very infrequently in living organisms.

δ chain

The heavy chain of immunoglobulin D (IgD).

D exon

A DNA sequence that encodes a portion of the immunoglobulin heavy chain's third hypervariable region. It is situated on the 5′ side of J exons. An intron lies between them. During lymphocyte differentiation, V-D-J sequences are produced that encode the complete variable region of the heavy chain.

D gene region

Diversity region of the genome that encodes heavy chain sequences in the immunoglobulin H chain hypervariable region.

D gene segment

The DNA region that codes for the D or diversity portion of an immunoglobulin heavy chain or a T lymphocyte receptor β or δ chain. It is the segment that encodes the third hypervariable region situated between the chain regions which the V gene segment and J gene segment encode. This part of the heavy chain variable region is frequently significant in determining antibody specificity.

D region

A segment of an immunoglobulin heavy chain variable region or the β or δ chain of the T lymphocyte receptor coded for by a D gene segment. A few residues constitute the D region in the third hypervariable region in most heavy chains of immunoglobulins. The D or diversity region governs antibody specificity and probably T cell receptor specificity as well.

DAF

Refer to decay-accelerating factor.

Dale, Henry Hallett (1875–1968)

British investigator who made a wide range of scientific contributions including work on the chemistry of nerve impulse transmissions, the discovery of histamine, and the development of the Schultz-Dale test for anaphylaxis. He received a Nobel Prize in 1935.

Dalen-Fuchs nodule

A hemispherical granulomatous nodule composed of epithelioid cells and retinal epithelial cells in the choroid of the eye in sympathetic ophthalmia patients and in some other diseases.

Dameshek, William (1900–1969)

Noted Russian–American hematologist who was among the first to understand autoimmune hemolytic anemias. He spent many years as Editor-in-Chief of the journal *Blood*.

dander antigen

A combination of debris such as desquamated epithelial cells, microorganisms, hair, and other materials trapped in perspiration and sebum that are constantly deleted from the skin. Dander antigens may induce immediate, IgE-mediated, type I hypersensitivity reactions in atopic individuals.

DANE particle

A 42-nm structure identified by electron microscopy in hepatitis B patients in the acute infective stage. The DANE particle has a 27-nm diameter icosahedral core that contains DNA polymerase.

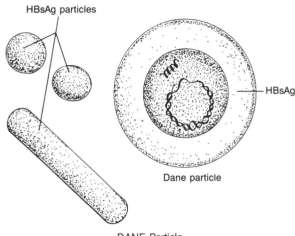

DANE Particle

Danysz, Jan (1860–1928)

Polish investigator who worked at the Pasteur Institute, Paris. The Danysz neutralization phenomenon was named for him. He had wide-ranging interests including viruses that are pathogenic specifically for rodents. He subsequently investigated chemotherapeutic agents.

Danysz phenomenon (Danysz effect)

The addition of toxin to the homologous antitoxin in several fractions with appropriate time intervals between them, resulting in a greater toxicity of the mixture than would occur if all the sample of toxin were added at once. Therefore, a greater amount of antitoxin is required for neutralization if the toxin is added in divided doses than if all toxin is added at one time, or less toxin is required to neutralize the given quantity of antitoxin if all toxin is added at one time than if it is added in divided doses with time intervals between. This form of reaction has been called the Danysz phenomenon or Danysz effect. Neutralization in the above instances is tested by injection of the toxin-antitoxin mixture into experimental animals. This phenomenon is attributed to the combination of toxin and antitoxin in multiple proportions. The addition of one fraction of toxin to excess antitoxin leads to maximal binding of antitoxin by toxin molecules. When a second fraction of toxin is added, insufficient antitoxin is available to bring about neutralization. Therefore, the mixture is toxic due to uncombined excess toxin. Equilibrium is reached after an appropriate time interval. The interaction between toxin and antitoxin is considered to occur in two steps: (1) rapid combination of toxin and antitoxin and (2) slower aggregation of the molecules.

dapsone

Diaminodiphenyl sulfone. A sulfa drug that has been used in the treatment of leprosy. It has also shown efficacy for prophylaxis of malaria and for therapy of dermatitis herpetiformis.

DAT

Abbreviation for direct antiglobulin test. Refer to direct Coombs' test.

Structure of heavy chain gene from IgM-producing cell

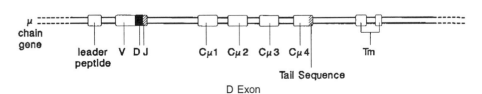

D Exon

Dausset, Jean Baptiste Gabriel (1916–)

French physician and investigator. He pioneered research on the HLA system and the immunogenetics of histocompatibility. For this work he shared a Nobel prize with Benacerraf and Snell in 1980. *Immunohematologie, Biologique et Clinique*, 1956; *HLA and Disease (with Svejaard)*, 1977.

ddC (dideoxycytidine)

An inhibitor of reverse transcriptase used in AIDS treatment. It resembles ddI.

ddI (2′,3′-dideoxyinosine)

A purine analog that blocks HIV-1 *in vivo*. It is transformed into a triphosphorylated substance, ddATP, which blocks HIV reverse transcriptase and suppresses the replication of HIV by inhibiting viral DNA synthesis. Administration of ddI may be followed by an elevation in the $CD4^+$ T helper cells and a significant decrease in p24 antigen, an indicator of HIV activity in the blood. AIDS patients tolerate ddI better than they do zidovudine.

DDS syndrome

A hypersensitivity reaction that occurs in 1 in 5000 leprosy patients who have been treated with dapsone (DDS, 4,4′-diaminodiphenyl sulfone), a drug that prevents folate synthesis by inhibiting the *p*-aminobenzoic acid condensation reaction. Patients develop hemolysis, agranulocytosis, and hypoalbuminemia, as well as exfoliative dermatitis and life-threatening hepatitis.

dead vaccine

Refer to inactivated vaccine.

Dean and Webb titration

Historically important assay for the measurement of antibody. While the quantity of antiserum is held constant, varying dilutions of antigen are added, and the tube contents are mixed. The tube in which flocculation occurs first represents the endpoint. It is in this tube that the ratio of antigen to antibody is in optimal proportions.

decay-accelerating factor (DAF)

A 70-kD membrane glycoprotein of normal human erythrocytes, leukocytes, and platelets, but it is absent from the red blood cells of paroxysmal nocturnal hemaglobulinuria patients. It facilitates dissociation of classical complement pathway C3 convertase (C4b2a) into C4b and C2a. It also promotes the association of alternative complement pathway C3 convertase (C3bBb) into C3b and Bb. DAF is found on selected mucosal epithelial cells and endothelial cells. It prevents complement cascade amplification on the surfaces of cells to protect them from injury by autologous complement. DAF's physiologic function may be to protect cells from lysis by serum. DAF competes with C2 for linkage with C4b to block C3 convertase synthesis in the classical pathway. The DAF molecule consists of a single chain bound to the cell membrane by phosphatidyl inositol. Paroxysmal nocturnal hemoglobulinuria develops as a consequence of DAF deficiency.

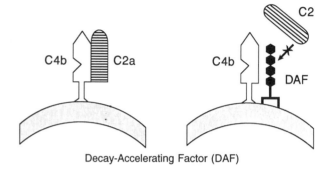

Decay-Accelerating Factor (DAF)

decomplementation

Deliberate inactivation of complement either *in vitro* or *in vivo*. To decomplement serum to remove hemolytic action, the specimen may be heated to 56°C for 30 min. Other methods for inactivation of complement include the addition of cobra venom factor, zymosan, or other substances which take up complement from the medium in which they are placed. Removal of complement activity in living animals may be accomplished through the injection of cobra venom factor or other substances to use up or inactivate the complement system.

decorate

Term used by immunologists to describe the reaction of tissue antigens with monoclonal antibodies, described as "staining", in the immunoperoxidase reaction. Thus, a tissue antigen stained with a particular antibody is said to be decorated with that monoclonal antibody. Immunoperoxidase techniques give a reddish-brown color to the reaction product that is read by light microscopic observation.

defensins

Widely reactive antimicrobial cationic proteins present in polymorphonuclear neutrophilic leukocyte granules. They block cell transport activities and are lethal for Gram-positive and Gram-negative microorganisms.

deficiency of secondary granules

A rare disorder in which neutrophils are bereft of secondary granules. This has an autosomal recessive mode of inheritance. Affected individuals show an increased incidence of infection by pyogenic microorganisms.

degranulation

A mechanism whereby cytoplasmic granules in cells fuse with the cell membrane to discharge the contents from the cell. A classic example is degranulation of the mast cell or basophil in immediate (type I) hypersensitivity. In phagocytic cells, cytoplasmic granules combine with phagosomes and release their contents into the phagolysosome formed by their union.

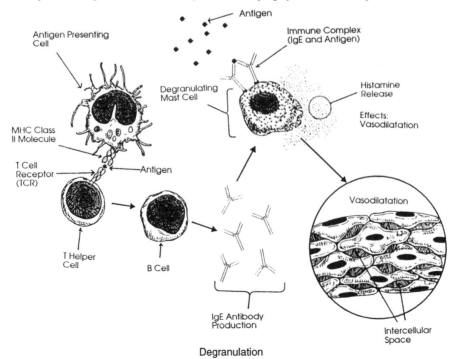

Degranulation

delayed-type hypersensitivity (DTH)

Cell-mediated immunity, or hypersensitivity mediated by sensitized T lymphocytes. Although originally described as a skin reaction that requires 24 to 48 h to develop following challenge with antigen, current usage emphasizes the mechanism, which is T cell mediated, as opposed to emphasis on the temporal relationship of antigen injection and host response. The CD4+ T lymphocyte is the principal cell that mediates delayed-type hypersensitivity reactions. To induce a DTH reaction, antigen is injected intradermally in a primed individual. If the reaction is positive, an area of erythema and induration develops 24 to 48 h following antigen challenge. Edema and infiltration by lymphocytes and macrophages occurs at the local site. The CD4+ T lymphocytes identify antigen on Ia positive macrophages and release lymphokines which entice more macrophages to enter the area where they become activated.

Skin tests are used clinically to reveal delayed-type hypersensitivity to infectious disease agents. Skin test antigens include such substances as tuberculin, histoplasmin, candidin, etc. Tuberculin or purified protein derivative (PPD), which are extracts of the tubercle bacillus, have long been used to determine whether or not a patient has had previous contact with the organism from which the test antigen was derived. Delayed-type hypersensitivity reactions are always cell mediated. Thus, they have a mechanism strikingly different from anaphylaxis or the Arthus reaction, which occur within minutes to hours following exposure of the host to antigen and are examples of antibody-mediated reactions. DTH is classified as type IV hypersensitivity (Coombs and Gell classification).

delta agent [hepatitis D virus (HDV)]

A viral etiologic agent of hepatitis that is a circular single-stranded incomplete RNA virus without an envelope. It is a 1.7-kilobase virus and consists of a small, highly conserved domain and a larger domain manifesting epitopes. HDV is a subviral satellite of the hepatitis B virus (HBV), on which it depends to fit its genome into virions. Thus, the patient must first be infected with HBV to have HDV. Individuals with the delta agent in their blood are positive for HBsAg, anti-HBC, and often HBe. This agent is frequently present in IV drug abusers and may appear in AIDS patients and hemophiliacs.

denaturation

Changing a protein's secondary and tertiary structure (coiling and folding) to produce a configuration that is either uncoiled or coiled more randomly. Whereas storage causes slow denaturation, heating or chemical treatment may induce more rapid denaturation of native protein molecules. Denaturation diminishes protein solubility and often abrogates the molecule's biologic activity. New or previously unexposed epitopes may be revealed as a consequence of denaturation.

dendritic cells

Mononuclear phagocytic cells found in the skin as Langerhans cells, in the lymph nodes as interdigitating cells, in the paracortex as veiled cells in the marginal sinus of afferent lymphatics, and as mononuclear phagocytes in the spleen where they present antigen to T lymphocytes. Dendritic reticular cells may have nonspecific esterase, Birbeck granules, endogenous peroxidase, possibly CD1, complement receptors CR1 and CR3, and Fc receptors.

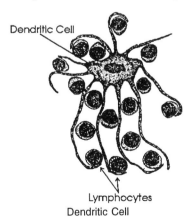

Dendritic Cell

Lymphocytes
Dendritic Cell

dendritic epidermal cell

Mouse epidermal cells that are Thy-1+, MHC class II molecule negative, and possess γδ T cell receptor associated with CD3. It is

believed to be a variety of bone marrow-derived T lymphocyte that is separate from Langerhans cells in the skin.

Dengue

An infection produced by the group B arbovirus, flavivirus, which the mosquito (*Stegomyia Aedes aegypti*) transmits. Dengue fever that occurs in the tropical regions of Africa and America may either be benign or produce malignant Dengue hemorrhagic shock syndrome, in which patients experience severe bone pain (break bone fever). They have myalgia, biphasic fever, headache, lymphadenopathy, and a morbilliform maculopapular rash on the trunk. They also manifest thrombocytopenia and lymphocytopenia.

dense-deposit disease

Type II membranoproliferative glomerulonephritis characterized by the deposition of electron-dense material, often containing C3, in the peripheral capillary basement membrane of the glomerulus. C3 is decreased in the serum as a consequence of alternate complement pathway activation. C4 is normal. There is an increase in sialic acid-rich glomerular basement membrane glycoproteins. Patients may possess a serum factor termed nephritic factor that activates the alternate complement pathway. This factor is an immunoglobulin molecule that reacts with alternate complement pathway-activated components such as the bimolecular C3b and activated factor B complex. Nephritic factor stabilizes alternate pathway C3 convertase.

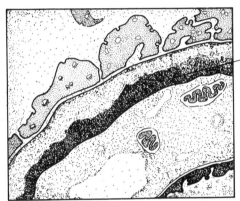

Dense deposits in glomerular capillary basement membrane

Dense-Deposit Disease

density gradient centrifugation

The centrifugation of relatively large molecules, such as in a solution of DNA, with a density gradient substance such as cesium chloride. This method also permits the separation of different types of cells as they are centrifuged through a density gradient produced by a substance to which they are impermeable. A commonly used material is Ficoll-Hypaque. Separation of cells is according to size as they progress through the gradient. When they reach the level where their specific gravity is the same as that of the medium, cell bands of different density are produced. This technique is widely employed to separate hematopoietic cells.

deoxyribonuclease

An endonuclease that catalyzes DNA hydrolysis.

deoxyribonuclease I

An enzyme that catalyzes DNA hydrolysis to a mono- and oligonucleotides mixture comprised of fragments terminating in a 5'-phosphorphyl nucleotide.

deoxyribonuclease II

An enzyme that catalyzes DNA hydrolysis to amono- and oligonucleotides mixture comprised of fragments terminating in a 3'-phosphoryl nucleotide.

deoxyribonucleoprotein antibodies

Antibodies reactive with insoluble deoxyribonucleoprotein (DNP) that occur in 60 to 70% of patients with active systemic lupus erythematosus. IgG DNP antibodies, that fix complement, cause the LE cell phenomenon and yield a homogeneous staining pattern.

depot-forming adjuvants

Substances that facilitate an immune response by holding an antigen at the injection site following inoculation. They facilitate the slow release of antigen over an extended period and help attract macrophages to the site of antigen deposition. To be effective, they must be administered together with the antigen. Water-in-oil emulsion of the Freund type, as well as aluminum salt (aluminum hydroxide) adjuvants, are examples of depot-forming adjuvants. In the past, depot-forming and centrally acting adjuvants were distinguished. However, adjuvant action depends upon far more complicated cellular and molecular mechanisms than the simplistic views of depot formation advanced in the past.

dermatitis herpetiformis (DH)

A skin disease with grouped vesicles and urticaria that is related to celiac disease. Dietary gluten exacerbates the condition and should be avoided to help control it. Most patients (70%) manifest no bowel disease symptoms. IgA and C3 granular immune deposits occur along dermal papillae at the dermal-epidermal junction. Groups of papules, plaques, or vesicles appear in a symmetrical distribution on knees, elbows, buttocks, posterior scalp, neck, and superior back region. The disease is chronic, unless gluten is deleted from the diet. Neutrophils (PMNs) and fibrin collect at dermal papillae tips, producing microabscesses. Microscopic blisters, which ultimately develop into subepidermal blisters, may develop at the tips of these papillae. These lesions must be distinguished from those of bullous pemphigoid. There is an association of DH with HLA-B8, HLA-DR3 and HLA-B44, and HLA-DR7 haplotypes.

dermatitis venenata

Refer to contact dermatitis.

dermatographism

Wheal and flare reaction of the immediate hypersensitivity type induced by scratching the skin. Thus, minor physical trauma induces degranulation of mast cells with the release of the pharmacological mediators of immediate hypersensitivity through physical stimulation. It is an example of an anaphylactoid reaction.

dermatomyositis

A connective tissue or collagen disease characterized by skin rash and muscle inflammation. It is a type of polymyositis presenting with a purple-tinged skin rash that is prominent on the superior eyelids, extensor joint surfaces, and base of the neck. There is weakness, muscle pain, edema, and calcium deposits in the subcutaneous tissue, especially prominent late in the disease. Blood vessels reveal lymphocyte cuffing, and autoantibodies against tRNA synthetases appear in the serum.

dermatopathic lymphadenitis

Benign lymph node hyperplasia that follows skin inflammation or infection.

Dermatophagoides

The house mite genus. Constituents of house mites represent the main allergen in house dust.

Dermatophagoides pteronyssinus

A house dust mite whose antigens may be responsible for house dust allergy in atopic individuals. It is a common cause of asthma.

dermatophytid reaction

Refer to id reaction.

desensitization

A method of treatment used by allergists to diminish the effects of IgE-mediated type I hypersensitivity. The allergen to which an individual has been sensitized is repeatedly injected in a form that favors the generation of IgG (blocking) antibodies rather than IgE antibodies that mediate type I hypersensitivity in humans. This

method has been used for many years to diminish the symptoms of atopy, such as asthma and allergic rhinitis, and to prevent anaphylaxis produced by bee venom. IgG antibodies are believed to prevent antigen interaction with IgE antibodies anchored to mast cell surfaces by intercepting the antigen molecules before they reach the cell-bound IgE. Thus, a type I hypersensitivity reaction of the anaphylactic type is prevented.

desetope

A term that originates from "determinant selection". It describes that region of class II histocompatibility molecules that reacts with antigen during antigen presentation. Allelic variation permits these contact residues to vary, which is one of the factors in histocompatibility molecule selection of a particular epitope that is being presented.

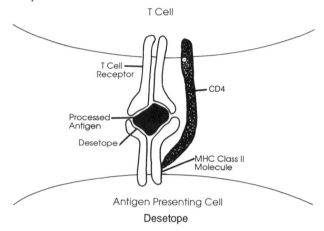

Desetope

designer antibody

A genetically engineered immunoglobulin needed for a specific purpose. The term has been used to refer to chimeric antibodies produced by linking mouse gene segments that encode the variable region of immunoglobulin with those that encode the constant region of a human immunoglobulin. This technique provides the antigen specificity obtained from the mouse antibody, while substituting the less immunogenic Fc region of the molecule from a human source. This greatly diminishes the likelihood of an immune response in humans receiving the hybrid immunoglobulin molecules, since most of the mouse immunoglobulin Fc region epitopes have been eliminated through the human Fc substitution.

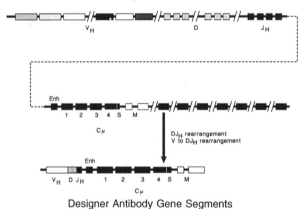

Designer Antibody Gene Segments

designer lymphocytes

Lymphocytes into which genes have been introduced to increase the cell's ability to lyse tumor cells. Tumor-infiltrating lymphocytes transfected with these types of genes have been used in experimental adoptive immunotherapy.

desmin

A 55-kD intermediate filament molecule found in mesenchymal cells that include both smooth and skeletal muscle, endothelial cells of the vessels, and probably myofibroblasts. In surgical pathologic diagnosis, monoclonal antibodies against desmin are useful to identify muscle tumors.

Antibody levels

3 6 12 24

IgG antibody

IgE antibody

Desensitization

desmoglein

A transmembrane glycoprotein that is one of the three components that make up a complex of epidermal polypeptides formed from the immunoprecipitate of pemphigus foliaceous autoantibodies.

despecification

A method to reduce the antigenicity of therapeutic antisera prepared in one species and used in another. To render immunoglobulin molecules less immunogenic in the heterologous recipient, they may be treated with pepsin to remove the molecules' most immunogenic portion, i.e., the Fc fragment, but leave intact the antigen bonding regions, i.e., F(ab′)₂ fragments which retain their antitoxic properties. Had such a treatment been available earlier in the 20th century, serum sickness induced by horse antitoxin administered to human patients as a treatment for diphtheria would have been greatly reduced.

determinant groups

Chemical mosaics found on macromolecular antigens that induce an immune response. Also called epitopes.

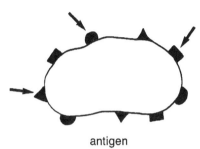

antigen
Determinant Groups

dextran

Polysaccharides of high mol wt comprised of D-glucohomopolymers with α glycoside linkages, principally α-1,6 bonds. Dextrans serve as murine B lymphocyte mitogens. Some dextrans may also serve as thymus-independent antigens. Dextrans of relatively low mol wt have been used as plasma expanders.

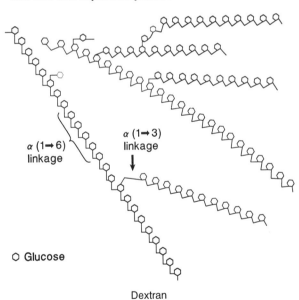

α (1→6)
linkage

α (1→3)
linkage

○ Glucose

Dextran

dhobi itch

Contact hypersensitivity (type IV hypersensitivity) induced by using a laundry marking ink made from Indian ral tree nuts. It occurs in subjects sensitized by wearing garments marked with such ink. It induces a dermatitis at sites of contact with the laundry marking ink.

diabetes mellitus, insulin dependent (type 1)

In type I (autoimmune) diabetes mellitus, autoantibodies against islet cells (and insulin) may be identified. Among the three to six genes governing susceptibility to type I diabetes are those encoding the MHC. Understanding human diabetes has been greatly facili-

tated by both immunologic and genetic studies in experimental animal models including nonobese diabetic mice (NOD mice) and biobreeding rats (BB rats).

Human type I diabetes mellitus results from autoimmune injury of pancreatic β cells. Specific autoantibodies signal pancreatic β cell destruction. The autoantibodies are against islet cell cytoplasmic or surface antigens or insulin. Antiidiotypic antibodies may also develop against antiinsulin antibodies, possibly leading to antibody blockade of insulin receptors and thereby inducing insulin resistance and β cell exhaustion. Autoantibodies have also been demonstrated against a 64-kD third islet cell antigen which could represent a primary target of autoimmune reactivity in type I diabetes and have been found in the sera of the diabetic before clinical onset of the disease. HLA typing is also useful. DNA sequence analysis has revealed that alleles of HLA-DQ β chain govern diabetes susceptibility and resistance. The amino acid at position 57 has a critical role in disease susceptibility and resistance. Although pancreatic β cells fail to express MHC class II antigens under normal circumstances, they become Ia MHC class II antigen positive following stimulation by IFN-γ and TNF or lymphotoxin. Class II positive pancreatic β cells may present islet cell autoantigens to T lymphocytes inaugurating an autoimmune response.

Patients at risk for diabetes or prediabetes might benefit from immunosuppressive therapy, such as by cyclosporine, although most are still treated with insulin. Transplantation of pancreatic islet cells remains a bright possibility in future treatment strategies.

dialysis

The separation of a solution of molecules that differ in mol wt by employing a semipermeable membrane. Antibody affinity is measured by equilibrium dialysis.

diapedesis (see also other figure, page 90)

Cell migration from the interior of small vessels into tissue spaces as a consequence of constriction of endothelial cells in the wall.

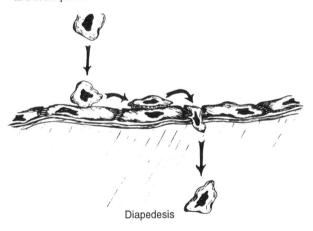

Diapedesis

diathelic immunization

Protective immunity induced by injecting antigen into the nipple or teat of a mammary gland.

diazo salt

A diazonium salt prepared by diazotization from an arylamine to yield a product with a diazo group ($-\overset{+}{N} \equiv N-$). Diazotization has been widely used in the preparation of hapten-carrier conjugates for use in experimental immunology.

diazotization (see figure, page 90)

A method to introduce the diazo group ($-\overset{+}{N} \equiv N-$) into a molecule. Karl Landsteiner used this technique extensively in coupling low mol wt chemicals acting as haptens to protein macromolecules serving as carriers. Aromatic amine derivatives can be coupled to side chains of selected amino acid residues to prepare protein-hapten conjugates, which when used to immunize experimental animals such as rabbits stimulate the synthesis of antibodies. Some of these antibodies are specific for the hapten, which by itself is unable to stimulate an immune response. First an aromatic amine reacts with nitrous acid generated through the combination of sodium nitrite with HCL. The diazonium salt is then combined with the protein at a pH that is slightly alkaline. The reaction products include monosubstituted tyrosine and histidine and also lysine residues that are disubstituted.

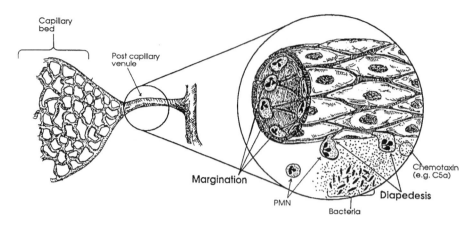

Diapedesis

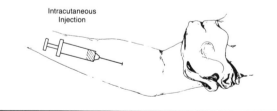

Diazotization

DIC

Abbreviation for disseminated intravascular coagulation.

Dick test

A skin test to signify susceptibility to scarlet fever in subjects lacking protective antibody against the erythrogenic toxin of *Streptococcus pyogenes*. A minute quantity of diluted erythrogenic toxin is inoculated intradermally in the individual to be tested. An area of redness (erythema) occurs at the injection site 6 to 12 h following inoculation of the diluted toxin in individuals who do not have neutralizing antibodies specific for the erythrogenic toxin and who are therefore susceptible to scarlet fever. A heat-inactivated preparation of the same diluted toxin is also injected intradermally in the

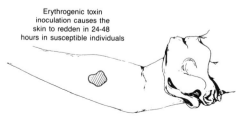

Dick Test

same individual as a control against nonspecific hypersensitivity to other products of the preparation.

differentiation antigen

Epitope that appears at various stages of development or in separate tissues.

differentiation factors

Substances that facilitate maturation of cells, such as the ability of interleukin-2 to promote the growth of T cells.

diffusion coefficient

Mathematical representation of a protein's diffusion rate in gel. The diffusion coefficient is useful in determining antigen mol wt. It is the diffusion rate to concentration gradient ratio.

DiGeorge syndrome

A T cell immunodeficiency in which there is failure of T cell development, but normal maturation of stem cells and B lymphocytes. This is attributable to failure in the development of the thymus, depriving the individual of the mechanism for T lymphocyte development. Maldevelopment of the thymus gland is associated with thymic hypoplasia. Anatomical structures derived from the third and fourth pharyngeal pouches during embryogenesis fail to develop. This leads to a defect in the function of both the thymus and parathyroid glands. DiGeorge syndrome is believed to be a consequence of intrauterine malfunction. It is not familial. Tetany and hypocalcemia, both characteristics of hypoparathyroidism, are observed in DiGeorge syndrome in addition to the defects in T cell immunity. Peripheral lymphoid tissues exhibit a deficiency of lymphocytes in thymic-dependent areas. By contrast, the B or bursa equivalent-dependent areas, such as lymphoid follicles, show normal numbers of B lymphocytes and plasma cells. Serum immunoglobulin levels are within normal limits, and there is a normal immune response following immunization with commonly employed immunogens. A defect in delayed-type hypersensitivity is demonstrated by the failure of affected patients to develop positive skin tests to commonly employed antigens such as candidin or streptokinase and the inability to develop an allograft response. Defective cell-mediated immunity may increase susceptibility to opportunistic infections and render the individual vulnerable to a graft-vs.-host reaction in blood transfusion recipients. There is also minimal or absent *in vitro* responsiveness to T cell antigens or mitogens. Considerable success in treatment has been achieved with fetal thymic transplants and by the passive administration of thymic humoral factors.

dilution end point

A value expressed as the titer that reflects the lowest amount of an antibody giving a reaction. It is determined by serial dilution of the antibody in serum or other body fluid, while maintaining a constant amount of antigen.

dinitrochlorobenzene (DNCB)

A substance employed to test an individual's capacity to develop a cell-mediated immune reaction. A solution of DNCB is applied to the skin of an individual not previously sensitized against this chemical, where it acts as a hapten, interacting with proteins of the skin. Reexposure of this same individual to a second application of DNCB 2 weeks after the first challenge results in a T cell-mediated delayed-type hypersensitivity (contact dermatitis) reaction. Persons with impaired delayed-type hypersensitivity or cell-mediated immunity might reveal an impaired response. The 2,4-dinitro-1-chlorobenzene interacts with free α amino terminal groups in polypeptide chains, as well as with side chains of lysine, tyrosine, histidine, cysteine, or other amino acid residues.

1-chloro-2,4-dinitrobenzene

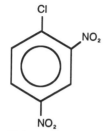

Dinitrochlorobenzene

dinitrofluorobenzene (2,4-dinitro-1-fluorobenzene) (DNFB)

A chemical employed to prepare hapten-carrier conjugates. It inserts the 2,4-dinitrophenyl group into molecules containing free $-NH_2$ groups. When placed on the skin, it leads to contact hypersensitivity.

dinitrophenyl (DNP) group

Designation for 2,4-dinitrophenyl groups that become haptens after they are chemically linked to $-NH_2$ groups of proteins that interact with chlorodinitrobenzene, 2,4-dinitrobenzene sulphonic acid, or dinitrofluorobenzene. These protein

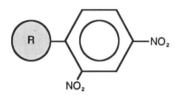

Dinitrophenyl Group

carrier-DNP hapten antigens are useful as experimental immunogens. Antibodies specific for the DNP hapten, which are generated through immunization with the carrier-hapten complex, interact with low mol wt substances that contain the DNP groups.

diphtheria antitoxin

An antibody generated by the hyperimmunization of horses against *Corynebacterium diphtheriae* exotoxin with injections of diphtheria toxoid and diphtheria toxins. When used earlier in the 20th century to treat children with diphtheria, many of the recipients developed serum sickness. It may be employed for passive immunization to treat diphtheria or for short-term protection during epidemics. Presently pepsin digestion of the serum globulin fraction of the antitoxin yields F(ab')₂ fragments of antibodies that retain their antigen-binding property, but lose the highly antigenic Fc region. This process diminishes the development of serum sickness-type reactions and is called despecification.

diphtheria immunization

The repeated administration of diphtheria toxoids [as alum precipitated toxoids (APT)]. Toxoid-antitoxin floccules (TAF) are an alternate form for adults who show adverse reaction to APT. Besides this active immunization procedure, diphtheria antitoxin can also be given for passive immunization in the treatment of diphtheria.

diphtheria toxin

A 62-kD protein exotoxin synthesized and secreted by *Corynebacterium diphtheriae*. The exotoxin, which is distributed in the blood, induces neuropathy and myocarditis in humans. Tryptic enzymes nick the single chain diphtheria toxin. Thiols reduce the toxin to produce two fragments. The 40-kD B fragment gains access to cells through their membranes, permitting the 21-kD A fragment to enter. Whereas the B fragment is not toxic, the A fragment is toxic and it inactivates elongation factor-2, thereby blocking eukaryocytic protein synthesis. Guinea pigs are especially sensitive to diphtheria toxin, which causes necrosis at injection sites, hemorrhage of the adrenals, and other pathologic consequences. Animal tests developed earlier in the century consisted of intradermal inoculation of *C. diphtheriae* suspensions into the skin of guinea pigs that were unprotected, compared to a control guinea pig that had been pretreated with passive administration of diphtheria antitoxin for protection. In later years, toxin

generation was demonstrated *in vitro* by placing filter paper impregnated with antitoxin at right angles to streaks of *C. diphtheriae* microorganisms growing on media in Petri plates. Formalin treatment or storage converts the labile diphtheria toxin into toxoid.

diphtheria toxoid

An immunizing preparation generated by formalin inactivation of *Corynebacterium diphtheriae* exotoxins. This toxoid, which is used in the active immunization of children against diphtheria, is usually administered as a triple vaccine, together with pertussis microorganisms and tetanus toxoid; as purified toxoid which has been absorbed to hydrated aluminum phosphate (PTAP); or as alum-precipitated toxoid (APT). Infants immunized with one of these preparations develop active immunity against diphtheria. Toxoid-antitoxin floccules (TAF) may be administered to adults who demonstrate adverse hypersensitivity reactions to toxoids.

diploid

A descriptor to indicate dual copies of each autosome and two sex chromosomes in a cell nucleus. The diploid cell has twice the number of chromosomes in a haploid cell.

direct agglutination

The aggregation of particulate antigens such as microorganisms, red cells, or antigen-coated latex particles when they react with specific antibody.

direct antiglobulin test

An assay in which washed erythrocytes are combined with antiglobulin antibody. If the red cells had been coated with nonagglutinating (incomplete) antibody *in vivo*, agglutination would occur. Examples of this in humans include hemolytic disease of the newborn, in which maternal antibodies coat the infant's erythrocytes, and autoimmune hemolytic anemia, in which the subject's red cells are coated with autoantibodies. This is the basis of the direct Coombs' test.

direct Coombs' test

Refer to direct antiglobulin test.

direct fluorescence antibody method

The use of antibodies, either polyclonal or monoclonal, labeled with a fluorochrome such as fluorescein isothiocyanate, which

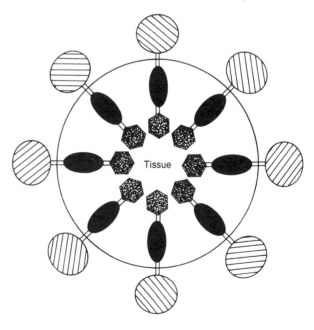

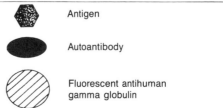

Antigen

Autoantibody

Fluorescent antihuman gamma globulin

Direct Fluorescent Antibody Method

yields an apple green color by immunofluorescence microscopy, or rhodamine isothiocyanate, which yields a reddish-orange color, to identify a specific antigen. This technique is routinely used in immunofluorescence evaluation of renal biopsy specimens as well as skin biopsy preparations to detect immune complexes comprised of the various immunoglobulin classes or complement components.

direct immunofluorescence

The use of fluorochrome-labeled antibody to identify antigens, especially those of tissues and cells. An example is the immuno-fluorescence evaluation of renal biopsy specimens.

direct reaction

Skin reaction caused by the intra-cutaneous injection of viable or nonviable lymphocytes into a host that has been sensitized against donor tissue antigens. This represents a type IV hypersensitivity reaction, which is classified as a delayed-type reaction mediated by T cells. Reactivity is against lymphocyte surface epitopes.

discoid lupus erythematosus

A type of lupus erythematosus that involves only the skin which manifests a characteristic rash. The viscera are not involved, but the skin manifests erythematous plaques and telangiectasis with plugging of the follicles. Also called cutaneous lupus erythematosus.

Discoid Lupus
Erythematous (LE)

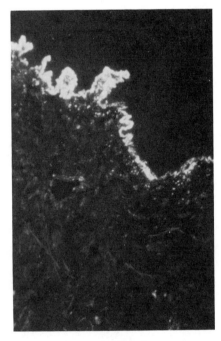

immune deposits
at dermal-epidermal
junction

Discoid Lupus Erythematous (LE)

disodium cromoglycate

A drug that is valuable in the treatment of immediate (type I) hypersensitivity reactions, especially allergic asthma. Although commonly used as an inhalant, the drug may also be administered orally or applied topically to the nose and eyes. Mechanisms of action that have been postulated include mast cell membrane stabilization and bridging of IgE on mast cell surfaces, thereby blocking bridging by antigen.

disseminated intravascular coagulation (DIC)

A tendency to favor coagulation over fibrinolysis in the blood circulation as a consequence of various factors. In DIC, 30 to 65% of

the cases are due to infection. It may occur as fast DIC or slow DIC. The fast variety is characterized by acute, fulminant, consumptive coagulopathy with bleeding as a result of Gram-negative substances, massive tissue injury, burns, etc. Deficient or consumed coagulation factors must be replaced in fast DIC. Slow DIC accompanies chronic diseases characterized by thrombosis, microcirculatory ischemia, and end-organ function. Examples include acute promyelocytic leukemia, neoplasia, vasculitis, and other conditions. The pathogenesis includes endothelial cell damage produced by endotoxins or other agents, activation of platelets, and activation of the intrinsic coagulation pathway. Tissue thromboplastin may be released by trauma or neoplasia, leading to the activation of the extrinsic pathway. Patients with DIC have elevated partial thromboplastin time, prothrombin time, fibrinopeptide A, and fibrinogen degradation products and have decreased fibrinogen, factor V, platelets, antithrombin III, factor VIII, and plasminogen. There is diffuse cortical necrosis of the kidneys.

dissociation constant

The equilibrium constant for dissociation. This is usually described in enzyme-substrate interactions. If the interaction of enzyme with substrate attains equilibrium prior to catalysis, the Michaelis constant (K_M) represents the dissociation constant. The Michaelis constant is equivalent to the concentration substrate when the reaction velocity is half maximal.

distemper vaccine

An attenuated canine distemper virus vaccine prepared from virus grown in tissue culture or chick embryos.

distribution ratio

The plasma immunoglobulin to whole body immunoglobulin ratio.

disulfide bonds

The –S–S– chemical bonds between amino acids that link polypeptide chains together. Chemical reduction may break these bonds. Disulfide bonds in immunoglobulin molecules are either intrachain or interchain, i.e., linking heavy to light chains as well as heavy to heavy chains. The different types of bonds in immunoglobulin molecules differ in their ease of chemical reduction.

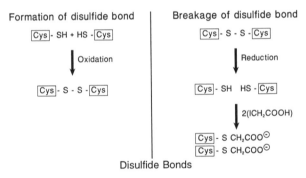

Disulfide Bonds

Dixon, Frank James (1920–)

American physician noted for his fundamental contributions to immunopathology that include the role of immune complexes in the production of disease. He is also known for his work on antibody formation. Dixon was the founding Director of the Research Institute of Scripps Clinic.

DNA dependent RNA polymerase

An enzyme that participates in DNA transcription. With DNA as a template, it catalyzes RNA synthesis from the ribonucleoside-5′-triphosphates.

DNA library

A gene library or clone library comprised of multiple nucleotide sequences that are representative of all sections of the DNA in a particular genome. It is a random assemblage of DNA fragments from one organism, linked to vectors and cloned in an appropriate host. This prevents any individual sequence from being systematically excluded. Adjacent clones will overlap, and cloning large fragments helps to insure that the library will contain all sequences. The DNA to be investigated is reduced to fragments by enzymatic or mechanical treatment, and the fragments are linked to appropriate vectors such as plasmids or viruses. The altered vectors are then introduced into host cells. This is followed by cloning. Transcribed DNA fragments termed exons and nontranscribed DNA fragments termed introns or spacers are part of the gene library. A probe may be used to screen a gene library to locate specific DNA sequences.

DNA Fingerprinting

DNA Sequence	Protein Sequence

DRI   CAG CT|T AAG TTT GAA TGT CAT TTC| TTC AAT   Glu Leu Lys Phe Glu Cys His Phe Phe Asn

1001

DR2(15 and 16)   --- |-c- ---   AGG --G --- ---| --- --- ---   - Pro - Trp Val -   -   -   -   -

1002

DNA Library

DNA ligase

An enzyme that joins DNA strands during repair and replication. It serves as a catalyst in phosphodiester, binding between the 3′-OH and the 5′-PO_4 of the phosphate backbone of DNA.

DNA nucleotidylexotransferase [terminal deoxynucleotidyl-transferase (TdT)]

DNA polymerase that randomly catalyzes deoxynucleotide addition to the 3′-OH end of a DNA strand in the absence of a template. It can also be employed to add homopolymer tails. Immature T and B lymphocytes contain TdT. The thymus is rich in TdT, which is also present in the bone marrow. TdT inserts a few nucleotides in T cell receptor gene and immunoglobulin gene segments at the V-D, D-J, and V-J junctions. This enhances sequence diversity.

DNA polymerase

An enzyme that catalyzes DNA synthesis from deoxyribonucle-otide triphosphate by employing a template of either single- or double-stranded DNA. This is termed DNA-dependent (direct) DNA polymerase in contrast to RNA-dependent (direct) DNA polymerase which employs an RNA template for DNA synthesis.

DNA polymerase I

DNA-dependent DNA polymerase whose principal function is in DNA repair and synthesis. It catalyzes DNA synthesis in the 5′ to 3′ sense. It also has a proofreading function (3′ → 5′ exonuclease) and a 5′ → 3′ exonuclease.

DNA polymerase II

DNA-dependent DNA polymerase in prokaryotes. It catalyzes DNA synthesis in the 5′ to 3′ sense, has a proofreading function (3′ → 5′ exonuclease), and is thought to play a role in DNA repair.

DNA polymerase III

DNA-dependent DNA polymerase in prokaryotes that catalyzes DNA synthesis in the 5′ to 3′ sense. It is the principal synthetic enzyme in DNA replication. It has a proofreading function (3′ → 5′ exonuclease) and 5′ → 3′ exonuclease activity.

DNBS (2,4-dinitrobenzene sulfonate)

A substance employed to generate dinitrophenylated proteins that are used as experimental antigens. Chemically, DNBS reacts principally with lysine residue free ε amino groups if an alkaline pH is maintained.

DNCB

Refer to dinitrochlorobenzene.

DNP

Refer to dinitrophenyl group.

DO and DM

MHC class II-like molecules. DO gene expression occurs exclusively in the thymus and on B lymphocytes. DNα and DOβ chains pair to produce the DO molecule. Information related to DM gene products awaits further investigation.

domain

A region of a protein or polypeptide chain that is globular and folded with 40 to 400 amino acid residues. The domain may have a spatially distinct "signature" which permits it to interact specifically with receptors or other proteins. In immunology, it refers to the loops in polypeptide chains that are linked by disulfide bonds on constant and variable regions of immunoglobulin molecule light and heavy polypeptide chains.

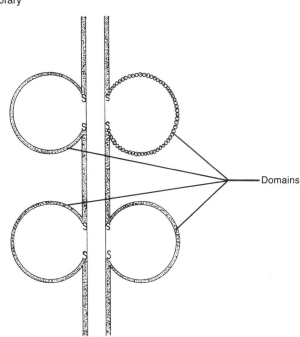

Domains

domesticated mouse

A mouse that has adjusted to a captive existence.

dominant phenotype

Trait manifested in an individual who is heterozygous at the gene locus of interest.

Donath-Landsteiner antibody

An immunoglobulin specific for P blood group antigens on human erythrocytes. This antibody binds to the patient's red blood cells at cold temperatures and induces hemolysis on warming. It occurs in subjects with paroxysmal cold hemaglobulinemia (PCH). Also called Donath-Landsteiner cold autoantibody.

donor

A descriptor for one who offers whole blood, blood products, bone marrow, or an organ to be given to another individual. Individuals who are drug addicts or test positively for certain diseases such as HIV-1 infection or hepatitis B, for example, are not suitable as donors. There are various other reasons for donor rejection not listed here. To be a blood donor, an individual must meet certain criteria which include blood pressure, temperature, hematocrit, pulse, and history. There are many reasons for donor rejection, including low hematocrit, skin lesions, surgery, drugs, or positive donor blood tests.

dot blot

A rapid hybridization method to partially quantify a specific RNA or DNA fragment found in a specimen without the need for a northern or Southern blot. After serially diluting DNA, it is "spotted" on a nylon or nitrocellulose membrane and then denatured with NaOH. It is then exposed to a heat-denatured DNA fragment probe

Dot-Blot

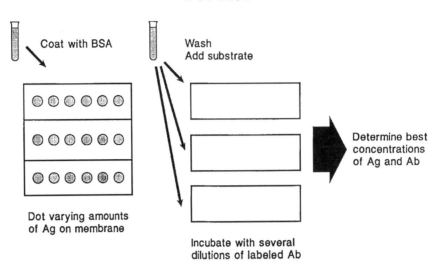

Coat with BSA

Wash
Add substrate

Dot varying amounts
of Ag on membrane

Incubate with several
dilutions of labeled Ab

Determine best
concentrations
of Ag and Ab

that is believed to be complementary to the nucleic acid fragment whose identity is being sought. The probe is labeled with ^{32}P or ^{35}S. When the two strands are complementary, hybridization takes place. This is detected by autoradiography of the radiolabeled probe. Enzymatic, nonradioactive labels may also be employed.

dot DAT

A variation of the Coombs' test known as a dot blot direct antiglobulin test. IgG is fixed on a solid phase support or nitrocellulose membrane. The patient's erythrocytes are incubated on the membrane. This technique eliminates subjective interpretation of results, which diminishes the number of false positives and false negatives.

double diffusion test

A test in which soluble antigen and antibody are placed in separate wells of a gel containing electrolyte. The antigen and antibody diffuse toward one another until their molecules meet at the point of equivalence and precipitate, forming a line of precipitation in the gel. In addition to the two-dimensional technique, double immunodiffusion may be accomplished in a tube as a one-dimensional technique. The Oakley-Fulthorpe test is an example of this type of reaction. This technique may be employed to detect whether antigens are similar or different or share epitopes. It may also be used to investigate antigen and antibody purity. Refer also to reaction of identity, reaction of nonidentity, and reaction of partial identity.

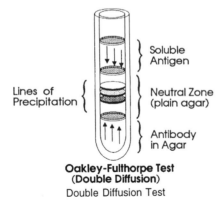

Lines of
Precipitation

Soluble
Antigen

Neutral Zone
(plain agar)

Antibody
in Agar

**Oakley-Fulthorpe Test
(Double Diffusion)**
Double Diffusion Test

double-emulsion adjuvant

Water-in-oil-in-water emulsion adjuvant.

double-layer fluorescent antibody technique

An immunofluorescence method to identify antigen in a tissue section or cell preparation on a slide by first covering and incubating it with antibody or serum containing antibody that is not labeled with a fluorochrome. After appropriate time for interaction, the preparation is washed and a second application of fluorochrome-labeled antibody such as goat or rabbit anti-human immunoglobulin

is applied to the tissue or cell preparation and again incubated. This technique has greater sensitivity than does the single layer immunofluorescent method. Examples include the application of serum from a patient with Goodpasture's syndrome to a normal kidney section acting as substrate followed by incubation and washing, and then covering with fluorochrome-labeled goat anti-human IgG to detect antiglomerular basement membrane antibodies in the patient's serum. A similar procedure is used in detecting antibodies against intercellular substance antigens in the serum of patients with pemphigus vulgaris.

double-negative thymocytes

CD4⁻CD8⁻ thymocytes that are few in number and serve as progenitors for all other thymocytes. They represent an intermediate step between pluripotent bone marrow stem cells and immature cells destined to follow T cell development. Significant heterogeneity is present in this cell population. Peripheral extrathymic CD4⁻CD8⁻ T cells have been examined in both skin and spleens of mice, and like their corresponding cells in the thymus, CD4⁻CD8⁻ T cells express T cell receptor (TCR) $\gamma\delta$ proteins. These double-negative cell populations are greatly expanded in certain autoimmune mouse strains such as those expressing the *lpr* or *gld* genes. Available evidence reveals two thymic populations of CD4⁻CD8⁻ cells. Immature double-negative bone marrow graft cells have stem cell features. However, double-negative cells of greater maturity and without stem cell functions quickly repopulate the thymus.

double-stranded DNA antibodies

Antibodies associated with systemic lupus erythematosus that may be studied serially using the Farr (ammonium sulfate precipitation) assay to predict activity of the disease and to monitor treatment. dsDNA antibody levels are interpreted in conjunction with serum C3 or C4 concentrations. If the dsDNA antibody level doubles or exceeds 30 IU/ml earlier than 10 weeks, an exacerbation of SLE is likely, especially if there is an associated decrease in the serum C4 concentration. This reflects selective stimulation of B cell stimulation known to occur in SLE patients. The Farr assay is more reliable than is the EIA method, since the Farr assay measures high avidity antibodies to dsDNA.

doubling dilution

A technique used in serology to prepare serial dilutions of serum. A fixed quantity (one volume) of physiologic saline is added to each of a row of serological tubes, except for the first tube in the row which receives two volumes of serum. One volume of serum from the first tube is added to one volume of saline in the second tube. After thoroughly mixing the contents with the transfer pipette, one volume of the second tube is transferred to the third, and the procedure is repeated down the row. This same volume is then discarded from the final tube after the contents have been thoroughly mixed. Thus, the serum dilution in each tube is double that in the preceding tube. The first tube is undiluted, the second contains a 1:2 dilution, the third a 1:4, the fourth a 1:8, etc.

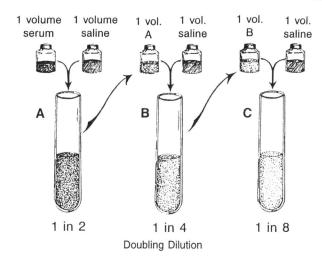

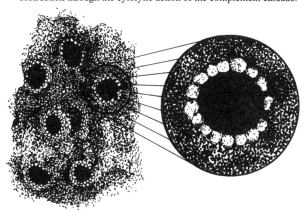

Doubling Dilution

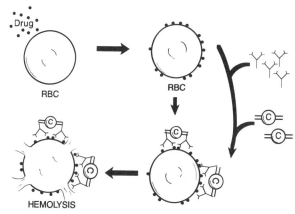

a bystander type of complement-mediated lysis. Another form of drug-induced autoimmunity is seen with nitrofurantoin, in which the autoimmunity involves cell-mediated phenomena without evidence of autoantibodies.

Drug-Induced Hemolysis

doughnut structure

The assembly and insertion of complement C9 protein monomers into a cell membrane to produce a transmembrane pore which leads to cell destruction through the cytolytic action of the complement cascade.

Doughnut Structure

DPT vaccine

Refer to DTP vaccine.

Drakeol 6VR®

A purified light mineral oil employed to prepare water-in-oil emulsion adjuvants.

drift

Refer to antigenic drift.

drug allergy

An immunopathologic or hypersensitivity reaction to a drug. Some drugs are notorious for acting as haptens that bind to proteins in the skin or other tissues that act as carriers. This hapten-carrier complex elicits an immune response manifested as either antibodies or T lymphocytes. Any of the four types of hypersensitivity in the Coombs' and Gell classification may be mediated by drug allergy. One of the best known allergies is hypersensitivity to penicillin. Antibodies to a drug linked to carrier molecules in the host may occur in autoimmune hemolytic anemia or thrombocytopenia, anaphylaxis, urticaria, and serum sickness. Skin eruptions are frequent manifestations of a T cell-mediated drug allergy.

drug-induced autoimmunity

Some mechanisms in drug-induced autoimmunity are similar to those induced by viruses. Autoantibodies may appear as a result of the helper determinant effect. With some drugs such as hydantoin, the mechanism resembles that of the Epstein-Barr virus (EBV). The generalized lymphoid hyperplasia also involves clones specific for autoantigens. A third form is that seen with α-methyldopa. This drug induces the production of specific antibodies. The drug attaches to cells *in vivo* without changing the surface antigenic makeup. The antibodies, which often have anti-e (Rh series) specificity, combine with the drug on cells, fix complement, and induce

DTaP vaccine

A cellular form of DTP vaccine used for the final in a series of four injections. It may be used also as a pre-school booster.

DTH

Delayed-type hypersensitivity.

DTH T cell

CD4+ T lymphocyte sensitized against a delayed-type hypersensitivity antigen.

DTP vaccine

A preparation used for protective immunization that is comprised of diphtheria and tetanus toxoids and pertussis vaccine. Children should receive a DPT vaccine at the ages of 2, 4, 6, and 15 months, with a booster given at 4 to 6 years of age. The tetanus and diphtheria toxoids should be repeated at 14 to 16 years of age. The vaccine is contraindicated in individuals who have shown prior allergic reactions to DPT or in subjects with acute or developing neurologic disease. DPT vaccine is effective in preventing most cases of the diseases it addresses. Because of the occasional production of neurologic complications by the pertussis part of the vaccine, a genetically altered mutant is being developed for use in the vaccine.

Duffy blood group

Human erythrocyte epitopes encoded by *Fya* and *Fyb* genes, located on chromosome 1. Since these epitopes are receptors for *Plasmodium vivax,* African-Americans who often express the Fy(a-b-) phenotype are not susceptible to the type of malaria induced by this species. Mothers immunized through exposure to fetal red cells bearing the Duffy antigens, which she does not possess, may synthesize antibodies that cross the placenta and induce hemolytic disease of the newborn.

Duncan's syndrome

Marked lymphoproliferation and agammaglobulinemia that are associated with Epstein-Barr virus infection. The rapidly proliferating B lymphocytes produce neoplasms which may rupture the spleen. This defect in the immune response with susceptibility to infection is inherited as an X-linked recessive disorder. The individual is not able to successfully resist infection by the Epstein-Barr virus.

dust, nuclear (leukocytoclasis)

Extensive basophilic granular material representing karyolytic nuclear debris that accompanies areas of inflammation and necrosis, as in leukocytoclastic vasculitis.

dye exclusion test

An assay for the viability of cells *in vitro*. Vital dyes such as eosin and trypan blue are excluded by living cells; however, the loss of cell membrane integrity by dead cells admits the dye that stains the cell. The dye exclusion principle is used in the microlymphocytotoxicity test employed for HLA typing in organ transplantation.

dye test

An assay to determine whether an individual has become infected with *Toxoplasma.* Antibody in an infected patient's serum prevents living toxoplasma organisms, obtained from an infected mouse's peritoneum, from taking up methylene blue. Therefore, the organisms do not stain blue if antitoxoplasma antibody is present in the serum.

dysgammaglobulinemia

A pathologic condition associated with selective immunoglobulin deficiency, i.e., depression of one or two classes of serum immunoglobulins and normal or elevated levels of the other immunoglobulin classes. The term has also described antibody-deficient patients whose antibody response to immunogenic challenge is impaired even though their immunoglobulin levels are normal. Use of the term is discouraged, as it is imprecise. Also called dysimmunoglobulinemia.

E

e allotype

A rabbit IgG heavy chain allotype. e allotypes are determined by an amino acid substitution at position 309 in C_H2. The e14 allotype heavy chains possess threonine at position 309, whereas e15 allotype heavy chains possess alanine at that location. They are determined by alleles at the *de* locus that encodes rabbit gamma chain constant regions.

E antigen

A hepatitis B virus antigen present in the blood sera of chronic active hepatitis patients.

E rosette

Human T lymphocyte encircled by rings of sheep red blood cells. It was used previously as a method to enumerate T lymphocytes. Refer also to E rosette-forming cell.

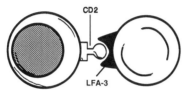

T lymphocyte   Sheep Erythrocyte

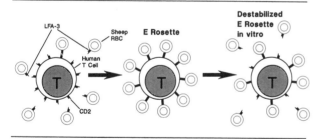

Erythrocyte Rosetting

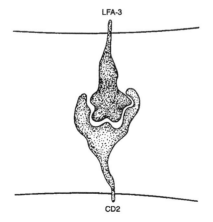

T cell

E rosette-forming cell

The formation of a complex of sheep red cells encircling a T lymphocyte to form a rosette. This was one of the first methods to enumerate human T cells, as the sheep red cells did not form spontaneous rosettes with human B lymphocytes. This is now known to be due to the presence of the CD2 marker, which is a cell adhesion (LFA-2) molecule, on T cells. CD2 positive T cells are now enumerated by the use of monoclonal antibodies and flow cytometry.

E-selectin [or endothelial leukocyte adhesion molecule 1 (ELAM-1)]

A molecule found on activated endothelial cells that recognizes sialylated Lewis X and related glycans. Its expression is associated with acute cytokine mediated inflammation. Also called CD62E.

E32

A protein formed early in development of B lymphocytes that has a role in immunoglobulin heavy chain transcription.

E5

Murine monoclonal IgM antibody to endotoxin that has proven safe and capable of diminishing mortality and helping to reverse organ failure in patients with Gram-negative sepsis who are not in shock during therapy. E5 may recognize and combine with lipid A epitope.

EA

An abbreviation for erythrocyte (E) coated with specific antibody (A). This is a technique to measure the activity of Fcγ receptors. Sheep red blood cells with subagglutinating quantities of IgG antibodies are placed in contact with cells at room temperature. IgG Fc receptor-bearing cells will combine with the EA, resulting in rosette formation.

EAC

Abbreviation for erythrocyte (E), antibody (A), and complement (C), as designated in studies involving the complement cascade. Historically, sheep red blood cells have been combined with antibody specific for them, and lysis is induced only after the addition of complement. This interaction of the three reagents has served as a useful mechanism to study the reaction sequence of the multiple component complement system, which is often written as EAC1423, etc. EAC may also be employed to reveal the presence of complement receptors 1 through 4. The combination of IgM-coated sheep red blood cells with sublytic quantities of complement to produce EAC provides a mechanism whereby rosettes are formed by cells bearing complement receptors once the EAC are layered onto test cells at room temperature. EAC rosette erythrocytes such as sheep red blood cells, coated or synthesized with antibody and complement, encircle human B lymphocytes to form a rosette. This is based on erythrocyte binding to C3b receptors. This technique has been used in the past to identify B lymphocytes. However, phagocytic cells bearing C3b receptors may also form EAC rosettes. This type of rosette is in contrast to the E rosette, which identifies T lymphocytes. B lymphocytes are now enumerated by monoclonal antibodies against the B lymphocyte CD markers using flow cytometry.

EAE

Abbreviation for experimental allergic encephalomyelitis.

EBNA (Epstein-Barr virus nuclear antigen)

Refer to Epstein-Barr nuclear antigen.

ECF-A (eosinophil chemotactic factor of anaphylaxis)

A 500-D acidic polypeptide that attracts eosinophils. Interaction of antigen with IgE antibody molecules on the surface of mast cells causes ECF-A to be released from the mast cells.

ECHO virus (enteric cytopathogenic human orphan virus)

Comprised of 30 types within the picornavirus family. It is cytopathic in cell culture and produces clinical manifestations in patients that include upper respiratory tract infections, diarrhea, exanthema, viremia, and sometimes poliomyelitis and viral meningitis.

eclipsed antigen

An antigen, such as one from a parasite, that so closely resembles host antigens that it fails to stimulate an immune response.

ecto-5'-nucleotidase deficiency

A purine metabolism alteration that produces immunodeficiencies of B lymphocytes.

eczema

A skin lesion that is characterized as a weeping eruption consisting of erythema, pruritus with edema, papules, vesicles, pustules, scaling, and possible exudation. It occurs in individuals who are atopic, such as those with atopic dermatitis. Application to the skin or the ingestion of drugs that may themselves act as haptens may induce this type of hypersensitivity. It may be seen in young children who subsequently develop asthma in later life.

eczema vaccinatum

On occasion, in subjects receiving smallpox vaccination in the past, the virus in the vaccine superinfected areas of skin affected by atopic dermatitis. This led to generalized vaccinia, which was severe and frequently fatal. It was also referred to as Kaposi's varicelliform eruption.

eczematoid skin reaction

The appearance of erythematous, vesicular, and pruritic lesions on the skin that resemble eczema, but are not due to atopy.

ED$_{50}$

The 50% effective dose. For example, 50% hemolysis can be determined more accurately than can a 100% endpoint.

Edelman, Gerald Maurice (1929–)

American investigator who was Professor at Rockefeller University and shared the Nobel Prize in 1972, with Porter, for their work on antibody structure. Edelman was the first to demonstrate that immunoglobulins are composed of light and heavy chains. He also did pioneering work with Bence-Jones protein, cell adhesion molecules, immunoglobulin amino acid sequence, and neurobiology.

edema

Tissue swelling as a result of fluid extravasation from the intravascular space.

edge artifact

In immunoperoxidase staining of paraffin-embedded tissues, tissue drying may produce nonspecific coloring at the periphery which is an artifact.

effector function

The nonantigen-binding functions of an antibody molecule that are mediated by the constant region of heavy chains. These include Fc receptor binding, complement fixation, binding to mast cells, etc. Effector function generally results in removal of antigen from the body, such as in phagocytosis or complement-mediated lysis.

effector lymphocyte

A lymphocyte activated through either specific or nonspecific mechanisms to carry out a certain function in the immune response. Examples of effector lymphocytes include the NK cell, the tumor-infiltrating lymphocyte (TIF), the lymphokine activated killer (LAK) cell, cytotoxic T lymphocyte, helper T lymphocyte, and suppressor T cell. Most commonly, the term signifies a T lymphocyte capable of mediating cytotoxicity, suppression, or helper function.

Ehrlich, Paul (1854–1915)

German immunologist and bacteriologist who shared the Nobel Prize in 1908, with Metchnikoff, for "their work on immunity". Ehrlich developed stains for mycobacteria and leukocytes early in his career. In 1897, he published a paper describing a practical, standard method for preparing diphtheria toxin and antitoxin. He presented his side chain theory of antibody formation in the Croonian Lecture before the Royal Society of Great Britain in 1900. His later chemotherapy research was directed toward treatments for syphilis and trypanosomiasis. He also made fundamental contributions to cancer research. *Collected Studies on Immunity*, 1906; *Collected Papers of Paul Ehrlich*, 3 Vols., 1957.

Ehrlich phenomenon (historical)

The demonstration that the L_0 dose of diphtheria toxin differs from the L_+ dose not by one minimal lethal dose (MLD) as one might anticipate, but by 10 to 100 MLD, depending on the particular sample. This represents the difference between the dose that causes minimal reactivity and the dose that produces death.

Ehrlich side chain theory

The first selective theory of antibody synthesis developed by Paul Ehrlich in 1900. Although elaborate in detail, the essential feature of the theory was that cells of the immune system possess the genetic capability to react to all known antigens and that each cell on the surface bears receptors with surface haptophore side chains. On combination with antigen, the side chains would be cast off into the circulation, and new receptors would replace the old ones. These cast-off receptors represented antibody molecules in the circulation. Although far more complex than this explanation, the importance of the theory was in the amount of research stimulated to try to disprove it. Nevertheless, it was the first effort to account for the importance of genetics in immune responsiveness at a time when Mendel's basic studies had not even yet been "rediscovered" by De Vries.

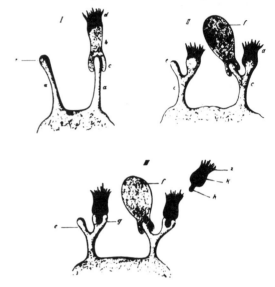

Diagram of Ehrlich's 'Side-Chain Theory' published in connection with the 'Anæmia' of Ehrlich and Lazarus in Nothnagel's 'Special Pathology and Therapy', Vol. 8, 1898–1901, as an appendix to Ehrlich's 'Schlussbetrachtungen', pages 163–185

Figure I. RECEPTOR OF FIRST ORDER. (*e*) haptophore complex; (*b*) adsorbed toxin molecule with (*c*) haptophore group; (*d*) toxophore group

Figure II. RECEPTOR OF SECOND ORDER with (*e*) haptophore group; (*d*) zymophore group and (*f*) adsorbed nutritive molecule

Figure III. RECEPTOR OF THIRD ORDER. (*e*) haptophore; (*g*) complementophile group; (*k*) complement with (*h*) haptophore; (*z*) zymotoxic group; (*f*) nutritive molecule

Ehrlich Side Chain Theory

EIA

Abbreviation for exercise-induced anaphylaxis.

EIA

Abbreviation for enzyme immunoassay.

eicosanoid

An arachidonic acid-derived 20-carbon cyclic fatty acid. It is produced from membrane phospholipids. Eicosanoids, as well as

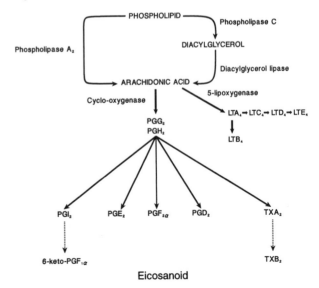

Eicosanoid

other arachidonic acid metabolites, are elevated during shock and following injury and are site specific. They produce various effects, including bronchodilation, bronchoconstriction, vasoconstriction, and vasodilation. Eicosanoids include leukotrienes, prostaglandins, thromboxanes, and prostacyclin.

Eisen, Herman Nathaniel (1918–)
American physician whose research contributions range from equilibrium dialysis (with Karush) to the mechanism of contact dermatitis.

ELAM-1 (endothelial leukocyte adhesion molecule 1)
A glycoprotein of the endothelium that facilitates adhesion of neutrophils. Structurally, it has an epidermal growth factor-like domain, a lectin-like domain, amino acid sequence homology with complement-regulating proteins, and six tandem-repeated motifs. Tumor necrosis factor, interleukin-1, and substance P induce its synthesis. Its immunoregulatory activities include attraction of neutrophils to inflammatory sites and mediating cell adhesion by sialyl-Lewis X, a carbohydrate ligand. It acts as an adhesion molecule or addressin for T lymphocytes that home to the skin.

electroimmunodiffusion
A double-diffusion in-gel method in which antigen and antibody are forced toward one another in an electrical field. Precipitation occurs at the site of their interaction. Refer to Laurell rocket test and rocket immunoelectrophoresis. Also called counter immunoelectrophoresis.

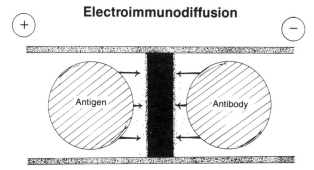

Electroimmunodiffusion

electrophoresis
A method for separating a mixture of proteins based on their different rates of migration in an electrical field. Zone electrophoresis represents a technical improvement in which a stabilizing medium such as cellulose acetate serves as a matrix for buffer and as a structure to which proteins can remain attached following fixation. By this technique, plasma proteins are resolved into five or six major peaks. Zone electrophoresis permits a gross evaluation of the levels of immunoglobulins and other proteins in the serum. In cases of increased levels, electrophoresis indicates whether this involves a general proliferation and hypersecretion by lymphocytes derived from multiple individual cells (polyclonal origin; the proteins are heterogeneous) or involves proliferation and hypersecretion by lymphoid cells derived from a limited number of individual cells (monoclonal origin; the proteins are homogeneous).

electroporation
A technique to insert molecules into cells through use of brief high voltage electric pulses. It can be used to insert DNA into animal cells. The electrical discharge produces tiny pores that are nanometers in diameter in the plasma membrane. These pores admit supercoiled or linear DNA.

Elek plate
A method to show toxin production by *Corynebacterium diphtheriae* colonies growing on an agar plate. Diphtheria antitoxin impregnated into a strip of filter paper is placed at a right angle to a streak of the microorganisms on the agar plate. Toxin formation by the growing microbes interacts with antitoxin in the filter paper to form a line of precipitation.

Plasma - agar plate showing staphylocoagulase effect inhibited by commercial antitoxin

Elek Plate

elephantiasis
Enlargement of extremities by lymphedema caused by lymphatic obstruction during granulomatous reactivity in filariasis.

elevated IgE, defective chemotaxis, recurrent infection, and eczema
Both men and women with the third disorder may manifest several bacterial infections, eczema, abscesses, and pneumonia associated with group A ß hemolytic streptococci and *Staphylococcus aureus*. They demonstrate normal function of both B and T cell limbs of the immune response, although their IgE levels are strikingly elevated, reaching values that are tenfold those of normal individuals. Only chemotactic function is defective, with other parameters of phagocytic cell function within normal limits. No treatment other than antibiotic therapy is available.

ELISA (enzyme-linked immunosorbent assay)
Refer to enzyme-linked immunosorbent assay.

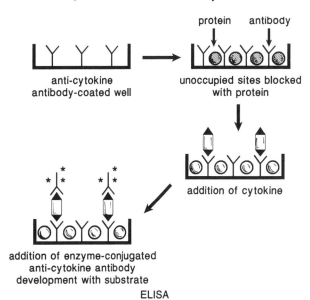

anti-cytokine antibody-coated well

unoccupied sites blocked with protein

addition of cytokine

addition of enzyme-conjugated anti-cytokine antibody development with substrate

ELISA

ELISPOT assay
A modification of the enzyme-linked immunosorbent assay (ELISA) which involves the capture of products secreted from cells placed in contact with antigen or antibody fixed to a plastic surface. An enzyme-linked antibody is then used to identify the captured products by cleaving a colorless substrate to yield a colored spot.

eluate
In immunology, a substance such as an antibody obtained by physical or chemical treatment of an antigen-antibody complex. By allowing particulate antigens such as erythrocytes to interact with antibody followed by heating the antibody particle or cell complex to 56°C, the antibody can be dissociated from the particulate antigen and is present in the eluate.

EMIT
Abbreviation for enzyme-multiplied immunoassay technique.

emperipolesis
The intrusion or penetration of a lymphocyte into the cytoplasm of another cell followed by passage through the cell. Emperipolesis also describes the movement of one cell within another cell's cytoplasm.

ENA antibodies
Antibodies against extractable nuclear antigens. This category includes antibodies to ribonucleoprotein (RNP), presently termed U1 snRNP (small nuclear ribonucleoproteins) or U1 RNP in addition to Sm antibodies which have specificity for Smith antigen. Sm antibodies are associated with systemic lupus erythematosus, whereas U1 snRNP antibodies in high titer are detected in patients with mixed connective tissue disease. U4/U6 snRNP antibodies are detectable in patients with systemic sclerosis.

encapsulation
The reaction of leukocytes to foreign material that cannot be phagocytized because of its large size. Multiple layers of flattened leukocytes form a wall surrounding the foreign body and isolate it within the tissues. This type of reaction occurs in invertebrates

including annelids, mollusks, and arthropods. In vertebrates, macrophages surround the foreign body, a granuloma is formed, and fibroblasts subsequently appear. A fibrous capsule is formed.

encephalitogenic factors

Myelin basic protein or related molecules found in the brain that can induce experimental allergic encephalomyelitis if administered to experimental animals together with Freund's complete adjuvant. The smallest constituent of myelin that is capable of inducing experimental encephalomyelitis is a nonapeptide (Phe-Leu-Trp-Ala-Glu-Gly-Gln-Lys).

end cell

A cell such as a mature plasma cell that is at the termination point in that cell line's maturation pattern. End cells do not further divide. They represent the final product of maturation.

end piece (historical)

In the early complement literature, complement activity in the pseudoglobulin fraction of serum is called the end piece, in contrast to the activity in the euglobulin fraction which is called mid-piece of complement. Current information reveals that what was referred to as the end piece did not contain C1, but did contain all of the C2 and some other complement components.

end point

The greatest dilution of an antibody in solution that will still yield an identifiable reaction when combined with antigen. The reciprocal of this dilution represents the titer.

end-stage renal disease (ESRD)

Chronic renal failure. Approximately one third of cases are linked to diabetes mellitus. Kidneys of patients on chronic dialysis ultimately develop ESRD. There is proliferation of intravascular smooth muscle induced by ischemia. There may also be venous thrombosis.

endocytosis

A mechanism whereby substances are taken into a cell from the extracellular fluid through plasma membrane vesicles. This is accomplished by either pinocytosis or receptor-facilitated endocytosis. In pinocytosis, extracellular fluid is captured within a plasma membrane vesicle. In receptor-facilitated endocytosis, extracellular ligands bind to receptors, and coated pits and coated vesicles facilitate internalization. Clathrin-coated vesicles become uncoated and fuse to form endosomes. Ligand and receptor dissociate within the endosomes, and the receptor returns to the cell surface. Endosomes fused with lysosomes form secondary lysosomes where ligand degradation occurs. Low-density lipoproteins are handled in this manner.

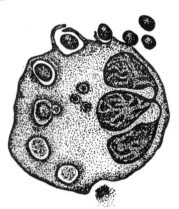

Endocytosis

endogenous pyrogen

A substance such as interleukin-1 or tumor necrosis factor synthesized by macrophages or other cells that induces fever and is found in the blood plasma in infections or is associated with cell-mediated immune reactivity. Diminishing prostaglandins in the hypothalamic region produce fever. Interleukin-6 is also an endogenous pyrogen.

endometrial antibodies

IgG autoantibodies present in two-thirds of women with endometriosis. They react with the epithelial glandular portion but not the stromal component of endometrium. They have been suggested as a possible cause for infertility that occurs in approximately 30 to 40% of women with endometriosis.

endomysial antibodies

IgA subclass (IgA EmA) antibodies that are specific for reticulin in smooth muscle endomysium. They are present in essentially all celiac disease (gluten-sensitive enteropathy) patients with villous atrophy and in 60 to 70% of individuals with dermatitis herpetiformis who are receiving a diet with normal gluten content.

endophthalmitis phacoanaphylactica

A condition resulting from the accidental release of lens protein into the blood circulation during cataract removal in humans. This interaction of a normally sequestered antigen, i.e., lens protein, with the host immune system activates an autoimmune response that results in inflammation of the eye concerned.

endoplasmic reticulum

A structure in the cytoplasm comprised of parallel membranes that are connected to the nuclear membranes. Lipids and selected proteins are synthesized in this organelle. The membrane is continuous and convoluted. Electron microscopy reveals rough endoplasmic reticulum, which contains ribosomes on the side exposed to the cytoplasm, and smooth endoplasmic reticulum without ribosomes. Fatty acids and phospholipids are synthesized and metabolized in smooth endoplasmic reticulum. Selected membrane and organelle proteins, as well as secreted proteins, are synthesized in the rough endoplasmic reticulum. Cells such as plasma cells that produce antibodies or other specialized secretory proteins have abundant rough endoplasmic reticulum in the cytoplasm. Following formation, proteins move from the rough endoplasmic reticulum to the Golgi complex. They may be transported in vesicles that form from the endoplasmic reticulum and fuse with Golgi complex membranes. Once secreted protein reaches the endoplasmic reticulum lumen, it does not have to cross any further barriers prior to exit from the cell.

endorphin

An endogenous opioid peptide which links to a cognate receptor. α, β, and γ endorphins exist. β Endorphin is associated with secretion by the pituitary gland and is possibly associated with pain perception.

endosome

A 0.1- to 0.2-μm intracellular vesicle produced by endocytosis.

endothelial cell antibodies (ECA)

IgG antibodies present in the blood sera of systemic lupus erythematosus patients that may mediate immunologic injury to blood vessel walls. They may be involved in the pathogenesis of rheumatoid vasculitis. Besides systemic lupus erythematosus, cytotoxic antibodies reactive with vascular endothelial cells are present in the sera of cardiac allograft recipients undergoing hyperacute rejection even though the cross match was negative. They may also be found in patients with hemolytic uremic syndrome, in Kawasaki syndrome and in renal allograft recipients who have rejected the transplant. Selected patients with Wegener's granulomatosis and with micropolyarteritis may reveal non-cytolytic ECA, which are also demonstrable in 44% of dermatomyositis patients, particularly those who also have interstitial lung disease. About 33% of IgA nephropathy patients possess antibodies that show specificity for vascular endothelial cells and for HLA class I antigens.

endothelial leukocyte adhesion molecule 1 (ELAM-1)

Facilitates focal adhesion of leukocyte to blood vessel walls. It is induced by endotoxins and cytokines and belongs to the adhesion molecule family. It is considered to play a significant role in the pathogenesis of atherosclerosis and infectious and autoimmune diseases. Neutrophil and monocyte adherence to endothelial cells occurs during inflammation *in vivo* where there is leukocyte margination and migration to areas of inflammation. Endothelial cells activated by IL-1 and TNF synthesize ELAM-1, at least in culture. A 115-kD chain and a 100-kD chain comprise the ELAM-1 molecule.

endothelin

A peptide comprised of 21 amino acid residues that is derived from aortic endothelial cells and is a powerful vasoconstrictor. A gene on chromosome 6 encodes the molecule. It produces an extended pressor response, stimulates release of aldosterone, inhibits release of renin, and impairs renal excretion. It is elevated in myocardial infarction and cardiogenic shock, major abdominal surgery, pulmonary hypertension, and uremia. It may have a role in the development of congestive heart failure.

endotoxin

A Gram-negative bacterial cell wall lipopolysaccharide (LPS) that is heat stable and causes neutrophils to release pyrogens. It may produce endotoxin or hemorrhagic shock and modify resistance against infection. Endotoxins comprise an integral constituent of the outer membrane of Gram-negative microorganisms. They are significant virulence factors and induce injury in a number of ways. Toxicity is associated with the molecule's lipid A fraction, which is comprised of a β-1,6-glucosaminyl-glucosamine disaccharide substituted with phosphate groups and fatty acids. Lipopolysaccharide (LPS) has multiple biological properties that include ability to induce fever, lethal action, initiation of both complement and blood coagulation cascades, mitogenic effect on B lymphocytes; the ability to stimulate production of such cytokines as tumor necrosis factor and interleukin-1; and the ability to clot *Limulus* amebocyte lysate. Relatively large amounts of lipopolysaccharide released from Gram-negative bacteria during Gram-negative septicemia may produce endotoxin shock.

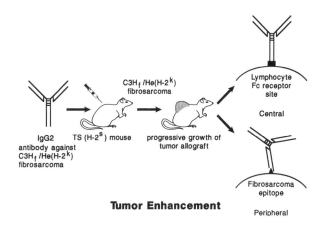

Tumor Enhancement

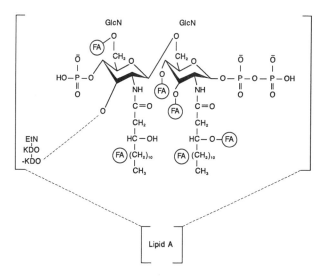

Lipid A

endotoxin shock

Falling blood pressure and disseminated intravascular coagulation (DIC) following exposure to relatively large amounts of endotoxin produced during bacterial sepsis with *Escherichia coli, Pseudomonas aeruginosa,* or meningococci. DIC leads to the formation of thrombi in small blood vessels, leading to such devastating consequences as bilateral cortical necrosis of the kidneys and blockage of the blood supply to the brain, lungs, and adrenals. When DIC affects the adrenal glands, as in certain meningococcal infections, infarction leads to adrenal insufficiency and death. This is the Waterhouse-Friderichsen syndrome.

enhancement

The prolonged survival, conversely the delayed rejection, of tumor or skin allografts in individuals previously immunized or conditioned by passive injection of antibody specific for graft antigens.

This is termed immunological enhancement and is believed to be due to a blocking effect by the antibody.

enhancer

A segment of DNA that can elevate the amount of RNA a cell produces. This DNA sequence activates the beginning of RNA polymerase II transcription from a promoter. Although initially described in the DNA tumor virus, SV40, enhancers have now been demonstrated in immunoglobulin μ and κ genes' J-C intron. Immunoglobulin enhancers function well in B cells, presumably due to precise regulatory proteins that communicate with the enhancer region.

enhancing antibodies

Blocking antibodies that favor survival of tumor or normal tissue allografts.

enkephalin(s)

A related group of endogenous opioids that are pentamers synthesized in the central nervous system and gastrointestinal tract. They share the initial four amino acids (H-Tyr-Gly-Gly-Phe-).

entactin/nidogen

Glomerular basement membrane antigens that might have a possible role in glomerular basement membrane nephritis. Entactin is a 150 kD glomerular membrane sulphated glycoprotein that is probably synthesized by endothelial and epithelial cells in the developing kidney. It has a role in cellular adhesion to extracellular material. It is complexed with laminin *in vivo*. Nidogen, also a basement membrane constitutent, is a 150 kD glycoprotein that has binding affinity for laminin. Entactin and nidogen closely resemble one another and may be identical. Serum IgG, IgM and/or IgA antibodies specific for entactin/nidogen have been found in over 40% of patients with glomerulonephritis in one study.

enterotoxin

A bacterial toxin that is heat stable and causes intestinal injury.

env

A gene of retroviruses that codes for env envelope glycoprotein. Refer to HIV-1 genes.

Retroviral Genome

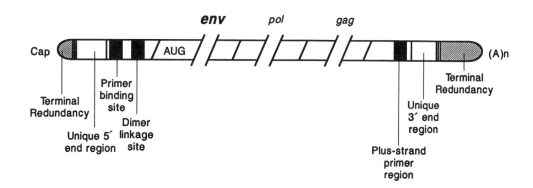

enzyme labeling

A method such as the immunoperoxidase technique that permits detection of antigens or antibodies in tissue sections by chemically conjugating them to an enzyme. By then staining the preparation for the enzyme, antigen or antibody molecules can be located. Refer to immunoperoxidase method.

enzyme-linked immunosorbent assay (ELISA)

An immunoassay which employs an enzyme linked to either antiimmunoglobulin or antibody specific for antigen and detects either antibody or antigen. This method is based on the sandwich or double-layer technique, in which an enzyme rather than a fluorochrome is used as the label. In this method, antibody is attached to the plastic tube, well, or bead surface to which the antigen-containing test sample is added. If antibody is being sought in the test sample, then antigen should be attached to the plastic surface. Following antigen-antibody interaction, the enzyme-antiimmunoglobulin conjugate is added. The ELISA test is read by incubating the reactants with an appropriate substrate to yield a colored product that is measured in a spectrophotometer. Alkaline phosphatase and horseradish peroxidase are enzymes that are often employed. ELISA methods have replaced many radioimmunoassays because of their lower cost, safety, speed, and simplicity in performing.

enzyme-multiplied immunoassay technique

An immunoassay used to monitor therapeutic drugs such as antitumor, antiepileptic, antiasthmatic, and metabolites of cocaine and of other agents subject to abuse. It is a one phase, competitive enzyme-labeled immunoassay.

eosinophil

16 μm.
Eosinophil with
Segmented Nucleus

A polymorphonuclear leukocyte identified in Wright- or Giemsa-stained preparations by staining of secondary granules in the leukocyte cytoplasm as brilliant reddish-or-ange refractile granules. Cationic peptides are released from these secondary granules when an eosinophil interacts with a target cell and may lead to death of the target. Eosinophils make up 2 to 5% of the total white blood cells in man. After a brief residence in the circulation, eosinophils migrate into tissues by passing between the lining endothelial cells. It is believed that they do not return to the circulation. The distribution corresponds mainly to areas exposed to external environment, such as skin, mucosa of the bronchi, and gastrointestinal tract. Eosinophils are elevated during allergic reactions, especially type I immediate hypersensitivity responses, and are also elevated in individuals with parasitic infestations.

eosinophil chemotactic factor

Mast cell granule peptides that induce eosinophil chemotaxis. These include two tetrapeptides: Val-Gly-Ser-Glu and Ala-Gly-Ser-Glu. Histamine also induces eosinophil chemotaxis.

eosinophilia

Elevated number of eosinophil in the blood. It occurs in immediate, type I hypersensitivity reactions, including anaphylaxis and atopy, and is observed in patients with parasitic infestations, especially by nematodes.

eosinophilic granuloma

A subtype of a macrophage lineage (histiocytosis X) tumor that contains eosinophils, especially in bone.

eosinophilic myalgia syndrome (EMA)

An intoxication syndrome observed in persons in the U.S. that appeared linked to the consumption of L-tryptophan, proposed by some health advocates as an effective treatment for various disorders such as insomnia, premenstrual syndrome, etc. It was associated with a strain of *Bacillus amyloliquefaciens* employed to produce tryptophan commercially. The inducing agent was apparently an altered amino acid, DTAA (ditryptophan aminal acetaldehyde), a contaminant introduced during manufacture. Clinical manifestations of the syndrome include arthralgia, myopathy, angioedema, alopecia, mobileform rash, oral ulcers, sclerodermoid lesions, restricted lung disease, fever, lymphadenopathy, and dyspnea, among other features. There was a significant eosinophilia. IL-5 was believed to have a role in injury to tissues. Histopathologic examination revealed arteriolitis and sclerosing skin lesions.

epibody

An antiidiotypic antibody reactive with an idiotype of a monoclonal, human anti-IgG autoantibody as well as with human IgG Fc region. These antibodies identify an antigenic determinant associated with the sequence Ser-Ser-Ser. The ability of an epibody to identify an epitope shared by a rheumatoid factor idiotope and an Fc γ epitope demonstrates that this variety of antiidiotypic antibody may function as a rheumatoid factor.

epidermal growth factor (EGF)

A trisulfated polypeptide consisting of 53 residues. It is a member of the tyrosine kinase family and is related to the *erb* oncogene. EGF has multiple functions that include stimulation of the mitogenic response, facilitation of wound healing, and many other functions. It is present in the saliva of rodents.

epidermal growth factor receptor (EGFR)

A 400-amino acid protein found in T cell carcinomas, neurons, cornea, fibroblasts, T lymphocytes, liver, vascular endothelium, and placenta. EGFR measurement is used to judge the aggressiveness of such neoplasia as breast cancer.

epithelial membrane antigen (EMA)

A marker that identifies, by immunoperoxidase staining, most epithelial cells and tumors derived from them, such as breast carcinomas. However, various nonepithelial neoplasms, such as selected lymphomas and sarcomas, may express EMA also. Thus, it must be used in conjunction with other markers in tumor identification and/or classification.

epithelial thymic-activating factor (ETAF)

An epithelial cell-culture product capable of facilitating thymocyte growth. The activity is apparently attributable to interleukin-1.

epithelioid cells

The epithelioid cell is a particular type of cell characteristic of some types of granulomas such as in tuberculosis, sarcoidosis, leprosy, etc. The cell has poorly defined cellular outlines; cloudy, abundant eosinophilic cytoplasm; and an elongated and pale nucleus. By electron microscopy, the cell shows a few short and slender pseudopodia and well-developed cellular organelles. Mitochondria are generally elongated, the Golgi complex is prominent, and lysosomal dense bodies are scattered throughout the cytoplasm. Strands of endoplasmic reticulum, free ribosomes, and fibrils are present in the ground substance.

The epithelioid cell derives from the monocyte-macrophage system. Peripheral blood monocytes made adherent to cellophane strips and implanted into the subcutaneous tissue of an experimental animal develop into epithelioid cells. Conversion of the macrophage to an epithelioid cell is not preceded by a mitotic division of the macrophage. On the contrary, epithelioid cells are able to divide, resulting in round, small daughter cells which mature in 2 to 4 days, gaining structural and functional characteristics of young macrophages. Material that is taken up by macrophages, but cannot be further processed prevents the conversion of epithelioid cells. The lifespan of the epithelioid cell is from 1 to 4 weeks.

epitope

An antigenic determinant. It is the simplest form or smallest structural area on a complex antigen molecule that can combine

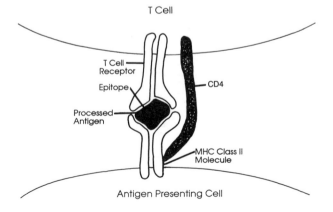

with an antibody or T lymphocyte receptor. It must be at least 1 kD to elicit an antibody response. A smaller molecule such as a hapten may induce an immune response if combined with a carrier protein molecule. Multiple epitopes may be found on large nonpolymeric molecules. Based on X-ray crystallography, epitopes consist of prominently exposed "hill and ridge" regions that manifest surface rigidity. Antigenicity is diminished in more flexible sites.

epitype
A family or group of related epitopes.

EPO
Refer to erythropoietin.

Epstein-Barr immunodeficiency syndrome
Duncan's X-linked immunodeficiency. This is an X-linked or autosomal recessive condition associated with congenital cardiovascular and central nervous system defects. Patients may develop infectious mononucleosis that is fatal. There is aplasia of the bone marrow, agammaglobulinemia, and agranulocytosis, and the response to mitogens and antigens by B cells is greatly diminished. Natural killer cell activity is decreased, and T cells are abnormal. Patients may develop hepatitis, B cell lymphomas, and immune suppression.

Epstein-Barr nuclear antigen
A molecule that occurs in B cells before virus-directed protein can be found in nuclei of infected cells. Thus, it is the earliest evidence of Epstein-Barr virus infection and can be found in patients with conditions such as infectious mononucleosis and Burkitt's lymphoma.

Epstein-Barr virus (EBV)
A DNA herpes virus linked to aplastic anemia, chronic fatigue syndrome, Burkitt's lymphoma, histiocytic sarcoma, hairy cell leukemia, and immunocompromised patients. EBV may promote the appearance of such lymphoid proliferative disorders as Hodgkin's and non-Hodgkin's lymphoma, infectious mononucleosis, nasopharyngeal carcinoma, and thymic carcinoma. It readily transforms B lymphocytes and is used in the laboratory for this purpose to develop long-term B lymphocyte cultures. Antibodies produced in patients with EBV infections include those that appear early and are referred to as EA, antibodies against viral capsid antigen (VCA), and antibodies against nuclear antigens (EBNA).

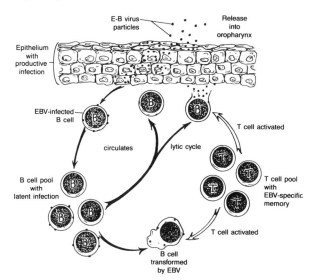

Epstein-Barr Virus-Host Interactions

equilibrium constant
A constant that expresses the state of equilibrium reached by molecules in a reversible reaction such as A + B $\rightleftharpoons$ AB. The equilibrium constant may be expressed as a dissociation constant, $K_D = [A][B]/[AB]$, or an association constant, $K_A = [AB]/[A][B]$.

equilibrium dialysis
Equilibrium dialysis was developed for the study of primary antibody-hapten interactions. The basis for the technique is as follows. Two cells are separated by a semipermeable membrane, allowing the free passage of hapten molecules, but not larger antibody molecules. At time zero (t_0), there is a known concentration of hapten in cell A and antibody in cell B. Hapten from cell A will then diffuse across the membrane into cell B until, at equilibrium, the concentration of free hapten is the same in both cells A and B; that is, the rate of diffusion of hapten from cell A to B is the same as that from cell B to A. Though the concentrations of free hapten are the same in both cells, the total amount of hapten in cell B is greater because some of the hapten is bound to the antibody molecules. A series of experiments are preformed varying the starting amount of hapten concentration, while keeping antibody concentration constant.

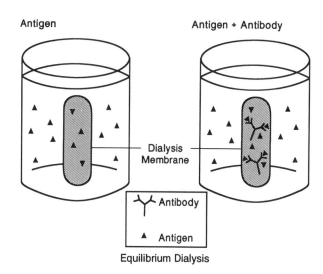

Antigen   Antigen + Antibody

Dialysis Membrane

Antibody

▲ Antigen

Equilibrium Dialysis

equivalence (or equivalence point)
In a precipitation reaction *in vitro*, the antigen to antibody ratio where maximal precipitation takes place. The supernatant should not contain free antigen or free antibody, as all of the antigen and antibody molecules react with one another at equivalence.

erbA, erbB
Oncogenes expressing tyrosine kinase activity. They are similar in structure to the avian erythroblastosis retrovirus. They code for cell membrane proteins. *erbB* is expressed in breast and salivary gland carcinomas and is a truncated version of epidermal growth factor receptor. Increased copy numbers of the c-*erbB*-2 (HER-2/neu) gene suggest an unfavorable prognosis for carcinoma of the breast.

ergotype
A T lymphocyte being activated. The injection of antiergotype T cells blocks full-scale activation of T lymphocytes and may prevent development of experimental autoimmune disease in animal models. An example is experimental allergic encephalomyelitis (EAE), in which antiergotype T lymphocytes may prevent full T lymphocyte activation.

erythema
Redness of the skin caused by dilatation of blood vessels lying near the surface.

erythema marginatum
Immune complex-induced vasculitis in the subcutaneous tissues associated with rheumatic fever.

erythema multiforme
Skin lesions resulting from subcutaneous vasculitis produced by immune complexes. They are frequently linked to drug reactions. The lesions are identified by a red center encircled by an area of pale edema which is encircled by a red or erythematous ring. This gives it a target appearance. Erythema multiforme usually signifies a drug allergy or may be linked to systemic infection. Lymphocytes and ma-crophages infiltrate the lesions. When there is involvement and sloughing of the mucous membranes, the lesion is considered quite severe and even life threatening. This form is called the Stevens-Johnson syndrome.

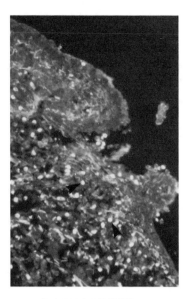

immunocytes in dermis

Erythema Multiforme

erythema nodosum

Slightly elevated erythematous nodules that develop on the shins and sometimes the forearm and head and are quite painful. These nodular lesions represent subcutaneous vasculitis involving small arteries. The phenomenon is associated with infection and is produced by antigen-antibody complexes. Erythema nodosum may be an indicator of inflammatory bowel disease, histoplasmosis, tuberculosis, sarcoidosis, or leprosy. It can follow the use of certain drugs. Although claimed in the past to be due to antigen-antibody deposits in the walls of small venules, the immunologic mechanism may involve type IV (delayed-type) hypersensitivity in the small venules. Neutrophils, macrophages, and lymphocytes infiltrate the subcutaneous fat.

erythroblastosis fetalis

A human fetal disease induced by IgG antibodies passed across the placenta from mother to fetus that are specific for fetal red blood cells, leading to their destruction. Although not often a serious problem until the third pregnancy, the escape of fetal red blood cells into the maternal circulation, especially at the time of parturition, produces a booster response in the mother of the IgG antibody that produces an even more severe reaction in the second and third fetus. The basis for this reaction is an isoantigen such as RhD antigen not present in the mother, but present in the fetal red cells and inherited from the father. Clinical consequences of this maternal-fetal blood group incompatibility include anemia, jaundice, kernicterus, hydrops fetalis, and even stillbirth. Preventive therapy now includes administration of anti-D antiserum (RhIG) within 72 h following parturition. This antibody combines with the fetal red cells dumped into the mother's circulation at parturition and dampens production of a booster response.

erythrocyte agglutination test

Refer to hemagglutination test.

erythrocyte autoantibodies

Autoantibodies against erythrocytes. They are of significance in the autoimmune form of hemolytic anemia and are usually classified into cold and warm varieties by the thermal range of their activity.

erythropoietin

A 46-kD glycoprotein produced by the kidney, more specifically by cells adjacent to the proximal renal tubules, based on the presence of substances such as heme in the kidneys which are oxygen sensitive. It stimulates red blood cell production. It combines with erythroid precursor receptors to promote mature red cell development. Erythropoietin formation is increased by hypoxia. It is useful in the treatment of various types of anemia.

essential mixed cryoglobulinemia

A condition that is identified by purpura (skin hemorrhages), joint pains, impaired circulation in the extremities on exposure to cold (Raynaud's phenomenon), and glomerulonephritis. Renal failure may result. Polyclonal IgG, IgM, and complement are detectable as granular deposits in the glomerular basement membranes. The cryoprecipitates containing IgG and IgM may also contain hepatitis B antigen, as this condition is frequently a sequela of hepatitis B.

estradiol

In diagnostic immunology, a marker identifiable in breast carcinoma tissue by monoclonal antibody and the immunoperoxidase technique that correlates, to a limited degree, with estrogen receptor activity in cytosols from the same preparation.

estrogen/progesterone receptor protein

Monoclonal antibodies against estrogen receptor protein and against progesterone receptor protein permit identification of tumor cells by their preferential immunoperoxidase staining for these markers, whereas stromal cells remain unstained. This method is claimed by some to be superior to cytosol assays in evaluating the clinical response to hormones.

euglobulin

A type of globulin that is insoluble in water, but dissolves in salt solutions. In the past, it was used to designate that part of the serum proteins that could be precipitated by 33% saturated ammonium sulfate at 4°C or by 14.2% sodium sulfate at room temperature. Euglobulin is precipitated from the serum proteins at low ionic strength.

eukaryote

A cell or organism with a real nucleus containing chromosomes encircled by a nuclear membrane.

EVI antibodies

Autoantibodies found in Chaga's disease to endocardium, vascular structures, and interstitium of striated muscle. The target is laminin but the relevance is in doubt, because other diseases produce anti-laminin antibodies that do not produce the unique pathology seen in Chagas disease.

exchange transfusion

A method that involves replacing the entire blood volume of a patient with donor blood. This is done to remove toxic substances such as those formed in kernicterus in infants with erythroblastosis fetalis or may be employed to remove anti-Rh antibodies causing hemolytic disease of the newborn.

excitation filter

A filter in fluorescent microscopes that permits only light of a specific excitation wavelength, such as ultraviolet light, to pass through.

exercise-induced asthma

An attack of asthma brought on by exercise.

exoantigen

Released antigen.

exocytosis

The release of intracellular vesicle contents to the exterior of the cell. The vesicles make their way to the plasma membrane, with which they fuse to permit the contents to be released to the external environment. Examples include immunoglobulin released from plasma cells and mast cell degranulation, which releases histamine and other pharmacological mediators of anaphylaxis to the exterior of the cell.

exon

That segment of a strand of DNA responsible for coding. This continuous DNA sequence in a gene encodes the amino acid sequence of the gene product. Exons are buttressed on both ends by introns, which are noncoding regions of DNA. The coding sequence is transcribed in mature mRNA and subsequently translated into proteins. Exons produce folding regions, functional regions, domains, and subdomains. Introns, which are junk DNA, are spliced out. They constitute the turns or edges of secondary structures.

exotoxins

An extracellular product of pathogenic microorganisms. Exotoxins are 3- to 500-kD polypeptides produced by such microorganisms as *Corynebacterium diphtheria, Clostridium tetani,* and *C. botulinum. Vibrio cholerae* produces exotoxins that elevate cAMP levels in intestinal mucosa cells and increase the flow of water and ions into the intestinal lumen, producing diarrhea. Exotoxins are polypeptides released from bacterial cells and are diffusible, thermolabile, and able to be converted to toxoids that are immunogenic, but not toxic. Bacterial exotoxins are either cytolytic, acting on cell membranes, or bipartite (A-B toxins), linking to a cell surface through the B segment of the toxin and releasing the A segment only after the molecule reaches the cytoplasm where it produces injury.

experimental allergic encephalomyelitis

An autoimmune disease induced by immunization of experimental animals with preparations of brain or spinal cord incorporated into Freund's complete adjuvant. After 10 to 12 days, perivascular accumulations of lymphocytes and mononuclear phagocytes surround the vasculature of the brain and spinal cord white matter. Demyelination may also be present, worsening as the disease becomes chronic. The animals often develop paralysis. The disease can be passively transferred from a sick animal to a healthy one of the same strain with T lymphocytes, but not with antibodies. The mechanism involves T cell receptor interaction with an 18-kD myelin basic protein molecule, which is an organ-specific antigen of nervous system tissue. The CD4+ T lymphocyte represents the phenotype that is reactive with myelin basic protein. The immune reaction induces myelinolysis, wasting, and paralysis. Peptides derived from myelin basic protein (MBP) itself may be used to induce experimental allergic encephalomyelitis in animals. This experimental autoimmune disease is an animal model for multiple sclerosis and postvaccination encephalitis in man.

experimental allergic neuritis

An experimental disease induced by injecting rats with peripheral nerve incorporated into Freund's complete adjuvant. P2 antigen is involved. Lymphocytes and macrophages infiltrate the sciatic nerve, and paralysis may develop.

experimental allergic orchitis

An experimental autoimmune disease induced by injecting experimental animals with isogeneic or allogeneic testicular tissue incorporated into Freund's complete adjuvant.

experimental allergic thyroiditis

An autoimmune disease produced by injecting experimental animals with thyroid tissue or extract or thyroglobulin incorporated into Freund's complete adjuvant. It represents an animal model of Hashimoto's thyroiditis in humans, with mixed extensive lymphocytic infiltrate. Another animal model is the spontaneous occurrence of this disease in the obese strain of chickens as well as in Buffalo rats.

OS Chicken   Normal Chicken
Experimental Allergic Thyroiditis (EAT)

experimental autoimmune myasthenia gravis (EAMG)

Myasthenia gravis (MG) is an autoantibody-mediated autoimmune disease. Experimental forms of MG were made possible through the ready availability of AChR from electric fish. Monoclonal antibodies were developed, followed by molecular cloning techniques that permitted definition of the AChR structure. EAMG can be induced in more than one species of animals by immunizing them with purified AChR from the electric ray (*Torpedo californica*). The autoantigen, nicotinic AChR, is T cell dependent. The *in vivo* synthesis of anti-AChR antibodies requires helper T cell activity. Antibodies specific for the nicotinic AChR of skeletal muscle react with the postsynaptic membrane at the neuromuscular junction.

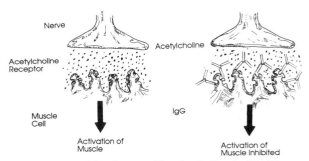

Experimental Autoimmune Myasthenia Gravis (EAMG)

experimental autoimmune thyroiditis (EAT)

A murine model for Hashimoto's thyroiditis. There is a strong MHC genetic component in susceptibility to Hashimoto's thyroiditis, which has been shown to reside in the *IA* subregion of the murine MHC (H-2), governing the immune response (*Ir*) genes to mouse thyroglobulin (MTg). Following induction of EAT with MTg, autoantibodies against MTg appear, and mononuclear cells infiltrate the thyroid. Repeated administration of soluble, syngeneic MTg without adjuvant leads to thyroiditis only in the murine haplotype susceptible to EAT. Autoreactive T cells proliferate *in vitro* following stimulation with MTg. Passively transfer the disease to naive recipients by adoptive immunization and differentiate into cytotoxic T lymphocytes (Tc) *in vitro*. Thus, lymphoid cells rather than antibodies represent the primary mediator of the disease. *In vitro* proliferation of murine-autoreactive T cells was found to show a good correlation with susceptibility to EAT and was dependent on the presence of Thy-1+, Lyt-1+, Ia+, and L3T4+ lymphocytes. Effector T lymphocytes (T$_E$) in EAT comprise various T cell subsets and Lyt-1 (L3T4) and Lyt-2 phenotypes. T lymphocytes cloned from thyroid infiltrates of Hashimoto's thyroiditis patients reveal numerous cytotoxic T lymphocytes and clones synthesizing IL-2 and IFN-γ. While the T cell subsets participate in the pathogenesis of Hashimoto's thyroiditis, autoantibody synthesis appears to aid perpetuation of the disease or result from it.

extended haplotype

Linked alleles in positive linkage disequilibrium situated between and including HLA-DR and HLA-B of the major histocompatibility complex of man. Examples of extended haplotypes include the association of B8/DR3/SCO1/GLO2 with membranoproliferative glomerulonephritis and of A25/B18/DR2 with complement C2 deficiency. Extended haplotypes may be a consequence of crossover suppression through environmental influences together with selected HLA types, leading to autoimmune conditions. The B27 relationship to *Klebsiella* is an example.

extrinsic allergic alveolitis

Inflammation in the lung produced by immune reactivity, mainly of the granulomatous type, to inhaled antigens such as dust, bacteria, mold, grains, or other substances. Also called farmer's lung.

extrinsic asthma

Caused by antigen-antibody reactions, with two, mutually nonexclusive, forms:

1. atopic, involving IgE antibodies
2. nonatopic, involving either antibodies other than IgE or immune complexes

The atopic form usually begins in childhood and is preceded by other atopic manifestations such as paroxysmal rhinitis, seasonal hay fever, or infantile eczema. A familial history is usually obtainable. Other environmental agents may sometimes be recognized in the nonatopic form. The patients with extrinsic asthma respond favorably to bronchodilators.

exudation

The passage of blood cells and fluid containing serum proteins from the blood into the tissues during inflammation.

F

F-actin

Actin molecules in a dual stranded helical polymer. Together with the tropomyosin-tropinin regulatory complex, it constitutes the thin filaments of skeletal muscle.

f allotype

A rabbit IgA subclass α heavy chain allotype. Allelic genes at the *f* locus encode five f allotypes. These are designated f69 through f73. More than one allotypic determinant is associated with each allotype.

f-Met peptides

Bacterial tripeptides such as formyl-Met-Leu-Phe that are chemotactic for inflammatory cells, inducing leukocyte migration.

F protein

A 42-kD protein in the cytoplasm of murine hepatocytes. It occurs in F.1 and F.2 allelic forms in separate inbred murine strains.

F_1 hybrid

The first generation of offspring after a mating between genetically different parents, such as from two different inbred strains.

F_1 hybrid disease

A graft-vs.-host reaction that takes place after the F_1 hybrid is injected with immunologically competent lymphoid cells from a parent. Whereas the hybrid is immunologically tolerant to the histocompatibility antigens of the parent, the injected T lymphocytes from the parent react against antigens of the other parent that are expressed in tissues of the F_1 hybrid, producing graft-vs.-host-like reactivity. The severity of the reaction depends on the histocompatibility differences between the two parents of the F_1 hybrid.

F_2 generation

The second generation of offspring produced following a specific mating.

Fab fragment

A product of papain digestion of an IgG molecule. It is comprised of one light chain and the segment of heavy chain on the N-terminal side of the central disulfide bond. The light and heavy chain segments are linked by interchain disulfide bonds. It is 47 kD and has a sedimentation coefficient of 3.5 S. The Fab fragment has a single antigen-binding site. There are two Fab regions in each IgG molecule.

Fab′ fragment

A product of reduction of an F(ab′)₂ fragment that results from pepsin digestion of IgG. It is comprised of one light chain linked by disulfide bonds to the N-terminal segment of heavy chain. The Fab′ fragment has a single antigen-binding site. There are two Fab′ fragments in each F(ab′)₂ fragment.

F(ab′)₂ fragment

A product of pepsin digestion of an IgG molecule. This 95-kD immunoglobulin fragment has a valence, or antigen-binding capacity, of two, which renders it capable of inducing aggulination or precipitation with homologous antigen. However, the functions associated with the intact IgG molecule's Fc region, such as complement fixation and attachment to Fc receptors on cell surfaces, are missing. Pepsin digestion occurs on the carboxy terminal side of the central disulfide bond at the hinge region of the molecule which leaves the central disulfide bond intact. The C_H2 domain is converted to minute peptides, yet the C_H3 domain is left whole, and the two C_H3 domains comprise the pFc′ fragment.

Fab″ fragment

Refer to F(ab′)₂ fragment.

Fabc fragment

A 5-S intermediate fragment produced by partial digestion of IgG by papain in which only one Fab fragment is cleaved from the parent

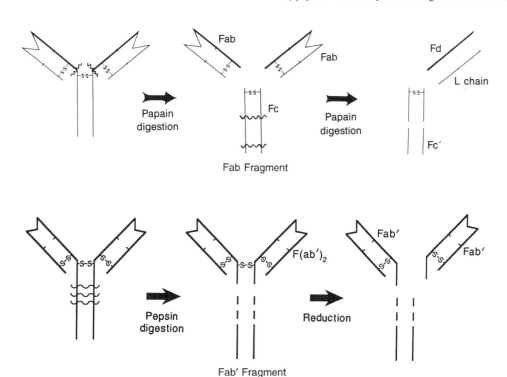

Fab Fragment

Fab′ Fragment

molecule in the hinge region. This leaves the Fabc fragment, which is comprised of a Fab region bound covalently to an Fc region and is functionally univalent.

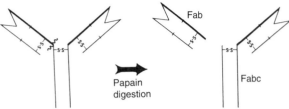

Papain
digestion

Fab

Fabc

Fabc Fragment

Facb fragment

Abbreviation for fragment antigen and complement binding. The action of plasmin on IgG molecules denatured by acid cleaves C_H3 domains from both heavy chain constituents of the Fc region. This yields a bivalent fragment functionally capable of precipitation and agglutination with an Fc remnant still capable of fixing complement.

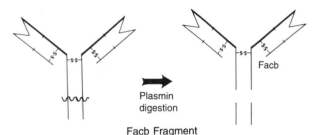

Plasmin
digestion

Facb

Facb Fragment

Factor B

An alternative complement pathway component. It is a 739-amino acid residue single polypeptide chain which combines with C3b and is cleaved by factor D to produce alternative pathway C3 convertase. Cleavage by factor D is at an arginine-lysine bond at position 234 to 235 to yield an amino terminal fragment Ba. The carboxy terminal fragment termed Bb remains attached to C3b. C3bBb is C3 convertase, and C3bBb3b is C5 convertase of the alternative complement pathway. The Bb fragment is the enzyme's active site. There are three short homologous, 60-amino acid residue repeats in factor B, and it possesses four attachment sites for N-linked oligosaccharides. Alleles for human factor B include *BfS* and *BfF*. The factor B gene is located in the major histocompatibility complex situated on the short arm of chromosome 6 in man and on chromosome 17 in mice. Also called C3 proactivator.

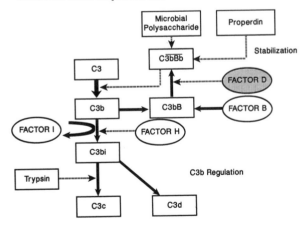

Alternative Complement Pathway
Factor B

Factor D

An alternative complement activation pathway serine esterase. It splits factor B to produce Ba and Bb fragments. It is also called C3 activator convertase.

Factor D deficiency

An extremely rare genetic deficiency of factor D which has an X-linked or autosomal recessive pattern of inheritance. There is only 1% of physiologic amounts of factor D in the serum of affected patients, which renders them susceptible to repeated infection by *Neisseria* microorganisms. There are half the physiologic levels of factor D in the serum of heterozygotes which have no clinical symptoms related to this deficiency.

Factor H

A regulator of complement in the blood under physiologic conditions. Factor H is a glycoprotein in serum that unites with C3b and facilitates dissociation of alternative complement pathway C3 convertase, designated C3bBb, into C3b and Bb. Factor I splits C3b if factor H is present. In man, factor H is a 1231-amino acid residue single polypeptide chain. It is comprised of 20 short homologous repeats which are comprised of about 60 residues present in proteins that interact with C3 or C4. Factor H is an inhibitor of the alternative complement pathway. Previously called β-1H globulin.

Factor H deficiency

Extremely rare genetic deficiency of factor H which has an autosomal recessive mode of inheritance. Only 1% of the physiologic level of factor H is present in the serum of affected individuals, which renders them susceptible to recurrent infections by pyogenic microorganisms. People who are heterozygotes contain 50% of normal levels of factor H in their serum and show no clinical effects.

Factor H receptor (fH-R)

A receptor that initial studies have shown to be comprised of a 170-kD protein expressed by Raji cells and tonsil B cells. Neutrophils, B lymphocytes, and monocytes express fH-R activity.

Factor I

A serine protease that splits the α chain of C3b to produce C3bi and the α chain of C4b to yield C4bi. Factor I splits a 17-amino acid residue peptide termed C3f, if factor H or complement receptor 1 are present, from the C3b α chain to yield C3bi. Factor I splits the C3bi, if complement receptor I or factor H are present, to yield C3c and C3dg. Factor I splits the C4b α chain, if C4-binding protein is present, to yield C4bi. C4c and C4d are produced by a second splitting of the α chain of C4bi. Factor I is a heterodimeric molecule. It is also called C3b/C4b inactivator.

Factor I deficiency

Very uncommon genetic deficiency of C3b inactivator. It has an autosomal recessive pattern of inheritance. There is less than 1% of the physiologic level of factor I in the serum of affected subjects, which renders them susceptible to repeated infections by pyogenic microorganisms. These individuals also reveal deficiencies of factor B and C3 in their serum, since these components are normally split *in vivo* by alternative pathway C3 convertase (C3bBb), which factors I and H inhibit under physiologic conditions. These patients may develop urticaria because of the formation of C3a which induces the release of histamine.

Factor P (properdin)

A key participant in the alternative pathway of complement activation. It is a γ globulin, but not an immunoglobulin, that combines with C3b and stabilizes alternative pathway C3 convertase (C3bB) to produce C3bBbP. Factor P is a 3 or 4 polypeptide chain structure.

Factor VIII

A coagulation protein produced by endothelial cells, which makes it a useful marker for vascular tumors. It is demonstrable by immunoperoxidase staining. Megakaryocytes and platelets also stain for factor VIII.

facultative phagocytes

Cells such as fibroblasts that may show phagocytic properties under special circumstantces.

Fagraeus-Wallbom, Astrid Elsa (1913–)

Swedish investigator noted for her doctoral thesis which provided the first clear evidence that immunoglobulins are made in plasma cells. *Antibody Production in Relation to the Development of Plasma Cells*, 1948.

FANA

Fluorescent antinuclear antibody.

farmer's lung

A pulmonary disease of farm workers who have been exposed repeatedly to organic dust and fungi such as the *Aspergillus* species. It occurs as an extrinsic allergic alveolitis or hypersensitivity pneumonitis in nonatopic subjects. It is mediated by IgG1.

Antibodies specific for moldy hay, in which a number of fungi grow readily, are manifest in 90% of individuals. These include *Microspora vulgaris* and *Micropolyspora faeni*. The pathogenesis is believed to involve a type III hypersensitivity mechanism with the deposition of immune complexes in the lung. Patients become breathless within hours after inhaling the dust and may develop interstitial pneumonitis with cellular infiltration of the alveolar walls where monocytes and lymphocytes are prominent. This may lead to pulmonary fibrosis following chronic inflammation, peribronchiolar granulomatous reaction, and foreign body-type giant cell reactions. Corticosteroids are used for treatment.

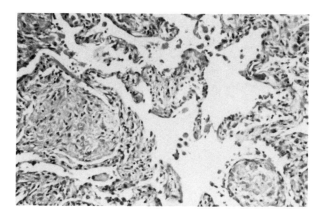

Farmer's Lung

Farr technique

An assay to measure primary binding of antibody with antigen as opposed to secondary manifestations of antibody-antigen interac-

tions such as precipitation, agglutination, etc. It is a quantitation of an antiserum's antigen-binding properties. It is appropriate for antibodies of all immunoglobulin classes and subclasses. The technique is limited to the assay of antibody against antigens soluble in 40% saturated ammonium sulfate solution in which antibodies precipitate. Following interaction of antibody with radiolabeled antigen, precipitation in ammonium sulfate separates the bound antigen from the free antigen. The quantity of radiolabeled antigen that reacted with antibody can be measured in the precipitate. The antibody dilution that precipitates part of the ligand reflects the antigen-binding ability.

Fb fragment

IgG fragment that is the product of subtilisin digestion. It is comprised of the Fab fragment's C_H1 and C_L (constant) domains.

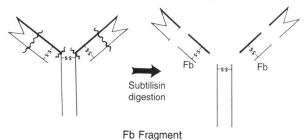

Fb Fragment

Fc fragment (fragment crystallizable)

A product of papain digestion of an IgG molecule. It is comprised of two C-terminal heavy chain segments (C_H2, C_H3) and a portion of the hinge region linked by the central disulfide bond and noncovalent forces. This 50-kD fragment is unable to bind antigen, but it has multiple other biological functions, including complement fixation, interaction with Fc receptors on the cell surfaces and placental transmission of IgG. One Fc fragment is produced by papain digestion of each IgG molecule.

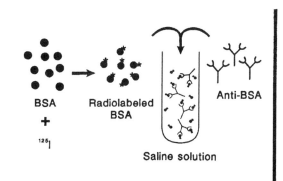

Farr Technique

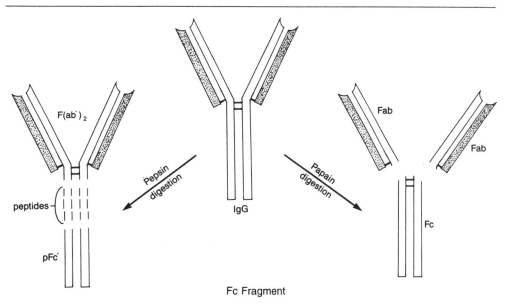

Fc Fragment

Fc piece
Refer to Fc fragment.

Fc receptor
A structure on the surface of some lymphocytes, macrophages, or mast cells that specifically binds the Fc region of immunoglobulin, often when the Fc is aggregated. The Fc receptors for IgG are designated FcγR. Those for IgE are designated FcεR. IgM, IgD, and IgA Fc receptors have yet to be defined. Neutrophils, eosinophils, mononuclear phagocytes, B lymphocytes, selected T lymphocytes, and accessory cells bear Fc receptors for IgG on their surfaces. When the Fc region of immunoglobulin binds to the cation permease Fc receptor, there is an influx of Na^+ or K^+ that activates phagocytosis, H_2O_2 formation, and cell movement by macrophages.

Fcγ Receptors

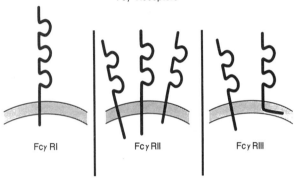

Fcα and Fcε Receptors

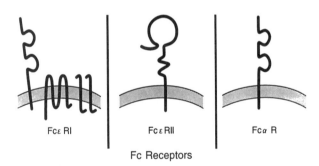

Fc Receptors

Fc receptors on human T cells
Fc receptors are carried on 95% of human peripheral blood T lymphocytes. On about 75% of the cells, the FcR are specific for IgM; the remaining 20% are specific for IgG. The FcR-bearing T cells are also designated T_M and T_G or T_μ and T_γ. The T_M cells act as helpers in B cell function. They are required for the B cell responses to pokeweed mitogen (PWM). In cultures of B cells with PWM, the T_M cells also proliferate, supporting the views on their helper effects. Binding of IgM to FcR of T_M cells is not a prerequisite for helper activity. In contrast to the T_M cells, the T_G cells effectively suppress B cell differentiation. They act on the T_M cells, and their suppressive effect requires prior binding of their Fc-IgG receptors by IgG immune complexes. There are a number of other differences between T_M and T_G cells. Circulating T_G cells may be present in increased numbers, often accompanied by a reduction in the circulating number of T_M cells. Increased numbers of T_G cells are seen in cord blood and in some patients with hypogammaglobulinemia, sex-linked agammaglobulinemia, IgA deficiency, Hodgkin's disease, and thymoma, to mention only a few.

Fc′ fragment
A product of papain digestion of IgG. It is comprised of two noncovalently bonded C_H3 domains that lack the terminal 13 amino acids. This 24-kD dimer consists of the region between the heavy chain amino acid residues 14 through 105 from the carboxy terminal end. Normal human urine contains minute quantities of Fc′ fragment.

Fcε receptor (FcεR)
Mast cell and leukocyte receptor for the Fc region of IgE. When immune complexes bind to Fcε receptors, the cell may respond by releasing the mediators of immediate hypersensitivity, such as histamine and serotonin. Modulation of antibody synthesis may also occur. There are two varieties of Fcε receptors, designated FcεRI and FcεRII (CD23). FcεRI represents a high-affinity receptor

FcεRI

found on mast cells and basophils. It anchors monomeric IgE to the cell surface. It possesses 1α, 1β, and 2γ chains. FcεRII represents a low-affinity receptor. It is found on mononuclear phagocytes, B lymphocytes, eosinophils, and platelets. Subjects with increased IgE in the serum have elevated numbers of FcεRII on their cells. It is a 321-amino acid single polypeptide chain that is homologous with asialoglycoprotein receptor.

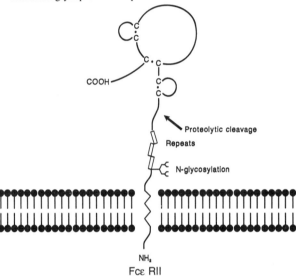

Fcε RII

Fcγ receptor (FcγR).
Receptor for the Fc region of IgG. B and T lymphocytes, natural killer cells, polymorphonuclear leukocytes, mononuclear phagocytes, and platelets contain FcγR. When these receptors bind immune complexes, the cell may produce leukotrienes, prostaglandins, modulate antibody synthesis, increase consumption of oxygen, activate oxygen metabolites, and become phagocytic. The three types of Fcγ receptors include FcγRI, or FcγRII (CD32), and FcγRIII (CD16). FcγRI represents a high-affinity receptor found on mononuclear phagocytes. In humans, it binds IgG1 and IgG3. FcγRII and FcγRIII represent low-affinity IgG receptors. Neutrophils, monocytes, eosinophils, platelets, and B lymphocytes express FcγRII on their membranes. Neutrophils, natural killer cells, eosinophils, macrophages, and selected T lymphocytes express FcγRIII on their membranes in humans and bind IgG1 and IgG3.

Fcγ Receptors

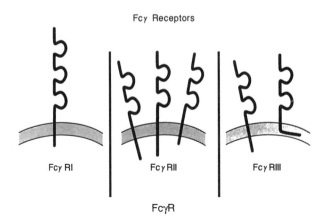

FcγR

Paroxysmal nocturnal hemoglobulinuria patients have deficient FcγRIII on their neutrophil membranes.

Fd fragment

The heavy chain portion of a Fab fragment produced by papain digestion of an IgG molecule. It is on the N-terminal side of the papain digestion site.

Fd piece

Refer to Fd fragment.

Fd' fragment

The heavy chain portion of an Fab' fragment produced by reduction of the F(ab')$_2$ fragment that results from pepsin digestion of IgG. It is comprised of V_H1, C_H1, and the heavy chain hinge region. Fd' contains 235 amino acid residues.

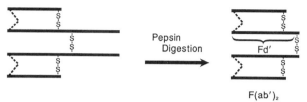

Fd' fragment

Fd' piece

Refer to Fd' fragment.

FDNB (1-fluoro-2-4-dinitrobenzene)

Refer to dinitrofluorobenzene (DNFB).

Felton phenomenon

Specific immunologic unresponsiveness or paralysis induced by the inoculation of relatively large quantities of pneumococcal polysaccharide into mice.

Felty's syndrome

A type of rheumatoid arthritis in which patients develop profound leukopenia and splenomegaly.

feral mouse

A mouse that has returned to the wild from its former commensal existence.

Fernandez reaction

An early (24- to 48-hour) tuberculin-like delayed-type hypersensitivity reaction to lepromin observed in tuberculoid leprosy; a skin test for leprosy.

ferritin

An iron-containing protein that is electron dense and serves as a source of stored iron until it is needed for the synthesis of hemoglobulin. It is an excellent antigen and is found in abundant quantities in horse spleen. Ferritin's electron-dense quality makes it useful to label antibodies or antigens to be identified or localized in electron microscopic preparations.

ferritin labeling

The conjugation of ferritin to antibody molecules to render them visible in histologic or cytologic specimens observed by electron microscopy. Antibodies may be labeled with ferritin by use of a cross-linking reagent such as toluene-2,4-diisocyanate. The

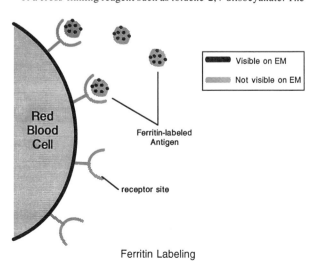

Ferritin Labeling

ferritin-labeled antibody may be reacted directly with the specimen, or ferritin-labeled antiimmunoglobulin may be used to react with unlabeled specific antibody attached to the target tissue antigen.

fertilizin

A glycoprotein present as a jelly-like substance surrounding sea urchin eggs. It behaves as an antigen-like substance from the standpoint of valence. Sperm, which contain proteinaceous antifertilizin, are agglutinated into clumps when combined with soluble fertilizin, which is dissolved from eggs by acidified sea water.

fetal antigen

A fetal or oncofetal antigen that is expressed as a normal constituent of embryos and not in adult tissues. It is reexpressed in neoplasms of adult tissues, apparently as a result of derepression of the gene responsible for its formation.

Feulgen reaction

A standard method that detects DNA in tissues.

fibrillarin antibodies

Antibodies against fibrillarin found in about 8% of diffuse and limited scleroderma patients. Fibrillarin is a 34 kD protein constituent of U3 ribonucleoprotein (U3-RNP). Mercuric chloride induces autoantibodies against U3 small nuclear ribonucleoprotein in susceptible mice.

fibrin

Protein responsible for the coagulation of blood. It is formed through the degradation of fibrinogen into fibrin monomers. Polymerization of the nascent fibrin molecules (comprising the α, β, and γ chains) occurs by end-to-end as well as lateral interactions. The fibrin polymer is envisaged as having two chains of the triad structure lying side by side in a staggered fashion in such a way that two terminal nodules are associated with the central nodule of a third molecule. The chains may also be twisted around each other. The fibrin polymer thus formed is stabilized under the action of a fibrin-stabilizing factor, another component of the coagulation system. Fibrinogen may also be degraded by plasmin. In this process, a number of intermediates, designated as fragments X, Y, D, and E, are formed. These fragments interfere with polymerization of fibrin by binding to nascent intact fibrin molecules, thus causing a defective and unstable polymerization. Fibrin itself is also cleaved by plasmin into similar, but shorter fragments, collectively designated fibrin degradation products. Of course, any excess of such fragments will impair the normal coagulation process, an event with serious clinical significance. Abzymes, such as thromboplastin activator linked to an antibody specific for antigens in fibrin that are not present in fibrinogen, are used clinically to lyse fibrin clots obstructing coronary arteries in myocardial infarction patients.

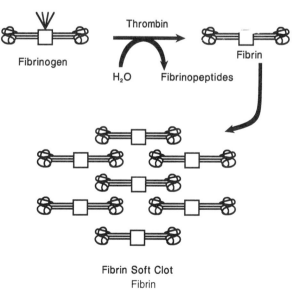

Fibrin Soft Clot

Fibrin

fibrinogen

Fibrinogen is one of the largest plasma proteins and has a mol wt of 330 to 340 kD, comprising more than 3000 amino acid residues. The concentration in the plasma ranges between 200 to 500mg/100 ml. The molecule contains 3% carbohydrate, about 28 to 29 disulfide linkages, and one free sulfhydryl group. Fibrinogen exists as a dimer and can be split into two identical sets comprising three different polypeptide chains. Fibrinogen is susceptible to enzymatic cleavage by a variety of enzymes. The three polypeptide chains of fibrinogen are designated Aα, Bβ, and γ. By electron microscopy, the dried fibrinogen molecule shows a linear arrangement of three nodules 50 to 70 Å in diameter connected by a strand about 15 Å thick.

fibrinoid necrosis
Tissue death in which there is a smudgy eosinophilic deposit that resembles fibrin microscopically and camouflages cellular detail. It is induced by proteases released from neutrophils that digest the tissue and cause fibrin deposition. Fibrinoid necrosis is seen in tissues in a number of connective tissue diseases with immune mechanisms. An example is systemic lupus erythematosus. Fibrinoid necrosis is classically seen in the walls of small vessels in immune complex vasculitis such as occurs in the Arthus reaction.

fibrinopeptides
Peptides released by the conversion of fibrinogen into fibrin. Thrombin splits fragments from the N-terminal region of Aα and Bβ chains of fibrinogen. The split fragments are called fibrinopeptide A and B, respectively, and are released in the fluid phase. They may be further degraded and apparently have vasoactive functions. The release rate of fibrinopeptide A exceeds that of fibrinopeptide B, and this differential release may play a role in the propensity of nascent fibrin to polymerize.

fibronectin
An adhesion-promoting dimeric glycoprotein found abundantly in the connective tissue and basement membrane. The tetrapeptide, Arg-Gly-Asp-Ser, facilitates cell adhesion to fibrin; Clq; collagens; heparin; and type I-, II-, III-, V-, and VI-sulfated proteoglycans. Fibronectin is also present in plasma and on normal cell surfaces. Approximately 20 separate fibronectin chains are known. They are produced from the fibronectin gene by alternative splicing of the RNA transcript. Fibronectin is comprised of two 250-kD subunits joined near their carboxy terminal ends by disulfide bonds. The amino acid residues in the subunits vary in number from 2145 to 2445. Fibronectin is important in contact inhibition, cell movement in embryos, cell-substrate adhesion, inflammation, and wound healing. It may also serve as an opsonin.

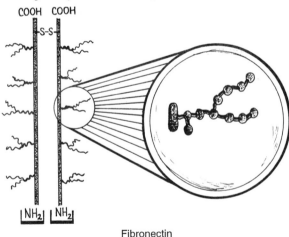

Fibronectin

FICA (fluoroimmunocytoadherence)
The use of column chromatography to capture antigen-binding cells.

ficin
A substance employed to delete sialic acid from cell surfaces which is especially useful in blood grouping to decrease the zeta potential and facilitate otherwise poorly agglutinating antibodies. Erythrocytes treated with ficin reveal enhanced expression of Kidd, Ii, Rh, and Lewis antigens. The treatment destroys MNSs, Lutheran, Duffy, Chido, Rogers, and Tn, among other antigens.

Ficoll
A 400-kD water-soluble polymer comprised of sucrose and epichlorohydrin. It is employed in the manufacture of Ficoll-Hypaque,

a density gradient substance used to separate and purify mononuclear cells by centrifugation following removal of the buffy coat.

Ficoll-Hypaque
A density gradient medium used to separate and purify mononuclear cells by centrifugation.

FIGE
Field inversion gel electrophoresis. Refer to pulsed-field gradient gel electrophoresis.

Cell Separation

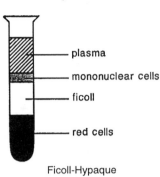

Ficoll-Hypaque

filariasis
Infection by filaria such as *Wuchereria bancrofti.*

fimbria
Hair-like filaments. They are adhesins that facilitate attachment to host cells.

fimbrial antigens
Epitopes of hair-like structures termed fimbriae or pili on Gram-negative bacteria.

final serum dilution
A serological term to designate the titration end point. It is the precise dilution of serum reached following combination with all components needed for the reaction, i.e., the addition of antigen and complement to the diluted serum.

fingerprinting, DNA
A method to demonstrate short, tandem-repeated, highly specific genomic sequences known as minisatellites. There is only a 1 in 30 billion probability that two persons would have the identical DNA fingerprint. It has greater specificity than restriction fragment length polymorphism (RFLP) analysis. Each individual has a different number of repeats. The insert-free wild-type M13 bacteriophage identifies the hypervariable minisatellites. The sequence of DNA that identifies the differences is confined to two clusters of 15-base pair repeats in the protein III gene of the bacteriophage. The specificity of this probe, known as the Jeffries probe, renders it applicable to human genome mapping, parentage testing, and forensic science.

RNA may also be split into fragments by enzymatic digestion followed by electrophoresis. A characteristic pattern for that molecule is produced and aids in identifying it.

first-set rejection (see figure, page 113)
Acute form of allograft rejection in a nonsensitized recipient. It is usually completed in 12 to 14 days. It is mediated by type IV (delayed-type) hypersensitivity to graft antigens.

first-use syndrome
An anaphylactoid reaction that may occur in some hemodialysis patients during initial use of a dialyzer. It may be produced by dialyzer material or by residual ethylene oxide used for sterilization.

FISH (fluorescence *in situ* hybridization)
A method to determine ploidy by examining interphase (nondividing) nuclei in cytogenetic and cytologic samples.

FITC (fluorescein isothiocyanate)
A widely used fluorochrome for labeling antibody molecules. It may also be used to label other proteins. Fluorescein-labeled antibodies are popular because they appear apple-green under ultraviolet irradiation, permitting easy detection of antigens of interest in tissues and cells. FITC fluoresces at 490 and 520 nm. FITC-labeled antibodies are useful for the demonstration of immune deposits in both skin and kidney biopsies.

fixed drug eruption
A hypersensitivity reaction to a drug that appears at the same local site on the body surface regardless of the route by which the drug is administered. The lesion is a clearly circumscribed plaque that is reddish-brown or purple and edematous. It may be covered by a bulla. Common sites of occurrence include the extremities, hands, and glans penis. Drugs that may induce this reaction include suflonamides, barbiturates, quinine, and tetracycline, among other substances. There is hydropic degeneration of the basal layer.

Types of Skin Graft Rejection

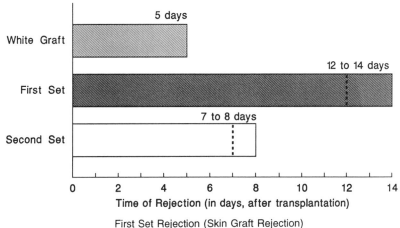

First Set Rejection (Skin Graft Rejection)

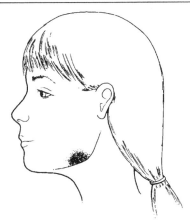

Fixed Drug Eruption

FK506

A powerful immunosuppressive agent synthesized by *Streptomyces tsukubaensis*. Its principal use is for immunosuppression to prevent transplant rejection. It has been used experimentally in liver transplant recipients. It interferes with the synthesis and binding of interleukin-2 and resembles cyclosporin, with which it may be used synergistically. Its immunosuppressive properties are 50 times greater than those of cyclosporin. It has been used in renal allotransplantation, but like cyclosporin also produces nephrotoxicity. It also has some neurotoxic effects and may have diabetogenic potential. The drug continues to be under investigation and is awaiting FDA approval.

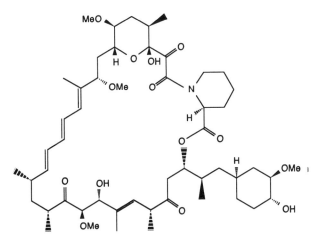

FKBP

A protein that binds FK506. A rotamase enzyme with an amino acid sequence that closely resembles that of protein kinase C. It serves as a receptor for both FK506 and rapamycin.

flagellar antigen

Epitopes of flagella in an organelle that renders some bacteria motile. Also called H antigens.

Bacterial Cell

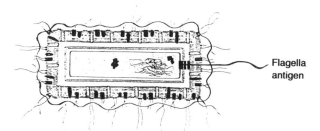

Flagellar Antigen

flagellin

A protein that is a principal constituent of bacterial flagella. It consist of 25- to 60-kD monomers that are arranged into helical chains that wind around a central hollow core. Polymeric flagellin is an excellent thymus-independent antigen. Mutations may occur in the central part of a flagellin monomer which has a variable sequence.

flame cells

Plasma cells whose cytoplasm stains intensely eosinophilic and contains glycoprotein globules. They occur in IgA myelomas especially, but also in Waldenström's macroglobulinemia and in the leptomeninges of African trypanosomiasis patients near neutrophil aggregates.

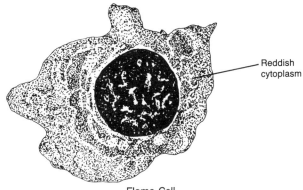

Flame Cell

Flocculation

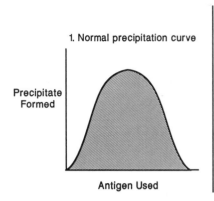

1. Normal precipitation curve

Precipitate Formed

Antigen Used

2. Flocculation curve

Precipitate Formed

Antigen Used

flocculation

A variant of the precipitin reaction in which soluble antigens interact with antibody to produce precipitation over a relatively narrow range of antigen to antibody ratios; also used in the past to refer to flocculation aggregation of such lipids as cardiolipin or other lipids in serological tests for syphilis, but these reactions are more correctly referred to as agglutination rather than flocculation. Flocculation differs from the classic precipitin reaction in that insoluble aggregates are not formed until a greater amount of antigen is added than would be required in a typical precipitin reaction. If the antibody (or total protein) precipitated is plotted vs. antigen added, the plot does not extrapolate to the origin. In flocculation reactions, excess antibody, as well as excess antigen, inhibits precipitation. Thus, precipitation occurs only over a narrow range of antibody to antigen ratios. Soluble antigen-antibody complexes are formed in both antigen and antibody excess.

Horse antisera commonly give flocculation reactions (for example, antisera to diphtheria toxin and certain streptococcal toxins). The peculiar aspects of the flocculation reaction must be attributed to the reacting antibodies as opposed to the antigen, which gives a typical precipitin reaction with rabbit antisera. For many years, this reaction was known as the toxin-antitoxin type of curve because it was observed with horse antibodies against diphtheria and tetanus toxins. In recent years, it has been observed with blood sera from some patients with Hashimoto's thyroiditis. These patients develop autoantibodies against human thyroglobulin. This antithyroglobulin antibody may give a classic precipitin curve, but some individuals develop a flocculation type of antibody response against the antigen.

fluid mosaic model

A fluid lipid molecular bilayer in the plasma membrane and organelle membranes of cells. This structure permits membrane proteins and glycoproteins to float. The lipid molecules are situated in a manner that arranges the polar heads toward outer surfaces and their hydrophobic side chains projecting into the interior. There can be lateral movement of molecules in the bilayer plain or they may rotate on their long axes. This is the Singer-Nicholson "fluid mosaic". The bilayer consists of glycolipids and phospholipids. Amphipathic lipids and globular proteins are spaced throughout the membrane. The fluid consistency permits movement of the receptors, proteins, and glycoprotein laterally.

Fluid Mosaic Model

fluorescein

A yellow dye that stains with brilliant apple-green fluorescence. Its isothiocyanate derivative is widely employed to label proteins such as immunoglobulins that are useful in diagnostic medicine as well as in basic science research.

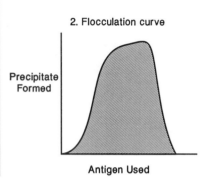

Fluorescein Isothiocyanate

fluorescein-labeled antibody

An antibody tagged with a fluorescein derivative such as fluorescein isothiocyanate. These antibodies are useful to localize antigens in tissues and cells by their brilliant apple-green fluorescence under ultraviolet light.

fluorescence

Emission of light of one wavelength by a substance irradiated with light of a different wavelength.

fluorescence-activated cell sorter (FACS)

An instrument designed to analyze and sort cells such as lymphocytes or other blood cells on the basis of their fluorescence and

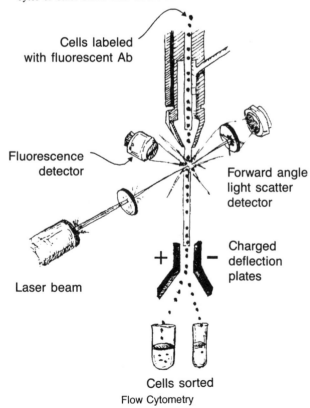

Cells labeled with fluorescent Ab

Fluorescence detector

Forward angle light scatter detector

Laser beam

Charged deflection plates

Cells sorted

Flow Cytometry

light-scattering abilities. The cell population to be analyzed is labeled with a fluorochrome and passed through a rapidly vibrating nozzle where minute droplets are formed. Some of the droplets contain one cell each. A laser focused on the stream immediately preceding its conversion to droplets permits measurement of each cell's fluorescence intensity by a photocell. Scattered light, which indicates the cell size, is measured by a second photocell. If the fluorescence and light scatter signals meet the appropriate standards, each droplet is given either a positive or a negative charge which permits it to be deflected into one receptacle or another as it descends between a positively charged and a negatively charged metal plate. Thus, cells can be separated and retrieved for use in other experiments based on their charge. Results may be displayed as either a dot plot that presents fluorescence intensity in cell size on two axes, or they may be presented as a histogram in which fluorescence intensity or light scatter is on one axis vs. the cell number on the other axis. Among the many uses of this technology is the enumeration of CD4+ lymphocytes that are first reacted with a monoclonal antibody specific for CD4 then followed by interaction with fluorescein-labeled anti-Ig. Besides immunophenotyping, ploidy analyses are another frequent use in the investigation of tumors.

fluorescence enhancement

Increased fluorescence of certain substances after their combination with antibody. This is attributable to changing the substance from an aqueous milieu to the antibody combining site's hydrophobic surroundings.

fluorescence microscopy

A special microscope which uses ultraviolet light to illuminate a tissue or cell stained with a fluorochrome-labeled substance such as an antibody against an antigen of interest in the tissue. When returning from an excited state to a ground state, the tissue emits fluorescent light, which permits the observer to localize an antigen of interest in the tissue or cell.

fluorescence quenching

A method to ascertain association constants of antibody molecules interacting with ligands. Fluorescence quenching results from excitation energy transfer, where certain electronically excited residues in protein molecules, such as tryptophan and tyrosine, transfer energy to a second molecule which is bound to the protein. Maximum emission is a wavelength of approximately 345 nm. The attachment of the acceptor molecule need not be covalent. This transfer of energy occurs when the absorbence spectrum of the acceptor molecule overlaps with that of the emission spectrum of the donor and takes place via resonance interaction. There is no need for direct contact between the two molecules for energy transfer. If the acceptor molecule is nonfluorescent, diminution of energy occurs through nonradiation processes. On the other hand, if the acceptor molecule is fluorescent, the transfer of radiation to it results in its own fluorescence (sensitized fluorescence). Fluorescence quenching techniques can provide very sensitive quantitative data on antibody-hapten interactions.

fluorescence treponemal antibody test

Refer to FTA-ABS test.

fluorescent antibody

An antibody molecule to which a fluorochrome has been conjugated, such as fluorescein isothiocyanate.

fluorescent antibody technique

An immunofluorescence method in which antibody labeled with a fluorochrome such as fluorescein isothiocyanate (FITC) is used to identify antigen in tissues or cells when examined by ultraviolet light used in fluorescence microscopy. Besides the direct technique,

antigens in tissue sections treated with unlabeled antibody can be "counterstained" with fluorescein-labeled antiimmunoglobulin to localize antigen in tissues by the indirect immunofluorescence method.

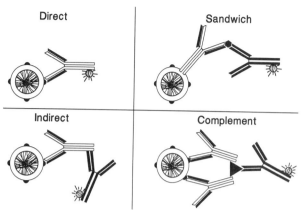

Fluorescent Antibody Techniques

fluorescent protein tracing

Fluorescent dyes are used in place of nonfluorescent dyes because they are detectable in a much lower concentration. Radioactive labeling is employed usually if the substance to be detected is present in minute amounts. Fluorescent labeling, however, provides simplicity of technique and precise microscopic observation of fluorescence. Fluorescent microscopic preparations require several hours and permit localization at the cellular level, whereas autoradiograms require a longer period and are localized at the tissue level. Either fluorescein (apple-green fluorescence) or rhodamine (reddish-orange fluorescence) compounds may be used for tracing.

fluorochrome

A label such as fluorescein isothiocyanate or rhodamine isothiocyanate, which is used to label antibody molecules or other substances. A fluorochrome emits visible light of a defined wavelength upon irradiation with light of a shorter wavelength such as ultraviolet light.

fluorodinitrobenzene

Refer to dinitrofluorobenzene (DNFB).

fluorography

A method to identify radiolabeled proteins following their separation by gel electrophoresis. A fluor such as diphenyl oxazole is incorporated into the gel where it emits photons on exposure to a radioisotope. After drying, the gel is placed on X-ray film in the dark.

fog fever

An episode of acute respiratory distress in cows approximately 7 days after their removal to a pasture where hay has been recently cut. They may die within 1 day, developing pulmonary edema with extensive emphysema. This disease may present as an atopic allergy in sensitized animals who are exposed to grass proteins, pollen, and fungal spores. A nonimmunologic intoxication has also been suggested as a cause. Cattle may also have a similar reaction, which resembles farmer's lung in man, if they had been fed hay containing *Micropolyspora faeni* spores. Precipitating antibodies are present in their blood sera.

Rhodamine disulphonic acid   Fluorescein

Fluorochrome

follicle

Circular or oval areas of lymphocytes in lymphoid tissues rich in B cells. They are present in the cortex of lymph nodes and in the splenic white pulp. Primary follicles contain B lymphocytes that are small and medium sized. They are demonstrable in lymph nodes prior to antigenic stimulation. Once a lymph node is stimulated by antigen, secondary follicles develop. They contain large B lymphocytes in the germinal centers where tingible body macrophages (those phagocytizing nuclear particles) and follicular dendritic cells are present.

follicular center cells

B lymphocytes in germinal centers (secondary follicles).

follicular dendritic cell

Cells manifesting narrow cytoplasmic processes that interdigitate between densely populated areas of B lymphocytes in lymph node follicles and in spleen. Antigen-antibody complexes adhere to the surfaces of follicular dendritic cells and are not generally endocytosed, but are associated with the formation of germinal centers. These cells are bereft of class II histocompatibility molecules, although Fc receptors, complement receptor 1, and complement receptor 2 molecules are demonstrable on their surfaces.

Follicular Dendritic Cell

follicular hyperplasia

Lymph node enlargement associated with an increase in follicle size and number. Germinal centers are usually present in the follicles. Follicular hyperplasia is often a postinfection reactive process in lymph nodes.

food allergy

Type I (anaphylactic) or type III (antigen-antibody complex) hypersensitivity mechanism responses to allergens or antigens in foods that have been ingested may lead to intestinal distress, producing nausea, vomiting, and diarrhea. There may be edema of the buccal mucosa, generalized urticaria, or eczema. Food categories associated with food allergy in some individuals include eggs, fish, or nuts. Both skin tests and RAST tests using the appropriate allergen or antigen may identify individuals with a particular food allergy.

footprinting

A method to ascertain the DNA segment (or segments) which binds to a protein. Radiolabeled double-stranded DNA is combined with the binding protein to yield a complex that is exposed to an endonuclease that cuts the molecules once and at random. The digested DNA is electrophoresed in polyacrylamide gel together with a control DNA sample (which has been treated similarly, but without added protein) to permit separation of fragments differing in length by one nucleotide. Autoradiography of the material reveals a series of bands representing the DNA fragments. In the area of protein binding, the DNA is spared from digestion, and no corresponding bands appear compared to the control. The protected area's specific location can be ascertained by running a DNA sequencing gel in parallel.

footprints

Macrophages filled with *Mycobacterium leprae* without caseation necrosis. A similar situation may be observed in anergic Hodgkin's disease patients and in AIDS patients infected with *M. avium-intercellulare.*

forbidden clone theory

According to this hypothesis, self-reactive lymphocyte clones are eliminated in the thymus during embryonic life, but mutation later on may permit the reappearance of self-reactive clones of lymphocytes that induce autoimmunity.

formol toxoids

A toxoid generated by the treatment of an exotoxin such as diphtheria toxin with formalin. Although first used nearly a century ago, it was subsequently modified to contain an adjuvant such as an aluminum compound to boost immune responsiveness to the toxoid. It was later replaced with the so-called triple vaccine of diphtheria, pertussis, and toxoid vaccine.

formyl-methionyl-leucyl-phenylalanine (F-Met-Leu-Phe)

A synthetic peptide that is a powerful chemotactic attractant for leukocytes, facilitating their migration. It may also induce neutrophil degranulation. This peptide resembles chemotactic factors released from bacteria. Following interaction with neutrophils, leukocyte migration is enhanced, and complement receptor 3 molecules are increased in the cell membrane.

Forssman antibody

An antibody specific for the Forssman (heterogenetic) antigen. Human serum may contain Forssman antibody as a natural antibody.

Forssman antigen

A heterophil or heterogenetic glycolipid antigen that stimulates the synthesis of anti-sheep hemolysin in rabbits. Its broad phylogenetic distribution spans both animal and plant kingdoms. The antigen is present in guinea pig and horse organs, but not in their red blood cells. In sheep, it is found exclusively in erythrocytes. Forssman antigen occurs in both red blood cells and organs in chickens. It is also present in goats, ostriches, mice, dogs, cats, spinach, *Bacillus anthracis,* and on the gastrointestinal mucosa of a limited number of people. It is absent in rabbits, rats, cows, pigs, cuckoos, beans, and *Salmonella typhi.* Forssman substance is ceramide tetrasaccharide. The Forssman antigen contains *N*-acylsphingosine (ceramide), galactose, and *N*-acetylgalactosamine. As originally defined, it is present in guinea pig kidney, heat stable, and alcohol soluble. Forssman antigen-containing tissue is effective in absorbing the homologous antibody from serum. Antibodies to the Forssman antigen occur in the sera of patients recovering from infectious mononucleosis.

Forssman, Magnus John Karl August (1868–1947)

Swedish professor for whom the heterophile antigen was named, following his discovery and work with that protein.

forward angle light scatter (FALS)

In flow cytometry, particle size is measured by the amount of scattered light (0.5 to 2 degrees) detected as the focused laser beam encounters the cell and continues as scattered light to the forward angle light scatter detector.

foscarnet

An investigational drug used to combat cytomegalovirus-induced pneumonia, hepatitis, colitis, and retinitis in AIDS patients rendered nonresponsive to gancyclovir, which is a frequently used treatment for cytomegalovirus infection.

Foscarnet

F:P ratio

Fluorescence to protein ratio that expresses the ratio of fluorochrome to protein in an antibody preparation labeled with the fluorochrome.

Fracastoro, Girolamo (1478–1553)

A physician who was born at Verona and educated at Padua. His interests ranged from poetry to geography. He proposed the theory of acquired immunity and was a leader in the early theories of contagion. *Syphilis sive Morbus Gallicus,* 1530; *De Sympathia et Antipathia Rerum,* 1546; *De Contagione,* 1546.

fractional catabolic rate

The total plasma immunoglobulin percentage that is catabolized each day. Predicted from the half-life of plasma or from the excretion rate of catabolized immunoglobulin products in urine.

fragmentins

Serine esterases present in cytotoxic T cell and natural killer cell cytoplasmic granules. The introduction of fragmentins into the cytosol of a cell causes apoptosis as the DNA in the nucleus is fragmented into 200 base pair multimers.

framework regions (FR)

Amino acid sequences in variable regions of heavy or light immunoglobulin chains other than the hypervariable sequences. There is much less variability in the framework region than in the hypervariable region. Two β-pleated sheets opposing one another comprise the structural features of an antibody domain's framework regions. Polypeptide chain loops join the β-pleated sheet strands. The framework regions contribute to the secondary and tertiary structure of the variable region domain, although they are less significant than the hypervariable regions for the antigen-binding site.

The framework region forms the folding part of the immunoglobulin molecule. Light chain FRs are found at amino acid residues 1 to 28, 38 to 50, 56 to 89, and 97 to 107. Heavy chain FRs are present at amino acid residues 1 to 31, 35 to 49, 66 to 101, and 110 to 117.

freemartin

The female member of dizygotic cattle twins where the other twin is a male. Their placentas are fused *in utero,* causing them to be exposed to each other's cells *in utero* prior to the development of immunologic maturity. This renders the animals immunologically tolerant of each other's cells and prevents them from rejecting grafts from the other twin. The female twin has reproductive abnormalities and is sterile.

Frei test

A tuberculin type of delayed hypersensitivity skin test employed to reveal delayed-type hypersensitivity in lymphogranuloma venereum patients. Following intradermal injection of lymphogranuloma venereum virus, an erythematous and indurated papule develops after 4 days.

Freund, Jules (1890–1960)

Hungarian physician who later worked in the United States. He made many contributions to immunology, including work on antibody formation, studies on allergic encephalomyelitis, and the development of Freund's adjuvant. He received the Lasker award in 1959.

Freund's adjuvant

A water-in-oil emulsion that facilitates or enhances an immune response to antigen that has been incorporated into the adjuvant. There are two forms. Freund's complete adjuvant (CFA) consists of light-weight mineral oil that contains killed, dried mycobacteria. Antigen in an aqueous phase is incorporated into the oil phase containing mycobacteria with the aid of an emulsifying agent such as Arlacel A. This emulsion is then used as the immunogen. Freund's incomplete adjuvant (IFA) differs from the complete form only in that it does not contain mycobacteria. In both cases, the augmenting effect depends on and parallels the magnitude of the local inflammatory lesion, essentially a nonnecrotic monocytic reaction with fibrous encapsulation. Whereas the complete form facilitates stimulation of both T and B limbs of the immune response, the incomplete variety enhances antibody formation, but does not stimulate cell-mediated immunity except for transient Jones-Mote reactivity. The adjuvant principle in mycobacteria is the cell wall wax D fraction. CFA does not potentiate the immune response to the so-called thymus-independent antigens such as pneumococcal polysaccharide or polyvinylpyrrolidone. CFA may be combined with normal tissues and injected into animals of the type supplying the tissue to induce autoimmune diseases such as thyroiditis, allergic encephalomyelitis, or adjuvant arthritis.

front typing

In blood typing for transfusion, antibodies of known specificity are used to identify erythrocyte ABO antigens. Differences between front and back typing might be attributable to acquired group B or B subtypes, diminished immunoglobulins, anti-B and anti-A_1 antibody polyagglutination, rouleaux formation, cold agglutinins, Wharton's jelly, or two separate cell populations.

FTA-ABS. (Fluorescence treponema antibody absorption)

A serological test for syphilis that is very sensitive, i.e., approaching 100% in the diagnosis of secondary, tertiary, congenital, and neurosyphilis.

FTA-ABS test

An assay for specific antibodies to *Treponema pallidum* in the serum of patients suspected of having syphilis. Before combining the patient's serum with killed *T. pallidum* microorganisms fixed on a slide, the serum is first absorbed with an extract of Reiter's treponemes to remove group-specific antibodies. After washing, the specimen is covered with fluorescein-labeled anti-human globulin. This is followed by examination by fluorescence microscopy equipped with ultraviolet light. Demonstration of positive fluorescence of the target microorganisms reveals specific antibody present in the patient's serum. The greater specificity and sensitivity of this test makes it preferable to the previously used FTA-200 assay.

functional affinity

Association constant for a bivalent or multivalent antibody's interaction with a bivalent or multivalent ligand. The multivalent reactivity may enhance the affinity of multiple antigen-antibody reactions. Avidity has a similar connotation, but it is a less precise term than is functional affinity.

functional antigen

Refer to protective antigen.

functional immunity

Refer to protective immunity.

Fv fragment

An antibody fragment consisting of the N-terminal variable segments of both heavy (V_H) and light (V_L) chain domains that are joined by noncovalent forces. The fragment has one antigen-binding site.

Fv region

The N-terminal variable segments of both the heavy and light chains in each Fab region of an immunoglobulin molecule with a four-chain unit structure.

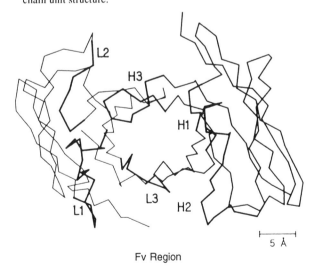

Fv Region

fyn

Refer to lck, fyn, ZAP (phosphotyrosine kinases in T cells).

G

G-CSF

Granulocyte colony-stimulating factor. A biological response modifier that facilitates formation of granulocytes in the bone marrow. It was first licensed by the FDA in 1991 and may be useful to reactivate granulocyte production in the marrow of irradiated or chemotherapy-treated patients. The genes for G-CSF are found on chromosome 17. Endothelial cells, macrophages, and fibroblasts produce G-CSF, which functions synergistically with IL-3 in stimulating bone marrow cells.

γδ T cell receptor (TCR)

Double-negative CD4⁻CD8⁻ T cells may express CD3-associated γδ TCRs. γδ TCRs may be capable of reacting with polymorphic ligand(s). Activation properties of cells expressing γδ TCRs closely resemble those of cells expressing αβ TCRs. The initial receptor to be expressed during thymic ontogeny is the γδ TCR. Cells lacking γδ TCR expression may subsequently rearrange α and β chains, resulting in αβ TCR expression. Thus, a single cell never normally expresses both receptors. Cells expressing the αβ TCR mediate helper T cell and cytotoxic T cell functions. Whereas γδ TCRs are expressed on double-negative cells only, αβ TCRs may be present on either CD4⁺ or CD8⁺. The γδ TCR apparently can function in the absence of MHC molecules, in contrast to the association of CD4⁺ with MHC class II or CD8⁺ with MHC class I recognition by cells expressing αβ TCRs.

γδ TCR-expressing cells might protect against microorganisms entering through the epithelium in the skin, lung, intestines, etc. γδ TCR-bearing cells constitute a majority of T cells during thymic ontogeny and in the epidermis of the mouse. It is not known whether the epidermis/epithelium in the skin can function as a site for T cell education and maturation. γδ TCR represents an evolutionary precursor of the αβ TCR, as reflected by the relatively low percentages of cells expressing the γδ TCRs in adults and the fact that cells expressing the αβ TCR carry out the principal immunologic functions. Diversity of the γδ TCR and lymphokine synthesis by cells expressing the γδ TCR attest to the significance of the cells in the immune system.

γδ T cells

Early T lymphocytes that express γ and δ chains comprising the T cell receptor of the cell surface. They comprise only 5% of the normal circulating T cells in healthy adults. γδ T cells home to the lamina propria of the gut. Their function is not fully understood.

γ globulin

Obsolete designation for immunoglobulin. Serum proteins that show the lowest mobility toward the anode during electrophoresis when the pH is neutral. The γ globulin fraction contains immunoglobulins. It is the most cationic of the serum globulins.

THE GAMMA GLOBULIN REGION

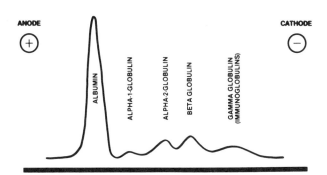

γ heavy chain disease

γ heavy chain disease, also called Franklin's disease, is a very rare syndrome in which the myeloma cells synthesize γ heavy chains only. Clinically, this disease affects mostly older individuals who have either a gradual or sudden onset. The patients may be weak, have fever and malaise, and demonstrate lymphadenopathy over a period of time. Swelling of the uvula and edema of the palate may be a consequence of lymphoid tissue involvement of the nasopharynx and Waldeyer's ring. Lymph nodes affected may include those of the axilla, mediastinum, tracheobronchial tree, and abdomen. Fever and enlargement of the spleen and liver may follow infection of these patients. The disease may last from several months to 5 years and usually leads to death, although remission has been described in occasional patients. The blood serum contains proteins with an electrophoretic peak that correspond in mobility to the homogeneous (Bence-Jones negative) protein present in the urine. There is an elevated sedimentation rate, mild anemia, thrombocytopenia, leukopenia, and sometimes eosinophilia. Abnormal plasma cells and lymphocytes may appear in the blood and occasionally be manifested as plasma cell leukemia. To confirm the diagnosis of γ heavy chain disease, one must demonstrate a spike with the electrophoretic mobility of a fast γ or β globulin reactive with antiserum against γ heavy chains, but not with κ or λ light chains. There are numerous deletions in γ chains that vary in location, but often include most of the variable region and the total C_H1 segment, with continuation of the normal sequence where the hinge region begins. Whereas patients with Franklin's disease demonstrate heavy γ chains in both the blood serum and urine, the other immunoglobulin levels are diminished. There is total failure in light chain synthesis in all individuals. Thus, γ heavy chain disease may mimic lymphomas of one type or another (as well as multiple myeloma, toxoplasmosis, and histoplasmosis) and may be associated occasionally with tuberculosis, rheumatoid arthritis, and various other conditions.

γ interferon

Refer to interferon γ.

γ macroglobulin

Obsolete term for IgM.

gag (see figure, page 120)

The retroviral HIV-1 gene that encodes the heterogeneous p24 protein of the virus core.

gammopathy

An abnormal increase in immunoglobulin synthesis. Gammopathies that are monoclonal usually signify malignancy such as multiple myeloma, Waldenström's disease, heavy chains disease, or chronic lymphocytic leukemia. Benign gammopathies occur in amyloidosis and monoclonal gammopathy of undetermined etiology. Inflammatory disorders are often accompanied by benign polyclonal gammopathies. These include rheumatoid arthritis, lupus erythematosus, tuberculosis, cirrhosis, and angioimmunoblastic lymphadenopathy.

gancyclovir (9-[2-hydroxy-1(hydroxymethyl) ethoxymethyl] guanine)

An antiviral drug used for the therapy of immunocompromised patients infected with cytomegalic inclusion virus. Five days of gancyclovir therapy has proven effective for clearing CMV from the blood, urine, and respiratory secretions. It has been used successfully to treat CMV retinitis, gastroenteritis, and hepatitis. Drug resistance may develop. The drug may induce neutropenia and thrombocytopenia as side effects. It has not proven very effective in AIDS or bone marrow transplant patients.

gas gangrene antitoxin

Antibodies found in antisera against exotoxins produced by *Clostridium perfringens, C. septicum,* and *C. oedematiens,* bacteria which may cause gas gangrene. In the past, this antiserum was used together with antibiotics and surgical intervention in the treatment of wounds where gas gangrene was possible.

Retroviral Genome

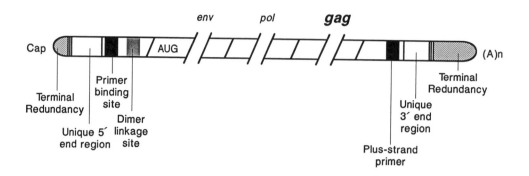

gastrin-producing cell antibodies (GPCA)

Antibodies present in 8 to 16% of antral (type B) chronic atrophic gastritis patients. GPCA may lead to diminished gastrin secretion and fewer gastrin-producing cells in antral gastritis. There are no parietal cell antibodies present, and GPCA are not linked to pernicious anemia. GPCA in type B gastritis are different from gastric fundus parietal cell antibodies that have been linked to fundal mucosal atrophy in type A gastritis.

gastrin receptor antibodies

Antibodies claimed to be present in 30% of pernicious anemia patients by some investigators but unconfirmed by others. Gastric receptor antibodies may inhibit gastrin binding and have been demonstrated to bind gastric parietal cells.

gatekeeper effect

Contraction of endothelium mediated by IgE, permitting components of the blood to gain access to the extravascular space as a consequence of increased vascular permeability.

gay bowel syndrome

A constellation of gastrointestinal symptoms in homosexual males related to both infectious and noninfectious etiologies before the AIDS epidemic. Clinical features include alterations in bowel habits, condyloma acuminata, bloating, flatulence, diarrhea, nausea, vomiting, adenomatous polyps, fistulas, fissures, hemorrhoids, and perirectal abscesses, among many other features. Associated sexually transmitted infections include syphilis, herpes simplex, gonorrhea, and *Chlamydia trachomatis*. Numerous other microbial species identified include human papilloma virus, *Campylobacter* organisms, hepatitis A and B, cytomegalovirus and parasites such as *Entamoeba histolytica*. Treatment varies with the etiology of various gay bowel syndrome manifestations.

gel

Agar gel is a semisolid substance prepared from seaweed agar that has been widely used in the past in bacteriology, but is used in immunology for diffusion of antigen and antibody in Ouchterlony-type techniques, electrophoresis, immunoelectrophoresis, and related methods.

gel diffusion

A method to evaluate antibodies and antigens based upon their diffusion in gels toward one another and their reaction at the point of contact in the gel.

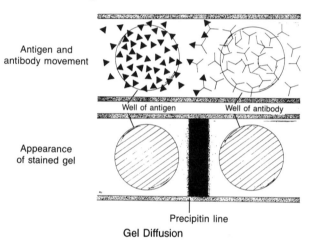

Antigen and antibody movement

Appearance of stained gel

Precipitin line

Gel Diffusion

gel filtration chromatography

This method permits the separation of molecules on the basis of size. Porous beads are allowed to swell in buffer, water, or other solutions and are packed into a column. The molecular pores of the beads will permit the entry of some molecules into them, but exclude others on

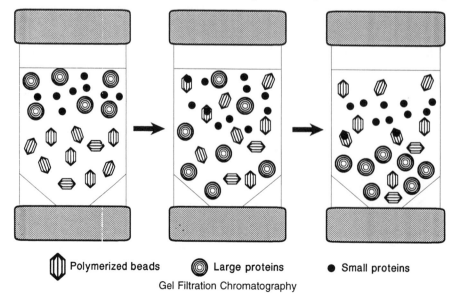

◇ Polymerized beads　◎ Large proteins　● Small proteins

Gel Filtration Chromatography

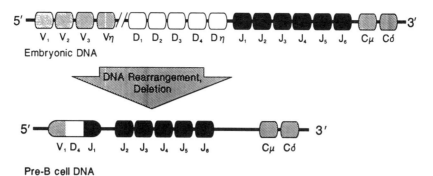

Embryonic DNA

DNA Rearrangement, Deletion

Pre-B cell DNA

Gene Rearrangement

the basis of size. Molecules larger than the pores will pass through the column and emerge with the void volume. Since the solute molecules within the beads maintain a concentration equilibrium with solutes in the liquid phase outside the beads, molecular species of a given weight, shape, and degree of hydration move as a band. Gel chromatography using spherical agarose gel particles is useful in the exclusion of IgM, which is present in the first peak. Of course, other molecules of similar size, such as α macroglobulins, are also present in this peak. IgG is present in the second peak, but the fractions of the leading side are contaminated by IgA and IgD.

gene bank

A synonym for DNA library.

gene cloning

The use of recombinant DNA technology to replicate genes or their fragments.

gene conversion hypothesis

Describes the method by which alternative sequences can be introduced into the MHC genes without reciprocal crossover events. This mechanism could account for the incredible polymorphism of alleles. The alternative sequences may include those found in the class I-like genes and pseudogenes present on chromosome 6. Hypothetically, gene conversion was an evolutionary event as well as an ongoing one, giving rise to new mutations, therefore new alleles, within a population.

gene diversity

Determination of the extent of an immune response to a particular antigen or immunogen as determined by mixing and matching of exons from variable, joining, diversity, and constant region gene segments. S. Tonegawa received the Nobel prize for revealing the mechanism of the generation of diversity in antibody formation.

gene mapping

Refers to gene localization or gene order. Gene localization can be in relationship to other genes or to a chromosomal band. The term may also refer to the ordering of gene segments.

gene rearrangement (see figure, above)

Genetic shuffling that results in elimination of introns and the joining of exons to produce mRNA. Gene rearrangement within a lymphocyte signifies its dedication to the formation of a single cell type, which may be immunoglobulin synthesis by B lymphocytes or production of a β chain receptor by T lymphocytes. Neoplastic transformation of lymphocytes may be followed by the expansion of a single clone of cells, which is detectable by Southern blotting.

generalized anaphylaxis

The signs and symptoms of anaphylactic shock manifest within seconds to minutes following the administration of an antigen or allergen that interacts with specific IgE antibodies bound to mast cell or basophil surfaces, causing the release of pharmacologically active mediators that include vasoactive amines from their granules. Symptoms may vary from transient respiratory difficulties (due to contraction of the smooth muscle and terminal bronchioles) to even death.

generalized vaccinia

A condition observed in some children being vaccinated against smallpox with vaccinia virus. There were numerous vaccinia skin lesions that occurred in these children who had a primary immunodeficiency in antibody synthesis. Although usually self-limited, children who also had atopic dermatitis in addition to the generalized vaccinia often died.

genetic code

The codons, i.e., nucleotide triplets, correlating with amino acid residues in protein synthesis. The nucleotide linear sequence in mRNA is translated into the amino acid residue sequence.

genetic switch hypothesis

A concept that predicts a switch in the gene governing heavy chain synthesis by plasma cells during immune response ontogeny.

genome

All genetic information that is contained in a cell or in a gamete.

genomic DNA

DNA found in the chromosomes. Refer to DNA.

genotype

An organism's genetic makeup.

germ-free animal

An animal such as a laboratory mouse raised under sterile conditions, where it is free from exposure to microorganisms and is not exposed to larger organisms. Germ-free animals have decreased serum immunoglobulin and lymphoid tissues that are not fully developed. The diet may also be controlled to avoid exposure to food antigens. Most difficult is the ability to maintain a virus-free environment for these animals.

germinal center

Germinal (or follicular) centers in lymph node and lymphoid aggregates within primary follicles of lymphoid tissues following antigenic stimulation. The mixed-cell population in the germinal center is comprised of B lymphoblasts (both cleaved and transformed lymphocytes), follicular dendritic cells, and numerous tingible

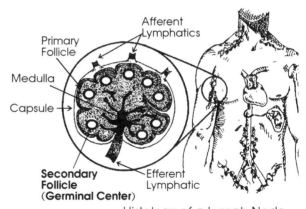

Histology of a Lymph Node

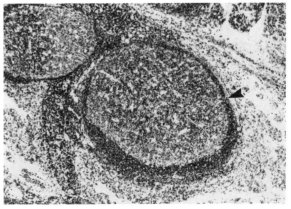

Germinal Center

body-containing macrophages. Germinal centers seen in various pathologic states include "burned out" germinal centers comprised of accumulations of pale histiocytes and scattered immunoblasts; "progressively transformed" centers that show a "starry sky" pattern containing epithelioid histiocytes, dendritic reticulum cells, increased T lymphocytes, and mantle zone lymphocytes; and "regressively transformed" germinal centers that are relatively small with few lymphocytes and reveal an onion-skin layering of dendritic reticulum cells, vascular endothelial cells, and fibroblasts.

germinal follicle
Refer to germinal center.

Ghon complex
The combination of a pleural surface-healed granuloma or scar on the middle lobe of the lung, together with hilar lymph node granulomas. The Ghon complex signifies healed primary tuberculosis.

giant cell arteritis
Inflammation of the temporal artery in middle-age subjects that may become a systemic arteritis in 10 to 15% of affected subjects. Blindness may occur eventually in some of them. There is nodular swelling of the entire artery wall. Neutrophils, mononuclear cells, and eosinophils may infiltrate the wall with production of giant cell granulomas.

gld gene
Murine mutant gene on chromosome 1. When homozygous the *gld* gene produces progressive lymphadenopathy and lupus-like immunopathology.

glial fibrillary acidic protein (GFAP)
An intermediate filament protein constituent of astrocytes which is also abundant in glial cell tumors. The immunoperoxidase technique employing monoclonal antibodies against the GFAP is used in surgical pathologic diagnosis to identify tumors based on their histogenetic origin.

globulin
Serum proteins comprised of α, β, and γ globulins and classified on the basis of their electrophoretic mobility. All three globulin fractions demonstrate anodic mobility that is less than that of albumin. α Globulins have the greatest negative charge, whereas γ globulins have the least negative charge. Originally, globulins were characterized based on their insolubility in water, i.e., the euglobulins, or sparing solubility in water, i.e., the pseudoglobulins. Globulins are precipitated in half-saturated ammonium sulfate solution.

glomerulonephritis (GN)
A group of diseases characterized by glomerular injury. Immune mechanisms are responsible for most cases of primary glomerulonephritis and many of the secondary glomerulonephritis group. Over 70% of glomerulonephritis patients have glomerular deposits of immunoglobulins, frequently with complement components. Antibody-associated injury may result from the deposition of soluble circulating antigen-antibody complexes in the glomerulus or by antibodies reacting in the glomerulus either with antigens intrinsic to the glomerulus or with molecules planted within the glomerulus. Cytotoxic antibodies may also cause glomerular injury. Goodpasture's syndrome is an example of a disease in which antibodies react directly with the glomerular basement membrane, interrupting its integrity and permitting red blood cells to pass into the urine. Antigen-antibody complexes such as those produced in systemic lupus erythematosus may be deposited in the walls of the peripheral capillary loops, especially in a subendothelial location, leading to various manifestations of glomerulonephritis depending

Types of Glomerulonephritis

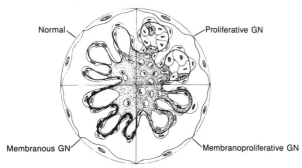

Glomerulonephritis (GN)

on the stage of the disease. In membranoproliferative glomerulonephritis type I, IgG, and C3 are found deposited in the glomerulus, whereas in dense deposit disease (membranoproliferative glomerulonephritis type II), C3 alone is demonstrable in the dense deposit within the capillary wall. The types of primary glomerulonephritis (GN) include acute diffuse proliferative GN, rapidly progressive (crescentic) GN, membranous GN, lipoid nephrosis, focal segmental glomerulosclerosis, membranoproliferative GN, IgA nephropathy, and chronic GN. Secondary diseases affecting the glomeruli include systemic lupus erythematosus, diabetes mellitus, amyloidosis, Goodpasture's syndrome, polyarteritis nodosa, Wegener's granulomatosis, Henoch-Schönlein purpura, and bacterial endocarditis.

glucocorticoids
Glucocorticoids are powerful immunosuppressive and anti-inflammatory drugs. They reduce circulating lymphocytes and monocytes, in addition to suppressing interleukin-1 (IL-1) and IL-2 production. However, their chronic use produces adverse effects, including increased susceptibility to infection, bone fractures, diabetes, and cataracts.

glucose-6-phosphate dehydrogenase deficiency
Occasionally, individuals of both sexes have been shown to have no glucose-6-phosphate dehydrogenase in their leukocytes. This could be associated with deficient NADPH with diminished hexose monophosphate shunt activity and reduced formation of hydrogen peroxide in leukocytes, which are unable to kill microorganisms intracellularly as was described in chronic granulomatous disease (CGD). Clinical aspects resemble those in CGD, except that glucose-6-phosphate dehydrogenase deficiency occurs later and affects both sexes, in addition to being associated with hemolytic anemia. Although the NBT test is within normal limits, glucose-6-phosphate dehydrogenase activity is deficient, the killing curve is altered, there is abnormal formation of H_2O_2, and oxygen consumption is inadequate. No effective treatment is available.

glutamic acid decarboxylase autoantibody
An autoantibody specific for a 64-kD antigen found in patients with insulin-dependent diabetes mellitus (type I) and in those with the "stiff man" syndrome.

gluten-sensitive enteropathy (celiac sprue, nontropical sprue)
A disease that results from defective gastrointestinal absorption due to hypersensitivity to cereal grain storage proteins, including gluten or its product gliadin, present in wheat, barley, and oats. There is diarrhea, weight loss, and steatorrhea. It is characterized by villous atrophy and malabsorption in the small intestine. It occurs mostly in Caucasians and occasionally in African-Americans, but not in Asians and is associated with HLA-DR3 and DR7. The disease may be limited to the intestines or associated with dermatitis herpetiformis, a vesicular skin eruption. Antigliadin antibodies, which are IgA are formed, and lymphocytes and plasma cells appear in the lamina propria in association with villous atrophy. Diagnosis is made by showing villous atrophy in a small intestinal biopsy. Administering a gluten-free diet leads to resolution of the disease.

GlyCAM-1
A molecule resembling mucin that is present on high endothelial venules in lymphoid tissues. L-selectin molecules on lymphocytes in the peripheral blood bind GlyCAM-1 molecules causing the lymphocytes to exit the blood circulation and circulate into the lymphoid tissues.

glycosylphophatidylinositol (GPI) linked membrane antigens
A class of cell surface antigens attached to the membrane by glycosylphophatidylinositol. Monoclonal antibody studies indicate that in both human and murine subjects GTI-linked antigens are capable of stimulating T cells and sometimes B cells. Structurally they are diverse. See also Ly-5 and Qa-2.

Gm allotype
A genetic variant determinant of the human IgG heavy chain. Allelic genes that encode the $\gamma1$, $\gamma2$, and $\gamma3$ heavy chain constant regions encode the Gm allotypes. They were recognized by the ability of blood sera from rheumatoid arthritis patients, which contain anti-IgG rheumatoid factor, to react with them. Gm allotypic determinants are associated with specific amino acid substitution in different γ chain constant regions in man. IgG subclasses are associated with certain Gm determinants. For example, IgG1 is associated with G1m(1) and G1m(4), and IgG3 is associated with

G3m(5). Although the great majority of Gm allotypes are restricted to the IgG-γ chain Fc region, a substitution at position 214 of C_H1 of arginine yields the G1m(4) allotype, and a substitution at this same site of lysine yields G1m(17). For Gm expression, the light chain part of the molecule must be intact.

GM-CSF

Granulocyte macrophage-colony stimulating factor. A growth factor for hemopoietic cells that is synthesized by lymphocytes, monocytes, fibroblasts, and endothelial cells. It has been prepared in recombinant form to stimulate production of leukocytes in AIDS patients and to initiate hematopoiesis following chemotherapy of bone marrow recipients. Patients with anemia and malignant neoplasms might also derive benefit from GM-CSF administration.

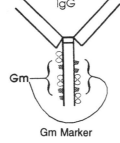

Gm Marker

γM globulin

Obsolete term for IgM.

Gm marker

Refer to Gm allotype.

gnotobiotic

An adjective that describes an animal or an environment where all the microorganisms are known, i.e., either a germ-free animal or an animal contaminated with one microorganism can be considered as gnotobiotic. Besides animal models, the so-called "bubble boy" who suffered severe combined immunodeficiency survived for 8 years in a plastic bubble that enclosed his gnotobiotic, germ-free environment.

gold therapy

A treatment for arthritis that inhibits oxidative degradation of membrane proteins and lipids and counteracts singlet oxygen produced as free radicals. It is administered in such forms as aurothioglucose and gold sodium thiomalate. Gold may induce gastrointestinal symptoms such as nausea and vomiting, diarrhea, and abdominal pain and renal symptoms such as nephrotic syndrome and proteinuria, as well as skin rashes, hepatitis, or blood dyscrasias. Renal biopsy may reveal IgG and C3 in a "moth-eaten" pattern in the glomerular basement membrane, as well as "feathery crystals" in the renal tubules.

Golgi complex

Tubular cytoplasmic structures that participate in protein secretion. The complex consists of flattened membranous sacs on top of each other termed cisternae. These are also associated with spherical vesicles. Proteins arriving from the rough endoplasmic reticulum are processed in the Golgi complex and sent elsewhere in the cell. Proteins handled in this manner include those that are secreted constitutive 'y such as immunoglobulins, those of the membrane, those that are stored in secretory granules to be released on command, and lysosomal enzymes.

gonococcal complement fixation test

A complement fixation test that uses as antigen an extract of *Neisseria gonorrhoea*. It is of little value in diagnosing early cases of gonorrhea that appear before the generation of an antibody response, but may be used to identify late manifestations in untreated individuals.

Good, Robert Alan (1922–)

American immunologist and pediatrician who has made major contributions to studies on the ontogeny and phylogeny of the immune response. Much of his work focused on the role of the thymus and the bursa of Fabricius in immunity. *The Thymus in Immunobiology,* 1964; *Phylogeny of Immunity,* 1966.

Goodpasture's antigen

An antigen found in the noncollagenous part of type IV collagen. It is present in human glomerular and alveolar basement membranes, making them a target for injury-inducing anti-GBM antibodies in blood sera of Goodpasture's syndrome patients. Interestingly, individuals with Alport's (hereditary) nephritis *do not* have the Goodpasture antigen in their basement membranes. Thus, renal transplants stimulate anti-GBM antibodies in Alport's patients.

Goodpasture's syndrome

A disease with pulmonary hemorrhage (with coughing up blood) and glomerulonephritis (with blood in the urine) induced by antiglomerular basement membrane autoantibodies that also interact with alveolar basement membrane antigens. A linear pattern of immunofluorescent staining confirms interaction of the IgG antibodies with basement membrane antigens in the kidney and lung, leading to membrane injury with pulmonary hemorrhage and acute (rapidly progressive or crescentic) proliferative glomerulonephritis. Pulmonary hemorrhage may precede hematuria. In addition to linear IgG, membranes may reveal linear staining for C3.

Goodpasture's Syndrome

Gorer, Peter Alfred (1907–1961)

British pathologist. With Snell he discovered the H-2 murine histocompatibility complex. Most of his work was in transplantation genetics. He identified antigen II and described its association with tumor rejection. *The Gorer Symposium,* 1985.

gp120

A surface 120-kD glycoprotein of human immunodeficiency virus type 1 (HIV-1) that combines with the CD4 receptor on T lymphocytes and macrophages. Synthetic soluble CD4 molecules have been used to block gp120 antigens and spare CD4+ lymphocytes from becoming infected. The *env* gene, which mutates frequently, encodes gp120, thereby interfering with host efforts to manufacture effective or protective antibodies.

gp160 Vaccine

A vaccine that contains a cloned segment of the envelope protein of HIV-1. It activates both humoral and cellular immunity against HIV products during early infection with HIV-1. It diminishes the rate at which CD4+ T lymphocytes are lost.

GPLA

The guinea pig major histocompatibility complex (MHC). There are two loci that encode class I MHC molecules (*GPLA-B* and *GPLA-S*). The *GPLA-Ia* locus encodes class II MHC molecules. Complement protein factor B, C2, and C4 are encoded by other loci. The genes are *BF* (factor B), C2, C4, *Ia, B,* and *S*.

Grabar, Pierre (1898–1986)

French educated immunologist, born in Kiev, who served as Chef de Service at the Institut Pasteur and as Director, National Center for Scientific Research, Paris. Best known for his work with Williams in the development of immunoelectrophoresis. He studied antigen-antibody interactions and developed the "carrier" theory of antibody function. He was instrumental in reviving European immunology in the era after World War II.

graft

The transplantation of a tissue or organ from one site to another within the same individual or between individuals of the same or a different species.

graft facilitation

Prolonged graft survival attributable to conditioning of the recipient with IgG antibody, which is believed to act as a blocking factor. It also decreases cell-mediated immunity. This phenomenon is related to immunologic enhancement of tumors by antibody and has been referred to as immunological facilitation (*facilitation immunologique*).

graft rejection

The immunologic destruction of transplanted tissues or organs between two members or strains of a species differing at the major histocompatibility complex for that species (i.e., HLA in man and H-2 in the mouse). The rejection is based upon both cell-mediated and antibody-mediated immunity against cells of the graft by the histoincompatible recipient. First-set rejection usually occurs within

2 weeks after transplantation. The placement of a second graft with the same antigenic specificity as the first in the same host leads to rejection within 1 week and is termed second-set rejection. This demonstrates the presence of immunological memory learned from the first experience with the histocompatibility antigens of the graft. When the donor and recipient differ only at minor histocompatibility loci, rejection of the transplanted tissue may be delayed, depending upon the relative strength of the minor loci in which they differ. Grafts placed in a hyperimmune individual, such as those with preformed antibodies, may undergo hyperacute or accelerated rejection. Hyperacute rejection of a kidney allograft by preformed antibodies in the recipient is characterized by formation of fibrin plugs in the vasculature as a consequence of the antibodies reacting against endothelial cells lining vessels, complement fixation, polymorphonuclear neutrophil attraction, and denuding of the vessel wall, followed by platelet accumulation and fibrin plugging. As the blood supply to the organ is interrupted, the tissue undergoes infarction and must be removed.

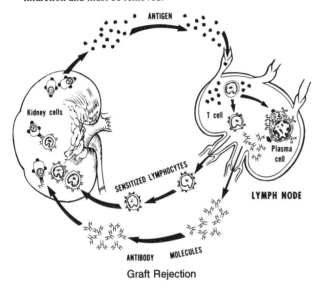

Graft Rejection

graft-vs.-host disease (GVHD)

Disease produced by the reaction of immunocompetent T lymphocytes of the donor graft that are histoincompatible with the tissues of the recipient into which they have been transplanted. For the disease to occur, the recipient must be either immunologically immature, immunosuppressed by irradiation or drug therapy, or tolerant to the administered cells. Also, the grafted cells must be immunocompetent. Patients develop skin rash, fever, diarrhea, weight loss, hepatosplenomegaly, and aplasia of the bone marrow. The donor lymphocytes infiltrate the skin, gastrointestinal tract, and liver. The disease may be either acute or chronic. Murine GVH disease is called "runt disease", "secondary disease", or wasting disease. Both allo- and autoimmunity associated with GVHD may follow bone marrow transplantation. Of patients receiving HLA-identical bone marrow transplants, 20 to 50% still manifest GVHD with associated weight loss, skin rash, fever, diarrhea, liver disease, and immunodeficiency. GVHD may be either acute, which is an alloimmune disease, or chronic, which consists of both allo- and autoimmune components. The conditions requisite for the GVH reaction include genetic differences between immunocompetent cells in the marrow graft and host tissues, immunoincompetence of the host, and alloimmune differences that promote proliferation of donor cells that react with host tissues. In addition to allogenic marrow grafts, the transfusion of unirradiated blood products to an immunosuppressed patient or intrauterine transfusion from mother to fetus may lead to GVHD.

graft-vs.-host reaction (GVHR)

The reaction of a graft containing immunocompetent cells against the genetically dissimilar tissues of an immunosuppressed recipient. Criteria requisite for a GVHR include (1) histoincompatibility between the donor and recipient, (2) passively transferred immunologically reactive cells, and (3) a recipient host who has been either naturally immunosuppressed because of immaturity or genetic

defect or deliberately immunosuppresed by irradiation or drugs. The immunocompetent grafted cells are especially reactive against rapidly dividing cells. Target organs include the skin, gastrointestinal tract (including the gastric mucosa), and liver, as well as the lymphoid tissues. Patients often develop skin rashes and hepatosplenomegaly and may have aplasia of the bone marrow. GVHR usually develops within 7 to 30 days following the transplant or infusion of the lymphocytes. Prevention of the GVHR is an important procedural step in several forms of transplantation and may be accomplished by irradiating the transplant. The clinical course of GVHR may take a hyperacute, acute, or chronic form as seen in graft rejection.

granulocyte

Refers to the three types of polymorphonuclear leukocytes that differ mainly because of the staining properties of their cytoplasmic granules. The three types are classified as neutrophils, eosinophils, and basophils. They are all mature myeloid-series cells and have different functions. Granulocytes constitute 58 to 71% of the leukocytes in the blood circulation. Refer to the individual cells for details.

granulocyte antibodies

IgG and/or IgM antibodies present in approximately 33% of adult patients with idiopathic neutropenia. They are also implicated in the pathogenesis of drug-induced neutropenia, febrile transfusion reactions, isoimmune neonatal neutropenia, Evans syndrome, primary autoimmune neutropenia of early childhood, systemic lupus erythematosus, Graves disease and the neutropenia of Felty syndrome and selected other autoimmune diseases.

granulocyte-specific antinuclear antibodies

Autoantibodies that react with the nuclei of neutrophils to produce a homogeneous "staining" pattern by immunofluorescence. They do not react with Hep-2 cells or with liver substrates. These antibodies have been claimed to be present in patients with active rheumatoid arthritis with vasculitis and/or neutropenia and in 17% of juvenile rheumatoid arthritis patients. Granulocyte-specific antinuclear antibodies may be correlated with erosive joint disease.

granulocytopenia

An abnormally low number of blood granulocytes.

granuloma

A tissue reaction characterized by altered macrophages (epithelioid cells), lymphocytes, and fibroblasts. These cells form microscopic masses of mononuclear cells. Giant cells form from some of these fused cells. Granulomas may be of the foreign body type, such as those surrounding silica or carbon particles, or of the immune type that encircle particulate antigens derived from microorganisms. Activated macrophages trap antigen, which may cause T cells to release lymphokines, causing more macrophages to accumulate. This process isolates the microorganism. Granulomas appear in cases of tuberculosis and develop under the influence of helper T cells that react against *Mycobacterium tuberculosis*. Some macrophages and epithelioid cells fuse to form multinucleated giant cells in immune granulomas. There may also be occasional neutrophils and eosinophils. Necrosis may develop.

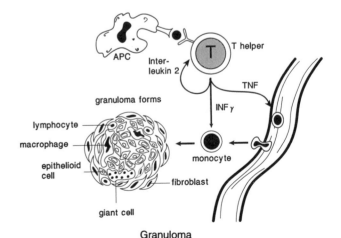

Granuloma

granulomatous hepatitis

Granulomatous inflammation of the liver.

granulopoietin

A 45-kD glycoprotein produced by monocytes that governs granulocyte formation in the bone marrow. Also called colony-stimulating factor.

granzymes

Proteases released from large granular lymphocyte (LGL) granules that contribute to fatal injury of target cells subjected to the cytotoxic action of perforin. Antigranzyme antibodies inhibit target cell lysis. Also called fragmentins.

Graves' disease (hyperthyroidism)

Thyroid gland hyperplasia with increased thyroid hormone secretion produces signs and symptoms of hyperthyroidism in the patient. Patients may develop IgG autoantibodies against thyroid-stimulating hormone (TSH) receptors. This autoantibody is termed long-acting thyroid stimulator (LATS). When the LATS IgG antibody binds to the TSH receptor, it has a stimulatory effect on the thyroid promoting hyperactivity. This IgG autoantibody can cross the placenta and produce transient hyperthyroidism in a newborn infant. The disease has a female predominance.

gross cystic disease fluid protein 15 (GCDFP-15) antigen

A 15-kD glycoprotein that is demonstrable with immunoperoxidase staining and expressed by primary and metastatic breast carcinomas with apocrine features and extramammary Paget's disease. Normal apocrine sweat glands, eccrine glands (variable), minor salivary glands, bronchial glands, metaplastic breast epithelium, benign sweat gland tumors of skin, and submandibular serous cells express GCDFP-15 antigen.

group agglutination

In the serologic classification of microorganisms, the identification of group-specific antigens rather than species-specific antigens by the antibody used for serotyping.

growth factors

Cytokines that facilitate the growth and proliferation of cells. Examples include platelet-derived growth factor, erythropoietin, interleukin-2 (T cell growth factor), and many others.

Guillain-Barre' syndrome

A type of idiopathic polyneuritis in which autoimmunity to peripheral nerve myelin leads to a condition characterized by chronic demyelination of the spinal cord and peripheral nerves.

gut-associated lymphoid tissue (GALT)

Lymphoid tissue situated in the gastrointestinal mucosa and submucosa. It constitutes the gastrointestinal immune system. GALT is present in the appendix, in the tonsils, and in the Peyer's patches subjacent to the mucosa. GALT represents the counterpart of BALT and consists of radially arranged and closely packed lymphoid follicles which impinge upon the intestinal epithelium, forming dome-like structures. In GALT, specialized epithelial cells overlie the lymphoid follicles, forming a membrane between the lymphoid cells and the lumen. These cells are called M cells. They are believed to be "gatekeepers" for molecules passing across. Other GALT components include IgA-synthesizing B cells and intraepithelial lymphocytes such as CD8+ T cells, as well as the lymphocytes in the lamina propria that include CD4+ T lymphocytes, B lymphocytes which synthesize IgA, and null cells.

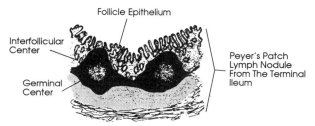

GALT (Gut Associated Lymphoid Tissue)

GVH

Refer to graft-vs.-host reaction and graft-vs.-host disease.

GVH disease

Refer to graft-vs.-host disease.

H

H-2 histocompatibility system

The major histocompatibility complex in the mouse. H-2 genes are located on chromosome 17. They encode somatic cell surface antigens as well as the host immune response (*Ir* genes). Each of these has a length of 600 base pairs. There are four regions in the H-2 complex designated K, I, S, and D. K region genes encode class I histocompatibility molecules designated K. I region genes encode class II histocompatibility molecules designated I-A and I-E. S region genes encode class III molecules designated C2, C4, factor B, and P-450 cytochrome (21-hydroxylase). D region genes encode class I histocompatibility molecules designated D and L. Antigens that represent the H-2 type of a particular inbred strain of mice are encoded by H-2 alleles. Thus, differences in the antigenic structure between inbred mice of differing H-2 alleles is of critical importance in the acceptance or rejection of tissue grafts exchanged between them. K, D, and L subregions of H-2 correspond to A, B, and C subregions of HLA in humans. The I-A and I-E regions are equivalent to the human HLA-D region.

H-2 locus

The mouse major histocompatibility region on chromosome 17.

H-2 restriction

MHC restriction involving the murine H-2 MHC.

H-2D

A murine H-2 MHC locus whose products are class I antigens. *H-2D* gene has multiple alleles.

H-2I region

A murine H-2 MHC region where the genes encoding class II molecules are found.

H-2K

A murine H-2 MHC locus. Genes at this locus, which has multiple alleles, encode MHC class I antigens.

H-2L

Murine class I histocompatibility antigen found on spleen cells that serves as a target epitope in graft rejection.

H65-RTA

An immunosuppressant used clinically in the treatment of acute graft-vs.-host disease and rheumatoid arthritis. H65-RTA is an immunoconjugate made up of anti-CD5 monoclonal antibody coupled to a cytotoxic enzyme. It leads to inhibition of protein synthesis, cell depletion, and lymphocyte activation.

H7

A pharmacological agent used to study T-cell activation. Its target is protein kinase C.

H antigen

(1) Epitopes on flagella of enteric bacteria that are motile and Gram-negative. H is from the German word *"Hauch",* which means breath, and refers to the production of a film on agar plates that resembles breathing on glass. In the Kaufmann-White classification scheme for *Salmonella,* H antigens serve as the basis for the division of microorganisms into phase I and phase II, depending on the flagellins they contain. A single cell synthesizes only one type

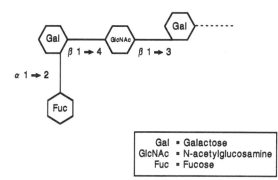

H Antigen

H-2 Complex (on Chromosome 17)

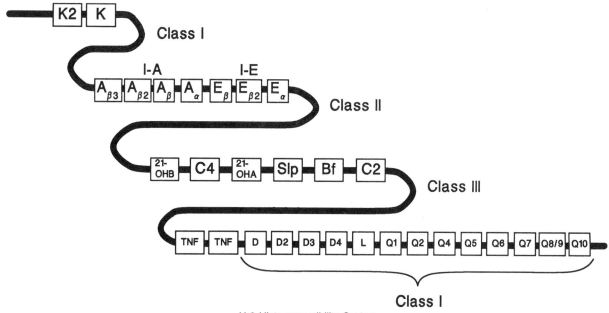

H-2 Histocompatibility System

of flagellin. Phase variation may result in a switch to production of the other type which is genetically controlled. (2) An ABO blood group system antigen that is also called H substance. (3) Histocompatibility antigen.

H chain (heavy chain)
A principal constituent of immunoglobulin molecules. Each immunoglobulin is comprised of at least one four-polypeptide chain monomer, which consists of two heavy and two light polypeptide chains. The two heavy chains are identical in any one molecule as are the two light chains.

H chain disease
A variant of multiple myeloma in which immunoglobulin free heavy chains are demonstrable in the patient's serum. This condition, characterized by free H chains in the serum, but not free light chains, is called heavy chain disease.

H substance
A basic carbohydrate of the ABO blood group system structure of man. Most people express this ABO related antigen. "Secretors" have soluble H substance in their body fluids.

H-Y
A Y chromosome-encoded minor histocompatibility antigen that may induce male skin graft rejection by females or destruction of lymphoid cells from males by effector cytotoxic T lymphocytes from females.

H_1, H_2 blocking agents
Refer to antihistamine.

H1 receptors
Histamine receptors on vascular smooth muscle cells through which histamine mediates vasodilation.

H2 receptors
Histamine receptors on different types of tissue cells through which histamine mediates bronchial constriction in asthma, gastrointestinal constriction in diarrhea, and endothelial constriction resulting in edema.

HA-1A
Human monoclonal IgM antibody specific for the lipid A domain of endotoxin. It can prevent death of laboratory animals with Gram-negative bacteremia and endotoxemia. In clinical trials, HA-1A has proven safe and effective for the treatment of patients with sepsis and Gram-negative bacteremia, whether or not they are in shock, but not in those with focal Gram-negative infection.

Hageman factor (HF)
A zymogen in plasma that is activated by contact with a surface or by the kallikrein system at the beginning of the intrinsic pathway of blood coagulation. This is an 80-kD plasma glycoprotein which, following activation, is split into an α and β chain. When activated, this substance is a serine protease that transforms prekallikrein into kallikrein. HF is coagulation factor XII.

hairpin loop
The looped structure of hairpin DNA.

hairy cell leukemia
A B lymphocyte leukemia in which the B cells have characteristic cytoplasmic filopodia.

half-life ($T_{1/2}$)
The time required for a substance to be diminished to one half of its earlier level by degradation or decay; by catabolism, as in biological half-life; or by elimination. In immunology, $T_{1/2}$ refers to the time in which an immunoglobulin remains in the blood circulation. For IgG, the half-life is 20 to 25 days; for IgA, 6 days; for IgM, 5 days; for IgD, 2 to 8 days; and for IgE, 1 to 5 days.

halogenation
Halogen binding to the cell wall of a microorganism with resulting injury to the microbe.

HALV
Human AIDS-lymphotrophic virus. A designation proposed to replace Montagnier's LAV designation and Gallo's HTLV-III designation for the AIDS virus. However, the designation HIV for human immunodeficiency virus was subsequently chosen instead.

HAM
HTLV-1-associated myelopathy.

HAM-1 and HAM-2
Histocompatibility antigen modifier. Two murine genes that determine formation of permeases that are antigen transporters (oligopeptides) from the cytoplasm to a membrane-bound compartment where antigen complexes with MHC class I and class II molecules. In man, the equivalence of HAM-1 and -2 are termed ATP-binding cassette transporters.

HAM test
Refer to paroxysmal nocturnal hemoglobulinuria (PNH).

HAMA
Human anti-mouse antibody. A group of murine monoclonal antibodies administered to selected cancer patients in research treatment protocols stimulate synthesis of anti-mouse antibody in the human recipient. The murine monoclonal antibodies are specific for the human tumor cell epitopes. The anti-mouse response affects the continued administration of the monoclonal antibody preparation. The hypersensitivity induced can be expressed as anaphylaxis, subacute allergic reactions, delayed-type hypersensitivity, rash, urticaria, flu-like symptoms, gastrointestinal disorders, dyspnea, hypotension, and renal failure.

hand-mirror cell
A lymphocyte in which the nucleus represents the mirror and the cytoplasm has extended into a uropod that resembles the handle of a hand mirror. Cells with this morphology may be found in both benign and neoplastic hematopoietic conditions. They may represent immature T lymphocytes, large granular lymphocytes such as natural killer cells, or atypical lymphocytes such as those seen in infectious mononucleosis. However, hand-mirror cells are most frequently found in acute lymphocytic leukemia of FAB L1 or L2 subtypes. Hand-mirror cells may also be 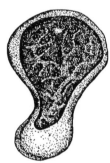 seen in cases of multiple myeloma, lymphosarcoma, Hodgkin's disease, acute myelogenous leukemia of the FAB M5a subtype, and acute and chronic lymphocytic leukemia. Hand-mirror cells in lymphocytic leukemia may be associated with defective immune function.

Hand Mirror Cell

Hand-Schuller-Christian disease
A macrophage lineage neoplasm (histiocytosis X) that is mainly in bone.

HANE (hereditary angioneurotic edema)
A condition induced by C1q esterase inhibitor (Clq-INH) deficiency in which immune complexes induce uptake of activated C1, C4, and C2. This may be activated by trauma, cold, vibration or other physical stimuli, histamine release, or menstruation. HANE induces nonpitting swelling that is not pruritic and not urticarial that reaches a peak within 12 to 18 h. It occurs on the face, extremities, toes, fingers, elbows, knees, gastrointestinal mucosa, and oral pharynx. It can produce edema of the epiglottis, which is fatal in approximately one third of the cases. Patients may have nausea and abdominal pain with vomiting. Four types have been described. Two are congenital. In type I, which accounts for 85% of cases, Clq-INH is diminished to 30% of normal. In type II variant, the product of the gene is present, but does not function properly. Type III may be acquired or autoimmune. The acquired type is associated with certain lymphoproliferative disorders such as Waldenström's macroglobulinemia, IgA myeloma, chronic lymphocytic leukemia, or other types of B cell proliferation. The autoimmune type is linked to IgG1 autoantibodies. There is unregulated Cls activation.

Hanganitziu-Deicher antigen
An altered ganglioside present in certain human neoplasms (CD3, GM1, and terminal 4NAcNeu).

haploid
Refers to a single copy of each autosome and one sex chromosome. This constitutes one set of unpaired chromosomes in a nucleus. The adjective may also refer to a cell containing this number of chromosomes.

haplotype
Those phenotypic characteristics encoded by closely linked genes on one chromosome inherited from one parent. It frequently describes several major histocompatibility complex (MHC) alleles on a single chromosome. Selected haplotypes are in strong linkage disequilibrium between alleles of different loci. According to Mendelian genetics, 25% of siblings will share both haplotypes.

hapten
A relatively small molecule that by itself is unable to elicit an immune response when injected into an animal, but is capable of reacting in vitro with an antibody specific for it. Only if complexed to a carrier molecule prior to administration can a hapten induce an immune response. Haptens usually bear only one epitope. Pneumococcal polysaccharide is an example of a larger molecule that may act as a hapten in rabbits, but as a complete antigen in humans.

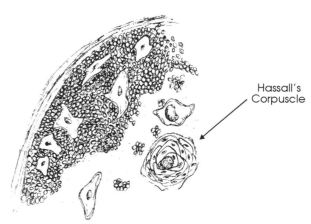

Hapten Conjugates

hapten conjugate response

The response to hapten conjugates requires two populations of lymphocytes: T and B cells. The cells producing the antibodies are derived from B cells. T cells act as helpers in this process. B cell preparations, depleted of T cells, cannot respond to hapten conjugates.

The T cells are responsive to the carrier portion of the conjugate, although in some cases they also recognize the hapten. The influence of the carrier on the ensuing response is called carrier effect. The experimental design for demonstrating the carrier effect involves adoptive transfer of hapten-sensitive B cells and of T cells primed with one or another carrier. The primed cells are those which have already had a past opportunity to encounter the antigen.

hapten inhibition test

An assay for serological characterization or elucidation of the molecular structure of an epitope by blocking the antigen-binding site of an antibody specific for the epitope with a defined hapten.

hapten X

Refer to CD15.

Harderian gland

A tear-secreting gland in the orbit of the eye in mammals and birds. IgG, IgM, and secretory IgA may originate from this location in birds.

Hasek, Milan (1925–1985)

Czechoslovakian scientist whose contributions to immunology include: investigations of immunologic tolerance and the development of chick embryo parabiosis. Hasek also made fundamental contributions to transplantation biology.

Hashimoto's disease (Hashimoto's thyroiditis)

Autoimmune thyroiditis that usually occurs in women past the age of 40. Thyroid gland inflammation is characterized by lymphocyte, plasma cell, and macrophage infiltration and often formation of lymphoid germinal centers. The thyroid becomes firm and enlarged. There is lymphocytic infiltration and oxyphilic alteration of the thyroid follicular epithelium. Thyroid follicles exhibit permanent germinal centers, and there is atrophy of the thyroid epithelium. Autoantibodies against thyroid peroxidase (microsomal antibodies), thyroglobulin, and colloid are demonstrable. Even though the thyroid gland increases in size, the acinar epithelium is continually destroyed. Thus, the principal pathogenesis is by autoimmune mechanisms. The disease leads to a decrease in thyroid function. This is considered to be an organ-specific autoimmune disease. Hormonal replacement therapy is used for treatment. Also called chronic autoimmune thyroiditis.

Hassall's corpuscles

Epithelial cell whorls in the medulla of the thymus. They are thought to produce thymic hormones. Whereas the center may exhibit degeneration, cells at the periphery may reveal endocrine secretion granules.

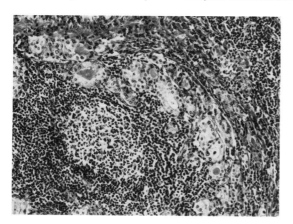

Hashimoto's Thyroiditis

Histology of the Thymus
Hassall's Corpuscle

HAT medium

A growth medium used in tissue culture of animal cells. It contains hypoxanthine, aminopterin, and thymidine and is used to selectively isolate hybrid cells. Aminopterin, a folic acid antagonist, blocks nucleic acid *de novo* synthesis, but does not have this effect for synthesis by the salvage pathway mechanism. Hypoxanthine and thymidine are substances from which nucleic acid can be synthesized by the second mechanism. The enzymes hypoxanthine-guanine phosphoribosyl transferase (HGPRT) and thymidine kinase are requisite for the salvage pathway. Thus, HAT medium cannot sustain cells lacking these enzymes, as nucleic acid synthesis is totally inhibited. Yet hybrid cells produced from each of the two cell lines whose enzyme defects are different can survive and grow in HAT medium. This makes it a useful selective medium for hybrid cells. It is widely employed in hybridoma production, as it permits myeloma cell-lymphocyte hybrids to survive. Nonfused myeloma cells die for lack of HGPRT, and unfused lymphocytes die because of their very limited proliferative capacity. The hybrid cell which survives in HAT medium receives the missing enzymes from the lymphocyte and proliferative capacity from the myeloma cell.

Haurowitz, Felix (1896–1988)

A noted protein chemist from Prague, who later came to the United States. He investigated the chemistry of hemoglobins. In 1930 (with Breinl) he advanced the instruction theory of antibody formation. *Chemistry and Biology of Proteins,* 1950; *Immunochemistry and Biosynthesis of Antibodies,* 1968.

HAV

Abbreviation for hepatitis A virus.

hay fever

Allergic rhinitis, recurrent asthma, rhinitis, and conjunctivitis in atopic individuals who inhale allergenic (antigenic) materials such as pollens, animal dander, house dust, etc. These substances do not induce allergic reactions in nonatopic (normal) individuals. Hay fever is a type I immediate hypersensitivity reaction mediated by homocytotrophic IgE antibodies specific for the allergen for which the individual is hypersensitive. Hay fever is worse during seasons when airborne environmental allergens are most concentrated.

HBcAg

Hepatitis B core antigen. This 27-nm core can be detected in hepatocyte nuclei.

HBeAg

Hepatitis B nucleocapsid constituent of relatively low mol wt which signifies an infectious state when it appears in the serum.

HBIG (hepatitis B immunoglobulin)

A preparation of donor pool-derived antibodies against hepatitis B virus (HBV). It is heat treated and shown not to contain human immunodeficiency virus. HBIG is given at the time when individuals are exposed to HBV and 1 month thereafter. Low-titer (1:128) and high-titer (1:100,000) preparations are available.

HBLV (human B lymphotropic virus)

Herpes virus 6.

HBsAg

Hepatitis B virus envelope or surface antigen.

HBV

Abbreviation for hepatitis B virus.

HBx

A regulatory gene of hepatitis B that codes for the production of an HBx protein which is a transcriptional transactivator of viral genes. This modifies expression of the host gene. In transgenic mouse-induced hepatomas, it may promote development of hepatocellular carcinoma.

hCG (human choriogonadotrophic hormone)

A glycoprotein comprised of lactose and hexosamine that is synthesized by syncytiotrophoblast, fetal kidney, and liver-selected tumors. It may be measured by radioimmunoassay or enzyme-linked immunosorbent assay (ELISA). It is elevated in patients with various types of tumors such as carcinoma of the liver, stomach, breast, pancreas, lungs, kidneys, and renal cortex, as well as conditions such as lymphoma, leukemia, melanoma, and seminoma.

HD$_{50}$

An uncommon designation for CH$_{50}$, which refers to hemolytic complement activity of a serum sample.

HDN

Hemolytic disease of the newborn.

heaf test

A type of tuberculin test in which an automatic multiple puncture device with six needles is used to administer the test material by intradermal inoculation. The multiple needles advance 2 to 3 mm into the skin. Also called Tine test.

heart-lung transplantation

A procedure that has proven effective for the treatment of primary respiratory disease with dysfunction of gas exchange and alveolar mechanics, together with a secondary elevation in pulmonary vascular resistance, and in primary high-resistance circulatory disorder associated with pulmonary vascular disease.

heat-aggregated protein antigen

Partial denaturation of a protein antigen by mild heating. This diminishes the protein's solubility, but causes it to express new epitopes. An example is the greater reactivity of rheumatoid factor (e.g., IgM anti-IgG autoantibody) with heat-aggregated γ globulin than with unheated γ globulin.

heat inactivation

Loss of biological activity through heating. Heating a serum sample at 56°C in a water bath for 30 min destroys complement activity through inactivation of C1, C2, and factor B. By diminishing the heat to 52°C for 30 min, only factor B is destroyed, whereas C1 and C2 remain intact. This inactivates the alternative complement pathway, but not the classical pathway. IgG, IgM, and IgA are resistant to incubation in a 56°C water bath for 30 min, whereas IgE antibodies are destroyed by this temperature.

heat-labile antibody

An antigen-specific immunoglobulin that loses its ability to interact with antigen following exposure to heating at 56°C for 30 min.

heat shock proteins (hsp)

A restricted number of highly conserved cellular proteins that increase during metabolic stress such as exposure to heat. Heat shock proteins affect protein assembly into protein complexes, proper protein folding, protein uptake into cellular organelles, and protein sorting. The main group of hsps are 70-kD proteins.

heat shock protein antibodies

Antibodies of the IgM, IgG and IgA classes specific for a 73 kD chaperonin that belongs to the hsp70 family present in the sera of approximately 40% of systemic lupus erythematosus patients, and in 10 to 20% of individuals with rheumatoid arthritis. Antibodies specific for the 65 kD heat shock protein derived from mycobacteria shows specificity for rheumatoid synovium. RA synovial fluid T cells specific for a 65 kD mycobacterial heat shock protein have been reported to be inversely proportional to the disease duration.

heavy chain

An immunoglobulin polypeptide chain that designates the class of immunoglobulin. The five immunoglobulin classes are based upon the heavy chains they possess and are IgM, IgG, IgA, IgD, and IgE. Each four-chain immunoglobulin molecule or each four-chain monomeric unit of IgM contains two heavy chains and two light chains. These are fastened together by disulfide bonds. At the amino terminus is the heavy chain's variable region designated V$_H$. Adjacent to this is the first constant region designated C$_H$1 through C$_H$3 or C$_H$4 domains based on immunoglobulin class. Heavy chain antigenic determinants determine not only the immunoglobulin class, but the subclass as well.

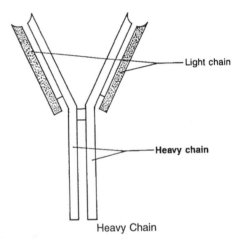

heavy chain class

Immunoglobulin heavy polypeptide chain primary (antigenic) structure present in all members of a species that is different from the other heavy chain classes. Primary structural features governing immunoglobulin heavy chain class are located in the constant region. Lowercase Greek letters such as μ, γ, α, δ, and ε designate heavy chain class.

heavy chain diseases

Monoclonal gammopathies or paraproteinemias associated with lymphoproliferative disease and characterized by excess synthesis of Fc fragments of immunoglobulins that appear in the serum and/or urine. The most common is the α heavy chain disease (Seligmann's disease) in which patients produce excess incomplete α chains of IgA1 molecules. It mainly affects Sephardic Jews, Arabs, and other Mediterranean residents, as well as subjects in South America and Asia. It may appear in childhood or adolescence as a lymphoproliferative disorder associated with the respiratory tract or gastrointestinal tract. Patients may manifest malabsorption, diarrhea, steatorrhea, hepatic dysfunction, weight loss, lymphadenopathy, hypocalcemia, and extensive mononuclear infiltration. α Heavy chain disease may either spontaneously remit or respond favorably to treatment with antibiotics. It may require chemotherapy in some cases. δ Heavy chain disease has been reported in an elderly male demonstrating abnormal plasma cell infiltration of the bone marrow together with osteolytic lesions. γ Heavy chain disease (Franklin's disease) occurs in older men and may either induce death rapidly (within weeks) or last for years. Death usually takes place within 1 year because of infection. Patients develop lymphoproliferation with fever, anemia, fatigue, angioimmunoblastic lymphadenopathy, hepatosplenomegaly, eosinophilic infiltrates, leukopenia, lymphoma, autoimmune disease, or tuberculosis. There is elevated IgG1 in the serum. It has been treated with cyclophosphamide, vincristine, and prednisone. µ Chain disease is a rare condition of middle-aged to older individuals. Some patients may ultimately develop chronic lymphocytic leukemia. µ Chain disease is characterized by lymphadenopathy, hepatosplenomegaly, infiltration of the bone marrow by vacuolated plasma cells, and frequently by elevated synthesis of κ light chain.

heavy chain subclass

Within an immunoglobulin heavy chain class, differences in primary structure associated with the constant region that can further distinguish these heavy chains of the same class are designated as subclasses. These differences are based on primary or antigenic structure. Heavy chain subclasses are designated as γ1, γ2, γ3, etc.

Heidelberger, Michael (1888–1991)

American. Father of immunochemistry. He began his career as an organic chemist. His contributions to immunology include the perfection of quantitative immunochemical methods and the immunochemical characterization of pneumococcal polysaccharides. His contributions to immunologic research are legion. During his career he received the Lasker Award, the National Medal of Science, the Behring Award, the Pasteur Medal, and the French Legion of Honor. *Lectures on Immunochemistry,* 1956.

helper/suppressor ratio

The ratio of CD4+ helper/inducer T lymphocytes to CD8+ suppressor/ cytotoxic T lymphocytes. This value normally ranges between 1.5 and 2.0. However, in certain virus infections, notably AIDS, the ratio becomes inverted as a consequence of greatly diminished CD4+ lymphocytes and either stationary or elevated levels of CD8+ lymphocytes. Inversion of the ratio continues as the clinical situation deteriorates. Eventually, the CD4+ cells may completely disappear in the AIDS patient. Other conditions in which the helper/suppressor (CD4/CD8) ratio is decreased include other viral infections such as herpes, cytomegalovirus, Epstein-Barr virus infection, and measles, as well as in graft-vs.-host disease, lupus erythematosus with renal involvement, severe burns, exercise, myelodysplasia, acute lymphocytic leukemia in remission, severe sunburn, exercise, and loss of sleep. Elevated ratios may occur in such conditions as atopic dermatitis, Sezary syndrome, psoriasis, rheumatoid arthritis, primary biliary cirrhosis, lupus erythematosus without renal involvement, chronic autoimmune hepatitis, and type I insulin-dependent diabetes mellitus.

helper T cells

CD4+ helper/inducer T lymphocytes. They represent a subset of T cells which are critical to induction of an immune response to a foreign antigen. Antigen is presented by an antigen-presenting cell such as the macrophage in the context of self MHC class II antigen and IL-1. Once activated, the CD4+ T cells express IL-2 receptors and produce IL-2 molecules which can act in an autocrine fashion by combining with the IL-2 receptors and mediating the CD4+ cells to proliferate. Differentiated CD4+ lymphocytes synthesize and secrete lymphokines that affect the function of other cells of the immune system, such as CD8+ cells, B cells, and NK cells. B cells differentiate into plasma cells that synthesize antibody. Activated macrophages participate in delayed-type hypersensitivity (type IV) reactions. Cytotoxic T cells also develop. Murine monoclonal antibodies are used to enumerate CD4+ T lymphocytes by flow cytometry.

hemadsorption inhibition test

A red blood cell suspension is added to a tissue culture infected with a hemagglutinating virus. Viral hemagglutinin, expressed at the tissue culture cell surfaces, facilitates the hemadsorption of erythrocyte aggregates to the tissue culture surfaces. Antiviral antibody added to the culture prevents this hemadsorption, which serves as the basis for testing for antiviral antibody.

hemagglutination

The aggregation of red blood cells by antibodies, viruses, lectin, or other substances.

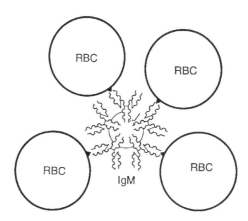

hemagglutination inhibition reaction

A serological test based upon inhibiting the aggregation of erythrocytes bearing antigen. The technique may be employed for diagnosis of such viral infections as rubella, variola-vaccinia, rubeola, herpes zoster, herpes simplex types I and II, cytomegalovirus, and Epstein-Barr virus. It is also being used in the diagnosis of adenovirus, influenza, coronavirus, parainfluenza, mumps, and the viral diseases that include St. Louis, Eastern, Venezuelan, and Western equine encephalitides. It has also been used in diagnosing various bacterial and parasitic diseases.

hemagglutination inhibition test

An assay for antibody or antigen based on the ability to interfere with red blood cell aggregation. Certain viruses are able to agglutinate red blood cells. In the presence of antiviral antibody, the ability to agglutinate erythrocytes is inhibited. Thus, this serves as a basis to assay the antibody.

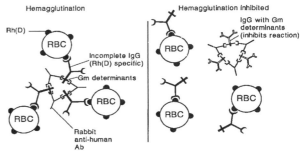

Hemagglutination Inhibition Test

hemagglutination test

An assay based upon the aggregation of red blood cells into clusters either through the action of antibody specific for their surface epitopes or through the action of a virus that possesses a hemagglutinin as part of its structure and which does not involve antibody.

hemagglutinin

A red blood cell agglutinating substance. Antibodies, lectins, and some viral glycoproteins may induce erythrocyte agglutination. In immunology, hemagglutinin usually refers to an antibody that causes red blood cell aggregation in physiological salt solution either at 37°C, in which case they are termed warm hemagglutinins, or at 4°C, in which case they are referred to as cold hemagglutinins.

hematopoietic stem cell

A bone marrow cell that is undifferentiated and serves as a precursor for multiple cell lineages. These cells are also demonstrable in the yolk sac and later in the liver in the fetus.

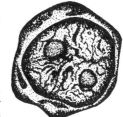

Hematopoietic Stem Cell

Hemolytic Disease of Newborn

1. First Pregnancy (Initial Sensitization)   2. Second and Third Pregnancies (2° and 3° Sensitizations)

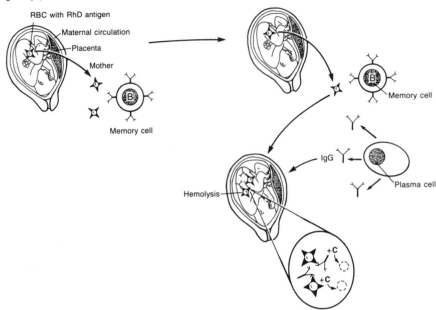

hematopoietic system
Those tissues and cells that generate the peripheral blood cells.

hematoxylin bodies
Nuclear aggregates of irregular shape found in areas of fibrinoid change or fibrinoid necrosis in systemic lupus erythematosus patients. These homogeneous-staining nuclear masses contain nuclear protein and DNA, as well as anti-DNA. They are probably formed from injured cell nuclei that have interacted with antinuclear antibodies *in vivo*. Hematoxylin staining imparts a bluish-purple color to hematoxylin bodies. They may be viewed in the kidney, lymph nodes, spleen, lungs, atrial endocardium, synovium, and serous membranes. Hematoxylin bodies of the tissue correspond to LE cells of the peripheral blood.

hematuria
Either macroscopic or microscopic blood in the urine from any cause, whether glomerular basement membrane injury or renal stones, for example.

hemocyanin
A blood pigment that transports oxygen in invertebrates. In immunology, hemocyanin of the keyhole limpet has been widely used as an experimental antigen.

hemocytoblast
A bone marrow stem cell.

hemolysin
A complement-fixing antibody that lyses erythrocytes in the presence of complement. An antibody which acts together with complement to produce an interruption in the membrane integrity of red blood cells, causing disruption. Historically, the term refers to rabbit anti-sheep erythrocyte antibody used in the visible part of a complement fixation test. Microbial products such as streptolysin O may disrupt (lyse) red blood cells in agar medium.

hemolysis
Interruption of the cellular integrity of red blood cells that may be either immune or nonimmune mediated. Clinically, immune hemolysis may be IgM-mediated when immunoglobulins combine with red blood cell surfaces for which they are specific, such as the ABO blood groups, and activate complement to produce lysis. This results in the release of free hemoglobin in the intravascular space with serious consequences. By contrast, hemolysis mediated by IgG in the extravascular space may be less severe. There is an elevation of indirect bilirubin, since the liver may not be able to conjugate the bilirubin in case of massive hemolysis. There is an elevation of lactate dehydrogenase, and hemoglobin appears in the blood and urine. There is elevated urobilinogen in both urine and feces. Hemolysis may be also attributable to the action of enzymes or other

chemicals acting on the cell membrane. It can also be induced by such mechanisms as placing the red cells in a hypotonic solution.

hemolytic anemia
A disease characterized by diminished circulating erythrocytes as a consequence of their destruction based either on an intrinsic abnormality, as in sickle cell anemia and thalassemia, or as a consequence of membrane-specific antibody and complement. Certain infections such as malaria are also associated with hemolytic anemia. Free hemoglobin in serum may lead to renal problems.

hemolytic disease of the newborn (HDN)
A fetus with RhD+ red blood cells can stimulate an RhD− mother to produce anti-RhD IgG antibodies that cross the placenta and destroy the fetal red blood cells when a sufficient titer is obtained. This is usually not until the third pregnancy with an Rh+ fetus. At parturition, the RhD+ red blood cells enter the maternal circulation and subsequent pregnancies provide a booster to this response. With the third pregnancy, a sufficient quantity of high-titer antibody crosses the placenta to produce considerable lysis of fetal red blood cells. This may lead to erythroblastosis fetalis (hemolytic disease of the newborn). In HDN cases, 70% are due to RhD incompatibility between the mother and fetus. Exchange transfusions may be required for treatment. Two other antibodies against erythrocytes that may likewise be a cause for transfusion exchange include anti-Fy^a and Kell. As bilirubin levels rise, the immature blood-brain barrier permits bilirubin to penetrate and deposit on the basal ganglia. Anti-D antibody passively injected into the mother following parturition unites with the RhD+ red cells, leading to their elimination by the mononuclear phagocyte system.

hemolytic plaque assay
A technique to identify and enumerate cells synthesizing antibodies. Typically, spleen cells from a mouse immunized against sheep red blood cells are combined with melted agar or agarose in which sheep erythrocytes are suspended. After gentle mixing, the suspension is distributed into Petri plates where it gels. This is followed by incubation at 37°C, after which complement is added to the dish from a pipette. Thus, the sheep erythrocytes (SRBC) surrounding cells secreting IgM antibody against SRBC are lysed by the added complement, producing a clear zone of hemolysis resembling the effect produced by β hemolytic streptococci on blood agar. IgG antibody against sheep erythrocytes can be identified by adding anti-IgG antibody to aid lysis by complement. Whereas modifications of this method have been used to identify cells producing antibodies against a variety of antigens or haptens conjugated to the sheep red cells, it can also be used to ascertain the immunoglobulin class being secreted.

hemolytic system

The visible phase of a complement fixation test in which sheep red blood cells combine with their homologous antibody and are added to a system where antigen is mixed with patient serum, presumed to contain specific antibody, followed by complement. If antibody is present in the patient's serum and combines with the antigen, the added complement is fixed and is no longer available to react with antibody-coated sheep red cells, which are added as the second phase or visible reaction. Lysis indicates the presence of free complement from the first phase of the reaction, indicating that antibody against the antigen in question was not present in the patient's serum. By contrast, no hemolysis indicates the presence of the antibody in the patient's serum.

The hemolytic system may also be used to describe a reaction in which erythrocytes are combined with their homologous antibody. The addition of all nine components of complement together with calcium and magnesium ions followed by incubation at 37°C results in immune lysis (hemolysis).

hemophilia

An inherited coagulation defect attributable to blood clotting factor VIII, factor IX, or factor XI deficiency. Hemophilia A patients are successfully maintained by the administration of exogenous factor VIII, which is now safe. Before mid-1985, factor products were a source of several cases of AIDS transmission when factor VIII was extracted from the blood of HIV positive subjects by accident. Hemophilia B patients are treated with factor IX. Hemophilia A and B are X-linked, but hemophilia C is autosomal.

Hemophilus influenzae vaccine

Refer to Hib.

hemopoietic resistance (HR)

Transplantation of allogeneic, parental, or xenogeneic bone marrow or leukemia cells into animals exposed to total body irradiation often results in the destruction of the transplanted cells. The mechanism causing the failure of the transplant appears similar with all three types of cells. This phenomenon, designated hemopoietic resistance (HR), has a genetic basis and mechanism different from conventional transplantation reactions against solid tumor allografts. It does not require prior sensitization and apparently involves the cooperation between natural killer (NK) cells and macrophages, both resistant to irradiation. The NK cells have the characteristics of null cells; macrophages play an accessory cell role. The cooperative activity seems to represent *in vivo* surveillance against leukemogenesis.

hemostatic plug

The immediate response to a vessel wall injury is the adhesion of platelets to the injury site and the growth by further aggregation of platelets, of a mass which tends to obstruct (often incompletely) the lumen of the damaged vessel. This platelet mass is called a hemostatic plug. The exposed basement membranes at the sites of injury are the substrate for platelet adhesion, but deeper tissue components may have a similar effect. Far from being static, the hemostatic plug has a continuous tendency to break up with new masses reformed immediately at the original site.

Henoch-Schoenlein purpura

A systemic form of small vessel vasculitis that is characterized by arthralgias, nonthrombocytopenic purpuric skin lesions, abdominal pain with bleeding, and renal disease. Immunologically, immune complexes containing IgA activate the alternate pathway of complement. Patients may present with upper respiratory infections preceding onset of the disease. Certain drugs, food, and immunizations have also been suspected as etiologic agents. The disease usually occurs in children 4 to 7 years of age, although it can occur in adults. Histopathologically, there is a diffuse leukocytoclastic vasculitis involving small vessels. The submucosa or subserosa of the bowel may be sites of hemorrhage. There may be focal or diffuse glomerulonephritis in the kidneys. Children may manifest lesions associated with the skin, gastrointestinal tract, or joints, whereas in adults the disease is usually associated with skin findings. The skin lesions begin as a pruritic urticarial lesion that develops into a pink maculopapular spot which matures into a raised and darkened lesion. The maculopapular lesion may ultimately resolve in 2 weeks without leaving a scar. Patients may also have arthralgias associated with the large joints of the lower extremities. Skin biopsy reveals the vasculitis, and immunofluorescence examination shows IgA deposits in vessel walls which is in accord with a diagnosis of Henoch-Schoenlein purpura.

HEP

Abbreviation for "high egg passage", which signifies multiple passages of rabies virus through eggs to achieve attenuation for preparation of a vaccine appropriate for use in immunizing cattle.

heparan sulfate

A glycosaminoglycan that resembles heparin and is comprised of the same disaccharide repeating unit. Yet, it is a smaller molecule and less sulfated than heparin. An extracellular substance, heparan sulfate is present in the lungs, arterial walls, and on numerous cell surfaces.

heparin

A glycosaminoglycan comprised of two types of disaccharide repeating units. One is comprised of D-glucosamine and D-glucuronic acid, whereas the other is comprised of D-glucosamine and L-iduronic acid. Heparin is extensively sulfated and is an anticoagulant. It unites with an antithrombin III, which can unite with and block numerous coagulation factors. It is produced by mast cells and endothelial cells. It is found in the lungs, liver, skin, and gastrointestinal mucosa. Based on its anticoagulant properties, heparin is useful for treatment of thrombosis and phlebitis.

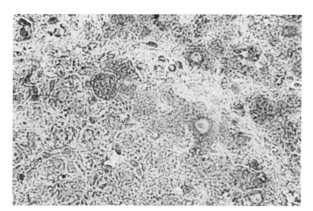

Heparin

hepatitis (immunopathology panel)

A profile of assays that are very useful to establish the clinical and immune status of a patient believed to have hepatitis. The panel for acute hepatitis may include hepatitis B surface antigen (HBsAg), antibody to hepatitis B core antigen (anti-HBc), antihepatitis B surface antigen (anti-ABs), antihepatitis A (IgM), anti-HBe, and antihepatitis C. The panel for chronic hepatitis (carrier) includes all of these except for the antihepatitis A test.

hepatitis B

A DNA virus that is relatively small and has four open reading frames. The S gene codes for HBsAg. The P gene codes for a DNA polymerase. There is an X gene and a core gene that code for HBcAg and the precore area that codes for HBeAg.

HBs Antigen in Liver Cells

hepatitis B vaccine

Human plasma-derived hepatitis B vaccine (Heptavax-B), which was developed in the 1980s, was unpopular because of the fear of AIDS related to any product for injection derived from human plasma. It was replaced by a recombinant DNA vaccine (Recombivax®) prepared in yeast (*Saccharomyces cervesiae*). It is very effective in inducing protective antibodies in most recipients. Nonresponders are often successfully immunized by intracutaneous vaccination.

hepatitis B virus protein X

Refer to HBx

hepatitis D virus

Refer to δ agent.

hepatitis E virus (HEV)

Enteric non-A, non-B hepatitis. A single-stranded RNA virus that has an oral-fecal route of transmission and can introduce epidemics under poor sanitary conditions where drinking water is contaminated and the population is poorly nourished.

hepatitis, non-A, non-B (C) (NANBH)

The principal cause of hepatitis that is transfusion related. Risk factors include intravenous drug abuse (42%), unknown risk factors (40%), sexual contact (6%), blood transfusion (6%), household contact (3%), and health professionals (2%). There are 150,000 cases per year in the U.S. Of those cases, 30 to 50% become chronic carriers, and one fifth develop cirrhosis. Parenteral NANBH is usually hepatitis C, and enteric NANBH is usually hepatitis E.

hepatitis serology

This refers to hepatitis B antigens and antibodies against them. Core antigen is designated HBc. The HBc particle is comprised of double-stranded DNA and DNA polymerase. It has an association with HBe antigens. Core antigen signifies persistence of replicating hepatitis B virus. Anti-HBc antibody is a serologic indicator of hepatitis B. It is an IgM antibody that increases early and is still detectable 20 years post-infection. The IgM anti-HBc antibody assay is the one best antibody assay for acute hepatitis B. The HBe antigen (HBe) follows the same pattern as HBsAg antigen. When found, it signifies a carrier state. The anti-HBe increases as HBe decreases. It appears in patients who are recovering and may last for years after the hepatitis has been resolved. The first antigen that is detectable following hepatitis B infection is surface antigen (HBs). It is detectable a few weeks before clinical disease and is highest with the first appearance of symptoms. This antigen disappears 6 months from infection. Antibody to HBs increases as the HBsAg levels diminish. Anti-HBs often is detectable for the lifetime of the individual.

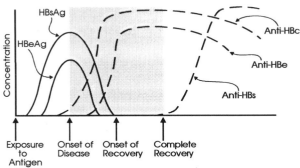

Hepatis Serology

hepatitis vaccines

The vaccine used to actively immunize subjects against hepatitis B virus contains purified hepatitis B surface antigen. Current practice uses an immunogen prepared by recombinant DNA technology referred to as Recombivax®. The antigen preparation is administered in three sequential intramuscular injections to individuals such as physicians, nurses, and other medical personnel who are at risk. Temporary protection against hepatitis A is induced by the passive administration of pooled normal human serum immunoglobulin which protects against hepatitis A virus for a brief time. Antibody for passive protection against hepatitis must be derived from the blood sera of specifically immune individuals.

hepatocyte-stimulating factor

A substance indistinguishable from interleukin-6, classified as a monokine, that stimulates hepatocytes to produce acute-phase reactants.

herbimycin A

An inhibitor of T cell activation. Its target is the Sre-family of protein tyrosine kinases.

herd immunity

Nonspecific factors, as well as specific immunity, may have a significant role in resistance of a group (herd) of humans or other animals against an infectious disease agent. Elimination of reservoirs of the disease agent may be as important as specific immunity in diminishing disease incidence among individuals. Herd immu-

nity also means that an epidemic will not follow infection of a single member of the herd or group if other members are immune to that particular infectious agent.

hereditary angioedema (HAE)

A disorder in which recurrent attacks of edema, persisting for 48 to 72 h, occur in the skin and gastrointestinal and respiratory tracts. It is nonpitting and life-threatening if laryngeal edema becomes severe enough to obstruct the airway. Edema in the jejunum may be associated with abdominal cramps and bilious vomiting. Edema of the colon may lead to watery diarrhea. There is no redness or itching associated with edema of the skin. Tissue trauma or no apparent initiating cause may induce an attack. It is due to decreased or absent C1 inhibitor (C1 INH). It is inherited in an autosomal dominant fashion. Heterozygotes for the defect develop the disorder. Greatly diminished C1 INH levels (5 to 30% of normal) are found in affected individuals. Activation of C1 leads to increased cleavage of C4 and C2, decreasing their serum levels during an attack. C1 INH is also a kinin system inactivator. The C1 INH deficiency in HAE permits a kinin-like peptide produced from C2b to increase vascular permeability leading to manifestations of HAE. Some have proposed that bradykinin may represent the vasopermeability factor. Hereditary angioedema has been treated with ε aminocaproic acid and transexamic acid, but they do not elevate C1 INH or C4 levels. Anabolic steroids such as danazol and stanozolol, which activate C1 INH synthesis in affected individuals, represent the treatment of choice.

hereditary ataxia telangiectasia

Refer to ataxic telangiectasia.

herpes gestationis

A very pruritic blistering condition of the skin characterized by vesicles and bullae that develop in rare cases of pregnancy. It is rare in black subjects and is associated with HLA-DR3 and DR4. By light microscopy, a subepidermal bulla with eosinophils may be identified in a few cases. There is edema of the papillary dermis, lymphohistiocytic infiltrate with eosinophils in perivascular regions, spongiosis, liquefactive degeneration, and possible necrosis in the epidermis. The eruption occurs usually in the second or third trimester of pregnancy and appears as hive-like plaques, blisters, or vesicles. Immunofluorescence reveals C3 at the dermal-epidermal junction in nearly all cases, and IgG in 25% of biopsies. Circulating IgG antibody binds to the herpes gestationis antigen, a 180 kD epidermal protein in the skin basement membrane.

herpes gestationis antibodies

Antibodies present in 89% of herpes gestationis sera that are IFA-positive and in blood sera of 47% of patients with bullous pemphigoid. These antibodies are not found in physiologic pregnancy. Direct immunofluorescence may be used to demonstrate IgG antibodies against the basement membrane zone in 30 to 50% of herpes gestationis cases. By contrast, C3 is demonstrable at the skin basement membrane zone in almost 100% of cases. Use of the sensitive immunoblotting method permits identification of antibodies against the hemidesmosome constituents of the basement membrane zone in the skin of 90% of herpes gestationis subjects.

herpes virus

A DNA virus family that contains a central icosahedral core of double-stranded DNA. There is a lipoprotein envelope that is

Herpesvirus

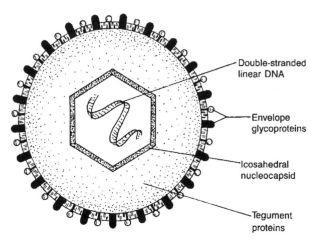

Double-stranded linear DNA

Envelope glycoproteins

Icosahedral nucleocapsid

Tegument proteins

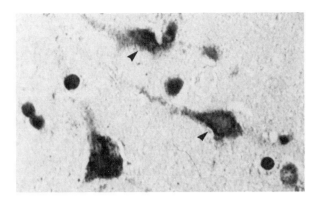

Herpes simplex
in brain

trilaminar and 100 nm in diameter and a nucleus that is 30 to 43 nm in diameter. Herpes viruses may persist for years in a dormant state. Six types have been described. HSV-1 (herpes simplex virus-1) can account for oral lesions such as fever blisters. HHV-2 (human herpes virus - 2) produces lesions below the waistline and is sexually transmitted. It may produce venereal disease of the vagina and vulva, as well as herpetic ulcers of the penis. HHV-3 (herpes varicella-zoster) occurs clinically as either an acute form known as chickenpox or a chronic form termed shingles. HHV-4 (Epstein-Barr virus), HHV-5 (cytomegalovirus), and HHV-6 (human B cell lymphotrophic virus) are the other types of herpes virus.

Herxheimer reaction
A serum sickness (type III) form of hypersensitivity that occurs following the treatment of selected chronic infectious diseases with an effective drug. When the microorganisms are destroyed in large numbers in the blood circulation, a significant amount of antigen is released from the disrupted microbes that tend to react with preformed antibodies in the circulation. This type of reaction has been described following the use of effective drugs to treat syphilis, trypanosomiasis, and brucellosis.

heteroantibody
An autoantibody with the ability to crossreact with an antigen of a different species.

heteroantigen
An antigen that induces an immune response in a species other than the one from which it was derived.

heteroclitic antibody
An antibody with greater affinity for a heterologous epitope than for the homologous one that stimulated its synthesis.

heterocliticity
The preferential binding by an antibody to an epitope other than the one that generated synthesis of the antibody.

heteroconjugate antibodies
Antibodies against a tumor antigen coupled covalently to an antibody specific for a natural killer cell or cytotoxic T lymphocyte surface antigen. These antibodies facilitate binding of cytotoxic effector cells to tumor target cells. Antibodies against effector cell

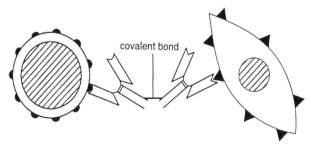

Cytotoxic C Cell   Tumor Cell
or NK Cell   (target)
Heteroconjugate Antibody

surface markers may also be coupled covalently with hormones that bind to receptors on tumor cells.

heterocytotropic antibody
An antibody that has a greater affinity when fixed to mast cells of a species other than the one in which the antibody is produced. Frequently assayed by skin-fixing ability, as revealed through the passive cutaneous anaphylaxis test. Interaction with the antigen for which these "fixed" antibodies are specific may lead to local heterocytotrophic anaphylaxis.

heterodimer
A molecule comprised of two components that are different, but closely joined structures, such as a protein comprised of two separate chains. Examples include the T cell receptor comprised of either α and β chains or of γ and δ chains, and class I as well as class II histocompatibility molecules.

heterogeneic
Refer to xenogeneic.

heterogenetic antibody
Refer to heterophile antibody.

heterogenetic antigen
Refer to heterophile antigen.

heterograft
Refer to xenograft.

heterokaryon
The formation of a hybrid cell by fusion of two separate cells in suspension leading to a cell form with two nuclei. Cell fusion may be accomplished through the use of polyethylene glycol or ultraviolet light-inactivated Sendai virus.

heterologous
In transplantation biology, refers to an organ or tissue transplant from one species to a recipient belonging to another species, i.e., a xenogeneic graft. It also refers to something from a foreign source.

heterologous antigen
A cross-reacting antigen.

heterologous vaccine
A vaccine that induces protective immunity against pathogenic microorganisms which the vaccine does not contain. Thus, the microorganisms that are present in the heterologous vaccine possess antigens that crossreact with those of the pathogenic agent absent from the vaccine. Measles vaccine can stimulate protection against canine distemper. Vaccinia virus was used in the past to induce immunity against smallpox because the agents of vaccinia and variola share antigens in common.

heterophile antibody
An antibody found in an animal of one species that can react with erythrocytes of a different and phylogenetically unrelated species. These are often IgM agglutinins. Heterophile antibodies are detected in infectious mononucleosis patients who demonstrate antibodies reactive with sheep erythrocytes. To differentiate this condition from serum sickness, which also is associated with a high titer of heterophile antibodies, the serum sample is absorbed with beef erythrocytes which contain Forssmann antigen. This treatment removes the heterophile antibody reactivity from the serum of infectious mononucleosis patients.

heterophile antigen
An antigen (epitope) present in divergent animal species, plants, and bacteria that manifest broad crossreactivity with antibodies of the heterophile group. Heterophile antigens induce the formation of heterophile antibodies when introduced into a species where they are absent. Heterophile antigens are often carbohydrates.

heterospecific
Showing specificity for a species other than the one from which the substance in question was derived.

heterotopic
An adjective that describes the placement of an organ or tissue graft in an anatomic site other than the one where it is normally located.

heterotopic graft
A tissue or organ transplanted to an anatomic site other than the one where it is usually found under natural conditions. For example, the anastomosis of the renal vasculature at an anatomical site that would situate the kidney in a place other than the renal fossa where it is customarily found.

heterotypic vaccine
Refer to heterologous vaccine.

HEV
Refer to high endothelial venules.

Heymann Nephritis

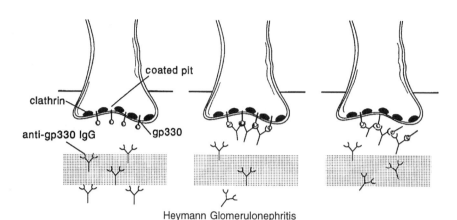

Heymann Glomerulonephritis

Heymann antigen

A 330-kD glycoprotein (GP330) present on visceral epithelial cell basal surfaces in coated pits, as well as on tubular brush borders. It participates in production of experimental nephritis in rats. See also Heymann glomerulonephritis.

Heymann glomerulonephritis

An experimental model of membranous glomerulonephritis induced by immunizing rats with proximal tubule brush border preparations containing subepithelial antigen or Heymann factor, a 330-kD protein incorporated in Freund's adjuvant. The rats produce antibodies against brush border antigens, and membranous glomerulonephritis is induced. Autoantibodies combine with shed epithelial cell antigen. The union of antibody with Heymann antigen, distributed in an interrupted manner along visceral epithelial cell surfaces, leads to subepithelial electron-dense, granular deposits. Immunoglobulins and complement are deposited in a granular rather than linear pattern along the glomerular basement membrane, as revealed by immunofluorescence. The glomerulonephritis results from interaction of antibrush border antibody with the 330-kD glycoprotein that is fixed, but discontinuously distributed on the base of visceral epithelial cells and is crossreactive with brush border antigens. Heymann nephritis closely resembles human membranous glomerulonephritis.

HGG

Abbreviation for human γ globulin.

HGP-30

An experimental vaccine for AIDS that employs a synthetic HIV core protein, p17.

HHV

Human herpes virus.

Hib (*Hemophilus influenzae* type b)

A microorganism that induces infection mostly in infants less than 5 years of age. Approximately 1000 deaths out of 20,000 annual cases are recorded. A polysaccharide vaccine (Hib Vac) was of only marginal efficacy and poorly immunogenic. By contrast, anti-Hib vaccine which contains capsular polysaccharide of Hib bound covalently to a carrier protein such as polyribosylribitol-diphtheria toxoid (PRP-D) induces a very high level of protection that reached 94% in one cohort of Finnish infants. PRP-tetanus toxoid has induced 75% protection. PRP-diphtheria toxoid vaccine has been claimed to be 88% effective.

hidden determinant

An epitope on a cell or molecule that is unavailable for interaction with either lymphocyte receptors or the antigen-binding region of antibody molecules because of stereochemical factors. These hidden or cryptic determinants neither react with lymphocyte or antibody receptors nor induce an immune response unless an alteration in the molecule's steric configuration causes the epitope to be exposed.

HIG

Abbreviation for human immune globulin.

high-dose tolerance

Specific immunologic unresponsiveness induced by the repeated administration of large doses of antigen (tolerogen), if the substance is a protein, or of a massive single dose, if the substance is a

polysaccharide, to immunocompetent adult animals. Although no precise inducing dose of antigen can be defined, usually in high-dose tolerance the antigen level exceeds 10^{-4} mol Ag per kilogram of body weight. Also called high-zone tolerance.

high endothelial postcapillary venules

Lymphoid organ vessels that are especially designed for circulating lymphocytes to gain access into the parenchyma of the organ. They contain cuboidal endothelium which permits lymphocytes to pass between the cells into the tissues. Lymphocyte recirculation from the blood to the lymph occurs through these vessel walls.

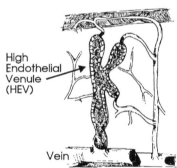

High Endothelial Postcapillary Venule

high endothelial venules (HEV)

Postcapillary venules of lymph node paracortical areas. They also occur in Peyer's patches, which are part of the gut-associated lymphoid tissue (GALT). Their specialized columnar cells bear receptors for antigen-primed lymphocytes. They signal lymphocytes to leave the peripheral blood circulation. A homing receptor for circulating lymphocytes is found in lymph nodes.

high responder

A descriptor for mouse, guinea pig, or other inbred strains that mount pronounced immune responses to selected antigens in comparison to the responses of other strains including so called "low responder" strains. Immune response (IR) genes encode this capability.

high-titer, low-avidity antibodies (HTLA)

Antibodies that induce erythrocyte agglutination at high dilutions in the Coombs' antiglobulin test. These antibodies cause only weak agglutination and are almost never linked to hemolysis of clinical importance. Examples of HTLA antibodies are anti-Bgᵃ, -Cs, -Ch, -Kna, -JMH, -Rg, and -Yk, among others.

high-zone tolerance

Antigen-induced specific immunosuppression with relatively large doses of protein antigens (tolerogens). B cell tolerance usually requires high antigen doses. High-zone tolerance is relatively short lived. Called also high-dose tolerance. Refer to high-dose tolerance.

hinge region

An area of an immunoglobulin heavy chain situated between the first constant domain and the second constant domain (C_H1 and

C_H2) in an immunoglobulin polypeptide chain. The high content of proline residues in this region provides considerable flexibility to this area, which enables the Fab region of an immunoglobulin molecule to combine with cell surface epitopes that it might not otherwise reach. Fab regions of an Ig molecule can rotate on the hinge region. There can be an angle up to 180° between the two Fab regions of an IgG molecule. In addition to the proline residues, there may be one or several half cysteines associated with the interchain disulfide bonds. Enzyme action by papain or pepsin occurs near the hinge region. Whereas, γ, α, and δ chains each contain a hinge region, μ and ε chains do not. The 5′ part of the C_H2 exon encodes the human and mouse a chain hinge region. Four exons encode the γ-3 chains of humans and two exons encode human δ chains.

histaminase
A common tissue enzyme, termed diamine oxidase, that transforms histamine into imidazoleacetic acid, an inactive substance.

histamine
A biologically active amine, i.e., β-aminoethylimidazole, of 111 mol wt that induces contraction of the smooth muscle of human bronchioles and small blood vessels, increased capillary permeability, and increased secretion by the mucous glands of the nose and bronchial tree. It is a principal pharmacological mediator of immediate (type I) hypersensitivity (anaphylaxis) in man and guinea pigs. Although found in many tissues, it is especially concentrated in mast cells of the tissues and basophils of the blood. It is stored in their cytoplasmic granules and is released following crosslinking of IgE antibodies by a specific antigen on their surfaces. It is produced by the decarboxylation of histidine through the action of histidine decarboxylase. When histamine combines with H_1 receptors, smooth muscle contraction and increased vascular permeability may result. Combination with H_2 receptors induces gastric secretion and blocks mediator release from mast cells and basophils. It may interfere with suppressor T cell function. Histamine attracts eosinophils that produce histaminase which degrades histomine.

histamine-releasing factor
A lymphokine produced from antigen-stimulated lymphocytes that induces the release of histamine from basophils.

histiocyte
A tissue macrophage that is fixed in the tissues, such as the connective tissue. Histiocytes are frequently around blood vessels and are actively phagocytic. They may be derived from monocytes in the circulating blood.

histiocytic lymphoma
A misnomer for large cell lymphomas, principally B cell tumors. The term histiocytic lymphoma more accurately describes a lymphoma of macrophage lineage.

histiocytosis X
A descriptor for neoplasms of macrophage lineage. Included in this category are Letterer-Siwe disease, Hand-Schüller-Christian disease, and eosinophilic granuloma of bone.

histocompatibility
Tissue compatibility as in the transplantation of tissues or organs from one member to another of the same species (an allograft) or from one species to another (a xenograft). The genes that encode antigens which should match if a tissue or organ graft is to survive in the recipient are located in the major histocompatibility complex (MHC) region. This is located on the short arm of chromosome 6 in man and of chromosome 17 in the mouse. Class I and class II MHC antigens are important in tissue transplantation. The greater the match between donor and recipient, the more likely the transplant is to survive. For example, a six-antigen match implies sharing of two HLA-A antigens, two HLA-B antigens, and two HLA-DR antigens between donor and recipient. Even though antigenically dissimilar grafts may survive when a powerful immunosuppressive drug such as cyclosporine is used, the longevity of the graft is still improved by having as many antigens match as possible.

histocompatibility antigen
One of a group of genetically encoded antigens present on tissue cells of an animal that provoke a rejection response if the tissue containing them is transplanted to a genetically dissimilar recipient. These antigens are detected by typing lymphocytes on which they are expressed. These antigens are encoded in man by genes at the HLA locus on the short arm of chromosome 6. In the mouse, they are encoded by genes at the H-2 locus on chromosome 17.

Histamine Synthesis

histocompatibility locus

The specific site on a chromosome where the histocompatibility genes that encode histocompatibility antigens are located. There are major histocompatibility loci such as HLA in man and H-2 in the mouse across which incompatible grafts are rejected within 1 to 2 weeks. There are also several minor histocompatibility loci, with more subtle antigenic differences, across which only slow, low level graft rejection reactions occur.

histocompatibility testing

Determination of the MHC class I and class II tissue type of both donor and recipient prior to organ or tissue transplantation. In man HLA-A, HLA-B, and HLA-DR types are determined, followed by crossmatching donor lymphocytes with recipient serum prior to transplantation. A mixed lymphocyte culture (MLC) is necessary in bone marrow transplantation, in conjunction with molecular DNA typing. The MLC may also be requested in living related organ transplants. As in renal allotransplantation, organ recipients have their serum samples tested for percent reactive antibodies, which reveals whether or not they have been presensitized against HLA antigens of an organ for which they may be the recipient.

histone antibodies

Antibodies of the IgG class against H2A-H2B histones detectable in essentially all procainamide-induced lupus patients manifesting symptoms. They are also present in approximately one-fifth of systemic lupus erythematosus patients and in procainamide-treated persons who do not manifest symptoms. Antibodies against H2A, H2B and H2A-H2B complex react well with histone fragments resistant to trypsin. By contrast, antibodies in the sera of SLE patients manifest reactivity for intact histones but not with their fragments. In lupus induced by hydralazine, antihistone antibodies react mainly with H3 and H4 and their fragments that are resistant to trypsin.

histoplasmin

An extract from cultures of *Histoplasma capsulatum* that is injected intradermally, in the same manner as the tuberculin test, to evaluate whether an individual has cell-mediated immunity against this microorganism.

histoplasmin test

A skin test analogous to the tuberculin skin test, which determines whether or not an individual manifests delayed-type hypersensitivity (cell-mediated) immunity to *Histoplasma capsulatum*, the causative agent of histoplasmosis in man. A positive skin test implies an earlier or a current infection with *H. capsulatum*.

histotope

That portion of an MHC class II histocompatibility molecule that reacts with a T lymphocyte receptor.

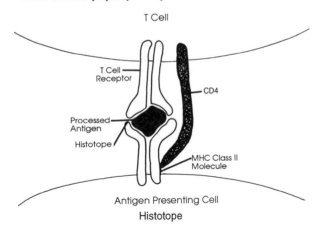

Histotope

HIV-1 genes

The *gag* gene encodes the structural core proteins p17, p24, p15, and p55 precursor. The *pol* gene encodes a protease that cleaves *gag* precursors. It also encodes reverse transcriptase that produces proviral DNA from RNA and encodes an integrase that is necessary for proviral insertion. The *env* gene encodes gp160 precursor, gp120, and gp41 in mature proteins; gp120 binds CD4 molecules, and gp41 is needed for fusion of the virus with the cell; vpr's function is unknown; *vif* encodes a 23-kD product that is necessary for infection of cells by free virus and is not needed for infection

from cell to cell; *tat* encodes a p14 product that binds to viral long-terminal repeat (LTR) sequence and activates viral gene transcription; *rev* encodes a 20-kD protein product that is needed for posttranscriptional expression of *gag* and *env* genes; *nef* encodes a 27-kD protein that inhibits HIV transcription and slows viral replication; and *vpu* encodes a 16-kD protein product that may be required for assembly and packaging of new virus particles.

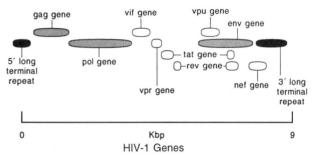

HIV-1 Genes

HIV-1 virus structure

HIV-1 is comprised of two identical RNA strands which constitute the viral genome. These are associated with reverse transcriptase and p17 and p24, which are core polypeptides. These components are all enclosed in a phospholipid membrane envelope that is derived from the host cell. Proteins gp120 and gp41 encoded by the virus are anchored to the envelope.

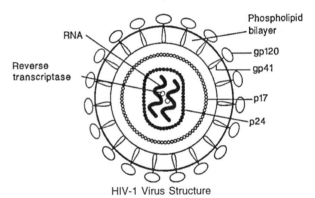

HIV-1 Virus Structure

HIV-2

Abbreviation for human immunodeficiency virus-2. Previously referred to as HTLV-IV, LAV-2, and SIV/AGM. This virus was first discovered in west African individuals who showed aberrant reactions to HIV-1 and simian immunodeficiency virus (SIV). It shows greater sequence homology (70%) with SIV/MAC than with HIV-1 (40% sequence homology). It has only 50% conservation for *gag* and *pol*. The remaining HIV genes are even less conserved than this. It has p24, gp36, and gp140 structural antigens. Its clinical course resembles that of AIDS produced by HIV-1, but it is confined principally to western Africa and is transmitted principally through heterosexual promiscuity.

HIV-2V

Possesses *X-ORF* and *VPX*, which are unique to it.

HIV infection

Recognition of infection by the human immunodeficiency virus (HIV) is through seroconversion. Following conversion to positive reactivity in an antibody screening test, a Western blot analysis is performed to confirm the result of positive testing for HIV. HIV mainly affects the immune system and the brain. It affects primarily the CD4+ lymphocytes which are necessary to initiate an immune response by interaction with antigen-presenting cells. This also deprives other cells of the immune system from receiving a supply of interleukin-2 through CD4+ lymphocyte stimulation, leading to a progressive decline in immune system function. HIV transmission is by either sexual contact, through blood products, or horizontally from mother to young. Although first observed in male homosexuals, it later became a major problem of intravenous drug abusers and ultimately has become more serious in the heterosexual population, affecting an increasing number of women as well as men. Clinically, individuals may develop acute HIV mononucleosis that usually occurs 2 to 6 weeks following infection, although it may occur later. The main symptoms include headache, fever, malaise, sore throat, and rash. Patients may develop pharyngitis; generalized lymphadenopathy; a macular or urticarial rash on the face, trunk, and limbs; and hepatosplenomegaly. The severity of the symptoms may vary from one individual to another. Acute HIV infection may also induce neurologic diseases, including meningitis, encephalitis, and other neurologic manifestations. Some individuals may not develop symptoms or illness for years. Other individuals develop AIDS-related complex (ARC), which represents progressive immune dysfunction. Symptoms include fever, night sweats, weight loss, chronic diarrhea, generalized lymphadenopathy, herpes zoster, and oral lesions. Individuals with ARC may progress to AIDS or death may occur in the ARC stage. ARC patients do not revert to an asymptomatic condition. Other individuals may develop persistent generalized lymphadenopathy (PGL) characterized by enlarged lymph nodes in the neck, axilla, and groin. The Centers for Disease Control (CDC) has set up criteria for the diagnosis of AIDS. These include the individuals who develop certain opportunistic infections and neoplasms, HIV-related encephalopathy, and HIV-induced wasting syndrome. The most frequent opportunistic infections in AIDS patients include *Pneumocystis carinii,* which produces pneumonia, and *Mycobacterium avium-intracellulare,* among other microorganisms. The most frequent tumor in AIDS patients is Kaposi's sarcoma. At the present time, AIDS is 100% fatal.

HIV Infection

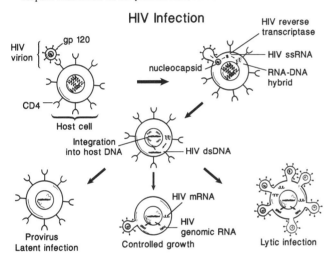

hives

A wheal and flare reaction of the anaphylactic type produced in the skin as a consequence of histamine produced by activated mast cells. There is edema, erythema, and pleuritis. This is a synonym for urticaria.

HLA

Abbreviation for human leukocyte antigen. The HLA histocompatibility system in humans represents a complex of MHC class I molecules distributed on essentially all nucleated cells of the body and MHC class II molecules that are distributed on B lymphocytes, macrophages, and a few other cell types. These are encoded by genes at the major histocompatibility complex. The HLA locus in humans is found on the short arm of chromosome 6. This has now been well defined, and in addition to encoding surface isoantigens, genes at the HLA locus also encode immune response (IR) genes. The class I region consists of HLA-A, HLA-B, and HLA-C loci and the class II region consists of the D region which is subdivided into HLA-DP, HLA-DQ, and HLA-DR subregions. Class II molecules play an important role in the induction of an immune response, since antigen-presenting cells must complex an antigen with class II molecules to present it in the presence of interleukin-1 to CD4$^+$ T lymphocytes. Class I molecules are important in presentation of intracellular antigen to CD8$^+$ T lymphocytes as well as for effector functions of target cells. Class III molecules encoded by genes located between those that encode class I and class II molecules include C2, BF, C4a, and C4b. Class I and class II molecules play an important role in the transplantation of organs and tissues. The microlymphocytotoxicity assay is used for HLA-A, -B, -C, -DR, and -DQ typing. The primed lymphocyte test is used for DP typing.

HLA-A

A class I histocompatibility antigen in humans. It is expressed on nucleated cells of the body. Tissue typing to identify an individual's HLA-A antigens employs lymphocytes.

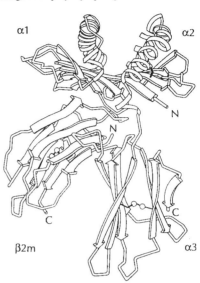

Polypeptide Chain Structure of HLA-A2

HLA allelic variation

Genomic analysis has identified specific individual allelic variants to explain HLA associations with rheumatoid arthritis, type I diabetes mellitus, multiple sclerosis, and celiac disease. There is a minimum of six α and eight β genes in distinct clusters, termed HLA-DR, -DQ, and -DP within the HLA class II genes. DO and DN class II genes are related, but map outside DR, DQ, and DP regions. There are two types of dimers along the HLA cell surface HLA-DR class II molecules. The dimers are made up of either DR α polypeptide associated with DRβ$_1$ polypeptide or DR with DRβ$_2$ polypeptide. Structural variation in class II gene products is linked to functional features of immune recognition leading to individual variations in histocompatibility, immune recognition, and susceptibility to disease. There are two types of structural variations which include variation among DP, DQ, and DR products in primary amino acid sequence by as much as 35% and individual variation attributable to different allelic forms of class II genes. The class II polypeptide chain possesses domains which are specific structural subunits containing variable sequences that distinguish among class II α genes or class II β genes. These allelic variation sites have been suggested to form epitopes, which represent individual structural differences in immune recognition.

HLA-B

A class I histocompatibility antigen in humans. It is expressed on the nucleated cells of the body. Tissue typing to define an individual's HLA-B antigens employs lymphocytes.

HLA-B27-related arthropathies

Joint diseases that occur with increased frequency in individuals who are HLA-B27 antigen positive. Juvenile rheumatoid arthritis, ankylosing spondylitis, Reiter's syndrome, *Salmonella*-related arthritis, psoriatic arthritis, and *Yersinia* arthritis belong to this group.

HLA-C

A class I histocompatibility antigen in human. It is expressed on nucleated cells of the body. Lymphocytes are employed for tissue typing to determine HLA-C antigens. HLA-C antigens play little or no role in graft rejection.

HLA-D region

The human MHC class II region comprised of three subregions designated DR, DQ, and DP. Multiple genetic loci are present in each of these. DN (previously DZ) and DO subregions are each comprised of one genetic locus. Each class II HLA molecule is comprised of one α and one β chain that constitute a heterodimer.

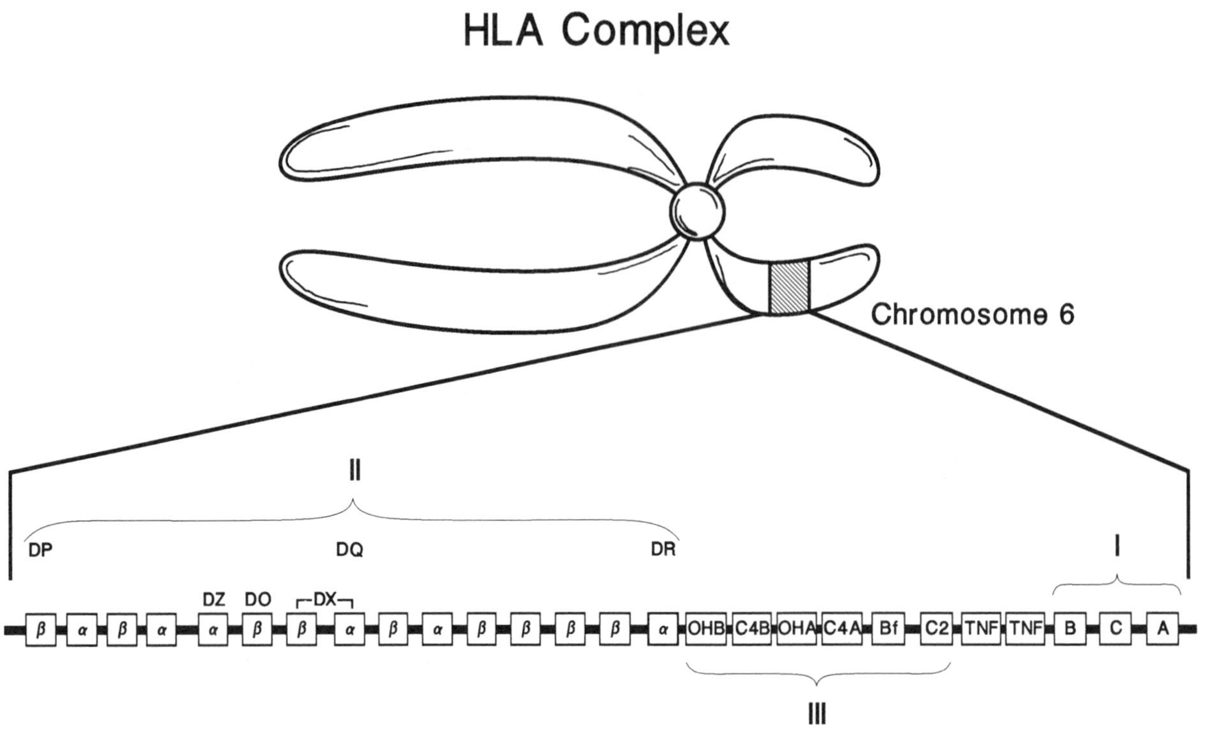

HLA Complex

Chromosome 6

DP DQ DR II I

DZ DO ┌DX┐

β α β α α β β α β α β β β β α OHB C4B OHA C4A Bf C2 TNF TNF B C A

III

Genes within each subregion encode a particular class II molecule's α and β chains. Class II genes that encode α chains are designated A, whereas class II genes that encode β chain are designated B. A number is used following A or B if a particular subregion contains two or more A or B genes.

HLA disease association

Certain HLA alleles occur in a higher frequency in individuals with particular diseases than in the general population. This type of data permits estimation of the "relative risk" of developing a disease with every known HLA allele. For example, there is a strong association between ankylosing spondylitis, which is an autoimmune disorder involving the vertebral joints, and the class I MHC allele, HLA-B27. There is a strong association between products of the polymorphic class II alleles HLA-DR and -DQ and certain autoimmune diseases, since class II MHC molecules are of great importance in the selection and activation of CD4+ T lymphocytes which regulate the immune responses against protein antigens. For example, 95% of Caucasians with insulin-dependent (type I) diabetes mellitus have HLA-DR3 or HLA-DR4 or both. There is also a strong association of HLA-DR4 with rheumatoid arthritis. Numerous other examples exist and are the targets of current investigations, especially in extended studies employing DNA probes. Calculation of the relative risk (RR) and absolute risk (AR) can be found elsewhere in this dictionary under definitions for those terms.

HLA-DP subregion

The site of two sets of genes designated HLA-DPA1 and HLA-DPB1 and the pseudogenes HLA-DPA2 and HLA-DPB2. DP α and DP β chains encoded by the corresponding genes DPA1 and DPB1 unite to produce the DP αβ molecule. DP antigen or type is determined principally by the very polymorphic DP β chain in contrast to the much less polymorphic DP α chain. DP molecules carry DPw1-DPw6 antigens.

HLA-DQ subregion

Two sets of genes, designated DQA1 and DQB1 and DQA2 and DQB2, are found in this region. DQA2 and DRB2 are pseudogenes. DQ α and DQ β chains, encoded by DQA1 and DQB1 genes, unite to produce the DQ αβ molecule. Although both DQ α and DQ β chains are polymorphic, the DQ β chain is the principal factor in determining the DQ antigen or type. DQ αβ molecules carry DQw1-DQw9 specificities.

HLA-DR antigenic specificities

Epitopes on DR gene products. Selected specificities have been mapped to defined loci. HLA serologic typing requires the identification of a prescribed antigenic determinant on a particular HLA molecular product. One typing specificity can be present on many different molecules. Different alleles at the same locus may encode these various HLA molecules. Monoclonal antibodies are now used to recognize certain antigenic determinants shared by various molecules bearing the same HLA typing specificity. Monoclonal antibodies have been employed to recognize specific class II alleles with disease associations.

HLA-DR subregion

The site of one HLA-DRA gene. Although DRB gene number varies with DR type, there are usually three DRB genes, termed DRB1, DRB2, and DRB3 (or DRB4). The DRB2 pseudogene is not expressed. The DR α chain, encoded by the DRA gene, can unite with products of DRB1 and DRB3 (or DRB4) genes which are the DR β-1 and DR β-3 (or DR β-4) chains. This yields two separate DR molecules, DR αβ-1 and DR αβ-3 (or DR αβ-4). The DRβ chain determines the DR antigen (DR type) since it is very polymorphic, whereas the DRα chain is not. DR αβ-1 molecules carry DR specificities DR1-DRw18. Yet, DR αβ-3 molecules carry the DRw52 and the DR αβ-4 molecules carry the DRw53 specificity.

HLA-E

HLA class I nonclassical molecule.

HLA-F

HLA class I nonclassical molecule.

HLA-G

A polymorphic class I HLA antigen with extensive variability in the α-2 domain. It is found on trophoblast, i.e., placenta cells and trophoblastic neoplasms. HLA-G is expressed only on cells such as placental extravillous cytotrophoblasts and choriocarcinoma that fail to express HLA-A, -B, and -C antigens. HLA-G expression is most pronounced during the first trimester of pregnancy. Trophoblast cells expressing HLA-G at the maternal-fetal junction may protect the semiallogeneic fetus from "rejection". Prominent HLA-G expression suggests maternal immune tolerance.

HLA-H

A pseudogene found in the MHC class I region that is structurally similar to HLA-A but is nonfunctional due to the absence of a cysteine residue at position 164 in its protein product and the deletion of the codon 227 nucleotide.

HLA locus

The major histocompatibility locus in man.

HLA nonclassical class I genes

Genes located within the MHC class I region that encode products that can associate with β2 microglobulin. However, their function and tissue distribution are different from those of HLA-A,-B,-C molecules. Examples include HLA-E, -F, and -G. Of these only HLA-G is expressed on the cell surface. It is uncertain whether or not these HLA molecules are involved in peptide binding and presentation like classical class I molecules.

HLA oligotyping

A recently developed method using oligonucleotide probes to supplement other histocompatibility testing techniques. Whereas serological and cellular methods identify phenotypic characteristics of HLA proteins, oligotyping defines the genotype of the DNA that encodes HLA protein structure and specificity. Thus, oligotyping can identify the DNA type even when there is a failure of expression of HLA genes which render serological techniques ineffective.

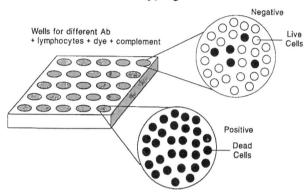

HLA Typing

Wells for different Ab + lymphocytes + dye + complement

Negative
Live Cells

Positive
Dead Cells

Hm-1

Designation for the Syrian hamster's major histocompatibility complex (MHC). Class II MHC genes have been recognized.

Hodgkin's disease

A type of lymphoma that involves the lymph nodes and spleen, causing a replacement of the lymph node architecture with binucleated giant cells known as Reed-Sternberg cells, reticular cells, neutrophils, eosinophils, and lymphocytes. There is both lymphadenopathy and splenomegaly. Patients manifest a deficiency of cell-mediated immunity which causes skin tests of the tuberculin type to be negative. By contrast, there is no alteration in their B cell function. They may have an increase in suppressor cell activity. There is also increased susceptibility to opportunistic infections.

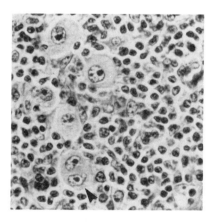

Hodgkin cell and Reed-Sternberg Cell

Hodgkin's Disease

Hof

A German word for courtyard which refers to the perinuclear clear zone adjacent to the nucleus in plasma cells. Lymphoblasts and Reed-Sternberg cells may also exhibit a hof.

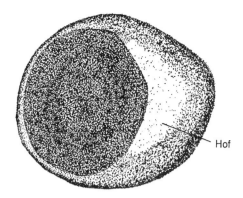

Hof

holoxenic

An adjective that describes an animal raised under ordinary conditions as opposed to an axenic or gnotobiotic animal.

homing receptors

Molecules on a cell surface that direct traffic of that cell to a precise location in other tissues or organs. For example, lymphocytes bear surface receptors that facilitate their attachment to high endothelial cells of postcapillary venules in lymph nodes.

homobody

An idiotypic determinant of an antibody molecule whose three-dimensional structure resembles antigen. Also called internal image of antigen. For example, antiidiotypic antibodies, of this type, to insulin receptor may partially mimic the action of insulin.

homocytotrophic antibody

An antibody that attaches better to animal cells of the same species in which it is produced than it attaches to animal cells of a different species. The term usually refers to an antibody that becomes fixed to mast cells of an animal of the same species that results in anaphylaxis with the release of pharmacological mediators of immediate hypersensitivity. These include histamines and other vasoactive amines when the mast cells degranulate.

homodimer

A protein comprised of dual peptide chains that are identical.

homograft

Allograft, i.e., an organ or tissue graft from a donor to a recipient of the same species.

homograft reaction

An immune reaction generated by a homograft (allograft) recipient against the graft alloantigens. Also called an allograft reaction.

homograft rejection

Allograft rejection. An immune response induced by histocompatibility antigens in the donor graft that are not present in the recipient. This is principally a cell-mediated type of immune response.

homologous

An adjective that describes something from the same source. For example, an organ allotransplant from one member to a recipient member of the same species, i.e., renal allotransplantation in humans.

homologous antigen

An antigen (immunogen) that stimulates the synthesis of an antibody and reacts specifically with it.

homologous disease

Refer to allogeneic disease and graft-vs.-host disease (GVHD).

homologous restriction factor (HRF)

An erythrocyte surface protein that prevents cell lysis by homologous complement on its surface. It bears a structural resemblance to C8 and C9.

homology region

A 105- to 115-amino acid residue sequence of heavy or light chains of immunoglobulins which have a primary structure that resembles other corresponding sequences of the same size. A homology region has a globular shape and an intrachain disulfide bond. The exons that encode homology regions are separated by introns. Light polypeptide chain homology regions are termed V_L and C_L. Heavy chain homology regions are designated V_H, C_H1, C_H2, and C_H3.

homology unit

A structural feature of an immunoglobulin domain.

homopolymer

A molecule comprised of repeating units of only one amino acid.

homotransplantation

Homograft, i.e., allograft transplantation.

homozygous typing cell (HTC) technique

An assay that employs a stimulator cell that is homozygous at the HLA-D locus. An HTC incorporates only a minute amount of tritiated thymidine when combined with a homozygous cell in the MLR. This implies that the HTC shares HLA-D determinants with the other cell type. By contrast, when an HTC is combined with a nonhomozygous cell, much larger amounts of tritiated thymidine are incorporated. Many variations between these two extremes are noted in actual practice. Homozygous typing cells are frequently obtained from the progeny of marriages between cousins.

homozygous typing cells (HTCs)

Cells obtained from a subject who is homozygous at the HLA-D locus. HTCs facilitate MLR typing of the human D locus.

hook effect

An artifact that may be seen in the immunoradiometric assay (IRMA) that occurs when a hormone being assayed is in very high concentration. The excess amount cannot be measured by the detector system since it will have obtained a theoretical limit. The diminished counts with the labeled antibody at the elevated hormone concentration yield spuriously low results. Thus, IRMA is not an appropriate method to assay hormones present in relatively high concentrations, such as gastrin, prolactin, or hCG. The hook effect requires measurement of two separate concentrations to establish linearity.

hookworm vaccine

A live vaccine to protect dogs against the hookworm, *Ancylostoma caninum.* The vaccine is comprised of X-irradiated larvae to halt their development to adult forms.

hormone immunoassays

Multiple hormones, including thyroid-stimulating hormone, human growth hormone, insulin, glucagon and many others, may be measured by an immunoassay using either radioactively labeled reagents or through enzyme color reactions using the ELISA technique. Labeled and unlabeled hormone are allowed to compete for binding sites with antihormone antibody. This is followed by the separation of bound from unbound hormone by one of several techniques.

horror autotoxicus (historical)

The term coined by Paul Ehrlich (*circa* 1900) to account for an individual's failure to produce autoantibodies against his own self constituents even though they are excellent antigens or immunogens in other species. This lack of immune reactivity against self was believed to protect against autoimmune disease. This lack of self-reactivity was postulated to be a fear of poisoning or destroying one's self. Abrogation of horror autotoxicus leads to autoimmune disease. Horror autotoxicus was later (1959) referred to as self-tolerance by F.M. Burnet.

horse serum sensitivity

An allergic or hypersensitive reaction in a human or other animal receiving antitoxin or antithymocyte globulin generated by immunization of horses whose immune serum is used for therapeutic purposes. Classic serum sickness is an example of this type of hypersensitivity which first appeared in children receiving diphtheria antitoxin early in the 20th century.

hot antigen suicide

The labeling of an antigen with a powerful radioisotope such as ^{131}I proves lethal upon contact with antigen-binding cells that have receptors specific for it. This leads to a failure to synthesize antibodies specific for that antigen, provided the antigen-binding and antibody-synthesizing cells are one and the same. Hot antigen suicide supports the clonal selection theory of antibody formation.

hot spot

A hypervariable region in DNA that encodes the variable region of an immunoglobulin molecule's heavy (V_H) and light (V_L) polypeptide chains. These are also designated complimentarity-determining regions (CDR). These are the areas for specific antigen binding, and they also determine the idiotype of an immunoglobulin molecule. The remaining background support structures of the heavy and light polypeptide chains are termed framework regions (FR). The κ and λ light chain hot spots are situated near amino acid residues 30, 50, and 95. Also called hypervariable regions.

house dust allergy

Type I immediate hypersensitivity reaction in atopic individuals exposed to house dust in which the principal allergen is *Dermatophagoides pteronyssinus,* the house dust mite. The condition is expressed as a respiratory allergy with the atopic subject manifesting either asthma or allergic rhinitis.

HPV

Human papilloma virus. A human virus that has the potential to be oncogenic and occurs most frequently in individuals with multiple sexual partners. There are 46 HPV genotypes. It can be demonstrated by *in situ* hybridization in proliferations of epithelial cells that are benign, such as condyloma accuminatum, or malignant, such as squamous cell carcinoma of the uterine cervix. Whereas HPV types 6 and 11 are not usually premalignant, HPV types 16, 18, 31, 33, and 35 are linked to cervical intraepithelial neoplasia (CIN), cervical dysplasia, and anogenital cancer. HPV is predicted to induce derepression as a neoplastic mechanism. HPV encodes E6, a viral protein that combines with the tumor suppressor protein p53.

HSA

Abbreviation for human serum albumin, an antigen commonly used in experimental immunology.

HSV

Herpes simplex virus.

HTLA (human T lymphocyte antigen)

An obsolete term for human T lymphocyte antigen. This is now replaced by cluster of differentiation (CD) designations.

HTLA antibody

High-titer, low-avidity antibody.

HTLV

Human T cell leukemia virus. Viruses linked to human T cell leukemias.

HTLV-IV

A human retrovirus isolated from west Africa that is related to HIV-1 and HIV-2, but appears nonpathogenic.

human chorionic gonadotropin (hCG)

A hormone synthesized by placental syncytiotrophoblasts. It serves as a marker demonstrable by immunoperoxidase staining in trophoblastic neoplasms, e.g., choriocarcinomas. Germ cell and nontrophoblastic neoplasms, such as various carcinomas, may express hCG.

human immune globulin (HIG)

A pooled globulin preparation from the plasma of donors who are negative for HIV that is used in the treatment of primary immunodeficiency, such as severe combined immunodeficiency, Burton's disease, and combined variable immunodeficiency, and in cases of idiopathic thrombocytopenic purpura. The method of production is extraction by cold ethanol fractionation at acid pH. Viruses are inactivated, which permits the safe administration of the HIG to patients without risk of HIV, HAV, HBV, or non-A, non-B hepatitis.

human immunodeficiency virus (HIV)

A retrovirus that induces acquired immune deficiency syndrome (AIDS) and associated disorders. It was previously designated as HTLV-III, LAV, or ARV. It infects CD4+ T lymphocytes, mononuclear phagocytes carrying CD4 molecules on their surface, follicular dendritic cells, and Langerhans cells. It produces profound immunodeficiency affecting both humoral and cell-mediated immunity. There is a progressive decrease in CD4+ helper/inducer T lymphocytes until they are finally depleted in many patients. There may be polyclonal activation of B lymphocytes with elevated synthesis of immunoglobulins. The immune response to the virus is not protective and does not improve the patient's condition. The virus is comprised of an envelope glycoprotein (gp160) which is its principal antigen. It has a gp120 external segment and a gp41 transmembrane segment. CD4 molecules on CD4+ lymphocytes and macrophages serve as receptors for gp120 of HIV. It has an inner core that contains RNA and is encircled by a lipid envelope. It contains structural genes designated *env, gag,* and *pol* that encode the envelope protein, core protein, and reverse transcriptase, respectively. HIV also possesses at least six additional genes, i.e., *tat,* that regulate HIV replication. It can increase production of viral protein several thousandfold. The *rev* gene encodes proteins that block transcription of regulatory genes; *vif (sor)* is the virus infectivity gene whose product increases viral infectivity and may promote cell to cell transmission; *nef* is a negative regulatory factor that encodes a product that blocks replication of the virus; and *vpr* (viral protein R) and *vpu* (viral protein U) genes have also been described. No successful vaccine has yet been developed, although several types are under investigation.

Genetically Engineered Antibodies

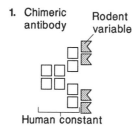

1. Chimeric antibody — Rodent variable / Human constant

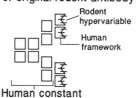

2. Reshaped humanized antibody with specificity of original rodent antibody — Rodent hypervariable / Human framework / Human constant

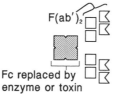

3. Antibody-toxin/enzyme chimeric molecule — F(ab')₂ / Fc replaced by enzyme or toxin

4. Single domain antibody — [M]–V

5. Single chain Fv molecule created by linked V domains from an L and an H chain — V$_H$ V$_L$ / Linker peptide

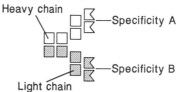

6. Bi-specific antibody created by protein engineering — Heavy chain / Specificity A / Specificity B / Light chain

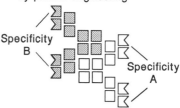

7. Bi-specific antibodies created by protein engineering — Specificity B / Specificity A

human milk-fat globulin (HMFG)

Human milk glycoprotein on secretory breast cell surfaces. Many breast and ovarian carcinomas are positive for HMFG.

humanized antibody (see figure, above)

An engineered antibody produced through recombinant DNA technology. A humanized antibody contains the antigen-binding specificity of an antibody developed in a rat, whereas the remainder of the molecule is of human origin. To accomplish this, hypervariable genes that encode the antigen-binding regions of a rat antibody are transferred to the normal human gene which encodes an immunoglobulin molecule that is mostly human, but expresses the antigen-binding specificity of the rat antibody in the variable region of the molecule. This greatly diminishes any immune response to the antibody molecule itself as a foreign protein by the human host, while retaining the desired functional capacity of reacting with the specific antigen.

humoral

In immunology, this term refers to the antibody limb of the immune response, in contrast to the cell-mediated limb, together with the action of complement. Thus, immunity based on antibodies or antibodies and complement is produced and referred to as humoral immunity. Humoral immunity of the antibody type represents the products of the B cell system.

humoral antibody

Antibody found in the blood plasma, lymph, and other body fluids. Humoral antibody, together with complement, mediates humoral immunity which is based upon soluble effector molecules.

humoral immunity

Immunity attributable to specific immunoglobulin antibody and present in the blood plasma, lymph, other body fluids, or tissues. The antibody may also adhere to cells in the form of cytophilic antibody. Antibody or immunoglobulin-mediated immunity acts in conjunction with complement proteins to produce either beneficial (protective) or pathogenic (hypersensitivity-tissue injuring) reactions. Antibodies that are the messengers of humoral immunity are derived from B cells. For purposes of discussion, it is separated from so-called cellular or T cell-mediated immunity. However, the two cannot be clearly distinguished since antibodies and T cells often participate in immune reactions together. However, the classification of humoral separate from cellular immunity is useful in understanding and explaining biological mechanisms.

hump

Immune deposits containing IgG and C3, as well as the alternate complement pathway components, properdin and factor B, that occur in postinfectious glomerulonephritis on the subepithelial side of peripheral capillary basement membranes. They resolve within 4 to 8 weeks of the infection in most individuals. They may also occur in selected other nonstreptococcal postinfectious glomerulonephritides.

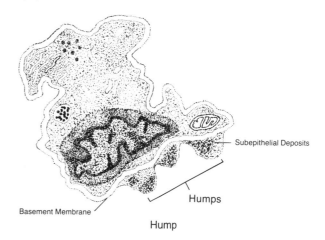

Subepithelial Deposits / Humps / Basement Membrane / Hump

HUT 78

The original designation for a cell line derived from a patient with mycosis fungoides, that is now termed H-9, that was susceptible to infection with HIV-1 virus and has greatly aided in HIV-1 culture *in vitro.*

hybrid antibody

An immunoglobulin molecule that may be prepared artificially, but may never occur in nature. Each of two antigen-bindings sites are of different specificity. If IgG whole molecules or F(ab')₂ fragments prepared from them are subjected to mild reduction, the central disulfide bond is converted to sulfhydryl groups. A preparation of half molecules or F(ab') fragments is produced. If either of these is reoxidized, hybrid molecules will constitute some of the products. These can be purified by passing over immunoadsorbents containing bound antigen of the appropriate specificity. Hybrid antibodies are monovalent and do not induce precipitation. They may be used in labeling cell surface antigens in which one antigen-binding site of the molecule is specific for cell surface

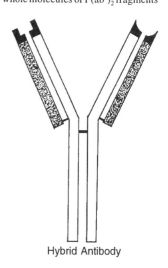

Hybrid Antibody

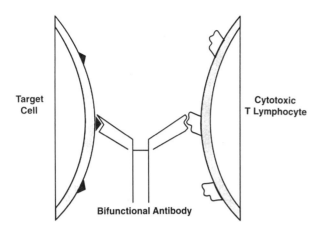

Target
Cell

Cytotoxic
T Lymphocyte

Bifunctional Antibody

epitopes and the other combines with a marker that renders the
reaction product visible.

hybrid cell

A cell produced when two cells fuse and their two nuclei fuse to
form a heterokaryon. Although hybrid cell lines can be established
from clones of hybrid cells, they lack stability and delete chromosomes, which is nevertheless useful for gene mapping. Hybrid cell
lines can be isolated by using HAT as a selective medium.

hybrid hapten

A hydrophobic type of hapten that lies within the folds of a protein
carrier away from the aqueous solvent, creating a new spatial
structure.

hybrid resistance

The resistance of members of an F_1 generation of animals to growth
of a transplantable neoplasm from either one of the parent strains.

hybridoma

A hybrid cell produced by the fusion of an antibody-secreting cell
isolated from the spleen of an animal immunized against that
particular antigen with a mutant myeloma cell of the same species
which is no longer secreting its own protein product. Polyethylene
glycol is used to effect cell fusion. Antibody-synthesizing cells
provide the ability to produce a specific monoclonal antibody. The
mutant myeloma cell line confers immortality upon the hybridoma.
If the nucleotide synthesis pathway is inhibited, the myeloma cells
become dependent on hypoxanthine guanine phosphoribosyl transferase (HGRPT) and the salvage pathway. The antibody-synthesizing cells provide the HGPRT, and the mutant myeloma cell enables
endless reproduction. Once isolated through use of a selective
medium such as HAT, hybridoma cell lines can be maintained for
relatively long periods. Hybridomas produce specific monoclonal
antibodies that may be collected in great quantities for use in
diagnosis and selected types of therapy.

hydrophilic

An adjective that describes a water-soluble substance. A cell membrane or protein that contains hydrophilic groups on its surface that
attract water molecules.

hydrophobic

A descriptor for a substance that is insoluble in water. Protein or
membrane hydrophobic groups are situated inside these structures
away from water.

hydrops fetalis

A hydropic condition that occurs in newborns who may appear
puffy and plethoric and that may be induced by either immune or
nonimmune mechanisms. In the immune type, the mother synthesizes IgG antibodies specific for antigens of the offspring, such as
anti-RhD erythrocyte antigen. These IgG antibodies pass across the
placenta into the fetal circulation causing hemolysis. Nonimmune
hydrops results from various etiologies not discussed here.

hydroxychloroquine

(2-[[4-[7-chloro-4-quinolyl]amno]ethyl-amino]ethanol sulfate). An
antimalarial drug that has also been used in therapy of the connective tissue diseases, such as systemic lupus erythematosus and
rheumatoid arthritis.

17-hydroxycorticosteroids (17-OHCS)

Adrenal steroid hormones synthesized by the action of 17-hydroxy-
lase, including cortisone, cortisol, 11-deoxycortisol, and tetrahydro
derivities of 17-hydroxylase. 17-OHCSs in the urine gives an
indication of the adrenal gland's functional status and catabolic

rates. 17-OHCSs are elevated in Cushing's disease, obesity, pregnancy, and pancreatitis, but decreased in hypopituitarism and
Addison's disease.

hyperacute rejection

Accelerated allograft rejection attributable to preformed antibodies
in the circulation of the recipient that are specific for antigens of the
donor. These antibodies react with antigens of endothelial cells
lining capillaries of the donor organ. It sets in motion a process that
culminates in fibrin plugging of the donor organ vessels, resulting
in ischemia and loss of function and necessitating removal of the
transplanted organ.

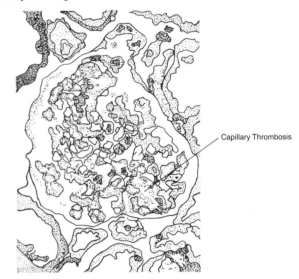

Capillary Thrombosis

Renal Histology Showing Hyperacute Graft Rejection

hypergammaglobulinemia

Elevated serum γ globulin (immunoglobulin) levels. A polyclonal
increase in immunoglobulins in the serum occurs in any condition
where there is continuous stimulation of the immune system, such
as chronic infection, autoimmune disease, systemic lupus
erythematosus, etc. Hypergammaglobulinemia may also result
from a monoclonal increase in immunoglobulin production, as in
multiple myeloma, Waldenström's macroglobulinemia, or other
conditions associated with the formation of monoclonal immunoglobulins. Repeated immunization may also induce hypergammaglobulinemia.

hypergammaglobulinemic purpura

Purpura hyperglobulinemia.

hyperimmune

A descriptor for an animal with a high level of immunity that is
induced by repeated immunization of the animal to generate large
amounts of functionally effective antibodies, in comparison to
animals subjected to routine immunization protocols, perhaps with
fewer boosters.

hyperimmunization

Successive administration of an immunogen to an animal to induce
the synthesis of antibody in relatively large amounts. This procedure is followed in the preparation of therapeutic antisera by
repeatedly immunizing animals to render them "hyperimmune".

hyperimmunoglobulin E syndrome (HIE)

A condition characterized by markedly elevated IgE levels, i.e.,
greater than 5000 IU/ml. The patients have early eczema and
repeated abscesses of the skin, sinuses, lungs, eyes, and ears.
*Staphylococcus aureus, Candida albicans, Hemophilus influenzae,
Streptococcus pneuomoniae*, and group A hemolytic streptococci
are among the more common infectious agents. The principal
infection produced by *Staphylococcus aureus* and *C. albicans* is a
"cold abscess" in the skin. The failure of IgE to fix complement and
therefore cause inflammation at the infection site is characteristic.
IgG antibodies against IgE form complexes that bind to mononuclear phagocytes, resulting in monokine release that induces
calcium resorption from bone. As calcium is lost from the bone,
osteoporosis results, leading to bone fractures. Patients with HIE
have diminished antibody responses to vaccines and to major
histocompatibility antigens. They may be anergic, and *in vitro*

challenge of their lymphocytes with mitogens or antigens leads to diminished responsiveness. There is also a decrease in the CD8⁺ T lymphocyte population in the peripheral blood. The disease becomes manifest in young infants, shows no predilection for males vs. females, and is not hereditary. Also called Job's syndrome.

hyperimmunoglobulin M syndrome

An immunodeficiency disorder in which the serum IgM level is normal or elevated. By contrast, the serum IgG and IgA levels are strikingly diminished or absent. These patients have repeated infections and may develop neoplasms in childhood. This syndrome may be transmitted in an X-linked or autosomal dominant fashion. It may also be related to congenital rubella. The condition is produced by failure of the T lymphocytes to signal IgM-synthesizing B cells to switch to IgG- and IgA-producing cells.

hyperplasia

An increase in the cell number of an organ that leads to an increase in the organ size. It is often linked to a physiological reaction to a stimulus and is reversible.

hypersensitivity

Increased reactivity or increased sensitivity by the animal body to an antigen to which it has been previously exposed. The term is often used as a synonym for allergy, which describes a state of altered reactivity to an antigen. Hypersensitivity has been divided into categories based upon whether it can be passively transferred by antibodies or by specifically immune lymphoid cells. The most widely adopted current classification is that of Coombs and Gell that designates immunoglobulin-mediated hypersensitivity reactions as types I, II, and III and lymphoid cell-mediated (delayed-type) hypersensitivity/cell-mediated immunity as a type IV reaction. *Hypersensitivity* generally represents the "dark side", signifying the undesirable aspects of an immune reaction, whereas the term *immunity* implies a desirable effect.

hypersensitivity angiitis

Small vessel inflammation most frequently induced by drugs.

hypersensitivity pneumonitis

Lung inflammation induced by antibodies specific for substances that have been inhaled. Within hours of inhaling the causative agent, dyspnea, chills, fever, and coughing occur. Histopathology of the lung reveals inflammation of alveoli in the interstitium with obliterating bronchiolitis. Immunofluorescence examination reveals deposits of C3. Hyperactivity of the lungs to air-borne immuogens or allergens may ultimately lead to interstitial lung disease. An example is farmer's lung, which is characterized by malaise, coughing, fever, tightness in the chest, and myalgias. Of the numerous syndromes and associated antigens that may induce hypersensitivity pneumonitis, humidifier lung (thermophilic actinomycetes), bagassosis (*Thermoactinomyces vulgaris),* and bird fancier's lung (bird droppings) are well known.

hypersensitivity vasculitis

An allergic response to drugs, microbial antigens, or antigens from other sources, leading to an inflammatory reaction involving small arterioles, venules, and capillaries.

hyperthyroidism

A metabolic disorder attributable to thyroid hyperplasia with an elevation in thyroid hormone secretion. Also called Graves' disease.

hypervariable regions

A minimum of four sites of great variability which are present throughout the H&L chain V regions. They govern the antigen-binding site of an antibody molecule. Thus, grouping of these hypervariable residues into areas govern both conformation and specificity of the antigen-binding site upon folding of the protein molecule. Hypervariable residues are also responsible for variations in idiotypes between immunoglobulins produced by separate cell clones. Those parts of the variable region that are not hypervariable are termed the framework regions. Hypervariable

The Six Hypervariable Regions of an Antibody

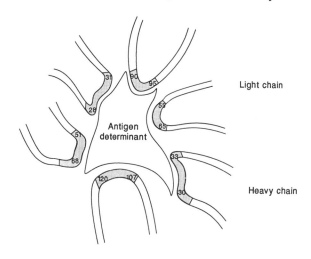

Light chain

Antigen determinant

Heavy chain

regions are also called complementarity-determining regions. Refer to hot spot.

hypocomplementemia

Literally, diminished complement in the blood. This can occur in a number of diseases in which immune complexes fix complement *in vivo,* leading to a decrease in complement protein. Examples include active cases of systemic lupus erythematosus, proliferative glomerulonephritis, and serum sickness. Protein deficient patients may also have diminished plasma complement protein levels.

hypocomplementemic glomerulonephritis

Decreased complement in the blood during the course of chronic progressive glomerulonephritis, in which C3 is deposited in the glomerular basement membrane of the kidney.

hypocomplementemic vasculitis urticarial syndrome

A type of systemic inflammation with leukocytoclastic vasculitis. There is diminished serum complement levels and urticaria.

hypogammaglobulinemia

Deficient levels of IgG, IgM, and IgA serum immunoglobulins. This may be attributable to either decreased synthesis or increased loss. Hypogammaglobulinemia can be physiologic in neonates. It may be a manifestation of either congenital or acquired antibody deficiency syndromes. Several types are described and include Bruton's disease and a congenital type, as well as an acquired type such as in chronic lymphocytic leukemia. Human γ globulin is used for treatment.

hypogammaglobulinemia of infancy

A temporary delay in immunoglobulin synthesis during the first 12 months or even 24 months of life. This leads only to a transient, physiologic immunodeficiency following catabolism of maternal immunoglobulins passed to the infant.

hyposensitization

A technique to decrease responsiveness to antigens acting as allergens in individuals with immediate (type I) hypersensitivity to them. Since the reaction is mediated by IgE antibodies specific for the allergen becoming fixed to the surface of the patient's mast cells, the aim of this mode of therapy is to stimulate the production of IgG-blocking antibodies that will combine with the allergen before it reaches the mast cell-bound IgE antibodies. This is accomplished by graded administration of an altered form of the allergen that favors the production of IgG rather than IgE antibodies. In addition to intercepting allergen prior to its interaction with IgE, the IgG antibodies may also inhibit further production of IgE.

I

I invariant (Ii)
Refer to invariant chain.

I-K
Abbreviation for immunoconglutinin.

I region
In the mouse, the DNA segment of the major histocompatibility complex where the gene that encodes MHC class II molecules are located. The 250-kb I region consists of I-A and I-E subregions. The genes designated pseudo $A_\beta 3$, $A_\beta 2$, $A_\beta 1$ and A_α are located in the 175-kb I-A subregion. The genes designated $E_\beta 2$ and E_α are located in the 75-kb I-E subregion. $E_\beta 1$ is located where the I-A and I-E subregions join. The S region contains the $E_\beta 3$ gene.

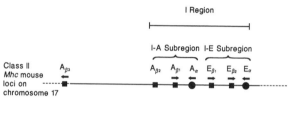

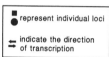

Ia antigens (immune-associated antigen)
Products of major histocompatibility complex (MHC) I region genes that encode murine cellular antigens. In man, the equivalent antigens are designated HLA-DR, which are encoded by MHC class II genes. B lymphocytes, monocytes, and activated T lymphocytes express Ia antigens.

iatrogenic
A physician-induced disease or injury that occurs as a result of therapy by a medical practitioner.

IBD
Abbreviation for inflammatory bowel disease.

Ibuprofen
((+)-2-(p-isobutylphenyl)propionic acid). An antiinflammatory drug used in the therapy of patients with rheumatoid arthritis, ankylosing spondylitis, and juvenile rheumatoid arthritis.

$$(CH_3)_2CHCH_2 - \underset{\text{Ibuprofen}}{\bigcirc} - \underset{CH_3}{\overset{|}{CHCOOH}}$$

iC3B (inactivated C3b)
Also designated as C3bi.

iC4b
Synonym for C4bi.

ICAM-1 (intercellular adhesion molecule 1)
A γ interferon-induced protein which is needed for the migration of polymorphonuclear neutrophils into areas of inflammation.

ICAM-2
Refer to intercellular adhesion molecule 2.

ICAM-3
Refer to intercellular adhesion molecule 3.

Homotype Adhesion

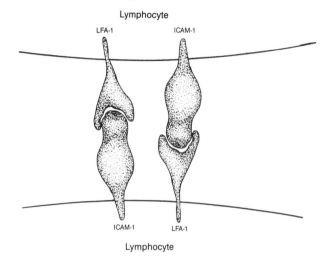

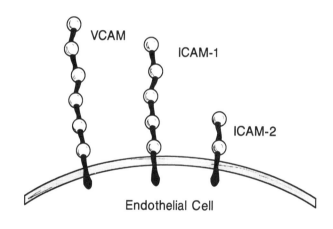

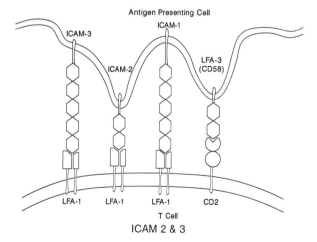

ICAM 2 & 3

id reaction

Dermatophytid reaction. A sudden rash linked to, but anatomically separated from an inflammatory reaction of the skin in a sensitized individual with the same types of lesions elsewhere. The hands and arms are usual sites of id reactions that are expressed as sterile papulovesicular pustules. They may be linked with dermatophytosis such as tinea capitis or tenia pedis. They may also be associated with stasis dermatitis, contact dermatitis, and eczema.

identity, reaction of

Refer to reaction of identity.

identity testing

See paternity testing.

idiosyncrasy

The term idiosyncrasy, mostly used with respect to drugs, is a quantitatively abnormal response usually not corresponding to the pharmacological action of the drug. It does not have an immunological mechanism.

idiotope

An epitope or antigenic determinant in the hypervariable region of the N-terminus of an immunoglobulin molecule. An idiotope is an epitope or antigenic determinant of an idiotype on an antibody molecule's V region. This type of antigenic determinant is present on immunoglobulin molecules synthesized by one clone or a few clones of antibody-producing cells.

idiotype

The idiotype refers to that segment of an immunoglobulin or antibody molecule that determines its specificity for antigen and is based upon the multiple combinations of variable (V), diversity (D), and joining (J) exons. The idiotype is located in the Fab region, and its expression usually requires participation of the variable regions of both heavy and light chains, namely the Fv fragment which contains the antigen-combining site.

The antigen-binding specificity of the combining site may imply that all antibodies produced by an animal in response to a given immunogen have the same idiotype. This is not true since the antibody response is heterogeneous. There will usually be a major idiotype representing 20 to 70% of the specific antibody response. The remainder carry different idiotypes that may crossreact with the major idiotype. Cross-reacting idiotypy represents the extent of heterogeneity among the antibodies of a given specificity.

The unique antigenic determinants that govern the idiotype (Id) of an immunoglobulin molecule occur on the products of either a single or several clones of cells synthesizing immunoglobulins. This unique idiotypic determinant is sometimes called a private idiotype that appears in all V regions of immunoglobulin molecules whose amino acid sequences are the same. Shared idiotypes are also known as public idiotype determinants. These may appear in a relatively large number of immunoglobulin molecules produced by inbred strains of mice or other genetically identical animals in response to a specific antigen. The localization of idiotypes in the antigen-binding site of the molecule's V region is illustrated by the ability of haptens to block or inhibit the interaction of antiidiotypic antibodies with their homologous antigenic markers or determinants in the antigen-binding region of antibody molecules.

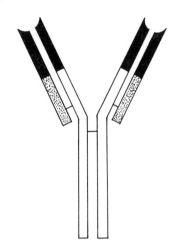

idiotype network (see figure, bottom of this page)

The interaction of idiotypes and antiidiotypes that lead to the regulation of antibody synthesis or of lymphoid cells bearing receptors that express these idiotypic specificities.

idiotype network theory

Refer to network theory.

idiotype suppression

The inhibition of idiotype antibody production by suppressor T lymphocytes activated by antiidiotype antibodies.

idiotype vaccine

Antibody preparations that mimic antigens at the molecular level. They induce immunity specific for the antigens they mimic. Such vaccines are not infectious to the recipient, are physiologic, and can be used in place of many antigens, e.g., idiotype vaccine related to *Plasmodium falciparum* circumsporozote (CS) protein.

idiotypic determinant

Refer to epitope.

idiotypic specificity

Characteristic folding of the antigen-binding site, thereby exposing various groups which by their arrangement confer antigenic properties to the immunoglobulin V region itself. Antibodies against a specific antigen usually carry both a few predominant idiotypes and other similar, but not identical, cross-reacting idiotypes. The proportion between the two indicates the degree of heterogeneity of the antibody response in a given individual.

Idiotype and Anti-Idiotype Antibodies

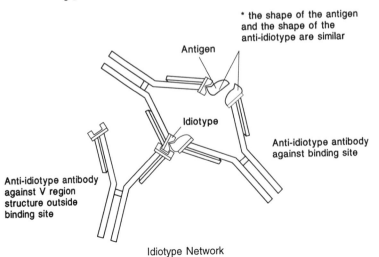

* the shape of the antigen and the shape of the anti-idiotype are similar

Antigen

Idiotype

Anti-idiotype antibody against binding site

Anti-idiotype antibody against V region structure outside binding site

Idiotype Network

IE
Abbreviation for immunoelectrophoresis.
IEF
Abbreviation for isoelectric focusing.
IEP
Abbreviation for immunoelectrophoresis.
IFE
Abbreviation for immunofixation electrophoresis.
IFN
Abbreviation for interferon.
Ig
Abbreviation for immunoglobulin.
Igα and Igβ
Proteins on the B cell surface that are noncovalently associated with cell surface IgM and IgD. They link the B cell antigen receptor complex to intracellular tyrosine kinases. Anti-Ig binding leads to their phosphorylation. Igα and Igβ are requisite for expression of IgM and IgD on the B cell surface.

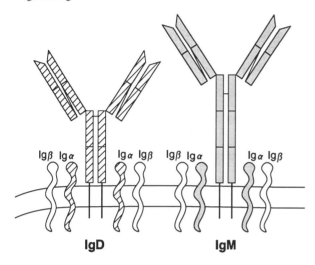

IgA deficiency
Refer to immunoglobulin A deficiency.
IgA nephropathy (Berger's disease)
A type of glomerulonephritis in which prominent IgA-containing immune deposits are present in mesangial areas. Patients usually present with gross or microscopic hematuria and often mild proteinuria. By light microscopy, mesangial widening or proliferation may be observed. However, immunofluorescence microscopy demonstrating IgA and C3, fixed by the alternative pathway, is requisite for diagnosis. Electron microscopy confirms the presence of electron-dense deposits in mesangial areas. Half of the cases progress to chronic renal failure over a 20-year course.

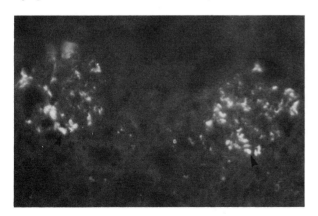

granular immune
deposits in mesangial
areas

IgA Nephropathy

IgA paraproteinemia
Myeloma in which the paraimmunoglobulin belonging to the IgA class occurs in only about one fifth of affected patients. IgA myeloma patients are reputed to have lower survival rates than those patients affected with IgG myeloma, but this claim awaits confirmation. IgA myeloma may be associated with myeloma cell infiltration of the liver leading to jaundice and altered liver function tests. It is frequently associated with hyperviscosity syndrome, which may be related to the ability of IgA molecules to polymerize as well as to form complexes with such substances as haptoglobin, α-1 antitrypsin, β lipoprotein, antihemophilic factor, and albumin.
IgG index
The ratio of IgG and albumin synthesis in the brain and in peripheral tissues. It is increased in multiple central nervous system infections, inflammatory disorders, and neoplasms.
IgG myeloma subclasses
The most frequent myeloma paraimmunoglobulin is IgG-κ, which belongs to the IgG1 subclass most commonly and to the IgG4 subclass only rarely. IgG3 paraimmunoglobulins are associated with increased serum viscosity, which leads to aggregates that are related to concentration and may be associated with serum cryoglobulin in an individual patient. In IgG paraimmunoglobulin myeioma, the serum concentration of IgG may reach 2000 mg/dl. Concomitantly there is a reduction in other immunoglobulins. Since the IgG myeloma paraimmunoglobulin belongs to one heavy chain subtype, the Gm allotypes are restricted in a manner similar to the restriction of light chains to either the κ or λ type.
IgG subclass deficiency
Decreased or absent IgG2, IgG3, or IgG4 subclasses. Total serum IgG is unaffected because it is 65 to 70% IgG1. Deletion of constant heavy chain genes or defects in isotype switching may lead to IgG subclass deficiency. IgG1 and IgG3 subclasses mature quicker than do IgG2 or IgG4. Patients have recurrent respiratory infections and recurring pyogenic sinopulmonary infections with *Hemophilus influenzae, Staphylococcus aureus,* and *Streptococcus pneumoniae.* Some patients may manifest features of autoimmune disease, such as systemic lupus erythematosus. IgG2-IgG4 deficiency is often associated with recurrent infections or autoimmune disease. IgG2 deficiency is reflected as recurrent sinopulmonary infections and nonresponsiveness to polysaccharide antigens such as those of the pneumococcus. A few individuals have IgG3 deficiency and develop recurrent infections. In the IgG4-deficient patients, recurrent respiratory infections, as well as autoimmune manifestations, also occur. The diagnosis is established by the demonstration of significantly lower levels of at least one IgG subclass in the patient compared with IgG subclass levels in normal age-matched controls. γ Globulin is the treatment choice. Very infrequently, C-gene segment deletions may lead to IgG subclass deficiency.
7 S IgM
Monomeric IgM.
IgM deficiency syndrome
An infrequent condition characterized by diminished activation of complement and decreased B lymphocyte membrane 5-ecto nucleotidase. Patients manifest heightened susceptibility to pulmonary infections, septicemia, and tumors.
IgM index
Reflects total IgM formation in the blood-brain barrier. This is elevated in infectious meningoencephalomyelitis and may also be increased in central nervous system lupus erythematosus and multiple sclerosis.
IgM paraproteinemia
Occasional cases of IgM myeloma occur and are characterized by infiltration of the bone marrow by plasmacytes, numerous osteolytic lesions, and occasionally bleeding diathesis. IgM myeloma is distinct from true Waldenström's macroglobulinemia, but is rare compared with the IgG and IgA types. Myelomas associated with the minor classes of immunoglobulin IgD and IgE occur with even less frequency than do the IgM myelomas.
IgT
An obsolete term for the T lymphocyte antigen receptor which was incorrectly thought of as an immunioglobulin of a special class. This should not be confused with IgG (t).
Ii antigens
Two nonallelic carbohydrate antigens (epitopes) on the surface membrane of erythrocytes of man. They may also occur on some

nonhematopoietic cells. The i epitope is found on fetal erythrocytes and red cell blood precursors. The I antigen is formed when aliphatic galactose-*N*-acetyl-glucosamine is converted to a complex branched structure. I represents the mature form and i the immature form. Mature erythrocytes express I. Antibodies against i antigen are hemolytic in cases of infectious mononucleosis.

IJ

An area of the murine I region that has been predicted to encode for a suppressor cell antigen.

IL

Abbreviation for interleukins.

IL-1

Abbreviation for interleukin-1.

IL-2

Abbreviation for interleukin-2.

IL-2/LAK cells

Interleukin-2/lymphokine-activated killer cells. NK cells, which express only the p70 and not the p55 receptor for IL-2, are incubated with IL-2, converting them into an activated form referred to as LAK cells. The IL-2/LAK combination has been used to treat cancer patients through adoptive immunotherapy, which has been successful in inducing transient regression of tumors in selected cases of melanoma, colorectal carcinoma, non-Hodgkin's lymphoma, and renal cell carcinoma, as well as regression of metastases in the liver and lung of some patients. There may be transient defective chemotaxis of neutrophils, and patients often develop "capillary leak syndrome", producing pulmonary edema. Patients may also develop congestive heart failure.

IL-2 receptor (CD25)

Refer to interleukin-2 receptor (IL-2R).

IL-3

Abbreviation for interleukin-3.

IL-4

Abbreviation for interleukin-4.

IL-5

Abbreviation for interleukin-5.

IL-6

Abbreviation for interleukin-6.

IL-6 receptor

A 468-amino acid residue structure on the membrane that shows homology with an immunoglobulin domain.

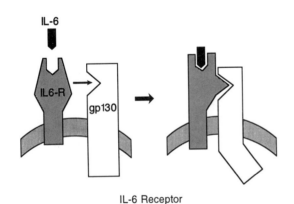

IL-6 Receptor

IL-7

Abbreviation for interleukin-7.

IL-8

Abbreviation for interleukin-8.

IL-9

Abbreviation for interleukin-9.

IL-10

Abbreviation for interleukin-10.

IL-11

Abbreviation for interleukin-11.

IL-12

Abbreviation for interleukin-12.

IL-13

Abbreviation for interleukin-13.

IL-14

Abbreviation for interleukin-14.

IL-15

Abbreviation for interleukin-15.

ImD$_{50}$

The antigen (or vaccine) dose capable of successfully immunizing 50% of a particular animal test population.

immediate hypersensitivity

Antibody-mediated hypersensitivity. In humans, this homocytotrophic antibody is IgE and in certain other species IgG1. The IgE antibodies are attached to mast cells through their Fc receptors. Once the Fab regions of the mast cell-bound IgE molecules interact with specific antigen, vasoactive amines are released from the cytoplasmic granules as described under type I hypersensitivity reactions. The term "immediate" is used to indicate that this type of reaction occurs within seconds to minutes following contact of a cell-fixed IgE antibody with antigen. Skin tests and RAST are useful to detect immediate hypersensitivity in humans, and passive cutaneous anaphylaxis reveals immediate hypersensitivity in selected other species. Examples of immediate hypersensitivity in humans include the classic anaphylactic reaction to penicillin administration, hay fever, and environmental allergens such as tree and grass pollens, bee stings, etc.

immediate-spin crossmatch

A test for incompatibility between donor erythrocytes and the recipient patient's serum. This assay reveals ABO incompatibility in practically all cases, but is unable to identify IgG alloantibodies against erythrocyte antigens.

immobilization test

A method for the identification of antibodies specific for motile microorganisms by determining the ability of antibody to inhibit motility. This may be attributable to adhesion or agglutination of the microorganisms' flagella or cell wall injury when complement is present.

immune

Either natural or acquired resistance to a disease. Either a subclinical infection with the causative agent or deliberate immunization with antigenic substances prepared from it may render a host immune. Because of immunological memory, the immune state is heightened upon second exposure of individuals to an immunogen. A subject may become immune as a consequence of having experienced and recovered from an infectious disease.

immune adherence

Attachment of antigen-antibody complexes or of antibody-coated bacteria or other particles carrying C3b or C4b to complement receptor 1 (CR1) expressing cells. Erythrocytes of primates, B cells, T cells, phagocytic cells, and glomerular epithelial cells all express CR1. It is absent on the erythrocytes of other mammals, but is found on their thrombocytes. Immune adherence facilitates the elimination of antigen-antibody complexes from the blood circulation, especially through their attachment to red blood cells and platelets followed by uptake by phagocytic cells through complement receptor 3.

immune adherence receptor

Synonym for complement receptor 1.

immune antibody

A term used to distinguish an antibody induced by transfusion or other immunogenic challenge in contrast to a natural antibody such as the isohemagglutinins against ABO blood group substances found in humans.

immune clearance

Refer to immune elimination.

immune complex disease (ICD)

As described under type III hypersensitivity reaction, antigen-antibody complexes have a fate that depends in part on the size. The larger insoluble immune complexes are removed by the mononuclear phagocyte system. The smaller immune complexes become lodged in the microvasculature such as the renal glomeruli. They may activate the complement system and attract polymorphonuclear neutrophils, initiating an inflammatory reaction. The antigen of an immune complex may be from microorganisms such as streptococci leading to subepithelial deposits in renal glomeruli in poststreptococcal glomerulonephritis. Or they may be endogenous such as DNA or nuclear antigens in systemic lupus erythematosus. Diphtheria antitoxin prepared in horses induced serum sickness when the foreign horse serum proteins stimulated antibodies in human recipients. Immune complex disease is characterized clinically by fever, joint pain, lymphadenopathy, eosinophilia, hypocomplementemia, proteinuria, purpura, and urticaria, among other features. Laboratory techniques to detect immune complexes include the solid-phase Clq assay, the Clq-binding assay, the Raji cell technique, and the staphylococcal protein assay, among other methods. Most autoimmune diseases have a type III (antigen-antibody complex)-mediated mechanism. The connective tissue diseases such as systemic lupus erythematosus, polyarteritis nodosa, progressive systemic sclerosis, dermatomyositis, rheumatoid arthritis, and others fall within this category. Viral infections such as hepatitis B, cytomegalovirus, infectious mononucleosis, and Dengue, as well as neoplasia such as carcinomas and melanomas, may be associated with immune complex formation. Refer to type III hypersensitivity reaction.

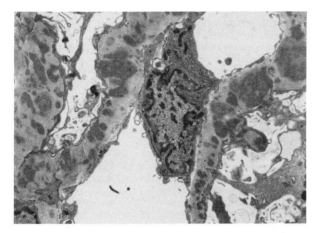

electron dense deposits

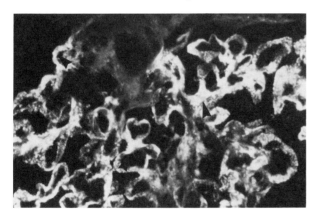

immune complexes
in renal glomerulus

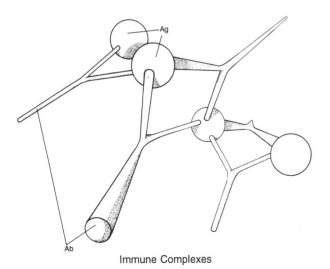

Immune Complexes

immune complex pneumonitis

A type III, immune complex Arthus-type reaction in the pulmonary alveoli.

immune complex reactions

Type III hypersensitivity in which antigen-antibody complexes fix complement and induce inflammation in tissues such as capillary walls.

immune complexes

Antigen-antibody complexes.

immune cytolysis

Complement-induced destruction or lysis of antibody-coated target cells such as tumor cells. Immune cytolysis of red blood cells is referred to as immune hemolysis. Immune cytolysis refers also to the destruction of target cells by cytotoxic T lymphocytes or natural killer (NK) cells (including K cells) through the release of perforins.

immune deviation

Antigen-mediated suppression of the immune response may selectively affect delayed-type hypersensitivity, leaving certain types of immunoglobulin responses relatively intact and unaltered. This selective suppression of certain phases of the immune response to an antigen without alteration of others has been termed immune deviation. Thus, "split tolerance" or immune deviation offers an experimental model for dissection of the immune response into its component parts. It is necessary to use an antigen capable of inducing formation of humoral antibody and development of delayed-type hypersensitivity to induce immune deviation. Since it is essential that both humoral and cellular phases of the immune response be directed to the same antigenic determinant group, defined antigens are required. Immune deviation selectively suppresses delayed-type hypersensitivity and IgG_2 antibody production. By contrast, immunologic tolerance affects both IgG_{1-} and IgG_2 antibody production and delayed-type hypersensitivity. For example, prior administration of certain protein antigens to guinea pigs may lead to antibody production. However, the subsequent injection of antigen incorporated into Freund's complete adjuvant leads to deviation from the expected heightened delayed-type hypersensitivity and formation of IgG2 antibodies to result in little of either, i.e., negligible delayed-type hypersensitivity and suppression of IgG2 formation.

immune elimination

Accelerated removal of an antigen from the blood circulation following its interaction with specific antibody and elimination of the antigen-antibody complexes through the mononuclear phagocyte system. A few days following antigen administration, antibodies appear in the circulation and eliminate the antigen at a much more rapid rate than that occuring in nonimmune individuals. Splenic and liver macrophages express Fc receptors that bind

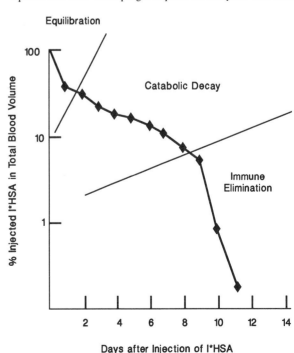

antigen-antibody complexes and also complement receptors which bind those immune complexes that have already fixed complement. This is followed by removal of immune complexes through the phagocytic action of mononuclear phagocytes. Immune elimination also describes an assay to evaluate the antibody response by monitoring the rate at which a radiolabeled antigen is eliminated in an animal with specific (homologous) antibodies in the circulation.

immune exclusion
Prevention of antigen entry into the body by the products of a specific immune response, such as the blocking of an antigen's access to the body by mucosal surfaces when secretory IgA specific for the antigen is present.

immune hemolysis
The lysis of erythrocytes through the action of specific antibody and complement.

immune interferon
Synonym for interferon γ.

immune modulation
Refer to immunomodulation.

immune network hypothesis of Jerne
The antibody molecules' antigen-binding sites (paratopes), which are encoded by variable region genes, have idiotopes as phenotypic markers. Each paratope recognizes idiotopes on a different antibody molecule. Interaction of idiotypes with antiidiotypes is physiologic idiotypy and is shared among Ig classes. This comprises antibodies produced in response to the same antigen. The idiotypic network consists of the interaction of idiotypes involving free molecules as well as B and T lymphocyte receptors. Idiotypes are considered central in immunoregulation involving autoantigens.

immune-neuroendocrine axis
There is a bidirectional regulatory circuit between the immune and neuroendocrine systems. The neuroendocrine and immune systems affect each other. Receptors for neurally active polypeptides, neurotransmitters, and hormones are present on immune system cells, whereas receptors reactive with products of the immune system may be identified on nervous system cells. Neuroendocrine hormones have variable immunoregulatory effects mediated through specific receptors. Neurotransmitter influence on cell function is determined in part by the receptor-linked signal amplification associated with second messenger systems. Neuroimmunoregulation is mediated either through a neural pathway involving pituitary peptides and adrenal steroid hormones or through a second pathway consisting of direct innervation of immune system tissues. The thymus, spleen, bone marrow, and perhaps other lymphoid organs contain afferent and efferent nerve fibers. ACTH, endorphins, enkephalins, and adrenal cortical steroids derived from the pituitary represent one direction for modulating the immune response. For example, prolactin (PRL) regulates lymphocyte function. Stimulated lymphocytes or nonstimulated macrophages may produce neuroendocrine hormone-related peptides, ACTH, and endorphins. The participation of these substances in a stress response represents the opposite direction of regulation. Thymic hormones may also induce an endocrine response.

Neuroendocrine hormones may exert either a positive or negative regulatory effect on the macrophage, which plays a key role in both inflammation and immune responsiveness. Leukocyte mediators are known to alter both central nervous system and immune system functions. IL-1 acts on the hypothalamus to produce fever and participates in antigen-induced activation. IL-1 is synthesized by macrophages and has a major role in inflammation and immune responsiveness.

Substance P (SP) has been postulated to have an effect in hypersensitivity diseases including arthritis and asthma. The nervous system may release SP into joints in arthritis and into the respiratory tract in asthma, perpetuating inflammation. SP has also been found to participate in immune system functions such as the induction of monocyte chemotaxis in *in vitro* experiments. Through their effect on mast cells, enkephalins influence hypersensitivity reactions. Enkephalins have been shown to diminish antibody formation against cellular and soluble immunogens and to diminish passive cutaneous anaphylaxis. Mast cells, which can be stimulated by either immunologic or nonimmunologic mechanisms, may be significant in immune regulation of neural function.

immune neutropenia
Neutrophil degradation by antibodies, termed leukoagglutinins, which are specific for neutrophil epitopes. Penicillin or other drugs, as well as blood transfusions, may induce immune neutropenia. The condition may be associated with such autoimmune disorders as systemic lupus erythematosus and may occur in neonates through passage of leukoagglutinins from mother to young.

immune paralysis
Refer to immunologic paralysis.

immune response
Reaction of the animal body to challenge by an immunogen. This is expressed as antibody production and/or cell-mediated immunity or immunologic tolerance. Immune response may follow stimulation by a wide variety of agents such as pathogenic microorganisms, tissue transplants, or other antigenic substances deliberately introduced for one purpose or another. Infectious agents may also induce inflammatory reactions characterized by the production of chemical mediators at the site of injury.

immune response (*Ir*) genes
Genes that regulate immune responsiveness to synthetic polypeptide and protein antigens as demonstrated in guinea pigs and mice. This property is transmitted as an autosomal dominant trait that maps to the major histocompatibility complex (MHC) region. Ir genes control helper T lymphocyte activation, which is requisite for the generation of antibodies against protein antigens. T lymphocytes with specific receptors for antigen fail to recognize free or soluble antigens, but they do recognize protein antigens noncovalently linked to products of MHC genes termed class I and class II MHC molecules. The failure of certain animal strains to respond may be due to ineffective antigen presentation in which processed antigen fails to bind properly to class II MHC molecules or to an ineffective interaction between the T cell receptor and the MHC class II-antigen complex.

immune serum
An antiserum containing antibodies specific for a particular antigen or immunogen. Such antibodies may confer protective immunity.

immune serum globulin
Injectable immunoglobulin that consists mainly of IgG extracted by cold ethanol fractionation from pooled plasma from up to 1000 human donors. It is administered as a sterile $16.5 \pm 1.5\%$ solution to patients with immunodeficiencies and as a preventive against certain viral infections including measles and hepatitis A.

immune stimulatory complexes
Refer to ISCOMs.

immune suppression
Decreased immune responsiveness as a consequence of therapeutic intervention with drugs or irradiation or as a consequence of a disease process that adversely affects the immune response, such as acquired immune deficiency syndrome (AIDS).

immune tolerance
Refer to immunologic tolerance.

immunity
A state of resistance or protection from a pathogenic microorganism or the effect of toxic substances such as snake or insect venom.

immunity deficiency syndrome
Synonym for immunodeficiency.

immunize
The deliberate administration of an antigen or immunogen for the purpose of inducing active immunity that is often protective, as in the case of immunization against antigenic products of infectious disease agents. As a consequence of contact between the antigen or immunogen and immunologically competent cells of the host, specific antibodies and specifically reactive immune lymphoid cells are induced to confer a state of immunity.

immunizing dose (ImD$_{50}$)
In vaccine standardization, the ImD$_{50}$ is that amount of the immunogen requisite to immunize 50% of the test animal population, as determined by appropriate immunoassay.

immunoabsorbent
A gel or other inert substance employed to absorb antibodies from a solution or to purify them.

immunoabsorption
Removal of a selected group of antibodies by antigen or the removal of antigen by interaction with specific antibody.

immunoassay
A test that measures antigen or antibody. When choosing an immunoassay technique, one should keep in mind the differing levels of sensitivity of various methods. Whereas immunoelectrophoresis is relatively insensitive, requiring 5 to 10,000 ng/ml for detection, the enzyme-linked immunoabsorbent assay (ELISA), radioimmunoas-

say (RIA), and immunofluorescence may detect less than 0.001 ng/ml. Between these two extremes are agglutination that detects 1 to 10,000 ng/ml and complement fixation which detects 5 ng/ml.

immunoaugmentive therapy (IAT)

An approach to tumor therapy practiced by selected individuals in Mexico, Germany, and the Bahama Islands that is not based on scientific fact and is of unproven efficacy or safety. Concoctions of tumor cell lysate and blood serum from tumor-bearing patients as well as from normal individuals is injected into cancer patients ostensively for the two to provide tumor antibody, "blocking" and "deblocking" proteins. The method is not based upon any scientifically approved protocol and is of very doubtful value. The treatment preparations are often contaminated with microorganisms.

immunobeads

Minute plastic spheres with a coating of antigen (or antibody) that may be aggregated or agglutinated in the presence of the homologous antibody. Immunobeads are used also for the isolation of specific cell subpopulations such as the separation of B cells from T cells that is useful in class II MHC typing for tissue transplantation.

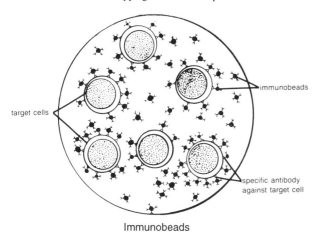

Immunobeads

immunoblast

Lymphoblast.

immunoblastic lymphadenopathy

Refer to angioimmunoblastic lymphadenopathy (AILA).

immunoblastic sarcoma

A lymphoma comprised of immunoblast-like cells that are divided into B and T cell immunoblastic sarcomas. Both are malignant lymphomas. B cell immunoblastic sarcoma is characterized by large immunoblastic plasmacytoid cells and Reed-Sternberg cells. This is the most frequent lymphoma that occurs in individuals with natural immunodeficiency, with suppression of the immune system, or who manifest immunoproliferative disorders, such as angioimmunoblastic lymphadenopathy, or autoimmune diseases, such as Hashimoto's thyroiditis, α chain disease, etc. The disease has a poor prognosis. T cell immunoblastic sarcoma is less frequent than the B cell variety. The tumor cells are large and have a clear cytoplasm containing round to oval nuclei with fine chromatin. One to three nucleoli are present amidst lobulation and nuclear folding. Other cells present include plasma cells and histiocytes. Patients with mycosis fungoides who develop polyclonal gammopathy and general lymphadenopathy may develop a T cell immunoblastic sarcoma.

immunoblot (Western blot)

The interaction between labeled antibodies and proteins that have been absorbed on nitrocellulose paper. Refer to Western blot.

immunoblotting

A method to identify antigen(s) by the polyacrylamide gel electrophoresis (PAGE) of a protein mixture containing the antigen. PAGE separates the components according to their electrophoretic mobility. After transfer to a nitrocellulose filter by electroblotting, antibodies labeled with enzyme or radioisotope and which are specific for the antigen in question are incubated with the cellulose membrane. After washing to remove excess antibody that does not bind, substrate can be added if an enzyme was used, or autoradiography can be used if a radioisotope was used to determine where the labeled antibodies were bound to homologous antigen. Also called Western blotting.

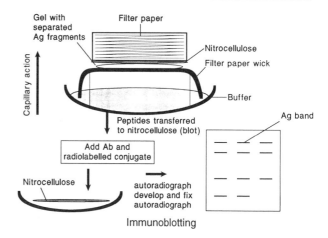

Immunoblotting

immunochemistry

That branch of immunology that is concerned with the properties of antigens, antibodies, complement, T cell receptors, MHC molecules, and of all the molecules and receptors that participate in immune interactions *in vivo* and *in vitro*. Immunochemistry aims to identify active sites in immune responses and define the forces that govern antigen-antibody interaction. It is also concerned with the design of new molecules such as catalytic antibodies and other biological catalysts. Also called molecular immunology.

immunoconglutination

Aggregation of C3bi- or C4bi-coated erythrocytes or bacteria by antibodies to C3bi or C4bi, produced by immunizing animals with erythrocytes or bacteria that have interacted with antibody and complement. The antibody is termed immunoconglutinin, which resembles conglutinin in its activity, but should not be confused with it. Autoantibodies specific for C3bi or C4bi may develop as a result of an infection and produce immunoconglutination.

immunoconglutinin

An autoantibody, usually of the IgM class, that is specific for neoantigens in C3bi or C4bi. It may be stimulated during acute and chronic infections caused by bacteria, viruses, or parasites and in chronic inflammatory disorders. The level is also increased in many autoimmune diseases, as well as following immunization with various immunogens. C3 nephritic factor is an example of an immunoconglutinin. Also called *immune conglutinin,* but it should not be confused with *conglutinin.* Refer also to immunoconglutination.

immunocyte

Literally "immune cell". A term sometimes used by pathologists to describe plasma cells in stained tissue sections, e.g., in the papillary or reticular dermis in erythema multiforme.

immunocytes in
reticular dermis

immunocytoadherence

A method to detect cells with surface immunoglobulin, either synthesized or attached through Fc receptors. Red blood cells coated with the homologous antigen are mixed with the immuno-globulin-bearing cells and result in rosette formation.

immunodeficiency

A failure in humoral antibody or cell-mediated limbs of the immune response. If attributable to intrinsic defects in T and/or B lymphocytes, the condition is termed a primary immunodeficiency. If the defect results from loss of antibody and/or lymphocytes, the condition is termed a secondary immunodeficiency.

immunodeficiency, acquired

A decrease in the immune response to immunogenic (antigenic) challenge as a consequence of numerous diseases or conditions that include acquired immune deficiency syndrome (AIDS), chemotherapy, immunosuppressive drugs such as corticosteroids, psychological depression, burns, nonsteroidal antiinflammatory drugs, radiation, Alzheimer's disease, coeliac disease, sarcoidosis, lymphoproliferative disease, Waldenström's macroglobulinemia, multiple myeloma, aplastic anemia, sickle cell disease, malnutrition, aging, neoplasia, diabetes mellitus, and numerous other conditions.

immunodeficiency associated with hereditary defective response to Epstein-Barr virus

An immunodeficiency that develops in previously healthy subjects with a normal immune system who have developed a primary Epstein-Barr virus infection. They develop elevated numbers of natural killer cells in the presence of a lymphopenia. The condition is serious, and its acute stage may lead to B cell lymphoma or failure of the bone marrow or agammaglobulinemia. The disease may be fatal. The condition was first considered to have an X-linked recessive mode of inheritance recurring only in males; it has now been found in occasional females. The term Duncan's syndrome is often used to describe the X-linked variety of this condition.

immunodeficiency, congenital

A varied group of unusual disorders with associated autoimmune manifestations, increased incidence of malignancy, allergy, and gastrointestinal abnormalities. These include defects in stem cells, B cells, T cells, phagocytic defects, and complement defects. An example is severe combined immunodeficiency due to various causes. The congenital immunodeficiencies are described under the separate disease categories.

immunodeficiency from hypercatabolism of immunoglobulin

Serum levels of immunoglobulins fluctuate according to their rates of synthesis and catabolism. Although many immunological deficiencies result from defective synthesis of immunoglobulins and lymphocytes, immunoglobulin levels in serum can decline as a consequence of either increased catabolism or loss into the gastrointestinal tract or other areas. Defective catabolism may affect one to several immunoglobulin classes. For example, in myotonic dystrophy, only IgG is hypercatabolized. In contrast to the normal levels of IgM, IgA, IgD, IgE, and albumin in the serum, the IgG concentration is markedly diminished. Synthesis of IgG in these individuals is normal, but the half-life of IgG molecules is reduced as a consequence of increased catabolism. Patients with ataxia telangiectasia and those with selective IgA deficiency have antibodies directed against IgA which remove this class of immunoglobulin. Patients with the rare condition known as familial hypercatabolic hypoproteinemia demonstrate reduced IgG and albumin levels in the serum and slightly lower IgM levels, but the IgA and IgE concentrations are either normal or barely increased. Although synthesis of IgG and albumin in such patients is within normal limits, the catabolism of these two proteins is greatly accelerated.

immunodeficiency from severe loss of immunoglobulins and lymphocytes

The gastrointestinal and urinary tracts are two sources of serious protein loss in disease processes. The loss of integrity of the renal glomerular basement membrane, renal tubular disease, or both may result in loss of immunoglobulin molecules into the urine. Since the small IgG molecules would pass through in many situations leaving larger IgA molecules in the intravascular space, all immunoglobulins are not lost from the serum at the same rate. More than 90 diseases that affect the gastrointestinal tract have been associated with protein-losing gastroenteropathy. This may be secondary to inflammatory or allergic disorders or disease processes involving the lymphatics. In intestinal lymphangiectasia associated with lym-

phatic blockage, lymphocytes as well as protein are lost. Lymphatics in the small intestine are dilated.

Intestinal lymphangiectasia patients show defects in both humoral and cellular immune mechanisms. The major immunoglobulins are diminished to less than half of normal. IgG is affected more than IgA and IgM, which are more affected than is IgE.

immunodeficiency with B cell neoplasms

B cell leukemias can be classified as pre-B cell, B cell, or plasma cell neoplasms. They include Burkitt's lymphoma, Hodgkin's disease, and chronic lymphocytic leukemia. Neoplasms of plasma cells are associated with multiple myeloma and Waldenström's macroglobulinemia. Many of these conditions are associated with hypogammaglobulinemia and a diminished capacity to form antibodies in response to the administration of an immunogen. In chronic lymphocytic leukemia (CLL), more than 95% of individuals have malignant leukemic cells that are identifiable as B lymphocytes expressing surface immunoglobulin. These patients frequently develop infections and have autoimmune manifestations such as autoimmune hemolytic anemia. CLL patients may have secondary immunodeficiency, which affects both B and T limbs of the immune response. Diminished immunoglobulin levels are due primarily to diminished synthesis.

immunodeficiency with partial albinism

A type of combined immunodeficiency characterized by decreased cell-mediated immunity and deficient natural killer cells. Patients develop cerebral atrophy and aggregation of pigment in melanocytes. This disease, which leads to death, has an autosomal recessive mode of inheritance.

immunodeficiency with T cell neoplasms

Almost one third of patients with acute lymphocytic leukemia (ALL), individuals with Sezary syndrome, and a very few chronic lymphocytic leukemia (CLL) patients develop a malignant type proliferation of lymphoid cells. Sezary cells are poor mediators of T cell cytotoxicity, but they can produce migration inhibitory factor (MIF)-like lymphokine. They produce neither immunoglobulin nor suppressor substances, but they do have a helper effect for immunoglobulin synthesis by B cell. In mycosis fungoides, skin lesions contain T lymphocytes, and there is an increased number of null cells in the blood with a simultaneous decrease in the numbers of B and T cells. T cell immunity is decreased in this condition, but IgA and IgE may be elevated. Whereas ALL patients show major defects in cell-mediated or in humoral (antibody) immunity, a few of them manifest profound reduction in their serum immunoglobulin concentration. This has been suggested to be due to malignant expansion of their T suppressor lymphocytes.

immunodeficiency with thrombocytopenia

Synonym for Wiskott-Aldrich syndrome.

immunodeficiency with thrombocytopenia and eczema (Wiskott-Aldrich syndrome)

An X-linked recessive disease characterized by thrombocytopenia, eczema, and susceptibility to recurrent infections, sometimes leading to early death. The lifespan of young boys is diminished as a consequence of extensive infection, hemorrhage, and sometimes malignant disease of the lymphoreticular system. Infectious agents affecting these individuals include the Gram-negative and Gram-positive bacteria, fungi, and viruses. The thymus is normal morphologically, but there is a variable decline in cellular immunity which is thymus dependent. The lymph node architecture may be altered as paracortical areas become depleted of cells with progression of the disease. IgM levels in serum are low, but IgA and IgG may be increased. Isohemagglutinins are usually undetectable in the serum. Patients may respond normally to protein antigens, but show defective responsiveness to polysaccharide antigens. There is decreased immune responsiveness to lipopolysaccharides from enteric bacteria and B blood group substances.

immunodeficiency with thymoma

An abnormality in B cell development with a striking decrease in B cell numbers and in immunoglobulins in selected patients with thymoma. There is a progressive decrease in their cell-mediated immunity as the disease continues. Most patients develop chronic sinopulmonary infections. Approximately 20% develop thrombocytopenia, and about 25% have splenomegaly. Immunoglobulin class alterations are variable. Skin test reactivity and responsiveness to skin allografts are decreased. Patients have few or no lymphocytes expressing surface immunoglobulin. The disease is due to a stem cell defect. The preferred treatment is γ globulin administration.

immunodiffusion

A method in which antigen and antibody are placed in wells at different sites in agar gel and are permitted to diffuse toward each other in the gel. The formation of precipitin lines at their point of contact in the agar gel signifies a positive reaction, showing that the antibody is specific for the antigen in question. Multiple variations of this technique have been described.

immunodominance

Descriptor for the immune response-generating capacity of that part of an epitope on an antigen molecule that serves as an immunodominant tip or area that provides the principal binding energy for reaction with a paratope on an antibody molecule or with a T cell receptor for antigen. The hapten portion of a hapten-carrier complex is often the immunodominant part of the molecule. Immunodominance refers to the region of an antigenic determinant that is the principal binding site for antibody.

immunodominant epitope

The antigenic determinant on an antigen molecule that binds or fits best with the antibody or T cell receptor specific for it.

immunodominant site

Refer to immunodominant epitope.

immunoelectroadsorption

A quantitative assay for antibody using a metal-treated glass slide to adsorb antigen followed by antibody from serum. Adsorption is facilitated by an electric current. Measurement of the antibody layer's thickness reflects the serum concentration.

immunoelectron microscopy

Traditionally, the use of antibodies labeled with ferritin to study the ultrastructure of subcellular organelles and, more recently, the use of immunogold labeling and related procedures for the identification and localization of antigens by electron microscopy.

immunoelectroosmophoresis

Refer to counter immunoelectrophoresis.

immunoelectrophoresis (IEP or IE)

A method to identify antigens on the basis of their electrophoretic mobility, diffusion in gel, and formation of precipitation arcs with specific antibody. Electrophoresis in gel is combined with diffusion of a specific antibody in a gel medium containing electrolyte to identify separated antigenic substances. The presence or absence of immunoglobulin molecules of various classes in a serum sample may be identified in this way. One percent agar containing electrolyte is layered onto microscope slides, allowed to gel, and patterns of appropriate troughs and wells are cut in the solidified medium. Antigen to be identified is placed in the circular wells cut into the agar medium. This is followed by electrophoresis, which permits separation of the antigenic components according to their electrophoretic mobility, and antiserum is placed in a long trough in the center of the slide. After antibody has diffused through the agar toward each separated antigen, precipitin arcs form where the antigen and antibody interact. Abnormal amounts of immunoglobulins result in changes in the shape and position of precipitin arcs when compared with the arcs formed by antibody against normal human serum components. With monoclonal gammapathies, the arcs become broad, bulged, and displaced. The absence of immunoglobulin classes such as those found in certain immunodeficiencies can also be detected with IEP.

immunoenhancement

The process of increasing or contributing to the level of immune response by various specific and nonspecific means such as immunization.

immunoferritin method

A technique to aid detection by electron microscopy of sites where antibody interacts with antigen of cells and tissues. Immunoglobulin may be conjugated with ferritin, an electron-dense marker, without altering its immunological reactivity. These ferritin-labeled antibodies localize molecules of antigen in subcellular areas. Electron-dense ferritin permits visualization of antibody binding to homologous antigen in cells and tissues by electron microscopy. In addition to ferritin, horseradish peroxidase-labeled antibodies may also be adapted for use in immunoelectron microscopy.

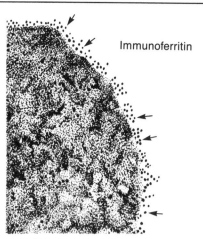

A ferritin-conjugated anti-A globulin treated group A red cell. The binding of ferritin particles to the surface of the erythrocyte denotes the site of the antigen.

immunoferritin technique

A method that involves the labeling of antibody molecules with an electron-dense material, i.e., ferritin, rendering their site of attachment to cell or tissue antigens visible when observed by electron microscopy. In addition to this direct method, the use of ferritin-labeled antiimmunoglobulin can detect unlabeled antibody bound to tissue antigen by indirect "sandwich" immunoferritin methodology.

immunofixation

A method to identify antigens in a protein mixture that has been electrophoresed in agarose gel. Antibodies are applied to the gel, which is then rinsed to remove excess reactants, and the preparation is stained to reveal precipitates of antigen and antibody.

immunofixation electrophoresis (IFE)

The application of specific antibodies to a gel in which the antigen has been electrophoresed. Precipitation occurs at the location in the gel where antibodies interact with the electrophoresed homologous antigen. The method is used to identify specific antigenic components that might otherwise be camouflaged because of close similarity of molecules that might not be otherwise discernible by conventional immunoelectrophoresis.

immunofluorescence

A method for the detection of antigen or antibody in cells or tissue sections through the use of fluorescent labels, termed fluorochromes, by fluorescent light microscopic examination. The most commonly used fluorochromes are fluorescein isothiocyanate, which imparts an apple-green fluorescence, and rhodamine B isothiocyanate, which imparts a reddish-orange tint. This method, developed by Albert Coons in the 1940s, has a wide application in diagnostic medicine and research. In addition to antigens and antibodies, complement and other immune mediators may also be detected by this method. It is based on the principal that following adsorption of light by molecules, they dispose of their increased energy by various means, such as emission of light of longer wavelength. Fluorescence is the process whereby emission is of relatively short duration (10^{-6} to 10^{-9} s) for return of the excited molecules to the ground state. The active groups in protein that allow them to attach fluorochromes include free amino and carboxyl groups at the ends of each polypeptide chain, many free amino groups and lysine side chains, many free carboxyl groups in asparatic and glutamic acid residues, the guanidino group of arginine, the phenolic group of tyrosine, and the amino groups of histidine and tryptophan. Labeling antibody molecules with fluorochromes does not alter their antigen-binding specificity. Several immunofluorescence techniques are available. In the direct test, smears of the substance to be examined are fixed with heat or methanol and followed by flooding with a fluorochrome-antibody conjugate. This is followed by incubating in a moist chamber for 30 to 60 min at 37°C, after which the smear is washed first in buffered saline for 5 to 10 min and second in tap water for another 5 to 10 min. These washing procedures remove uncombined conjugated globulin. After adding a small drop of buffered glycerol and the cover slip, the smear may be examined with the fluorescence light microscope. In the indirect test, which is more sensitive than the direct, a smear or tissue section is first flooded with unlabeled antibody specific for the antigen being sought. After washing, fluorescein-labeled antiimmunoglobulin of the species of the primary antibody is layered over the section. After appropriate incubation and washing, the section is cover slipped and examined as in the direct method. Other variations, such as complement staining, are also available. The indirect method is more sensitive than the direct method and considerably less expensive than one fluorochrome-labeled antiimmunoglobulin may be used with multiple primary antibodies specific for a battery of antigens. The technique is widely used to diagnose and classify renal diseases, bulbous skin diseases, and the study of cells and tissues in connective tissue disorders such as systemic lupus erythematosus.

immunogen
A substance that is able to induce a humoral antibody and/or cell-mediated immune response rather than immunological tolerance. The term "immunogen" is sometimes used interchangeably with "antigen", yet the term signifies the ability to stimulate an immune response as well as react with the products of it, e.g., antibody. By contrast, "antigen" is reserved by some to mean a substance that reacts with antibody. The principal immunogens are proteins and polysaccharides, whereas lipids may serve as haptens.

immunogenic
An adjective that denotes the capacity to induce humoral antibody and/or cell-mediated immune responsiveness, but not immunological tolerance. Immunogenicity depends on characteristics of the immunogen and on the injected animal's genetic capacity to respond to the immunogen. To be immunogenic, a substance must be foreign to the recipient. An immunogen that is of significant molecular size and complexity, as well as host factors such as previous exposure to the immunogen and immunocompetence, are all critical factors in immunogenicity.

immunogenicity
The ability of an antigen serving as an immunogen to induce an immune response in a particular species of recipient. Immunogenicity depends on a number of physical and chemical characteristics of the immunogen (antigen), as well as on the genetic capacity of the host response.

immunoglobulin
A mature B cell product synthesized in response to stimulation by an antigen. Antibody molecules are immunoglobulins of defined specificity produced by plasma cells. The immunoglobulin molecule consists of heavy (H) and light (L) chains fastened together by disulfide bonds. The molecules are subdivided into classes and subclasses based on the antigenic specificity of the heavy chains. Heavy chains are designated by lower case Greek letters (μ, γ, α, δ, and ϵ), and the immunoglobulins are designated IgM, IgG, IgA, IgD, and IgE, respectively. The three major classes are IgG, IgM, and IgA and the two minor classes are IgD and IgE, which together comprise less than 1% of the total immunoglobulins. The two types of light chains (termed κ and λ) are present in all five immunoglobulin classes, although only one type is present in an individual molecule.

IgG, IgD, and IgE have two H and two L polypeptide chains, whereas IgM and IgA consist of multimers of this basic chain structure. Disulfide bridges and noncovalent forces stabilize immunoglobulin structure. The basic monomeric unit is Y shaped, with a hinge region rich in proline and susceptible to cleavage by proteolytic enzymes. Both H and L chains have a constant region at the carboxy terminus and a variable region at the amino terminus. The two heavy chains are alike, as are the two light chains in any individual immunoglobulin molecule. Approximately 60% of human immunoglobulin molecules have κ light chains and 40% have λ light chains. The five immunoglobulin classes are termed isotypes based on the heavy chain specificity of each immunoglobulin class. Two immunoglobulin classes, IgA and IgG, have been further subdivided into subclasses based on H chain differences. There are four IgG subclasses, designated as IgG1 through IgG4, and two IgA subclasses, designated IgA1 and IgA2.

Digestion of IgG molecules with papain yields two Fab and one Fc fragments. Each Fab fragment has one antigen-binding site. By contrast, the Fc fragment has no antigen-binding site, but is responsible for fixation of complement and attachment of the molecule to a cell surface. Pepsin cleaves the molecule toward the carboxy terminal end of the central disulfide bond, yielding an $F(ab')_2$ fragment and a pFc' fragment. $F(ab')_2$ fragments have two antigen-

binding sites. L chains have a single variable and constant domain, whereas H chains possess one variable and three to four constant domains.

Secretory IgA is found in body secretions such as saliva, milk, and intestinal and bronchial secretions. IgD and IgM are present as membrane-bound immunoglobulins on B cells, where they interact with antigen to activate B cells. IgE is associated with anaphylaxis, and IgG, which is the only immunoglobulin capable of crossing the placenta, is the major human immunoglobulin.

immunoglobulin A (IgA)
IgA comprises 5 to 15% of the serum immunoglobulins and has a half-life of 6 days. It has a mol wt of 160 kD and a basic four-chain monomeric structure. However, it can occur as monomers, dimers, trimers, and multimers. It contains α heavy chains and κ or λ light chains. There are two subclasses of IgA designated as IgA1 and IgA2. In addition to serum IgA, a secretory or exocrine variety appears in body secretions and provides local immunity. For example, the Sabin oral polio vaccine stimulates secretory IgA antibodies in the gut, which provides effective immunity against poliomyelitis. IgA-deficient individuals have an increased incidence of respiratory infections associated with a lack of secretory IgA in the respiratory system. Secretory or exocrine IgA appears in colostrum, intestinal, and respiratory secretions; saliva; tears; and other secretions.

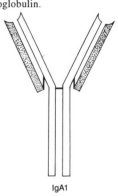

IgA1

IgA2

immunoglobulin α chain
A 58-kD, 470-amino acid residue heavy polypeptide chain that confers class specificity on immunoglobulin A molecules. The chain is divisible into three constant domains, designated C_H1, C_H2, and CH_H3, and one variable domain, designated V_H. A hinge region is situated between C_H1 and C_H2 domains. An additional segment of 18 amino acid residues at the penultimate position of the chain contains a cysteine residue where the J chain can be linked through a disulfide bond. The IgA subclass is divisible into IgA1 and IgA2 subclasses, reflecting two separate α chain isotypes. The α-2 chain has two allotypes designated A2m(1) and A2m(2) and does not have disulfide bonds linking H to L chains. Residues that are subclass specific are found in a number of positions in C_H1, the hinge region, and C_H2, where α-1 and α-2 chains differ, but α-2 chains are the same. Differences in the two α chains are found in two C_H1 and five C_H3 positions. Thus, there are three varieties of α heavy chain in man.

immunoglobulin A deficiency
The most frequent human immunodeficiency that affects 1 in 600 persons in the United States. Even though the B cells of these individuals have IgA on the surface, they do not differentiate into plasma cells that secrete the IgA. IgA levels are decreased from a normal value of 76 to 390 mg/dl to a value below 5 mg/dl. Almost half develop anti-IgA antibodies that are subclass specific. Those IgA-deficient individuals with heightened susceptibility to infection by pyogenic microorganisms also are deficient in IgG2 and often IgG4. The administration of a blood transfusion to IgA-deficient patients possessing anti-IgA antibodies can lead to anaphylactic shock due to IgE antibodies specific for IgA or to a fatal hemolytic transfusion reaction. Patients may also have intestinal lymphangiectasia, arthritis, gluten-sensitive enteropathy, allergies, and myotonic dystrophy. They may also develop low mol wt IgM antibodies against food substances such as milk. Other clinical features may include sinopulmonary infections, cirrhosis, and autoimmune disease. IgA deficiency is detectable in approximately three fourths of ataxia telangiectasia patients. Intravenous immune globulin with only minute quantities of IgA may be beneficial to these patients.

immunoglobulin class
The subdivision of immunoglobulin molecules based on antigenic and structural differences in the Fc regions of their heavy polypep-

tide chains. Immunoglobulin molecules belonging to a particular class have at least one constant region isotypic determinant in common. The different classes, such as IgG, IgM, and IgA, designate separate isotypes. Since the light chains of immunoglobulin molecules are one of two types, the heavy chains determine immunoglobulin class. There is about 30% homology of amino acid sequence among the five immunoglobulin heavy chain constant regions in man. Heavy chains (or isotypes) also differ in carbohydrate content. Immunization of a nonhuman species with human immunoglobulin provides antisera that may be used for class or isotype determination. Ig G is divided into four subclasses and IgA is divided into two subclasses.

immunoglobulin class switching

The mechanism whereby an IgM-producing B cell switches isotype to begin producing IgG molecules instead. Further differentiation may lead to a B cell producing IgA. However, the antigen-binding specificity of the antibody molecules with a different isotype remains unchanged.

immunoglobulin D (IgD)

IgD, which has a mol wt of 185 kD, comprises less than 1% of serum immunoglobulins. It has the basic four-chain monomeric structure with two δ heavy chains (mol wt 63,000 D each) and either two κ or two λ light chains (mol wt 22,000 D each). The half-life of IgD is only 2 to 3 days, and the role of IgD in immunity remains elusive. Membrane IgD serves with IgM as an antigen receptor on B cell membranes.

immunoglobulin δ chain

A 64-kD, 500-amino acid residue heavy polypeptide chain consisting of one variable region, designated V_H, and a three-domain constant region, designated C_H1, C_H2, and C_H3. There is also a 58-residue amino acid residue hinge region in human δ chains. Two exons encode the hinge region. IgD is very susceptible to the action of proteolytic enzymes at its hinge region. Two separate exons encode the membrane component of δ chain. A distinct exon encodes the carboxy terminal portion of the human δ chain that is secreted. The human δ chain contains three N-linked oligosaccharides. Two δ chains and two light chains, either κ or λ, fastened together by disulfide bonds constitute an IgD molecule.

immunoglobulin deficiency with elevated IgM

Antibody deficiency characterized by heightened susceptibility to infection by pyogenic microorganisms. Whereas IgM and IgD levels are elevated in the serum, the IgA and IgG concentrations are greatly diminished or not detectable. Patients often manifest IgM autoantibodies against neutrophils and platelets. The only immunoglobulins secreted by B cells in this X-linked recessive immunodeficiency syndrome are IgM and IgD.

immunoglobulin domain

An immunoglobulin heavy or light polypeptide chain structural unit that is comprised of approximately 110 amino acid residues. Immunoglobulin functions may be linked to certain domains. There is much primary and three-dimensional structural homology among immunoglobulin domains. A particular exon may encode an immunoglobulin domain.

immunoglobulin E (IgE)

IgE constitutes less than 1% of the total immunoglobulins and has a half-life of approximately 2.5 days. This antibody has a four-chain unit structure with two ε heavy chains (mol wt 75,000 D each) and either two κ or two λ light chains per molecule (total mol wt 190 kD). IgE does not precipitate with antigen *in vitro* and is heat labile. IgE is responsible for anaphylactic hypersensitivity in man. It is elevated and plays a beneficial role in parasitic infections.

immunoglobulin ε chain

A 72-kD, 550-amino acid residue heavy polypeptide chain comprised

of one variable region, designated V_H, and a four-domain constant region, designated C_H1, C_H2, C_H3, and C_H4. This heavy chain does not possess a hinge region. In man, the ε heavy chain has 428 amino acid residues in the constant region. There is no carboxy terminal portion of the ε chains. Two ε heavy polypeptide chains, together with two κ or two λ light chains, fastened together by disulfide bonds, comprise an IgE molecule.

immunoglobulin fold

An immunoglobulin domain's three-dimensional configuration. An immunoglobulin fold has a sandwich-like structure comprised of two β-pleated sheets that are nearly parallel. There are four antiparallel chain segments in one sheet and three in the other. Approximately 50% of the domain's amino acid residues are in the β-pleated sheets. The other 50% of the amino acid residues are situated in polypeptide chain loops and in terminal segments. The turns are sites of invariant glycine residues. Hydrophobic amino acid side chains are situated between the sheets.

immunoglobulin fragment

A term reserved for products that result from the action of proteolytic enzymes on immunoglobulin molecules. Intrachain disulfide bonds can be severed by reduction in the presence of denaturing agents such as urea, guanidine, or detergents. Peptide bonds in intact domains are not easily split by proteolytic enzymes. Light chains can be cleaved at the V-C junction, giving rise to large segments that correspond to the V_L and C_L domains. Similar cleavage of the heavy chain is more difficult to achieve. Papain cleaves H chains at the N-terminus of the H-H disulfide bonds, giving two individual portions of the terminus of the molecule, called Fab, and the fragment of the C-terminus region, Fc, which is crystallizable. In contrast, pepsin cleaves H chains at the C-terminus of the H-H disulfide bonds. Thus, the two Fab fragments will remain joined and are called $F(ab')_2$. It degrades the C_H2 domains, but splits the C_H3 domains, which remain noncovalently bonded in dimeric form and are called pFc'. Further digestion of the pFc' with papain results in smaller dimeric fragments called Fc'. Plasmin has been found to cleave the immunoglobulin molecule between C_H2 and C_H3 giving rise to a fragment designated Facb. The heavy chain portion of the Fab, designated as Fd, and the heavy chain portion of the Fab' fragment, designated Fd', results from the breakdown of an $F(ab')_2$ fragment produced by pepsin digestion of the IgG molecule. The Fv fragment consists of the variable domain of heavy and light chains on an immunoglobulin molecule where antigen binding occurs.

immunoglobulin G (IgG)

IgG comprises approximately 85% of the immunoglobulins in adults. It has a mol wt of 154 kD based on two L chains of 22,000

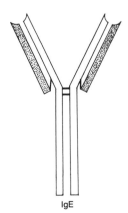

IgD

IgE

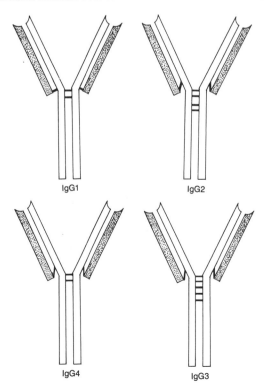

IgG1   IgG2

IgG4   IgG3

The Immunoglobulins

Ig	IgG	IgM	IgA	IgD	IgE
Serum concentration (mg/dl)	800-1700	50-190	140-420	0.3-0.40	<0.001
Total Ig (%)	85	5-10	5-15	<1	<1
Complement fixation	+	++++	-	-	-
Principal biological effect	Resistance-opsonin; secondary response	Resistance-precipitin; primary response	Resistance- prevents movement across mucous membranes	?	Anaphylaxis
Principal site of action	Serum	Serum	Secretions	?; Receptor for B cells	Mast cells
Molecular weight (kd)	154	900	160 (+ dimer)	185	190
Serum half-life (days)	23	5	6	2-3	2-3
Antibacterial lysis	+	+++	+	?	?
Antiviral lysis	+	+	+++	?	?
H-chain class	γ	μ	α	δ	ε
Subclass	$\gamma_1\ \gamma_2\ \gamma_3\ \gamma_4$	$\mu_1,\ \mu_2$	$\alpha_1,\ \alpha_2$		

D each and two H chains of 55,000 D each. It has the longest half-life (23 days) of the five immunoglobulin classes, crosses the placenta, and is the principal antibody in the anamnestic or booster response. IgG shows high avidity or binding capacity for antigen, fixes complement, stimulates chemotaxis, and acts as an opsonin to facilitate phagocytosis.

immunoglobulin γ chain

A 51-kD, 450-amino acid residue heavy polypeptide chain comprised of one variable V_H domain and a constant region with three domains designated C_H1, C_H2, and C_H3. The hinge region is situated between C_H1 and C_H2. There are four subclasses of IgG in man with four corresponding γ chain isotypes designated γ-1, γ-2, γ-3, and γ-4. IgG1, IgG2, IgG3, and IgG4 have differences in their hinge regions and differ in the number and position of disulfide bonds that link the two γ chains in each IgG molecule. There is only a 5% difference in amino acid sequence among human γ chain isotypes, exclusive of the hinge region. Cysteine residues, which make it possible for inter-heavy (γ) chain disulfide bonds to form, are found in the hinge area. IgG1 and IgG4 have 2 inter-heavy chain disulfide bonds, IgG2 has 4, and IgG3 has 11. Proteolytic enzymes, such as papain and pepsin, cleave an IgG molecule in the hinge region to produce Fab and F(ab')₂ and Fc fragments. Four murine isotypes have also been described. Two exons encode the carboxy terminal region of membrane γ chain. Two γ chains, together with two κ or γ light chains, fastened together by disulfide bonds comprise an IgG molecule.

immunoglobulin genes

Genes that encode heavy and light polypeptide chains of antibody molecules are found on different chromosomes, i.e., chromosome 14 for heavy chain, chromosome 2 for κ light chain, and chromosome 22 for λ light chain. The DNA of the majority of cells does not contain one gene that encodes a complete immunoglobulin heavy or light polypeptide chain. Separate gene segments that are widely distributed in somatic cells and germ cells come together to form these genes. In B cells, gene rearrangement leads to the creation of an antibody gene that codes for a specific protein. Somatic gene rearrangement also occurs with the genes that encode T cell antigen receptors. Gene rearrangement of this type permits the great versatility of the immune system in recognizing a vast array of epitopes. Three forms of gene segments join to form an immunoglobulin light chain gene. The three types include light chain variable region (V_L), joining (J_L) and constant region (C_L) gene segments. V_H, J_H, and C_H as well as D (diversity), gene segments assemble to encode the heavy chain. Heavy and light chain genes have a closely similar organizational structure. There are 100 to 300 V_κ genes, 5 J_κ genes, and 1 C_κ gene on chromome 2's κ locus. There are 100 V_H genes, 30 D genes, six J_H genes, and 11 C_H genes on chromosome 14's heavy chain locus. Several V_λ, 6 J_λ, and 6 C_λ genes are present on chromosome 22's λ locus in man. V_H and V_L genes are classified as V gene families, depending on the sequence homology of their nucleotides or amino acids.

Immunoglobulin Gene Rearrangement

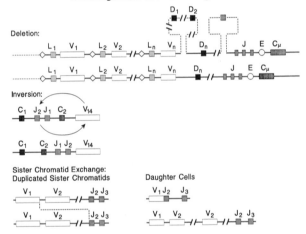

Immunoglobulin Gene

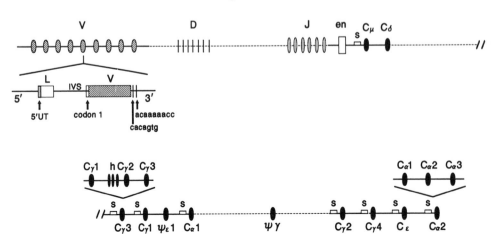

immunoglobulin heavy chain

A 51- to 71-kD polypeptide chain present in immunoglobulin molecules that serves as the basis for dividing immunoglobulins into classes. The heavy chain is comprised of three to four constant domains, depending upon class, and one variable domain. In addition, a hinge region is present in some chains. There is approximately 30% homology, with respect to amino acid sequence, among the five classes of immunoglobulin heavy chain in man. The heavy chain of IgM is μ, of IgG is γ, of IgA is α, of IgD is δ, and of IgE is ε.

immunoglobulin heavy chain-binding protein (BiP)

A 77-kD protein that combines with selected membrane and secretory proteins. It is believed to facilitate their passage through the endoplasmic reticulum.

immunoglobulin κ chain

A 23-kD 214-amino acid residue polypeptide chain that is comprised of a single variable region and a single constant region. It is one of the two types of light polypeptide chain present in all five immunoglobulin classes. Approximately 60% of light immunoglobulin chains in man are κ with wide variations of their percentages in other species. Whereas κ chains are virtually absent in immunoglobulins of dogs, they comprise the vast majority of murine immunoglobulin light chains. κ light chain allotypes in man are termed Km1, Km1,2, and Km3.

immunoglobulin λ chain

A 23-kD 214-amino acid residue polypeptide chain with a single variable region and a single constant region. λ chains represent one of two light polypeptide chains comprising all five classes of immunoglobulin molecules. Approximately 40% of immunoglobulin light chains in man are λ. Wide variations in percentages are observed in other species. For example, the great majority of immunoglobulin light chains in horses and dogs are λ, whereas they constitute only 5% of murine light chains. Constant region differences among λ light chains of mice and man distinguish the molecules into four isotypes in humans. A different C gene segment encodes the separate constant regions defining each λ light chain isotype. The human λ light chain isotypes are designated Kern⁻Oz⁺, Kern⁺Oz⁻, and Mcg.

immunoglobulin light chain

A 23-kD, 214-amino acid polypeptide chain comprised of a single constant region and a single variable region that is present in all five classes of immunoglobulin molecules. The two types of light chains are designated κ and λ. They are found in association with heavy polypeptide chains and immunoglobulin molecules and are fastened to these structures through disulfide bonds.

immunoglobulin-like domain

A 100-amino acid residue structure found in selected β sheet-rich proteins with intrachain disulfide bonds. It is found in immunoglobulins, interleukins 1 and 6, the T cell receptor, and platelet-derived growth factor.

immunoglobulin M (IgM)

IgM comprises 5 to 10% of the total immunoglobulins in adults and has a half-life of 5 days. It is a pentameric molecule with 5 four-chain monomers joined by disulfide bonds and the J chain, with a total mol wt of 900 kD. Theoretically, this immunoglobulin has ten antigen-binding sites. IgM is the most efficient immunoglobulin in fixing complement. A single IgM pentamer can activate the classic pathway. Monomeric IgM is found with IgD on the B lymphocyte cell surface, where it serves as the receptor for antigen. Because IgM is relatively large, it is confined to intravascular locations. IgM is particularly important for immunity against polysaccharide antigens on the exterior of pathogenic microorganisms. It also promotes phagocytosis and bacteriolysis through its complement activation activity.

Monomeric
IgM

Pentameric
IgM

immunoglobulin μ chain

A 72-kD, 570-amino acid heavy polypeptide chain comprised of one variable region, designated V_H, and a four-domain constant region, designated C_H1, C_H2, C_H3, and C_H4. The μ chain does not have a hinge region. A "tail piece" is located at the carboxy terminal end of the chain. It is comprised of 18 amino acid residues. A cysteine residue at the penultimate position of a carboxy terminal

region of the μ chain forms a disulfide bond that joins to the J chain. There are five N-linked oligosaccharides in the μ chain of man. Secreted IgM ($μ_s$) and membrane IgM ($μ_m$) and μ chain differ only in the final 20 amino acid residues at the carboxy terminal end. The membrane form of IgM has 41 different residues substituted for the final 20 residues in the secreted form. A 26-residue region of this carboxy terminal section in the membrane form of IgM apparently represents the hydrophobic transmembrane part of the chain.

immunoglobulin staining

Demonstrable by immunoperoxidase staining of plasma cell and B lymphocyte cytoplasm in frozen or paraffin-embedded sections. B-5 fixative is preferable to formalin for demonstration of intracellular IgG or light chains in paraffin sections. Monoclonal cytoplasmic staining for either κ or λ light chains aids the diagnosis of B cell lymphomas.

immunoglobulin subclass

The subdivision of immunoglobulin classes according to structural and antigenic differences in the constant regions of their heavy polypeptide chains. All molecules in an immunoglobulin subclass must express the isotypic antigenic determinants unique to that class, but they also express other epitopes that render that subclass different from others. IgG has four subclasses designated as IgG1, IgG2, IgG3, and IgG4. Whereas there is only 30% identity among the five immunoglobulin classes, there is three times that similarity among IgG subclasses. The IgA class is divisible into two subclasses, whereas the remaining three immunoglobulin classes have not been further subdivided into subclasses. The structural differences in subclasses are exemplified by the variations and number of inter-heavy chain disulfide bonds which the four IgG subclasses possess. The function of immunoglobulin molecules differs from one subclass to another, as exemplified by the inability of IgG4 to fix complement.

Ig	IgG	IgM	IgA	IgE	IgD
H Chain Class	γ	μ	α	ε	δ
Subclass	γ₁ γ₂ γ₃ γ₄	μ₁ μ₂	α₁ α₂		

Immunoglobulin Subclass

Immunoglobulin Superfamily
Adhesion Receptors

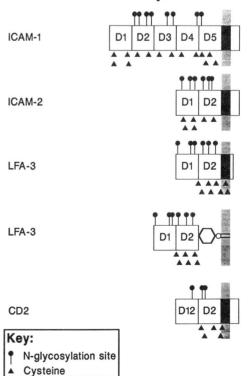

Key:
• N-glycosylation site
▲ Cysteine

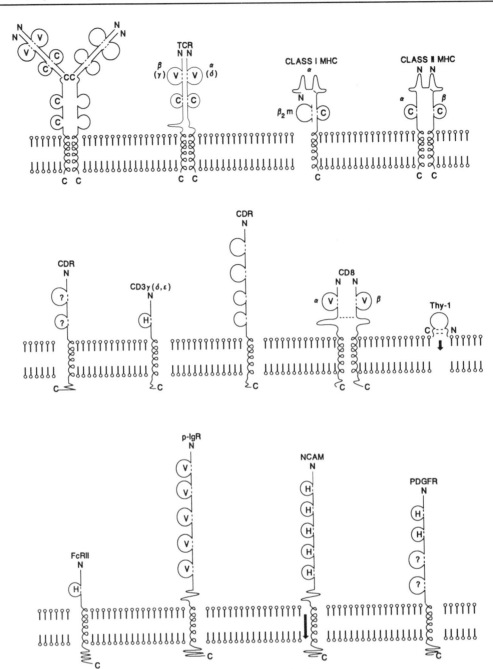

Immunoglobulin Superfamily

immunoglobulin superfamily (see also figure, page 159)

Several molecules that participate in the immune response and show similarities in structure, causing them to be named the immunoglobulin supergene family. Included are CD2, CD3, CD4, CD7, CD8, CD28, T cell receptor (TCR), MHC class I and MHC class II molecules, leukocyte function associated antigen 3 (LFA-3), the IgG receptor, and a dozen other proteins. These molecules share in common with each other an immunoglobulin-like domain, with a length of approximately 100 amino acid residues and a central disulfide bond that anchors and stabilizes antiparallel β strands into a folded structure resembling immunoglobulin. Immunoglobulin superfamily members may share homology with constant or variable immunoglobulin domain regions. Various molecules of the cell surface with polypeptide chains whose folded structures are involved in cell to cell interactions belong in this category. Single gene and multigene members are included.

immunogold labeling

A technique to identify antigens in tissue preparations by electron microscopy. Sections are incubated with primary antibody and

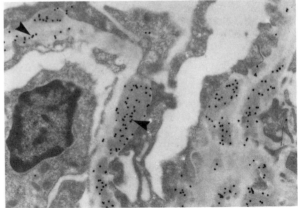

Immunogold

followed by treatment with colloidal gold-labeled anti-IgG antibody. Electron-dense particles are localized at sites of antigen-antibody interactions.

immunogold silver staining (IGSS)

An immunohistochemical technique to detect antigens in tissues and cells by light microscopy. IGSS offers higher labeling intensity than that of most other methods when examined in a bright field or in conjunction with polarized light. The technique successfully stains tissue sections from paraffin wax, resin, or cryostat preparations. It is also effective for cell suspensions or smears, cytospin preparations, cell cultures, or tissue sections. Both l- and 5-nm gold conjugates are used for light microscopy. The l-nm particles are advantageous in studies of cell penetration. In immunogold silver staining, primary antibody is incubated with tissues or cells to localize antigens which are identified with gold-labeled secondary antibodies and silver enhanced.

immunohematology

The study of blood group antigens and antibodies and their interactions in health and disease. Both the cellular elements and the serum constituents of the blood have distinct profiles of antigens. There are multiple systems of blood cell groups, all of which may stimulate antibodies and interact with them. These may be associated with erythrocytes, leukocytes, or platelets.

immunoincompetence

Inability to produce a physiologic immune response. For example, patients with acquired immune deficiency (AIDS) become immunoincompetent as a consequence of destruction of their helper/inducer (CD4$^+$) T lymphocyte population. Infants born without a thymus or experimental animals thymectomized at birth are immunologically incompetent. Children born with severe combined immunodeficiency due to one or several causes are unable to mount an appropriate immune response. Immunoincompetence may involve either the B cell limb, as in Bruton's hypogammaglobulinemia, or the T cell limb, as in patients with DiGeorge's syndrome.

immunoinhibitory genes

Selected HLA genes that appear to protect against immunological diseases. Their mechanisms of action are in dispute.

immunoisolation

The enclosure of allogeneic tissues such as pancreatic islet cell allografts within a membrane that is semipermeable, but does not itself induce an immune response. Substances of relatively low mol wt can reach the graft through the membrane, while it remains protected from immunologic rejection by the host.

immunologic (or immunological)

Adjective referring to those aspects of a subject that fall under the purview of the scientific discipline of immunology.

immunologic adjuvant

A substance that enhances an immune response, either humoral or cellular or both, to an immunogen (antigen).

immunologic barrier

An anatomical site that diminishes or protects against an immune response. This refers principally to immunologically privileged sites where grafts of tissue may survive for prolonged periods without undergoing immunologic rejection. This is based mainly on the lack of adequate lymphatic drainage in these areas. Examples include prolonged survival of foreign grafts in the brain.

immunologic competence

The capability to mount an immune response.

immunologic enhancement

The prolonged survival, conversely the delayed rejection, of a tumor allograft in a host as a consequence of contact with specific antibody. Both the peripheral and central mechanisms have been postulated. In the past, coating of tumor cells with antibody was presumed to interfere with the ability of specifically reactive lymphocytes to destroy them, but today a central effect in suppressing cell-mediated immunity, perhaps through suppressor T lymphocytes, is also possible.

immunologic facilitation (*facilitation immunologique*)

Slightly prolonged survival of certain normal tissue allografts, e.g., skin, in mice conditioned with isoantiserum specific for the graft.

immunologic paralysis

Immunologic unresponsiveness induced by the injection of large doses of pneumococcal polysaccharide into mice where it is metabolized slowly. Any antibody that is formed is consumed and not detectable. The pneumococcal polysaccharide antigen remains in tissues of the recipient for months, during which time the animals produce no immune response to the antigen. Immunologic paralysis is much easier to induce with polysaccharide than with protein antigens. It is highly specific for the antigen used for its induction. Felton's first observation of immunologic paralysis preceded the demonstration of acquired immunologic tolerance by Medawar et al.

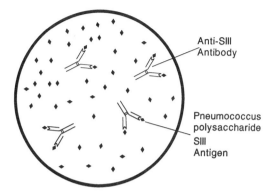

Anti-SIII Antibody

Pneumococcus polysaccharide SIII Antigen

Schematic View of Antigen Excess
Immunologic Paralysis

immunologic tolerance

An active, but carefully regulated response of lymphocytes to self antigens. Autoantibodies are formed against a variety of self antigens. Maintenance of self-tolerance is a quantitative process. When comparing the ease with which T and B cell tolerance may be induced, it was found that T cell tolerance is induced more rapidly and is longer lasting than B cell tolerance. For example, T cell tolerance may be induced in a single day, whereas B cells may require 10 days for induction. In addition, 100 times more tolerogen may be required for B cell tolerance than for T cell tolerance. The duration of tolerance is much greater in T cells, e.g., 150 days, compared to that in B cells, which is only 50 to 60 days. T suppressor cells are also very important in maintaining natural tolerance to self antigens. For example, they may suppress T helper cell activity. Maintenance of tolerance is considered to require the continued presence of specific antigens. Low antigen doses may be effective in inducing tolerance in immature B cells leading to clonal abortion, whereas T cell tolerance does not depend upon the level of maturation. Another mechanism of B cell tolerance is cloning exhaustion, in which the immunogen activates all of the B lymphocytes specific for it, leading to maturation of cells and transient antibody synthesis and thereby exhausting and diluting the B cell response. Another mechanism of B cell tolerance is antibody-forming cell blockade. Antibody-expressing B cells are coated with excess antigen, rendering them unresponsive to the antigen.

immunological deficiency state

Immunodeficiency.

immunological ignorance

A type of tolerance to self in which a target antigen and lymphocytes capable of reacting with it are both present simultaneously in an individual without an autoimmune reaction occurring. The abrogation of immunologic ignorance may lead to autoimmune disease.

immunological inertia

Specific immunosuppression related to paternal histocompatibility antigens during pregnancy, such as suppression of maternal immune reactivity against fetal histocompatibility antigens.

immunological reaction

An *in vivo* or *in vitro* response of lymphoid cells to an antigen they have never previously encountered or to an antigen for which they are already primed or sensitized. An immunological reaction may consist of antibody formation, cell-mediated immunity, or immunological tolerance. The humoral antibody and cell-mediated immune reactions may mediate either protective immunity or hypersensitivity, depending on various conditions.

immunological rejection

The destruction of an allograft or even a xenograft in a recipient host whose immune system has been activated to respond to the foreign tissue antigens.

immunological unresponsiveness

Failure to form antibodies or develop a lymphoid cell-mediated response following exposure to immunogen (antigen). Immunosuppression that is specific for only one antigen with no interference with the response to all other antigens is termed immunological tolerance. By contrast, the administration of powerful immunosuppressive agents such as azathioprine, cyclosporine, or total body irradiation causes generalized immunological unresponsiveness to essentially all immunogens to which the host is exposed.

immunologically activated cell

An immunologically competent cell following its interaction with antigen. This response may be expressed either as lymphocyte transformation, immunological memory, cell-mediated immunity, immunologic tolerance, or antibody synthesis.

immunologically competent cell

A lymphocyte, such as a B cell or T cell, that can recognize and respond to a specific antigen.

immunologically privileged sites

Certain anatomical sites within the animal body provide an immunologically privileged environment which favors the prolonged survival of alien grafts. The potential for development of a blood and lymphatic vascular supply connecting graft and host may be a determining factor in the qualification of an anatomical site as an area which provides an environment favorable to the prolonged survival of a foreign graft. Immunologically privileged areas include (1) the anterior chamber of the eye, (2) the substantia propria of the cornea, (3) the meninges of the brain, (4) the testis, and (5) the cheek pouch of the Syrian hamster. Foreign grafts implanted in these sites show a diminished ability to induce transplantation immunity in the host. These immunologically privileged sites usually fail to protect alien grafts from the immune rejection mechanism in hosts previously or simultaneously sensitized with donor tissues.

immunologist

A person who makes a special study of immunology.

immunology

That branch of biomedical science concerned with the response of the organism to immunogenic (antigenic) challenge, the recognition of self from nonself, and all the biological (in vivo), serological (in vitro), physical, and chemical aspects of immune phenomena.

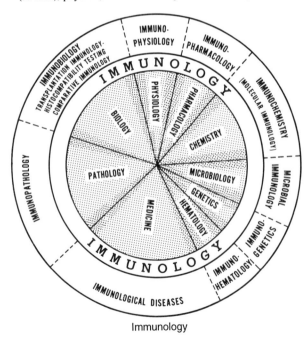

Immunology

immunolymphoscintigraphy

A method to determine the presence of tumor metastasis to lymph nodes. Antibody fragments or monoclonal antibodies against specific tumor antigens are radiolabeled and then detected by scintigraphy.

immunomodulation

Therapeutic alteration of the immune system by the administration of biological response modifiers such as lymphokines or antibodies against cell surface markers bound to a toxin such as ricin.

immunonephelometry

A test that measures light which is scattered at a 90° angle to a laser or light source as it is passed through a suspension of minute complexes of antigen and antibody. Measurement is made at 340 to 360 nm using a spectrophotometer.

immunoosmoelectropheresis

Refer to counter immunoelectrophoresis.

immunoparasitology

Immunologic aspects of the interaction between animal parasites and their hosts.

immunopathic

Injury to cells, tissues, or organs induced by either humoral (antibodies) or cellular products of an immune response.

immunopathology

The study of disease processes that have an immunological etiology or pathogenesis involving either humoral antibody (from B cells) and complement or T cell-mediated mechanisms. Immunologic injury of tissues and cells may be mediated by any of the four types of hypersensitivity (described separately).

immunoperoxidase method

Nakene and Pierce in 1966 first proposed that enzymes be used in the place of fluorochromes as labels for antibodies. Horseradish peroxidase (HRP) is the enzyme label most widely employed. The immunoperoxidase technique permits the demonstration of antigens in various types of cells and fixed tissues. This method has certain advantages that include the following: (1) the use of conventional light microscopy, (2) the stained preparations may be kept permanently, (3) the method may be adapted for use with electron microscopy of tissues, and (4) counterstains may be employed. The disadvantages include the following: (1) the demonstration of relatively minute positively staining areas is limited by the light microscope's

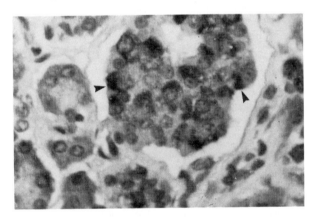

insulin in β cells

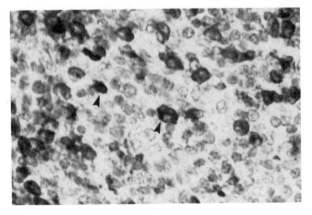

plasma cells
decorated with
antibody
(immunoperoxidase)

Immunoperoxidase Method

A B C TECHNIQUE

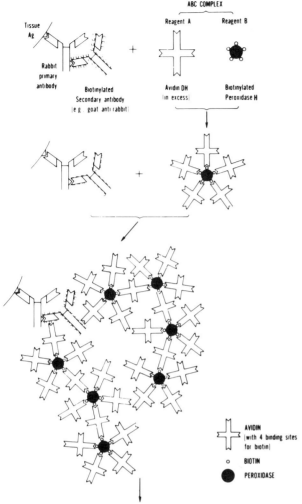

Development in chromogenic hydrogen donor and hydrogen peroxide (The reaction product is seen as a reddish brown or brown granular deposit depending upon the chromogenic hydrogen donor used)

Immunoperoxidase Method

resolution, (2) endogenous peroxidase may not have been completely eliminated from the tissue under investigation, and (3) diffusion of products resulting from the enzyme reaction away from the area where antigen is localized. The peroxidase-antiperoxidase (PAP) technique employs unlabeled antibodies and a PAP reagent. This has proven highly successful for the demonstration of antigens in paraffin-embedded tissues as an aid in surgical pathologic diagnosis. Tissue sections preserved in paraffin are first treated with xylene, and after deparaffinization they are exposed to a hydrogen peroxide solution which destroys the endogenous peroxidase activity in tissue. The sections are next incubated with normal swine serum which suppresses nonspecific binding of immunoglobulin molecules to tissues containing collagen. Thereafter, the primary rabbit antibody against the antigen to be identified is reacted with the tissue section. Primary antibody that is unbound is removed by rinsing the sections which are then covered with swine antibody against rabbit immunoglobulin. This so-called linking antibody will combine with any primary rabbit antibody in the tissue. It is added in excess, which will result in one of its antigen-binding sites remaining free. After washing, the PAP reagent is placed on the section, and the antibody portion of this complex, which is raised in rabbits, will be bound to the free antigen-binding site of the linking antibody on the sections. The unbound PAP complex is then washed away by rinsing. To read the sections microscopically, it is necessary to add a substrate of hydrogen peroxide and aminoethylcarbazole (AEC) which permits the formation of a visible product that may be detected with the light microscope. The AEC is oxidized to produce a reddish-brown pigment that is not water soluble. Peroxidase catalyzes the reaction. Because peroxidase occurs only at sites where the PAP is bound via

linking antibody and primary antibody to antigen molecules, the antigen is identified by the reddish-brown pigment. The tissue sections can then be counterstained with hematoxylin or other suitable dye, covered with mounting medium and cover slips, and read by conventional light microscopy. The PAP technique has been replaced, in part, by the avidin-biotin complex (ABC) technique.

immunophenotyping
The use of monoclonal antibodies and flow cytometry to reveal cell surface or cytoplasmic antigens that yield information that may reflect clonality and cell lineage classification. This type of data is valuable clinically in aiding the diagnosis of leukemias and lymphomas through the use of a battery of B cell, T cell, and myeloid markers. However, immunophenotyping results must be used only in conjunction with morphologic criteria when reaching a diagnosis of leukemia or lymphoma.

immunophilins
High-affinity receptor proteins in the cytoplasm that combine with such immunosuppressants as cyclosporin A, FK506, and rapamycin. They prevent the activity of rotamase by blocking conversion between cis- and trans-rotamers of the peptide and protein substrate peptidyl-prolylamide bond. Immunophilins are important in transducing signals from the cell surface to the nucleus. Immunosuppressants have been postulated to prevent signal transduction mediated by T lymphocyte receptors, which blocks nuclear factor activation in activated T lymphocytes. Cyclophilin- and FK506-binding proteins represent immunophilins. Drug-immunophilin complexes are implicated in the mechanism of action of the immunosuppressant drugs, cyclosporin, FK506, and rapamycin.

immunophysiology
The physiologic basis of immunologic processes.

immunopotency
The capability of a part of an antigen molecule to function as an epitope and induce the synthesis of specific antibodies.

immunopotentiation
Facilitation of the immune response usually with the aid of adjuvants such as muramyl dipeptide, Freund's adjuvant, synthetic polynucleotides, or other agents.

immunoproliferative small intestinal disease (IPSID)
(Mediterranean lymphoma α heavy chain disease.) A varied group of disorders in which there is monoclonal synthesis of immunoglobulin heavy chain (often α). Light chains are not produced. The variable region and often the C_H1 constant region may be missing. Whereas the monoclonal protein usually elevated is the α chain, some cases may manifest γ or μ chain elevations. Patients experience weight loss, pain in the abdomen, diarrhea, and malabsorption. There is expansion of the mesenteric lymphoid tissue and of the proximal small intestinal. There is also clubbing of the fingers. Tetracyclines are the recommended treatment.

immunoprophylaxis
Disease prevention through the use of vaccines to induce active immunization or antisera to induce passive immunization.

immunoradiometric assay (IRMA)
A quantitative method to assay certain plasma proteins based on a "sandwich" technique using radiolabeled antibody, rather than

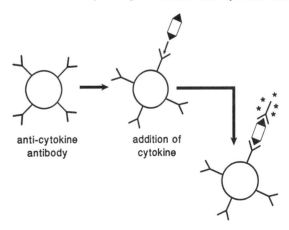

anti-cytokine antibody   addition of cytokine

addition of iodinated monoclonal antibody

IRMA (Immunoradiometric Assay)

radiolabeled hormone competing with hormone from a patient in the radioimmunoassay (RIA).

immunoradiometry

A radioimmunoassay method in which the antibody rather than the antigen is radiolabeled.

immunoscintigraphy

The formation of two-dimensional images of the distribution of radioactivity in tissues following the administration of antibodies labeled with a radionuclide that are specific for tissue antigens. A scintillation camera is used to record the images.

immunosuppression

(1) The deliberate administration of drugs, such as cyclosporine, azathioprine, corticosteroids, FK506, or rapamycin, or the administration of specific antibody or the use of irradiation to depress immune reactivity in recipients of organ or bone marrow allotransplants. (2) Profound depression of the immune response that occurs in patients with certain diseases such as acquired immune deficiency syndrome in which the helper-inducer (CD4+) T lymphocytes are destroyed by the HIV-1 virus.

immunosuppressive agent

A drug, such as cyclosporine, FK506, rapamycin, azathioprine, or corticosteroids, or an antibody, such as antilymphocyte serum, or irradiation that produces mild to profound depression of a host's ability to respond to an immunogen (antigen), as in the conditioning of an organ allotransplant recipient.

immunotactoid glomerulopathy

A renal malady characterized by glomerular deposits of fibrillar material comprised of 10- to 48.9-nm microfibrils or microtubules as viewed by electron microscopy. These glomerular fibrillar deposits are not birefringent when stained with Congo red and examined by polarizing light microscopy. This differentiates them from amyloid. Usually, no extraglomerular fibrillar deposits are present, differentiating the condition from amyloid or light chain deposition. It is not associated with concomitant systemic disease such as cryoglobulinemia or systemic lupus erythematosus. Typically affects middle-aged males who may manifest hypertension, nephrotic range proteinuria, and microscopic hematuria.

immunotherapy

The use of immunologic mechanisms to combat disease. These include the hyposensitization treatment used by allergists in raising blocking antibodies against common allergens, nonspecific stimulation of the immune response with BCG immunotherapy in treating certain types of cancer, and the IL-2/LAK cell adoptive immunotherapy technique for treating certain tumors.

immunotoxin

The linkage of an antibody specific for target cell antigens with a cytotoxic substance such as the toxin ricin yields an immunotoxin. Upon parenteral injection, its antibody portion directs the immunotoxin to the target and its toxic portion destroys target cells on contact. Among its uses is the purging of T cells from hematopoietic cell preparations used for bone marrow transplantation.

Substance produced by the union of a monoclonal antibody or one of its fractions to a toxic molecule such as a radioisotope, a bacterial or plant toxin, or a chemotherapeutic agent. The antibody portion is intended to direct the molecule to antigens on a target cell, such as those of a malignant tumor, and the toxic portion of the molecule is for the purpose of destroying the target cell. Contemporary methods of recombinant DNA technology have permitted the preparation of specific hybrid molecules for use in immunotoxin therapy. Immunotoxins may have difficulty reaching the intended target tumor, may be quickly metabolized, and may stimulate the development of antiimmunotoxin antibodies. Crosslinking proteins may likewise be unstable.

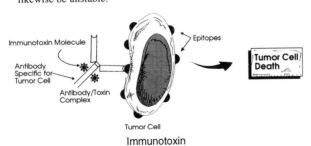

Immunotoxin

immunotyping

Refer to immunophenotyping.

***in situ* hybridization**

A technique to identify specific DNA or RNA segments in cells or tissues, viral plaques, or colonies of microorganisms. DNA in cells or tissue fixed on glass slides must be denatured with formamide before hybridization with a radiolabeled or biotinylated DNA or RNA probe that is complementary to the tissue mRNA being sought. Proof that the probe has hybridized to its complementary strand in the tissue or cell under study must be by auto-radiography or enzyme-labeled probes, depending on the technique being used.

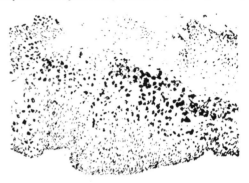

in situ hybridization
HPV DNA 16 and 18 probe group
from a cervical biopsy specimen

***in situ* transcription**

A method in which mRNA acts as a template for complementary DNA for reverse transcription in tissues that have been fixed.

inaccessible antigens

Refer to hidden determinant (epitopes).

inactivated vaccine

An immunizing preparation that contains microorganisms such as bacteria or viruses that have been killed to stop their replication while preserving their protection-inducing antigens. Formaldehyde, phenol, and β-propiolactone have been used to inactivate viruses, whereas formaldehyde, acetone, phenol, or heating have been methods used to kill bacteria to be used in vaccines.

inactivation

A term used mostly by immunologists to signify loss of complement activity in a serum sample that has been heated to 56°C for 30 min or by hydrazine treatment. Inactivation also applies to chemical or heat treatment of pathogenic microorganisms in a manner that preserves their antigenicity for use in inactivated vaccines.

inbred strain

Laboratory animals developed by sequential brother-sister matings. After 20 generations, the animals, e.g., mice, are said to be inbred. They are homozygous at approximately 98% of genetic loci. This homogeneity at the histocompatibility loci permits successful grafting, without rejection, among members of the inbred stain. Recessive deleterious genes may become homozygous during inbreeding, leading them to manifest their negative effects with respect to such factors as growth rate, susceptibility to disease, or fertility, and thus, limiting the number of possible inbred generations. An additional problem is the development of sublines of an inbred strain caused by mutations and evolutionary factors.

inbreeding

The mating of animals of a species that are genetically more similar to one another than to members of that same species selected by chance or at random in the population. Deliberate inbreeding of experimental animals is carried out to induce genetic uniformity or homozygosity. Raising inbred strains of mice for laboratory investigation involves brother-sister matings for 20 or more generations. Thereafter, the progeny are said to be "inbred".

incompatibility

Dissimilarity between the antigens of a donor and recipient as in tissue allotransplantation or blood transfusions. The transplantation of a histoincompatible organ or the transfusion of incompatible blood into a recipient may induce an immune response against the antigens not shared by the recipient in injurious consequences.

incomplete antibody

A nonagglutinating antibody that must have a linking agent such as anti-IgG to reveal its presence in an agglutination reaction. Refer to Coombs' test.

incomplete Freund's adjuvant (IFA)

Light-weight mineral oil, without mycobacteria, which when combined with aqueous-phase antigen, as a water-in-oil emulsion, enhances the humoral or antibody (B cell) limb of the immune response. It does not facilitate T cell-mediated immune responsiveness.

index of variability

Refer to Wu-Kabat plot.

indirect agglutination (passive agglutination)

The aggregation or agglutination of a specific antibody with carrier particles such as latex particles or tanned red blood cells to which antigens have been adsorbed or with bis-diazotized red blood cells to which antigens have been linked chemically. Refer to passive agglutination.

indirect antiglobulin test

A method to detect incomplete (nonagglutinating) antibody in a patient's serum. Following incubation of red blood cells or other cells possessing the antigen for which the incomplete antibodies of interest are specific, rabbit anti-human globulin is added to the antibody-coated cells which have been first washed. If agglutination results, incomplete agglutinating antibody is present in the serum with which the antigen-bearing red cells have been incubated.

indirect complement fixation test

A complement fixation assay for antibodies that are unable to fix guinea pig complement. It involves the addition of a rabbit antibody of established guinea pig complement-fixing capacity to an antigen-(avian) antibody-guinea pig complement complex. This is followed by the addition of a visible hemolytic system. Cell lysis indicates the initial presence of avian antibody.

indirect Coombs' test

Refer to indirect antiglobulin test.

indirect fluorescence antibody technique

A method to identify antibody or antigen using a fluorochrome-labeled antibody which combines with an intermediate antibody or antigen rather than directly with the antibody or antigen being sought. The indirect test has a greater sensitivity than those of the direct fluorescence antibody technique. It is often referred to as the sandwich or double-layer method.

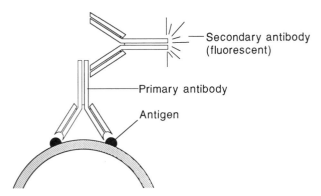

Indirect Fluorescent Antibody Technique

indirect hemagglutination test

Refer to passive agglutination test.

indirect immunofluorescence

The interaction of unlabeled antibody with cells or tissues expressing antigen for which the antibody is specific, followed by treatment of this antigen-antibody complex with fluorochrome-labeled antiimmunoglobulin that interacts with the first antibody, forming a so-called sandwich.

indirect template theory (historical)

A variation of the template hypothesis which postulated that instructions for antibody synthesis were copied from the antigen configuration into the DNA encoding the specific antibody. This was later shown to be untenable and is of historical interest only.

indomethacin

[1-(4-chlorobenzylyl)-5-methoxy-2-methyl-1H-indole-3-acetic acid]. A drug that blocks synthesis of prostaglandin. It is used for therapy of rheumatoid arthritis and ankylosing spondylitis.

Indomethacin

inducer

A substance that promotes cellular differentiation, such as colony-stimulating factors.

inducer determinant

Hapten determinant.

inducer T lymphocyte

A cell required for the initiation of an immune response. The inducer T lymphocyte recognizes antigens in the context of MHC class II histocompatibility molecules. It stimulates helper, cytotoxic, and suppressor T lymphocytes, whereas helper T cells activate B cells. The human leukocyte common antigen variant, termed 2H4, occurs on the inducer T cell surface, and 4B4 surface molecules are present on CD4+ helper T cells. CD4+ T lymphocytes must be positive for either 4B4 or for 2H4.

inductive phase

The time between antigen administration and detection of immune reactivity.

inert particle agglutination tests

Assays that employ particles of latex, bentonite, or other inert materials to adsorb soluble antigen on their surface to test for the presence of specific antibody in the passive agglutination test. The particles coated with adsorbed antigen agglutinate if antibody is present. An example is the RA test in which pooled human IgG is adsorbed to latex particles which agglutinate if combined with a serum sample containing rheumatoid factor, i.e., anti-IgG autoantibody.

infantile agammaglobulinemia

Synonym for X-linked agammaglobulinemia.

infantile sex-linked hypogammaglobulinemia

Antibody deficiency syndrome that is sex linked and occurs in males following the disappearance of passively transferred antibodies from the mother following birth. Serum immunoglobulin concentrations are relatively low, and there is defective antibody synthesis giving rise to recurring bacterial infections. Cell-mediated immunity remains intact.

infection allergy

T cell-mediated delayed-type hypersensitivity associated with infection by selected microorganisms such as *Mycobacterium tuberculosis*. This represents type IV hypersensitivity to antigenic products of microorganisms inducing a particular infection. It also develops in *Brucellosis, Lymphogranuloma venereum,* mumps, and vaccinia. Also called infection hypersensitivity.

infection hypersensitivity

Tuberculin-type sensitivity that is more evident in some infections than others. It develops with great facility in tuberculosis, brucellosis, lymphogranuloma venereum, mumps, and vaccinia. The sensitizing component of the antigen molecule is usually protein, although polysaccharides may induce delayed reactivity in cases of systemic fungal infections such as those caused by *Blastomyces, Histoplasma,* and *Coccidioides.*

infectious bursal agent

A virus infection of cells in the bursa of Fabricius in young chicks, leading to destruction of the bursa and most of the antibody-producing capability of the animal.

infectious mononucleosis

A disease of teenagers and young adults who have a sore throat, fever, and enlarged lymph nodes. Atypical large lymphocytes with increased cytoplasm which is also vacuolated, are found in the peripheral blood and have been shown by immunophenotyping to be T cells. They are apparently responding to Epstein-Barr virus-infected B lymphocytes. There is also lymphocytosis, neutropenia, and thrombocytopenia. Patients also develop heterophile antibodies, which agglutinate horse, ox, and sheep red blood cells as revealed by the Paul-Bunnell test. Infectious mononucleosis is the most common condition that EBV causes. There may be splenomegaly and chemical hepatitis.

infectious mononucleosis syndrome(s)

Conditions induced by viruses that produce an acute and striking peripheral blood monocytosis and lead to symptoms resembling those of infectious mononucleosis induced by Epstein-Barr virus. Examples include herpes virus, cytomegalovirus, HIV-1, HHV-6, and *Toxoplasma gondii* infections.

infectious tolerance

Infectious tolerance was described in the 1970s. Animals rendered tolerant to foreign antigens were found to possess suppressor T lymphocytes associated with the induced unresponsiveness. Thus, self-tolerance was postulated to be based on the induction of suppressor T cells. Rose has referred to this concept as clonal balance rather than clonal deletion. Self antigens are considered to normally induce mostly suppressor rather than helper T cells, leading to a negative suppressor balance in the animal body. Three factors with the potential to suppress immune reactivity against self include nonantigen-specific suppressor T cells, antigen-specific suppressor T cells, and antiidiotypic antibodies. Noel Rose suggested that suppressor T lymphocytes leave the thymus slightly before the corresponding helper T cells. Suppressor T cells specific for self antigens are postulated to be continuously stimulated and usually in greater numbers than the corresponding helper T cells.

inflammation

A defense reaction of living tissue to injury. The literal meaning of the word is burning and originates from the cardinal symptoms of rubor, calor, tumor, and dolor, the Latin terms equivalent to redness, heat, swelling, and pain, respectively. It is beneficial for the host and essential for survival of the species, although in some cases the response is exaggerated and may be itself injurious.

Inflammation is the result of multiple interactions which have as a first objective localization of the process and removal of the irritant. This is followed by a period of repair. Inflammation is not necessarily of immunologic nature, although immunologic reactions are among the immediate causes inducing inflammation, and the immunologic status of the host determines the intensity of the inflammatory response. Inflammation tends to be less intense in infants whose immune system is not fully mature.

The causes of inflammation are numerous and include living microorganisms such as pathogenic bacteria and animal parasites which act mainly by the chemical poisons they produce and less by mechanical irritation; viruses which become offenders after they have multiplied in the host and cause cell damage; and fungi which grow at the surface of the skin but produce little or no inflammation in the dermis. Other causes of inflammation include physical agents such as trauma, thermal and radiant energy, and chemical agents which represent a large group of exogenous or endogenous causes which include immunologic offenders. Refer to inflammatory response.

inflammatory bowel disease (IBD)

A general term that applies to ulcerative colitis and Crohn's disease, as well as to idiopathic inflammatory bowel disease that resembles the other two. There is an hereditary predisposition to IBD. Intestinal epithelial cells express HLA-DR (MHC class II) antigens in Crohn's disease, ulcerative colitis, and infectious colitis patients which might render them capable of becoming autoantigen-presenting cells. Inflammatory bowel disease patients may become sensitized to cow's milk antigens, developing IgG and IgM antibodies against this protein. Leukotrienes have been shown to be of greater significance than prostaglandins in mediating inflammation in ulcerative colitis. Mucocutaneous conditions such as oral ulcers, epidermolysis bullosa acquisita, erythema nodosa, etc., as well as eye diseases such as uveitis and iridocyclitis, may be associated with IBD. Some IBD patients may also have cirrhosis, chronic active hepatitis, or joint involvement such as ankylosing spondylitis. IBD patients may have abdominal pain, fever, and diarrhea. Granulomas may develop in the gut wall, and lymphocytes that stain for IgA in the cytoplasm are often abundant. Autoantibodies reactive with fetal colon antigens may be present.

inflammatory cell

Cells of the blood and tissues that participate in acute and chronic inflammatory reactions. These include polymorphonuclear neutrophils, eosinophils, and macrophages.

inflammatory macrophage

A macrophage in a peritoneal exudate induced by thioglycolate broth or mineral oil injection into the peritoneal cavity of an experimental animal.

inflammatory mediator

A substance that participates in an inflammatory reaction.

inflammatory response

Acute inflammation represents an early defense mechanism to contain an infection and prevent its spread from the initial focus. When microbes multiply in host tissues, two principal defense mechanisms mounted against them are antibodies and leukocytes. The three major events in acute inflammation are (1) dilation of capillaries to increase blood flow; (2) changes in the microvasculature structure, leading to escape of plasma proteins and leukocytes from the circulation; and (3) leukocyte emigration from the capillaries and accumulation at the site of injury. Widening of interendothelial cell junctions of venules or injury of endothelial cells facilitates the escape of plasma proteins from the vessels. Neutrophils attached to the endothelium through adhesion molecules escape the microvasculature and are attracted to sites of injury by chemotactic agents. This is followed by phagocytosis of microorganisms that may lead to their intracellular destruction. Activated leukocytes may produce toxic metabolites and proteases that injure endothelium and tissues when they are released. Activation of the third complement component (C3) is also a critical step in inflammation.

Multiple chemical mediators of inflammation derived from either plasma or cells have been described. Mediators and plasma proteins such as complement are present as precursors that require activation to become biologically active. Mediators derived from cells are present as precursors in intracellular granules, such as histamine and mast cells. Following activation, these substances are secreted. Other mediators such as prostaglandins may be synthesized following stimulation. These mediators are quickly activated by enzymes or other substances such as antioxidants. A chemical mediator may also cause a target cell to release a secondary mediator with a similar or opposing action.

Besides histamine, other preformed chemical mediators in cells include serotonin and lysosomal enzymes. Those that are newly synthesized include prostaglandins, leukotrienes, platelet-activating factors, cytokines, and nitric oxide. Chemical mediators in plasma include complement fragments C3a and C5a and the C5b-g sequence. Three plasma-derived factors including kinins, complement, and clotting factors are involved in inflammation. Bradykinin is produced by activation of the kinin system. It induces arteriolar dilation and increased venule permeability through contraction of endothelial cells and extravascular smooth muscle. Activation of bradykinin precursors involves activated factor XII (Hageman factor) generated by its contact with injured tissues.

During clotting, fibrinopeptides produced during the conversion of fibrinogen to fibrin increase vascular permeability and are chemotactic for leukocytes. The fibrinolytic system participates in inflammation through the kinin system. Products produced during arachidonic acid metabolism also affect inflammation. These include prostaglandins and leukotrienes, which can mediate essentially every aspect of acute inflammation.

influenza vaccine

Purified and inactivated immunizing preparation made from viruses grown in eggs. It cannot lead to infection. It contains (H1N1) and (H3N2) type A strains and one type B strain. These are the strains considered most likely to cause influenza in the United States. Whole virus and split virus preparations are available. Children tolerate the split virus preparation better than the whole virus vaccine.

influenza viruses

Infectious agents that induce an acute, febrile respiratory illness often associated with myalgia and headache. The classification of influenza A viruses into subtypes is based on their hemagglutinin (H) and neuraminidase (N) antigens. Three hemagglutinin (H1, H2, and H3) and two neuraminidase (N1 and N2) subtypes are the principal influenza A virus antigenic subtypes that produce disease in man. Due to antigenic change (antigenic drift), infection or vaccination by one strain provides little or no protection against subsequent infection by a distantly related strain of the same subtype. Influenza B viruses undergo less frequent antigenic variation.

inhibition test

(1) Blocking an established serological test such as agglutination or precipitation through the addition of an antigen for which the

antibody in the test system is specific. It shows the specificity of the reactants. (2) Inhibition of an antigen-antibody interaction through the addition of a hapten for which the antibody is specific. Refer to hapten inhibition test. (3) Preventing the action of a virus through addition of and antibody specific for the virus.

inhibition zone
Refer to prozone.

innate immunity
Natural or native immunity that is present from birth and is designed to protect the host from injury or infection without previous contact with the infectious agent. It includes such factors as protection by the skin, mucous membranes, lysozyme in tears, stomach acid, and numerous other factors. Phagocytes, natural killer cells, and complement represent key participants in natural innate immunity.

innocent bystander
A cell that is fatally injured during an immune response specific for a different cell type.

innocent bystander hemolysis
Drugs acting as haptens induce immune hemolysis of "innocent" red blood cells. The reaction is drug specific and involves complement through activation of the alternate complement pathway. The direct antiglobulin (Coombs') test identifies split products that are membrane associated, yet the indirect Coombs' test remains negative.

inoculation
The introduction of an immunogen into an animal to induce immunity, usually to protect against an infectious disease agent.

instructive theory
A hypothesis that postulated acquisition of antibody specificity after contact with a specific antigen. According to one template theory of antibody formation, it was necessary that the antigen be present during the process of antibody synthesis. According to the refolding template theory, uncommitted and specific globulins could become refolded upon the antigen, serving as a template for it. The cell released the complementary antibodies, which rigidly retained their shape through disulfide bonding. This theory had to be abandoned when it was shown that the specificity of antibodies in all cases is due to the particular arrangement of their primary amino acid sequence. *De novo* synthesis template theories that recognized the necessity for antibodies to be synthesized by amino acids, in the proper and predetermined order, still had to contend with the serious objection that proteins cannot serve as informational models for the synthesis of proteins. Instructive theories were abandoned when immunologic tolerance was demonstrated and when antigen was shown not to be necessary for antibody synthesis to occur. The template theories had never explained the anamnestic (memory) immune response. Antibody specificity depends upon the variable region amino acid sequence, especially the complementarity-determining or hypervariable regions.

instructive theory of antibody formation
Refer to instructive theory.

insulin-dependent (type I) diabetes mellitus (IDDM)
Juvenile-onset diabetes caused by diminished capacity to produce insulin. Genetic factors play a major role, as the disease is more common in HLA-DR3 and HLA-DR4 positive individuals. There are significant autoimmune features that include IgG autoantibodies against glucose transport proteins and anticytoplasmic and antimembrane antibodies directed to antigens in the pancreatic islets of Langerhans. β Cells are destroyed, and the pancreatic islets become infiltrated by T lymphocytes and monocytes in the initial period of the disease. Experimental models of IDDM include the NOD mouse and the BB rat.

insulin receptor antibodies
Antibodies that lead to insulin resistance and may also be associated with acanthosis nigricans and manifestations of autoimmune disease. Insulin receptor antibodies may lead to hypoglycemia that may be associated with autoimmune disease and Hodgkin's disease. Selected patients with insulin-dependent diabetes mellitus and noninsulin-dependent diabetes mellitus may have insulin receptor antibodies that react at either the site that binds insulin or at some other binding region.

insulin resistance
Diminished responsiveness to insulin as revealed by its decreased capacity to induce hypoglycemia may or may not have an immunologic basis. The administration of bovine or porcine insulin to humans can lead to antibody production that may contribute to insulin resistance.

Intal®
Commercial preparation of disodium cromoglycate.

integrin family of leukocyte adhesive proteins
The CD11/CD18 family of molecules.

integrins
A family of cell membrane glycoproteins that are heterodimers comprised of α and β chain subunits. They serve as extracellular matrix glycoprotein receptors. They identify the RGD sequence of the β subunit, which consists of the arginine-glycine-aspartic acid tripeptide that occasionally also includes serine. The RGD sequence serves as a receptor recognition signal. Extracellular matrix glycoproteins, for which integrins serve as receptors, include fibronectin, C3, and lymphocyte function-associated antigen 1 (LFA-1), among other proteins. Differences in the β chain serve as the basis for division of integrins into three categories. Each category has distinctive α chains. The β chain provides specificity. The same 95-kD β chain is found in one category of integrins that includes lymphocyte function-associated antigen 1 (LFA-1), p150,95, and complement receptor 3 (CR3). The same 130-kD β chain is shared among VLA-1, VLA-2, VLA-3, VLA-4, VLA-5,

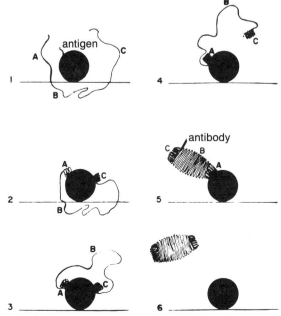

Instructive Theory of Antibody Formation

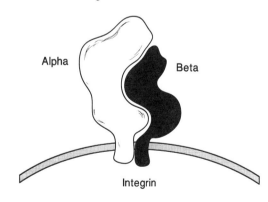

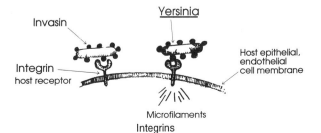

Integrins

Representative Integrin Family Adhesion Receptors

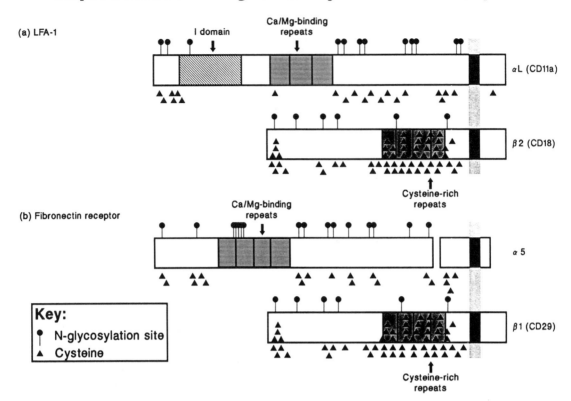

VLA-6, and integrins found in chickens. A 110-kD β chain is shared in common by another category that includes the vitronectin receptor and platelet glycoprotein IIb/IIIa. There are four repeats of 40 amino acid residues in the β chain extracellular domains. There are 45 amino acid residues in the β chain intracellular domains. The principal function of integrins is to link the cytoskeleton to extracellular ligands. They also participate in wound healing, cell migration, killing of target cells, and in phagocytosis. Leukocyte adhesion deficiency syndrome occurs when the β subunit of LFA-1 and Mac-1 are missing. VLA proteins facilitate binding of cells to collagen (VLA-1, -2, and -3), laminin (VLA-1, -2, and -6), and fibronectin (VLA-3, -4, and -5).

intercellular adhesion molecule 1 (ICAM-1)
A 90-kD cellular membrane glycoprotein that occurs in multiple cell types including dendritic cells and endothelial cells. It is the lymphocyte function-associated antigen 1 (LFA-1) ligand. The LFA-1 molecules on cytotoxic T lymphocytes (CTL) interact with ICAM-1 molecules found on CTL target cells. Interferon γ, tumor necrosis factor, and IL-1 can elevate ICAM-1 expression.

intercellular adhesion molecule 2 (ICAM-2)
A protein that is a member of the immunoglobulin superfamily that is important in cellular interactions. It is a cell surface molecule that serves as a ligand for leukocyte integrins. ICAM-2 facilitates lymphocytes binding to antigen-presenting cells or to endothelial cells. ICAM-2 binds to LFA-1, a T lymphocyte integrin.

intercellular adhesion molecule 3 (ICAM-3)
A leukocyte cell surface molecule that plays a critical role in the interaction of T lymphocytes with antigen presenting cells. The interaction of the T lymphocyte with an antigen presenting cell through union of ICAM-1, ICAM-2, and ICAM-3 with LFA-1 molecules is also facilitated by the interaction of the T cell surface molecule CD2 with LFA-3 present on antigen-presenting cells.

intercrine cytokines
A family comprised of a minimum of 8- to 10-kD cytokines that share 20 to 45% amino acid sequence homology. All are believed to be basic heparin-binding polypeptides with proinflammatory and reparative properties. Their cDNA has conserved single open reading frames, 5′ region typical signal sequences, and 3′ untranslated regions that are rich in AP sequences. Human cytokines that include interleukin-8, platelet factor 4, β thromboglobulin, IP-10, and

melanoma growth-stimulating factor or GRO comprise a subfamily encoded by genes on chromosome 4. They possess a unique structure. LD78, ACT-2, I-309, RANTES, and macrophage chemotactic and activating factor (MCAF) comprise a second subset and are encoded by genes on chromosome 17 of man. Human chromosome 4 bears the intercrine α genes, and human chromosome 17 bears the intercrine β genes. Four cysteines are found in the intercrine family. Adjacent cysteines are present in the intercrine β subfamily, which includes huMCAF, huBLD-78, huACT-II, huRANTES, muTCA-III, muJE, muMIP-1 α, and muMIP-1 β. One amino acid separates cysteines of the intercrines α subfamily, which is comprised of huPF-4, hubetaTG, h:IL-8, ch9E3, huGRO, huIP-10, and muMIP-2. The cysteines are significant for tertiary structure and for intercrine binding to receptors.

interdigitating cell
Antigen-processing macrophages in the paracortical, thymus-dependent regions of lymph nodes.

interfacial test
Refer to ring test.

interferon(s) α (IFN-α)
At least 13 immunomodulatory 189-amino acid residue glycoproteins synthesized by macrophages and B cells that are able to prevent the replication of viruses, are antiproliferative, and are pyrogenic, inducing fever. IFN-α stimulates natural killer cells and induces expression of class I MHC antigens. It also has an immunoregulatory effect through alteration of antibody responsiveness. The 14 genes that encode IFN-α are positioned on the short arm of chromosome 9 in man. Polyribonucleotides, as well as RNA or DNA viruses, may induce IFN-α secretion. Recombinant IFN-α has been prepared and used in the treatment of hairy cell leukemia, Kaposi's sarcoma, chronic myeloid leukemia, human papilloma virus-related lesions, renal cell carcinoma, chronic hepatitis, and selected other conditions. Patients may experience severe flu-like symptoms as long as the drug is administered. They also have malaise, headache, depression, and supraventricular tachycardia and may possibly develop congestive heart failure. Bone marrow suppression has been reported in some patients.

interferon β (IFN-β)
An antiviral, 20-kD protein comprised of 187 amino acid residues. It is produced by fibroblasts and prevents replication of viruses. It

has 30% amino acid sequence homology with interferon α. RNA or DNA viruses or polyribonucleotides can induce its secretion. The gene encoding it is located on chromosome 9 in man.

Interferons

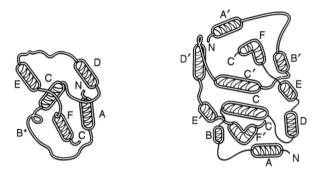

Interferon-β   Interferon-γ

interferon γ (IFN-γ)

A glycoprotein that is a 21- to 24-kD homodimer synthesized by activated T lymphocytes and natural killer (NK) cells, causing it to be classified as a lymphokine. IFN-γ has antiproliferative and antiviral properties. It is a powerful activator of mononuclear phagocytes, increasing their ability to destroy intracellular microorganisms and tumor cells. It causes many types of cells to express class II MHC molecules and can also increase expression of class I. It facilitates differentiation of both B and T lymphocytes. IFN-γ is a powerful activator of NK cells and also activates neutrophils and vascular endothelial cells. It is decreased in chronic lymphocytic leukemia, lymphoma, and IgA deficiency, as well as those infected with rubella, Epstein-Barr virus, and cytomegalovirus. Recombinant IFN-γ has been used for treatment of a variety of conditions including chronic lymphocytic leukemia, mycosis fungoides, Hodgkin's disease, and various other disorders. It has been found effective in decreasing synthesis of collagen by fibroblasts and might have potential in the treatment of connective tissue diseases. People receiving it may develop headache, chills, rash, or even acute renal failure. The one gene that encodes IFN-γ in man is found on the long arm of chromosome 12.

interferon γ receptor

A 90-kD glycoprotein receptor comprised of one polypeptide chain. The only cells found lacking this receptor are erythrocytes. It is encoded by a gene on chromosome 6q in man.

IFN-γ Receptor

Human IFN-γ
receptor

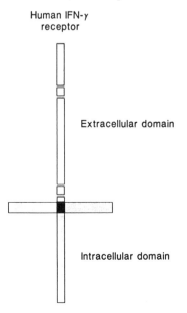

Extracellular domain

Intracellular domain

interferons (IFNs)

A group of immunoregulatory proteins synthesized by T lymphocytes, fibroblasts, and other types of cells following stimulation with viruses, antigens, mitogens, double-stranded DNA, or lectins. Interferons are classified as α and β, which have antiviral properties, and as γ which is known as immune interferon. α and β Interferons share a common receptor, but γ has its own. Interferons have immunomodulatory functions. They enhance the ability of macrophages to destroy tumor cells, viruses, and bacteria. Interferons α and β were formerly classified as type I interferons. They are acid stable and synthesized mainly by leukocytes and fibroblasts. Interferon γ is acid labile and is formed mainly by T lymphocytes stimulated by antigen or mitogen. This immune interferon has been termed type II interferon in the past. Whereas the ability of interferon to prevent infection of noninfected cells is species specific, it is not virus specific. Essentially, all viruses are subject to its inhibitory action. Interferons induce formation of a second inhibitory protein that prevents viral messenger RNA translation. In addition to γ interferon formation by T cells activated with mitogen, natural killer cells also secrete it. Interferons are not themselves viricidal.

interleukin(s) (IL)

A group of cytokines synthesized by lymphocytes, monocytes, and selected other cells that promote growth of T cells, B cells, and hematopoietic stem cells and have various other biological functions. Refer to IL-1 through IL-13.

interleukin-1 (IL-1)

A cytokine synthesized by activated mononuclear phagocytes that have been stimulated by ribopolysaccharide or by interaction with CD4⁺ T lymphocytes. It is a monokine and is a mediator of inflammation, sharing many properties in common with tumor necrosis factors (TNF). IL-1 is comprised of two principal polypeptides of 17 kD, each with isoelectric points of 5.0 and 7.0. They are designated IL-1α and IL-1β, respectively. Genes found on chromosome 2 encode these two molecular species. They have the same biological activities and bind to the same receptor on cell surfaces. Both IL-1α and IL-1β are derived by proteolytic cleavage of 33-kD precursor molecules. IL-1α acts as a membrane-associated substance, whereas IL-1β is found free in the circulation. IL-1 receptors are present on numerous cell types. IL-1 may either activate adenylate cyclase, elevating cAMP levels and then activating protein kinase A, or it may induce nuclear factors that serve as cellular gene transcriptional activators. IL-1 may induce synthesis of enzymes that generate prostaglandins, which may in turn induce fever, a well-known action of IL-1. IL-1's actions differ according to whether it is produced in lower or in higher concentrations. At low concentrations, the effects are mainly immunoregulatory. IL-1 acts with polyclonal activators to facilitate CD4⁺ T lymphocyte proliferation, as well as B lymphocyte growth and differentiation. IL-1 stimulates multiple cells to act as immune or inflammatory response effector cells. It also induces further synthesis of itself, as well as of IL-6, by mononuclear phagocytes and vascular endothelium. It resembles tumor necrosis factor (TNF) in inflammatory properties. IL-1 secreted in greater amounts produces endocrine effects as it courses through the peripheral blood circulation. For example, it produces fever and promotes the formation of acute-phase plasma proteins in the liver. It also induces cachexia. Natural inhibitors of IL-1 may be produced by mononuclear phagocytes activated by immune complexes in humans. The inhibitor is biologically inactive and prevents the action of IL-1 by binding with its receptor, serving as a competitive inhibitor. Corticosteroids and prostaglandins suppress IL-1 secretion. IL-1 was formerly called lymphocyte-activating factor (LAF).

interleukin-1 receptor (IL-1R)

An 80-kD receptor on T lymphocytes, chondrocytes, osteoblasts, and fibroblasts that binds IL-1α and IL-1β. Helper/inducer CD4⁺ T lymphocytes are richer in IL-1R than are suppressor/cytotoxic (CD8⁺) T cells. The IL-1R has an extracellular portion that binds ligand and contains all N-linked glycosylation sites. A 217-amino acid segment, apparently confined to the cytoplasm, could be involved in signal transduction. Further studies of ligand-binding have been facilitated through the development of a soluble form of the cloned IL-1R molecule which contains the extracellular part, but not the transmembrane cytoplasmic region of the molecule. IL-1α and IL-1β molecules bind with equivalent affinities. The IL-1R has been claimed to have more than one subunit. This is based on the demonstration of bands such as a 100-kD band, in addition to that characteristic of the receptor which is 80-kD. The recombinant IL-1R functions in signal transduction. When the cytoplasmic part of the IL-1R is depleted, the molecule does not function. The human T cell IL-1R has now been cloned and found to be quite similar to its murine counterpart. Two affinity classes of binding sites for IL-1 have been described.

interleukin-1 receptor antagonist protein (IRAP)

A substance on T lymphocytes and endothelial cells that inhibits IL-1 activity.

interleukin-1 receptor deficiency

CD4+ T cells deficient in IL-1 receptors in affected individuals fail to undergo mitosis when stimulated and fail to generate IL-2. This leads to a lack of immune responsiveness and constitutes a type of combined immunodeficiency. Opportunistic infections are increased in affected children who have inherited the condition as an autosomal recessive trait.

interleukin-2 (IL-2)

A 15.5-kD glycoprotein synthesized by CD4+ T helper lymphocytes. It was formerly called T cell growth factor. IL-2 has an autocrine effect acting on the CD4+ T cells that produce it. Although mainly produced by CD4+ T cells, a small amount is produced by CD8+ T cells. Physiologic amounts of IL-2 do not have an endocrine effect, since it acts on the cells producing it or on those nearby acting as a paracrine growth factor. IL-2's main effects are on lymphocytes. The amount of IL-2 which CD4+ T lymphocytes synthesize is a principal factor in determining the strength of an immune response. It also facilitates formation of other cytokines produced by T lymphocytes, including interferon γ and lymphotoxins. Inadequate IL-2 synthesis can lead to antigen-specific T lymphocyte anergy. IL-2 interacts with T lymphocytes by reacting with IL-2 receptors. IL-2 also promotes NK cell growth and potentiates the cytolytic action of NK cells through generation of lymphokine-activated killer (LAK) cells. Although NK cells do not have the p55 lower affinity receptor, they do express the high-affinity p70 receptor and thus require high IL-2 concentrations for their activation. IL-2 is a human B cell growth factor and promotes synthesis of antibody by these cells. However, IL-2 does not induce isotype switching. IL-2 promotes the improved responsiveness of immature bone marrow cells to other cytokines. In the thymus, it may promote immature T cell growth. The IL-2 gene is located on chromosome 4 in man. Corticosteroids, cyclosporin A, and prostaglandins inhibit IL-2 synthesis and secretions.

interleukin-2 receptor (IL-2R)

Also known as CD25. IL-2R is a structure on the surface of T lymphocytes, natural killer, and B lymphocytes characterized by the presence of a 55-kD polypeptide, p55, and a 70-kD polypep-

IL-2 Receptor

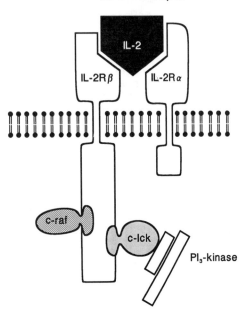

tide termed p70, which interacts with IL-2 molecules at the cell surface. The p55 polypeptide chain is referred to as Tac antigen, an abbreviation for T activation. The expression of both p55 and p70 permits a cell to bind IL-2 securely with a K_d of about 10^{-8}. p55, the low-affinity receptor, apparently complexes with p70, the high-affinity receptor, to accentuate the p70 receptor's affinity for IL-2. This permits increased binding in cells expressing both receptors. In addition, lesser quantities of IL-2 than would otherwise be required for stimulation are effective when both receptors are present on the cell surface. Antibodies against p55 or p70 can block IL-2 binding. Powerful antigenic stimulation such as in transplant rejection may lead to the shedding of p55 IL-2 receptors into the serum. The gene encoding the p55 chain is located on chromosome 10p14 in man. IL-2, IL-1, IL-6, IL-4, and TNF may induce IL-2R expression.

interleukin-2 receptor α subunit (IL-2Rα)

The 55-kD polypeptide subunit of the IL-2R with a kD of 10^{-8} M. The α subunit is responsible for increasing the affinity between cytokine and receptor, however it has no role in signal transduction. It is expressed only on antigen stimulation of T cells, usually within 2 hours. Following long-term T cell activation the α subunit is shed, making it a potential candidate for a serum marker of strong or prolonged antigen stimulation. The gene encoding IL-2Rα is located on chromosome 10p14 in man.

interleukin-2 receptor αβγ subunit (IL-α2Rβγ)

The complete IL-2 receptor consists of two distinct polypeptides, IL-2α, which is induced upon activation, and IL-2βγ, which is present on resting T cells. Upon expression of all three proteins affinity increases to 10^{-11} M and very low (physiologic) levels of IL-2 are capable of stimulating the cell. IL-2R is found on T lymphocytes, natural killer cells, and B lymphocytes although natural killer cells do not express IL-2Rα. IL-2, IL-1, IL-6, IL-4, and TNF may induce IL-2R expression.

Activated T Cell

IL-2 Receptor Protein

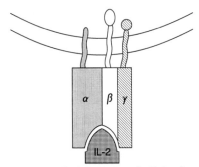

Interleukin-2 Receptor αβγ Subunit

interleukin-2 receptor β subunit (IL-2Rβ)

The 70- to 74-kD subunit of IL-2R with a kD of 10^{-9} M. The β subunit is a member of the cytokine receptor family Type I due to its tryptophan serine-x-tryptophan-serine (WSXWS) domain. It is a constitutive membrane protein coordinately expressed with IL-2Rγ.

interleukin-2 receptor βγ subunit (IL-2Rβγ)

A heterodimer found on resting T cells. Only those T cells expressing IL-2Rβγ are capable of growth in response to IL-2, as this is the portion of the receptor responsible for signal transduction.

interleukin-2 receptor γ subunit (IL-2Rγ)

This subunit is also a type I (WSXWS) receptor that is associated with IL-4 and IL-7 receptors as well as IL-2Rγ. Mutations in the γ subunit have been found in some SCIDS cases with X-linked inheritance, resulting in decreased proliferation of B and T cells.

interleukin-3 (IL-3)

A 20-kD lymphokine synthesized by activated CD4+ T helper lymphocytes that acts as a colony-stimulating factor by facilitating proliferation of some hematopoietic cells and promoting proliferation and differentiation of other lymphocytes. It acts by binding to high- and low-affinity receptors, inducing tyrosine phosphorylation and colony formation of erythroid, myeloid, megakaryocytic, and lymphoid hematopoietic cells. It also facilitates mast cell

proliferation and the release of histamine. It facilitates T lymphocyte maturation through induction of 20α-hydroxysteroid dehydrogenase. The gene encoding IL-3 is situated on the long arm of
chromosome 5.

interleukin-4 (IL-4) (B cell growth factor)

A 20-kD cytokine produced mainly by CD4+ T lymphocytes, but
also by activated mast cells. Most studies of IL-4 have been in mice,
where it serves as a growth and differentiation factor for B cells and
is a switch factor for synthesis of IgE. It also promotes growth of a
cloned CD4+ T cell subset. Further properties of murine IL-4
include its function as a growth factor for mast cells and activation
factor for macrophages. It also causes resting B lymphocytes to
enlarge and enhances class II MHC molecule expression. IL-4 was
previously termed B cell growth factor I (BCGF-I) and also B cell-
stimulating factor 1 (BSF-1). In man, CD4+ T lymphocytes also
produce IL-4, but the human variety has not been shown to serve as
a B cell or mast cell growth factor. Human IL-4 also fails to activate
macrophages. Both murine and human IL-4 induce switching of B
lymphocytes to synthesize IgE. Thus, IL-4 may be significant in
allergies. Human IL-4 also induces CD23 expression by B lymphocytes and macrophages in man. IL-4 may have some role in cell-
mediated immunity.

interleukin-5 (IL-5) (eosinophil differentiation factor)

A 20-kD cytokine synthesized by some activated CD4+ T lymphocytes and by activated mast cells. Formerly, it was called T cell
replacing factor or B cell growth factor II. It facilitates B cell growth
and differentiation into cells that secrete IgA. It is a costimulator
with IL-2 and IL-4 of B cell growth and differentiation. IL-5 also
stimulates eosinophil growth and differentiation. It activates mature eosinophils to render them capable of killing helminths. Through
IL-5, T lymphocytes exert a regulatory effect on inflammation
mediated by eosinophils. Because of its action in promoting eosinophil differentiation, it has been called eosinophil differentiation
factor (EDF). IL-5 can facilitate B cell differentiation into plaque-
forming cells of the IgM and IgG classes. In parasitic diseases, IL-
5 leads to eosinophilia.

interleukin-6 (IL-6)

A 26-kD cytokine produced by vascular endothelial cells, mononuclear phagocytes, fibroblasts, activated T lymphocytes, and
various neoplasms such as cardiac myxomas, bladder cancer, and
cervical cancer. It is secreted in response to IL-1 or TNF. Its main
actions are on hepatocytes and B cells. Although it acts on many
types of cells, a significant function is its ability to cause B
lymphocytes to differentiate into cells that synthesize antibodies.
IL-6 induces hepatocytes to form acute-phase proteins that include fibrinogen. It is the main growth factor for activated B
lymphocytes late in B cell differentiation. IL-6 is a growth factor
for plasmacytoma cells which produce it. IL-6 also acts as a
costimulator of T lymphocytes and of thymocytes. It acts in
concert with other cytokines that promote the growth of early bone
marrow hematopoietic stem cells. It acts together with IL-1 to
costimulate activation of T_H cells. IL-6 was formerly termed B cell
differentiation factor (BCDF) and B cell-stimulating factor 2
(BSF-2). It has also been implicated in the pathogenesis of plaques
in psoriasis.

interleukin-7 (IL-7)

Facilitates lymphoid stem cell differentiation into progenitor B cell.
Principally, a T lymphocyte growth factor synthesized by bone
marrow stromal cells. It promotes lymphopoiesis, governing stem
cell differentiation into early pre-T and B cells. It is also formed by
thymic stroma and promotes the growth and activation of T cells
and activates macrophages. It also enhances fetal and adult thymocyte proliferation.

interleukin-8 (IL-8) (neutrophil-activating protein 1)

It is an 8-kD protein of 72 residues produced by macrophages and
endothelial cells. It has a powerful chemotactic effect on T lymphocytes and neutrophils and upregulates the binding properties of
leukocyte adhesion receptor CD11b/CD18. IL-8 regulates expression of its own receptor on neutrophils and has antiviral,
immunomodulatory, and antiproliferative properties. It prevents
adhesion of neutrophils to endothelial cells activated by cytokines,
thereby blocking neutrophil-mediated injury. It participates in inflammation and the migration of cells. It facilitates neutrophil
adherence to endothelial cells. It accomplishes this through the
induction of $β_2$ integrins by neutrophils.

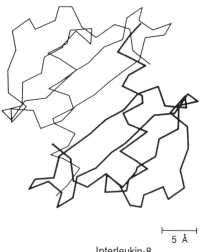

|—— 5 Å ——|

Interleukin-8

interleukin-9 (murine growth factor P40, T cell growth factor III)

A cytokine that facilitates the growth of some T helper cell clones,
but not cytolytic T lymphocyte clones. It is encoded by genes
comprised of 5 exons in a 4-kb segment of DNA in both mice and
humans. The gene encoding IL-9 is located on chromosomes 5 and
13. This hematopoietic growth factor glycoprotein can be derived
from a megakaryoblastic leukemia. Selected human T lymphocyte
lines and peripheral lymphocytes activated by mitogen express it.
IL-9 is related to mast cell growth-enhancing activity both
structurally and functionally. In the presence of erythropoietin, IL-
9 supports erythroid colony formation. In conjunction with IL-2,
IL-3, IL-4, and erythropoietin, IL-9 may enhance hematopoiesis *in
vivo*. It may facilitate bone marrow-derived mast cell growth
stimulated by IL-3 and fetal thymocyte growth in response to IL-2.
T_H2 cells preferentially express IL-9 following stimulation with con
A or by antigen presented on syngeneic antigen presenting cells.

interleukin-10 (IL-10) (cytokine synthesis inhibitory factor)

An 18-kD polypeptide devoid of carbohydrate, in humans, that acts
as a cytokine synthesis inhibitory factor. It is expressed by CD4+
and CD8+ T lymphocytes, monocytes, macrophages, activated B
lymphocytes, B lymphoma cells, and keratinocytes. It inhibits some
immune responses and facilitates others. It inhibits cytokine synthesis by T_H1 cells, blocks antigen presentation, and the formation
of interferon γ. It also inhibits the macrophage's ability to present
antigen and to form IL-1, IL-6, and TNF-α. It also participates in
IgE regulation. Although IL-10 suppresses cell-mediated immunity, it stimulates B lymphocytes, IL-2 and IL-4 T lymphocyte
responsiveness *in vitro*, and murine mast cells exposed to IL-3 and
IL-4. IL-10 might have future value in suppressing T lymphocyte
autoimmunity in multiple sclerosis and type I diabetes mellitus and
in facilitating allograft survival. Murine T_H2 cells secrete IL-10,
which suppresses synthesis of cytokines by T_H1 cells.

interleukin-11 (IL-11)

A cytokine produced by stromal cells derived from the bone marrow
of primates. It is a growth factor that induces IL-6-dependent
murine plasmacytoma cells to proliferate. IL-11 has several biological actions that include its hematopoietic effect. In man, the
genomic sequence and gene encoding IL-11 is comprised of five
exons and four introns. The gene is located at band 19q13.3-13.4 on
the long arm of chromosome 19. It may facilitate plasmacytoma
establishment, possibly representing an important role for IL-11 in
tumorigenesis. In combination with IL-3, IL-11 can potentiate
megakaryocyte growth, producing increased numbers, size, and
ploidy values. It may be important in the formation of platelets. In
the presence of functional T lymphocytes, IL-11 can stimulate the
production of B cells that secrete IgG. It has a synergistic effect in
primitive hematopoietic cell proliferation that is IL-3 dependent.

interleukin-12 (IL-12)

A heterodimeric molecule comprised of 35- and 40-kD chains
linked by disulfide bonds. It acts on T cells as a cytotoxic lymphocyte maturation factor (CLMF). It also serves as a natural killer

(NK) cell stimulatory factor (NKSF). IL-12 is a growth factor for activated CD4+ and CD8+ T lymphocytes and for NK cells. It facilitates NK cell and LAK cell lytic action exclusive of IL-2. It can induce resting peripheral blood mononuclear cells to form interferon γ *in vitro*. IL-12 may act synergistically with IL-2 to increase responses by cytotoxic lymphocytes. It may have potential as a therapeutic agent in the treatment of tumors or infections, especially if used in combination with IL-2.

interleukin-13 (IL-13)

A cytokine expressed by activated T cells that inhibits inflammatory cytokine production by lipopolysaccharide in human peripheral blood monocytes. It synergizes with interleukin-2 to regulate the synthesis of interferon γ in large granular lymphocytes. Mapping reveals that the IL-13 gene is linked closely to the IL-4 gene. IL-13 could be considered a modulator of B cell responses. It stimulates B cell proliferation with anti-Ig and anti-CD40 antibodies as well as promotes IgE synthesis. It also induces resting B cells to express CD23/ FcεRII and class II MHC antigens. The biological activities of IL-13 are similar to those of IL-4. Therefore, both IL-13 and IL-4 may contribute much to the development of allergies. Although IL-13 and IL-4 affect B cells similarly, they have different functions. IL-13 induces less IgE and IgG4 production than does IL-4. IL-13 does not act on T cells or T cell clones; it has no T cell growth-promoting activity.

IL-13 suppresses cell-mediated immunity. It suppresses the cytotoxic functions of monocytes/macrophages and the generation of proinflammatory cytokines. IL-13 has a pleiotropic action on monocytes/macrophages, neutrophils, and B cells. It usually produces an inhibitory effect on monocytes/macrophages, and down-regulates Fcγ receptors and secretion of inflammatory cytokines and nitric oxide induced by lipopolysaccharide. IL-13 inhibits macrophage function in cell-mediated immunity. It induces neutrophils to form IL-1 receptor antagonist and IL-1 receptor II. This leads to inhibition of IL-1, a central inflammatory mediator, which is in accord with the antiinflammatory function of IL-13.

interleukin-14 (IL-14)

Formerly known as high-molecular-weight B cell growth factor (HMW-BCGF), IL-14 is a cytokine produced by follicular dendritic cells, germinal center T cells, and some malignant B cells. Normal and malignant B cells, notably germinal center B cells and NHL-B cells, respectively, express receptors for IL-14. Its predominant activity is to enhance the proliferation of B cells and to induce memory B cell production and maintenance. Work with NHL-B cell lines has shown that inhibition of the expression of the IL-14 gene results in diminished cell growth and eventual cell death.

interleukin-15 (IL-15)

A T cell growth factor that shares many of the biological properties of IL-2. IL-15 enhances peripheral blood T cell proliferation, and *in vitro* studies demonstrate its ability to induce cytotoxic T cells. IL-15 mRNA has not been found in activated peripheral T cells, but monocyte-enriched peripheral blood cell lines as well as placental and skeletal muscle tissues do express IL-15. The IL-2 receptor and IL-15 receptor also share a common component for successful signal transduction.

intermediate filaments

These are 7- to 11-nm diameter intracellular filaments observed by electron microscopy that are lineage specific. They are intermediate in size between actin microfilaments, which are 6 nm in diameter, and microtubules, which are 25 nm in diameter. They are detected in cell and tissue preparations by monoclonal antibodies specific for the filaments and are identified by the immunoperoxidase method. The detection of various types of intermediate filaments in tumors is of great assistance in determining the histogenetic origin of many types of neoplasms.

internal image

According to the Jerne network theory, antibodies are produced against the antibodies induced by an external antigen. Some of the antiantibodies produced will bear idiotopes that precisely fit the paratope or antigen-binding site of the original antibody against the external antigen. Since they bear close structural similarity to the epitope on the antigen molecule that was originally administered, they are termed the internal image of the antigen.

International Unit of Immunological Activity

The use of an international reference standard of a biological preparation of antiserum or antigen of a precise weight and strength. The potency or strength of biological preparations such as antitoxins, vaccines, and test antigens derived from microbial products and antibody preparations may be compared against such standards to reflect their strength or potency.

interstitial nephritis

Inflammation characterized by mononuclear cell (lymphocytic) infiltrate in the interstitium surrounding renal tubules. This occurs following autoantibody reaction with tubular basement membrane in the kidney. Other etiologies include analgesic abuse.

intervening sequence

Refer to intron.

intestinal lymphangiectasia

Escape of immunoglobulins, as well as other proteins and lymphocytes, into the intestinal tract as a consequence of lymphatic dilation in the intestinal villi. The loss of immunoglobulin leads to secondary immunodeficiency. In addition to primary intestinal telangiectasia, obstruction of lymphatic drainage of the intestine produced by a lymphoma represents a secondary type.

intolerance

Adverse reactivity following administration of normal doses of a drug.

intracellular immunization

A recent term used to describe interference with wild-type virus replication by a dominant negative mutant viral gene. This has been suggested to be of possible use in protecting cells against HIV-1 infection because of the easy accessibility of CD4+ cells. By using *tat, gag,* and *rev* mutant genes and a mutant CD4 cell which bears the KDEL sequence, HIV envelope protein transport to the cell surfaces is inhibited.

intraepidermal lymphocytes

Primarily CD8+ T cells within the dermis. Murine intraepidermal lymphocytes express mainly the γδ receptor. It is believed that these cells have a more restricted repertoire of antigen receptors than those homing to extracutaneous sites. Most skin associated lymphocytes are found in the dermis with only about 2% in the epidermis. Intraepidermal lymphocytes express CLA-1, which may play a role in homing.

intraepithelial lymphocyte

A lymphocyte present in the intestinal epithelium or other specialized epithelial layer.

intravenous immune globulin (IVIG)

An immunoglobulin preparation comprised principally of IgG derived from the blood plasma of about 1000 donors. This preparation may be effective against hepatitis A and B, cytomegalovirus, rubella, varicella-zoster, tetanus, and various other agents. IVIG is administered to children with common variable immunodeficiency, X-linked agammaglobulinemia, or other defects of the antibody limb of the immune system. Even though AIDS is a CD4+ T cell defect, IVIG may help to protect against microbial infection in HIV-I-infected children. It may also be effective in autoimmune or idiopathic thrombocytopenia, autoimmune neutropenia, or Kawasaki's disease. IVIG may induce anaphylactic reactions and fever, as well as headache, muscle aches, and cardiovascular effects on blood pressure and heart rate.

intrinsic affinity

Synonym for intrinsic association constant.

intrinsic association constant

The association constant that describes univalent ligand binding to a special site on a protein macromolecule if all sites of this type are identical and noninteracting when found on the same molecule. Equilibrium dialysis and other techniques that determine bound and free ligand concentration evaluate the intrinsic association constant.

intrinsic asthma

Nonallergic or idiopathic asthma that usually occurs first during adulthood and follows a respiratory infection. Patients experience chronic or recurrent obstruction of bronchi associated with exposure to pollen or other allergens. This is in marked contrast to patients with extrinsic (allergic) asthma mediated by immune (IgE) mechanisms in the bronchi. Intrinsic asthma patients have negative skin tests to ordinary allergens when the IgE content of their serum is normal. They do manifest eosinophilia. There is no family history of atopic diseases.

intrinsic factor

A glycoprotein produced by parietal cells of the gastric mucosa that is necessary for vitamin B_{12} absorption. A lack of intrinsic factor leads to vitamin B_{12} deficiency and pernicious anemia.

intrinsic factor antibodies

Antibodies found in three fourths of pernicious anemia patients that are specific for either the binding site (type I or "blocking antibodies") or some other intrinsic factor antigenic determinant (type II antibodies).

intron

Structural gene segment that is not transcribed into RNA. Introns have no known function and are believed to be derived from "junk" DNA.

inulin

A homopolysaccharide of D-fructose that is found in selected plants such as dahlias. Although it is used to measure renal clearance, soluble inulin is known to activate the alternative complement pathway.

Inv

Former designation for human κ light polypeptide chain allotype. Allotypic epitopes in the immunoglobulin κ light chain constant region. Km replaced Inv.

Inv allotypes

Original terminology for Km allotypes, which is now the preferred nomenclature.

Inv allotypic determinant

Refer to Km allotypic determinant.

Inv marker

Refer to Km allotypic determinant.

invariant (Ii) chain

A nonpolymorphic, 31-kD glycoprotein that associates with class II histocompatibility molecules in the endoplasmic reticulum. It inhibits the linking of endogenous peptides with the class II molecule, conveying it to appropriate intracellular compartments. Truncation of the invariant chain stimulates a second signal that may function in the trans-Golgi network, prior to the conveyance of MHC class II molecules to the cell surface.

inverted repeat

Complementary sequence segments on a single strand of DNA. A palindrome when an inverted repeat's halves are placed side by side.

Complementary segments of sequence on a single strand of DNA

e.g., in duplex DNA

Inverted Repeat

ion exchange chromatography

A method that permits the separation of proteins in a solution, taking advantage of the net charge differences among them. It involves the electrostatic binding of proteins onto a charged resin suspended in buffer and packed into a column. Since serum proteins vary in their charge, binding to or elution from the column is possible by gradually increasing or decreasing the salt concentration (with or without changes in pH). This affects the type of proteins binding to the resin. With buffers of low molarity and pH greater than 6.5, the IgG in solution is not adsorbed on the column and passes through with the first buffer volume.

Ir genes

Immune response genes. Class II MHC genes that control immune responses are found in the Ir region. These genes govern the ability of an animal to respond immunologically to any particular antigen.

irradiation chimera

An animal or human whose lymphoid and myeloid tissues have been destroyed by lethal irradiation and successfully repopulated with donor bone marrow cells that are genetically different.

ISCOMs

Immune stimulatory complexes of antigen bound in a matrix of lipid that serves as an adjuvant and facilitates uptake of antigen into the cytoplasm of a cell following plasma membrane-lipid fusion.

islet cell transplantation

An experimental method aimed at treatment of type I diabetes mellitus. The technique has been successful in rats, but less so in man. It requires sufficient functioning islets from a minimum of two cadaveric donors that have been purified, cultured, and shown to produce insulin. The islet cells are administered into the portal vein.

The liver serves as the host organ in the recipient who is treated with FK506 or other immunosuppressant drugs.

islet of Langerhans

Groups of endocrine cells within the exocrine pancreas that consist of α cells that secrete glucagon, β cells that secrete insulin, and δ cells that secrete somatostatin.

isoagglutinin

An alloantibody present in some individuals of a species that is capable of agglutinating cells of other members of the same species.

isoallergens

Allergenic determinants with similar size, amino acid composition, peptide fingerprint, and other characteristics. They are present in a given material, but each is able individually to sensitize a susceptible subject. They are molecular variants of the same allergen.

isoallotypic determinant (nonmarkers)

An antigenic determinant present as an allelic variant on one immunoglobulin class or immunoglobulin subclass heavy chain that occurs on every molecule of a different immunoglobulin class or subclass heavy chain.

isoantibody

An antibody that is specific for an antigen present in other members of the species in which it occurs. Thus, it is an antibody against an isoantigen. Also called alloantibody.

isoantigen

An antigen found in a member of a species that induces an immune response if injected into a genetically dissimilar member of the same species. These are antigens carrying identical determinants in a given individual. Isoantigens of two individuals may or may not have identical determinants. In the latter case they are allogeneic with respect to each other and called alloantigens.

Since the individual red blood cell antigens have the same molecular structure and are identical in different individuals, they have been referred to in the past as isoantigens. This is only a descriptive term and should not be used, because two individuals may be allogeneic by virtue of the assortment of the antigens present on their red blood cells. An isoantigen is an antigen of an isograft.

isoelectric focusing (IEF)

An electrophoretic method to fractionate amphoteric molecules, especially proteins, according to their distribution in a pH gradient in an electric field created across the gradient. Molecular distribution is according to isoelectric pH values. The anode repels proteins that are positively charged and the cathode repels proteins that are negatively charged. Thus, each protein migrates in the pH gradient and bands at a position where the gradient pH is equivalent to the isoelectric pH of the protein. A chromatographic column is used to prepare a pH gradient by the electrolysis of amphoteric substances. A density gradient or a gel is used to stabilize the pH gradient. Proteins or peptides focus into distinct bands at that part of the gradient that is equivalent to their isoelectric point. Isoelectric focusing is a technique that permits the separation of protein substances on the basis of their isoelectric characteristics. Thus, this technique can be employed to define heterogeneous antibodies. It may also be employed to purify homogeneous immunoglobulins from heterogeneous pools of antibody.

isoelectric point (pI)

The pH at which a molecule has no charge, as the number of positive and negative charges are equal. At the isoelectric pH, the molecule does not migrate in an electric field. The solubility of most substances is minimal at their isoelectric point.

isogeneic (isogenic)

An adjective implying genetic identity such as identical twins. Although used as a synonym for syngeneic when referring to the genetic relationship between members of an inbred strain (of mice), the inbred animals never show the absolute identity, i.e., identical genotypes, observed in identical twins.

isograft

A tissue transplant from a donor to an isogenic recipient. Grafts exchanged between members of an inbred strain of laboratory animals such as mice are syngeneic rather than isogenic.

isohemagglutinins

An antibody in some members of a species that recognizes erythrocyte isoantigens on the surfaces of red blood cells from other members of the same species. In the ABO blood group system, the anti-A antibodies in the blood sera of group B individuals and the anti-B antibodies in the blood sera of group A individuals are examples of isohemagglutinins.

isoimmunization

An immune response induced in the recipient of a blood transfusion in which the donor red blood cells express isoantigens not present in the recipient. The term also refers to maternal immunization by fetal red blood cells bearing isoantigens the mother does not possess.

isoleukoagglutinins

An antibody in the blood sera of multiparous females and of patients receiving multiple blood transfusions that recognizes surface isoantigens of leukocytes and leads to their agglutination.

isologous

Derived from the same species. Also called isogeneic or syngeneic.

isophile antibody

An antibody induced by and specifically reactive with erythrocytes, but it is not reactive with other species' red blood cells. These antibodies are against antigens of red blood cells unique to the species from which they were derived.

isophile antigen

Antigen that is species specific; often refers to erythrocyte antigens.

isoproterenol

(dl-β-[3,4-dihydroxyphenyl]-α-isopropylaminoethanol). A β adrenergic amine that is used to treat patients with asthma. It is able to relax the bronchial smooth muscle constriction that occurs in asthmatics.

isoschizomer

One of several restriction endonucleases derived from different organisms that identify the same DNA base sequence for cleavage but do not always cleave DNA at the same location in the sequence. Target sequence methylation affects the action of isoschizomers, which are valuable in investigations of DNA methylation.

isotope

An isotypic determinant or epitope of an isotype.

isotopic labeling (radionuclide labeling)

The introduction of a radioactive isotope into a molecule by either external labeling through tagging molecules with ^{125}I or other appropriate isotope or by internal labeling in which ^{14}C- or ^{3}H-labeled amino acids are added to tissue culture which allows the cells to incorporate the isotope. Once labeled, molecules can be easily traced and their fate monitored by measuring radioactivity.

isotype

Antigens that determine the class or subclass of heavy chains or the type and subtype of light chains of immunoglobulin molecules. Every normal member of a species expresses each isotype. An immunoglobulin subtype is found in all normal individuals. Among the immunoglobulin classes, IgG and IgA have subclasses that are

designated with Arabic numerals. They are distinguished according to domain number and size, as well as the constant region's number of both intrachain and interchain disulfide bonds. The four isotypes of IgG are designated IgG1, IgG2, IgG3, and IgG4. The two IgA isotypes are designated IgA1 and IgA2. The μ, δ, and ε heavy chains and the κ and λ light chains each have one isotype.

Isotype

isotype switching

The mechanism whereby a cell changes from synthesizing a heavy polypeptide chain of one isotype to that of another, as from μ chain to γ chain formation in B cells that have received a switch signal from a T cell.

isotypic determinant

Immunoglobulin epitope present in all normal individuals of a species. Isotypic determinants of immunoglobulin heavy and light chains determine immunoglobulin class and subclass and light chain type.

isotypic specificities

Species-specific variability of antibody (human, mouse, rabbit, etc.). Examples of isotypes include IgG, IgM, and κ light chains.

isotypic variation

Differences among antigens found in members of a species such as the epitopes that differentiate immunoglobulin classes and subclasses and light chain types among immunoglobulin chains.

ITP

Abbreviation for idiopathic thrombocytopenic purpura.

IVIG

Abbreviation for intravenous immunoglobulin.

J

J chain

A 17.6-kD polypeptide chain present in polymeric immunoglobulins that include both IgM and IgA. It links four-chain immunoglobulin monomers to produce the polymeric immunoglobulin structure. J chains are produced in plasma cells and are incorporated into IgM or IgA molecules prior to their secretion. Incorporation of the J chain appears essential for transcytosis of these immunoglobulin molecules to external secretions. The J chain comprises 2 to 4% of an IgM pentamer or a secretory IgA dimer. Tryptophan is absent from both mouse and human J chains. J chains are comprised of 137 amino acid residues and a single complex N-linked oligosaccharide on asparagine. Human J chain contains three forms of the oligosaccharide which differ in sialic acid content. The J chain is fastened through disulfide bonds to penultimate cysteine residues of μ or α heavy chains. The human J chain gene is located on chromosome 4q21, whereas the mouse J chain gene is located on chromosome 5.

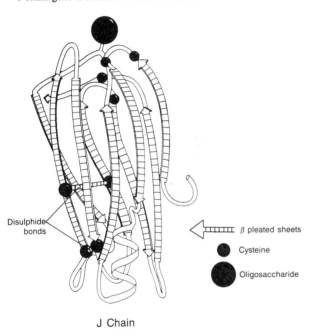

Disulphide bonds

β pleated sheets

Cysteine

Oligosaccharide

J Chain

J exon

A DNA sequence that encodes part of the third hypervariable region of a light or heavy chain located near the 5′ end of the κ, λ, and γ constant region genes. An intron separates the J exon from them. The J exon should not be confused with the J chain. The H constant region gene is associated with several J exons. The V region gene is translocated to a site just 5′ to one of the J exons during stem cell differentiation to a lymphocyte.

J gene segment

DNA sequence that codes for the carboxy terminal 12 to 21 residues of T lymphocyte receptor or immunoglobulin polypeptide chain variable regions. Through gene rearrangement, a J gene segment unites either a V or a D gene segment to intron 5′ of the C gene segment.

J region

The variable part of a polypeptide chain, comprising a T lymphocyte receptor or immunoglobulin that a J gene segment encodes. An immunoglobulin light chain's J region is comprised of the third hypervariable region carboxy terminal (1 or 2 residues) and the fourth framework region (12 to 13 residues). An immunoglobulin heavy chain's J region is comprised of the third hypervariable region carboxy terminal portion and the fourth framework region (15 to 20 residues). The heavy chain's J region is slightly longer than that of the light chain. The variable region carboxy terminal portion represents the J region of the T cell receptor.

Jarisch-Herxheimer reaction

A systemic reaction associated with fever, lymphadenopathy, skin rash, and headaches that follows the injection of penicillin into patients with syphilis. It is apparently produced by the release of significant quantities of toxic or antigenic substances from multiple *Treponema pallidum* microorganisms.

Jenner, Edward (1749–1823)

English physician best known for his use of vaccinia (cowpox-virus) for preventive vaccination against smallpox. Jenner was also interested in animal behavior and zoology. *Inquiry into the Cause and Effects of the Variolae Vaccinal,* 1798.

Jerne network theory

Niels Jerne's hypothesis that antibodies produced in response to a specific antigen would themselves induce a second group of antibodies which would in turn downregulate the original antibody-producing cells. The second antigen (Ab-2) would recognize epitopes of antibody 1's antibody-binding region. These would be antiidiotypic antibodies. Such antiidiotypic antibodies would also be reactive with the antigen-binding region of T cell receptors for which they were specific. Thus, a network of antiantibodies would produce a homeostatic effect on the immune response to a particular antigen. This theory was subsequently proven and confirmed by numerous investigators.

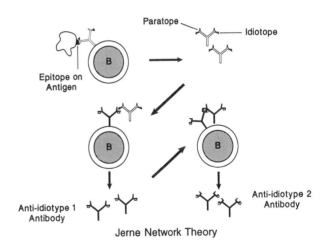

Paratope

Idiotope

Epitope on Antigen

Anti-idiotype 1 Antibody

Anti-idiotype 2 Antibody

Jerne Network Theory

Jerne, Niels Kaj (1911–)

Immunologist, born in London and educated at Leiden and Copenhagen, who received the Nobel Prize in 1984 for his contributions to immune system theory. He studied antibody synthesis and avidity, perfected the hemolytic plaque assay, developed natural selection theory of antibody formation, and formulated the idiotypic network theory.

Jerne plaque assay (see figure, page 176)

Refer to hemolytic plaque assay.

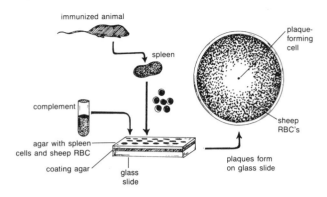

Jerne Plaque Assay

Jo1 syndrome

A clinical condition in which anti-Jo1 antigen (histidyl-tRNA synthetase) antibodies are produced. Arthritis and myositis, as well as interstitial lung disease, may be present. One quarter of myositis patients may manifest anti-Jo1 antibodies.

Job's syndrome

Cold staphylococcal abscesses or infections by other agents that recur. There is associated eczema, elevated levels of IgE in the serum, and phagocytic dysfunction associated with glutathione reductase and glucose-6-phosphatase deficiencies. The syndrome has an autosomal recessive mode of inheritance.

bronchus

Job's syndrome
with bronchogenic abscess

Johnin

An extract from culture medium in which *Mycobacterium johnei* is growing. It can be used in a skin test of cattle for the diagnosis of Johne's disease. Its preparation parallels the extraction of old tuberculin or purified protein derivative (PPD) used in the tuberculin test.

Jones criteria

Signs and symptoms used in the diagnosis of acute rheumatic fever.

Jones-Mote reaction

Delayed-type hypersensitivity to protein antigens associated with basophil infiltration, which gives the reaction the additional name of *cutaneous basophil hypersensitivity*. Compared to the other forms of delayed-type hypersensitivity, it is relatively weak and appears on challenge several days following sensitization with minute quantities of protein antigens in aqueous medium or in incomplete Freund's adjuvant. No necrosis is produced. Jones-Mote hypersensitivity can be produced in laboratory animals such as guinea pigs appropriately exposed to protein antigens in aqueous media or in incomplete Freund's adjuvant. It can be passively transferred by T lymphocytes.

jugular bodies

Nodules ventral to external jugular veins that contain lobules of lymphoid cells separated by sinusoids paved with phagocytic cells. Jugular bodies filter the blood, but are not a part of the lymphoid system. They may be found in selected amphibian species.

junctional diversity (see figure, bottom of page)

When gene segments join imprecisely, the amino acid sequence may vary and affect variable region expression. This can alter codons at gene segment junctions. These include the V-J junction of the genes encoding immunoglobulin κ and λ light chains and the V-D, D-J, and D-D junctions of genes encoding immunoglobulin heavy chains or the genes encoding T cell receptor β and δ chains.

juvenile onset diabetes

Synonym for type I insulin-dependent diabetes mellitus.

juvenile rheumatoid arthritis

When a child has inflammation in one or more joints that persists for a minimum of 3 months with no other explanation for arthritis. It may be associated with uveitis, pericarditis, rheumatoid nodules, fever, and rash; however, rheumatoid factor (RF) is found in the serum in less than 10% of these patients. Subgroups include (1) polyarticular disease divided into group 1, which is HLA-DR4 associated, and group 2, which is HLA-DR5 and DR8 associated; (2) systemic Still's disease with hepatosplenomegaly, enlarged lymph nodes, pericardial inflammation, rash, and fever; and (3) Pauciarticular disease in which one to nine joints are affected and associated with uveitis and anti-DNA antibodies, principally in young women.

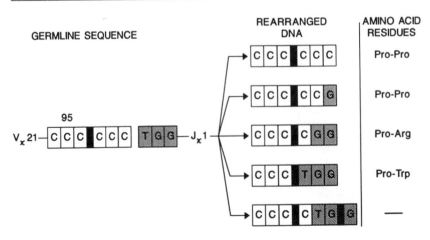

Junctional Diversity

K

K antigens

Surface epitopes of Gram-negative microorganisms. They are either proteins (fimbriae) or acid polysaccharides found on the surface of *Klebsiella* and *Escherichia coli* microorganisms. K antigens are exterior to somatic O antigens. They are labile to heat and crossreact with the capsules of other microorganisms such as *Hemophilus influenzae, Streptococcus pneumoniae,* and *Neisseria meningitidis.* K antigens may be linked to virulent strains of microorganisms that induce urinary tract infections. Anti-K antibodies are only weakly protective.

K cell

The K (killer) cells, also called null cells, have lymphocyte-like morphology, but functional characteristics different from those of B and T cells. They are involved in a particular form of immune response, the antibody-dependent cellular cytotoxicity

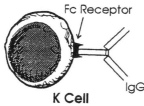

K Cell

(*ADCC*), killing target cells coated with IgG antibodies. A K cell is an Fc-bearing killer cell that has an effector function in mediating antibody-dependent cell-mediated cytotoxicity. An IgG antibody molecule binds through its Fc region to the K cell's Fc receptor. Following contact with a target cell bearing antigenic determinants on its surface for which the Fab regions of the antibody molecule attached to the K cell are specific, the lymphocyte-like K cell releases lymphokines that destroy the target. This represents a type of immune effector function in which cells and antibody participate. Besides K cells, other cells that mediate antibody-dependent cell-mediated cytotoxicity include natural killer (NK) cells, cytotoxic T cells, neutrophils, and macrophages.

κ chain

One of two types of light polypeptide chains present in immunoglobulin molecules of human and other species. κ Light chains are found in approximately 60% of human immunoglobulin, whereas λ light chains are present in approximately 40% of human immunoglobulin molecules. A single immunoglobulin molecule contains either κ or λ light chains, not one of each.

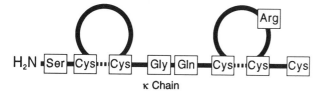

κ Chain

κ light chain deficiency

A rare condition in which point mutations in the $C_κ$ gene at chromosome 2p11 cause an absence of κ light chains in the serum and generate B lymphocytes whose surfaces are bereft of κ light chains.

K region

The K, as well as the D, region refer to class I MHC segments in the murine genome.

K562 cells

A chronic myelogenous leukemia cell line that serves as a target cell in a ^{51}Cr release assay of natural killer (NK) cells. Following incubation of NK cells with ^{51}Cr-labeled target K562 cells, the amount of chromium released into the supernatant is measured, and the cytotoxicity is determined by use of a formula.

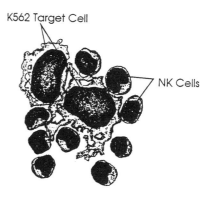

Kabat, Elvin Abraham (1914–)

American immunochemist. With Tiselius he was the first to separate immunoglobulins electrophoretically. He also demonstrated that γ globulins can be distinguished as 7S or 19S. Other contributions include research on antibodies to carbohydrates, the antibody combining site, and the discovery of immunoglobulin chain variable regions. He received the National Medal of Science. *Experimental Immunochemistry (with Mayer),* 1948; *Blood Group Substances: Their Chemistry and Immunochemistry; Structural Concepts in Immunology and Immunochemistry,* 1956; 1968.

kallikrein

An enzyme that splits kininogens to generate bradykinin which has an effect on pain receptors and smooth muscle and exerts a chemotactic effect on neutrophils. Bradykinin is a nonapeptide that induces vasodilation and increases capillary permeability. Kallikreins, also known as kininogenases, are present both in the plasma and in tissue and also in glandular secretions such as saliva, pancreatic juice, tears, urine, and others. Trypsin, pepsin, proteases of snake venoms, and bacterial products are also able to hydrolyze kininogens, but the substrate specificity and potency of kallikreins is greater. Plasma and tissue kallikreins are physically and immunologically different. It is not known whether plasma contains more than one form of kallikrein.

kallikrein-kinin system

Vasopressive peptides that control blood pressure through maintenance of regional blood flow and the excretion of water and electrolytes. Kallikrein causes the release of renin and the synthesis of kinins which interact with the immune system and increase urinary sodium excretion, as well as act as powerful vasodilators.

kallikrein inhibitors

Natural inhibitors of kallikreins belong to the group of natural inhibitors of proteolysis. They also inhibit other proteolytic enzymes, but each has its own preference for one protease or another. Apronitin, also known by its registered name of Trasylol®, is particularly active on tissue kallikreins.

karyotype

The number and shape of chromosomes within a cell. The karyotype may be characteristic for a particular species.

Kawasaki's disease

Mucocutaneous lymph node syndrome that occurs in children under 5 years of age. The incubation period may be 1 to 2 weeks. It is an acute febrile disease characterized by erythema of the conjunctiva and oral cavity, skin rash, and swollen (especially cervical) lymph nodes. It occurs mostly in Japan, with some cases in the U.S. Cardiac lesions such as coronary artery aneurysms may be found in 70% of the patients. Coronary arteritis causes death in 1 to 2% of patients. There is necrosis and inflammation of the vessel wall. The etiology is unknown, but it has been suggested to be infection by a retrovirus. Associated immunoregulatory disorders include T and B lymphocyte activation, circulating immune complexes, and autoantibodies cytolytic for endothelial cells activated by cytokines.

keratinocyte growth factor (KGF)

A 19-kD protein comprised of 163 amino acid residues that binds heparin and facilitates the growth of keratinocytes and other epithelial cells that comprise 95% of the epidermis.

kernicterus

Deposition in the skin, leading to yellowish discoloration, as well as deposition in the central nervous system of erythrocyte breakdown products in the blood of infants with erythroblastosis fetalis. It may lead to neurologic dysfunction.

keyhole limpet hemocyanin (KLH)

A traditional antigen widely used as a carrier in studies on immune responsiveness to haptens. This respiratory pigment containing copper is found in mollusks and crustaceans. It is usually immunogenic in vertebrate animals.

Ki-1 (CD30 antigen)

A marker of Reed-Sternberg cells found in Hodgkin's disease of the mixed-cellularity, nodular-sclerosing, and lymphocyte-depleted types and in selected cases of large cell non-Hodgkin's lymphomas.

Ki-67 or -780

Nuclear antigens expressed by both normal and neoplastic-proliferating cells. They are demonstrable by immunoperoxidase staining. A relatively high percentage of positive cells in a neoplasm implies an unfavorable prognosis.

killed vaccine

An immunizing preparation comprised of microorganisms, either bacterial or viral, that are dead, but retain their antigenicity, making them capable of inducing a protective immune response with the formation of antibodies and/or stimulation of cell-mediated immunity. Killed vaccines do not induce even a mild case of the disease which is sometimes observed with attenuated (greatly weakened, but still living) vaccines. Although the first killed vaccines contained intact dead microorganisms, some modern preparations contain subunits or parts of microorganisms to be used for immunization. Killed microorganisms may be combined with toxoids, as in the case of the DPT (diphtheria-pertussis-tetanus) preparations administered to children.

killer cell (K cell)

A large granular lymphocyte bearing Fc receptors on its surface for IgG, which makes it capable of mediating antibody-dependent cell-mediated cytotoxicity. Complement is not involved in the reaction. Antibody may attach through its Fab regions to target cell epitopes and link to the killer cell through attachment of its Fc region to the K cell's Fc receptor, thereby facilitating cytolysis of the target by the killer cell, or an IgG antibody may first link via its Fc region to the Fc receptor on the killer cell surface and direct the K cell to its target. Cytolysis is induced by insertion of perforin polymer in the target cell membrane in a manner that resembles the insertion of C9 polymers in a cell membrane in complement-mediated lysis. Perforin is showered on the target cell membrane following release from the K cell.

kilobase (kb)

1000 DNA or RNA base pairs.

kininases

An enzyme in the blood that degrades kinins to inactive peptides. Inactivation occurs when any of the eight bonds in the kinin is cleaved. There are two kininases of plasma. Kininase I, or carboxy peptidase N, cleaves the C-terminal arginine of kinins and of anaphylatoxins. It differs from pancreatic carboxy peptidase B with respect to mol wt, subunit structure, carbohydrate content, antigenic properties, substrate specificity, and inhibition pattern. The purified enzyme has a mol wt of 280 kD. The site of synthesis of carboxy peptidase N is believed to be the liver. Kininase II, a peptidyl dipeptidase, cleaves the C-terminal Phe-Arg of kinins and also liberates angiotensin II from angiotensin I. (Angiotensin is another vasoactive substance.) The vascular endothelium of both lung and peripheral vascular bed are rich in this enzyme. Peptidyl dipeptidase is present also outside the circulatory system and has different functions at these sites. Other enzymes with kininase activity are present in the spleen and kidney (cathepsin) and the endothelial cells of the gastrointestinal tract.

kininogens

The precursors of kinins, kininogens are glycoproteins synthesized in the liver. Plasma kininogens comprise two, possibly three,

classes of compounds with species variation. (1) Low mol wt kininogens (LMK) are acidic proteins with a mol wt of about 57 kD and are susceptible to conversion into kinins by kininogen-converting enzymes (kallikreins) of tissue origin. LMK has two forms, I and II, which represent the main plasma kininogens. (2) High mol wt kininogens (HMK) are α glycoproteins with a mol wt of 97 kD which can be converted into kinins both by plasma and tissue kallikreins. HMK exists in two forms, a and b, which differ both in enzyme sensitivity and generated kinin.

kinins

A family of straight chain polypeptides generated by enzymatic hydrolysis of plasma α-2 globulin precursors, collectively called kininogens. They exert potent vasomotor effects, causing vasodilatation of most vessels in the body, but vasoconstriction of the pulmonary bed. They also increase vascular permeability and promote the diapedesis of leukocytes.

Kitasato, Shibasaburo (1892–1931)

Co-discoverer, with Emil von Behring, of antitoxin antibodies.

KLH

Abbreviation for keyhole limpet hemocyanin.

Km (formerly Inv)

The designation for the κ light chain allotype genetic markers.

Km allotypes

Three Km allotypes have been described in human immunoglobulin κ light chains. They are designated Km1, Km1,2, and Km3. They are encoded by alleles of the gene that codes for the human κ light chain constant regions. Allotype differences are based on the amino acid residue at positions 153 and 191, which are in proximity to one another in a folded immunoglobulin $C_κ$ domain. One person may have a maximum of two out of the three Km allotypes on their light chains. To fully express Km determinants, the heavy immunoglobulin chains should be present, probably to maintain appropriate three-dimensional configuration.

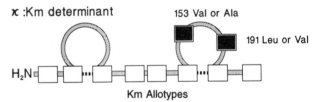

Km Allotypes

Koch, Heinrich Hermann Robert (1843–1910)

German bacteriologist awarded the Nobel Prize in 1905 for his work on tuberculosis. Koch made many contributions to the field of bacteriology. Along with his postulates for proof of etiology, Koch instituted strict isolation and culture methods in bacteriology. He studied the life cycle of anthrax and discovered both the cholera vibrio and the tubercle bacillus. The Koch phenomenon and Koch-Weeks bacillus both bear his name.

Koch phenomenon

Delayed hypersensitivity reaction in the skin of a guinea pig after it has have been infected with *Mycobacterium tuberculosis*. Robert Koch described the phenomenon in 1891 following the injection of either living or dead *M. tuberculosis* microorganisms into guinea pigs previously infected with the same microbes. He observed a

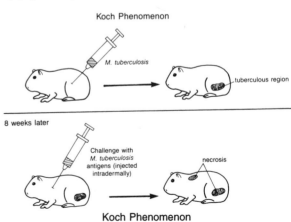

Koch Phenomenon

severe necrotic reaction at the site of inoculation, which occasionally became generalized and induced death. The injection of killed *M. tuberculosis* microorganisms into healthy guinea pigs caused no ill effects. This is a demonstration of cell-mediated immunity and is the basis for the tuberculin test.

Kohler, Georges J.F. (1946–)

German immunologist who shared the Nobel Prize in 1984 with Cesar Milstein for their work in the production of monoclonal antibodies by hybridizing mutant myeloma cells with antibody producing B cells (hybridoma technique).

Ku antibodies

Antibodies detectable in 15 to 50% of individuals with Sjögren's syndrome, systemic lupus erythematosus, mixed connective tissue disease and scleroderma. They are found in 5 to 15% of myositis patients in the USA.

Kunkel, Henry George (1916–1983)

American physician and immunologist. The primary focus of his work was immunoglobulins. He characterized myeloma proteins as immunoglobulins and rheumatoid factor as an autoantibody. He also discovered IgA and contributed to immunoglobulin structure and genetics. Kunkel received the Lasker Award and the Gairdner Award.

Kupffer cell

A liver macrophage that has become fixed as a mononuclear phagocytic cell in the liver sinusoids. It is an integral part of the reticuloendothelial system. Monocytes become attached to the interior surfaces of liver sinusoids where they develop into macrophages. They have CR1 and CR2 receptors, surface Fc receptors, and MHC class II molecules. They are actively phagocytic and remove foreign substances from the blood as it flows through the liver. Under certain disease conditions, they may phagocytize erythrocytes, leading to the deposition of hemosiderin particles derived from hemoglobulin breakdown products.

kuru

A slow virus disease of some native tribes of Guinea that practice cannibalism. Transmission is through skin lesions of individuals preparing infected brains for consumption. The virus accumulates in brain cell membranes.

Kveim reaction (historical)

A skin reaction for the diagnosis of sarcoidosis in which ground lymph node tissue of a known sarcoidosis patient is suspended in physiological salt solution and inoculated intracutaneously into a suspected sarcoidosis patient. A positive reaction, on histopathologic examination of an injection site biopsy 1 month to 6 weeks after inoculation, reveals a nodular epithelioid cell granuloma-like reaction. A positive Kveim test confirms the diagnosis of sarcoidosis. The danger of possibly transmitting hepatitis, AIDS, and other viruses precludes the use of this reaction.

L

L cell conditioned medium

A powerful growth factor for macrophages, termed macrophage colony-stimulating factor (MCSF), present in L cell cultures.

L chain

A 22-kD polypeptide chain found in all immunoglobulin molecules. There are two types designated κ or λ. Each four-chain immunoglobulin monomer contains either two κ or two λ light chains. The two types of light chain never occur in one molecule under natural conditions.

L-selectin

Molecule found on lymphocytes that is responsible for the homing of lymphocytes to lymph node high endothelial venules. L-selectin is also found on neutrophils where it acts to bind the cells to activated endothelium early in the inflammatory process. Also called CD62L.

λ5 B cell development

Refer to VpreB protein.

λ chain

One of the two light polypeptide chain types found in immunoglobulin molecules. The κ light chain is the other type. Each immunoglobulin molecule contains either two λ or two κ light chains. The κ to λ light chain ratio differs among species. Approximately 60% of IgG molecules in humans are κ and 40% are λ.

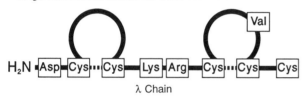

λ Chain

λ cloning vector

A genetically engineered λ phage that can accept foreign DNA and be used as a vector in recombinant DNA studies. Phage DNA is cleaved with restriction endonucleases, and foreign DNA is inserted. Insertion vectors are those with a single site where phage DNA is cleaved and foreign DNA inserted. Substitution or replacement vectors are those with two sites which span a DNA segment that can be excised and replaced with foreign DNA.

L$_+$ dose (historical)

The smallest amount of toxin which, when mixed with one unit of antitoxin and injected subcutaneously into a 250-g guinea pig, will kill the animal within 4 days. This is the unit used for standardization of antitoxin.

L2C leukemia

A B cell neoplasm of guinea pigs that is transplantable.

L3T4

A CD4 marker on mouse lymphocytes that signifies the T helper/inducer cell. It is detectable by specific monoclonal antibodies and is equivalent to the CD4$^+$ lymphocyte in man.

L3T4+ T lymphocytes

Murine CD4$^+$ T cells.

lactalbumin

A breast epithelial cell protein demonstrable by immunoperoxidase staining that is found in approximately one half to two thirds of breast carcinomas for which it is relatively specific. More than 50% of metastatic breast tumors and some salivary gland and skin appendage tumors stain positively for lactalbumin.

lacteals

Minute lymphatics that drain intestinal villi.

lactoferrin

A protein that combines with iron and competes with microorganisms for it. This represents a nonantibody humoral substance that contributes to the body's natural defenses against infection. It is present in polymorphonuclear neutrophil granules as well as in milk. By combining with iron molecules, it deprives bacterial cells of this needed substance.

lactoperoxidase

An enzyme present in milk and saliva that may be inhibitory to a number of microorganisms and serves as a nonantibody humoral substance that contributes to nonspecific immunity. Its mechanism of action resembles that of myeloperoxidase.

LAF (lymphocyte-activating factor)

Refer to interleukin-1.

LAK cells

Lymphokine-activated killer cells.

LAM-1 (leukocyte adhesion molecule 1)

A homing protein found on membranes that combines with target cell-specific glycoconjugates. It helps to regulate migration of leukocytes through lymphocyte binding to high endothelial venules and to regulate neutrophil adherence to endothelium at inflammatory sites.

lamin antibodies

Antibodies present in the sera of chronic autoimmune disease patients manifesting hepatitis, leukocytoclastic angiitis or brain vasculitis, cytopenia and circulating anticoagulant or cardiolipin antibodies. They form a rim-type antinuclear staining pattern in immunofluorescence assays. A minority of systemic lupus erythematosus patients develop antibodies to lamin.

lamina propria

The thin connective tissue layer that supports the epithelium of the gastrointestinal tract. The epithelium and lamina propria form the mucous membrane. The lamina propria may be the site of immunologic reactivity in the gastrointestinal tract, representing an area where lymphocytes, plasma cells, and mast cells congregate.

laminin

A relatively large (820 kD) basement membrane glycoprotein comprised of three polypeptide subunits. It belongs to the integrin

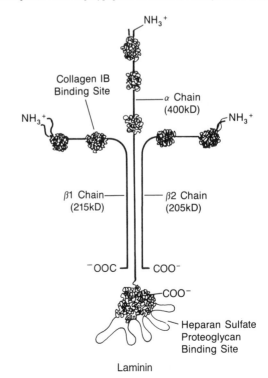

Laminin

receptor family which includes a 400-kD α heavy chain and two 200-kD light chains designated β-1 and β-2. By electron microscopy the molecule is arranged in the form of a cross. The domain structures of the α and β chains resemble one another. There are six primary domains. I and II have repeat sequences forming α helices. III and V are comprised of cysteine-rich repeating sequences. The globular regions are comprised of domains IV and VI. There is an additional short cysteine-rich α domain between the I and II domains in the β-1 chain. There is a relatively large globular segment linked to the C-terminal of domain I, designated the "foot" in the α chain. Five "toes" on the foot contain repeat sequences. Laminins have biological functions and characteristics that include facilitation of cellular adhesion and linkage to other basement membrane constituents such as collagen type IV, heparan, and glycosaminoglycans. Laminins also facilitate neurite regeneration, an activity associated with the foot of the molecule. There is more than one form of laminin, each representing different gene products, even though they possess a high degree of homology. S laminin describes a form found only in synaptic and nonmuscle basal lamina. This is a single 190-kD polypeptide (in the reduced form) and is greater than 1000 kD in the nonreduced form. It is associated with the development or stabilization of synapses. It is homologous to the β-1 chain of laminin. Laminin facilitates cell attachment and migration. It plays a role in differentiation and metastasis and is produced by macrophages. Macrophages, endothelial cells, epithelial cells, and Schwann cells produce it.

laminin receptor

A membrane protein comprised of two disulfide bond-linked subunits, one relatively large and one relatively small. Its function appears to be for attachment of cells and for the outgrowth of neurites. It may share structural similarities with fibronectin and vitronectin, both of which are also integrins.

Lancefield precipitation test

A ring precipitation test developed by Rebecca Lancefield to classify streptococci according to their group-specific polysaccharides. The polysaccharide antigen is derived by treatment of cultures of the microorganisms with HCl, formimide, or a *Streptomyces albus* enzyme. Antiserum is first placed into a serological tube, followed by layering the polysaccharide antigen over it. A positive reaction is indicated by precipitation at the interface.

Landsteiner, Karl (1868–1943)

Viennese pathologist and immunologist who later worked at the Rockefeller Institute for Medical Research in New York. Received the Nobel Prize in 1930 "for his discovery of the human blood groups". He was the first to infect monkeys with poliomyelitis and syphilis to allow controlled studies of those diseases. He established the immunochemical specificity of synthetic antigens and haptens. Landsteiner felt his most important contribution was in the area of antibody-hapten interactions. *Die Spezifizität der serologiochen Reaktionen,* 1933; *The Specificity of Serological Reactions,* 1945.

Landsteiner's rule (historical)

Landsteiner (1900) discovered that human red blood cells could be separated into four groups based on their antigenic characteristics. These were designated as blood groups O, A, B, and AB. He found naturally occurring isohemagglutinins in the sera of individuals specific for the (ABO) blood group antigen which they did not possess (i.e., anti-A and anti-B isohemagglutinin in group O subjects; anti-B in group A individuals; and anti-A in group B persons; neither anti-A nor anti-B in individuals of group AB). This principle became known as Landsteiner's rule.

lane

The path of migration of a molecule of interest from a well or point of application in gel electrophoresis. A substance is propelled within the confines of this path or corridor by an electric current which induces migration and separation of the molecules into bands according to size.

Langerhans cell

Dendritic-appearing accessory cells interspersed between cells of the upper layer of the epidermis. They can be visualized by gold chloride impregnation of unfixed sections and show dendritic processes, but no intercellular bridges. By electron microscopy, they lack tonofibrils or desmosomes; have indented nuclei; and contain tennis racket-shaped Birbeck granules, which are relatively small vacuoles, round to rectangular and measuring 10 nm. Following their formation from stem cells in the bone marrow, Langerhans cells migrate to the epidermis and then to the lymph nodes, where

they are described as dendritic cells based upon their thin cytoplasmic processes that course between adjacent cells. Langerhans cells express both class I and class II histocompatibility antigens, as well as C3b receptors and IgG Fc receptors on their surfaces. They function as antigen-presenting cells. Epidermal Langerhans cells express complement receptors 1 and 3, Fcγ receptors, and fluctuating quantities of CD1. Dendritic cells do not express Fcγ receptors or CD1. Langerhans cells in the lymph nodes are found in the deep cortex. Epidermal Langerhans cells are important in the development of delayed-type hypersensitivity through the uptake of antigen in the skin and transport of it to the lymph nodes. Veiled cells in the lymph are indistinguishable from Langerhans cells.

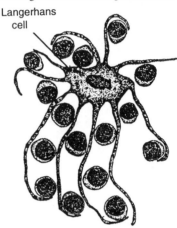

Langerhans cell

Langerhans Cell

lapinized vaccine

A preparation used for immunization that has been attenuated by passage through rabbits until its original virulence has been lost.

large granular lymphocytes (LGL)

Refer to natural killer cells.

large lymphocyte

A 12-μm or greater diameter lymphocyte.

large pyroninophilic blast cell

Cells that stain positively with methyl green pyronin stain. They are found in thymus-dependent areas of lymph nodes and other peripheral lymphoid tissues.

late-onset immune deficiency

A disease of unknown cause associated with gastric carcinoma, pernicious anemia, atrophic gastric autoimmunity, and several other conditions.

late-phase reaction (LPR)

An inflammatory response that begins approximately 5 to 8 h after exposure to antigen in IgE-mediated allergic diseases. In addition to inflammation, there is pruritus and minor cellular infiltration. Asthmatic patients may produce a delayed or secondary response following antigenic challenge which involves the release of histamines from neutrophils. This induces secondary degranulation of mast cells and basophils which stimulates bronchiole hyperreactivity. Whereas prostaglandin and then PGD₂ is produced in the primary response, it is not formed in the late-phase reaction. Cold air, ozone, viruses or other irritants may induce the late-phase reaction. This condition may be treated with β adrenergic aerosols.

latent allotype

The detection of an unexpected allotype in an animal's genetic constitution. This latent allotype is expressed as a temporary replacement of the nominal heterozygous allotype of the animal. This has been described among rabbit immunoglobulin allotypes. Whereas an F₁ rabbit would be expected to synthesize a1 or a2 heavy chain allotypes if one of its parents was homozygous for a1 and its other parent was homozygous for a2, the F₁ rabbit might under certain circumstances of stimulation produce immunoglobulin molecules of the a3 allotype, which would represent a latent allotype.

Latex fixation test

A technique in which latex particles are used as passive carriers of soluble antigens adsorbed to their surfaces. Antibodies specific for the adsorbed antigen then cause agglutination of the coated latex particles. This has been widely used and is the basis of a rheumatoid arthritis test in which pooled human IgG molecules are coated on the surface of latex particles, which are then agglutinated by antiimmuno-globulin antibodies in the sera of rheumatoid arthritis patients.

Latex Fixation Test

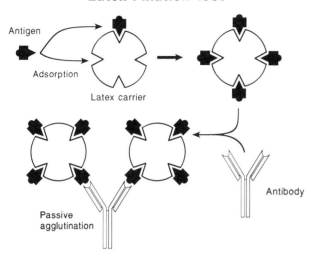

latex particles

Inert particles of defined size that are used as carriers of either antigens or antibodies in latex agglutination immunoassays. An example is the rheumatoid arthritis (RA) test in which latex particles are coated with pooled human IgG that serves as antigen. These IgG-coated particles are agglutinated by rheumatoid factor (antiimmunoglobulin antibody) that may be detected in a rheumatoid arthritis patient's serum.

LATS (long-acting thyroid stimulator)

See long-acting thyroid stimulator.

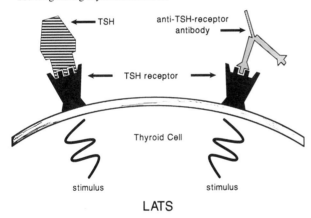

LATS

LATS protector

An antibody found in Graves' disease patients that inhibits LATS neutralization *in vitro*. This forms the basis for a LATS protector assay in which serum from Graves' patients is tested for the ability to "protect" a known LATS serum from being neutralized by binding to human thyroid antigen.

lattice theory

The concept that soluble antigen and antibody combine with each other in the precipitation reaction to produce an interconnecting structure of molecules. This structure has been likened to a criss-cross pattern of wooden strips fastened together to reveal a series of diamond-shaped structures. Lattice formation requires interaction of bivalent antibodies with multivalent antigens to produce a connecting linkage of many molecules to produce a complex whose density becomes sufficient to settle out of solution. The more epitopes recognized by the antibody molecules present, the more extensive is the complex formation.

Laurell crossed immunoelectrophoresis

Refer to crossed immunoelectrophoresis.

Laurell rocket test

A method to quantify protein antigens by rapid immunoelectro-phoresis. Antiserum is incorporated in agarose into which wells are cut and protein antigen samples are distributed. The application of an electric current at 90° angles to the antigen row drives antigen into the agar. Dual lines of immune precipitate emanate from each well and merge to form a point where no more antigen is present, producing a structure which resembles a rocket. The amount of antigen can be determined by measuring the rocket length from the well to the point of precipitate. This length is proportional to the total amount of antigen in the preparation.

LAV

Lymphadenopathy-associated virus (refer to HIV-1).

Lawrence, Henry Sherwood (1916–)

American immunologist. While studying type IV hypersensitivity and contact dermatitis, he discovered transfer factor. *Cellular and Humoral Aspects of Delayed Hypersensitivity,* 1959.

lazy leukocyte syndrome

A disease of unknown cause in which patients experience an increased incidence of pyogenic infections such as abscess formation, pneumonia, and gingivitis which is linked to defective neutrophil chemotaxis in combination with neutropenia. Random locomotion of neutrophils is also diminished and abnormal. This is demonstrated by the vertical migration of leukocytes in capillary tubes. There is also impaired exodus of neutrophils from the bone marrow.

LCA (leukocyte common antigen)

See leukocyte common antigen.

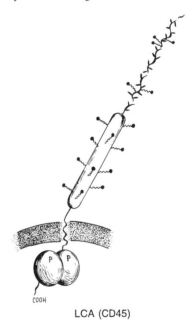

LCA (CD45)

LCAM

Leukocyte cell adhesion molecule.

lck, fyn, ZAP (phosphotyrosine kinases in T cells)

The PTKs associated with early signal transduction in T cell activation. Lck is a src type PTK. It is found on T cells, in physical association with CD4 and CD8 cytoplasmic regions. Deficiency of

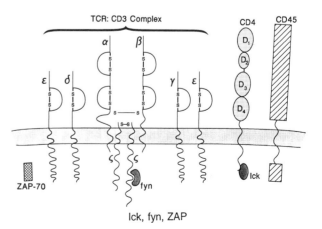

lck, fyn, ZAP

lck results in decreased stimulation of T cells and decreased T cell growth. Fyn is also a src PTK, however it is found on hematopoietic cells. Increased fyn results in enhanced T cell activation, but a deficiency of fyn has not been shown to decrease T cell growth. Fyn deficiency inhibits T cell activation in only some T cell subsets. ZAP or ζ associated protein kinase is like the syk PTK in B cells. It is found only on T cells and NK cells. Activity of the enzyme is dependent on association of ZAP with the ζ chain following TCR activation.

LCM

Refer to lymphocytic choriomeningitis.

LD$_{50}$

The dose of a substance such as a bacterial toxin or microbial suspension that leads to the death of 50% of a group of test animals within a certain period following administration. This has been employed to evaluate toxicity or virulence and to evaluate the protective qualities of vaccines administered to experimental animals.

LDCF (lymphocyte-derived chemotactic factor)

Lymphokines that are chemotactic factors, especially for mononuclear phagocytes.

LE cell

A neutrophil (PMN) in the peripheral blood or synovium of lupus erythematosus patients produced when the PMN phagocytizes a reddish-purple staining homogeneous lymphocyte nucleus with Wright's stain that has been coated with antinuclear antibody. In addition to lupus erythematosus, LE cells are seen also in scleroderma, drug-induced lupus erythematosus, and lupoid hepatitis.

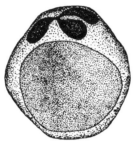

Lupus Erythematosus Cell (L.E. Cell)

Formation of an LE Cell

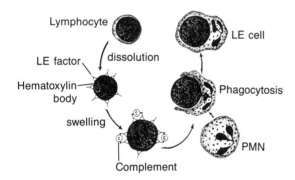

LE cell "prep"

Glass beads are added to heparinized blood samples, causing nuclei to be released from some blood cells which become coated with antinuclear antibody present in the serum. These opsonized nuclei are then phagocytized by polymorphonuclear neutrophils to produce LE cells. The LE cells produce homogeneous chromatin, which imparts a glassy appearance to the phagocytized nuclear material.

LE cell test

A diagnostic test no longer used which detects antinuclear antibodies in the blood sera of systemic lupus erythematosus patients. Antinuclear antibodies present in the serum react with nascent lymphocyte nuclei and serve as opsonins, enhancing phagocytosis of the nucleus-antibody complex by polymorphonuclear neutrophils. Thus, the appearance of a polymorphonuclear neutrophil with its own nucleus displaced to the periphery by an ingested lymphocyte nucleus which appears as a homogeneous mass and is coated with antinuclear antibody represents the so-called LE cell. These cells develop following incubation of blood containing the appropriate antibody for 1 h at 37°C. This earlier diagnostic test for the presence of antinuclear antibody in systemic lupus erythematosus has been replaced by a more sophisticated antinuclear antibody test.

LE cells are also present in other connective tissue diseases in addition to systemic lupus erythematosus.

LE factor

Antinuclear antibodies present in the blood sera of systemic lupus erythematosus patients. LE factor facilitates LE cell formation.

leader peptide

A 20-amino acid sequence situated at the N-terminus of free heavy and light polypeptide chains, but not present in secreted immunoglobulins. Once the light and heavy polypeptide chains reach the cisternal space of the endoplasmic reticulum, the peptide is split from the polypeptide chains. It is thought to facilitate vectorial release of the chains and their secretion.

leader sequence

Refer to leader peptide.

leading front technique

A method to assay chemotaxis or cell migration which evaluates differences in the migration of stimulated and nonstimulated cells.

lectin

Glycoproteins that bind to specific sugars and oligosaccharides and link to glycoproteins or glycolipids on the cell surface. They can be extracted from plants or seeds, as well as from other sources. They are able to agglutinate cells such as erythrocytes through recognition of specific oligosaccharides and occasionally will react with a specific monosaccharide. Many lectins also function as mitogens and induce lymphocyte transformation, during which a small resting lymphocyte becomes a large blast cell that may undergo mitosis. Well-known mitogens used in experimental immunology include phytohemagglutinin, pokeweed mitogen, and concanavalin A.

lectin-like receptors

Macrophage and monocyte surface structures that bind sugar residues. The ability of these receptors to anchor polysaccharides and glycoproteins facilitates attachment during phagocytosis of microorganisms. Steroid hormones elevate the number of these cell surface receptors.

Lederberg, Joshua (1925–)

American biochemist who made a significant contribution to immunology with his work on the clonal selection theory of antibody formation. He received a Nobel Prize in 1958 (with Beadle and Tatum) for genetic recombination and organization of genetic material in bacteria.

LEP (low egg passage)

A type of vaccine for rabies that has been employed for the immunization of dogs and cats.

lepra cells

Foamy macrophages that contain clusters of *Mycobacterium leprae* microorganisms that are not degraded because cell-mediated immunity has been lost. These are found in lepromatous leprosy, but are not observed in tuberculoid leprosy.

lepromin

A heat-inactivated extract of skin nodules laden with *Mycobacterium leprae*. This is used as a skin test antigen which upon intradermal injection induces formation of a nodular granuloma within 2 to 4 weeks in patients with tuberculoid leprosy, but not in those with lepromatous leprosy who are anergic. The development of a positive test signifies the presence of cell-mediated immunity against *M. leprae*.

lepromin test

A tuberculin-type skin test in which a *Mycobacterium leprae* suspension of microorganisms is used as the test material. This delayed-type (type IV) hypersensitivity reaction is negative in lepromatous leprosy, but positive in tuberculoid leprosy as well as in normal subjects. The reaction may occur either early and peak at 24 to 48 h (described as a Fernandez reaction) or it may occur late at approximately 4 weeks (referred to as a Mitsuda reaction). Although not useful in diagnosis, this test has been used in the past to classify leprosy and to give some idea of prognosis.

Lesch-Nyhan syndrome

A deficiency of hypoxanthine-guanine phosphoribosyl transferase that leads to neurological dysfunction and B cell immunodeficiency.

lethal dose

An amount of a toxin, virus, or any other material that produces death in all members of the species receiving it within a specified period of time following administration.

Letterer-Siwe disease

A macrophage lineage tumor (histiocytosis X) that may appear in the skin, lymph nodes, and spleen.

Leu-CAM

Leukocyte cell adhesion molecules.

Leu-M1

A granulocyte-associated antigen. Immunoperoxidase staining detects this marker on myeloid cells, but not on B or T cells, monocytes, erythrocytes, or platelets. It can be detected in Hodgkin's cells and Reed-Sternberg cells.

leukapheresis

A method that removes circulating leukocytes from the blood of healthy individuals for transfusion to recipients with decreased immunity or who are leukopenic. Leukapheresis is also used in leukemia patients who have too many white cells. The procedure leads to temporary relief of symptoms attributable to hyperleukocytosis.

leukemia

A malignant proliferation of circulating leukocytes. The disease often originates in the bone marrow and leads to the production of increased numbers of leukocytes in the peripheral blood circulation.

leukemia inhibitory factor (LIF)

A lymphoid factor that facilitates maintenance of embryonic stem cells through suppression of spontaneous generation. It also induces mitogenesis of selected cell lines, stimulation of bone remodeling, facilitation of megakaryocyte formation *in vivo*, and suppression of cellular differentiation in culture. The recombinant form is a 20-kD protein comprised of 180 amino acid residues.

leukoagglutinin

An antibody or other substance that induces the aggregation or agglutination of white blood cells into clumps.

leukocidin

A bacterial toxin produced especially by staphylococci that is cytolytic. It is toxic principally for polymorphonuclear leukocytes and, to a lesser extent, for monocytes. It contains an F and an S component which combine with the cell membrane, causing altered permeability. Less than toxic doses interfere with locomotion of polymorphonuclear neutrophils.

leukocyte

White blood cell. The principal types of leukocytes in the peripheral blood of man include polymorphonuclear neutrophils, eosinophils and basophils (granulocytes), and lymphocytes and monocytes.

leukocyte activation

The first step in activation is adhesion through surface receptors on the cell. Stimulus recognition is also mediated through membrane-bound receptors. An inducible endothelial-leukocyte adhesion molecule that provides a mechanism for leukocyte-vessel wall adhesion has been described. Surface adherent leukocytes undergo a large prolonged respiratory burst. NADPH oxidase, which utilizes hexose monophosphate shunt-generated NADPH, catalyzes the respiratory burst. Both Ca^{2+} and protein kinase C play a key role in the activation pathway. Complement receptor 3 (CR3) facilitates the ability of phagocytes to bind and ingest opsonized particles. Molecules found to be powerful stimulators of PMN activity include recombinant IFN-γ, granulocyte-macrophage colony-stimulating factor, TNF, and lymphotoxin.

leukocyte adhesion deficiency (LAD)

Recurrent bacteremia with staphylococci or *Pseudomonas* linked to defects in the leukocyte adhesion molecules known as integrins. These include the CD11/CD18 family of molecules. CD18 β chain gene mutations lead to a lack of complement receptors CR3 and CR4 to produce a congenital disease marked by recurring pyogenic infections. Deficiency of p150,95, LFA-1, and complement receptor 3 (CR3) membrane proteins leads to diminished adhesion properties and mobility of phagocytes and lymphocytes. There is a flaw in synthesis of the 95-kD β chain subunit which all three of these molecules share. The defect in mobility is manifested as altered chemotaxis, defective random migration, and faulty spreading. Particles coated with C3 are not phagocytized and therefore fail to activate a respiratory burst. The CR3 and p150,95 deficiency account for the defective phagocytic activity. LAD patients' T cells fail to respond normally to antigen or mitogen stimulation and are also unable to provide helper function for B cells producing immunoglobulin. They are ineffective in fatally injuring target cells, and they do not produce the lymphokine, γ interferon. LFA-1 deficiency accounts for the defective response of these T lymphocytes as well as all natural killer cells, which also have impaired ability to fatally injure target cells. Clinically, the principal manifestations are a consequence of defective phagocyte function rather than of defective T lymphocyte function. Patients may have recurrent severe infections, a defective inflammatory response, abscesses, gingivitis, and periodontitis. There are two forms of leukocyte adhesion deficiency. Those with the severe deficiency do not express the three α and four β chain complexes, whereas those with moderate deficiency express 2.5 to 6% of these complexes. There is an autosomal recessive mode of inheritance for leukocyte adhesion deficiency.

leukocyte adhesion proteins

Membrane-associated dimeric glycoproteins comprised of a unique α subunit and a shared 95-kD β subunit involved in cell-to-cell interactions. They include lymphocyte function-associated antigen (LFA-1), which is found on lymphocytes, neutrophils, and monocytes; Mac-1, which is found on neutrophils, eosinophils, NK cells, and monocytes; and p150,95, which is common to all leukocytes.

leukocyte chemotaxis inhibitors

Humoral factors that inhibit the chemotaxis of leukocytes. They play a role in the regulation of inflammatory responses of both immune and nonimmune origin.

leukocyte common antigen (LCA, CD45)

A family of high molecular weight glycoproteins (180 to 220 kD) densely expressed on lymphoid and myeloid cells including lymphocytes, monocytes, and granulocytes. Expression of LCA on leukocytes, but not on other cells makes this a valuable marker in immunophenotyping of tumors with respect to determination of histogenetic origin. LCA antigen function is unknown, but it has a high carbohydrate content and is believed to be associated with the cytoskeleton. LCA molecules are heterogeneous and appear on T and B lymphocytes as well as selected other hematopoietic cells. Some LCA epitopes are present in all LCA molecules, while others are confined to B lymphocyte LCA and still other epitopes are associated with B cell, CD8⁺ T cell, and most CD4⁺ T cell LCA molecules. About 30 exons are present in the gene that encodes LCA molecules. It is designated as CD45.

leukocyte groups

Leukocytes may be grouped according to their surface antigens such as MHC class I and class II histocompatibility antigens. These surface antigens may be detected by several techniques that include the microlymphotoxicity assay and DNA typing.

leukocyte inhibitory factor (LIF)

A lymphokine that prevents polymorphonuclear leukocyte migration. T lymphocytes activated *in vitro* may produce this lymphokine, which can interfere with the migration of polymorphonuclear neutrophils from a capillary tube, as observed in a special chamber devised for the laboratory demonstration of this substance. Serine esterase inhibitors inhibit LIF activity, although they do not have this effect on macrophage migration inhibitory factor (MIF). This inhibitor is released by normal lymphocytes stimulated with the lectin concanavalin A or by sensitized lymphocytes challenged with the specific antigen. LIF is a 65- to 70-kD protein.

leukocyte migration inhibitory factor

Refer to leukocyte inhibitory factor.

leukocyte transfer

Refer to adoptive transfer.

leukocytoclastic vasculitis

A type of vasculitis in which there is karyorrhexis of inflammatory cell nuclei. Fragments of neutrophil nuclei and immune complexes

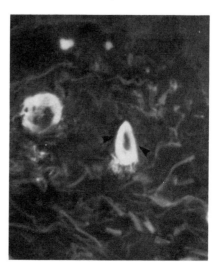

Leukocytoclastic Vasculitis

are deposited in vessels. Direct immunofluorescence reveals IgM, C3, and fibrin in vessel walls. There is nuclear dust, necrotic debris, and fibrin staining of the postcapillary venules. Leukocytoclastic vasculitis represents a type of allergic cutaneous arteriolitis or necrotizing angiitis. It is seen in a variety of diseases including Henoch-Schonlein purpura, rheumatoid arthritis, polyarteritis nodosa, and Wegener's granulomatosis, as well as other diseases.

leukocytosis
An increase above normal of the peripheral blood leukocytes as reflected by a total white blood cell count of greater than 11,000/mm³ of blood.

leukopenia
Reduction below normal of the number of white blood cells in the peripheral blood.

leukotaxis
Chemotaxis of leukocytes.

leukotriene
An arachidonic acid product generated during an anaphylactic reaction.

levamisole
An antihelminthic drug used extensively in domestic animals and birds that was found to also produce immunostimulant effects. The drug may potentiate or restore the function of T lymphocytes and other leukocytes.

Levine, Philip (1900–)
Russian–American immunohematologist. With Landsteiner, he conducted pioneering research on blood group antigens, including discovery of the MNP system. His work contributed much to transfusion medicine and transplantation immunobiology.

Lewis system
An erythrocyte antigen system that differs from other red cell groups in that the antigen is present in soluble form in the blood and saliva. Lewis antigens are adsorbed from the plasma onto the red cell membrane. The Lewis phenotype expressed is based on whether the individual is a secretor or a nonsecretor of the Lewis gene product. Expression of the Lewis phenotype is dependent also on the ABO phenotype. Lewis antigens are carbohydrates chemically. Lewis blood secretors have an increased likelihood of urinary tract infections induced by *Escherichia coli* or other microbes because of the linkage of carbohydrate residues of glycolipids and glycoproteins on urothelial cells.

Lewis Blood Group System

Genotype	Secretor Status	Phenotype
(a) Le, H, se	Non-secretor	Le a+b-
(b) Le, H, Se	Secretor	Le a-b+
(c) le, H, se	Non-secretor	Le a-b-
(d) le, H, Se	Secretor	Le a-b-

L$_f$ dose (historical)
The flocculating unit of diphtheria toxin is that amount of toxin which flocculates most rapidly with one unit of antitoxin in a series of mixtures containing constant amounts of toxin and varying amounts of antitoxin. This unit must be calculated.

L$_f$ flocculating unit (historical)
The flocculating unit of diphtheria toxin is that amount of toxin which flocculates most rapidly with one unit of antitoxin in a series of mixtures containing constant amounts of toxin and varying amounts of antitoxin. Historically, a unit of antitoxin was considered as the least quantity that would neutralize 100 minimal lethal doses of toxin administered to a guinea pig. Modern usage relates antitoxic activity to an international standard antitoxin.

LFA-1 deficiency
An immunodeficiency that is caused by a defect in lymphocyte function-associated antigen, a 95-kD β chain linked to CD11a which aids NK-binding, T helper cell reactivity, and cytotoxic T cell-mediated killing. This deficiency is associated with pyogenic mucocutaneous infections, pneumonia, diminished respiratory burst, and abnormal cell adherence in chemotaxis causing poor wound healing among other features.

LGL (large granular lymphocyte or null cell)
These lymphocytes do not express B and T cell markers, but they have Fc receptors for IgG on their surface. They comprise approximately 3.5% of lymphocytes and originate in the bone marrow. The LGLs include natural killer cells, which comprise 70% of LGLs, and killer cells, which mediate antibody-dependent cell-mediated cytotoxicity (ADCC).

Liacopoulos phenomenon (nonspecific tolerance)
The daily administration of 0.5 to 1.0 g of bovine γ globulin and bovine serum albumin to guinea pigs for at least 8 days suppresses their immune response to these antigens. If an unrelated antigen is injected and then continued for several days thereafter, the response to the unrelated antigen is reduced. This phenomenon has been demonstrated for circulating antibody, delayed hypersensitivity, and graft-vs.-host reaction. It describes the induction of nonspecific immunosuppression for one antigen by the administration of relatively large quantities of an unrelated antigen.

liberated CR1
A truncated complement receptor-1 (CR1) without transmembrane and intracytoplasmic domain that may help to limit the size of myocardial infarcts by diminishing complement activation. Liberated CR1 may also play a therapeutic role in other types of ischemia, burns, autoimmunity, and inflammation because it is a natural inhibitor of complement activation.

lichenification
Profound hyperkeratosis or skin thickening produced by chronic inflammation.

ligand
A molecule that binds or forms a complex with another molecule such as a cell surface receptor.

light chain
A 22-kD polypeptide chain found in all immunoglobulin molecules. Each four-chain immunoglobulin monomer contains two identical light polypeptide chains. They are joined to two like heavy chains by disulfide bonds. There are two types of light chains designated κ and λ. An individual immunoglobulin molecule possesses two light chains that are either κ or λ, but never a mixture of the two. The types of light polypeptide chains occur in all five of the immunoglobulin classes. Each light chain has an N-terminal V region which constitutes part of the antigen-binding site of the antibody molecule. The C region or constant terminal reveals no variation except for the Km and Oz allotype markers in humans.

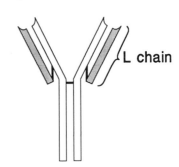

light chain deficiencies
In addition to deficiencies in heavy chains, one may observe light chain deficiencies. The ratio of κ to λ light chains may be altered in individuals with immunodeficiency. κ Chain deficiency has been associated with respiratory infections, megaloblastic anemia, and diarrhea. It has also been associated with achlorhydria and pernicious anemia and has even been seen in cases of malabsorption, diabetes, and cystic fibrosis. T cell function in all of these individuals was within normal limits, with only defective B cell immunity observed. Abnormal κ and λ light chain ratios are secondary findings in certain diseases, whereas in others they may be primary etiologic agents.

light chain disease
A paraproteinemia termed Bence-Jones myeloma which makes up one fifth of all myelomas. Excess monoclonal light chains are produced. These are linked to renal amyloidosis and renal failure as a consequence of blockage of tubules by certain Bence-Jones proteins. Four fifths of patients have monoclonal light chains in the blood circulation, and 60% have diminished γ globulin and lytic lesions of the bone. The λ type is usually more severe than the κ type of light chain disease. Light chain disease patients have a worse course than do patients with IgA or IgG myelomas.

light chain subtype

The subdivision of a type of light polypeptide chain based on its primary or antigenic structure which appears in all members of an individual species. Subtype differences distinguish light chains that share a common type. These relatively minor structural differences are located in the light chain constant region. Oz+, Oz−, Kern+, and Kern−, markers represent subtypes of λ light chains in man.

light chain type

Term for the classification of immunoglobulin light chains based on their primary or antigenic structure. Two types of light chain have been described and are designated as κ and λ. Two κ chains or two λ chains, never one of each, are present in each monomeric immunoglobulin subunit of vertebrate species.

light scatter

Light dispersion in any direction which can be useful in the study of cells by flow cytometry. A cell passing through a laser beam both absorbs and scatters light. Fluorochrome staining of cells permits absorbed light to be emitted as fluorescence. Forward angle light scatter permits identification of a cell in flow and determination of its size. If higher angle light scatter is added, some specific cell populations may also be identified. Light scatter measured at 90° to the laser beam and flow stream yields data on cell granularity or fine structure. Light scatter depends on such factors as cell size and

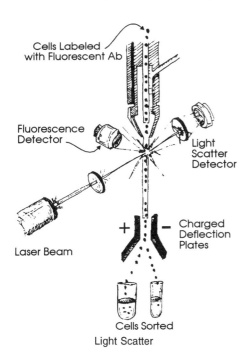

Light Scatter

shape, cell orientation in flow, cellular internal structure, laser beam shape and wavelength, and the angle of light collection.

limiting dilution (see figure, bottom of page)

A method of preparing aliquots that contain single cells through dilution to a point where each aliquot contains only one cell. The apportionment of cells by this method follows the Poissonian distribution, which yields 37% of aliquots without any cells and 63% with one or more cells. This technique can be used to estimate a certain cell's frequency in a population. For example, it may be employed to approximate the frequency of helper T lymphocytes, cytotoxic T lymphocytes, or B lymphocytes in a lymphoid cell suspension or to isolate cells for cloning in the production of monoclonal antibodies.

lineage infidelity

Cells that undergo neoplastic transformation may express molecules on their surface that are alien to the cell's lineage.

linear determinants

Antigenic determinants produced by adjacent amino acid residues in the covalent sequence in proteins. Linear determinants of six amino acids interact with specific antibody. Occasionally, linear determinants may be on the surface of a native folded protein, but they are more commonly unavailable in the native configuration and only become available for interaction with antibody upon denaturation of the protein.

linear staining

The interaction of IgG and possibly C3 on peripheral capillary loops of renal glomeruli in antiglomerular basement membrane diseases such as Goodpasture's syndrome. The use of fluorescein-labeled goat or rabbit antiimmunoglobulin preparations permits this smooth, thin, delicate, ribbon-like staining pattern to be recognized by immunofluorescence microscopy. It is in sharp contrast to the lumpy bumpy pattern of immunofluorescence staining seen in immune complex diseases.

linkage disequilibrium

Linkage disequilibrium refers to the appearance of HLA genes on the same chromosome with greater frequency than would be expected by chance. This has been demonstrated by detailed studies in both populations and families, employing outbred groups where numerous different haplotypes are present. With respect to the HLA-A, -B, and -C loci, a possible explanation for linkage disequilibrium is that there has not been sufficient time for the genes to reach equilibrium. However, this possibility is remote for HLA-A, -B, and -D linkage disequilibrium. Natural selection has been suggested to maintain linkage disequilibrium that is advantageous. If products of two histocompatibility loci play a role in the immune response and appear on the same chromosome, they might reinforce one another and represent an advantageous association. An example of linkage disequilibrium in the HLA system of man is the occurrence on the same chromosome of HLA-A3 and HLA-B7 in the Caucasian American population.

linked recognition

Refer to MHC restriction.

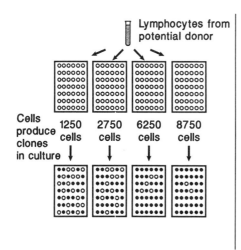

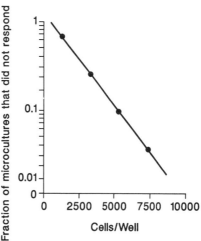

Limiting Dilution

Cross-section of bacterial cell wall

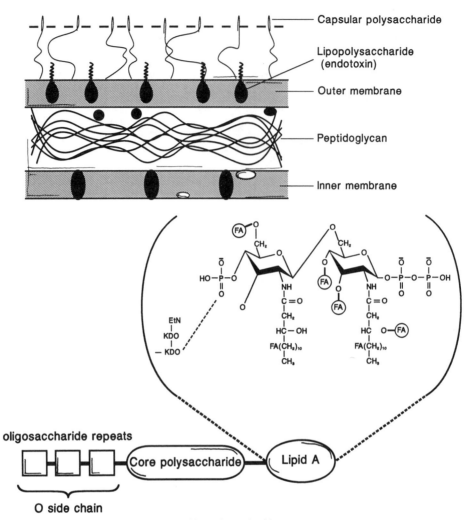

Lipopolysaccharide

lipopolysaccharide (LPS)
Refer to endotoxin.

liposome
A spherical lipid vesicle comprised of 5-nm phospholipid bilayers that enclose one or several aqueous units. These microspheres can be produced by dispersing phospholipid mixtures with or without sterols in aqueous solution. The liposomes produced consist of concentric phospholipid bilayers. Thus, they represent ideal models of cell membranes into which antigens may be embedded to induce an immune response. They also have been used to deliver drugs. They may or may not fuse with the cell membrane, and there are other problems associated with their use, such as whether or not they will leave the circulatory system or can be phagocytized by reticuloendothelial cells. Liposomes may act as immunologic adjuvants when antigens are incorporated into them. Liposomes serving as biological membrane models have also been used in studies on complement-mediated lysis.

lipoxygenase pathway
Enzymatic metabolism of arachidonic acid derived from the cell membrane which is the source of leukotrienes.

liquefactive degeneration
Dermal-epidermal interface liquefaction that is induced by immune mechanisms. This engages basal cells, leading to coalescing sub-epidermal vesicles in such skin diseases as dermatitis herpetiformis, erythema multiforme, fixed-drug reaction, lichen planus, lupus erythematosus, and many other skin conditions.

lissamine rhodamine (RB200)
A fluorochrome that produces orange fluorescence. Interaction with phosphorus pentachloride yields a reactive sulphonyl chloride

12- and 5-Lipoxygenase Pathway

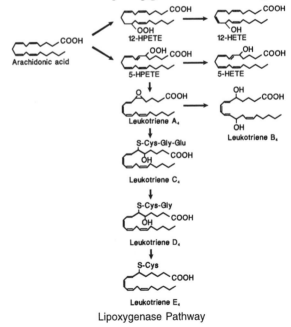

Lipoxygenase Pathway

that is useful for labeling protein molecules to be used in immunofluorescence staining methods.

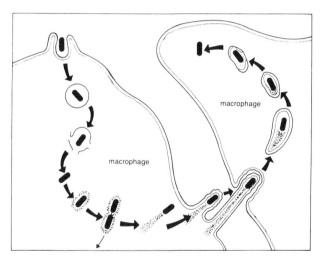

Lissamine Rhodamine B

Listeria

A genus of small Gram-positive motile bacilli that have a palisading pattern of growth. The best known is *L. monocytogenes*, which has a special affinity for monocytes and macrophages in which it takes up residence. It may be transmitted in contaminated milk and cheese. Approximately one third of the cases are in pregnant females, resulting in transplacental infection that may induce abortion or stillbirth. Infected infants may develop septicemia, vomiting, diarrhea, cardiorespiratory distress, and meningoencephalitis. Individuals with defective immune reactivity may develop endocarditis, meningoencephalitis, peritonitis, or other infectious processes. Treatment is with ampicillin, erythromycin, gentamycin, or chloramphenicol.

Listeria

live attenuated vaccine

An immunizing preparation consisting of microorganisms whose disease-producing capacity has been weakened deliberately in order that they may be used as immunizing agents. Response to a live attenuated vaccine more closely resembles a natural infection than does the immune response stimulated by killed vaccines. The

microorganisms in the live vaccine are actually dividing to increase the dose of immunogen, whereas the microorganisms in the killed vaccines are not reproducing, and the amount of injected immunogen remains unchanged. Thus, in general, the protective immunity conferred by the response to live attenuated vaccines is superior to that conferred by the response to killed vaccines. Examples of live attenuated vaccines include those used to protect against measles, mumps, polio, and rubella.

live vaccine

An immunogen for protective immunization that contains an attenuated strain of the causative agent, an attenuated strain of a related microorganism that cross-protects against the pathogen of interest, or the introduction of a disease agent through an avenue other than its normal portal of entry or in combination with an antiserum.

liver-kidney microsomal antibodies

Antibodies present in a subset of individuals with autoimmune chronic active hepatitis who are ANA-negative. By immunofluorescence, these antibodies can be shown to interact with hepatocyte and proximal renal tubule cell cytoplasm. These antibodies are not demonstrable in the sera of non-A, non-B chronic active hepatitis patients.

liver membrane antibodies

Antibodies specific for the 26 kD LM protein target antigen in the sera of 70% of autoimmune chronic active hepatitis patients who are HBsAg-negative. These antibodies are demonstrable by immunofluorescence. They may be demonstrated also in primary biliary cirrhosis, chronic hepatitis B, alcoholic liver disease and sometimes in Sjörgren's syndrome patients. "Lupoid" autoimmune chronic active hepatitis patients may develop antibodies not only against liver membrane but also against smooth muscle and nuclear constituents. Liver membrane antibodies are not useful for either diagnosis or prognosis.

LMP genes (see figure, bottom of page)

Two genes located in the MHC class II region in humans and mice that code for proteasome subunits. They are closely associated with the two *TAP* genes.

L₀ dose (historical)

This is the largest amount of toxin, which when mixed with one unit of antitoxin and injected subcutaneously into a 250-g guinea pig, will produce no toxic reaction.

local anaphylaxis (see figure, top of page 190)

A relatively common type I immediate hypersensitivity reaction. Local anaphylaxis is mediated by IgE crosslinked by allergen molecules at the surface of mast cells which then release histamines and other pharmacological mediators that produce signs and symptoms. The reaction occurs in a particular target organ such as the gastrointestinal tract, skin, or nasal mucosa. Hay fever and asthma represent examples.

local immunity

Immunologic reactivity confined principally to a particular anatomic site such as the respiratory or gastrointestinal tract. Local antibodies, as well as lymphoid cells, present in the area may mediate a specific immunologic effect. For example, secretory IgA produced in the gut may react to food or other ingested antigens.

locus

The precise location of a gene on a chromosome.

long-acting thyroid stimulator (LATS)

An IgG autoantibody that mimics the action of thyroid-stimulating hormone in its effect on the thyroid. The majority of patients with Graves disease, i.e., hyperthyroidism, produce LATS. This IgG

MHC Class II region

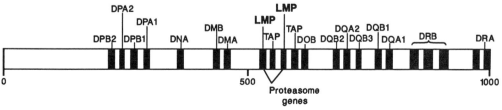

LMP Genes

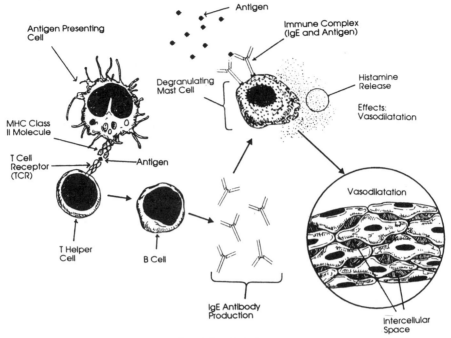

Local Anaphylaxis

autoantibody reacts with the receptors on thyroid cells that respond to thyroid-stimulating hormone. Thus, the antibody-receptor interaction results in the same biological consequence as does hormone interaction with the receptor. This represents a stimulatory type of hypersensitivity and is classified in the Gell and Coombs classification as one of the forms of type II hypersensitivity.

long homologous repeat
Refer to consensus sequence of C3/C4-binding proteins.

long-lived lymphocyte
Small lymphocyte derived principally from the thymus that survives for months to years without dividing. These are in contrast to short-lived lymphocytes.

low-dose tolerance
Antigen-specific immunosuppression induced by the administration of antigen in a suboptimal dose. Low-dose tolerance is achieved easily in the neonatal period, in which the lymphoid cells of the animal are not sufficiently mature to mount an antibody or cell-mediated immune response. This renders helper T lymphocytes tolerant, thereby inhibiting them from signaling B lymphocytes to respond to immunogenic challenge. Although no precise inducing dose of antigen can be defined, usually in low-dose tolerance 10^{-8} mol Ag/kg body weight is effective. Low-dose tolerance is relatively long lasting. Also called low-zone tolerance.

low responder mice
Inbred mouse strains that produce a poor immune response to selected antigens in comparison to the response by other inbred mouse strains. This is associated with the low responder's lack of appropriate *Ir* genes. Low responsiveness is governed by class II MHC genes.

low-zone tolerance
Refer to low-dose tolerance.

LPAM-1
A combination of α4 and β7 integrin chains that mediate the binding of lymphocytes to the high endothelial venules of Peyer's patch in mice. The addressin for LPAM-1 is MadCAM-1.

LPR
Abbreviation for late-phase reaction.

LPS
Abbreviation for lipopolysaccharide, which may serve as an endotoxin.

L$_r$ dose (historical)
This is the least amount of toxin which, after combining with one unit of antitoxin, will produce a minimal skin lesion when injected intracutaneously into a guinea pig.

LSGP (leukocyte sialoglycoprotein)
A richly glycosylated protein present on thymocytes and T lymphocytes. B lymphocytes are devoid of leukocyte sialoglycoprotein.

lupoid hepatitis
Autoimmune hepatitis that appears usually in young females who may produce antinuclear, antimitochondria, and antismooth muscle antibodies. Fifteen percent of these patients may show LE cells in the blood. This form of hepatitis has the histologic appearance of chronic active hepatitis, which generally responds well to corticosteroids.

lupus anticoagulant
IgG or IgM antibody that develops in lupus erythematosus patients, in certain individuals with neoplasia or drug reactions, in some normal persons, and was recently reported in AIDS patients who have active opportunistic infections. These antibodies are specific for phospholipoproteins or phospholipid constituents of coagulation factors. *In vitro,* these antibodies inhibit coagulation dependent upon phospholipids.

lupus erythematosus
A connective tissue disease associated with the development of autoantibodies against DNA, RNA, and nucleoproteins. It is believed to be due to hyperactivity of the B cell limb of the immune response. Clinical manifestations include skin lesions (including the so-called butterfly rash on the cheeks and across the bridge of the nose) that are light sensitive. Patients may develop vasculitis, arthritis, and glomerulonephritis. When the disease is confined to the skin, it

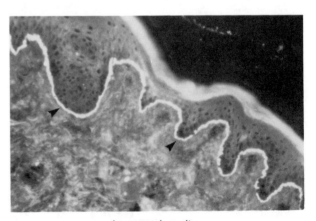

immune deposits
at dermal-epidermal
junction

Lupus Erythematosus

is referred to as discoid LE or cutaneous LE. Approximately 75% of lupus patients have renal involvement.

lupus erythematosus and pregnancy

Pregnant lupus erythematosus patients may experience fetal wasting caused by thromboses. Spontaneous abortion together with anticardiolipin antibody and anti-Rh antibodies may be linked to fertility failure and death of the fetus.

lupus erythematosus, drug induced

Certain drugs such as procainamide, hydralazine, D-penicillamine, phenytoin, isoniazid, and ergot substances may produce a condition that resembles lupus in patients receiving these substances. Most of these cases develop antinuclear and antihistone antibodies with only approximately one third of them developing clinical signs and symptoms of lupus, such as arthralgia, serositis, and fever. These cases do not usually develop the renal and CNS lesions seen in classic lupus. Nonacetylated metabolites accumulate in many of these individuals who are described as "slow acetylators". These nonacetylated metabolites act as haptens by combining with macromolecules. This may lead to an autoimmune response due to metabolic abnormality.

lupus inhibitor

Refer to lupus anticoagulant.

Lutheran blood group

Human erythrocyte epitopes recognized by alloantibodies against Lu^a and Lu^b products. Antibodies developed against Lutheran antigens during pregnancy may induce hemolytic disease of the newborn.

Lw antibody

An antibody that was first believed to be an anti-Rh specificity, but was subsequently shown to be directed against a separate red cell antigen closely linked to the Rh gene family. Its inheritance is separate from that of the Rh group. Lw is the designation given to recognize the research of Landsteiner and Wiener on the Rhesus system. The rare anti-Lw antibody reacts with Rh+ or Rh− erythrocytes and are nonreactive with Rh_{null} red cells.

Ly6

GPI linked murine cell surface alloantigens found most often on T and B cells but found also on nonlymphoid tissues such as brain, kidney, and heart. Monoclonal antibodies to these antigens indicate T cell receptor (TCR) dependence.

Ly antigen

A murine lymphocyte alloantigen that is expressed to different degrees on mouse T and B lymphocytes and thymocytes. Also referred to as Lyt antigen.

Ly1 B cell

A murine B lymphocyte that expresses CD5 (Ly1) epitope on its surface. This cell population is increased in inbred strains of mice susceptible to autoimmune diseases, such as the New Zealand mouse strain.

Lyb

A murine B lymphocyte surface alloantigen.

Lyb-3 antigen

Mature murine B cells express a surface marker designated Lyb-3. It is a single membrane-bound 68-kD polypeptide. On sodium dodecyl sulfate polyacrylamide gel electrophoresis (SDS-PAGE), it appears distinct from the SIg chains δ and μ. It does not contain disulfide bridges. The gene coding for Lyb-3 appears X-linked and recessive, and mutant mice lacking Lyb-3 antigens are known. Lyb-3 is involved in the cooperation between T and B cells in response to thymus-dependent antigens and seems to be manifested particularly when the amount of antigen used for immunization is suboptimal. The number of cells carrying Lyb-3 increases with the age of the animal.

Lyme disease

A condition that was first described in Lyme, CT, where an epidemic of juvenile rheumatoid arthritis (Still's disease) was found to be due to *Borrelia burgdorferi*. It is the most frequent zoonosis in the U.S. with concentration along the eastern coast. Insect vectors include the deer tick (*Ixodes dammini*), white-footed mouse tick (*I. pacificus*), wood tick (*I. ricinus*), and lone star tick (*Amblyomma americanum*). Deer and field mice are the hosts. In stage I a rash termed *erythema chronicum migrans* occurs. The rash begins as a single reddish papule and plaque that expands centrifugally to as much as 20 cm. This is accompanied by induration at the periphery with central clearing that may persist months to years. The vessels contain IgM and C3 deposits. Stage II is the cardiovascular stage,

which may be accompanied by pericarditis, myocarditis, transient atrial ventricular block, and ventricular dysfunction. Neurological symptoms also ensue and include Bell's palsy, meningoencephalitis, optic atrophy, and polyneuritis. Stage III is characterized by migratory polyarthritis. The diagnosis requires the demonstration of IgG antibodies against the causative agent by Western immunoblotting. Lyme disease is treated with the antibiotics tetracycline, penicillin, and erythromycin.

lymph

The fluid that circulates in the lymphatic system vessels. Its composition resembles that of tissue fluids, although there is less protein in lymph than in plasma. Lymph in the mesentery contains fat, and lymph draining the intestine and liver often possesses more protein than does other lymph. The principal cell type in the lymph is the small lymphocyte, with only rare large lymphocytes, monocytes, and macrophages. Occasional red cells and eosinophils are present. Coagulation factors are also present in lymph.

lymph gland

More correctly referred to as lymph node.

lymph node

A relatively small, i.e., 0.5 cm, secondary lymphoid organ that is a major site of immune reactivity. It is surrounded by a capsule and contains lymphocytes, macrophages, and dendritic cells in a loose reticulum environment. Lymph enters this organ from afferent lymphatics at the periphery, percolates through the node until it reaches the efferent lymphatics, exits at the hilus, and circulates to central lymph nodes and finally to the thoracic duct. The lymph node is divided into a cortex and medulla. The superficial cortex contains B lymphocytes in follicles, and the deep cortex is comprised of T lymphocytes. Differentiation of the specific cells continues in these areas and is driven by antigen and thymic hormones. Conversion of B cells into plasma cells occurs chiefly in the medullary region, where enclosed lymphocytes are protected from undesirable influences by a macrophage sleeve. The postcapillary venules from which lymphocytes exit the lymph node are also located in the medullary region. Macrophages and follicular dendritic cells interact with antigen molecules that are transported to the lymph node in the lymph. Reticulum cells form medullary cords and sinuses in the central region. T lymphocytes percolate through the lymph nodes. They enter from the blood at the postcapillary venules of the deep cortex. They then enter the medullary sinuses

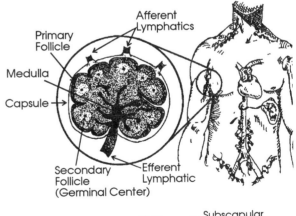

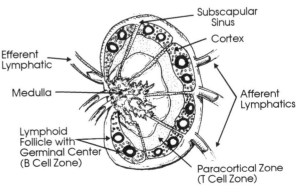

Lymph Node

and pass out of the node through the efferent lymphatics. T cells that interact with antigens are detained in the lymph node which may be a site of major immunologic reactivity. The lymph node is divided into B and T lymphocyte regions. Individuals with B cell or T cell immunodeficiencies may reveal an absence of one or the other lymphocyte type in the areas of the lymph node normally occupied by that cell population. The lymph node acts as a filter and may be an important site for phagocytosis and the initiation of immune responses.

lymphadenitis
Lymph node inflammation often caused by microbial (bacterial or viral) infection.

lymphadenoid goiter
Refer to Hashimoto's thyroiditis.

lymphadenopathy
Lymph node enlargement due to any of several causes. Lymphadenopathies are reactive processes in lymph nodes due to various exogenous and endogenous stimulants. Possible etiologies include microorganisms, autoimmune diseases, immunodeficiencies, foreign bodies, tumors, and medical procedures. "Lymphadenitis" is reserved for lymph node enlargement caused by microorganisms, whereas "lymphadenopathy" applies to all other etiologies of lymph node enlargement. Lymphadenopathies are divided into reactive lymphadenopathies, lymphadenopathies associated with clinical syndromes, vascular lymphadenopathies, foreign body lymphadenopathies, and lymph node inclusions. In benign lymphadenopathy, there is variability of germinal center size, no invasion of the capsule or fat, mitotic activity confined to germinal centers, and localization in the cortex and nonhomogenous follicle distribution.

lymphatics
Vessels that transport the interstitial fluid called lymph to lymph nodes and away from them, directing it to the thoracic duct from whence it reenters the blood stream.

lymphoblast
A relatively large cell of the lymphocyte lineage that bears a nucleus with fine chromatin and basophilic nucleoli. They frequently form following antigenic or mitogenic challenge of lymphoid cells, which leads to enlargement and division to produce effector lymphocytes that are active in immune reactions. The Epstein-Barr virus (EBV) is commonly used to transform B cells into B lymphoblasts in tissue culture to establish B lymphoblast cell lines.

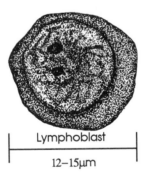

Lymphoblast

12–15μm

lymphocyte
A round cell that measures 7 to 12 μ and contains a round to ovoid nucleus that may be indented. The chromatin is densely packed and stains dark blue with Romanowsky stains. Small lymphocytes contain a thin rim of robin's egg blue cytoplasm, and a few azurophilic granules may be present. Large lymphocytes have more cytoplasm and a similar nucleus. Electron microscopy reveals villi

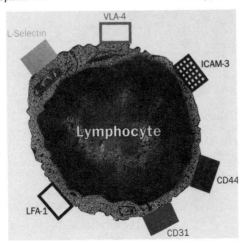

that cover most of the cell surface. Lymphocytes are divided into two principal groups termed B and T lymphocytes. They are distinguished not on morphology, but on the expression of distinctive surface molecules that have precise roles in immune reaction. In addition, natural killer cells, which are large granular lymphocytes, comprise a small percentage of the lymphocyte population.

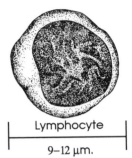

Lymphocyte

9–12 μm.

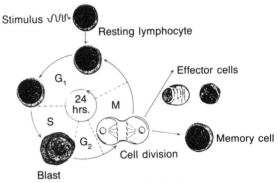

Lymphocyte

lymphocyte activation
The stimulation of lymphocytes *in vitro* by antigen or mitogen which renders them metabolically active. Activated lymphocytes may undergo transformation or blastogenesis.

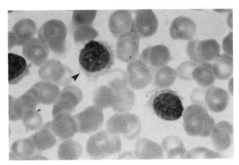

Lymphocyte Activation

lymphocyte activation factor (LAF)
Refer to interleukin-1.

lymphocyte defined (LD) antigens
Histocompatibility antigens on mammalian cells that induce reactivity in a mixed-lymphocyte culture (MLC) or mixed-lymphocyte reaction.

lymphocyte determinant
Target cell epitopes identified by lymphocytes rather than antibodies from a specifically immunized host.

lymphocyte function-associated antigen-1 (LFA-1) (see figure, facing pace)
A glycoprotein comprised of a 180-kD α chain and a 95-kD β chain expressed on lymphocyte and phagocytic cell membranes. LFA-1's ligand is the intercellular adhesion molecule 1 (ICAM-1). It facilitates natural killer cell and cytotoxic T cell interaction with target cells. Complement receptor 3 and p150,95 share the same specificity of the 769-amino acid residue β chain found in LFA-1. A gene on chromosome 16 encodes the α chain whereas a gene on chromosome 21 encodes the β chain. Refer to CD11a and CD18.

lymphocyte function-associated antigen-2 (LFA-2)
Refer to CD2.

lymphocyte function-associated antigen-3 (LFA-3) (see figure, facing page)
A 60-kD polypeptide chain expressed on the surfaces of B cells, T cells, monocytes, granulocytes, platelets, fibroblasts, and endothelial cells of vessels. LFA-3 is the ligand for CD2 and is encoded by genes on chromosome 1 in man.

Adhesion to Artificial Membranes

T cell

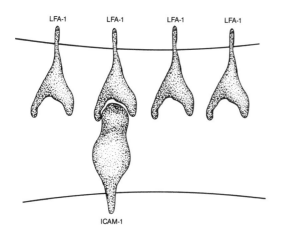

Lymphocyte Function Associated Antigen-1

Helper T cell

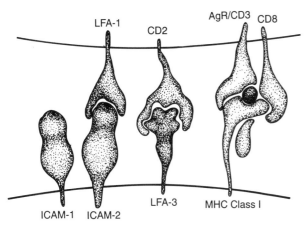

Antigen Presenting Cell

Lymphocyte Function-Associated Antigen-3

lymphocyte transfer reaction
Refer to normal lymphocyte transfer reaction.

lymphocyte transformation
An alteration in the morphology of a lymphocyte induced by an antigen, mitogen, or virus interacting with a small, resting lymphocyte. The transformed cell increases in size and amount of cytoplasm. Nucleoli develop in the nucleus, which becomes lighter staining as the cell becomes a blast. Epstein-Barr virus transforms B cells, and the human T cell leukemia virus transforms T cells.

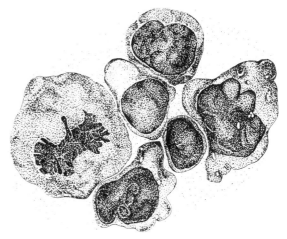

Lymphocyte Transformation

lymphocytes, circulating (or recirculating)
The lymphocytes present in the systemic circulation represent a mixture of cells derived from different sources: (1) B and T cells exiting from bone marrow and thymus on their way to seed the peripheral lymphoid organs; (2) lymphocytes exiting the lymph nodes via lymphatics, collected by the thoracic duct and discharged into the superior vena cava; and (3) lymphocytes derived from direct discharge into the vascular sinuses of the spleen. About 70% of cells in the circulating pool are recirculating; that is, they undergo a cycle during which they exit the systemic circulation to return back to lymphoid follicles, lymph nodes, and spleen and start the cycle again. The cells in this recirculating pool are mostly long-lived mature T cells. About 30% of the lymphocytes of the intravascular pool do not recirculate. They comprise mostly short-lived immature T cells which either live their lifespan intravascularly or are activated and exit the intravascular space. The exit of lymphocytes into the spleen occurs by direct discharge from the blood vessels. In lymph nodes and lymphoid follicles, the exit of lymphocytes occurs through

CD4+ T cell (Helper T)

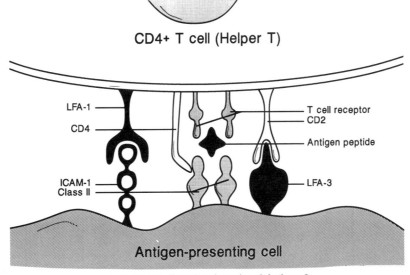

Lymphocyte Function-Associated Antigen-3

specialized structures, the postcapillary venules. These differ from other venules in that they have a tall endothelial covering. The exiting lymphocytes percolate through the endothelial cells, a mechanism whose real significance is not known. A number of agents such as cortisone or the bacterium *Bordetella pertussis* increase the extravascular exit of lymphocytes and prevent their return to circulation. The lymphocytes travel back and forth between the blood and lymph. They attach to and pass through the high endothelial cells of the postcapillary venules of lymph nodes or the spleen's marginal sinuses. Within 24 to 48 h they return via the lymphatics to the thoracic duct where they then reenter the blood.

lymphocytic choriomeningitis (LCM)

A murine viral disease that produces inflammatory brain lesions in the affected mouse as a result of delayed-type hypersensitivity to viral antigens on brain cells infected with the LCM virus. This infectious agent, which is classified as an arenavirus, is endemic in the mouse population and occasionally occurs in humans. Only adult mice that become infected develop the lesions, as those infected *in utero* are rendered immunologically tolerant to the viral antigens and fail to develop disease. An adult with an intact immune system exposed to LCM virus either becomes immune or succumbs to the acute infection, which is associated with lymphadenopathy, splenomegaly, and T lymphocyte perivascular infiltration of the viscera, especially the brain. A chronic carrier state can be induced in either neonatal mice or those with an impaired immune system through infection with the virus. Although carriers generate significant quantities of antiviral antibodies, the infection persists, and virus-antibody immune complexes become deposited in the renal glomeruli, walls of arteries, liver, lungs, and heart. The passive transfer of cytotoxic T lymphocytes from an immune animal to a carrier results in specific reactivity against LCM viral epitopes on cell membranes in the brain and meninges, which leads to profound inflammation with death of the animal. Transmission of this disease is through the excrement of rodents and is seen especially in winter when rodents enter dwellings. There is fever, headache, flu-like symptoms, and lymphocytosis in the cerebrospinal fluid. There may be associated leukopenia and thrombocytopenia. This disease must be distinguished from infectious mononucleosis, *Herpes zoster,* and enterovirus infection.

lymphocytic interstitial pneumonia (LIP)

A diffuse pulmonary disease of middle-aged females who may also have Sjögren's disease, hypergammaglobulinemia, or hypogammaglobulinemia. They develop shortness of breath, and reticulonodular infiltrates appear on chest films. Mature lymphocytes and plasma cells appear in the nodular interstitial changes in alveolar and interlobular septae with perivascular accumulation of round cells. LIP may resemble lymphoma based on the monotonous accumulation of small lymphocytes, and patients may ultimately develop end-stage lung disease or lymphoma.

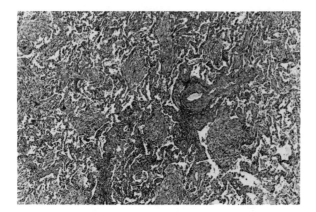

Interstitial Pneumonia
with organization

Lymphocytic Interstitial Pneumonia

lymphocytopenic center

Refer to germinal center.

lymphocytosis

An elevated number of peripheral blood lymphocytes.

lymphocytotoxin

Refer to lymphotoxin.

lymphocytotrophic

The property of possessing a special attraction or affinity for lymphocytes. Examples include the attraction of the Epstein-Barr virus for B lymphocytes and the affinity of human immunodeficiency virus (HIV) for the helper/inducer (CD4) T lymphocyte.

lymphogranuloma venereum (LGV)

A sexually transmitted disease induced by *Chlamydia trachomatis* that is divided into L_1, L_2, and L_3 immunotypes. It is rare in the U.S. but endemic in Africa, Asia, and South America. Clinically, patients develop papulo ulcers that heal spontaneously at the inoculation site. This is followed by development of inguinal and perirectal lymphadenopathy. There is skin sloughing, hemorrhagic proctocolitis, purulent draining, fever, headache, myalgia, aseptic meningitis, arthralgia, conjunctivitis, hepatitis, and erythema nodosum. Various antibody assays used in the diagnosis of LGV include complement fixation, with a titer of greater than 1:32, and immunofluorescence. Also, the Frei test, which consists of the intracutaneous inoculation of a crude antigen into the forearm, is used and can be read after 72 h. It is considered positive if the area of induration is greater than 6 mm.

lymphoid

An adjective that describes tissues such as the lymph node, thymus, and spleen, that contain a large population of lymphocytes.

lymphoid cell

A cell of the lymphoid system. The classic lymphoid cell is the lymphocyte.

lymphoid cell series

(1) Cell lineages whose members morphologically resemble lymphocytes, their progenitors, and their progeny. (2) Organized tissues of the body in which the predominant cell type is the lymphocyte or cells of the lymphoid cell lineage. These include the lymph nodes, thymus, spleen, and gut-associated lymphoid tissue, among others.

lymphoid follicle

Refer to the lymphoid nodules.

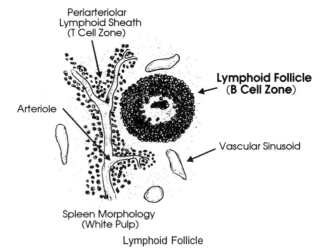

Spleen Morphology
(White Pulp)

Lymphoid Follicle

lymphoid nodules (or follicles)

Aggregates of lymphoid cells present in the loose connective tissue supporting the respiratory and digestive membranes. They are also present in the spleen and may develop beneath any mucous membrane as a result of antigenic stimulation. They are poorly defined at birth. Characteristic lymphoid nodules are round and nonencapsulated. They may occur as isolated structures or may be confluent, such as in the tonsils, pharynx, and naso-pharynx. In the tongue and pharynx, they form a characteristic structure referred to as Waldeyer's ring. In the terminal ileum, they form oblong patches termed Peyer's patches. The lymphoid nodules contain B and T cells and macrophages. Plasma cells in submucosal sites synthesize IgA that is released in secretions.

lymphoid system
The lymphoid organs and the lymphatic vessels.

lymphoid tissues
Tissues that include the lymph nodes, spleen, thymus, Peyer's patches, tonsils, bursa of Fabricius in birds, and other lymphoid organs in which the predominant cell type is the lymphocyte.

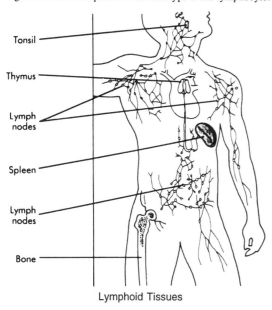

Tonsil

Thymus

Lymph nodes

Spleen

Lymph nodes

Bone

Lymphoid Tissues

lymphokine
A nonimmunoglobulin polypeptide substance synthesized mainly by T lymphocytes that affects the function of other cells. It may either enhance or suppress an immune response. Lymphokines may facilitate cell proliferation, growth, and differentiation, and they may act on gene transcription to regulate cell function. Lymphokines have either a paracrine or autocrine effect. Many lymphokines have now been described. Well-known examples of lymphokines include interleukin-2, interleukin-3, migration inhibitory factor (MIF), and γ interferon. The term cytokine includes lymphokines and soluble products produced by lymphocytes, as well as monokines and soluble products produced by monocytes.

lymphokine-activated killer (LAK) cells
Lymphoid cells derived from normal or tumor patients cultured in medium with recombinant IL-2 become capable of lysing NK-resistant tumor cells as revealed by ^{51}Cr-release cytotoxicity assays. These cells are also referred to as lymphokine-activated killer cells. Most LAK activity is derived from NK cells. The large granular lymphocytes (LGL) contain all LAK precursor activity and all active NK cells. In accord with the phenotype of precursor cells, LAK effector cells are also granular lymphocytes expressing markers associated with human NK cells. The asialo Gm_1+ population, known to be expressed by murine NK cells, contains most LAK precursor activity. Essentially all LAK activity resides in the LGL population in the rat. LAK cell and IL-2 immunotherapy has been employed in human cancer patients with a variety of histological tumor types when conventional therapy has been unsuccessful. Approximately one fourth of LAK- and IL-2-treated patients manifested significant responses, and some individuals experienced complete remission. Serious side effects include fluid retention and pulmonary edema attributable to the administered IL-2.

lymphoma
A malignant neoplasm of lymphoid cells. Hodgkin's disease, non-Hodgkin's lymphoma, and Burkitt's lymphoma are examples.

lymphoma belt
An area across Central Africa on either side of the equator where an increased incidence of Epstein-Barr virus-induced Burkitt's lymphoma occurs. Burkitt's lymphoma is a relatively common childhood cancer in Uganda.

lymphomatoid granulomatosis
Vasculitis in the lung of unknown etiology with an ominous prognosis. Atypical lymphocytes and plasma cells extensively infiltrate the pulmonary vasculature. Many of these lymphocytes are undergoing mitosis. The lungs may develop cavities, and occasionally, the nervous system, skin, and kidneys may be sites of nodular vasculitis.

lymphomatosis
Numerous lymphomas occurring in different parts of the body, such as those occurring in Hodgkin's disease.

lymphopenia
A decrease below normal in the number of lymphocytes in the peripheral blood.

lymphoreticular
Adjective that describes the system composed of lymphocytes and monocyte-macrophages, as well as the stromal elements that support them. The thymus, lymph nodes, spleen, tonsils, bone marrow, Peyer's patches, and avian bursa of Fabricius comprise the lymphoreticular tissues.

lymphorrhages
Accumulations of lymphocytes in inflamed muscle in selected muscle diseases such as myasthenia gravis.

lymphotoxin
A T lymphocyte lymphokine that is a heterodimeric glycoprotein comprised of a 5- and a 15-kD protein fragment. This cytokine is inhibitory to the growth of tumors either *in vitro* or *in vivo,* and it also blocks chemical-, carcinogen-, or ultraviolet light-induced transformation of cells. Lymphotoxin has cytolytic or cytostatic properties for tumor cells that are sensitive to it. Approximately three quarters of the amino acid sequence is identical between human and mouse lymphotoxin. Human lymphotoxin has 205 amino acid residues, whereas the mouse variety has 202 amino residues. Lymphotoxin does not produce membrane pores in its target cells, such as those produced by perforin or complement, but it is taken into cells after it is bound to their surface and subsequently interferes with metabolism. Lymphotoxin is also called tumor necrosis factor β (TNF-β).

lysin
Factors such as antibodies and complement or microbial toxins that induce cell lysis. For an antibody to demonstrate this capacity, it must be able to fix complement.

lysis
Disruption of cells due to interruption of their cell membrane integrity. This may be accomplished nonspecifically, as with hypotonic salt solution, or through the interaction of surface membrane epitopes with specific antibody and complement or with cytotoxic T lymphocytes.

lysosome
A cytoplasmic organelle enclosed by a membrane that contains multiple hydrolytic enzymes including ribonuclease, deoxyribonuclease, phosphatase, glycosidase, collagenase, arylsulfatase, and cathespins. Lysosomes occur in numerous cells, but are especially prominent in neutrophils and macrophages. The enzymes are critical for intracellular digestion. Refer also to phagosome and phagocytosis.

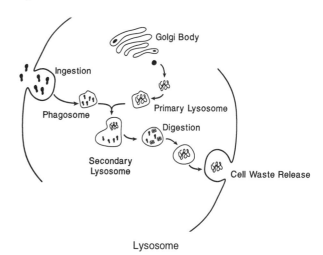

Golgi Body

Ingestion

Phagosome

Primary Lysosome

Digestion

Secondary Lysosome

Cell Waste Release

Lysosome

lysozyme (muraminidase)
A cationic low mol wt enzyme found in egg white, tears, nasal secretions, body fluids, lysosomal granules, on skin, and in lesser

amounts in serum that leads to hydrolysis of the β-1,4 glycosidic bond that joins N-acetylmuramic acid with N-acetylglucosamine in bacterial cell wall mucopeptide. This causes osmotic lysis of the bacterial cell. Lysozyme is effective principally against Gram-positive cocci, but it may facilitate the effect of antibodies and complement on Gram-negative microorganisms. This substance, which is found widely among vertebrates, invertebrates, plants, and bacteria, has been sequenced and its three-dimensional structure determined. It is widely distributed in such normal cells as histiocytes, leukocytes, and monocytes. By immunoperoxidase staining, this marker identifies histiocytes and neoplasms associated with them.

Lyt antigens

Murine T cell surface alloantigens that distinguish T lymphocyte subpopulations designated as helper (Lyt1) and suppressor (Lyt2 and Lyt3) antigens. Corresponding epitopes on B cells are termed Ly.

Lyt1,2,3

A category of murine T lymphocyte surface antigens that is used to subdivide them into helper T cells (Lyt1) and suppressor T cells (Lyt2 and Lyt3).

M

M cell

A gastrointestinal tract epithelial cell that conveys microorganisms and macromolecular substances from the gut lumen to Peyer's patches. M cells are non-antigen presenting cells found in the epithelial layer of the Peyer's patches, that, nevertheless may have an important role in antigen delivery. They have relatively large surfaces with microfolds that attach to microorganisms and macromolecular surfaces. The M cell cytoplasmic processes extend to CD4$^+$ T cells underneath them. Materials attached to microvilli are conveyed to coated pits and moved to the basolateral surface, which has pronounced invaginations rich in leukocytes and mononuclear phagocytes. Thus, materials gaining access by way of M cells come into contact with lymphoid cells as they reach the basolateral surface. This is believed to facilitate induction of immune responsiveness.

m chain

The IgM heavy polypeptide chain. Membrane μ chain is designated μ$_m$. Secreted μ chain is designated μ$_s$.

M component

A spike or defined peak observed on electrophoresis of serum proteins which suggests monoclonal proliferation of mature B lymphocytes synthesizing IgG, IgA, or IgM. M component can be seen in such diseases as multiple myeloma, heavy chain disease, and Waldenström's macroglobulinemia.

m heavy chain disease

A type of myelomatosis in which aberrant monoclonal immunoglobulin μ chains are present in the serum, but not in the urine and Bence-Jones proteins are present in the urine. Although very uncommon, this condition may be associated with chronic lymphocytic leukemia or reticulum cell sarcoma. Vacuolated plasma cells have been demonstrated in the bone marrow and are very suggestive of the diagnosis of μ heavy chain disease. The μ chains produced by bone marrow plasma cells have deletions in the variable region and involve the C$_H$1 domain, but have a normal sequence in the C$_H$2 domain. Light chains synthesized by these patients are not incorporated into molecular IgM; therefore, these individuals demonstrate a distinct failure in assembly of immunoglobulin molecules. Heavy and light chains have different electrophoretic mobilities, which becomes an important observation in establishing a diagnosis of μ heavy chain disease by electrophoresis.

M macroglobulin

An IgM paraprotein that occurs in Waldenström's macroglobulinemia.

M protein

(1) Monoclonal immunoglobulin or immunoglobulin components such as a myeloma protein. The M protein represents 3 to 10% of the total serum proteins. This level remains constant throughout life or decreases with age. (2) Group A β hemolytic streptococcal type-specific cell surface antigens, such as streptococcal M protein.

M13 bacteriophage

An *Escherichia coli* bacteriophage that contains single-stranded, circular DNA. It is male specific and infects *E. coli* by linking to the F-pilus on the bacterial cell surface. Viral DNA becomes double stranded after entering the host cell and replicates quickly. M13 bacteriophage has been popular as a cloning vector because of the ease of obtaining single-stranded or double-stranded DNA with it. Double-stranded DNA isolated from the bacterial cell can be employed to prepare recombinants *in vitro*. Single-stranded DNA from phage can be employed as a template for DNA sequencing. Due to M13's filamentous structure, it can house variable quantities of DNA.

MAB

Abbreviation for monoclonal antibody.

MAC

Abbreviation for membrane attack complex of the complement system.

MAC

Mycobacterium avium complex. A systemic infection that regularly affects subjects with AIDS, up to 66% of whom still have peripheral blood CD4$^+$ T lymphocytes. Infection with this complex is a clear indication of immunosuppression. MAC is successfully treated with clarithromycin.

MAC-1

A monoclonal antibody specific for macrophages.

macroglobulin

A relatively high mol wt serum protein. Macroglobulins have a sedimentation coefficient of 18 to 20 S and high carbohydrate content. Each type of macroglobulin belongs to a particular Ig class and is more homogeneous than the Igs produced in immune responses. Elevated levels appear on electrophoresis as a sharp peak in the migration area of the corresponding Ig class. Macroglobulins are monoclonal in origin and restricted to one κ or λ light chain type. The level of macroglobulins increases significantly in lymphocytic and plasmolytic disorders such as multiple myeloma or leukemia. It also increases in some collagen diseases, reticulosis, chronic infectious states, and carcinoma. The 820- and 900-kD IgM molecules are both α$_2$ macroglobulins.

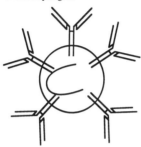

Pentameric IgM

Macroglobulin

macroglobulinemia

The presence of greater than normal levels of macroglobulins in the blood.

macroglobulinemia of Waldenström

A condition usually of older men in which monoclonal IgM is detected in the serum and elevated numbers of lymphoid cells and plasmacytoid lymphocytes expressing cytoplasmic IgM are found in the bone marrow. However, these subjects do not have the osteolytic lesions observed in multiple myeloma. Due to the high mol wt of the IgM and increased levels of this immunoglobulin, blood viscosity increases, leading to circulatory embarrassment. Patients often develop skin hemorrhages and anemia, as well as neurological problems. This condition is considered less severe than multiple myeloma.

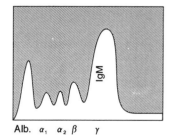

Alb. α$_1$ α$_2$ β γ

Waldenström's Macroglobulinemia with IgM spike

Alb. α$_1$ α$_2$ β γ

macrophage-activating factor (MAF)

A lymphokine such as γ interferon that accentuates the ability of macrophages to kill microbes and tumor cells. A lymphokine that enhances a macrophage's phagocytic activity and bactericidal and tumoricidal properties.

macrophage activation

Multiple processes are involved in stimulation of macrophages. These include an increase in size and number of cytoplasmic granules and a spread of membrane ruffling. Functional alterations include elevated metabolism and transport of amino acids and glucose; increased enzymatic activity; and an elevation in prostaglandins, cGMP, plasminogen activator, intracellular calcium ions, phagocytosis, pinocytosis, and the ability to lyse bacteria and tumor cells.

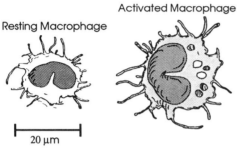

Resting Macrophage   Activated Macrophage

20 μm

Macrophage Activation

macrophage chemotactic factor (MCF)

Cytokines that act together with macrophages to induce facilitating migration. Among these substances are interleukins and interferons.

macrophage chemotactic and activating factor (MCAF)

A chemoattractant and activator of macrophages produced by fibroblasts, monocytes, and endothelial cells as a result of exogenous stimuli and endogenous cytokines such as TNF, IL-1, and PDGF. It also has a role in activating monocytes to release an enzyme that is cytostatic for some tumor cells. MCAF also has a role in ELAM-1 and CD11a&b surface expression in monocytes and is a potent degranulator of basophils.

macrophage cytophilic antibody

An antibody that becomes anchored to the Fc receptors on macrophage surfaces. This cytophilic antibody can be demonstrated by the immunocytoadherence test.

Macrophage Cytophilic Antibody

macrophage functional assays

Tests of macrophage function include (1) chemotaxis: macrophages are placed in one end of a Boyden chamber and a chemoattractant is added to the other end. Macrophage migration toward the chemoattractant is assayed. (2) Lysis: macrophages acting against radiolabeled tumor cells or bacterial cells in suspension can be measured after suitable incubation by measuring the radioactivity of the supernatant. (3) Phagocytosis: radioactivity of macrophages that have ingested a radiolabeled target can be assayed.

macrophage immunity

Cellular immunity.

macrophage inflammatory peptide-2 (MIP-2)

IL-8 type II receptor competitor and chemoattractant that is also involved in hemopoietic colony formation as a costimulator. It also degranulates murine neutrophils. The inflammatory activities of MIP-2 are very similar to IL-8.

macrophage inflammatory protein-1-α (MIP-1)

An endogeneous fever-inducing substance that binds heparin and is resistant to cyclooxygenase inhibition. Macrophages stimulated by endotoxin may secrete this protein, termed MIP-1, which differs from tumor necrosis factor (TNF) and IL-1, as well as other endogenous pyrogens because its action is not associated with prostaglandin synthesis. It appears indistinguishable from hematopoietic stem cell inhibitor and may function in growth regulation of hematopoietic cells.

macrophage migration inhibitory factor

Migration inhibitory factor.

macrophage migration test

An *in vitro* assay of cell-mediated immunity. Macrophages and lymphocytes from the individual to be tested are placed into segments of capillary tubes about the size of microhematocrit tubes and incubated in tissue culture medium containing the soluble antigen of interest, with maintenance of appropriate controls incubated in the same medium not containing the antigen. Lymphocytes from an animal or human sensitized to the antigen release a lymphokine called migration inhibitory factor that will block migration of macrophages from the end of the tube where the cells form an aggregated mass. The macrophages in the control preparation (that does not contain antigen) will migrate out of the tube into a fan-like pattern.

macrophages

Mononuclear phagocytic cells derived from monocytes in the blood that were produced from stem cells in the bone marrow. These cells have a powerful, although nonspecific role in immune defense. These intensely phagocytic cells contain lysosomes and exert microbicidal action against microbes which they ingest. They also have effective tumoricidal activity. They may take up and degrade both protein and polysaccharide antigens and present them to T lymphocytes in the context of major histocompatibility complex class II molecules. They interact with both T and B lymphocytes in immune reactions. They are frequently found in areas of epithelium, mesothelium, and blood vessels. Macrophages have been referred to as adherent cells since they readily adhere to glass and plastic and may spread on these surfaces and manifest chemotaxis. They have receptors for Fc and C3b on their surfaces, stain positively for nonspecific esterase and peroxidase, and are Ia antigen positive when acting as accessory cells that present antigen to CD4+ lymphocytes in the generation of an immune response. Monocytes, which may differentiate into macrophages when they migrate into the tissues, make up 3 to 5% of leukocytes in the peripheral blood. Macrophages that are tissue bound may be found in the lung alveoli, as microglial cells in the central nervous system, as Kupffer cells in the liver, as Langerhans cells in the skin, and as histiocytes in connective tissues, as well as macrophages in lymph nodes and peritoneum. Multiple substances are secreted by macrophages including complement components C1 through C5, factors B and D, properdin, C3b inactivators, and β-1H. They also produce monokines such as interleukin-1, acid hydrolase, proteases, lipases, and numerous other substances.

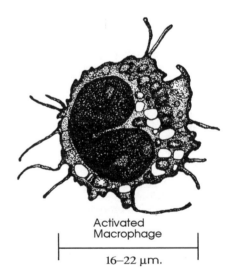

Activated Macrophage

16–22 μm.

MadCAM-1

Mucosal addressin cell adhesion molecule-1 which is an addressin in Peyer's patches of mice. This three Ig domain structure with a polypeptide backbone binds the α4β7 integrin. MadCAM-1 facilitates access of lymphocytes to the mucosal lymphoid tissues, as in the gastrointestinal tract.

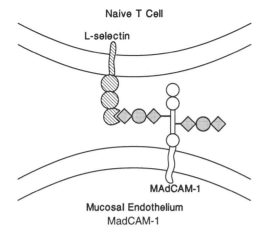

Naive T Cell

L-selectin

MAdCAM-1

Mucosal Endothelium
MadCAM-1

MAF
Macrophage-activating factor.

MAGE-1
Melanoma antigen-1 gene in humans that was derived from a malignant melanoma cell line. It codes for an epitope that a cytotoxic T lymphocyte clone specific for melanoma recognizes. This clone was isolated from a patient bearing melanoma.

MAGE-1 protein
A protein found on one-half of all melanomas and one-fourth of all breast carcinomas, but is not expressed on the majority of normal tissues. Even though MAGE-1 has not been shown to induce tumor rejection, cytotoxic T lymphocytes in melanoma patients manifest specific memory for MAGE-1 protein.

magic bullet
A term coined by Paul Ehrlich in 1900 to describe what he considered to be the affinity of a drug for a particular target. He developed "606" (salvarsan), an arsenical preparation, to treat syphilis. In immunology, it describes a substance that could be directed to a target by a specific antibody and injure the target once it arrives. Monoclonal antibodies have been linked to toxins such as diphtheria toxin or ricin as well as to cytokines for use as magic bullets.

MAIDS
Abbreviation for (1) murine acquired immunodeficiency syndrome and (2) monoclonal anti-idiotypic antibodies.

MAIS complex
Mycobacterium avium-intracellulare-scrofulaceum. Three species of mycobacteria that express the same antigens and lipids on their surfaces and also have the same biochemical reactions, antibiotic susceptibility, and pigment formation. They frequently occur together clinically. MAIS complex is relatively rare, but it occurs in 5 to 8% of AIDS patients when their CD4+ T lymphocyte levels diminish to less than 100 cells per cubic millimeter of blood. Affected patients have persistent diarrhea, night sweats and fever, abdominal pain, anemia, and extrahepatic obstruction. Ciproflozacin, clofazimine, ethambutol, and rifampicin, as well as rifabutin, clarithromycin, and azithromycin, have been used in treatment.

major basic protein (MBP)
A 10- to 15-kD protein present in eosinophilic granules. It has an isoelectric point that exceeds pH 10, thus the descriptor "basic". MBP induces injury to the bronchial epithelium and is linked to asthma. When inoculated intracutaneously, it can induce a wheal and flare response. Thus, this substance induces tissue injury in allergic and inflammatory diseases.

major histocompatibility complex (MHC)
A locus on a chromosome comprised of multiple genes encoding histocompatibility antigens that are cell surface glycoproteins. MHC genes encode both class I and class II MHC antigens. These antigens play critical roles in interactions among immune system cells, such as class II antigen participation in antigen presentation by macrophages to CD4+ lymphocytes and the participation of class I MHC antigens in cytotoxicity mediated by CD8+ T lymphocytes against target cells such as those infected by viruses, as well as various other immune reactions. MHC genes are very polymorphic and also encode a third category termed class III molecules that include complement proteins C2, C4, and factor B; P-450 cytochrome 21-hydroxylase; tumor necrosis factor; and lymphotoxin. The MHC locus in man is desig-

nated HLA, in the mouse as H2, in the chicken as B, in the dog as DLA, in the guinea pig as GPLA, and in the rat as RT1. The mouse and human MHC loci are the most widely studied. When organs are transplanted across major MHC locus differences between donor and recipient, graft rejection is prompt.

major histocompatibility complex restriction
Refer to MHC restriction.

major histocompatibility system
Refer to major histocompatibility complex.

Makari test
An assay no longer recommended that consisted of preparing an extract from a patient's tumor, incubating it with the subject's serum, and inoculating it into the skin, where it would induce an immediate skin reaction.

malaria vaccine
Although there is no effective vaccine against malaria, several vaccine candidates are under investigation, including an immunogenic, but nonpathogenic *Plasmodium* sporozoite that has been attenuated by radiation. Circumsporoite proteins combined with sporozoite surface protein 2(SSP-2) are immunogenic. Murine studies have shown the development of transmission-blocking antibodies following immunization with vaccinia into which has been inserted the *P. falciparum* surface 25-kD protein designated Pfs25. Attempts have been made to increase natural antibodies against circumsporozoite (CS) protein to prevent the prehepato-invasive stage. The high mutability of *P. falciparum* makes prospects for an effective vaccine dim.

malignolipin (historical)
A substance claimed in the past to be specific for cancer and to be detectable in the patient's blood early in the course of the disease. This is no longer considered valid. Malignolipin is comprised of fatty acids, phosphoric acid choline, and spermine. When injected into experimental animals, it can produce profound anemia, leukopoiesis, and cachexia.

MALT (mucosa-associated lymphoid tissue)
Extranodal lymphoid tissue associated with the mucosa at various anatomical sites, including the skin (SALT), bronchus, (BALT), gut (GALT), breast, and uterine cervix. The mucosa-associated lymphoid tissues provide localized or regional immune defense since they are in immediate contact with foreign antigenic substances, thereby differing from the lymphoid tissues associated with lymph nodes, spleen, and thymus. Secretory or exocrine IgA is associated with the MALT system of immunity.

Mancini test
Single radial diffusion test.

mannan-binding protein (Man-BP)
A substance that induces carbohydrate-mediated activation of complement, in contrast to complement activation mediated by immune complexes which is initiated by C1q. For example, the inability to opsonize *Saccharomyces cerevisiae* may be attributable to Man-BP deficiency. Alveolar macrophage mannose receptors facilitate ingestion of *Pneumocystis carinii.*

mantle
A dense zone of lymphocytes that encircles a germinal center.

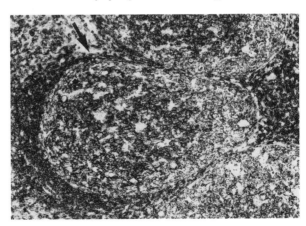

Mantle zone

mantle zone lymphoma

A type of follicular lymphoma of intermediate grade. There is small lymphocyte proliferation in the mantle zone encircling benign germinal centers, splenomegaly, and generalized lymphadenopathy. Histopathologically, there are B cells that vary in size from small to relatively large blasts containing clumped chromatin. IgM is usually present.

Mantoux test

A type of tuberculin reaction in which an intradermal injection of tuberculin tests for cell-mediated immunity. A positive test signifies delayed (type IV) hypersensitivity to *Mycobacterium tuberculosis*, which indicates previous or current infection with this microorganism.

Marek's disease

A lymphoproliferative disease of chickens induced by a herpes virus. Demyelination may occur as a consequence of autoimmune lymphocyte reactivity.

marginal zone

Exterior layer of lymphoid follicles of the spleen where T and B lymphocytes are loosely arranged encircling the periarterial lymphatic sheet. When antigens are injected intravenously, macrophages in this area actively phagocytize them.

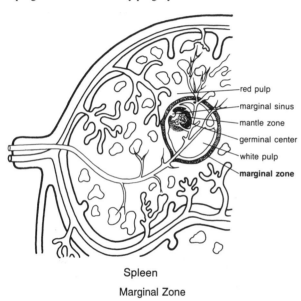

Spleen

Marginal Zone

margination

The adherence of leukocytes in the peripheral blood to the endothelium of vessel walls. Approximately 50% of polymorphonuclear neutrophils marginate at one time. During inflammation, there is margination of leukocytes, followed by their migration out of the vessels.

mast cell activation

Mast cells may be activated immunologically through cross-linking by antigen of surface IgE attached to FcεRI. They may be activated also by anti-IgE antibody, C5a, substance P, or local trauma.

mast cell-eosinophil axis

A term that characterizes the interactions between mast cells and eosinophils during inflammatory reactions recognized as immediate hypersensitivity. This involves the attraction of eosinophils and their activation by mast cell-derived ECF-A, as well as a dampening effect exerted by eosinophils upon mast cells. During this process, the released mediators influence the reactions in the microenvironment. When the causative agent is a parasite, the antiparasitic cytotoxic mechanisms of eosinophils reinforce the defense. The other effector cells attracted to the involved sites join forces in the defense activities. The inhibitory effects of eosinophils upon mast cells are exerted through a number of enzymes which inactivate or destroy some of the mast cell-derived mediators. Intact granules released from mast cells in the microenvironment are phagocytized by eosinophils and can be demonstrated in these cells by metachromatic staining. This represents an important detoxification mechanism, since even intact granules have been shown to exert proteolytic activity.

mast cell growth factor-1

Synonym for interleukin-3.

mast cell growth factor-2

Synonym for interleukin-4.

mast cells

A normal component of the connective tissue that plays an important role in immediate (type I) hypersensitivity and inflammatory reactions by secreting a large variety of chemical mediators from storage sites in their granules upon stimulation. Their anatomical location at mucosal and cutaneous surfaces and about venules in deeper tissues is related to this role. They can be identified easily by their characteristic granules which stain metachromatically. The size and shape of mast cells vary, i.e., 10 to 30 μm in diameter. In adventitia of large vessels, they are elongated; in loose connective tissue, they are round or oval; and the shape in fibrous connective tissue may be angular. On their surfaces, they have Fc receptors for IgE. Crosslinking by either antigen for which the IgE Fab regions are specific or by anti-IgE or antireceptor antibody leads to degranulation with the release of pharmacological mediators of immediate hypersensitivity from their storage sites in the mast cell granules. Leukotrienes, prostaglandins, and platelet-

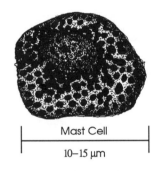

Mast Cell

10–15 μm

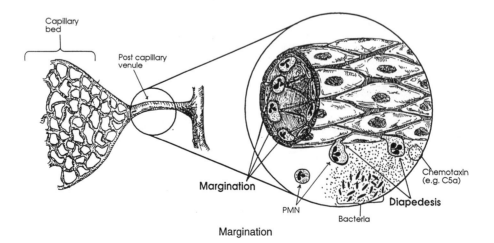

White blood cells in lumen of venule

Capillary bed

Post capillary venule

Margination

Margination

PMN

Bacteria

Diapedesis

Chemotaxin (e.g. C5a)

Margination

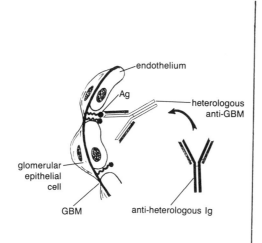

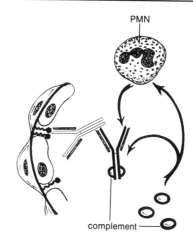

Masugi Nephritis

activating factor are also produced and released following Fcε receptor crosslinking. Mast cell granules are approximately 0.5 μ in diameter and are electron dense. They contain many biologically active compounds, of which the most important are heparin, histamine, serotonin, and a variety of enzymes. Histamine is stored in the granule as a complex with heparin or serotonin. Mast cells also contain proteolytic enzymes such as plasmin and also hydroxylase, β glucuronidase, phosphatase, and a high uronidase inhibitor, to mention only the most important. Zinc, iron, and calcium are also found. Some substances released from mast cells are not stored in a preformed state, but are synthesized following mast cell activation. These represent secondary mediators as opposed to the preformed primary mediators. Mast cell degranulation involves adenylate cyclase activation with rapid synthesis of cyclic AMP, protein kinase activation, phospholipid methylation, and serine esterase activation. Mast cells of the gastrointestinal and respiratory tracts that contain chondroitin sulfate produce leukotriene C_4, whereas connective tissue mast cells that contain heparin produce prostaglandin D_2.

Masugi nephritis

Experimental model of human antiglomerular basement membrane (anti-GBM) nephritis. The disease is induced by the injection of rabbit anti-rat glomerular basement membrane antibody into rats. The antiserum for passive transfer is raised in rabbits immunized with rat kidney basement membranes. The passively administered antibodies become bound to the glomerular basement membrane, fix complement, and induce glomerular basement membrane injury with increased permeability. Neutrophils and monocytes may infiltrate the area. Masugi nephritis is an experimental model of Goodpasture's syndrome in man. Also called nephrotoxic nephritis.

maternal immunity

Passive immunity conferred on the neonate by its mother. This is accomplished prepartum by active immunoglobulin transport across the placenta from the maternal to the fetal circulation in primate animals including humans. Other species such as ungulates transfer immunity from mother to young by antibodies in the colostrum, since the intestine can pass immunoglobulin molecules across its surface in the early neonatal period. The egg yolk of avian species is the mechanism through which immunity is passed from mother to young in birds.

maturation of affinity

During the course of immunization with a particular antigen, the antibodies formed show a progressive increase in their affinity.

MBP

Myelin basic protein or major basic protein.

MBSA

Methylated bovine serum albumin.

McCleod phenotype

Human erythrocytes without Kell or Cellano antigens. These red cells lack Kx, a precursor in the biosynthetic pathway of the Kell

blood group system. Kx is encoded by a gene on the X chromosome termed $X^l k$ and is normally found on granulocytes and fibroblasts. Red cells lacking Kx have decreased survival, diminished permeability to water, and are acanthocytic morphologically with spikes on their surface. They also have decreased expression of Kell system antigens. This group of erythrocyte abnormalities is termed the McCleod phenotype. Subjects with McCleod erythrocytes have a neuromuscular system abnormality characterized by elevated serum levels of creatine phosphokinase (CPK). Older individuals may have disordered muscular functions. The $X^l k$ gene maps to the short arm of the X chromosome where it is linked to the chronic granulomatous disease gene.

Mcg isotypic determinant

A human immunoglobulin λ chain epitope that occurs on some of every person's λ light polypeptide chains in immunoglobulin molecules. The Mcg isotypic determinant is characterized by asparagine at position 112, threonine at position 114, and lysine at position 163.

MCTD

Abbreviation for mixed-connective tissue disease.

MDP

Muramyl dipeptide.

measles vaccine

An attenuated virus vaccine administered as a single injection to children at 2 years of age or between 1 and 10 years old. Contraindications include a history of allergy or convulsions. Puppies may be protected against canine distemper in the neonatal period by the administration of attenuated measles virus which represents a heterologous vaccine. Passive immunity from the mother precludes early immunization of puppies with live canine distemper vaccine.

Mediterranean lymphoma

Refer to immunoproliferative small intestinal disease (IPSID).

medullary cord

A region of the lymph node medulla composed of macrophages as well as plasma cells that lies between the lymphatic sinusoids.

medullary sinus

Potential cavities in the lymph node medulla that receive lymph prior to its entering efferent lymphatics.

megakaryocyte (see figure, top of page 202)

Relatively large bone marrow giant cells that are multinuclear and from which blood platelets are derived by the breaking up of membrane-bound cytoplasm to produce the thrombocytes.

MEL-14

A selectin on the surface of lymphocytes significant in lymphocyte interaction with endothelial cells of peripheral lymph nodes. Selectins are important for adhesion despite shear forces associated with circulating blood. MEL-14 is lost from the surface of both granulocytes and T lymphocytes following their activation. Mel-14 combines with phosphorylated oligosaccharides.

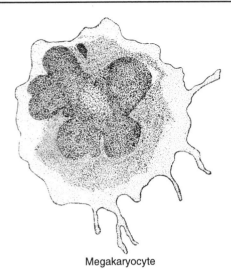

Megakaryocyte

Mel-14 antibody

Identifies a gp90 receptor that permits lymphocyte binding to peripheral lymph node high endothelial venules. Immature double-negative thymocytes comprise cells that vary from high to low in MEL-14 content. The gp90 MEL-14 epitope is a glycoprotein on murine lymph node lymphocyte surfaces. MEL-14 antibody prevents these lymphocytes from binding to postcapillary venules. The gp90 MEL-14 is apparently a lymphocyte homing receptor that directs these cells to lymph nodes in preference to lymphoid tissue associated with the gut.

melanoma-associated antigens (MAA)

Antigens associated with the aggressive, malignant, and metastatic tumors arising from melanocytes or melanocyte-associated nevus cells. Monoclonal antibodies have identified 40+ separate MAAs. They are classified as MHC molecules, cation binding proteins, growth factor receptors, gangliosides, high mol wt extracellular matrix-binding molecules, and nevomelanocyte differentiation antigens. Some of the antigens are expressed on normal cells, whereas others are expressed on tumor cells. Melanoma patient blood sera often contain anti-MAA antibodies, which are regrettably not protective. Monoclonal antibodies against MAAs aid studies on the biology of tumor progression, immunodiagnosis, and immunotherapy trials.

melphalan (L-phenylalanine mustard)

Nitrogen mustard that is employed for therapy of multiple myeloma patients.

1-phenylalanine mustard

membrane attack complex (MAC)

Five terminal proteins, i.e., C5, C6, C7, C8, and C9, associate into a membrane attack complex (MAC) on a target cell membrane to mediate injury. Initiation of MAC assembly begins with C5 cleavage into C5a and C5b fragments. A $(C5b678)_1(C9)_n$ complex then forms either on natural membranes or, in their absence, in combination with such plasma inhibitors as lipoproteins, antithrombin III, and S protein. C9 and C8 α proteins resemble each other not only structurally, but also in sequence homologies. Both bind calcium and furnish domains that bind lipid, enabling MAC to attach to the membrane. Mechanisms proposed for complement-mediated cytolysis include extrinsic protein channel incorporation into the plasma membrane or membrane deformation and

destruction. Central regions of C6, C7, C8α, C8β, and C9 have been postulated to contain amphiphilic structures which may be membrane anchors.

A single C9 molecule per C5b678 leads to erythrocyte lysis. Gram-negative bacteria, which have both outer and inner membranes, resist complement action by lengthening the O-antigen chain at the outer membrane or heightening surface carbohydrate content, which interferes with MAC binding. MAC assembly and insertion into the outer membrane is requisite for lysis of bacteria. Nucleated cells may rid their surfaces of MAC through endocytosis or exocytosis. Platelets have provided much data concerning sublytic actions of C5b-9 proteins.

Control proteins acting at different levels may inhibit killing of homologous cells mediated by MAC. Besides C8-binding protein or homologous restriction factor (HRF) found on human erythrocyte membranes, the functionally similar but smaller phosphatidyl inositol glycan (PIG)-tailed membrane protein harnesses complement-induced cell lysis. Sublytic actions of MAC may be of greater consequence for host cells than are its cytotoxic effects.

Membrane Attack Complex (MAC)

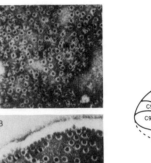
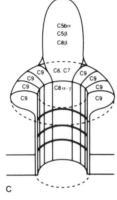

A. Electron micrograph of complement lesions (approximately 100 Å) in erythrocyte membranes formed by poly C9 tubular complexes

B. Electron micrograph of membrane lesions (approximately 160 Å) induced on a target cell by a clone cytolytic T lymphocyte (CTL) line

CTL and natural killer (NK) induced membrane lesions are formed by tubular complexes of perforin which is homologous to C9. Therefore, except for the larger internal diameter, the morphology of the lesions is similar to that of complement-mediated lesions.

C. Model of the MAC subunit arrangement

membrane cofactor protein (MCP)

A 50- to 70-kD protein, depending on the extent of glycosylation. MCP present on granulocytes, monocytes, and B and T lymphocytes acts together with factor I to produce C3b and C4b proteolysis.

membrane immunofluorescence

The reaction of a fluorochrome-labeled antibody with cell surface receptors of viable cells. This reaction of fluorescent antibody with surface antigens rather than internal antigens is the basis for many immunologic assays such as labeling of lymphocytes with such reagents for immunophenotyping by flow cytometry, patching, and capping and to detect changes in surface antigens through antigenic variation.

membrane immunoglobulin

Cell surface immunoglobulin that serves as an antigen receptor. Virgin B cells contain surface membrane IgM and IgD molecules. Following activation by antigen, the B cell differentiates into a plasma cell that secretes IgM molecules. Whereas membrane-bound IgM is a four-polypeptide chain monomer, the secreted IgM is a pentameric molecule containing five four-chain unit monomers and one J chain. Other immunoglobulin classes have membrane and secreted types. IgG and IgA membrane immunoglobulins probably serve as memory B cell antigen receptors. That segment of the immunoglobulin introduced into the cell membrane is a hydrophobic heavy chain region in the vicinity of the carboxy terminus. Within a particular isotype, the heavy chain is of greater length in the membrane form than in the secreted molecule. This greater length is at the carboxy terminal end of the membrane form. Separate mRNA molecules from one gene encode the membrane and secreted forms of heavy chain.

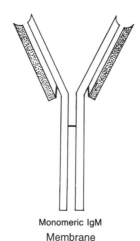

Monomeric IgM
Membrane
Immunoglobulin

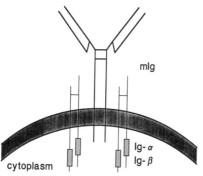

Membrane Immunoglobulin

membranoproliferative glomerulonephritis (MPGN)

A nephropathy in which the glomerular basement membrane is altered and glomerular cells proliferate, especially in mesangial areas, leading to the synonym mesangiocapillary GN. Patients may present with hematuria and/or proteinuria. In type I MPGN, IgG-, Clq-, C4-, and C3-containing subendothelial electron-dense deposits are present in approximately 66% of cases, as revealed by immunofluorescence and electron microscopy. Conventional light microscopy reveals splitting of basement membranes. In type II MPGN, also called dense deposit disease, the glomerular basement membrane's lamina densa appears as an electron-dense ribbon on either side of which C3 can be detected by immunofluorescence. Complement is fixed only by the alternate pathway. Type II patients have C3 nephritic factor (C3NeF) in their serum, which facilitates stabilization of alternate C3 convertase, thereby promoting C3 degradation and hypocomplementemia. Half of these patients develop chronic renal failure over a 10-year period.

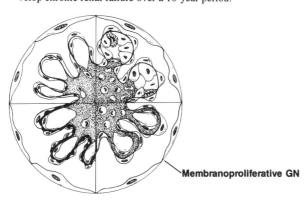

Membranoproliferative GN

membranous glomerulonephritis

A disease induced by deposition of electron-dense, immune (Ag-Ab) deposits in the glomerular basement membrane in a subepithelial location. This leads to progressive thickening of glomerular membranes. Most cases are idiopathic, but membranous glomerulonephritis may follow development of other diseases such as systemic lupus erythematosus; lung or colon carcinoma; and exposure to gold, mercury, penicillamine, or captopril. It can also be a sequela of certain infections, e.g., hepatitis B, or metabolic disorders, e.g., diabetes mellitus. Clinically, it is a principal cause of nephrotic syndrome in adults. The subepithelial immune deposits are shown by immunofluorescence to contain both immunoglobulins and complement. Proteinuria persists in 70 to 90% of cases, and half of the patients develop renal insufficiency over a period of years. A less severe course appears in 10 to 30% of cases.

Membranous GN

memory cells

Immunocompetent T and B lymphocytes that have the ability to mount an accentuated response to antigen, compared to that of virgin immunocompetent cells, because of their previous exposure to the antigen through immunization or infection.

2-mercaptoethanol agglutination test

A simple test to determine whether or not an agglutinating antibody is of the IgM class. If treatment of an antibody preparation, such as a serum sample, with 2-mercaptoethanol can abolish the serum's ability to produce agglutination of cells, then agglutination was due to IgM antibody. Agglutination induced by IgG antibody is unaffected by 2-mercaptoethanol treatment and just as effective after the treatment as it was before. Dithiothreitol (DTT) produces the same effect as 2-mercaptoethanol in this test.

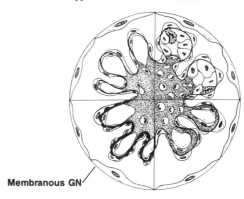

2-Mercaptoethanol

6-mercaptopurine (6-MP)

A powerful immunosuppressive drug used prior to the introduction of cyclosporine in organ transplantation. It is also an effective chemotherapeutic agent for the treatment of acute leukemia of childhood as well as other neoplastic conditions. 6-MP is a purine analog in which a thiol group replaces the 6-hydroxyl group. Hypoxanthine-guanine phosphoribosyl transferase (HGPRT) transforms 6-MP to 6-thioinosine-5'-phosphate. This reaction product blocks various critical purine metabolic reactions. 6-MP is also incorporated into DNA as thioguanine.

metaproterenol (dl-β-[3,5-dihydroxyphenyl]-α-isopropylamino-ethanol)

A β adrenergic amine that induces smooth muscle relaxation, especially in the bronchi. This substance has been used to treat asthma.

Metaproterenol

Metchnikoff, Elie (1845–1916)

Russian zoologist and immunologist who worked at the Pasteur Institute, Paris. For his work on cellular immunity he shared the 1908 Nobel Prize with Ehrlich. He was a lifetime proponent of the cellular (phagocytic) theory of immunity. *Lecons sur le Pathologie de l'Inflammation*, 1892; *L'Immunite dans les Maladies Infectieuses*, 1901; *Etudes sur la Nature Humaine*, 1903.

methotrexate (N-[p-[[2,4-diamino-6-pteridinyl-methyl]methylamino]benzoyl]glutamic acid)

A drug that blocks synthesis of DNA and thymidine in addition to its well-known use as a chemotherapeutic agent against neoplasia. It blocks dihydrofolate reductase, the enzyme required for folic acid conversion to tetrahydrofolate. Methotrexate has been used to treat cancer, psoriasis, rheumatoid arthritis, polymyositis, Reiter's syndrome, graft-vs.-host disease, and steroid-dependent bronchial asthma. It inhibits both humoral and cell-mediated immune responses. Its major toxicity is hepatic fibrosis, which is dose related. It may also produce hypersensitivity pneumonitis and megaloblastic anemia.

Methotrexate

methyl green pyronin stain

A stain used in histology or histopathology that renders DNA green and RNA red. It has been widely used to demonstrate plasma cells and lymphoblasts that contain multiple ribosomes containing RNA in their cytoplasm.

MGSA (melanoma growth stimulating activity)

Potent chemoattractant and neutrophil activator that competes for type II IL-8 receptors found on myelocytic cells. It is responsible for accelerated growth of melanoma cell lines.

MGUS

Refer to monoclonal gammopathy of undetermined significance.

MHC

Abbreviation for major histocompatibility complex.

MHC class I deficiency

A type of severe combined immunodeficiency in which class I MHC molecules are not expressed on the patient's lymphocyte membranes. The trait has an autosomal recessive mode of inheritance.

MHC class II deficiency

A type of combined immunodeficiency in which the patient's lymphocytes and monocytes fail to express class II MHC molecules on their surfaces. The cells also have diminished expression of class I MHC antigens. The condition, which appears principally in North African children, has an autosomal recessive mode of inheritance. Patients are able to synthesize the class I invariant chain β_2 microglobulin. Whereas the numbers of B cells and T lymphocytes in the circulating blood are normal, patients have agammaglobulinemia and diminished cell-mediated immunity. Malabsorption in the gastrointestinal tract and diarrhea are commonly associated with this deficiency.

MHC genes

Major histocompatility complex genes. Genes that encode the major, as opposed to minor, histocompatibility antigens that are expressed on cell membranes. MHC genes in the mouse are located at the H-2 locus on chromosome 17, whereas the MHC genes in man are located at the HLA locus on the short arm of chromosome 6.

MHC restriction (see facing page)

The recognition of antigen in the context of either class I or class II molecules by the T cell receptor for antigen. In the afferent limb of the immune response, when antigen is being presented at the surface of a macrophage or other antigen-presenting cell to CD4$^+$ T lymphocytes, this presentation must be in the context of MHC class II molecules for the CD4$^+$ lymphocyte to recognize the antigen and proliferate in response to it. By contrast, cytotoxic (CD8$^+$) T lymphocytes recognize foreign antigen, such as viral antigens on infected target cells, only in the context of class I MHC molecules. Once this recognition system is in place, the cytotoxic T cell can fatally injure the target cell through release of perforin molecules that penetrate the target cell surface.

MHD

Minimal hemolytic dose, which refers to an amount of complement used in a complement fixation test. MHD is also an abbreviation for minimal hemagglutinating dose of a virus, as in a hemagglutination inhibition test.

microfilaments

Cellular organelles that comprise a network of fibers of about 60 Å in diameter present beneath the membranes of round cells, occupying protrusions of the cells, or extending down microprojections such as microvilli. They are found as highly organized and prominent bundles of filaments, concentrated in regions of surface activity during motile processes or endocytosis. Microfilaments consist mainly of actin, a globular 42-kD protein. In media of appropriate ionic strength, actin polymerizes in a double array to form the microfilaments, which are critical for cell movement, phagocytosis, fusion of phagosome and lysosome, and other important functions of cells belonging to the immune and other systems.

microglial cell

Phagocytic cell in the central nervous system. It is a bone marrow-derived perivascular cell of the mononuclear phagocyte system. In the central nervous system, it may act as an antigen-presenting cell, functioning in an MHC class II-restricted manner.

microglobulin

Refer to β_2 microglobulin. A globulin molecule or its fragment with a mol wt of 40 kD or less. Bence-Jones proteins in the serum or urine would be an example. β_2 Microglobulin that is a constituent of MHC class I molecules is another example.

microlymphocytotoxicity

A widely used technique for HLA tissue typing. Lymphocytes are separated from heparinized blood samples by either layering over Ficoll-hypaque, centrifuging and removing lymphocytes from the interface or by using beads. After appropriate washing, these

Short Arm of Chromosome 6

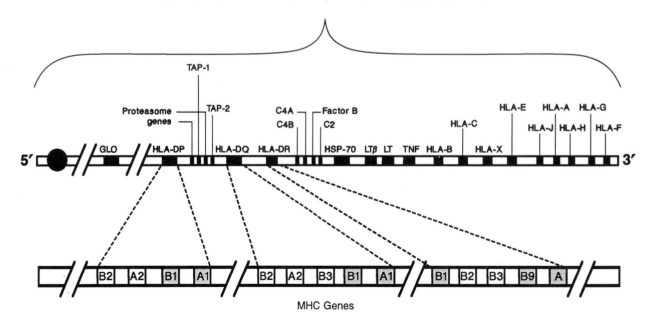

MHC Genes

MHC Restriction

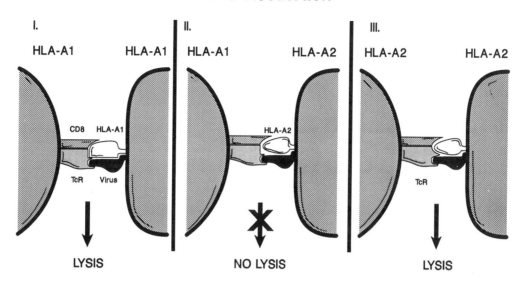

I.
HLA-A1   HLA-A1

CD8   HLA-A1

TcR   Virus

LYSIS

II.
HLA-A1   HLA-A2

HLA-A2

NO LYSIS

III.
HLA-A2   HLA-A2

TcR

LYSIS

purified lymphocyte preparations are counted, and aliquots are dispensed into microtiter plate wells containing predispensed quantities of antibody. When used for human histocompatibility (HLA) testing, antisera in the wells are specific for known HLA antigenic specificities. After incubation of the cells and antisera, rabbit complement is added, and the plates are again incubated. The extent of cytotoxicity induced is determined by incubating the cells with trypan blue, which enters dead cells and stains them blue, while leaving live cells unstained. The plates are read by using an inverted phase contrast microscope. A scoring system from 0 to 8 (where 8 implies >80% of target cells killed) is employed to indicate cytotoxicity. Most of the sera used to date are multispecific, as they are obtained from multiparous females who have been sensitized during pregnancy by HLA antigens determined by their spouse. Monoclonal antibodies are being used with increasing frequency in tissue

typing. This technique is useful to identify HLA-A, HLA-B, and HLA-C antigens. When purified B cell preparations and specific antibodies against B cell antigens are employed, HLA-DR and HLA-DQ antigens can be identified.

microtiter technique
Refer to Takatsy method.

microtubules
These organelles are hollow, cylindrical fibers of about 240 Å in diameter, radiating from the center of eukaryotic cells, including lymphocytes, phagocytes, and mast cells, in all directions toward the plasma membrane. The mitotic spindle is comprised of them. Microtubules form a sturdy cytoskeleton. They originate from the centriole, a structure occupying the concavity of the nucleus. Microtubules provide orientation of gross membrane activities, associate directly or indirectly with granules to enable their contact and fusion with endocytic vesicles, and direct reorganization of the cell membrane. Although not critical for the cell movement of chemotaxis, they are needed for "fine tuning" of cell locomotion. The major component of microtubules is tubulin, a dimeric protein.

mid-piece (historical)
A term used by investigators in the early 1900s. It refers to the components of complement present in the serum euglobulin fraction, which actually contains the entire C1 and selected other complement components, but no C2 component.

MIF
Macrophage/monocyte migration inhibitory factor. A substance synthesized by T lymphocytes in response to immunogenic chal-

Microlymphocytotoxicity
(trypan blue dye exclusion test)

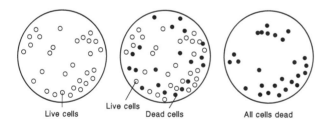

Live cells   Live cells   Dead cells   All cells dead

Lymphocytotoxicity Test

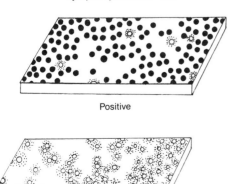

Positive

Negative

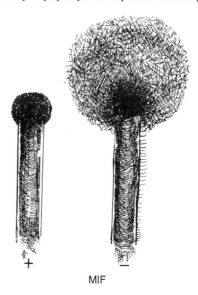

+   −

MIF

lenge that inhibits the migration of macrophages. MIF is a 25-kD lymphokine. Its mechanism of action is by elevating intracellular cAMP, polymerizing microtubules, and stopping macrophage migration. MIF may increase the adhesive properties of macrophages, thereby inhibiting their migration. The two types of the protein MIF include one that is 65 kD with a pI of 3 to 4 and another that is 25 kD with a pI of approximately 5.

Mikulicz's syndrome
Lymphocytic inflammation in the parotid gland. This condition represents a type of Sjögren's syndrome.

Milstein, Cesar (1927–)
Argentinian born immunologist who worked in the United Kingdom. He shared the 1984 Nobel Prize with G. F. Kohler for their production of monoclonal antibodies by hybridizing mutant myeloma cells with antibody producing B cells (hybridoma technique).

minimal hemagglutinating dose (MHD)
In the hemagglutination inhibition test for antiviral antibodies, the MHD is the least amount of hemagglutinating virus that will completely agglutinate the red cells in a single volume of a standard suspension.

minimal hemolytic dose (MHD)
The smallest amount of complement that can completely lyse a defined volume of a standardized suspension of red blood cells sensitized with antibody.

minimum lethal dose (MLD)
That dose of a substance or agent that will kill 100% of the population being tested.

minisatellite
DNA regions comprised of tandem repeats of DNA short sequences.

minor histocompatibility antigens
Molecules expressed on cell surfaces that are encoded by the minor histocompatibility loci, not the major histocompatibility locus. They represent weak transplantation antigens by comparison with the major histocompatibilty antigens. However, they are multiple, and their cumulative effect may contribute considerably to organ or tissue graft rejection. Graft rejection based on a minor histocompatibility difference between donor and recipient requires several weeks compared to the 7 to 10 days required for a major histocompatibility difference. Minor histocompatibility antigens may be difficult to identify by serological methods.

minor histocompatibility locus
A chromosomal site of genes that encode minor histocompatibility antigens that stimulate immune responses against grafts containing these antigens.

minor histocompatibility peptides
H antigens. Among minor antigens thus far identified are H-3 antigens, male-specific H-Y antigen, β_2 microglobulin, and numerous others that have not yet been firmly established.

minor lymphocyte-stimulating (Mls) determinants
Characterized by their activation of a marked primary mixed-lymphocyte reaction (MLR) between lymphocytes of mice sharing an identical major histocompatibility complex (MHC) haplotype. MHC class II molecules on various cell surfaces present Mls epitopes to naive T lymphocytes which mount a significant response. V β-specific monoclonal antibodies have facilitated the definition of Mls epitopes. Mls determinants activate T lymphocytes expressing selected β specificities. Refer also to Mls antigens.

minor lymphocyte-stimulating genes
See Mls genes.

MIP-1 (macrophage inflammatory protein-1-α)
See macrophage inflammatory protein.

MIRL (membrane inhibitor of reactive lysis)
Inhibitor of membrane attack complexes on self-tissue. Also known as CD59.

mitochondria
Cytoplasmic organelles in aerobic eukaryotic cells where respiration, electron transport, oxidative phosphorylation, and citric acid cycle reactions occur. Mitochondria possess DNA and ribosomes.

mitochondrial antibodies
IgG antibodies present in 90 to 95% of primary biliary cirrhosis (PBC) patients. These antibodies, which are of doubtful pathogenic significance, are specific for the pyruvate dehydrogenase enzyme complex E2 component situated at the inner mitochondrial membrane (M2). These antibodies are also specific for another E2 associated protein. Other antimitochondrial antibodies include the M1 antibodies of syphilis, the M3 antibodies found in pseudolupus, the M5 antibodies in collagen diseases, the M6 antibodies in hepatitis induced by iproniazid, the M7 antibodies associated with myocarditis and cardiomyopathy, the M8 antibodies that may be a marker for prognosis and the M9 antibodies that serve as a marker for beginning primary biliary cirrhosis. Mitochondrial antibody titer may be an indicator of primary biliary cirrhosis progression. A patient with a titer of 1 to 40 or more should be suspected of having primary biliary cirrhosis, whether or not symptoms are present and even if the alkaline phosphatase is normal.

mitogen
A substance, often derived from plants, that causes DNA synthesis and induces blast transformation and division by mitosis. Lectins, representing plant-derived mitogens or phytomitogens, have been widely used in both experimental and clinical immunology to evaluate T and B lymphocyte function *in vitro*. Phytohemagglutinin (PHA) is principally a human and mouse T cell mitogen, as is concanavalin A (Con A). By contrast, lipopolysaccharide (LPS) induces B lymphocyte transformation in mice, but not in humans. Staphylococcal protein A is the mitogen used to induce human B lymphocyte transformation. Pokeweed mitogen (PWM) transforms B cells of both humans and mice, as well as their T cells.

mitogenic factor
In immunology, a substance that induces cell division and proliferation of cells, such as phytohemagglutinin (PHA) which induces proliferation of lymphocytes. Many substances may serve as mitogenic factors.

Mitsuda reaction
A graduated response to an intracuteneous inoculation of lepromin, a substance used in the lepromin test. A nodule representing a subcutaneous granulomatous reaction to lepromin occurs 2 to 4 weeks after inoculation and is maximal at 4 weeks. It indicates granulomatous sensitization in a leprosy patient. Although not a diagnostic test, it can distinguish tuberculoid from lepromatous leprosy in that this test is positive in tuberculoid leprosy, as well as in normal adult controls, but is negative in lepromatous leprosy patients.

mixed agglutination
Aggregation (agglutination) produced when morphologically dissimilar cells that share a common antigen are reacted with antibody specific for this epitope. The technique is useful in demonstrating antigens on cells which by virtue of their size or irregular shape are not suitable for study by conventional agglutination tests. It is convenient to use an indicator, such as a red cell which possesses the antigen being sought. Thus, the demonstration of mixed agglutination in which the indicator cells are linked to the other cell type suspected of possessing the common antigen constitutes a positive test.

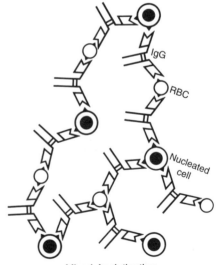

Mixed Agglutination

mixed-antiglobulin reaction
A test to demonstrate antibodies adsorbed to cell surfaces. The addition of antiglobulin-coated red cells to a suspension of cells

suspected of containing cell surface antibodies results in formation of erythrocyte-test cell aggregates if the test is positive. This is caused by linkage of the antiglobulin to the immunoglobulin on the surface of test cells.

mixed-connective tissue disease (MCTD)

A connective tissue disease that shares characteristics in common with systemic lupus erythematosus, rheumatoid arthritis, and dermatomyositis. By immunofluorescence, a speckled nuclear pattern attributable to antinuclear antibody in the circulation is revealed. There are high titers of antinuclear antibodies specific for nuclear ribonucleoproteins. Treatment with corticosteroids is quite effective. Also called Sharp syndrome or overlap syndrome.

mixed hemadsorption

The demonstration of antiviral antibody by the mixed-antiglobulin reaction.

mixed-lymphocyte culture (MLC)

The combination of lymphocytes from two members of a species in culture where they are maintained and incubated for 3 to 5 days. Lymphoblasts are formed as a consequence of histoincompatibility between the two individuals donating the lymphocytes. The lymphocyte antigens of these genetically dissimilar subjects each stimulate DNA synthesis by the other which is measured by tritiated thymidine uptake that is assayed in a scintillation counter. Refer to mixed-lymphocyte reaction (MLR).

mixed-lymphocyte reaction (MLR)

Lymphocytes from potential donor and recipient are combined in tissue culture. Each of these lymphoid cells has the ability to respond by proliferating following stimulation by antigens of the

One-Way MLR

Mixed Lymphocyte Reaction

other cell. In the one-way reaction, the donor cells are treated with mitomycin or irradiation to render them incapable of proliferation. Thus, the donor antigens stimulate the untreated responder cells. Antigenic specificities of the stimulator cells that are not present in the responder cells lead to blastogenesis of the responder lymphocytes. This leads to an increase in the synthesis of DNA and cell division. This process is followed by introduction of a measured amount of tritiated thymidine, which is incorporated into the newly synthesized DNA. The mixed-lymphocyte reaction usually measures a proliferative response and not an effector cell killing response. The test is important in bone marrow and organ transplantation to evaluate the degree of histoincompatibility between donor and recipient. Also called mixed-lymphocyte culture.

mixed vaccine

A preparation intended for protective immunization that contains antigens of more than one pathogenic microorganism. Thus, it induces immunity against those disease agents whose antigens are represented in the vaccine. It may also be called a polyvalent vaccine.

MK-571

A powerful synthetic antagonist of leukotriene D_4 receptor which prevents bronchoconsriction induced by exercise in patients with asthma.

MLC

Abbreviation for mixed-lymphocyte culture.

MLD (minimum lethal dose)

That dose of a substance or agent that will kill 100% of the population being tested. Ehrlich defined this as the least amount of a given toxin that will kill a 250-g guinea pig within 4 days after subcutaneous injection.

MLNS (mucocutaneous lymph node syndrome)

Refer to Kawasaki's disease.

Mls antigens

Minor lymphocyte stimulatory antigens. Cell surface molecules originally observed on mouse cells that stimulate previously unsensitized T lymphocytes. They occur in two stimulatory forms designated as Mls_a and Mls_c. The V β chain of the T cell receptor is encoded by the Mls genes. The anti-Mls response is linked to T lymphocyte receptor V β expression. The anti-Mls responses also link to intrathymic contact of CD8+ T lymphocytes and are critical in the induction of immunologic tolerance. As the immune system matures, clones of autoreactive T lymphocytes expressing the T cell receptor V β chain are deleted.

Mls genes

Minor lymphocyte stimulatory genes. Mouse mammary tumor retroviruses code for *Mls* genes. The proteins formed serve as superantigens that are powerful inducers of CD4+ T lymphocyte proliferation in mixed-lymphocyte cultures.

MMR vaccine

Measles-mumps-rubella vaccine. A live attenuated virus vaccine given at 15 months of age or earlier. A booster injection is given later. The vaccine is effective in stimulating protective immunity in most cases. It might prove ineffective in children younger than 15 months of age if they still have massively transferred antibodies from the mother. This vaccine should not be given to pregnant women, immunodeficient individuals undergoing immunosuppressive therapy, or individuals with acute febrile disease.

MNSs blood group system

Human erythrocyte glycophorin epitopes. There are four distinct sialoglycoproteins (SGP) on red cell membranes. These include α-SGP (glycophorin A, MN), β-SGP (glycophorin C), γ-SGP (glycophorin D), and δ-SGP (glycophorin B). MN antigens are present on α-SGP and δ-SGP. M and N antigens are present on α-SGP, with approximately 500,000 copies detectable on each erythrocyte. This is a 31-kD structure that is comprised of 131 amino acids, with about 60% of the total weight attributable to carbohydrate. This transmembrane molecule has a carboxy terminus that stretches into the cytoplasm of the erythrocyte with a 23-amino acid hydrophobic segment embedded in the lipid bilayer. The amino terminal segment extends to the extracellular compartment. Blood group antigen activity is in the external segment. In α-SGP with M antigen activity, the first amino acid is serine and the fifth is glycine. When it carries N antigen activity, leucine and glutamic acid replace serine and glycine at positions 1 and 5, respectively. The Ss antigens are encoded by allelic genes at a locus closely linked to the MN locus. The U antigen is also considered a part of the MNSs system.

Whereas anti-M and anti-N antibodies may occur without red cell stimulation, antibodies against Ss and U antigens generally follow erythrocyte stimulation. The MN and Ss alleles positioned on chromosome 4 are linked. Antigens of the MNSs system may provoke the formation of antibodies that can mediate hemolytic disease of the newborn.

Mo1

An adhesive glycoprotein present on neutrophils and monocytes. It is termed Leu-CAM, which is an iC3b receptor and thus helps mediate monocyte functions that are complement dependent.

modulation

Refer to antigenic modulation.

molecular hybridization probe

A molecule of nucleic acid that is labeled with a radionuclide or fluorochrome that can reveal the presence of complementary nucleic acid through molecular hybridization such as *in situ*.

molecular mimicry

The sharing of antigenic determinants or epitopes between cells of an immunocompetent host and a microorganism may lead to pathologic sequelae if antibodies produced against the microorganism combine with antigens of self and lead to immunologic injury. Ankylosing spondylitis and rheumatic fever are examples. Immunologic crossreactivity between a viral antigen and a self antigen or between a bacterial antigen such as streptococcal M protein and human myocardial sarcolemmal membranes may lead to tissue injury.

Moloney test

Diphtheria toxoid is injected intradermally, and the skin response is observed to determine whether or not the subject is hypersensitive to diphtheria prophylactic substances.

monoclonal

An adjective that refers to derivation from a single clone.

monoclonal antibody (MAb)

An antibody synthesized by a single clone of B lymphocytes or plasma cells. The first to be observed were produced by malignant plasma cells in patients with multiple myeloma and associated gammapathies. The identical copies of the antibody molecules produced contain only one class of heavy chain and one type of light chain. Kohler and Millstein in the mid-1970s developed B lymphocyte hybridomas by fusing an antibody-producing B lymphocyte with a mutant myeloma cell that was not secreting antibody. The B lymphocyte product provided the specificity, whereas

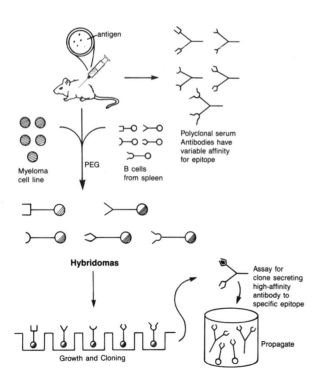

Monoclonal Antibody

the myeloma cell conferred immortality on the hybridoma clone. Today monoclonal antibodies (MAb) are produced in large quantities against a plethora of antigens for use in diagnosis and sometimes in treatment. MAb are homogeneous and are widely employed in immunoassays, single antigen identification in mixtures, delineation of cell surface molecules, and assay of hormones and drugs in serum, among many other uses. Since the response to some immunogens is inadequate in mice, monoclonal antibodies have also been generated using rabbit cells. Monoclonal antibodies have been radioactively labeled and used to detect tumor metastases; to differentiate subtypes of tumors, with monoclonal antibodies against membrane antigens or intermediate filaments; to identify microbes in body fluids; and for circulating hormone assays. MAb may be used to direct immunotoxins or radioisotopes to tumor targets with potential for tumor therapy.

monoclonal antibody HECA-452

An antibody that recognizes cutaneous lymphocyte antigen.

monoclonal gammopathy

A pathologic state characterized by a monoclonal immunoglobulin (M component) in the patient's serum. Examples include multiple myeloma, benign monoclonal gammopathy, and Waldenström's macroglobulinemia.

monoclonal gammopathy of undetermined significance (MGUS)

Benign monoclonal gammopathy. A condition in which the serum contains an M component, serum albumin is less than 2g/dl, and there is no Bence-Jones protein in the urine. There are no osteolytic lesions, and plasma cells comprise less than 5% of bone marrow constituents. Of these patients, 20 to 40% ultimately develop a monoclonal malignancy. When monoclonal spikes exceed 2g/dl with other immunoglobulins decreased and no Bence-Jones protein in the urine, the condition is usually malignant. One to two percent of myelomas are nonsecretory monoclonal gammapathies. More than half of the patients with elevated monoclonal IgM continue as MGUS, with other cases evolving into Waldenström's macroglobulinemia and a few progressing to lymphoma or chronic lymphocytic leukemia.

monoclonal immunoglobulin

A protein formed by an expanding clone of antibody-synthesizing cells in the body of a patient with either a tumor or a benign condition. The classic example is multiple myeloma or Waldenström's macroglobulinemia, in which an expanded clone of cells producing a homogeneous and uniform immunoglobulin product can be demonstrated. A number of tumors may stimulate monoclonal immunoglobulin synthesis. These include adenocarcinoma and carcinoma of the cervix, liver, and bladder, as well as Kaposi's sarcoma and angiosarcoma. Selected infections may also evoke a monoclonal immunoglobulin response. Among these are viral hepatitis, tuberculosis, and schistomiasis. Other conditions that stimulate this response include thalassemia, autoimmune hemolytic anemia, autoimmune diseases such as scleroderma and pemphigus vulgaris, and various other conditions.

monoclonal immunoglobulin deposition disease (MIDD)

A condition in which monotypic light or heavy chains are deposited in the tissues. The deposits may be fibrillar or nonfibrillar. The AL type of amyloidosis, characterized by light chains and amyloid P component, represent an example of fibrillar deposits, whereas light chain or light and heavy chain diseases represent nonfibrillar deposits. Patients may develop azotemia, albuminuria, hypogammaglobulinemia, cardiomyopathy or nephropathy. Some of them may develop multiple myeloma or another plasma cell neoplasm. Another variety of MIDD is amyloid H, which is comprised of V_H, D_D, J_H, and C_H3 domains.

monoclonal protein

A protein synthesized by a clone of identical cells derived from a single cell.

monocyte

Mononuclear phagocytic cells in the blood that are derived from promonocytes in the bone marrow. Following a relatively brief residence in the blood, they migrate into the tissues and are transformed into macrophages. They are less mature than macrophages, as suggested by fewer surface receptors, cytoplasmic organelles, and enzymes than the latter. Monocytes are larger than polymorphonuclear leukocytes, are actively phagocytic, and constitute 2 to 10% of the total white blood cell count in humans. The monocyte in the blood circulation is 15 to 25 μm in diameter. It has grayish-blue cytoplasm that contains lysosomes with enzymes such as acid phosphatase, arginase cachetepsin, collagenase, deoxyribonuclease, lipase, glucosidase, and plasminogen activator. The cell has a reniform nucleus with delicate lace-like chromatin. The monocyte has surface receptors such as the Fc receptor for IgG and a receptor for CR3. It is actively phagocytic and plays a significant role in antigen processing. Monocyte numbers are elevated in both benign and malignant conditions. Certain infections stimulate a reactive type of monocytosis, such as in tuberculosis, brucellosis, HIV-1 infection, and malaria.

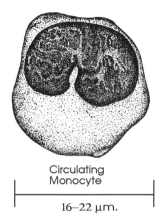

Circulating
Monocyte

16–22 µm.

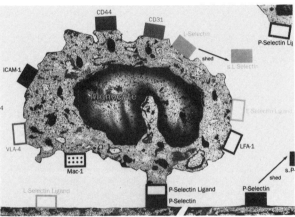

Monocyte

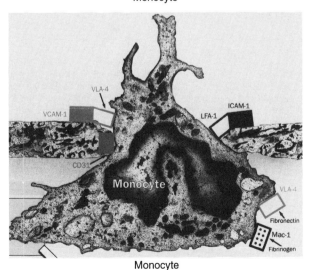

Monocyte

monocyte-derived neutrophil chemotactic factor

Refer to interleukin-8.

monocyte-phagocytic system

A system of cells that provides nonspecific immunity and is dependent on the activity of the monocyte/macrophage lineage cells, which are especially prominent in the spleen.

monogamous bivalency

The binding of a bivalent antibody molecule, such as IgG, with two identical antigenic determinants or epitopes on the same antigen molecule, in contrast to each Fab region of the IgG molecule uniting with an identical antigenic determinant on two separate antigen molecule. For this monogamous binding to take place, the epitopes must be positioned on the surface of the antigen molecule in such a manner that the binding of one Fab region to an epitope can position the remaining Fab of the IgG molecule for easy interaction with an adjacent identical epitope. Interaction of this type represents high affinity of binding, which lends a stability to the antigen-antibody complex. The combination of one IgM molecule to multiple epitopes on a single molecule of antigen would represent monogamous multivalency.

Monogamous Bivalency   Monogamous Multivalency

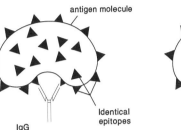 

antigen molecule

Identical
epitopes

IgG

IgM molecule

monokine

A cytokine produced by monocytes. Any one of a group of biologically active factors secreted by monocytes and macrophages that has a regulatory effect on the function of other cells such as lymphocytes. Examples include interleukin-1 and tumor necrosis factor.

monomers

(1) Subunits of an immunoglobulin molecule, *viz.,* two heavy and two light chains that comprise the four polypeptide chains of a typical IgG molecule. (2) Also used to refer to a basic four-chain monomer or monomeric unit of an immunoglobulin molecule. In contrast to one monomeric immunoglobulin unit for IgG, IgM possesses five, which are linked together by disulfide bonds.

mononuclear cell

Leukocytes with single, round nuclei such as lymphocytes and macrophages, in contrast to polymorphonuclear leukocytes. Thus, the term refers to the mononuclear phagocytic system or to lymphocytes.

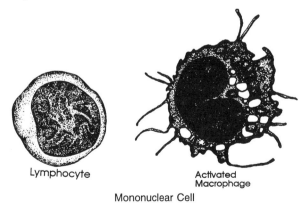

Lymphocyte   Activated
Macrophage

Mononuclear Cell

mononuclear phagocyte system

Mononuclear cells with pronounced phagocytic ability that are distributed extensively in lymphoid and other organs. "Mononuclear phagocyte system" should be used in place of the previously popular "reticulo-endothelial system" to describe this group of cells. Mononuclear phagocytes originate from stem cells in the bone marrow that first differentiate into monocytes that appear in the blood for approximately 24 h or more with final differentiation into macrophages in the tissues. Macrophages usually occupy perivascular areas. Liver macrophages are termed Kupffer cells, whereas those in the lung are alveolar macrophages. The microglia represent macrophages of the central nervous system, whereas histiocytes represent macrophages of connective tissue. Tissue stem cells are monocytes that have wandered from the blood into the tissues and may differentiate into macrophages. Mononuclear phagocytes have a variety of surface receptors that enable them to bind carbohydrates or such protein molecules as C3 via complement receptor 1 and complement receptor 3, and IgG and IgE through Fcγ and Fcε receptors. The surface expression of MHC class II molecules enables both monocytes and macrophages to serve as antigen-presenting cells to CD4+ T lymphocytes. Mononuclear phagocytes secrete a rich array of molecular substances with various functions. A few of these include interleukin-1; tumor necrosis factor α; interleukin-6; C2, C3, C4, and factor B complement proteins; prostaglandins; leukotrienes; and other substances.

Development of Mononuclear Phagocyte System

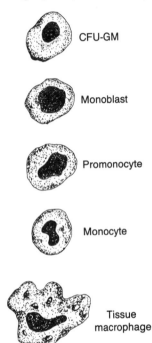

CFU-GM

Monoblast

Promonocyte

Monocyte

Tissue macrophage

monospecific antiserum

An antiserum against only a single antigen or epitope.

monovalent

Univalent. A combining power of one.

monovalent antiserum

A monospecific antiserum.

Montenegro test

A diagnostic assay for South American leishmaniasis induced by *Leishmania brasiliensis*. The intracutaneous injection of a polysaccharide antigen derived from the causative agent induces a delayed hypersensitivity response in the patient.

Montgu, Lady Mary Wortley (1689–1762)

British. Lived in Turkey, with her husband the British Ambassador, where she learned the practice of smallpox inoculation of well children in order to prevent their contracting a virulent form of the disease. She introduced the practice to England when she returned there.

Moro test

A variant of the tuberculin test in which tuberculin is incorporated into an ointment that is applied to the skin to permit the tuberculin to enter the body by inunction.

moth-eaten mouse

A C57Bl/6J mouse strain mutant designated (me/me). These animals are especially prone to develop autoimmune disease and infectious disease. They have areas of hair loss. Immunologically, they develop polyclonal hypergammaglobulinemia, defective cell-mediated immunity, and a type of autoimmune disease in which immune deposits are found on the kidneys.

MOTT (mycobacteria other than *Mycobacterium tuberculosis*)

An acronym for mycobacteria other than those that induce tuberculosis. Their recognition is increasing.

MOTT cell

A type of plasma cell in which refractile eosinophilic inclusion bodies that resemble Russell bodies are found. It is associated with African sleeping sickness. It is demonstrable in periarteriolar cuffs in the brain of patients in late stages of African trypanosomiasis.

mouse hepatitis virus (MHV)

A DNA virus that causes murine hepatitis and encephalitis. MHV infects oligodendrocytes and leads to demyelination without the presence of immune system cells.

MRL-lpr/lpr mice

A mouse strain genetically prone to develop lupus erythematosus-like disease spontaneously. Its congenic subline is MRL-+/+. The lymphoproliferation *(lpr)* gene in the former strain is associated with development of autoimmune disease, i.e., murine lupus. Although the MRL-+/+ mice are not normal immunologically, they develop autoimmune disease only late in life and without lymphadenopathy. MRL-lpr/lpr mice differ from New Zealand mice mainly in the development of striking lymphadenopathy in both males and females of the MRL-lpr/lpr strain between 8 and 16 weeks of age with a 100-fold increase in lymph node weight. Numerous Thy-1+, Ly-1+, Ly-2−, L3T4− lymphocytes that express and rearrange α and β genes of the T cell receptor, but fail to rearrange immunoglobulin genes are present in the lymph nodes. Multiple antinuclear antibodies, including anti-Sm, are among serological features of murine lupus in the MRL-lpr/lpr mouse model. These are associated with the development of immune complexes which mediate glomerulonephritis. Although the *lpr* gene is clearly significant in the pathogenesis of autoimmunity, the development of anti-DNA and anti-Sm even in low titers and late in life in the MRL-+/+ congenic line points to the role of factors other than the *lpr* gene in the development of autoimmunity in this strain.

mucocutaneous candidiasis

Cellular immunodeficiency is associated with this chronic *Candida* infection of the skin, mucous membranes, nails, and hair, with about 50% of patients manifesting endocrine abnormalities. Cell-mediated immunity to *Candida* antigens alone is absent or suppressed. The individual manifests anergy following the injection of *Candida* antigen into the skin. Immunity to other infectious agents, including other fungi, bacteria, and viruses, is not impaired. The B cell limb of the immune response, even to *Candida* antigens, does not appear to be affected. The antibody response to *Candida* and other antigens is within normal limits. The relative numbers of both T and B lymphocytes are normal, and immunoglobulins are at normal or elevated levels. Four clinical patterns have been described. The most severe is known as early chronic mucocutaneous candidiasis with granuloma and hyperkeratotic scales on the nails or face. These have an associated endocrinopathy in about 50% of the cases. The second type is late-onset chronic mucocutaneous candidiasis, which involves the oral cavity or occasionally the nails. The third form is transmitted as an autosomal recessive trait and is usually not associated with endocrine abnormalities. It is a mild to moderately severe disorder. The fourth form is known as juvenile familial polyendocrinopathy with candidiasis which may be associated with hypoparathyroidism with or without Addison's disease. Those individuals in whom endocrinopathy is associated with mucocutaneous candidiasis may demonstrate autoantibodies against the endocrine tissue involved. In addition to the immunologic abnormalities described above, there is diminished formation of lymphokines, e.g., macrophage migration inhibitory factor (MIF), directed against *Candida* antigens. Recommended treatment includes antimycotic agents and immunologic intervention designed to improve resistance of the host.

mucocutaneous lymph node syndrome

Refer to Kawasaki's disease.

mucosa homing

The selective return of immunologically reactive lymphoid cells that originated in mucosal follicles, migrated to other anatomical locations, and then returned to their site of origin in mucosal areas.

mucosal immune system

Aggregates of lymphoid tissues or lymphocytes near mucosal surfaces of the respiratory, gastrointestinal, and urogenital tracts. There is local synthesis of secretory IgA and T cell immunity at these sites. Research in this field is so great that a special society entitled The Society for Mucosal Immunology has been established to represent interest in this area.

Mucosal Colonization

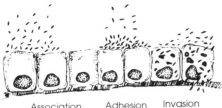

Association   Adhesion   Invasion

Mucosal Colonization

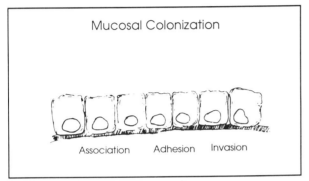

Association   Adhesion   Invasion

Mucosal Immune System

multilocus probes (MLPs)

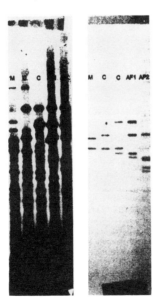

Multilocus Probes   **Single Locus Probes**

Probes used to identify multiple related sequences distributed throughout each person's genome. Multilocus probes may reveal as many as 20 separate alleles. Because of this multiplicity of alleles, there is only a remote possibility that two unrelated persons would share the same pattern, i.e., about 1 in 30 billion. There is, however, a problem in disciphering the multibanded arrangement of minisatellite RFLPs, as it is difficult to ascertain which bands are allelic. Mutation rates of minisatellite HVRs remain to be demonstrated, but are recognized occasionally. Used in resolving cases of disputed parentage.

multiple autoimmune disorders (MAD)

In type I MAD, a patient must manifest a minimum of two of the diseases designated Addison's disease, mucocutaneous candidiasis, or hypoparathyroidism. Type II MAD is known as Schmidt syndrome, in which patients manifest at least two conditions from a category that includes autoimmune thyroid disease, Addison's disease, mucocutaneous candaiasis, and insulin-dependent diabetes mellitus, with or without hypopituitarism.

multiple emulsion adjuvant

Water-in-oil-in-water emulsion adjuvant.

multiple myeloma

A plasmacytoma or plasma cell neoplasm associated with the production of a paraprotein that appears in the serum. The neoplastic plasma cells usually synthesize and secrete monoclonal, highly homogenous immunoglobulins. Serum electrophoresis reveals a narrow monoclonal band in 98 to 99% of patients. IgG paraimmunoglobulin manifests in 80% of myeloma patients, while 15% manifest monoclonal IgA. A few cases of the IgD and IgE types have been described. Homogeneous light chain dimers, which are identical to the corresponding light chain portion of immunoglobulin in the individual's blood, appear in the urine. These light chain dimers in the urine are called Bence-Jones proteins. These segments of light polypeptide chains do not represent degradation products of immunoglobulin, since they are synthesized separately from it. The disease affects 3 in 100,000 persons, usually men over 50 years of age. Patients develop anemia, anorexia, and weakness. The tumor infiltrates the bone marrow cavities, ultimately leading to erosion of the bone cortex. This may take years. Osteolytic lesions are the hallmark of multiple myeloma. The long bones, ribs, vertebrae, and skull manifest diffuse osteoporosis which leads to the appearance of punched-out areas and pathologic fractions. Tumor invasion of the marrow and erosion of the cortex as well as osteoclast-activating substances produce the bone lesions. Lung or renal infections may also occur. Hypogammaglobulinemia results from decreased functioning of normal plasma cells that leads to diminished antibody to combat infections. The malignant plasma cells produce an excess of nonsense paraimmunoglobulin which does not protect against infection. There is also defective phagocytic activity. Patients may have altered B cell function and increased susceptibility to pyogenic infections. Some patients may develop myeloma kidney, signified by proteinuria, followed by oliguria, kidney failure, and possibly death.

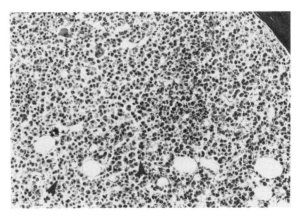

bone marrow
plasma cell
myeloma

Multiple Myeloma

multiple sclerosis (MS)

A demyelinating nervous system disease of unknown cause. It is most frequent in young adult females and has an incidence of 1 in 2500 individuals in the U.S. MS shows a disease association with HLA-A3, B7, and Dw2 haplotypes. Patients express multiple neurological symptoms that are worse at some times than others. They have paresthesias, muscle weakness, visual and gait disturbances, ataxia, and hyperactive tendon reflexes. There is infiltration of lymphocytes and macrophages in the nervous system which facilitates demyelination. Autoimmune mechanisms mediated by T cells, which constitute the majority of infiltrating lymphocytes, are involved. At least 20 viruses have been suggested to play a role in the etiology of MS. Infected oligodendrocytes are destroyed by the immune mechanism, and there may be also "innocent bystander" demyelination. Antibodies against HTLV-I GAG (p24) protein have been identified in the cerebrospinal fluid of MS patients. HTLV-I gene sequences have been identified in monocytes of MS patients. An oligoclonal increase in CSF IgG occurs in 90% of MS patients. There is inflammation, demyelination, and glial scarring. Paraventricular, frontal, and temporal areas of the brain are first involved, followed by regions of the brain stem, optic tracts, and white matter of the cortex with patchy lesions of the spinal cord. Attempts at treatment have included cop-1, a polypeptide mixture that resembles myelin basic protein, and numerous other agents.

Multiple sclerosis (100 x concentrate)

Zone electrophoresis patterns of cerebrospinal fluid

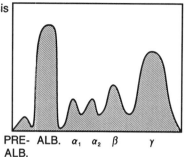

PRE- ALB. α_1 α_2 β γ
ALB.

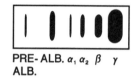

PRE- ALB. α_1 α_2 β γ
ALB.

Multiple Sclerosis

multivalent

In immunology, antibody or antigen molecules with a combining power greater than two.

multivalent antiserum

An immune serum preparation containing antibodies specific for more than two antigens. Multivalent means possessing more than two binding sites.

multivalent vaccine

Refer to polyvalent vaccine.

mumps vaccine

An attenuated virus vaccine prepared from virus generated in chick embryo cell cultures.

muramyl dipeptide (MDP)

N-Acetyl-muramyl-L-alanyl-D-isoglutamine. The active principle responsible for the immunologic adjuvant properties of complete Freund's adjuvant. It is an extract of the peptidoglycan of the cell walls of mycobacteria in complete Freund's adjuvant that has the immunopotentiating property of inducing delayed-type hypersensitivity and boosting antibody responses. It induces fever and lyses blood platelets and may produce a temporary leukopenia. However, purified derivatives without adverse side effects have been prepared for use as immunologic adjuvants and may prove useful for use in human vaccines.

N-acetylmuramyl-L-alanyl-D-isoglutamine

myasthenia gravis (MG)

An autoantibody-mediated autoimmune disease. Antibodies specific for the nicotinic acetylcholine receptor (AChR) of skeletal muscle react with the postsynaptic membrane at the neuromuscular junction and diminish the number of functional receptors. Patients develop muscular weakness and some voluntary muscle fatigue. Thus, MG represents a receptor disease mediated by antibodies. The nicotinic

AChR is the autoantigen. Contemporary research hopes to identify epitopes on the autoantigen(s) that interact with B and T cells in an autoimmune response. AChR is a four-subunit transmembrane protein. Most autoantibodies in humans are against the main immunogenic regions (MIR). Antibodies against the MIR crosslink AChR molecules, leading to their internalization and lysosomal degradation. This leads to a decreased number of postsynaptic membrane AChRs. Humans with MG and animals immunized against AChR develop circulating antibodies and clinical manifestations of MG. A subgroup of patients have seronegative MG; they resemble classic MG patients clinically, but have no anti-AChR antibodies in their circulation. Since the IgG anti-AChR autoantibodies cross the placenta from mother to fetus, newborns of mothers with this disease may also manifest signs and symptoms of the disease. Neonatal MG establishes the antibody-mediated autoimmune nature of the disease. The thymus of an MG patient may reveal either lymphofollicular hyperplasia (70%) or thymoma (10%). Anti-AChR synthesizing B cells and T helper lymphocytes may be found in hyperplastic follicles. These are often encircled by myoid cells that express AChR. Interdigitating follicular dendritic cells closely associated with myoid cells have been suggested to present AChR autoantigen to autoreactive T helper lymphocytes. Antiidiotypic antibodies have been used to suppress or enhance experimental autoimmune myasthenia gravis (EAMG), depending on the antibody concentration employed. Conjugate immunotoxins to anti-Id antibodies have been able to suppress autoimmunity to AChR. Both thymectomy and anticholinesterase drugs have proven useful in treatment.

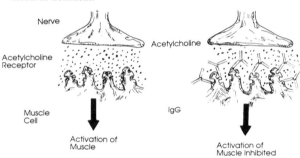

Myasthenia Gravis

mutant

An adjective that describes a mutation that may have occurred in a gene, protein, or cell.

mutation

A structural change in a gene that leads to a sudden and stable alteration in the genotype of a cell, virus, or organism. It is a heritable change in the genome of a cell, a virus, or an organism apart from that induced through the incorporation of "foreign" DNA. It represents an alteration in DNA's base sequence. Germ cell mutations may be inherited by future generations, whereas somatic cell mutations are inherited only by the progeny of that cell produced through mitotic division. A point mutation is an alteration in a single base pair. Mutations in chromosomes may be expressed as translocation, deletion, inversion, or duplication.

c-*myb*

A protooncogene that codes for the 75- to 89-kD phosphoprotein designated c-Myb in the nucleus, which immature hematopoietic cells express during differentiation. When casein kinase II phosphorylates Myb at an N-terminal site, this blocks the union of Myb to DNA and blocks continued activation. This phosphorylation site is deleted during oncogenic transformation, which permits Myb to combine with DNA.

myc

An oncogene designated v-*myc* when isolated from the avian myelocytomatosis retrovirus and designated c-*myc* when referring to the cellular homologue. Two others designated N-L-*myc* have been cloned. *myc* Genes are activated by overexpression either by upregulation, caused by transcriptional regulatory signal mutations in the first intron, or by gene amplification. Normal tissues contain c-*myc*. When c-*myc* is in its normal position on chromosome 8, it remains transcriptionally silent, but when it is translocated, as in Burkitt's lymphoma, it may become activated. The protooncogene c-*myc* is amplified in early carcinoma of the uterine cervix, lung cancer, and promyelocytic leukemia.

mycobacterial adjuvants

Substances that have long been used to enhance both humoral and cellular immune responses to antigen. Killed, dried mycobacteria, including *Mycobacterium tuberculosis,* among other strains, are ground and suspended in light-weight mineral oil. With the aid of an emulsifying agent such as arlacel A, added antigen in aqueous medium is incorporated to produce a water-in-oil emulsion. This mixture is used for immunization. The mycobacteria are especially effective in stimulating cell-mediated immunity to the antigen. The administration of this adjuvant without antigen may induce adjuvant arthritis in rats. Incorporation of normal tissues such as thyroid or adrenal into Freund's complete adjuvant may induce autoimmune disease if reinoculated into the animal of origin or other members of the strain with the same genetic background.

mycobacterial peptidoglycolipid

A constituent of the wax D fraction of mycobacteria. The wax D fraction of *Mycobacterium tuberculosis* var. *hominis* contains the adjuvant principle associated with such mycobacterial adjuvants as Freund's adjuvant. By electron microscopy, the peptidoglycolipid has a homogeneous, intertwined filamentous structure.

Mycoplasma-AIDS link

A mechanism postulated by L. Montagnier for AIDS development. HIV-1 virus binds to cells first activated by *Mycoplasma* infection.

mycosis fungoides

A chronic disorder involving the lympho-reticular system. There is major involvement in which the skin appears scaly with eczematous areas that are erythematous. There is infiltration by lichenified plaques. Finally, ulcers and neoplasms of the internal organs and lymph nodes develop. Cells in the skin lesion reveal markers that identify them as T lymphocytes. This disease usually appears after 50 years of age and is more frequent in males than females and in Blacks than Caucasians. There is an increase in the number of null cells in the blood circulation, where there is a simultaneous decrease in the numbers of B and T lymphocytes. T cell immunity is diminished both *in vitro* and *in vivo,* as revealed by diminished lymphocyte unresponsiveness to mitogens and by a poor response and skin test. Immunoglobulin A and E levels may be elevated in the serum.

myelin basic protein (MBP)

A principal constituent of the lipoprotein myelin that first appears during late embryogenesis. It is a 19-kD protein that is increased in multiple sclerosis patients who may generate T lymphocyte reactivity against MBP. T lymphocytes with the V β-17 variant of the T cell receptor are especially prone to react to MBP.

myelin basic protein antibodies

Antibodies against myelin proteins have been investigated for their possible role in the demyelination that accompanies multiple sclerosis, acute idiopathic optic neuritis, Guillain-Barré syndrome, chronic relapsing polyradiculoneuritis, carcinomatous polyneuropathy, and subacute sclerosing panencephalitis. Antibodies against myelin basic protein are well known to play a role in experimental allergic encephalomyelitis in laboratory animals, but the role they play in patients with multiple sclerosis or other neurological diseases has yet to be established.

myeloid antigen

A surface epitope of myeloid leukocytes. Examples include CD13, CD14, and CD33. A poor prognosis is indicated when the leukocytes of a patient with acute lymphocytic leukemia express myeloid antigens. The expression of these antigens is a better indicator of decreased survival than are other features of the disease.

myeloid cell series

An immature bone marrow cell (myeloblast) that is a precursor of the polymorphonuclear leukocyte series. This 18-μm diameter cell has a relatively large nucleus with finely distributed chromatin and two conspicuous nucleoli. The cytoplasm is basophilic when stained. During maturation, the cytoplasm becomes populated with large azurophilic primary granules, representing the promyelocyte stage. Later, the specific or secondary granules appear, representing the myelocyte stage. The nucleoli vanish as the nuclear chromatin forms dense aggregates. The chromatin in the nucleus condenses, and the cells no longer divide at this metamyelocyte stage of development. The nucleus assumes a sausage-like configuration known as a band. This subsequently develops into a three-lobed polymorphonuclear leukocyte, which develops into the neutrophils, eosinophils, and basophils that constitute myeloid cells. These latter three types are present in normal peripheral blood.

myeloma

A plasmacytoma that represents proliferation of a neoplastic plasma cell clone in bone marrow with the production of a monoclonal immunoglobulin paraprotein. Besides myeloma that occurs in humans, the experimental variety can be induced in certain inbred mouse strains such as BALB/c by intraperitoneal injection of mineral oil.

myeloma, IgD

A myeloma in which the monoclonal immunoglobulin is IgD. It constitutes 1 to 2% of myelomas and usually occurs in older males. There is lymphadenopathy and hepatosplenomegaly. Approximately one half of the cases develop dissemination beyond the bones. Patients develop osteolysis, hypercalcemia, anemia, azotemia, and aberrant plasma cells and plasma blasts.

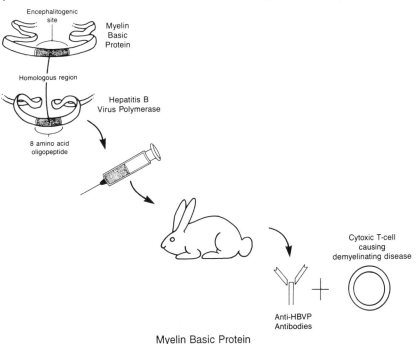

Myelin Basic Protein

myeloma protein

The immunoglobulin synthesized in excess in patients with multiple myeloma (plasmacytoma). Myeloma proteins are products of proliferating plasma cells of a malignant clone. The heavy and light chains are usually assembled to produce the homogeneous monoclonal paraimmunoglobulin, but if the synthesis of light chains is exclusive or exceeds that of heavy chains, a Bence-Jones protein may appear in addition to the paraimmunoglobulin or it may occur alone. A myeloma protein may be either a whole molecule of monoclonal immunoglobulin or part of the molecule synthesized by malignant plasma cells.

myelomatosis

A condition in which bone marrow plasma cells undergo malignant transformation and produce excessive homogeneous monoclonal immunoglobulin molecules that represent a paraprotein of a specific immunoglobulin isotype such as IgG or IgA. IgD and IgE myelomas also occur from time to time. Serum electrophoresis reveals a clearly demarcated band. Isoelectric focusing shows a classic monoclonal banding pattern. Some of the patients also have Bence-Jones protein in the urine. Patients are often males past 50 years of age and commonly present with spontaneous bone fracture or anemia due to replacement of bone marrow.

myeloperoxidase

An enzyme present in the azurophil granules of neutrophilic leukocytes that catalyzes peroxidation of many microorganisms. Myeloperoxidase, in conjunction with hydrogen peroxidase and halide, has a bactericidal effect.

myeloperoxidase (MPO) deficiency

A lack of 116-kD myeloperoxidase in both neutrophils and monocytes. This enzyme is located in the primary granules of neutrophils. It possesses a heme ring which imparts a dark-green tint to the molecule. MPO deficiency has an autosomal recessive mode of inheritance. Clinically, affected patients have a mild version of chronic granulomatous disease. *Candida albicans* infections are frequent in this condition.

myocardial antibodies

Antibodies against myocardium that have been demonstrated in two-thirds of coronary artery bypass patients and in Dressler syndrome. Acute rheumatic fever patients often manifest myocardial antibodies reactive with sarcolemmal, myofibrillar or intermyofibrillar targets. Dilated cardiomyopathy patients and patients with systemic hypertension/autoimmune polyendocrinopathy may develop autoantibodies against myocardium.

myoglobin

Oxygen-storing muscle protein that serves as a marker of muscle neoplasms, demonstrable by immunoperoxidase staining for surgical pathologic diagnosis.

myoid cell

A cell present in the neonatal thymus of humans and other species. It contains skeletal myofibrils.

N

N-linked oligosaccharide

An oligosaccharide that is covalently linked to asparagine residues in protein molecules. N-linked oligosaccharide manifests a core structure that is branched and comprised of two *N*-acetylglucosamine residues and three mannose residues. There are three types that differ on the basis of their exterior branches: (1) high-mannose oligosaccharide reveals two to six additional mannose residues linked to the polysaccharide core; (2) complex oligosaccharide comprised of two to five terminal branches that consist of *N*-acetylglucosamine, galactose, often *N*-acetylneuraminic acid, and occasionally fucose or another sugar; and (3) hybrid molecules that reveal characteristics of high-mannose and complex oligosaccharides.

N region

A brief segment of an immunoglobulin molecule's or T cell receptor chain's variable region that is not encoded by germline genes, but instead by brief nucleotide (N) insertions at recombinational junctions. These N nucleotides may be present both 3′ and 5′ to the rearranged immunoglobulin heavy chain's D gene segment as well as at the V-J, V-D-J, and D-D junctions of the variable region genes of the T lymphocyte receptor.

N-region diversification

In junction diversity, the addition at random of nucleotides that are not present in the genomic sequence at VD, DJ, and VJ junctions. Terminal deoxynucleotidyl transferase (TdT) catalyzes N-region diversification, which takes place in TCR α and β genes and in Ig heavy chain genes but not in Ig light chain genes.

N-terminus

The amino end of a polypeptide chain bearing a free amino-NH_2 group.

NAP

Neutrophil alkaline phosphatase.

NAP-1

Neutrophil attractant or activation protein-1. Refer to interleukin-8.

naproxen (2-naphthaleneacetic acid, 6-methoxy-α-methyl)

An antiinflammatory drug used in the treatment of arthritis, especially rheumatoid arthritis of children and adults, as well as ankylosing spondylitis.

Naproxen

native immunity

Genetically determined host responsiveness that prevents healthy humans from becoming infected under normal circumstances by selected microorganisms that usually infect animals. This may be altered in the case of profound immunosuppression of humans, as in the case of acquired immune deficiency syndrome in which humans become infected with microorganisms such as *Mycobacterium avium intracellulare*.

natural antibody

Antibodies found in the serum of an individual who has no known previous contact with that antigen, such as by previous immunization or infection with a microorganism containing that antigen. The anti-A and anti-B antibodies related to the ABO blood group system are natural antibodies. Natural antibodies may be a consequence of exposure to cross-reacting antigen(s), e.g., ABO blood group antibodies resulting from exposure to bacterial antigens in the gut.

natural fluorescence

Autofluorescence.

natural immunity

Innate immune mechanisms that do not depend upon previous exposure to an antigen. The skin, mucous membranes, and other barriers to infection; lysozyme in tears and other antibacterial molecules; and natural killer cells are among the numerous factors that contribute to natural resistance.

natural killer (NK) cells

Natural Killer (NK) Cell

9–12μm

Cells that attack and destroy certain virus-infected cells. They constitute an important part of the natural immune system, do not require prior contact with antigen, and are not MHC restricted by the major histocompatibility complex (MHC) antigens. NK cells are lymphoid cells of the natural immune system that express cytotoxicity against various nucleated cells including tumor cells and virus-infected cells. NK cells, killer (K) cells, or antibody-dependent cell-mediated cytotoxicity (ADCC) cells induce lysis through the action of antibody. Immunologic memory is not involved as previous contact with the antigen is not necessary for NK cell activity. The NK cell is approximately 15 μm in diameter and has a kidney-shaped nucleus with several, often three, large cytoplasmic granules. The cells are also called large granular lymphocytes (LGL). In addition to the ability to kill selected tumor cells and some virus-infected cells, they also participate in antibody-dependent cell-mediated cytotoxicity (ADCC) by anchoring antibody to the cell surface through an Fc γ receptor. Thus, they are able to destroy antibody-coated nucleated cells. NK cells are believed to represent a significant part of the natural immune defense against spontaneously developing neoplastic cells and against infection by viruses. NK cell activity is measured by a ^{51}Cr release assay employing the K562 erythroleukemia cell line as a target.

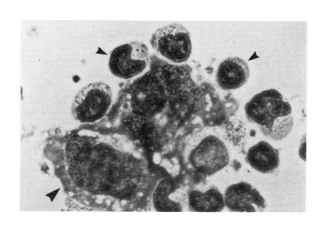

▲ NK cells
▲ K562 target cell

natural passive immunity
The transfer of IgG antibodies across the placenta from mother to child.

natural resistance
Refer to natural immunity.

natural suppressor cells
Lymphocytes that demonstrate nonspecific inhibition of some immune responses. These cells have been found following total irradiation and in neonatal animals. They may share a common lineage and function with NK cells.

naturally acquired immunity
Immunity that develops as a consequence of unplanned and coincidental contact with an antigen, as contrasted with immunity that is acquired through deliberate immunization.

necrosis
Tissue death.

necrotaxis
The attraction of leukocytes to dead or injured cells and tissues.

negative phase
The decrease in antibody titer immediately following injection of a second or booster dose of antigen to an animal previously given a primary injection of the same antigen. Following this initial drop of preformed antibody in the circulation, there is a rapid and pronounced rise in antibody titer, representing immunologic memory.

negative selection
The process whereby those thymocytes that recognize *self antigens* in the context of self MHC undergo clonal deletion (apoptosis) or clonal anergy (inactivation). The resulting cell population is self MHC restricted and self antigen tolerant. (Refer to positive selection.)

Neisser-Wechsberg phenomenon
The deviation of complement by antibody. Although complement does not react with antigen alone, it demonstrates weak affinity for unreacted antibody.

neoantigen
Modification of proteins by phosphorylation or specific proteolysis may change their covalent architecture to yield new antigenic determinants or epitopes termed neoantigens. An epitope that is newly expressed on cells during development or in neoplasia. Neoantigens include tumor-associated antigens. New antigenic determinants may also emerge when a protein changes conformation or when a molecule is split, exposing previously unexpressed epitopes.

Neoantigenic determinant

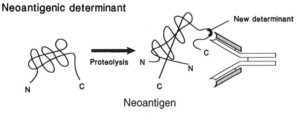

Neoantigen

neonatal thymectomy syndrome
Refer to wasting disease.

neopterin
A guanosine triphosphate metabolite which macrophages synthesize following their stimulation by γ interferon from activated T lymphocytes. Neopterin levels in both serum and urine of HIV-1 infected patients rise as the infection progresses. This together with diminishing CD4+ lymphocyte levels reflect progression of HIV-1 infection to clinical AIDS.

nephelometry
A technique to assay proteins and other biological materials through the formation of a precipitate of antigen and homologous antibody. The assay depends on the turbidity or cloudiness of a suspension. It is based on determination of the degree to which light is scattered when a helium-neon laser beam is directed through the suspension. The antigen concentration is ascertained using a standard curve devised from the light scatter produced by solutions of known antigen concentration. This method is used by many clinical immunology laboratories for the quantification of complement components and immunoglobulins in patients' sera or other body fluids.

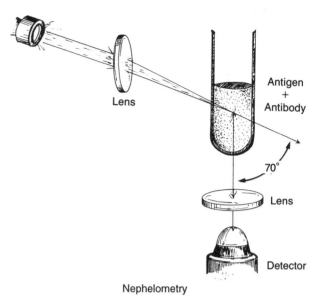

Nephelometry

nephritic factor
C3 nephritic factor (C3NeF) is a substance found in the serum of patients with type II membranoproliferative glomerulonephritis, i.e., dense deposit disease. This factor is able to activate the alternate complement pathway. It is an immunoglobulin that interacts with alternate complement pathway C3 convertase and stabilizes it. Thus, it activates the pathway and leads to the generation of complement fragments that are biologically active. C3NeF is an autoantibody against alternate pathway C3 convertase. Membranoproliferative glomerulonephritis patients have a genetic predisposition to develop the disease. The excessive C3 consumption leads to hypocomplementemia.

nephritic syndrome
A clinical complex of acute onset characterized by hematuria with red cell and hemoglobin casts in the urine, oliguria, azotemia, and hypertension. There may be limited proteinuria. Inflammatory reactions within glomeruli cause injury to capillary walls which permits the release of erythrocytes into the urine. Acute diffuse proliferative glomerulonephritis is an example of a primary glomerular disease associated with acute nephritic syndrome.

nephrotic syndrome
A clinical complex that consists of massive proteinuria with the loss of greater than 3.5 g of protein per day; generalized edema; hypoalbuminemia, i.e., less than 3 g/dl; hyperlipidemia; and lipiduria. Nephrotic syndrome in children often follows lipoid nephrosis, whereas in adults it is frequently associated with membranous glomerulopathy. Both are primary glomerular diseases.

network hypothesis
Niels Jerne's theory that antiidiotypic antibodies form in response to the antigen-binding regions of antibody molecules or of lymphocyte surface receptors. These in turn elicit anti-antiidiotypic antibodies, etc. Each new immune response stimulated in this network interrupts the finely tuned immune network balance as antiidiotype antibodies are produced, eventually downregulating the response and bringing it back to homeostasis.

network theory
An hypothesis proposed by Niels Jerne which explains immunoregulation through a network of idiotype-antiidiotype reactions involving T cell receptors and the antigen-binding regions of antibody molecules, i.e., the paratope regions. Exposure to antigens interrupts the delicate balance of the idiotype-antiidiotype network, leading to the increased synthesis of some idiotypes as well as of the corresponding antiidiotypes, leading to modulation of the response. Antiidiotypes occur following immunization against selected antigens and may prevent the response to the antigen. Selected antiidiotypic antibodies have a binding site that is closely similar to the immunizing epitope. It is referred to as the internal image of the epitope. Other antiidiotypic antibodies are directed to idiotopes of the antigen-binding region and are not internal images. Antiidiotypic antibodies with an internal image

may be substituted for an antigen, leading to specific antigen-binding antibodies. These are the basis for so-called idiotypic vaccines in which the individual never has to be exposed to the infecting agent. Antiidiotypes may also block T cell receptors for the corresponding antigen.

neuraminidase

An enzyme that cleaves the glycosidic bond between neuraminic acid and other sugars. Neuraminic acid is a critical constituent of multiple cell surface glycoproteins and confers a negative charge on the cells. Cells treated with neuraminidase agglutinate more readily than do normal cells because of the diminished coulombic forces between them. Cells treated with neuraminidase activate the alternate complement pathway. Neuraminidase is produced by myxoviruses, paramyxoviruses, and such bacteria as *Clostridium perfrigens* and *Vibrio cholerae*. Neuraminidase together with hemagglutinin is found on the spikes of the influenza virus.

neurofilament

A marker, demonstrable by immunoperoxidase staining, for neural-derived tumors as well as selected endocrine neoplasms with neural differentiation.

neuroleukin

A cytokine synthesized in the brain and in T lymphocytes as a 56-kD protein that shares sequence homology with the gp120 of HIV-1 and with phosphohexose isomerase. Thus, the ability of the AIDS virus' gp120 to compete with neuroleukin for the neuroleukin receptor might be associated with AIDS dementia.

neuron specific enolase (NSE)

An enzyme of neurons and neuroendocrine cells, as well as their derived tumors, e.g., oat cell carcinoma of lung, demonstrable by immunoperoxidase staining. NSE occurs also in some neoplasms not derived from neurons or endocrine cells.

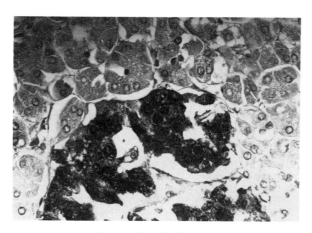

Neuron Specific Enolase

neuronal antibodies

Antibodies present in the cerebrospinal fluid of approximately three-fourths of systemic lupus erythematosus (SLE) patients with neuropsychiatric manifestations. 11% of SLE patients who do not manifest neuropsychiatric disease also develop them. Neuronal antibodies are identified by their interaction with human neuroblastoma cell lines. The presence of neuronal antibodies generally indicates CNS-SLE. The lack of neuronal antibodies in the serum and cerebrospinal fluid mitigates against CNS involvement in SLE patients.

neurotoxin

An exotoxin that can interfere with the conduction of a nerve impulse or block transmission at the synapse by linking to a voltage-gated Na^{2+} channel protein. *Corynebacterium diphtheriae*, *Clostridium tetani*, *C. botulinum*, and *Shigella dysenteriae* produce neurotoxin. An example of another neurotoxin is conotoxin.

neutralization

The inactivation of a microbial product such as a toxin by antibody or counteraction of a microorganism's infectivity, especially the neutralization of viruses.

Neutralizing Antibody Molecule

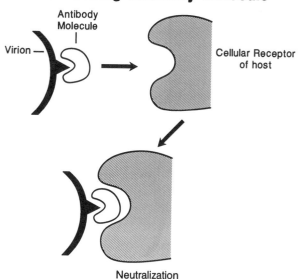

Neutralization

neutralization test

An assay based on the ability of antibody to inactivate the biological effects of an antigen or of a microorganism expressing it. Neutralization applies especially to inactivation of virus infectivity or of the biological activity of a microbial toxin.

neutropenia

A diminished number of polymorphonuclear neutrophilic leukocytes in the peripheral blood circulation.

neutrophil attracting peptide (NAP-2)

Chemoattractant of neutrophils to sites of platelet aggregation. NAP-2 competes weakly with IL-8 for the IL-8 type II receptor. However, since it can be found at much higher concentrations than IL-8 at platelet aggregation sites, NAP-2 is considered an active participant in the inflammatory process.

neutrophil-activating factor-1

Interleukin-8 (IL-8).

neutrophil-activating protein-1 (NAP-1)

Former term for IL-8.

neutrophil cytoplasmic antibodies

Autoantibodies specific for myeloid specific lysosomal enzymes. Anti-neutrophil cytoplasmic antibodies (ANCA) are of two types. The C-ANCA variety stains the cytoplasm where it reacts with alpha granule proteinase-3 (PR-3). By contrast, P-ANCA stains the perinuclear zone through its reaction with myeloperoxidase (MPO). 84 to 100% of Wegener's granulomatosis patients with generalized active disease develop C-ANCA, although fewer individuals with the limited form of Wegener's granulomatosis develop these antibodies. Organ involvement cannot be predicted from the identification of either C-ANCA or P-ANCA antibodies. P-ANCA antibodies reactive with myeloperoxidase may be found in patients with certain vasculitides that include Churg-Strauss syndrome, polyarteritis nodosa, microscopic polyarteritis and polyangiitis. A P-ANCA staining pattern unrelated to antibodies against PR-3 or to MPO has been described in inflammatory bowel disease (IBD). 59 to 84% of ulcerative colitis patients and 65 to 84% of primary sclerosing cholangitis patients are positive for P-ANCA. By contrast, only 10 to 20% of Crohn's disease patients are positive for it. These antibodies are classified as neutrophil nuclear antibodies (ANNA). Hep-2 cells are used to differentiate granulocyte-specific ANA from ANA that are not tissue specific.

neutrophil leukocyte

A peripheral blood polymorphonuclear leukocyte derived from the myeloid lineage. Neutrophils comprise 40 to 75% of the total white blood count numbering 2500 to 7500 cells per cubic millimeter. They are phagocytic cells and have a multilobed nucleus and azurophilic and specific granules that appear lilac following staining with Wright's or Giemsa stains. They may be attracted to a local site by such chemotactic factors as C5a. They are the principal cells of acute inflammation and actively phagocytize invading microorganisms. Besides serving as the first line of cellular defense in infection, they participate in such reactions as the uptake of antigen-antibody complexes in the Arthus reaction.

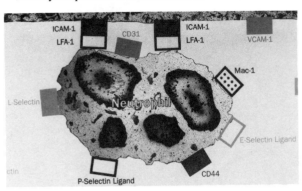

Neutrophil Leukocyte

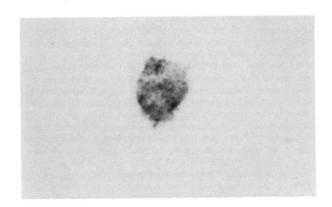

formazan crystals in PMN

Nitroblue Tetrazolium (NBT) Test

neutrophil microbicidal assay

A test that assesses the capacity of polymorphonuclear neutrophil leukocytes to kill intracellular bacteria.

New Zealand black (NZB) mice

An inbred strain of mice that serves as an animal model of autoimmune hemolytic anemia. They develop antinuclear antibodies in low titer, have defective T lymphocytes, have defects in DNA repair, and have B cells that are spontaneously activated.

New Zealand white (NZW) mice

An inbred strain of white mice that when mated with NZB strain, which develop autoimmunity, produce an F_1 generation of NZB/NZW mice that represent an animal model of autoimmune disease and especially of a lupus erythematosus-like condition.

Newcastle disease

Follicular conjunctivitis induced by an avian paramyxovirus that blocks the oxidative burst in phagocytes. Cytokines that produce fever are formed. There is recovery in approximately 1 to 2 weeks. In birds, the agent induces pneumoencephalitis, which is fatal.

Newcastle disease vaccines

(1) An inactivated virus raised in chick embryos that is incorporated into aluminum hydroxide gel adjuvant. (2) Live virus grown in chick embryos and attenuated in a graded manner. Strains with medium virulence are administered parenterally, and those that are less virulent are given to birds either in drinking water or as an aerosol.

Nezelof's syndrome

Hypoplasia of the thymus leading to a failure of the T lymphocyte compartment with no T cells and no T cell function. By contrast, B lymphocyte function remains intact. Thus, this is classified as a T lymphocyte immunodeficiency.

NF-AT

Nuclear factor of activated T lymphocytes. Refer to immunophilins.

NF-κB

A protein produced at the pre-B cell stage of B cell development. It is requisite for transcription of immunoglobulin light chains.

nick translation

A technique used to make a radioactive probe of a DNA segment. Nick translation signifies the movement of a nick, i.e., single-stranded break in the double-stranded helix, along a duplex DNA molecule.

NIP (4-hydroxy,5-iodo,3-nitrophenylacetyl)

Used as a hapten in experimental immunology.

nitroblue tetrazolium (NBT) test

An assay that evaluates the hexose monophosphate shunt in phagocytic cells. The soluble yellow dye, nitroblue tetrazolium, is taken up by neutrophils and monocytes during phagocytosis. In normal neutrophils, the NBT is reduced by enzymes to insoluble, dark blue formazan crystals within the cell. Neutrophils from patients with chronic granulomatous disease are unable to reduce the nitroblue tetrazolium. The ability to reduce NBT to the insoluble deep blue formazan crystals depends on the generation of superoxide in the neutrophil being tested.

NK 1.1

A natural killer (NK) cell alloantigen identified in selected inbred mouse strains such as C57BL/6 mice.

NK cell

Refer to natural killer cells.

NOD (nonobese diabetic) mouse

A mutant mouse strain that spontaneously develops type I, insulin-dependent diabetes mellitus, an autoimmune disease. There is an autosomal recessive pattern of inheritance for the NOD mutation. Lymphocytes infiltrate NOD mouse islets of Langerhans in the pancreas and kill β cells. There is a defect in the HLA-DQ part of the MHC class II region in humans with insulin-dependent diabetes mellitus and in the class II IA region of the mouse MHC class II. A major DNA segment is missing from the NOD mouse MHC IE region. When the IE segment is inserted or the IA defect is corrected in transgenic NOD mice, disease progression is halted.

non-A non-B hepatitis

Refer to hepatitis, non-A and non-B.

non-Hodgkin's lymphomas (NHL)

All lymphoid neoplasms without the characteristics of Hodgkin's disease constitute the non-Hodgkin's lymphomas. They are monoclonal lymphoid neoplasms that are very heterogeneous morphologically, antigenically, and with respect to kinetic phenotypes. They also differ greatly in clinical expression. They are divided into low, intermediate, and high grade. Low-grade lymphomas are treated conservatively, whereas intermediate- and high-grade lymphomas are aggressive and are appropriately treated. T lymphocyte, B lymphocyte, and NULL cell lymphomas make up the NHL group.

NON mouse

The normal control mouse for use in studies involving the NOD mouse strain that spontaneously develops type I (insulin-dependent) diabetes mellitus. The two strains differ only in genes associated with the development of diabetes.

nonadherent cell

A cell that fails to stick to a surface such as a culture flask. A lymphocyte is an example of a nonadherent cell, whereas macrophages readily adhere to the glass surface of a tissue culture flask.

nonidentity, reaction of

Refer to reaction of nonidentity.

nonimmunologic classic pathway activators

Selected microorganisms such as *Escherichia coli* and low virulence *Salmonella* strains as well as certain viruses such as parainfluenza react with C1q, leading to C1 activation without antibody. Thus, this represents classic pathway activation, which facilitates defense mechanisms. Various other substances such as myelin basic protein, denatured bacterial endotoxin, heparin, and urate crystal surfaces may also directly activate the classic complement pathway.

nonprecipitating antibodies

The addition of antigen in increments to an optimal amount of antibody precipitates only ~78% of the amount of antibody that would be precipitated by one step addition to the antigen. This demonstrates the presence of both precipitating and nonprecipitating antibodies. Although the nonprecipitating variety cannot lead to the formation of insoluble antigen-antibody complexes, they can be assimilated into precipitates that correspond to their specificity.

Rather than being univalent as was once believed, they may merely have a relatively low affinity for the homologous antigen. Monogamous bivalency, which describes the combination of high-affinity antibody with two antigenic determinants on the same antigen particle, represents an alternative explanation for the failure of these molecules to precipitate with their homologous antigen. The formation of nonprecipitating antibodies, which usually represents 10 to 15% of the antibody population produced, is dependent upon such variables as heterogeneity of the antigen, characteristics of the antibody, and animal species.

The equivalence zone is narrower with native proteins of 40 to 60 kD and their homologous antibodies than with polysaccharide antigens or aggregated denatured proteins and their specific antibody. The equivalence zone with synthetic polypeptide antigens varies with the individual compound used. The solubility of antibody-antigen complexes and the nature of the antigen are related to these variations at the equivalence zone. The extent of precipitation is dependent upon characteristics of both the antigen and antibody. At the equivalence zone, not all antigen and antibody molecules are present in the complexes. For example, rabbit anti-BSA (bovine serum albumin) precipitates only 46% of BSA at equivalence.

nonresponder

An animal that fails to generate an immune response following antigenic challenge. This may be genetically based, as in strain 13 guinea pigs which reveal unresponsiveness to selected antigens based upon genetic factors.

nonsecretor

An individual whose body secretions such as gastric juice, saliva, tears, and ovarian cyst mucin do not contain ABO blood group substances. Nonsecretors make up approximately one fifth of the population and are homozygous for the gene *se*.

nonspecific esterase (α naphthyl acetate esterase)

An enzyme of mononuclear phagocytes and lymphocytes that is demonstrable by cytochemical staining. There is diffuse granular staining of the cytoplasm of mononuclear phagocytes which may help to identify them. Some human T cells are positive for nonspecific esterase, but appear as one or several small localized dots within the T cell.

nonspecific fluorescence

Fluorescence emission that does not reflect antigen-antibody interaction and may confuse interpretation of immunofluorescence tests. Either free fluorochrome or fluorochrome tagging of proteins other than antibody such as serum albumin, α globulin, or β globulin may contribute to nonspecific fluorescence. Nonspecific staining is accounted for in appropriate controls.

nonspecific immunity

Mechanisms such as phagocytosis that nonspecifically remove invading microorganisms, as well as the action of chemical and physical barriers to infection such as acid in the stomach and the skin, respectively. Other nonspecific protective factors include lysozyme, β lysin, and interferon. Nonspecific or natural immunity does not depend on immunologic memory. Natural killer cells represent an important part of the natural immune cell system. Phagocytosis of invading microorganisms by polymorphonuclear neutrophils and monocytes represents another important aspect of nonspecific immunity.

nonspecific T cell suppressor factor

A CD8+ suppressor T lymphocyte soluble substance that nonspecifically suppresses the immune response.

nonspecific T lymphocyte helper factor

A soluble factor released by CD4+ helper T lymphocytes that nonspecifically activates other lymphocytes.

nonsquamous keratin (NSK)

A marker, demonstrable by immunoperoxidase staining, that is found in glandular epithelium and adenocarcinomas, but not in stratified squamous epithelium.

nonsterile immunity

Refer to premunition.

nonsteroidal anti-inflammatory drugs (NSAIDs)

A group of drugs used in the treatment of rheumatoid arthritis, gouty arthritis, ankylosing spondylitis, and osteoarthritis. The drugs are weak organic acids. They block prostaglandin synthesis by inhibiting cyclooxygenase and lipooxygenase. They also interrupt membrane-bound reactions such as NADPH oxidase in neutrophils, monocyte phospholipase C, and processes regulated by G protein. They also exert a number of other possible activities such as diminished generation of free radicals and superoxides which may alter intracellular cAMP levels, diminishing vasoactive mediator release from granulocytes, basophils, and mast cells.

nontissue-specific antigen

An antigen that is not confined to a single organ, but is distributed in more than one normal tissue or organ, such as nuclear antigens.

normal lymphocyte transfer reaction

The intracutaneous injection of an individual with peripheral blood lymphocytes from a genetically dissimilar, allogeneic member of the same species leads to the development of a local, erythematous reaction that becomes most pronounced after 48 h. The size of the reaction has been claimed to give some qualitative indication of histocompatibility or histoincompatibility between a donor and recipient. This test is not used in clinical practice.

Northern blotting (see below)

A method to identify specific mRNA molecules. Following denaturation of RNA in a particular preparation with formaldehyde to cause the molecule to unfold and become linear, the material is separated by size through gel electrophoresis and blotted onto a natural cellulose or nylon membrane. This is then exposed to a solution of labeled DNA "probe" for hybridization. This step is followed by autoradiography. Northern blotting corresponds to a similar method used for DNA fragments which is known as Southern blotting.

Nossal, Gustav Joseph Victor (1931–)

Australian immunologist whose seminal works have concentrated on antibodies and their formation. He is Director of the Walter and

Northern Blot

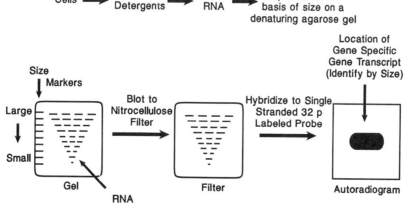

Northern Blot

Eliza Hall Institute of Medical Research in Melbourne. *Antibodies and Immunity,* 1969; *Antigens, Lymphoid Cells and the Immune Response (with Ada),* 1971.

NP (4-hydroxy,3-nitrophenylacetyl)

A hapten used in experimental immunology that exhibits limited crossreactivity with NIP.

NSAID (nonsteroidal anti-inflammatory drugs)

Used in the treatment of arthritis.

200-4 nuclear matrix protein

A marker expressed preferentially by malignant cells rather than by normal cells. Demonstrated by immunoperoxidase staining.

nucleoside phosphorylase

An enzyme that is only seldom decreased in immunodeficiency patients. It catalyzes inosine conversion into hypoxanthine.

nude mouse

A mouse strain that is hairless and has a congenital absence of the thymus and of T lymphocyte function. They serve as highly effective animal models to investigate immunologic consequences of not having a thymus. They fail to develop cell-mediated (T lymphocyte-mediated) immunity, are unable to reject allografts, and are unable to synthesize antibodies against the majority of antigens. Their B lymphocytes and natural killer cells are normal even though T lymphocytes are missing. Also called nu/nu mice. Valuable in the investigation of graft-vs.-host disease.

null cell

A lymphocyte that does not manifest any markers of T or B cells, including cluster of differentiation (CD) antigens or surface immunoglobulins. Approximately 20% of peripheral lymphocytes are null cells. They play a role in antibody-dependent cell-mediated cytotoxicity (ADCC). They may be the principal cell in certain malignancies such as acute lymphocytic leukemia of children. The three types of null cells include (1) undifferentiated stem cells that may mature into T or B lymphocytes, (2) cells with labile IgG and high-affinity Fc receptors that are resistant to trypsin, and (3) large granular lymphocytes that constitute NK and K cells.

null cell compartment

Null cells make up 37% of the bone marrow lymphocytes, i.e., they do not have any of the markers characteristic of B or T cell lineage. They may differentiate into either B or T cells upon appropriate induction, the mechanism of which is unknown. Some null cells differentiate into killer (K) cells by developing Fc and complement receptors. The NK cells are also present in this cell population. Null, K, and NK cells, like committed lymphocytes, also migrate to the peripheral lymphoid organs such as spleen and lymph nodes or to the thymus, but they represent only a very small fraction of the total cells present there. At all locations, the null cells are part of the rapidly renewed pool of immature cells with short lifespan (5 to 6 days). The null cells which have been committed to the T cell lineage migrate to the thymus to continue their differentiation.

null phenotype

The failure to express protein because the gene that encodes it is either defective or absent on both inherited haplotypes. Most of these involve erythrocytes, but are quite rare.

nylon wool

A material that has been used to fractionate T and B cells from a mixture of the two based upon the tendency for B cells to adhere to the nylon wool, whereas the T cells pass through. B cells are then eluted from the column. Previously, tissue typing laboratories used this technique to isolate B lymphocytes for MHC class II (B cell) typing. Magnetic beads have replaced nylon wool for lymphocyte T and B cell separation.

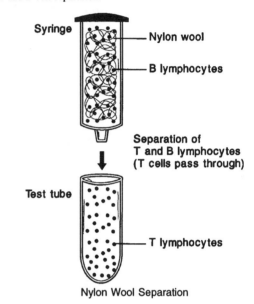

Nylon Wool Separation

NZB/NZW F₁ hybrid mice

A mouse strain genetically prone to develop lupus erythematosus-like disease spontaneously. This was the first murine lupus model, developed from the NZB/B₁ mouse model of autoimmune hemolytic anemia mated with the NZW mouse which develops a positive Coombs test, antinuclear antibodies, and glomerulonephritis. This F₁ hybrid develops various antinuclear autoantibodies and forms immune complexes that subsequently induce glomerulonephritis and shorten the lifespan due to renal disease. Hemolytic anemia is minimal in the F₁ strain compared to the NZB parent. NZB mice show greater lymphoid hyperplasia than do the F₁ hybrids, yet the latter show features that are remarkably similar to those observed in human lupus, such as major sex differences.

O

OA

Abbreviation for ovalbumin.

O125 (ovarian celomic)

Nonmucinous ovarian tumor antigen demonstrable with homologous antibody by immunoperoxidase staining. Selected mesotheliomas express this antigen as well.

O antigen

(1) A lipopolysaccharide-protein antigen of enteric microorganisms which is used for their serological classification. O antigens of the *Proteus* species serve as the basis for the Weil-Felix reaction which is employed to classify *Rickettsia*. O antigens of *Shigella* permit them to be subdivided into 40 serotypes. The exterior oligosaccharide repeating unit side chain is responsible for specificity and is joined to lipid A to form lipopolysaccharide and to lipid B. The O antigen is the most variable part of the lipopolysaccharide molecule.

(2) In the ABO blood group system, O antigen is an oligosaccharide precursor form of A and B antigens: a fucose-galactose-*N*-acetylglucosamine-glucose.

Blood Type	RBC surface antigen	Antibody in Serum
A	A antigen	Anti-B
B	B antigen	Anti-A
AB	AB antigens	no antibody
O	no A or B antigens	Both Anti-A and Anti-B

O Anitgen

O blood group

One of the groups described by Landsteiner in the ABO blood group system. Refer to ABO blood group system.

Oakley-Fulthorpe test

Double-diffusion type of precipitation test performed by incorporating antibody into agar which is placed in the tube followed by a layer of plain agar. A solution of antigen is placed on top of the plain agar in the tube and precipitation occurs where antigen and antibody meet in the plain agar layer.

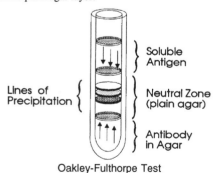

Oakley-Fulthorpe Test

Oct-2

A protein formed early in the development of B lymphocytes that has a role in immunoglobulin heavy chain transcription.

OKT monoclonal antibodies

Commercially available preparations used to enumerate human T cells according to their surface antigens to determine the immunophenotype. OKT designations have been replaced by CD designations.

OKT®3 (Orthoclone OKT®3)

A commercial mouse monoclonal antibody against the T cell surface marker CD3. It may be used, therapeutically, to diminish T cell reactivity in organ allotransplant recipients experiencing a rejection episode. OKT3 may act in concert with the complement system to induce T cell lysis, or it may act as an opsonin, rendering T cells susceptible to phagocytosis.

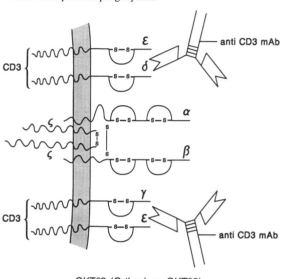

OKT®3 (Orthoclone OKT®3)

OKT4

Refer to CD4.

OKT8

Refer to CD8.

Old tuberculin (OT)

A broth culture, heat-concentrated filtrate of medium in which *Mycobacterium tuberculosis* microorganisms were grown. It was developed by Robert Koch for use in tuberculin skin tests nearly a century ago.

oligoclonal bands

When cerebrospinal fluid of some multiple sclerosis patients is electrophoresed in agarose gel, immunoglobulins with re-

Cerebrospinal Fluid Electrophoresis

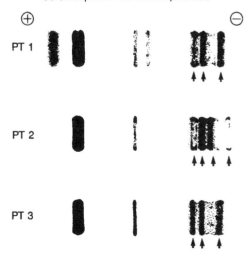

(gamma region)

↟ = oligoclonal IgG bands

stricted electrophoretic mobility may appear as multiple distinct bands in the γ region. Although nonspecific, 90 to 95% of multiple sclerosis patients show this. They may appear in selected other central nervous system diseases such as herpetic encephalitis, bacterial or viral meningitis, carcinomatosis, toxoplasmosis, neurosyphilis, progressive multifocal leukoencephalopathy, and subacute sclerosing panencephalitis and may appear briefly during the course of Guillian-Barre' disease, lupus erythematosus vasculitis, spinal cord compression, diabetes, and amyotrophic lateral sclerosis.

oligoclonal response

An immune response characterized by only a few separate clones of immunocompetent cells responding to yield a small number of immunoglobulin bands in agarose gel electrophoresis.

oligosaccharide determinant

An epitope or antigenic determinant of a polysaccharide hapten that consists of relatively few, i.e., two to seven, pentoses, hexoses, or heptoses united by glycoside linkages.

oncofetal antigens

Markers or epitopes present in fetal tissues during development, but not present, or found in minute quantities, in adult tissues. These cell-coded antigens may reappear in certain neoplasms of adults due to derepression of the gene responsible for their formation. Examples include carcinoembryonic antigen (CEA), which is found in the liver, intestine, and pancreas of the fetus, but also in both malignant and benign gastrointestinal conditions. Yet it is still useful to detect recurrence of adenocarcinoma of the colon based upon demonstration of CEA in the patient's serum; α-fetoprotein (AFP) is demonstrable in approximately 70% of hepatocellular carcinomas.

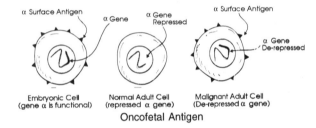

Oncofetal Antigen

oncogene(s)

Genes with the capacity to induce neoplastic transformation of cells. They are derived from either normal genes termed protooncogenes or from oncogenic RNA (oncorna) viruses. Their protein products are critical for regulation of gene expression or growth signal transduction. Translocation, gene amplification, and point mutation may lead to neoplastic transformation of protooncogene. Oncogenes may be revealed through use of viruses that induce tumors in animals or by derivation of tumor-causing genes from cancer cells. There are greater than 20 protooncogenes and cellular oncogenes in the human genome. An oncogene alone cannot produce cancer. It must be accompanied by malignant transformation which involves multiple genetic steps. Oncogenes encode four types of proteins that include growth factors, receptors, intracellular transducers, and nuclear transcription factors.

oncogene theory

A concept of carcinogenesis that assigns tumor development to latent retroviral gene activation through irradiation or carcinogens. These retroviral genes are considered to be normal constituents of the cell. Following activation, these oncogenes are presumed to govern the neoplasm through hormones that are synthesized and even the possible construction of a complete oncogenic virus. This concept states that all cells may potentially become malignant.

oncogenesis

The process whereby tumors develop.

oncogenic virus

Any virus, whether DNA or RNA, that can induce malignant transformation of cells. An example of a DNA virus would be human papillomavirus and an RNA virus would be retrovirus.

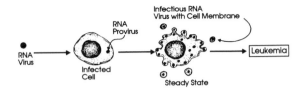

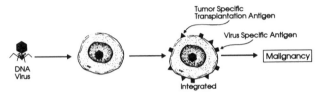

Oncogenic Virus

oncomouse

A commercially developed transgenic animal into which human genes have been introduced to make the mouse more susceptible to neoplasia. This transgenic mouse is used for both medical and pharmaceutical research.

one gene, one enzyme theory (historical)

An earlier hypothesis that proposed that one gene encodes one enzyme or other protein. Although basically true, it is now known that one gene encodes a single polypeptide chain, and it is necessary to splice out mRNA introns, comprised of junk DNA, before mRNA can be translated into a protein.

one-hit theory

A concept related to complement activation which states that only a single site of preparation is necessary for red cell lysis during complement-antibody-mediated injury to the cell membrane.

ontogeny

The development of an individual organism from conception to maturity.

open reading frame (ORF)

A length of RNA or DNA that encodes a protein and may signal the identification of a protein not described previously. An ORF begins with a start codon and does not contain a termination codon, but it ends at a stop codon.

opportunistic infection

An infection produced by microorganisms that are usually of relatively low virulence, but become more aggressive in subjects with altered or defective cell-mediated immunity. Susceptible individuals would include acquired immune deficiency (AIDS), severe combined immunodeficiency, and organ transplant recipients. Typical microorganisms producing opportunistic infections include *Pneumocystis carinii, Candida albicans, Mycobacterium avium-intracellulare, Cryptospiridium, Toxoplasma,* cytomegalovirus, and herpes virus.

opsonins

A substance that binds to bacteria, erythrocytes, or other particles to increase their susceptibility to phagocytosis. Opsonins include antibodies such as IgG3, IgG1, and IgG2 that are specific for epitopes on the particle surface. Following interaction, the Fc region of the antibody becomes anchored to Fc receptors on phagocyte surfaces, thereby facilitating phagocytosis of the particles. In contrast to these so-called heat-stable antibody opsonins are the heat-labile products of complement activation such as C3b or C3bi which are linked to particles by transacylation with the C3 thiolester. C3b combines with complement receptor 1 and C3bi combines with complement receptor 3 on phagocytic cells. Other substances that act as opsonins include the basement membrane constituent, fibronectin.

opsonization

Facilitation of the phagocytosis of microorganisms or other particles such as erythrocytes through the coating of their surface with either immune or nonimmune opsonins.

oral unresponsiveness

The mucosal immune system's selective ability to not react immunologically against antigens or food and intestinal microorganisms even though it responds vigorously to pathogenic microorganisms.

organ bank

A site where selected tissues for transplantation, such as acellular bone fragments, corneas, and bone marrow, may be stored for relatively long periods until needed for transplantation. Several hospitals often share such a facility. Organs such as kidneys, liver, heart, lung, and pancreatic islets must be transplanted within 48 to 72 h and are not suitable for storage in an organ bank.

organ brokerage

The selling of an organ such as a kidney from a living related donor to the transplant recipient is practiced in certain parts of the world, but is considered unethical and is illegal in the U.S. as it is in violation of the National Organ Transplant Act (Public Law 98-507,3 USC).

organ-specific antigen

An antigen that is unique to a particular organ even though it may be found in more than one species.

organism-specific antibody index (OSAI)

The ratio of organism-specific IgG to total IgG in cerebrospinal fluid compared to the ratio of organism-specific IgG in serum to the total serum IgG. This is illustrated in the following formula:

$$OSAI = \left(\frac{\text{Organism - specific IgG in CSF}}{\text{Total IgG in CSF}} \right) \div \left(\frac{\text{Organism - specific IgG in Serum}}{\text{Total IgG in Serum}} \right)$$

If the index is greater than 1, signifying a greater quantity of organism-specific immunoglobulin in CSF than in the blood serum, this implies that organism-specific IgG is being synthesized in the intra-blood-brain barrier (IBBB) and suggests that the specific organism of interest is producing an infection of the central nervous system. Similar indices can be calculated for IgM and IgA antibody classes.

original antigenic sin

When an individual is exposed to an antigen that is similar to, but not identical to an antigen to which he was previously exposed by either infection or immunization, the immune response to the second exposure is still directed against the first antigen. This was first noticed in the influenza virus infection. Due to antigenic drift and antigenic shift in influenza virus, reinfection with an antigenically altered strain generates a secondary immune response that is specific for the influenza virus strain that produced an earlier infection.

orthotopic

An adjective that describes an organ or tissue transplant that has been in the site usually occupied by that organ or tissue.

orthotopic graft

An organ or tissue transplant that is placed in the location that is usually occupied by that particular organ or tissue.

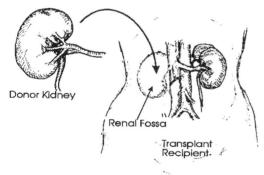

Orthotopic Graft (Transplant)

osteoclast-activating factor (OAF)

A lymphokine produced by antigen-activated lymphocytes that promotes bone resorption through activation of osteoclasts. Besides lymphotoxin produced by T cells, interleukin-1, tumor necrosis factor, and prostaglandins synthesized by macrophages also have OAF activity. Osteoclast-activating factors might be responsible for the bone resorption observed in multiple myeloma and T cell neoplasms.

OT (historical)

Old tuberculin.

Ouchterlony, Orjan Thomas Gunnersson (1914–)

Swedish bacteriologist who developed the antibody detection test that bears his name. Two-dimensional double diffusion with subsequent precipitation patterns is the basis of the assay. *Handbook of Immunodiffusion and Immunoelectrophoresis,* 1968.

Ouchterlony test

A double diffusion in a gel type of precipitation test. Antigen and antibody solutions are placed in separate wells that have been cut into an agar plate prepared with electrolyte. As the antigen and antibody diffuse through the gel medium, a line of precipitation forms at the point of contact between antigen and antibody. Results are expressed as reaction of identity, reaction of partial identity, or reaction of nonidentity. Refer to those entries for further details.

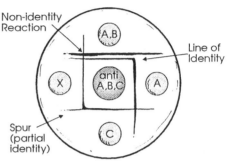

Ouchterlony Test (Double Diffusion)

Oudin, Jacques (1908–1986)

French immunologist who was Director of Analytical Immunology at the Pasteur Institute, Paris. His accomplishments include discovery of idiotypy and the agar single diffusion method antigen-antibody assay.

Oudin test

A type of precipitation in gel that involves single diffusion. Antiserum, incorporated into agar, is placed in a narrow test tube. This is overlaid with an antigen solution which diffuses into the agar to yield precipitation rings. Also called single radial diffusion test. A band of precipitation forms at the equivalence point.

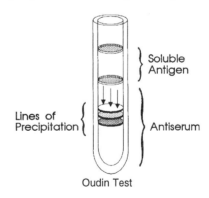

Oudin Test

outbreeding

Mating of subjects who showed greater genetic differences between themselves than randomly chosen individuals of a population. This process encourages genetic diversity. It is in contrast to inbreeding and random breeding.

ovalbumin (OA)

A protein derived from avian egg albumin (the egg white). It has been used extensively as an antigen in experimental immunology.

ovary antibodies (OA)

Antibodies present in 15 to 50% of premature ovarian failure patients in whom ovarian function failed after puberty but prior to

40 years of age. These individuals have elevated levels of gonadot-ropins in serum but diminished serum estradiol levels. Premature ovarian failure is associated with an increased incidence of autoim-mune disease, especially of the thyroid. They also develop autoan-tibodies of both the organ specific and non-organ-specific types. These women are often HLA-DR3 positive and have elevated CD4/CD8 ratios. Ovarian antibodies are specific for steroid cells, which causes them to cross-react with steroid-synthesizing cells in the placenta, adrenals and testes.

Owen, Ray David (1915–)

American geneticist who described erythrocyte mosaicism in dizygotic cattle twins. This discovery of re-ciprocal erythrocyte tolerance con-tributed to the concept of immuno-logic tolerance.

owl eye appearance

Inclusions found by light micros-copy in cytomegalovirus (CMV)

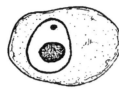

Owl Eye Appearance

infection. CMV-infected epithelial cells are enlarged and exhibit prominent eosinophilic intranuclear inclusions that are half the size of the nucleus and are encircled by a clear halo.

oxazolone (4-ethoxymethylene-2-phenyloxazol-5-one)

A substance used in experimental immunology to induce contact hypersensitivity in laboratory animals.

Oz isotypic determinant

Oz represents an isotypic marker. λ Light chains of human immu-noglobulin that are Oz$^+$ contain lysine at position 190, whereas those that are Oz$^-$ contain arginine at this position. A fraction of each person's λ light chains express Oz determinants.

λ :Oz determinant

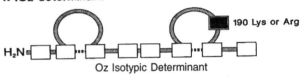

Oz Isotypic Determinant

P

P

Abbreviation for properdin. Also called factor P.

P1 kinase

A serine/threonine kinase activated by interferons α and β. It prevents translation by phosphorylating eIF2, the eukaryotic protein synthesis initiation factor. This facilitates inhibition of viral replicaton.

P1^{A1} antibodies

These antibodies, specific for the P1^{A1} antigen, are responsible for three fourths of the cases of neonatal alloimmune thrombocytopenic purpura and posttransfusion purpura. Anti-P1^{A1} antibodies prevent clot retraction and platelet aggregation.

p24 antigen

A human immunodeficiency virus type 1 (HIV-1), 24-kD core antigen that is the earliest indicator of infection with HIV-1. It is demonstrable days to weeks prior to seroconversion to antibody synthesis against HIV-1. Testing for the p24 antigen does not reveal anti-HIV-1 seronegative persons or those with inapparent infections who wish to donate blood.

P-80

An assay of an antiserum's ability to precipitate antigen. This test yields data equivalent to that obtained by the quantitative precipitation reaction. A constant quantity of radioisotope-labeled antigen is added to doubling dilutions of antiserum in a row of tubes. The tube in the zone of antigen excess in which precipitation of 80% of the antigen occurs is the end point.

P antigen

An ABO blood group-related antigen found on erythrocyte surfaces that is comprised of the three sugars, galactose, *N*-isoacetyl-galactosamine, and *N*-acetyl-glucosamine. The P antigens are designated P$_1$, P$_2$, P^k, and p. P$_2$ subjects rarely produce anti-P$_1$ antibody, which may lead to hemolysis in clinical situations. Paroxysmal cold hemoglobinuria patients develop a byphasic autoanti-P antibody that fixes complement in the cold (4°C) and lyses red blood cells at 37°C.

P-selectin

A molecule found in the storage granules of platelets and the Weibel-Palade bodies of endothelial cells. Ligands are sialylated Lewis X and related glycans. P-selectins are involved in the binding of leukocytes to endothelium and platelets to monocytes in areas of inflammation. Also called CD62P.

PAF

Abbreviation for platelet-activating factor.

palindrome

A DNA segment with dyad symmetrical structure. When read from 5' to the 3', it reveals an equivalent sequence whether read from forward or backward or from left or right. The base sequence in one strand is identical to the sequence in the second strand.

pan keratin antibodies

A "cocktail" of antibodies reactive with high mol wt cytokeratin and low mol wt keratin (AE1/AE3). By immunoperoxidase staining, these antibodies identify most epithelial cells and their derived neoplasms, irrespective of the site of origin or the level of differentiation.

pan-T cell markers

Surface epitopes found on all normal T lymphocytes. These include the 50-kD CD2 molecule that is the sheep erythrocyte rosette marker and is found exclusively on T lymphocytes, the 41-kD CD7 molecule, CD1 present on peripheral T lymphocytes and cortical thymocytes, the mature T lymphocyte marker CD3, and CD5.

panagglutination

Aggregation of cells with multiple antigenic specificities by certain blood sera, such as agglutination of normal red blood cells by a particular serum sample. It may also refer to an antibody that identifies an antigenic specificity held in common by a group of cells bearing a common antigenic specificity even though they differ in other antigenic specificities. Contamination of blood sera or of cells to be typed can result in aggregation of all the cells, leading to false-positive results as in blood grouping or cross-matching procedures.

pancreatic islet cell hormones

Immunoperoxidase staining of islet cell adenomas with antibodies to insulin, glucagon, somatostatin, and gastrin facilitates definition of their clinical phenotype.

pancreatic transplantation

A treatment for diabetes. Either a whole pancreas or a large segment of it, obtained from cadavers, may be transplanted together with kidneys into the same diabetic patient. It is important for the patient to be clinically stable and for there to be as close a tissue (HLA antigen) match as possible. Graft survival is 50 to 80% at 1 year.

panning (see figure, page 226)

A technique to isolate lymphocyte subsets through the use of petri plates coated with monoclonal antibodies specific for lymphocyte surface markers. Thus, only lymphocytes bearing the marker being sought bind to the petri plate surface.

pannus (see figure, page 226)

Granulation tissue reaction that is chronic and progressive and produces joint erosion in patients with rheumatoid arthritis. It is a structure that develops in synovial membranes during the chronic proliferative and destructive phase of rheumatoid arthritis. It is a membrane of granulation tissue induced by immune complexes that are deposited in the synovial membrane. They stimulate macrophages to release interleukin-1, fibroblast-activating factor, prostaglandins, substance P, and platelet-derived growth factor. This leads to extensive injury to chondroosseous tissues. The articular surface of the joint is covered by this synovitis. There is edema, swelling, and erythema in the joints. Palisades of histiocytes are present. This entire process can fill the joint space, leading to demineralization and cystic resorption.

PAP (peroxidase-antiperoxidase) technique

A method for immunoperoxidase staining of tissue to identify antigens with antibodies. This method employs unlabeled antibodies and a PAP reagent. The same PAP complex may be used for dozens of different unlabeled antibody specificities. If the primary antibody against the antigen being sought is made in the rabbit, then tissue sections treated with this reagent are exposed to sheep anti-rabbit immunoglobulin followed by the PAP complex. For human primary antibody, an additional step must link the human antibody into the rabbit sandwich technique. Paraffin-embedded tissue sections are first treated with xylene, and after deparaffinization, they are exposed to hydrogen peroxide to destroy the endogenous peroxidase. Sections are next incubated with normal sheep serum to suppress nonspecific binding of immunoglobulin to tissue collagen. Primary rabbit antibody against the antigen to be identified is combined with the tissue section. Unbound primary antibody is removed by rinsing the sections which are then covered with sheep antibody against rabbit immunoglobulin. This linking antibody will combine with any primary rabbit antibody in the tissue. It is added in excess, which results in one of its antigen-binding sites remaining free. After washing, the PAP is placed in the section, and the rabbit antibody part of this complex will be bound to the free antigen-binding site of the linking antibody. The unbound PAP complex is then washed away by rinsing. A substrate of hydrogen peroxide and aminoethyl-carbazole (AEC) is placed on the tissue section, leading to formation of a visible color reaction product which can be seen by light microscopy. Peroxidase is localized only at sites where the

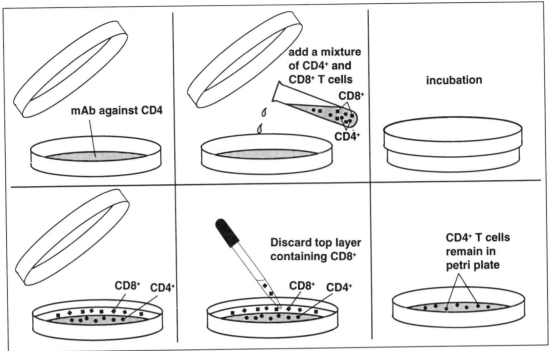

Panning

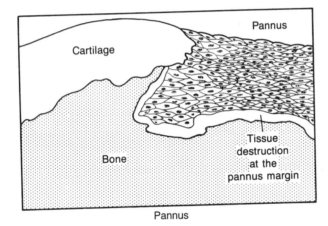

Pannus

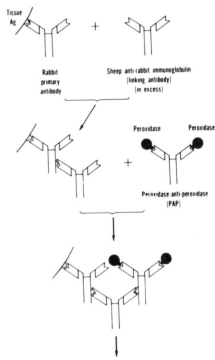

PAP is bound via linking antibody and primary antibody to antigen molecules, permitting the antigen to be identified as an area of reddish-brown pigment. Tissues may be counterstained with hematoxylin.

Development in chromogenic hydrogen donor and hydrogen peroxide. (The reaction product is seen as a reddish brown or brown granular deposit depending upon the chromogenic hydrogen donor used)

PAP (Peroxidate-Antiperoxidase) Technique

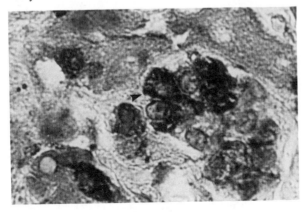

prolactin staining
in pituitary

PAP (Peroxidate-Antiperoxidase) Technique

papain

A proteolytic enzyme extracted from *Carica papaya* that is used to digest each IgG immunoglobulin molecule into two Fab fragments and one crystallizable Fc fragment. This aids efforts to reveal the molecular structure of immunoglobulins. Papain cleaves the immunoglobulin G molecule on the opposite side of the central disulfide bond from pepsin, which cleaves the molecule to the C-terminus side leading to the formation of 1 F(ab′)$_2$ fragment, which is bivalent in contrast to the Fab fragments which are univalent. The Fc

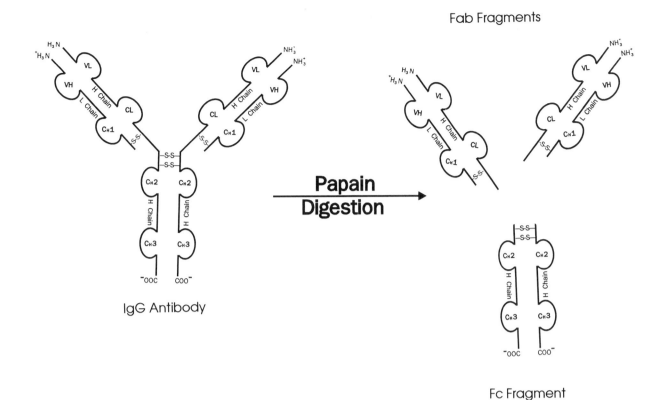

Fab Fragments

Papain
Digestion

IgG Antibody

Fc Fragment

Papain

fragment of papain digestion has no antigen-binding capacity, although it does have complement-fixing functions and attaches immunoglobulin molecules to Fc receptors on a cell membrane. The enzyme has also been used to render red blood cell surfaces susceptible to agglutination by incomplete antibody.

papain hydrolysis
Cleavage of IgG molecules into two Fab fragments and one Fc fragment. When the immunoglobulin is exposed to papain with cysteine present, papain cleaves a histidyl-threonine peptide bond of the heavy chain.

papovavirus
Minute tumor viruses that are icosahedral and contain double-stranded DNA. Included in the group are SV40 and polyomavirus that may cause malignant and benign tumors. Permissive or nonpermissive infections occur with papovavirus. Following permissive infection of monkey cells, papovavirus replicates, leading to lysis. T antigens, which are early papovavirus proteins that occur in nonpermissive rodent cells, can lead to transformation of the cells that is not reversible if the viral genome is integrated into the host genome. It is reversible if the cell can eliminate the viral genome.

para-Bombay phenotype
A variant Bombay phenotype that is of the ABO blood group system. Individuals expressing it have an Se secretor gene that encodes synthesis of blood groups A and B which are detectable in secretions. However, these subjects do not produce A and B erythrocytes as the H gene is absent. By comparison, Bombay phenotype individuals do not have the H gene or the enzyme it produces, i.e., fucosyl transferase, and do not have A or B blood group substances on their erythrocytes or in their secretions.

parabiotic intoxication
The result of a surgical union of allogeneic adult animals. The course of immune reactivity can be modified to take a single direction by uniting parental and F_1 animals. A hybrid recognizes parental cells as its own and does not mount an immune response against them, but alloantigens of F_1 hybrid cells stimulate the parental cells leading to graft-vs.-host disease.

paracortex
A T lymphocyte thymus-dependent area beneath and between lymph node cortex follicles.

paracrine
Local effects of a hormone acting on cells in its immediate vicinity.

paradoxical reaction
Death of experimental animals from anaphylaxis when administered a second injection of an antigen to which they had been previously immunized. Early workers administering repeated injections of tetanus toxoid observed the phenomenon. This term is no longer in use.

paraendocrine syndromes
Clinical signs and symptoms induced by hormones synthesized by neoplasms.

paraimmunoglobulins
(1) The physical characteristics of some immunoglobulins present in a variety of pathologic conditions or in others of unknown etiology, and (2) the secretory products of neoplastic lymphocytes. One form, the M protein of macroglobulin, is present in the normal serum, but increased levels may result in increased serum viscosity with sluggish blood flow and development of thrombi or in central nervous system lesions. Increased levels of M protein are considered as paraimmunoglobulinopathies.

paralysis
The masking of an immune response by the presence of excessive quantities of antigen. This mimics acquired immunologic tolerance and is considered a false tolerance state.

paraneoplastic autoantibodies
Autoantibodies that cross react with tumor and normal tissue in the same patient. Examples are Yo antibodies against cerebellar Purkinje cells that occur in paraneoplastic cerebellar degeneration, neuronal nuclear (Hu) antibodies in paraneoplastic subacute sensory neuronapathy and sensory neuropathies, anti-keratinocyte polypeptides in paraneoplastic pemphigus, antibodies against voltage-gated calcium channels in Lambert-Eaton syndrome, antibodies against retina in reginopathy associated with cancer and antibodies against myenteric and submucosal plexuses in pseudo-obstruction of the intestine.

paraneoplastic pemphigus
A rare autoimmune condition that may occasionally be seen in lymphoproliferative disorders. It is caused by autoantibodies against desmoplakin I and bullous pemphigoid antigen, as well as other

epithelial antigens. Clinically, there is erosion of the oropharynx and vermilion border, as well as pseudomembranous conjunctivitis and erythema of the upper trunk skin.

paraprotein

Homogeneous, monoclonal immunoglobulin molecules synthesized by an expanding clone of plasma cells, as observed in patients with multiple myeloma or with Waldenström's macroglobulinemia. The homogeneity of the paraprotein is reflected by all molecules belonging to the same immunoglobulin class and subclass as well as the same light chain type. On electrophoresis, a serum paraprotein appears as a distinct band. This is a consequence of a biologic event such as neoplastic transformation rather than antigenic stimulation.

paraproteinemias

Malignant diseases in which there is proliferation of a single clone of plasma cells producing monoclonal immunoglobulin. These are commonly grouped as paraproteinemias that may be manifested in several forms. Diseases associated with paraproteinemias include multiple myeloma, Waldenström's macroglobulinemia, cryoglobulinemia, plasmacytoma of soft tissues, amyloidosis, heavy chain disease, lymphomas, leukemia, sarcomas, gastrointestinal disorders associated with tumors, chronic infection, and some endocrine disorders.

paratope

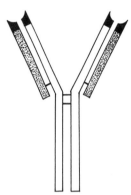

The antigen-binding site of an antibody molecule. It is the variable or Fv region of an antibody molecule and is the site for interaction with an epitope of an antigen molecule. It is complementary for the epitope for which it is specific.

A paratope is the portion of an antibody molecule where the hypervariable regions are located. There is less than 10% variability in the light and heavy chain amino acid positions in the variable regions. However, there is 20 to 60% variability in amino acid sequence in the so-called "hot spots" located at light chain amino acid positions 29-34, 49-52, and 91-95 and at heavy chain positions 30-34, 51-63, 84-90, and 101-110. Great specificity is associated with this variability and is the basis of an idiotype. This variability permits recognition of multiple antigenic determinants.

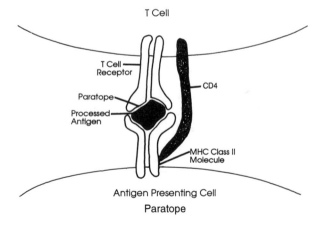

Paratope

parenteral

Administration or injection of a substance into the animal body by any route except for the alimentary tract.

parietal cell antibodies

Antibodies present in 50 to 100% of pernicious anemia (PA) patients. They are also found in 2% of normal individuals. Their frequency increases with aging and in subjects with insulin-dependent diabetes mellitus. The frequency of parietal cell antibodies diminishes with disease duration in pernicious anemia. Parietal cell autoantibodies in pernicious anemia and in autoimmune gastritis recognize the α and β subunits of the gastric proton pump (H^+/K^+

ATPase) which are the principal target antigens. Parietal cell antibodies react with the α and β subunits of the gastric proton pump and inhibit the gastric mucosa's acid-producing H^+/K^+ adenosine triphosphatase. Parietal cell antibodies relate to type A gastritis in which there is fundal mucosal atrophy, achlorhydria, development of pernicious anemia and autoimmune endocrine disease.

paroxysmal cold hemoglobinuria (PCH)

A rare type of disease that accounts for 10% of cold autoimmune hemolytic anemias. It may be either a primary idiopathic disease or secondary to syphilis or viral infection and is characterized by the passage of hemoglobin in the urine after exposure to cold. In addition to passing dark-brown urine, the patient may experience chills, fever, and pain in the back, legs, or abdomen. The disease is associated with a hemolysin termed the Donath-Landsteiner antibody which is a polyclonal IgG antibody. It sensitizes the patient's red blood cells in the cold, complement attaches to the erythrocyte surface, and hemolysis occurs on warming to 37°C. The antibody's specificity is for the P antigen of red blood cells.

paroxysmal nocturnal hemoglobinuria (PNH)

A rare form of hemolytic anemia in which the red blood cells, as well as neutrophils and platelets, manifest strikingly increased sensitivity to complement lysis. PNH red blood cell membranes are deficient in decay-accelerating factor (DAF), LFA-3, and FcµRIII. Without DAF, which protects the cell membranes from complement lysis by classic pathway C5 convertase and decreases membrane attack complex formation, the erythrocytes and lymphocytes are highly susceptible to lysis by complement. Interaction of these PNH erythrocytes with activated complement results in excessive C3b binding which leads to the formation of more C3b through the alternate complement pathway by way of factors B and D. Intravascular hemolysis follows activation of C5 convertase in the C5-9 membrane attack complex (MAC). The blood platelets and myelocytes in affected subjects are also DAF deficient and are readily lysed by complement. There is leukopenia, thrombocytopenia, iron deficiency, and diminished leukocyte alkaline phosphatase. The Coombs' test is negative, and there is also very low acetylcholine esterase activity in the red cell membrane. No antibody participating in this process has been found in either the serum or on the erythrocytes. The disease is suggested by episodes of intravascular hemolysis, iron deficiency, and hemosiderin in the urine. It is confirmed by hemolysis in acid medium, termed the HAM test.

partial identity

Refer to reaction of partial identity.

parvovirus

A minute icosahedral virus comprised of single-stranded DNA that may replicate in previously uninfected host cells or in those already infected with adenovirus.

PAS

(1) Abbreviation for periodic acid Schiff stain for polysaccharides. This technique identifies mucopolysaccharide, glycogen, and sialic acid among other chemicals containing 1,2-diol groups. (2) Abbreviation for para-aminosalicylic acid, which is used in the treatment of tuberculosis.

passive agglutination

The aggregation of particles with soluble antigens adsorbed to their surfaces by the homologous antibody. The soluble antigen may be linked to the particle surface through covalent bonds rather than by mere adsorption. Red blood cells, latex, bentonite, or collodion particles may be used as carriers for antigen molecules adsorbed to their surfaces. When the red blood cell is used as a carrier particle, its surface has to be altered in order to facilitate maximal adsorption of the antigen to its surface. Several techniques are employed to accomplish this. One is the tanned red blood cell technique which involves treating the red blood cells with a tannic acid solution which alters their surface in a manner

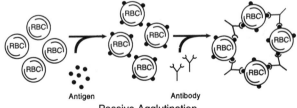

Passive Agglutination

that favors the adsorption of added soluble antigen. A second method is the treatment of red cell preparations with other chemicals such as bis-diazotized benzidine.

With this passive agglutination technique, even relatively minute quantities of soluble antigens may be detected by the homologous antibody agglutinating carrier cells on which they are adsorbed. Since red blood cells are the most commonly employed particle, the technique is referred to as passive hemagglutination. Latex particles are used in the rheumatoid arthritis (RA) test in which pooled IgG molecules are adsorbed to latex particles and reacted with sera of rheumatoid arthritis patients containing rheumatoid factor (IgM anti-IgG antibody) to produce agglutination. Polysaccharide antigens will stick to red blood cells without treatment. When proteins are used, however, covalent linkages are required.

passive agglutination test

An assay to recognize antibodies against soluble antigens which are attached to erythrocytes, latex, or other particles by either adsorption or chemical linkage. In the presence of antibodies specific for the antigen, aggregation of the passenger particles occurs. Examples of this technique include the RA latex agglutination test, the tanned red cell technique, the bentonite flocculation test, and the bis-diazotized benzidine test.

passive anaphylaxis

An anaphylactic reaction in an animal which has been administered an antigen after it has been conditioned by an inoculation of antibodies derived from an animal immunized against the antigen of interest.

passive Arthus reaction

An inflammatory vasculitis produced in experimental animals by the passive intravenous injection of significant amounts of precipitating IgG antibody, followed by the intracutaneous or subcutaneous injection of the homologous antigen for which the antibodies are specific. This permits microprecipitates to occur in the intercellular spaces between the intravascular precipitating antibody and antigen in the extravascular space. This is followed by interaction with complement, attraction of polymorphonuclear leukocytes, and an inflammatory response as described under Arthus reaction.

passive cutaneous anaphylaxis (PCA)

A skin test that involves the *in vivo* passive transfer of homocytotropic antibodies that mediates type I immediate hypersensitivity (e.g., IgE in man) from a sensitized to a previously nonsensitized individual by intradermally injecting the antibodies which become anchored to mast cells through their Fc receptors. This is followed hours or even days later by intravenous injection of antigen mixed with a dye such as Evans Blue. Crosslinking of the cell-fixed (e.g., IgE) antibody receptors by the injected antigen induces a type I immediate hypersensitivity reaction in which histamine and other pharmacological mediators of immediate hypersensitivity are released. Vascular permeability factors act on the vessels to permit plasma and dye to leak into the extravascular space, forming a blue area which can be measured with calipers. In humans, this is called the Prausnitz-Küstner (PK) reaction.

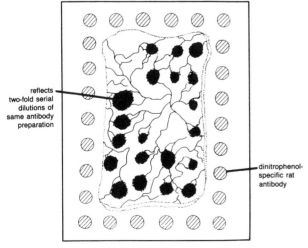

reflects two-fold serial dilutions of same antibody preparation

dinitrophenol-specific rat antibody

Passive Cutaneous Anaphylaxis in Rats

passive hemagglutination

Aggregation by antibodies of erythrocytes bearing adsorbed or covalently bound soluble antigen on their surface.

passive hemolysis

The lysis of erythrocytes used as carriers for soluble antigen bound to their surface. Following interaction with antibody, complement induces cell lysis. In passive hemolysis, the antigen is not a part of the cell surface structure, but is only attached to it.

passive immunity

A form of acquired immunity induced by the transfer of immune serum containing specific antibodies or of sensitized lymphoid cells from an immune to a nonimmune recipient host. Examples of passive immunity are the transfer of IgG antibodies across the placenta from mother to young or the ingestion of colostrum containing antibodies by an infant. Antitoxins generated to protect against diphtheria or tetanus toxins represent a second example of passive humoral immunity, as used in the past. Specifically sensitized lymphoid cells transferred from an immune to a previously nonimmune recipient is termed adoptive immunization. The passive transfer of antibodies in immune serum can be used for the temporary protection of individuals exposed to certain infectious disease agents. They may be injected with hyperimmune globulin.

passive immunization

The transfer of a specific antibody or of sensitized lymphoid cells from an immune to a previously nonimmune recipient host. Unlike active immunity which may be of a relatively long duration, passive immunity is relatively brief, lasting only until the injected immunoglobulin or lymphoid cells have disappeared. Examples of passive immunization include (1) the administration of γ globulin to immunodeficient individuals and (2) the transfer of immunity from mother to young, i.e., antibodies across the placenta or the ingestion of colostrum containing antibodies.

passive sensitization

The transfer of antibodies or primed lymphocytes from a donor previously exposed to antigen to a normal recipient for the purpose of conveying hypersensitivity from a sensitized to a nonsensitized individual. The Prausnitz-Küstner reaction is an example.

passive systemic anaphylaxis

Rendering a normal, previously unsensitized animal susceptible to anaphylaxis by a passive injection, often intravenously, of homocytotrophic antibody derived from a sensitized animal, followed by antigen administration. Anaphylactic shock occurs soon after the passively transferred antibody and antigen interact *in vivo,* releasing the mediators of immediate hypersensitivity from mast cells of the host.

passive transfer

The transfer of immunity or hypersensitivity from an immune or sensitized animal to a previously nonimmune or unsensitized (and preferably syngeneic) recipient animal by serum containing specific antibodies or by specifically immune lymphoid cells. The transfer of immunity by lymphoid cells is referred to as adoptive immunization. Humoral immunity and antibody-mediated hypersensitivity reactions are transferred with serum, whereas delayed-type hypersensitivity, including contact hypersensitivity, is transferred with lymphoid cells. Passive transfer was used to help delineate which immune and hypersensitivity reactions were mediated by cells and which were mediated by serum.

Pasteur, Louis (1822–1895)

French. Father of immunology. One of the most productive scientists of modern times, Pasteur's contributions included the crystalization of L- and O-tartaric acid, disproving the theory of spontaneous generation, studies of diseases in wine, beer, and silkworms, and the use of attenuated bacteria and viruses for vaccination. *Les Maladies des Vers a Soie,* 1865; *Etudes sur le Vin,* 1866; *Etudes sur la Biere,* 1876; *Oeuvres,* 1922–1939.

patch test

An assay to determine the cause of skin allergy, especially contact allergic (type IV) hypersensitivity. A small square of cotton, linen, or paper impregnated with the suspected allergen is applied to the skin for 24 to 48 h. The test is read by examining the site 1 to 2 days after applying the patch. The development of redness (erythema), edema, and formation of vesicles constitutes a positive test. The impregnation of tuberculin into a patch was used by Vollmer for a modified tuberculin test. There are multiple chemicals, toxins, and other allergens that may induce allergic contact dermatitis in exposed members of the population.

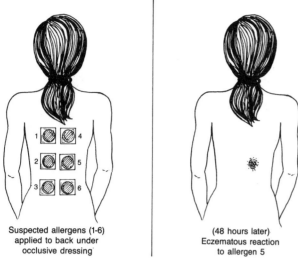

Suspected allergens (1-6)
applied to back under
occlusive dressing

(48 hours later)
Eczematous reaction
to allergen 5

PatchTest

patching

The accumulation of membrane receptor proteins crosslinked by antibodies or lectins on a lymphocyte surface prior to "capping". The antigen-antibody complexes are internalized following capping, permitting antigen processing and presentation in the context of MHC molecules. Membrane protein redistribution into patches is passive and does not require energy. The process depends on the lateral diffusion of membrane constituents in the plain of the membrane.

Patch

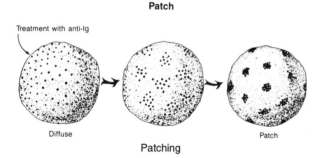

Treatment with anti-Ig

Diffuse

Patch

Patching

paternity testing

Tests performed to ascertain the biological (genetic) parentage of a child. In the past, these have included erythrocyte enzymes, red blood cell antigens, HLA antigens, immunoglobulin allotypes, nonimmunoglobulin serum proteins, and more recently DNA "fingerprinting" (typing). The demonstration of a genetic marker in a child that is not present in either the father or mother or in cases where none of the paternal antigens are present in the child is enough evidence for direct exclusion of paternity. Another case of direct exclusion of paternity is when a child fails to express a gene found in both the mother and putative father. When a child expresses a gene that only the man can transmit and which the putative father does not express, this is evidence for indirect exclusion of paternity. When a child is homozygous for a marker not present in the mother or putative father or if the parent is homozygous for a marker not found in the child, then paternity can be excluded as an indirect exclusion. Also called identity testing.

pathogenicity

The capacity of a microorganism to induce disease.

pathologic autoantibodies

Autoantibodies generated against self antigens that induce cell and tissue injury following interaction with the cells bearing epitopes for which they are specific. Many autoantibodies are physiologic, representing an epiphenomenon during autoimmune stimulation, whereas others contribute to the pathogenesis of tissue injury. Autoantibodies that lead to red blood cell destruction in autoimmune hemolytic anemia represent pathogenic autoantibodies,

whereas rheumatoid factors such as IgM anti-IgG autoantibodies have no proven pathogenic role in rheumatoid arthritis.

Paul-Bunnell test

An assay for heterophile antibodies in infectious mononucleosis patients. It is a hemagglutination test in which infectious mononucleosis patient serum induces sheep red blood cell agglutination. Absorption of the serum with guinea pig kidney tissue removes antibody to the Forssman antigen, but does not remove the sheep red blood cell agglutinin which can be absorbed with ox cells. This hemagglutinin is distinct from antibodies against the causative agent of infectious mononucleosis, i.e., the Epstein-Barr virus.

PBC

Abbreviation for primary biliary cirrhosis.

PCA

Abbreviation for passive cutaneous anaphylaxis.

PCH

Abbreviation for paroxysmal cold hemaglobinuria.

PCP

Abbreviation for *Pneumocystis carinii* pneumonia.

PCR

Abbreviation for polymerase chain reaction.

PDGF

Abbreviation for platelet-derived growth factor.

PECAM (CD31)

An immunoglobulin-like molecule present on leukocytes and at endothelial cell junctions. These molecules participate in leukocyte-endothelial cell interactions, as during an inflammatory response.

pediatric AIDS

Acquired immunodeficiency syndrome (AIDS) in infants who are infected vertically, i.e., from mother to young through intrauterine or intrapartum infection. They show symptoms usually between 3 weeks and 2 years of age. They develop lymphadenopathy; fever; increased numbers of B lymphocytes in the peripheral blood; thrombocytopenia; and increased levels of IgG, IgM, and IgD in the serum. They may develop lymphoid interstitial pneumonia, chronic otitis media encephalopathy, recurrent bacterial infections, and *Candida* esophagitis. Epstein-Barr virus infection may produce interstitial pneumonia and salivary gland inflammation. They may also have blood-borne infections with such microorganisms as *Hemophilus influenzae* or pneumococci. The adult age pattern of either opportunistic infections or Kaposi's sarcoma are rarely seen in HIV-1-infected infants. These children have a 50% 5-year mortality.

PEG (polyethylene glycol)

(1) A synthetic, inert molecule with a long chain structure which masks recognition of certain proteins by the immune system when PEG is attached to them. For example, PEG has been linked to adenosine deaminase (ADA), the absence of which may induce severe combined immunodeficiency. This PEG attachment permits extended survival of ADA in the body. (2) PEG is also used to promote cell fusion in hybridoma technology in which an antibody-secreting cell and a mutant myeloma cell are fused to yield a hybridoma which is immortal and continues to produce a monoclonal antibody product.

pemphigoid

A blistering disease of the skin in which bullae form at the dermal-epidermal junction, in contrast to the intraepidermal bullae of pemphigus vulgaris. Autoantibodies develop against the dermal basement membrane. By using fluorochrome-labeled goat or rabbit anti-human IgG, linear fluorescence can be demonstrated at the base of the subepidermal bullae by immunofluorescence microscopy. Dermal basement membrane IgG autoantibodies can also be demonstrated in the patient's serum. This disease occurs principally in elderly individuals. C3 linear fluorescence at the dermal-epidermal junction is often demonstrable as well.

pemphigus erythematosus (Senear-Usher syndrome)

A clinical condition with immunopathologic characteristics of both pemphigus and lupus erythematosus. Skin lesions may be on the seborrheic regions of the head and upper trunk, as seen in pemphigus foliaceus. However, immune deposits are also demonstrable at the dermal-epidermal junction and in skin biopsy specimens obtained from areas exposed to sunlight, reminiscent of lupus erythematosus. Light microscopic examination may reveal an intraepidermal bulla of the type seen in pemphigus foliaceus. Facial skin lesions may even include the "butterfly rash" seen in lupus. Immunofluorescence staining may reveal intercellular IgG and C3 in a "chickenwire" pattern in the epidermis with concomitant granular immune deposits containing immunoglobulins and complement at the dermal-epidermal junction. The serum may reveal both antinuclear antibodies and pemphigus antibodies. Pemphigus erythematosus has been reported in patients with neoplasms of internal organs, and in drug addicts, among other conditions. Indirect immunofluorescence using serum with both antibodies may reveal simultaneous staining for intercellular antibodies and peripheral (rim) nuclear fluorescence in the same specimen of monkey esophagus used as a substrate.

pemphigus vulgaris

A blistering lesion of the skin and mucous membranes. The bullae develop on normal appearing skin and rupture easily. The blisters are prominent on both the oral mucosa and anal/genital mucous membranes. The disorder may have an insidious onset appearing in middle-aged individuals, and tends to be chronic. It may be associated with autoimmune diseases, thymoma, and myasthenia gravis. Certain drugs may induce a pemphigus-like condition. By light microscopy, intraepidermal bullae are present. There is suprabasal epidermal acantholysis with only mild inflammatory reactivity in early pem-phigus. Suprabasal unilocular bullae develop, and there are autoantibodies to intercellular substance with activation of classic pathway-mediated immunologic injury. Acantholysis results as the epidermal cells become disengaged from one another as the bulla develops. Epidermal proteases activated by autoantibodies may actually cause the loss of intercellular bridges. Immunofluorescence staining reveals IgG, Clq, and C3 in the intercellular substance between epidermal cells. In pemphigus vulgaris patients, 80 to 90% have circulating pemphigus antibodies. Their titer usually correlates positively with clinical manifestations. Corticosteroids and immunosuppressive therapy, as well as plasma-pheresis, have been used with some success.

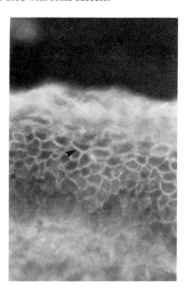

"chickenwire staining"
antibody to
intercellular antigen

Pemphigus Vulgaris

penicillin hypersensitivity

Allergic reactions to penicillin or its degradation products such as penicillinic acid may be either antibody mediated or cell mediated. Penicillin derivatives may act as haptens by conjugating to tissue proteins to yield penicilloyl derivatives. These conjugates may induce antibody-mediated hypersensitivity manifested as an anaphylactic reaction when the patient is subsequently exposed to penicillin, or it may be manifested as a serum sickness-type reaction with fever, urticaria, and joint pains. Penicillin hypersensitivity may also be manifested as hemolytic anemia in which the penicillin derivatives have become conjugated to the patient's red blood cells

or as allergic contact dermatitis, especially in pharmacists or nurses who come into contact with penicillin on a regular basis. Whereas the patch test using material impregnated with penicillin may be applied to the skin to detect cell-mediated (delayed-type, type IV) hypersensitivity, individuals who have developed anaphylactic hypersensitivity with IgE antibodies specific for penicilloyl-protein conjugates may be identified by injecting penicilloyl-polylysine into their skin. The development of a wheal and flare response signifies the presence of IgE antibodies which mediate anaphylactic reactivity in man.

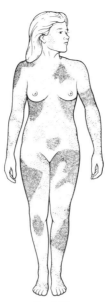

Penicillin
Hypersensitivity

pentamidine isoethionate

A substance useful in the treatment of *Pneumocystis carinii* pneumonia in AIDS patients who have failed to respond to trimethoprimsulfamethoxazole therapy. It is administered by aerosol and has diminished *P. carinii* pneumonia by 65%. Adverse effects include azothemia, arrythmia, hypotension, diabetes mellitus, pancreatitis, and severe hypoglycemia.

pentraxin family

A category of glycoproteins in the blood that have a cyclic pentameric symmetric structure. They include C-reactive protein, serum amyloid P, and complement C1.

pepsin

A proteolytic enzyme used to hydrolyze immunoglobulin molecules into F(ab')$_2$ fragments plus small peptides.

pepsin digestion (see figure, top of page 232)

A proteolytic enzyme used to cleave immunoglobulin molecules into F(ab')$_2$ fragments together with fragments of small peptides that represent what remains of the Fc fragment. Each immunoglobulin molecule yields only one F(ab')$_2$ fragment, which is bivalent and may manifest many of the same antibody characteristics as intact IgG molecules, such as antitoxic activity in neutralizing bacterial toxins. Cleaving the Fc region from an IgG molecule deprives it of its ability to fix complement and bind to Fc receptors on cell surfaces. Pepsin digestion is useful in diminishing the immunogenicity of antitoxins. It converts them to F(ab')$_2$ fragments which retain antitoxin activity.

pepsinogen antibodies

Antibodies against pepsinogen, which is produced by chief cells and mucous neck cells. Autoantibodies against pepsinogen develop in autoimmune atrophic gastritis patients with pernicious anemia. Three-fourths of peptic ulcer patients have pepsinogen antibodies.

peptide map

A fingerprint of peptides in two dimensions prepared by digestion of a protein with an enzyme such as pepsin. Thin layer chromotography in one direction and electrophoresis in the other direction at pH 6.5 yields the fingerprint pattern.

peptide T

A small HIV-1 envelope polypeptide that was first believed to have potential in treating AIDS, but was later withdrawn.

peptidergic endothelium-derived contrasting factor (EDCF)

A unique 21-amino acid peptide with two intrachain disulfide bonds. It is homologous to Surafotoxin S6b. Four isoforms of ET and two ET receptor subtypes have been described. Also called endothelin (ET).

peptidoglycolipid

A part of the wax D fraction extracted from mycobacteria that contains the adjuvant principle in Freund's adjuvant.

Percoll®

A density gradient centrifugation medium used to isolate certain cell populations such as natural killer (NK) cells. It is a colloidal suspension.

perforin

Perforin from cytolytic T lymphocyte and NK cell granules produces target cell lysis. Perforin isolation and sequence determination through cDNA cloning made possible studies on perforin

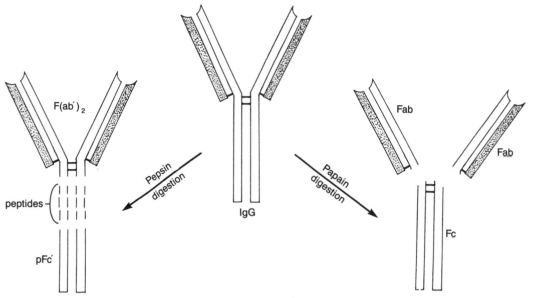

Pepsin Digestion

structure and function. This is a 70-kD glycoprotein. Granules containing perforin mediate lysis in a medium with calcium. A transmembrane polyperforin tubular channel is formed in the membrane's lipid bilayer. Perforin that has been released inserts into the membrane of the target cell to produce a 5- to 20-nm doughnut-shaped polymeric structure that serves as a stable conduit for intracellular ions to escape to the outside extracellular environment, promoting cell death. Both murine and human perforin cDNA have now been cloned and sequenced. They share 67% homology, and each is 534 amino acids long. Perforin and C9 molecules appear to be related in a number of aspects. This is confirmed by sequence comparison, functional studies, and morphologic and immunologic comparisons. Whereas lysis of host cells by complement is carefully regulated, cytotoxic T lymphocytes lyse virus-infected host cells. Decay-accelerating factor (DAF) and HRF guard host cells

against lysis by complement. By contrast, perforin is able to mediate lysis without such control factors. Using perforin cDNA as the hybridization probe, mRNA levels have been assayed for perforin expression. Thus far, all cytotoxic T lymphocytes have been shown to express the perforin message. Perforin expression is an *in vitro* phenomenon, but perforin expression has been shown *in vivo*. This was based upon cytotoxic CD8+ T cells containing perforin mRNA.

periarterial lymphatic sheath

The thymus-dependent region in the splenic white pulp.

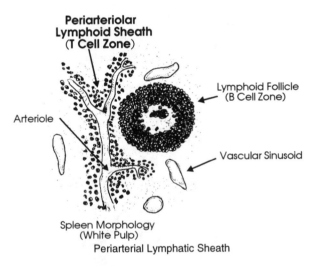

Spleen Morphology
(White Pulp)
Periarterial Lymphatic Sheath

periarteritis nodosa

A synonym for polyarteritis nodosa.

perinuclear antibodies

Antibodies against perinuclear granules in buccal mucosal cells in man. They are present in about 78% of patients with classical rheumatoid arthritis and in 40% of RA patients that are IgM rheumatoid factor negative. Their presence portends a poor prognosis in the rheumatoid factor negative group. Perinuclear antibodies may also be found in selected other rheumatic diseases and are often present in subjects infected with Epstein-Barr virus. They are also demonstrable in approximately one-fourth of primary biliary cirrhosis patients.

peripheral lymphoid organs

Lymphoid organs that are not required for ontogeny of the immune response. They include the lymph nodes, spleen, tonsils, and Peyer's patches.

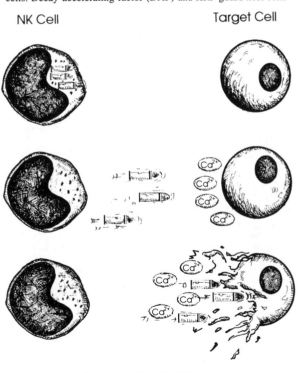

NK Cell   Target Cell

[⊏ ▯ ⊐] = Perforin

Perforin

Peripheral Lymphoid Tissues

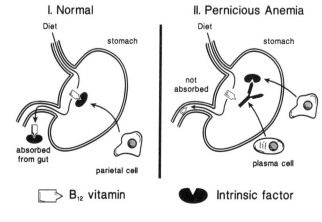

Peripheral Lymphoid Organs

antibodies, and at least half of them also develop antibodies against intrinsic factor, which is necessary for the absorption of B_{12}. The antiparietal cell antibodies are against a microsomal antigen found in gastric parietal cells. Intrinsic factor is a 60-kD substance that links to vitamin B_{12} and aids its uptake in the small intestine. PA may be a complication of common variable immunodeficiency or may be associated with autoimmune thyroiditis. PA is caused principally by injury to the stomach mediated by T lymphocytes. Patients may manifest megaloblastic anemia, deficiency of vitamin B_{12}, and increased gastrin in serum.

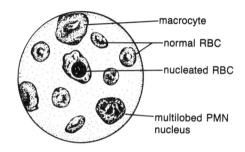

Macrocytic anemia

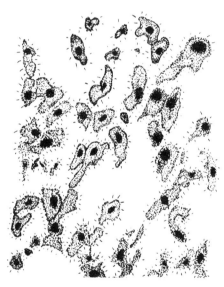

Parietal Cell Antibody
(FITC - labelled)
Pernicious Anemia

peripheral tolerance

Inhibition of expression of the immune response. The cells delivering the actual response are functionally impaired, but not defective. Peripheral tolerance affects the efferent limb of the immune response, which is concerned with the generation of effector cells. Refer to immunologic tolerance.

peripolesis

A gathering of cells of one type around another kind of cell, such as the accumulation of lymphocytes around macrophages. This facilitates cell-to-cell interaction in the induction of an immune response.

peritoneal exudate cells (PEC)

Inflammatory cells resident in the peritoneum of experimental animals following the interperitoneal injection of an inflammatory agent such as glycogen, peptone, or paraffin oil. The cell population varies with time after inoculation. The first cells to appear are polymorphonuclear neutrophilic leukocytes that are found several hours following injection. These are followed by lymphocytes and, within 72 h, by macrophages. The desired population of cells such as macrophages may be harvested by peritoneal lavage with an appropriate tissue culture medium.

permeability factors

Refer to vascular permeability factors.

permeability-increasing factor

A lymphokine that enhances the permeability of vessels.

pernicious anemia (PA)

An autoimmune disease characterized by the development of atrophic gastritis, achlorhydria, decreased synthesis of intrinsic factor, and malabsorption of vitamin B_{12}. Patients present with megaloblastic anemia caused by the vitamin B_{12} deficiency that develops. The majority of pernicious anemia patients develop antiparietal cell

persistent generalized lymphadenopathy (PGL)

A clinical stage of HIV infection.

pertussis adjuvant

Killed *Bordetella pertussis* microorganisms have been mixed with antigen to enhance antibody production. Pertussis adjuvant particularly facilitates IgE synthesis in rats or other animals. When used as a component in the triple vaccine, i.e., diphtheria-pertussis-tetanus (DPT) preparation used for childhood immunization, the killed *B. pertussis* microorganisms not only stimulate antibodies that protect against whooping cough, but they also facilitate antibody synthesis against both the diphtheria and tetanus toxoid preparations, thereby serving as an immunologic adjuvant.

pertussis vaccine

A preparation used for prophylactic immunization against whooping cough in children. It consists of virulent *Bordetella pertussis* microorganisms that have been killed by treatment with formalin. It is administered in conjunction with diphtheria toxoid and tetanus toxoid as a so-called triple vaccine. In addition to stimulating protective immunity against pertussis, the killed *B. pertussis* microorganisms act as an adjuvant and facilitate antibody production against the diphtheria and tetanus toxoid components in vaccine.

Rarely does a hypersensitivity reaction occur. To reduce the toxic effects of the vaccine, an acellular product is now in use.

Peyer's patches

Lymphoid tissue in the submucosa of the small intestine. It is comprised of lymphocytes, plasma cells, germinal centers, and thymus-dependent areas.

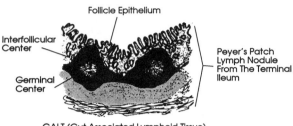

GALT (Gut Associated Lymphoid Tissue)
Peyer's Patches

PF4 (platelet factor 4)

An α granule platelet protein that is released when platelets aggregate. PF4 is a tetramer of high mol wt associated with chondroitin sulfate. It participates in immune modulation, chemotaxis, and inhibition of bone resorption and angiogenesis.

PFC (plaque-forming cell)

An *in vitro* technique in which antibody synthesizing cells derived from the spleen of an animal immunized with a specific antigen produce antibodies that lyse red blood cells coated with the corresponding antigen in the presence of complement in a gel medium. The reaction bears some resemblance to β hemolysis produced by streptococci on a blood agar plate. When examined microscopically, a single antibody-producing cell can be detected in the center of the plaque-forming unit.

pFc' fragment

A fragment of pepsin digestion of IgG or of the Fc fragment. Pepsin digestion of IgG or of the Fc fragment yields low mol wt peptides and a pFc' fragment that is still capable of binding to an Fc receptor on a macrophage or monocyte. It is a 27-kD dimer without a covalent bond comprised of two C_H3 domains, the carboxyl terminal 116 residues of each chain. Unlike the Fc' fragment, it has the basic N-terminal and C-terminal peptides of this immunoglobulin domain.

Pfeiffer phenomenon

The rapid lysis of *Vibrio cholerae* microorganisms that have been injected into the peritoneal cavity of guinea pigs immunized against them. The microorganisms are first rendered nonmotile, followed by complement-induced lysis in the presence of antibody. Immune bacteriolysis *in vivo* involving the cholera vibrio became known as the Pfeiffer phenomenon.

PFU

Abbreviation for plaque-forming unit. An assay of plaques that develop in the hemolytic plaque assay and related techniques.

PHA

Abbreviation for phytohemagglutinin. PHA is principally a T lymphocyte mitogen, producing a greater stimulatory effect on CD4+ helper/inducer T lymphocytes than on CD8 suppressor/cytotoxic T cells. It has a weaker mitogenic effect on B lymphocytes.

phacoanaphylactic endophthalmitis

Introduction of lens protein into the circulation following an acute injury of the eye involving the lens may result in chronic inflammation of the lens as a consequence of autoimmunity to lens protein.

phacoanaphylaxis

Hypersensitivity to lens protein of the eye following an injury that introduces lens protein, normally a sequestered antigen, into the circulation. The immune system does not recognize it as self and responds to it as it would any other foreign antigen.

phage neutralization assay

A laboratory test in which bacteriophage is combined with antibodies specific for it to diminish its capacity to infect a host bacterium. This neutralization of infectivity may be quantified by showing the decreased numbers of plaques produced when the phage which has been incubated with specific antibody is plated on appropriate bacteria. The technique is sensitive and can demonstrate even weak antibody activity.

phagocyte

Cells such as mononuclear phagocytes and polymorphonuclear neutrophils that ingest and frequently digest particles such as

bacteria, blood cells, and carbon particles, among many other particulate substances.

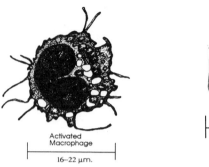

Phagocyte

phagocytic cell function deficiencies

Patients with this group of disorders frequently show an increased susceptibility to bacterial infections, but are generally able to successfully combat infections by viruses and protozoa. Phagocytic dysfunction can be considered as either extrinsic or intrinsic defects. Extrinsic factors include diminished opsonins that result from deficiencies in antibodies and complement, immunosuppressive drugs or agents that reduce phagocytic cell numbers, corticosteroids that alter phagocytic cell function, and autoantibodies against neutrophil antigens that diminish the number of PMNs in the blood circulation. Complement deficiencies or inadequate complement components may interfere with neutrophil chemotaxis to account for other extrinsic defects. By contrast, intrinsic defects affect the ability of phagocytic cells to kill bacteria. This is related to deficiencies of certain metabolic enzymes associated with the intracellular digestion of bacterial cells. Among the disorders are chronic granulomatous disease, myeloperoxidase deficiency, and defective glucose-6-phosphate-dehydrogenase.

phagocytic dysfunction

An altered ability of macrophages, neutrophils, or other phagocytic cells to ingest microorganisms or to digest them following ingestion. This represents a type of immunodeficiency involving phagocytic function.

phagocytic index (PI)

An *in vivo* measurement of the ability of the mononuclear phagocyte system to remove foreign particles. It may be represented by the rate at which injected carbon particles are cleared from the blood. The PI is increased in graft-vs.-host disease.

phagocytosis

An important clearance mechanism for the removal and disposition of foreign agents and particles or damaged cells. Macrophages, monocytes, and polymorphonuclears cells are phagocytic cells. In special circumstances, other cells such as fibroblasts may show phagocytic properties; these are called facultative phagocytes. Phagocytosis may involve nonimmunologic or immunologic mechanisms. Nonimmunologic phagocytosis refers to the ingestion of inert particles such as latex particles or of other particles that have been modified by chemical treatment or coated with protein. Damaged cells are also phagocytized by nonimmunologic mechanisms. Damaged cells may become coated with immunoglobulin or other proteins which facilitate their recognition.

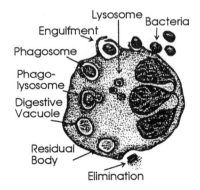

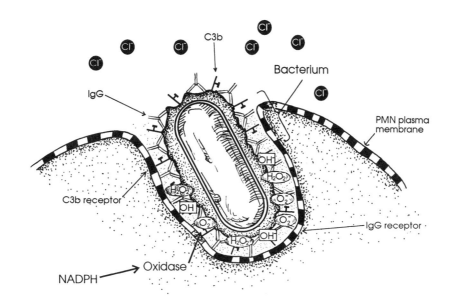

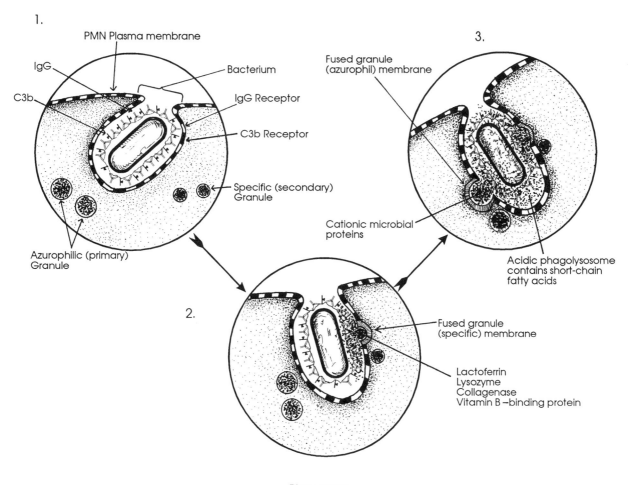

Phagocytosis

Phagocytosis of microorganisms involve several steps: attachment, internalization, and digestion. After attachment, the particle is engulfed within a membrane fragment and a phagocytic vacuole is formed. The vacuole fuses with the primary lysosome to form the phagolysosome, in which the lysosomal enzymes are discharged and the enclosed material is digested. Remnants of indigestible material can be recognized subsequently as residual bodies. Polymorphonuclear neutrophils (PMNs), eosinophils, and macrophages play an important role in defending the host against microbial infection. PMNs and occasional eosinophils appear first in response to acute inflammation, followed later by macrophages. Chemotactic factors are released by actively multiplying microbes. These chemotactic factors are powerful attractants for phagocytic cells which have specific membrane receptors for the factors. Certain pyogenic bacteria may be destroyed soon after phagocytosis as a result of oxidative reactions. However, certain intracellular microorganisms such as *Mycobacteria* or *Listeria* are not killed merely by ingestion and may remain viable unless there is adequate cell-mediated immunity induced by γ interferon activation of macrophages.

Phagocytic dysfunction may be due to either extrinsic or intrinsic defects. The extrinsic variety encompasses opsonin deficiencies secondary to antibody or complement factor deficiencies, suppression of phagocytic cell numbers by immunosuppressive agents, corticosteroid-induced interference with phagocytic function, neutropenia, or abnormal neutrophil chemotaxis. Intrinsic phagocytic dysfunction is related to deficiencies in enzymatic killing of engulfed microorganisms. Examples of the intrinsic disorders include chronic granulomatous disease, myeloperoxidase deficiency, and glucose-6-phosphate dehydrogenase deficiency. Consequences of phagocytic dysfunction include increased susceptibility to bacterial infections, but not to viral or protozoal infections. Selected phagocytic function disorders may be associated with severe fungal infections. Severe bacterial infections associated with phagocytic dysfunction range from mild skin infections to fatal systemic infections.

phagolysosome (see below)

A cytoplasmic vesicle with a limiting membrane produced by the fusion of a phagosome with a lysosome. Substances within a phagolysosome are digested by hydrolysis.

phagosome

A phagocytic membrane-limited vesicle in a phagocyte that contains phagocytized material which is digested by lysosomal enzymes that enter the vesicle after fusion with lysosomes in the cytoplasm.

pharyngeal pouch

Embryonic structure in the neck that provides the thymus, parathyroids, and other tissues with epithelial cells.

pharyngeal pouch syndrome

Thymic hypoplasia.

pharyngeal tonsils

Lymphoid follicles found in the roof and posterior wall of the nasopharynx. They are similar to Peyer's patches in the small intestine. Mucosal lymphoid follicles are rich in IgA producing B cells that may be found in germinal centers.

(Phe,G)AL

A poly-L-lysine backbone to which side chains of phenylalanine and glutamic acid short polymers are linked by alanine residues.

phenotype

Observable features of a cell or organism that are a consequence of interaction between the genotype and the environment. The phenotype represents those genetically encoded characteristics that are expressed. Phenotype may also refer to a group of organisms with the same physical appearance and the same detectable characteristics.

phenylbutazone (4-butyl-1,2-diphenyl-3,5-pyrazolidenedione)

A drug that prevents synthesis of prostaglandin and serves as a powerful antiinflammatory drug. It is used in therapy of rheumatoid arthritis and ankylosing spondylitis.

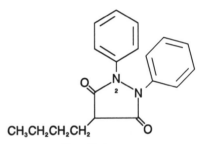

Phenylbutazone

Philadelphia chromosome

The translocation of one arm of chromosome 22 to chromosome 9 or 6 in chronic granulocytic leukemia cells in humans.

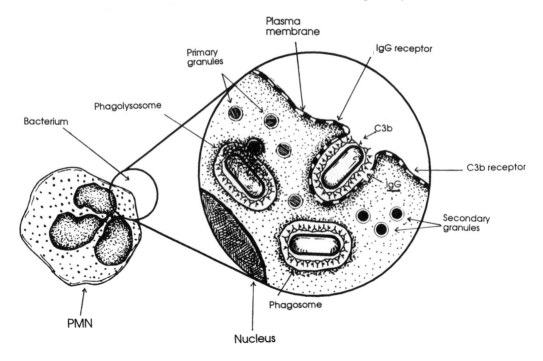

Phagolysosome

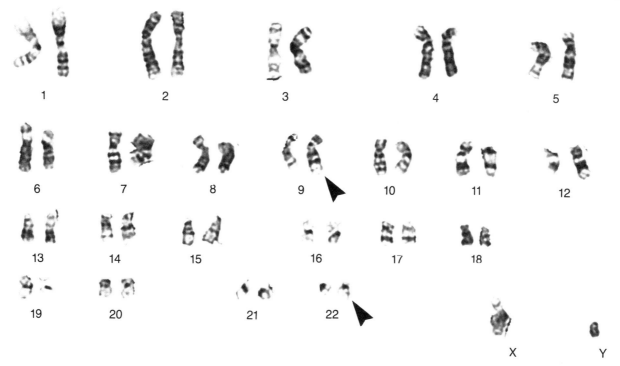

Philadelphia Chromosome

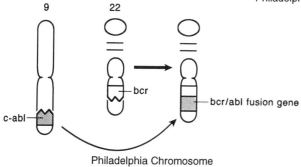

Philadelphia Chromosome

phorbol ester(s)

Esters of phorbol alcohol (4,9,12-β,13,20-pentahydroxy-1,6-tigliadien-3-on) found in croton oil and myristic acid. Phorbol myristate acetate (PMA), which is of interest to immunologists, is a phorbol ester that is 12-*O*-tetradecanoylphorbol-13-acetate (TPA). This is a powerful tumor promoter that also exerts pleotrophic effects on cells in culture, such as stimulation of macromolecular synthesis and cell proliferation, induction of prostaglandin formation, alteration in the morphology and permeability of cells, and disappearance of surface fibronectin. PMA also acts on leukocytes. It links to and stimulates protein kinase C. This leads to threonine

Phorbol 12,13-dibutyrate (PDBu)

Phorbol Ester

and serine residue phosphorylation in the transmembrane protein cytoplasmic domains such as in the CD2 and CD3 molecules. These events enhance interleukin-2 receptor expression on T cells and facilitate their proliferation in the presence of interleukin-1 as well as TPA. Mast cells, polymorphonuclear leukocytes, and platelets may all degranulate in the presence of TPA.

photoallergy

An anaphylactoid reaction induced by exposing an individual to light.

phylogenetic-associated residues

Amino acid residues in immunoglobulin variable regions at a specific position in immunoglobulin (or other protein) molecules in one or more species.

phylogeny

The evolutionary development of a species.

phytoalexins

Plant substances that are active in phytoimmunity.

phytoimmunity

Both active and passive immune-like phenomena in plants. Plant substances active in phytoimmunity include phytonicides and phytoalexins. Plant resistance to many diseases is associated with the presence of antibiotic substances in plant tissues. Antibiotic substances are inherent in both susceptible and resistant varieties of plant species. They may be constitutional inhibitors present in a plant before contact with a parasite or induced antibiotic substances that arise in plants after contact with a parasite. Defense reactions in plants that are associated with the formation and conversion of antibiotic substances include reactions to wounding and the necrotic reaction. Plant resistance to a specific disease is determined by the various antibiotic substances they contain and the synergistic action of these agents with differing roles in phytoimmunity. Plant varieties differ in the quantity of antibiotic substances present in intact tissues and the intensity of their generation in response to infection. They also differ in the nature of subsequent conversions that may produce a marked increase in antibiotic activity.

phytomitogens

Plant glycoproteins that activate lymphocytes through stimulation of DNA synthesis and induction of blast transformation.

phytonicides

Substances produced in both traumatized and nontraumatized plants that represent one of the factors active in plant immunity. Phytonicides have bactericidal, fungicidal, and protisticidal properties.

picryl chloride (1-chloro-2,4,6-trinitrobenzene)

A substance used to add picryl groups to proteins. When applied to the skin of an experimental animal such as a guinea pig, a solution of picryl chloride may conjugate with skin proteins, where it acts as a hapten and may induce contact (type IV) hypersensitivity.

piecemeal necrosis

Death of individual liver cells which are encircled by lymphocytes in chronic active hepatitis.

pigeon breeder's lung

Hypersensitivity pneumonitis. Also called pigeon fancier's lung.

pili

Structures that facilitate adhesion of bacteria to host cells and are therefore direct determinants of virulence.

pinocytosis

The uptake by a cell of small liquid droplets, minute particles, and solutes.

Pirquet von Cesenotics, Clemens Peter Freiherr von (1874–1929)

Viennese physician who coined the term "allergy" and described serum sickness and its pathogenesis. He also developed a skin test for tuberculosis. *Die Serumkrenkheit (with Schick)*, 1905; *Klinische Studien über Vakzination und Vakzinale Allergie*, 1907; *Allergy*, 1911.

pituitary hormones

Immunoperoxidase staining of pituitary adenomas with antibodies to ACTH, GH, prolactin, FSH, and LH facilitates definition of their clinical phenotype.

PK test

Abbreviation for Prausnitz-Küstner reaction.

plague vaccine

Yersinia pestis microorganisms killed by heat or formalin are injected intramuscularly to induce immunity against plague. It is administered in three doses 4 weeks or more apart. The duration of the immunity is approximately 6 months. A live attenuated vaccine, used mainly in Java, has also been found to induce protective immunity.

plantibodies

Antibodies derived from plants such as tobacco which may be used in research. They might prove less immunogenic than murine antibodies.

plaque-forming assay

Refer to hemolytic plaque assay.

plaque-forming cell (PFC) assay

Technique for demonstrating and enumerating cells forming antibodies against a specific antigen. Mice are immunized with sheep red blood cells (SRBC). After a specified period of time, a suspension of splenic cells from the immunized mouse is mixed with antigen (SRBC) and spread on a suitable semisolid gel medium. After or during incubation at 37°C, complement is added. The erythrocytes which have anti-SRBC antibody on their surface will be lysed. Circular areas of hemolysis appear in the gel medium. If viewed under a microscope, a single antibody-forming cell can be identified in the center of the lytic area. There are several modifications of this assay, since some antibodies other than IgM may fix complement less efficiently. In order to enhance the effects, an antiglobulin antibody called developing antiserum is added to the mixture. The latter technique is called indirect PFC assay.

plaque-forming cells

The antibody-producing cells in the center of areas of hemolysis observed microscopically when reading a hemolytic plaque assay. The antibodies they form are specific for red blood cells suspended in the gel medium surrounding them. Once complement is added,

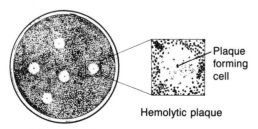

Top view of
Petri dish

Plaque forming cell

Hemolytic plaque

the antibody-coated erythrocytes lyse, producing clear areas of hemolysis surrounding the antibody-forming cell. The antibody produced may be specific not only for red blood cell surface antigens, but also for soluble antigens deliberately coated on their surfaces for assay purposes.

plaque technique

Refer to hemolytic plaque assay or phage neutralization assay.

plasma

A transparent yellow fluid that constitutes 50 to 55% of the blood volume. It is 92% fluid and 7% protein. Inorganic salts, hormones, sugars, lipids, and gases make up the remaining 1%. Plasma from which fibrinogen and clotting factors have been removed is known as serum.

plasma cell antigen

A murine plasmacyte membrane alloantigen. It may be designated PC-1, PC-2.

plasma cell dyscrasias

Lymphoproliferative disorders in which there is monoclonal plasma cell proliferation, leading to such conditions as multiple myeloma or to the less ominous extramedullary plas-macytoma.

plasma cell leukemia

A malignancy associated with plasma cells in the circulating blood that constitute greater than 20% of the leukocytes. The absolute plasma cell number is more than 2000 per cubic millimeter of blood. Advanced cases reveal extensive infiltration of the tissue with plasma cells and replacement of the marrow. Reactive plasmacytosis must be considered in the diagnosis.

plasma cells

Antibody-producing cells. Immuno-globulins are present in their cyto-plasm, and secretion of immunoglo-bulin by plasma cells has been directly demonstrated *in vitro*. Increased levels of immunoglobulins in some patho-logic conditions are associated with increased numbers of plasma cells, and conversely, their number at antibody producing sites increases following immunization. Plasma cells develop from B cells and are large, spherical or ellipsoidal cells, 10 to 20 μ in size. Mature plasma cells have abundant cytoplasm, which stain deep blue with Wright's stain, and have an eccentrically located, round or oval nucleus, usually surrounded by a well-defined perinuclear clear zone. The nucleus contains coarse and clumped masses of chromatin, often arranged in a cartwheel fashion. The nuclei of normal, mature plasma cells have no nucleoli, but those of neoplastic plasma cells such as those seen in multiple myeloma have conspicuous nucleoli. The cytoplasm of normal plasma cells has conspicuous Golgi complex and rough endoplasmic reticulum and frequently contains vacuoles. The nuclear to cytoplasmic ratio is 1:2. By electron microscopy, plasma cells show very abundant endoplasmic reticulum, indicating extensive and active protein synthesis. Plasma cells do not express surface immunoglobulin or complement receptors which distinguishes them from B lymphocytes.

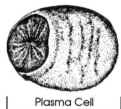

Plasma Cell

14–20 μm.

plasma half-life (T$^{1/2}$)

Determination of the catabolic rate of any component of the blood plasma. With respect to immunoglobulins, it is the time required for one half of the plasma immunoglobulins to be catabolized.

plasma pool

The amount of plasma immunoglobulin per unit of body weight. This may be designated as milligrams of immunoglobulin per kilogram of body weight.

plasmablast

An immature cell of the plasma cell lineage that reveals distinctive, clumped nuclear chromatin developing endoplasmic reticulum and a Golgi apparatus.

plasmacyte

Plasma cell.

plasmacytoma

A plasma cell neoplasm. Also termed myeloma or multiple myeloma. To induce an experimental plasmacytoma in laboratory mice or rats, paraffin oil is injected into the peritoneum. Plasmacytomas may occur spontaneously. These tumors, comprised of neoplastic plasma cells, synthesize and secrete monoclonal immunoglobulins, yielding a homogeneous product that forms a spike in electrophoretic analysis of the serum. Plasmacytomas were used extensively to generate monoclonal immunoglobulins prior to the development of B cell hybridoma technology to induce monoclonal antibody synthesis at will to multiple antigens.

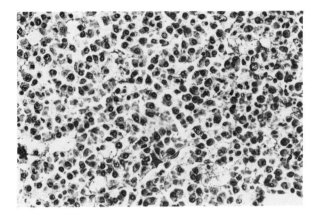

bone marrow plasmacytoma
pure plasma cell
infiltrate
Plasmacytoma

plasmapheresis

A technique in which blood is withdrawn from an individual, the desired constituent is separated by centrifugation, and the cells are reinjected into the patient. Thus, plasma components may be removed from the circulation of an individual by this method. The technique is also useful to obtain large amounts of antibodies from the plasma of an experimental animal.

Plasmapheresis is used therapeutically to rid the body of toxins or autoantibodies in the blood circulation. Blood taken from the patient is centrifuged, the cells are saved, and the plasma is removed. Cells are resuspended in albumin, fresh normal plasma, or albumin in saline and returned to the patient. The ill effects of a toxin or of an autoantibody may be reduced by 65% by removing approximately 2500 ml of plasma. Removal of twice this amount of plasma may diminish the level of a toxin or of an autoantibody by an additional 20%. This procedure has been used to treat patients with myasthenia gravis, Eaton-Lambert syndrome, Goodpasture's syndrome, hyperviscosity syndrome, posttransfusion purpura, and acute Guillain-Barre syndrome.

plasmid

Extrachromosomal genetic structure that consists of a circular, double-stranded DNA molecule that permits the host bacterial cell to resist antibiotics and produce other effects that favor its survival. Plasmid replication is independent of the bacterial chromosome. Plasmids have been used widely in recombinant DNA technology.

plasmin

A proteolytic enzyme in plasma that is generated from its inactive precursor plasminogen. It is a 90-kD enzyme that derives from cleavage of a single arginyl-valyl bond in the C-terminal region of plasminogen. It consists of two unequal chains, termed heavy (A) and light (B) chains, linked by a single disulfide bond. The A chain derives from the N-terminal region plasminogen. The B chain carries the serine active site. Plasmin catalyzes the hydrolysis of fibrin. Thus, it facilitates the dissolution of intravascular blood clots. In addition to its fibrinolytic activity, plasmin has numerous other functions associated with coagulation, fibrinolysis, and inflammation that include (1) enhancement of antibody responses to both thymus-dependent and thymus-independent antigens, (2) augmentation of agglutination by lectins, (3) facilitation of the escape of cells from contact inhibition and culture, (4) enhancement of cytotoxicity with or without participation of antibodies, and (5) stimulation of B cell proliferation.

plasminogen

The inactive precursor of the proteolytic enzyme plasmin. Several serine proteases such as urokinase convert it to active plasmin. It is a β globulin widely distributed in tissue, body fluids, and plasma. It is a single chain monomeric molecule. Plasminogen activation

occurs in two stages. The GLU-plasminogen activation begins with removal of two peptides at the N-terminus of the molecule and conversion to Lys-plasminogen. The second step involves the rapid conversion of Lys-plasminogen to Lys-plasmin.

Plasminogen

plasminogen activator

An enzyme produced by macrophages that converts plasminogen to plasmin which degrades fibrin.

platelet

A small (3 μ in diameter) round disk that is derived from bone marrow megakaryocytes, but is present in the blood. Platelets function in blood clotting by releasing thromboplastin. They also harbor serotonin and histamine which may be released during type I (anaphylactic hypersensitivity reactions). Complement receptor 1 (CR1) is present on the platelets of mammals other than primates and is significant for immune adherence.

platelet-activating factor (PAF)

A phospholipid with a mol wt of about 300 to 500 D formed by leukocytes, macrophages, and endothelial cells that induces aggregation of platelets and promotes amine secretion, aggregation of neurophils, release of enzymes, and an increase in vascular permeability. Its effect resembles that of IgE-mediated changes in anaphylaxis and cold urticaria. It may also participate in endotoxin shock and is derived from phosphorylcholine. The combination of antigens with the Fab regions of antibody molecules bound through Fc receptors to mast cells, polymorphonuclear leukocytes, and macrophages results in platelet-activating factor release. PAF release accompanies anaphylactic shock. It apparently mediates inflammation and allergic reactions. PAF induces a transient reduction in blood platelets, causes hypotension, and facilitates vascular permeability, but it has no effects on contracting smooth muscle and has no chemotactic activity. There is probably more than a single compound with this activity. PAF is resistant to arylsulfatase B, but is sensitive to phospholipases.

platelet antibodies

Refer to platelet antigens.

platelet antibodies, drug induced

Amphotericin B, cephalothin, methicillin, pentamidine, trimethoprim-sulfamethoxizole, and vancomycin may all induce the synthesis of antiplatelet antibodies.

platelet antigens

Surface epitopes on thrombocytes that may be immunogenic, leading to platelet antibody formation which causes such conditions as neonatal alloimmune thrombocytopenia and post-transfusion purpura. The PlA1 antigen may induce platelet antibody formation in PlA1 antigen negative individuals. Additional platelet antigens associated with purpura include PlA2, Baka, and HLA-A2.

platelet-associated immunoglobulin (PAIgG)

PAIgG is present in 10% of normal individuals, 50% of those with tumors, and 76% of septic patients and may be induced by graft-vs.-host disease. PAIgG is present in 71% of autologous marrow graft recipients and in 50% of allogeneic marrow graft recipients.

platelet autoantibodies

Platelets possess surface FcγRII that combine to IgG or immune complexes. The platelet surfaces can become saturated with immune complexes, as in autoimmune (or idiopathic) thrombocytopenic purpura (ITP) or AITP. Fab-mediated antibody binding to platelet antigens may be difficult to distinguish from Fc-mediated binding of immune complexes to the surface.

platelet-derived growth factor (PDGF)

A low mol wt protein derived from human platelets that acts as a powerful connective tissue mitogen, causing fibroblast and intimal smooth muscle proliferation. It also induces vasoconstriction and chemotaxis and activates intracellular enzymes. PDGF plays an important role in atherosclerosis and fibroproliferative lesions such as glomerulonephritis, pulmonary fibrosis, myelofibrosis, and other processes. It is comprised of a two chain, i.e., A or B, dimer. It can be an AA or BB homodimer or an AB heterodimer. Human PDGF-AA is a 26.5-kD A chain homodimeric protein comprised of 250 amino acid residues, whereas PDGF-BB is a 25-kD B chain homodimeric protein comprised of 218 amino acid residues.

platelet-derived growth factor receptor (PDGF-R)

A glycoprotein in the membrane that has five extracellular domains that resemble those of immunoglobulins. It also has a kinase insert in the cytoplasm. The receptor protein must undergo a conformational change for signal transduction. A gene on chromosome 4q11 encodes PDGF-R.

platelet transfusion

The administration of platelet concentrates prepared by centrifuging a unit of whole blood at low speed to provide 40 to 70 ml of plasma that contains 3 to 4×10^{11} platelets. This amount can increase an adult's platelet concentration by 10,000 per cubic millimeter of blood. It is best to store platelets at 20 to 24°C, subjecting them to mild agitation. They must be used within 5 days of collection.

PMN

Abbreviation for polymorphonuclear leukocyte.

pneumococcal polysaccharide

A polysaccharide found in the *Streptococcus pneumoniae* capsule that is a type-specific antigen. It is a virulence factor. Serotypes of this microorganism are based upon different specificities in the capsular polysaccharide which is comprised of oligosaccharide repeating units. Glucose and glucuronic acid are the repeating units in type III polysaccharide.

pneumococcal polysaccharide vaccine

A 23 valent vaccine containing capsular polysaccharides of *Streptococcus pneumoniae*. This vaccine counters 85 to 90% of the serotypes causing invasive pneumococcal infections. Elderly, immunocompromised, and chronically ill persons are advised to receive this vaccine every 3 years.

Pneumocystis carinii

A protozoan parasite that infects immunocompromised subjects such as AIDS patients, transplant recipients, lymphoma patients, leukemia patients, and others immunosuppressed for one reason or another. It is diagnosed in tissue sections stained with the Gomorimethenamine silver stain. A mannose receptor facilitates the organism's uptake by macrophages. Approximately one half of those hospitalized with a first infection by *Pneumocystis carini* pneumonia die.

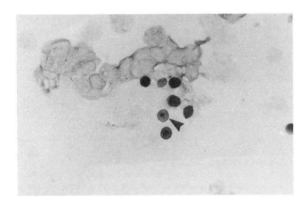

Pneumocystis carinii

PNH cells

Red blood cells of paroxysmal nocturnal hemoglobinuria patients. At weakly acid pH, PNH cells disrupt spontaneously. The ability of complement to lyse these cells is much more pronounced than its action on normal erythrocytes subjected to conditions that promote their lysis by complement.

POEMS

Abbreviation for polyneuropathy, organomegaly, endocrinopathy, monoclonal gammopathy, and skin alterations. Refer to POEMS syndrome.

POEMS syndrome

A condition that manifests polyneuropathy, organomegaly, endocrinopathy, monoclonal gammopathy, and skin alterations.

poison ivy

A plant containing the chemical urushiol which may induce severe contact hypersensitivity of the skin in individuals who have come into contact with it. Urushiol is also found in mango trees, Japanese lacquer trees, and cashew plants. Urushiol is present not only in *Toxicodendron radicans* (poison ivy) found in the eastern U.S. but also in *T. diversilobium* (poison oak) found in the western U.S. and *T. vernix* (poison sumac) found in the southern U.S. Setting fire to these plants is also hazardous in that the smoke containing the chemical may induce tracheitis and pulmonary edema in allergic individuals. The chemical may remain impregnated in unwashed clothing and cause reactions in people who may come into contact with it for long periods of time.

Rhus toxicodendron

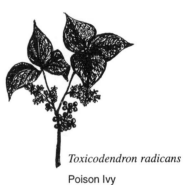

Toxicodendron radicans

Poison Ivy

poison ivy hypersensitivity

Hypersensitivity that is principally type IV contact hypersensitivity induced by urushiols which are chemical constituents of poison ivy (*Rhus toxicodendron);* when poison ivy plants containing this chemical come into contact with the skin, it becomes hypersensitive. The urushiol acts as a hapten by complexing with skin proteins to induce cellular (type IV) hypersensitivity on contact. Also called delayed-type hypersensitivity.

pokeweed mitogen (PWM)

A lectin extracted from the pokeweed *Phytolacca americana*. It is carbohydrate binding and may stimulate both human B and T cells. It has been used to promote B cell growth and proliferation in tissue culture, leading to immunoglobulin production.

pol (see facing page)

A retrovirus structural gene that codes for reverse transcriptase. The structural genes of HIV-1 also include *gag* and *env*.

polar cap

Refer to capping.

poliomyelitis vaccines

The three strains of poliomyelitis virus combined into a live attenuated oral poliomyelitis vaccine first introduced by Sabin. Replication in the gastrointestinal tract stimulates effective local immunity associated with IgA antibody synthesis. Individuals to be immunized receive three oral doses of the vaccine. This largely replaces the Salk vaccine which was introduced in the early 1950s as a vaccine comprised of the three strains of poliovirus that had been killed with formalin. This preparation must be administered subcutaneously.

pollen hypersensitivity

Immediate (type I) hypersensitivity which atopic individuals experience following inhalation of pollens such as ragweed in the U.S. This is an IgE-mediated reaction that results in respiratory symptoms expressed as hay fever or asthma. Sensitivity to certain pollens can be detected through skin tests with pollen extracts.

Retroviral Genome

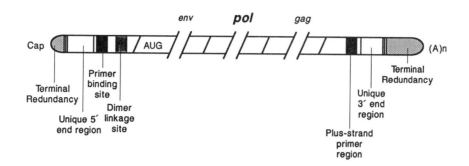

polyacrylamide gel

A cross-linked medium which is used for the SDS-PAGE and disk gel electrophoresis techniques for protein and nucleic acid separation. Varying the porosity of the polymerized and cross-linked acrylamide that forms a gel after being solubilized permits molecules of different size and charge to be separated.

polyacrylamide gel electrophoresis (PAGE)

Zone electrophoresis that employs a cross-linked polyacrylamide gel that is used in SDS-PAGE and disk gel electrophoretic methods. SDS-PAGE is a method to separate proteins by their mol wt. Proteins are combined with sodium dodecyl sulfate (SDS) at increased temperature and with a reducing agent to denature the protein. Electrophoresis is carried out in the transparent synthetic polyacrylamide gel which serves as a molecular sieve. Migration of the protein is inversely proportional to its mol wt. Coomassie blue is used to stain unlabeled proteins, and autoradiography is used to detect radiolabeled proteins. An unknown protein's mol wt can be predicted using standard proteins for comparison.

polyagglutination

The aggregation of erythrocytes by antibodies, autoagglutinins, or alloagglutinins in blood serum. Polyagglutination also refers to aggregation of normal red blood cells treated with neuraminidase and also to red blood cells with altered membranes that are improperly aggregated by anti-A or anti-B antibodies. This is linked to altered glycoproteins such as Tn, T, and Cad. Acquired B antigens may also lead to polyagglutination, as can bacterial infection of the patient. Serum contaminants including detergents, microbes, metal cations, or silica may also cause polyagglutination. Also called panagglutination.

polyarteritis nodosa

A necrotizing vasculitis of small- and medium-sized muscular arteries. It often involves renal and visceral arteries and spares the pulmonary circulation. The disease is often characterized by immune complex deposition in arteries, often associated with chronic hepatitis B virus infection. Early lesions of the vessels often reveal hepatitis B surface antigen, IgM, and complement components. This uncommon disease has a male to female ratio of 2.5:1. The mean age at onset is 45 years. Characteristically the kidneys, heart, abdominal organs, and both peripheral and central nervous systems are involved. Lesions of vessels are segmental and show preference for branching and bifurcation in small- and medium-sized muscular arteries. Usually, the venules and veins are unaffected and only rarely are granulomas formed. Characteristically, aneurysms form following destruction of the media and internal elastic lamina. There is proliferation of the endothelium with degeneration of the vessel wall and fibrinoid necrosis, thrombosis, ischemia, and infarction developing. Other than hepatitis B, polyarteritis nodosa is also associated with tuberculosis, streptococcal infections, and otitis media. Presenting signs and symptoms include weakness, abdominal pain, leg pain, fever, cough, and neurologic symptoms. There may be kidney involvement, arthritis, arthralgia, or myalgia, as well as hypertension. Of patients, 40% may have skin involvement manifested as a maculopapular rash. Laboratory findings include elevated erythrocyte sedimentation rate, leukocytosis, anemia, thrombocytosis, and cellular casts in the urinary sediment signifying renal glomerular disease. Angiography is important in revealing the presence of aneurysm and changes in vessel caliber.

There is no diagnostic immunologic test, but immune complexes, cryoglobulins, rheumatoid factor, and diminished complement component levels are often found. Biopsies may be taken from skeletal muscle or nerves for diagnostic purposes. Corticosteroids may be used, but cyclophosphamide is the treatment of choice in the severe progressive form.

polyarthritis

Multiple joint inflammation as occurs in rheumatic fever, systemic lupus erythematosus, and related diseases.

polyclonal

Originating from multiple clones.

polyclonal activation

The stimulation of multiple lymphocyte clones, thereby leading to a heterogeneous immune response.

polyclonal activator

Substances that activate multiple lymphocyte clones, including both T and B lymphocytes, regardless of antigen specificity. Phytohemagglutinin activates mainly T lymphocytes, whereas staphylococcal protein A stimulates human B lymphocytes and lipopolysaccharide derived from Gram-negative bacteria stimulates murine B lymphocytes. Con A stimulates T cells.

polyclonal antibodies

Multiple immunoglobulins responding to different epitopes on an antigen molecule. This multiple stimulation leads to the expansion of several antibody-forming clones whose products represent a mixture of immunoglobulins in contrast to proliferation of a single clone which would yield a homogeneous monoclonal antibody product.

Thus, polyclonal antibodies represent the natural consequence of an immune response in contrast to monoclonal antibodies which occur in vivo in pathologic conditions such as multiple myeloma or are produced artificially by hybridoma technology against one of a variety of antigens.

polyclonal hypergammaglobulinemia

An elevation in the blood plasma of γ globulin which is comprised of increased quantities of the different immunoglobulin classes rather than an increase of just one immunoglobulin class.

polyclone mitogens

Mitogens that stimulate multiple lymphocyte subpopulations.

polyclone proteins

Protein molecules from multiple cell clones.

polyendocrine deficiency syndrome (polyglandular autoimmune syndrome)

Two related endocrinopathies with gonadal failure that may be due to defects in the hypothalamus. There is vitiligo and autoimmune adrenal insufficiency. Four fifths of the patients have autoantibodies. Type I occurs in late childhood and is characterized by hypoparathyroidism, alopecia, mucocutaneous candidiasis, malabsorption, pernicious anemia, and chronic active hepatitis. It has an autosomal recessive mode of inheritance. Type II occurs in adults with Addison's disease and autoimmune thyroiditis or insulin-dependent diabetes mellitus.

polygenic inheritance

Phenotypic inheritance based on genetic variation at multiple loci. Numerous genetic loci may contribute to the inherited phenotype. Certain forms of immune responsiveness and susceptibility to selected diseases may be influenced by polygenic inheritance.

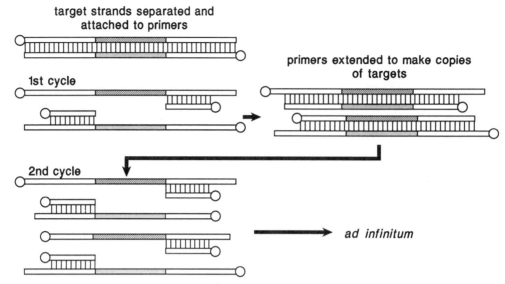

target strands separated and attached to primers

1st cycle

primers extended to make copies of targets

2nd cycle

ad infinitum

Polymerase Chain Reaction

polyimmunoglobulin receptor

An attachment site for polymeric immunoglobulins located on epithelial cell and hepatocyte surfaces that facilitate polymeric IgA and IgM transcytosis to the secretions. After binding, the receptor-immunoglobulin complex is endocytosed and enclosed within vesicles for transport. Exocytosis takes place at the cell surface where the immunoglobulin is discharged into the intestinal lumen. A similar mechanism in the liver facilitates IgA transport into the bile. The receptor-polymeric immunoglobulin complex is released from the cell following cleavage near the cell membrane. The receptor segment that is bound to the polymeric immunoglobulin is known as the secretory component which can only be used once in the transport process.

polymerase chain reaction (PCR)

A technique to amplify a small DNA segment beginning with as little as 1 μg. The segment of double-stranded DNA is placed between two oligonucleotide primers through many cycles of amplification. Amplification takes place in a thermal cycler, with one step occurring at a high temperature in the presence of DNA polymerase that is able to withstand the high temperature. Within a few hours, the original DNA segment is transformed into millions of copies. PCR methodology has been used for multiple purposes including detection of HIV-1, the prenatal diagnosis of sickle cell anemia, and gene rearrangements in lymphoproliferative disorders, among numerous other applications. The technique is used principally to prepare enough DNA for analysis by available DNA methods. The polymerase chain reaction has a 99.99% sensitivity. The technique is widely used in DNA diagnostic work.

polymers

Molecules comprised of more than one repeating unit. In immunology, immunoglobulins comprised of more than one basic monomeric four-polypeptide chain unit. IgA may exist as dimers with two units or as multimers. The IgM molecule is pentameric, containing five monomeric units.

polymorphism

The occurrence of two or more forms, such as ABO and Rh blood groups, in individuals of the same species. This is due to two or more variants at a certain genetic locus occurring with considerable frequency in a population. Polymorphisms are also expressed in the HLA system of human leukocyte antigens as well as in the allotypes of immunoglobulin γ and κ chains.

polymorphonuclear leukocyte (PMN)

White blood cells with lobulated nuclei that are often trilobed. These cells are of the myeloid cell lineage and, in the mature form, can be differentiated into neutrophils, eosinophils, and basophils. This distinction is based on the staining characteristics of their cytoplasmic specific or secondary granules. These cells, which measure approximately 13 μ in diameter, are active in acute inflammatory responses.

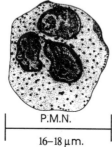

P.M.N.

16–18 μm.

Polymorphonuclear Leukocyte

polymyositis

An acute or chronic inflammatory disease of muscle that occurs in women twice as commonly as in men. Lymphocytes in polymyositis subjects produce a cytotoxin when incubated with autologous muscle. Biopsies of involved muscle reveal infiltration by lymphocytes and plasma cells. Antibodies can be demonstrated against the nuclear antigens Jo-1, PM-Scl, and RNP. Patients may develop polyclonal hypergammaglobulinemia. One fifth of the patients may develop rheumatoid factors and antinuclear antibodies. Cellular immunity appears important in the pathogenesis. This is exemplified by lymphocytes of patients with polymyositis responding to their own muscle antigens as if they were alien. Patients often complain of muscle weakness, especially in the proximal muscles of extremities. To diagnose polymyositis, a minimum of three of the following must be present: (1) shoulder or pelvic girdle weakness, (2) myositis as revealed by biopsy, (3) increased levels of muscle enzymes, and (4) electromyographic findings of myopathy. Corticosteroids have been used to decrease muscle inflammation and increase strength. Methotrexate or other cytotoxic agents may be used when steroids prove ineffective.

polynucleotides

A linear polymer comprised of greater than ten nucleotides joined by 3′,5′-phosphodiester bonds. Double-stranded DNA chains that may serve as adjuvants when inoculated with antigens.

polyspecific anti-human globulin (AHG)

This is known as the Coombs' reagent which consists of antibody against human IgG and C3d. It may also have anti-C3b, anti-C4b, and anti-C4d antibodies. Although it demonstrates only minimal reactivity with IgM and IgA heavy chains, it may interact with these molecules by reacting with their κ or λ light chains. It is used for the direct antiglobulin test.

polyvalent

Multivalent.

polyvalent antiserum

An antiserum comprised of antibodies specific for multiple antigens.

polyvalent vaccine

A vaccine comprised of multiple antigens from more than one strain of a pathogenic microorganism or from a mixture of immunogens such as the diphtheria, pertussis, and tetanus toxoid preparation.

Porter, Rodney Robert (1917–1985)

British biochemist who received the Nobel Prize in 1972, with Gerald Edelman, for their studies of antibodies and their chemical structure. Porter cleaved antibody molecules with the enzyme papain to yield Fab and Fc fragments. He suggested that antibodies have a four-chain structure.

Portier, Paul Jules (1866–1962)

French physiologist who, with Richet, was the first to describe anaphylaxis.

positive selection

The survival of those thymocytes that recognize *self MHC* as well as self or foreign antigen, and the death of those that do not recognize self MHC. The resulting cell population is self MHC restricted and capable of interacting with both self and foreign antigens. (Refer to negative selection.)

postcapillary venules

Relatively small blood vessels lined with cuboidal epithelium through which blood circulates after it exits the capillaries and before it enters the veins. It is a frequent site of migration of lymphocytes and inflammatory cells into tissues during inflammation. Recirculating lymphocytes migrate from the blood to the lymph through high endothelial venules of lymph nodes.

postcardiotomy syndrome

A condition that follows heart surgery or traumatic injury. Autoantibodies against heart antigens may be demonstrated by immunofluorescence within weeks of the surgery or trauma. Corticosteroid therapy represents an effective treatment. Myocardial infarct patients may also develop a similar condition.

postinfectious encephalomyelitis

Demyelinating disease following a virus infection that is mediated by autoimmune delayed-type (type IV) hypersensitivity to myelin.

postinfectious iridocyclitis

Inflammation of the iris and ciliary body of the eye. This may occur after a virus or bacterial infection and is postulated to result from an autoimmune reaction.

postrabies vaccination encephalomyelitis

A demyelinating disease produced in humans actively immunized with rabies vaccine containing nervous system tissue to protect against the development of rabies. Serial injections of rabbit brain tissue containing rabies virus killed by phenol could induce demyelinating encephalomyelitis in the recipient. This method of vaccination was later replaced with a vaccine developed in tissue culture that did not contain any nervous tissue.

poststreptococcal glomerulonephritis

An acute proliferative glomerulonephritis that may follow a streptococcal infection of the throat or skin by 1 to 2 weeks. It is usually seen in 6- to 10-year old children, but it may occur in adults as well. The onset is heralded by evidence of acute nephritis. Ninety percent of patients have been infected with Group A β hemolytic streptococci that are nephritogenic, specifically types 12, 4, and 1 which are revealed by their cell wall M protein. Poststreptococcal glomerulonephritis is mediated by antibodies induced by the streptococcal infection. Most patients show elevated antistreptolysin-O (ASO) titers. Serum complement levels are decreased. Immunofluorescence of renal biopsies demonstrates granular immune deposits that contain immunoglobulin and complement in the glomeruli. This is confirmed by electron microscopy. The precise streptococcal antigen has never been identified; however, a cytoplasmic antigen termed endostreptosin together with some cationic streptococcal antigens are found in glomeruli. Subepithelial immune deposits appear as "humps". They are antigen-antibody complexes that may also appear in the mesangium or occasionally in a subendothelial or intramembranous position. These immune deposits stain positively for IgG and complement by immunofluorescence. Affected children develop fever, nausea, oliguria, and hematuria within 2 weeks following a streptococcal sore throat or skin infection. Erythrocyte casts and mild proteinuria may be identified. There may be periorbital edema and hypertension upon examination. The BUN and ASO titer may also be elevated. More than 95% of children with poststreptococcal glomerulonephritis recover, although a few, less than 1%, develop rapidly progressive glomerulonephritis and a few others develop chronic glomerulonephritis.

posttransfusion graft-vs.-host disease

A condition that resembles postoperative erythroderma that occurs in immunocompetent recipients of blood. There is dermatitis, fever, marked diarrhea, pancytopenia, and liver dysfunction.

postvaccinal encephalomyelitis

A demyelinating encephalomyelitis that occurs approximately 2 weeks following vaccination of infants less than 1 year of age, as well as adults, with vaccinia virus to protect against smallpox. This rare complication of the smallpox vaccination frequently leads to death.

postzone

Lack of serologic reactivity as a consequence of high dilution of antibody, as in a serial dilution procedure. This is a zone of relative antigen excess.

PPD

Purified protein derivative of tuberculin.

PPLO (pleuropneumonia-like organisms)

Mycoplasma pneumoniae. A microorganism that causes asymptomatic respiratory tract infection or upper respiratory tract inflammation. It spreads in the air. Patients develop headache, muscle pain, chest tenderness, and a low-grade fever. They may manifest cold agglutinin-induced hemolysis.

Prausnitz-Giles, Carl (1876–1963)

German physician who conducted extensive research on allergies. He and Küstner successfully transferred food allergy with serum. This became the basis for the Prausnitz-Küstner test.

Prausnitz-Küstner (PK) reaction (historical)

A skin test for hypersensitivity in which serum containing IgE antibodies specific for a particular allergen is transferred from an allergic individual to a nonallergic recipient by intradermal injection. This is followed by injection of the antigen or allergen in question into the same site as the serum injection. Fixation of the IgE antibodies in the "allergic" serum to mast cells in the recipient results in local release of the pharmacological mediators of immediate hypersensitivity that include histamine. It results in a local anaphylactic reaction with a wheal and flare response.

pre-B cells

Pre-B cells develop from lymphoid stem cells in the bone marrow. These are large, immature lymphoid cells that express cytoplasmic μ chains, but no light chains or surface immunoglobulin and are found in fetal liver and adult bone marrow. They are the earliest cells of the B cell lineage. Antigen is not required for early differentiation of the B cell series. Pre-B cells differentiate into immature B cells, followed by mature B cells that express surface immunoglobulin. Pre-B cell immunoglobulin genes contain heavy chain V, D, and J gene segments that are contiguous. No rearrangement of light chain gene segments has yet occurred. In addition to their cytoplasmic IgM, pre-B cells are positive for CD10, CD19, and HLA-DR markers.

Pre-B

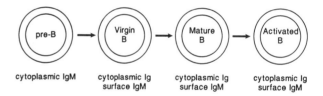

pre-T cell

Developed from pro-T cells through gene rearrangement. They give rise to γδ TCR-bearing cells through rearrangement and expression of γ and

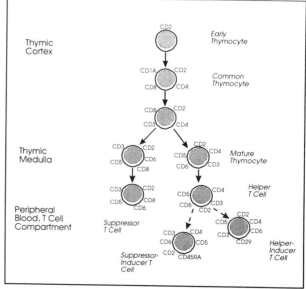

Pre-T Cell

δ TCR genes. Pre-T cells that give rise to T lymphocytes expressing the αβ T cell receptor rearrange α TRC genes, deleting the δ TCR genes situated on the chromosome between the Vα and Cα genes.

Pre-T

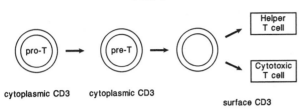

pre-T lymphocyte

Refer to prothymocyte.

precipitating antibody

A precipitin.

precipitation

Following the union of soluble macromolecular antigen with the homologous antibody in the presence of electrolytes *in vitro* and *in vivo,* which occurs within seconds after contact, complexes of increasing density form in a lattice arrangement and settle out of solution, as in the precipitation or precipitin reaction. Thus, the materials needed for a precipitin reaction include antigen, antibody, and electrolyte. The reaction of soluble antigen and antibody in the precipitin test may be observed in liquid or in gel media. The reaction in liquid media may be qualitative or quantitative. Following discovery of the precipitin reaction by Kraus, quantitative and semiquantitative measurements of antibody could be made. The term precipitinogen is sometimes employed to designate the antigen, and precipitin is the antibody in a precipitation reaction.

precipitation curve

The milligrams of antibody in the precipitate are plotted on the ordinate and the milligrams of antigen added are plotted on the abscissa of a graph. The precipitin curve contains an ascending and a descending limb and zones of antibody excess, equivalence, and antigen excess. By testing with the homologous reagents, unreacted antibodies and antigens can be detected in the supernatants. If antigen is homogeneous or if antibodies specific for only one of a mixture of antigens are studied by the precipitin reaction, none of the supernatants contain both unreacted antibodies and unreacted antigens that can be detected. The ascending limb of the precipitin curve represents the zone of antibody excess — where free antibody molecules are present in the supernatants. The descending limb represents the zone of antigen excess where free antigen is present in the supernatants. Precipitation is maximum in the zone of equivalence (or equivalence point) — neither antigen nor antibody can be detected in the supernatants. In contrast to the nonspecific system described above, the presence of more than one antigen-antibody system in the reaction medium may be revealed by the demonstration of unreacted antibody and antigen in certain supernatants. This occurs when there is an overlap between the zone of antigen excess in one antigen-antibody combination with the zone of antibody excess of a separate antigen-antibody system. The lattice theory proposed by Marrack explains how multivalent antigen molecules and bivalent antibodies can combine to yield antigen to antibody ratios that differ from one precipitate to another, depending upon the zone of the precipitin reaction in which they are formed. When the ratio of antibody to antigen is above 1.0, a visible precipitate forms. However, when the ratio is less than 1.0, soluble complexes result and remain in the supernatant. The soluble complexes are associated with the precipitin curve's descending limb. Also termed precipitin curve.

precipitation in gel media

Oudin in 1946 overlaid antibody incorporated in agar in a test tube with the homologous antigen. A band of precipitation appeared in the gel where the antigen-antibody interaction occurred. Mixtures of antibodies of several specificities were overlaid with a mixture of the homologous antigens, and a distinct band for each resulted. Oudin's technique involves simple or single diffusion in one dimension. In 1953, Oakley and Fulthorpe placed antiserum incorporated into agar in the bottom of a tube, covered this with a layer of plain agar which was permitted to solidify, and then added antigen. This was double diffusion in one dimension. Double diffusion in two dimensions was developed by Ouchterlony and independently by Elek in 1948. Agar is poured on a flat glass surface such as a microscope slide, glass plate, or Petri dish. Wells or troughs are cut in the agar, and these are filled with antigen and antibody solutions under study. Multiple component systems may be analyzed by use of this method, and crossreactivities may be detected. Double diffusion in agar is a useful method to demonstrate similarity among structurally related antigens. Equidistant holes are punched in agar gel containing electrolyte. Antigen is placed in one well and antiserum in another adjacent to it, and the plates are observed the following day for a precipitation line where antigen and antibody have migrated toward one another and reached equivalent concentrations. A single line implies a single antigen-antibody system. If agar plates containing one central well with others cut equidistant from it at the periphery are employed, a reaction of identity may be demonstrated by placing antibody in the central well and the homologous antigen in adjacent peripheral wells. A confluent line of precipitate is produced in the shape of an arc. This implies that the antigen preparations in adjacent peripheral wells are identical; that is, they have the same antigenic determinants. If antibodies against two unrelated antigen preparations are combined and placed in the central well and their homologous antigens are placed in separate adjacent peripheral wells, a line of precipitation is produced by each antigen-antibody reaction to give the appearance of crossed sword points. This constitutes a reaction of nonidentity. It implies that the antigenic determinants are different in each of the two samples of antigen. A third pattern known as a reaction of partial identity occurs when two antigen preparations that are related, but not the same are placed in separate adjacent wells and an antibody preparation that crossreacts with both of them is placed in a central well. The precipitation lines between each antigen-antibody system converge, but a spur or extension of one of the precipitation line occurs. This reaction of partial identity with spur formation implies that the antigen preparations are similar, but that one of them has an antigenic determinant(s) not present in the other. A reaction of identity and nonidentity may be observed simultaneously, implying that two separate antigen preparations have both common and different antigenic determinants.

precipitin

An antibody that interacts with a soluble antigen to yield an aggregate of antigen and antibody molecules in a lattice framework called a precipitate. Under appropriate conditions, the majority of antibodies can act as precipitins.

precipitin test (see top of facing page)

An assay in which antibody interacts with soluble antigen in the presence of electrolyte to produce a precipitate. Both qualitative and quantitative precipitin reactions have been described.

preemptive immunity

Resistance shown by virus-infected cells to superinfection with a different virus.

prekallikrein

A kallikrein precursor. The generation of kallikrein from prekallikrein can activate the intrinsic mechanism of blood coagulation.

premunition

A type of protective immunity characterized by the presence of a few pathogenic microorganisms of a particular species remaining in the body, apparently stimulating a protective immune response. Premunity is associated with certain protozoal diseases such as bovine babesiosis.

preprogenitor cells

This pool of cells represents a second step in the maturation of B cells and is induced by nonspecific environmental stimuli. They are the immature cells that are unable themselves to mount an immune

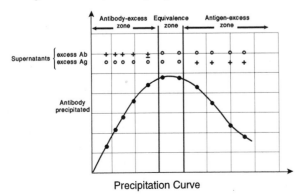

Precipitation Curve

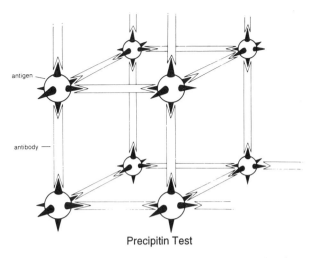

Precipitin Test

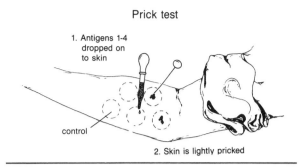

1. Antigens 1-4
dropped on
to skin

control

2. Skin is lightly pricked

(Several minutes later)

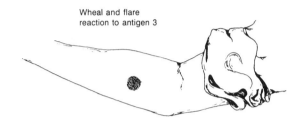

Wheal and flare
reaction to antigen 3

Prick Test

response, but are the pool from which the specific responsive clones will be selected by the specific antigen. They are present both in the bone marrow and peripheral lymphoid organs such as the spleen, but in the latter they form a minor population. They are characterized by the presence of some surface markers, frequently doublet or triplet surface immunoglobulins, and capable of being stimulated by selected activators. Sometimes termed B1 cells.

prick test (see facing column)
An assay for immediate (IgE-mediated) hypersensitivity in humans. The epidermal surface of the skin on which drops of diluted antigen (allergen) are placed is pricked by a sterile needle passed through the allergen. The reaction produced is compared with one induced by histamine or another mast cell secretogogue. This test is convenient, simple, rapid, and produces little discomfort for the patient in comparison with the intradermal test. It even may be used for infants.

primary agammaglobulinemia
Refer to antibody deficiency syndrome.

primary allergen
The antigenic material that sensitized a patient who subsequently shows cross-sensitivity to a related allergen.

primary biliary cirrhosis (PBC)
A chronic liver disease of unknown cause that affects middle-aged women in 90% of the cases. There is chronic intrahepatic cholestasis caused by chronic inflammation and necrosis of intrahepatic bile ducts with progression to biliary cirrhosis. It is believed to be an autoimmune disease based on its association with autoimmune conditions and the presence of autoantibodies. Patients develop pruritis, fatigue, steatorrhea, renal tubular acidosis, hepatic osteodystrophy,

and increased incidence of hepatocellular carcinoma and breast carcinoma. Four fifths of the patients also have a connective tissue or autoimmune disease such as rheumatoid arthritis, autoimmune thyroiditis, scleroderma, and Sjögren's syndrome. Most patients manifest a high titer of antimitochondrial antibodies. There is elevated IgM in the serum. Lymphocytic infiltration occurs together with intrahepatic bile duct destruction. Alkaline phosphatase is greatly increased, in addition to the elevation of IgM and antimitochondrial antibodies. The M2 antimitochondrial antibody is most frequently associated with PBC. In addition to hepatosplenomegaly and skin hyperpigmentation, patients may develop severe jaundice, petechiae, and purpura as the disease progresses. Liver transplantation is the only treatment for end-stage disease.

primary follicle
A densely packed accumulation of lymphocytes in the lymph node cortex or in the splenic white pulp where B cells develop into germinal centers when stimulated by antigen.

primary granule
Azurophil granule.

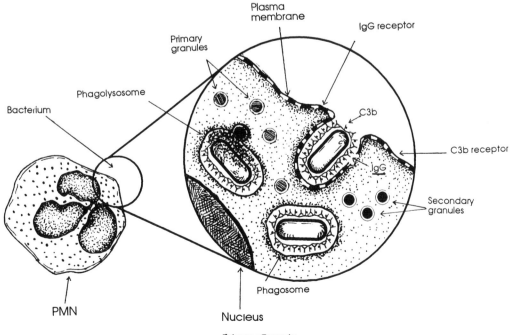

Primary Granule

primary immune response

The animal body's response to first contact with antigen. It is characterized by a lag period of a few days following antigen administration before antibodies or specific T lymphocytes are detectable. The antibody produced consists mostly of IgM and is of relatively low titer and low affinity. This is in contrast to the secondary immune response in which the latent period is relatively brief and IgG is the predominant antibody. The most important event that occurs in the primary response is the activation of memory cells that recognize antigen immediately on second encounter with antigen, leading to a secondary response. A similar pattern is followed in cell-mediated responses.

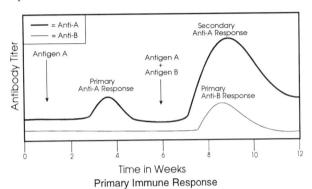

Primary Immune Response

primary immunodeficiency

Diminished immune reactivity attributable to an intrinsic abnormality of T or B lymphocytes.

primary interaction

Antigen and antibody binding that may or may not lead to a secondary visible reaction such as precipitation. Primary antigen-antibody interaction may be measured by equilibrium dialysis, the Farr assay, fluorescence polarization, fluorescence quenching, and selected radioimmunoassays such as radioimmunoelectrophoresis.

primary lymphoid organ

The site of maturation of B or T lymphocytes. The primary lymphoid organ for B lymphocytes in avian species is the bursa of Fabricius, whereas it is the bone marrow in adult mammals. By contrast, T cell development occurs in the thymus of all vertebrates. Stem cell maturation in primary lymphoid organs occurs without stimulation by antigen.

primary lysosome

A lysosome that has not yet fused with a phagosome.

primary nodule

Refer to primary follicle.

primary response

First response to an immunogen to which the recipient has not previously been exposed. It is principally an IgM antibody response.

primary structure

A polypeptide or protein molecule's linear amino acid sequence.

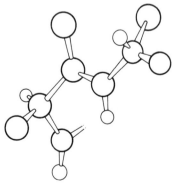

Primary Structure

primed

A lymphoid cell or an intact animal that has been exposed once to a specific antigen and which mounts a rapid and heightened response upon second exposure to this same antigen. Products of the reaction may be manifested as either increased antibody production or heightened cell-mediated immunity against that antigen.

primed lymphocyte

A lymphocyte that has interacted with an antigen *in vivo* or *in vitro*.

primed lymphocyte test (PLT)

Lymphocytes previously exposed or primed to a certain antigen in a primary mixed lymphocyte culture will divide rapidly when reexposed to the same antigen. Using a primed cell, one can determine whether or not an unknown cell possesses the original stimulating antigen. Cells previously exposed to MHC class II HLA antigens can be used in HLA typing for HLA-D region antigens. It is an assay for the detection of lymphocyte-associated determinants (LAD). For this procedure, lymphocytes donated by a normal person can serve as responder cells against the antigens of a known cell type. The test is based on the secondary stimulation of the primed or sensitized lymphocytes. The original stimulator serves as a positive control. The response of the sensitized cell to other cells measured by the incorporation of tritiated thymidine, by comparison with the control, may suggest sharing of HLA-D-associated antigens with the original stimulator cell if high stimulation values result. The HTC typing procedure, on the other hand, implies an antigenic determinant shared between the two cell types when there is little or no response.

The primed lymphocyte test (PLT) is a positive typing procedure and has the advantage that homozygous donor cells are not required. Primed or sensitized cells can be prepared whenever they are needed and frozen for future use. These cells can be used to type unknowns within a period of 24 h. This eliminates the 5 to 6 days needed for a homozygous cell typing procedure.

primed lymphocyte typing (PLT)

A method to type for HLA-D antigenic determinants. It is a type of mixed lymphocyte reaction in which cells previously exposed to allogeneic lymphocytes of known specificity can be reexposed to unknown lymphocytes to determine their HLA-DP type, for example.

priming dose

The initial dose of an immunogen administered to an animal for the purpose of inducing an immune response.

prion

An infectious particle comprised of a protein with appended carbohydrate. It is the most diminutive infectious agent known. The three human diseases in which prions have been implicated include kuru, Creutzfeldt-Jakob disease, and Gerstmann-Straussler syndrome. They have also been implicated in the following animal diseases: sheep and goat scrapie, bovine spongiform encephalopathy, chronic wasting of elk and mule deer, and transmissible mink encephalopathy. Prions do not induce inflammation nor do they stimulate antibody synthesis. They resist formalin, heat, ultraviolet radiation, and other agents that normally inactivate viruses. They possess a 28-kD,

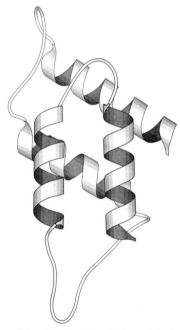

Prion Protein, Scrapie Associated

Paper Radioimmunosorbent Test (PRIST)

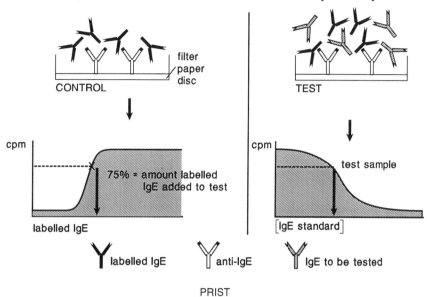

PRIST

hydrophobic, glycoprotein particle that polymerizes, forming an amyloid-like fibrillar structure.

PRIST (see above)

Abbreviation for paper radioimmunosorbent test. A technique to assay serum IgE levels. It resembles the radioimmunoabsorbent test except that filter paper disks impregnated with anti-human IgE is used in place of Sephadex discs.

private antigen

(1) An antigen confined to one major histocompatibility complex (MHC) molecule. (2) An antigenic specificity restricted to a few individuals. (3) A tumor antigen restricted to a specific chemically induced tumor. (4) A low-frequency epitope present on red blood cells of fewer than 0.1% of the population, i.e., Pt^a, By, Bp^a, etc. (5) HLA antigen encoded by one allele such as HLA-B27.

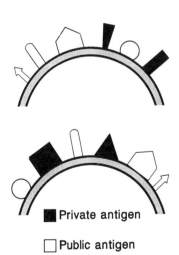

private idiotypic determinant

A determinant produced by a particular amino acid sequence in the immunoglobulin heavy or light chain hypervariable region of an antibody synthesized by only one individual.

private specificity

An epitope found on a protein encoded by a single allele. Thus, it is found only on one member of a group of proteins, such as alloanti-

gens of the major histocompatibility complex, even though it may also apply to other alloantigenic systems.

privileged site

Anatomical locations in the body that are protected from immune effector mechanisms because of the absence of normal lymphatic drainage. Antigenic substances such as tissue allografts may be placed in these sites without evoking an immune response. Privileged sites include the anterior chamber of the eye, the cheek-pouch of the Syrian hamster, and the central nervous system. Tissue allografts in these locations enjoy a period of protection from immunologic rejection, as the diffusion of antigen from graft sites to lymphoid tissues is delayed.

pro-C3

A polypeptide single chain that is split into C3 α and β (amino terminal) chains.

pro-C4

A polypeptide single chain that is split into C4 α, β (amino terminal), and γ (carboxy terminal) chains.

pro-C5

A polypeptide single chain that is split into C5 α and β (amino terminal) chains.

pro-T cell

The earliest identifiable thymocyte that is recognized by expression of cell surface antigens such as CD2 and CD7 and by CD3 ϵ protein in the cytoplasm, but is absent from the cell surface. Rearrangement of δ, γ, and β TCR genes accompanies differentiation of pro-T cells into pre-T cells.

procomplementary factors

Serum components of selected species such as swine that resemble complement in their action and confuse evaluation of complement fixation tests.

professional antigen-presenting cells

Macrophages, dendritic cells, and B cells that are capable of initiating T lymphocyte responsiveness to antigen. These cells display antigenic peptide fragments in association with the proper class of MHC molecules and also bear costimulatory surface molecules.

progressive multifocal leukoencephalopathy

A central nervous system disease characterized by demyelination, very little inflammation associated with patches of cortical degeneration, oligodendrogliocytes, intranuclear viral inclusions, aberrant large astrocytes, and reactive fibrillary astrocytes. This condition occurs in immunosuppressed individuals such as those with AIDS or latent virus infections such as measles. Papova (DNA type), often JC virus, usually causes it.

progressive systemic sclerosis

A connective tissue or collagen-vascular disease in which the skin and submucosal connective tissue become thickened and scarred. There is increased collagen deposition in the skin. It is slowly progressive and chronic and may involve internal organs. The female to male ratio is 2:1. Although the etiology is unknown, patients demonstrate antinuclear antibodies, rheumatoid factor, and polyclonal hypergammaglobulinemia. There is no demonstrable immunoglobulin at the dermal-epidermal junction. These patients may have altered cellular immunity. The epidermis is thin, dermal appendages atrophy, and rete pegs are lost. There is also a marked increase in collagen deposition in the reticular dermis together with fibrosis and hyalinization of arterioles. The GI tract may also reveal increased collagen deposition in the lamina propria, submucosa, and muscularis layers. At the onset, 90% of affected individuals experience Raynaud's phenomenon. Skin changes are usually the initial manifestation with involvement of the hands, feet, forearms, and face or possibly diffuse involvement of the trunk. A variation of the disease is called the CREST syndrome which consists of calcinosis, Raynaud's, esophageal dysmotility, sclerodactyly, and telangiectasia. This form of the disease may become stabilized for a number of years. The skin may exhibit a tight, smooth, waxy appearance in the sclerotic phase with no wrinkles or folds apparent. Ulcers may develop on the fingertips in many patients with a mask-like appearance of the face with thin lips. The skin may either become atrophic or return to a normal soft structure. The lungs may be involved, leading to dyspnea on exertion. Pulmonary fibrosis may lead to cor pulmonale. The principal immunologic findings include antinuclear antibodies with a speckled or nucleolar pattern, anticentromere antibodies that are found in individuals with known CREST syndrome, and the development of antibodies specific for acid-extractable nuclear antigen. Approximately one third of the individuals with diffuse involvement of the trunk reveal antibodies specific for topoisomerase (anti-Scl-70 antibodies).

progressive transformation of germinal centers (PTGC)
Germinal center enlargement in the presence of follicular hyperplasia and loss of the distinct boundary between the mantle zone and the germinal center. Transformed germinal centers contain small lymphocytes with diffuse immunoblasts and histiocytes and often occur in one enlarged lymph node in young men. It is not believed to be neoplastic or to portend development of lymphoma in the future, although it may be observed in some patients with nodular lymphocyte-predominant Hodgkin's disease.

progressive vaccinia
An adverse reaction to smallpox vaccination in children with primary cell-mediated immunodeficiency, such as severe combined immunodeficiency. The vaccination lesion would begin to spread from the site of inoculation and cover extensive areas of the body surface, leading to death.

prokaryote
An organism that contains one linear chromosome rather than a true nucleus.

promoter
(1) The DNA molecular site where RNA polymerase attaches and the point at which transcription is initiated. The promoter is frequently situated adjacent to the operator, and upstream from it is an operon. A TATA box and a promoter are both required for immunoglobulin gene transcription. (2) In tumor biology, a promoter mediates the second stage or promotion stage in the process of carcinogenesis. It may be a substance that can induce a tumor in an experimental animal that has been previously exposed to a tumor initiator. Yet the promoter alone is not carcinogenic.

properdin (factor P)
A globulin in normal serum that has a central role in activation of the alternative complement pathway activation. Additional factors such as magnesium ions are required for properdin activity. It is an alternative complement pathway protein that has a significant role in resistance against infection. It combines with and stabilizes the C3 convertase of the alternate pathway which is designated C3bBb. It is a 441-amino acid residue polypeptide chain with two points where N-linked oligosaccharides may become attached. Electron microscopy reveals it to have a cyclic oligomer conformation. Molecules are comprised of six repeating 60 residue motifs which are homologous to 60 amino acids at C7, C8 α, and C8 β amino carboxy terminal ends and the C9 amino terminal end.

properdin deficiency
An X-linked recessive disorder in which there is an increased susceptibility to infections by *Neisseria* microorganisms. Males with the deficiency reveal 2% or less of normal serum levels of properdin. In some heterozygous females, the serum properdin

level may be only 50%, while in others serum properdin levels are normal.

properdin pathway
An earlier synonym for alternative pathway of complement activation.

properdin system
An older term for alternative complement pathway. It consists of several proteins that are significant in resistance against infection. The first protein to be discovered was properdin. The properdin system also consists of factor B, a 95-kD β_2 globulin which is also termed C3 proactivator, glycine-rich β glycoprotein, or β_2 glycoprotein II. Properdin factor D is a 25-kD, α globulin which is also termed C3 proactivator convertase or glycine-rich β glycoproteinase. The properdin system, or alternative pathway, does not require the participation of antibody to activate complement. It may be activated by endotoxin or other substances. See alternative complement pathway.

prophylactic immunization
Prevention of disease through either active immunization or passive immunization. Active immunization usually induces longer lasting protection than does passive immunization.

prostacyclin (PC)
A derivative of ara-chidonic acid that is related to prostaglandins. It has a second five-membered ring. It inhibits aggregation of platelets and is a potent vasodilator. Prostacyclin's actions are the opposite of the actions of thromboxanes.

Prostacyclin

prostaglandin (PG)
A family of biologically active lipids derived from arachidonic acid through the effects of the enzyme cyclooxygenase. Although first described in the prostate gland, they are now recognized in practically all tissues of mammals. The hormonal effects of prostaglandin include decreasing blood pressure; stimulating contraction of smooth muscle; and regulation of inflammation, blood clotting, and the immune response. Prostaglandins are grouped on the basis of their substituted five-membered ring structure. During anaphylactic reactions mediated by IgE on mast cells, PGD_2 is released, producing small blood vessel dilation and constriction of bronchial and pulmonary blood vessels. Mononuclear phagocytes may release PGE_2 after binding of immune complexes to Fcγ receptors. Other effects of PGE_2 include blocking of MHC class II molecule expression in T cells and macrophages and inhibition of T cell growth. PGD_2 and PGE_2 both prevent aggregation of platelets. Antiinflammatory agents such as aspirin block prostaglandin synthesis.

Prostaglandin F$_2\alpha$

Prostaglandin D$_2$

Prostaglandin E$_2$

prostate specific antigen (PSA)
A marker in serum or tissue sections for adenocarcinoma of the prostate. PSA is a 34-kD glycoprotein found exclusively in benign and malignant epithelium of the prostate. Normal levels of PSA in males should be less than 4 ng/ml. The PSA molecule is smaller than prostatic acid phosphatase (PAP). In patients with prostate cancer, preoperative PSA serum levels are positively correlated with the disease. PSA is more stable and shows less diurnal variation than does PAP. PSA is increased in 95% of new cases of prostatic carcinoma compared with 60% for PAP. It is increased in 97% of recurrent cases compared with 66% for PAP. PSA may also be increased in selected cases of benign prostatic hypertrophy and prostatitis, but these elevations are less than those associated with

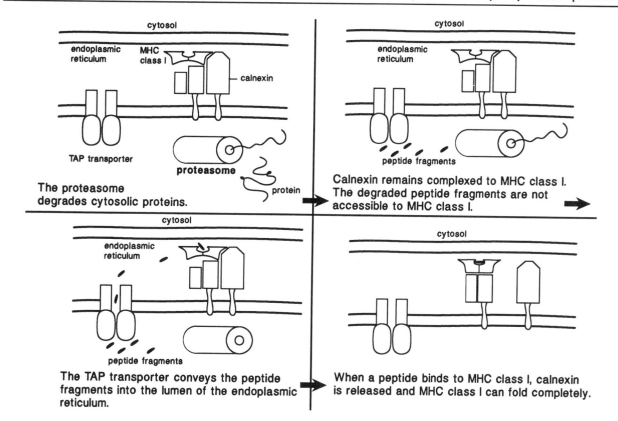

The proteasome degrades cytosolic proteins.

Calnexin remains complexed to MHC class I. The degraded peptide fragments are not accessible to MHC class I.

The TAP transporter conveys the peptide fragments into the lumen of the endoplasmic reticulum.

When a peptide binds to MHC class I, calnexin is released and MHC class I can fold completely.

adenocarcinoma of the prostate. It is inappropriate to use either PSA or PAP alone as a screen for asymptomatic males. TUR, urethral instrumentation, prostatic needle biopsy, prostatic infarct, or urinary retention may also result in increased PSA values. PSA is critical for the prediction of recurrent adenocarcinoma in postsurgical patients. PSA is also a useful immunocytochemical marker for primary and metastatic adenocarcinoma of the prostate.

prostatic acid phosphatase (PAP)/prostatic epithelial antigen

Prostate antigens, identifiable by immunoperoxidase staining, that are prostate specific and sensitive. Used together, they detect approximately 99% of prostatic adenocarcinomas.

proteasome (see above)

A 650-kD organelle in the cytoplasm termed the low molecular mass polypeptide complex. The proteasome is believed to generate peptides by degradation of proteins in the cytosol. It is a cylindrical structure comprised of as many as 24 protein subunits. The proteasome participates in degradation of proteins in the cytosol that are covalently linked to ubiquitin. It is requisite for some protein antigens to be ubiquinated prior to presentation to MHC class I-restricted T lymphocytes. Proteasomes that include MHC gene encoded subunits are especially adept at forming peptides that bind MHC class I molecules.

proteasome genes

Two genes in the MHC class II region that encode two proteasome subunits. The proteasome is a protease complex in the cytosol that may participate in the generation of peptides from proteins in the cytosol.

protective antigens

The antigenic determinants of a pathogenic microorganism that stimulate an immune response that can protect a host against an infection by that microorganism. Thus, these particular antigenic specificities can be used for prophylactic immunization in vaccines to immunize susceptible hosts against possible future infections.

protective immunity

Both natural, nonspecific immune mechanisms as well as actively acquired specific immunity that results in the defense of a host against a particular pathogenic microorganism. Protective immunity may be induced either by active immunization with a vaccine prepared from antigens of a pathogenic microorganism or by experiencing either a subclinical or clinical infection with the pathogenic microorganism.

protein A

A *Staphylococcus aureus* bacterial cell wall protein comprised of a solitary polypeptide chain whose binding sites manifest affinity for the Fc region of IgG. It combines with IgG1, IgG2, and IgG4, but not with IgG3 subclasses in humans, the IgG subclasses of mice, as well as the IgG of rabbits and certain other species. It has been used extensively for the isolation of IgG during protein purification and for the protection of IgG in immune (antigen-antibody) complexes. Protein A is antiphagocytic, a property that may be linked to its ability to bind an opsonizing antibody's Fc region. Protein A has

MHC Class II region

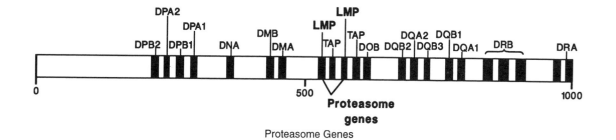

Proteasome Genes

been postulated to facilitate escape from the immune response by masking its epitopes with immunoglobulins. Protein A is used in immunology for mitogenic stimulation of human B lymphocytes, investigation of lymphocyte Fc receptors, and agglutination tests, as well as detection and purification of immunoglobulins by the ELISA technique.

protein AA
An 8.5-kD protein isolated by gel filtration and comprised of 76 amino acids. The sequence has no relationship to that of immunoglobulins or to other known sequences of human proteins. The sequence of protein AA extracted from various tissues appears identical, but several genetic variants are identified by sequence analysis. The molecule lacks cysteine and accordingly has no crosslinks. The N-terminal residue is arginine (in man), and in some, the second amino acid is phenylalanine. It also lacks carbohydrate or other attached small molecules. Amyloid protein AA is insoluble in ordinary aqueous solvents. Protein AA is the predominant component in the secondary form of amyloidosis.

protein B
A group B streptococcal protein capable of binding the Fc regions of IgA molecules. It is used in immunoassays and purification techniques for human serum and secretory IgA. It has the unique capacity to bind specifically to human IgA1 and IgA2 subclasses and shows no crossreactivity with other immunoglobulin classes or serum proteins.

protein blotting
Refer to immunoblotting.

protein F
Fusion protein. A 70-kD glycoprotein believed to be critical to the generation of an effective immune response against respiratory syncytial virus.

protein G
An antibody-binding protein that combines securely and specifically with certain species of antibody. (1) A group G streptococcus cell wall constituent that combines with IgG Fc regions of four IgG subclasses which may be significant in host resistance to this microorganism. (2) Protein G is a 90-kD attachment glycoprotein that is critical in the immune response to respiratory syncytial virus (RSV). G protein antigenic differences are associated with RSV strain differences.

protein kinase C (PKC)
An enzyme that Ca^{2+} activates in the cytoplasm of cells. It participates in cell activation and is a receptor for phorbol ester that acts by signal transduction, leading to hormone secretion, enzyme secretion, neurotransmitter release, and mediation of inflammation. It is also involved in lipogenesis and gluconeogenesis. PKC participates also in differentiation of cells and in tumor promotion.

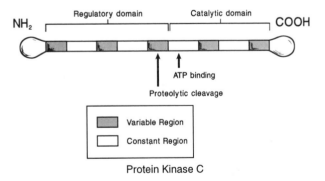

Protein Kinase C

protein M (M antigen)
Group A streptococcal protein found on fimbriae. It is antiphagocytic and facilitates virulence of streptococci.

protein P
A 23-kD pentameric protein detectable in amyloid deposits.

protein S
A 69-kD plasma protein that is vitamin K dependent and serves as a cofactor for activated protein C. It occurs as an active single chain protein or as a dimeric protein that is disulfide linked and inactive. Protein S, in the presence of phospholipid, facilitates protein C inactivation of factor Va and combines with C4b-binding proteins. Protein S deficiency, which is transmitted as an autosomal dominant, is characterized clinically by deep vein thrombosis, pulmo-

nary thrombosis, and thrombophlebitis. Laurell rocket electrophoresis is used to assay protein S.

protein SAA
A soluble precursor of AA which is present in minor quantities in the serum. It has a mol wt of 100 kD. SAA is present in the cord blood and during the first three decades of life and has an average concentration of 50 ± 40 ng/ml. In the following three decades of life it shows a slow, but steady increase in concentration, with doubling of the level in the eighth decade. Levels of the serum precursor are elevated in almost all patients with amyloidosis except in cases of extreme protein dysfunction due to severe nephrosis. Levels reaching 700 ng/ml are detectable in patients with tuberculosis, lymphoma, carcinoma, and leukemia, but SAA levels have no value in the diagnosis of amyloidosis. High levels of SAA are seen in the secondary form of the disease. The level of SAA may increase transiently during various infections.

proteinuria
Protein in the urine.

prothymocyte
Hematopoietic stem cells from the bone marrow migrate to the thymus by the blood circulation and enter through the epithelial cell lining of the cortex. Prothymocytes (pre-T cells) differentiate in the thymus microenvironment. The prothymocytes are educated in the thymus to function as T cells. There are four thymic peptide hormones termed thymulin, thymosin α, thymosin β, and thymopoietin. These hormones are significant in T lymphocyte proliferation and differentiation. Direct interaction with the thymus epithelium, which expresses HLA antigens, is necessary for forming functional T lymphocytes and for learning to recognize major histocompatibility complex (MHC) antigens. Prothymocytes proliferate and migrate from the cortex to the medulla. Some of them are short-lived and die. The long-lived cells acquire new characteristics and are called thymocytes. They exit the thymus as immature cells and seed specific areas of the peripheral lymphoid organs where they continue to differentiate through a process driven by an antigen. From these areas, they recirculate throughout the body.

protooncogene
A cellular gene that shows homology with a retroviral oncogene. It is found in normal mammalian DNA and governs normal proliferation and probably also differentiation of cells. Mutation or recombination with a viral genome may convert a protooncogene into an oncogene, signifying that it has become activated. Oncogenes may act in the induction and/or maintenance of a neoplasm. Protooncogenes united with control elements may induce transformation of normal fibroblasts into tumor cells. Examples of protooncogenes are c-*fos*, c-*myc*, c-*myb*, c-*ras*, etc. Alteration of the protooncogenes, leading to synthesis of an aberrant gene product, is believed to facilitate its becoming tumorigenic. An elevation in the quantity of gene product produced is also believed to be associated with protooncogenes becoming tumorigenic.

protoplast
A bacterial cell from which the cell wall has been removed. It includes the cell protoplasm and the cytoplasmic membrane. Lysozyme digestion of Gram-positive bacteria that contain a peptidoglycan cell wall yields protoplasts that require hypertonic media for survival, and they do not usually multiply. The hypertonic solution protects them from lysis. Protoplasts can be produced from Gram-positive bacteria also by treatment with penicillin or other antibiotics which inhibit synthesis of the cell wall. Gram-negative bacteria have a cell wall comprised of a thin peptidoglycan layer enclosed by an exterior membrane of lipopolysaccharide. Protoplasts prepared from Gram-negative bacteria are frequently termed spheroplasts.

protoplast fusion
A technique for DNA transfer from one group of bacteria to others, to myeloma cells, or other animal cells in culture. The exposure of plasmid-bearing *Escherichia coli* microorganisms to lysozyme and EDTA yields protoplasts that may be fused with myeloma cells by polyethylene glycol treatment.

provocation poliomyelitis
An uncommon consequence of attempted immunization against poliomyelitis in which paralysis followed soon after injection of vaccines such as those that contained *Bordetella pertussis* or alum.

prozone
In agglutination and precipitation reactions, the lack of agglutination or precipitation in tubes where the antibody concentration is greatest.

This is attributable to suboptimal agglutination or precipitation in the region of antibody excess. Agglutination or precipitation becomes readily apparent in the tubes where the same antibody is more dilute. This is known as a prozone or prozone phenomenon. It is attributable to either blocking antibody or antibody combining with only individual cells or molecules, as in antibody excess, or to some serum lipid or protein-induced nonspecific inhibition reaction. Soluble complexes of antigen and antibody may be present in the antibody excess zone of certain precipitation reactions. Excess antibody coating the surfaces of cells in an agglutination reaction, in the antibody excess zone, may prevent crosslinking, which is requisite for agglutination to become manifest. The prozone represents a false-negative reaction. When a serum sample is believed to contain a certain antibody that is being masked or is demonstrating a prozone phenomenon, the sample must be diluted serially to demonstrate reactivity in more dilute tubes to avoid reporting a false-negative result.

Prozone Effect

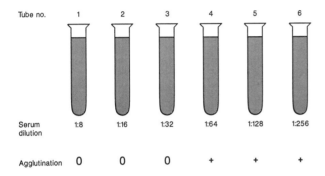

Tube no.	1	2	3	4	5	6
Serum dilution	1:8	1:16	1:32	1:64	1:128	1:256
Agglutination	0	0	0	+	+	+

PRP antigen

Polyribosyl-ribitol capsular polysaccharide. An antiphagocytic cell wall constituent of *Hemophilus influenzae* that provides this microorganism with an effective mechanism to induce disease. Type-specific antibodies that facilitate immunization are requisite for protective immunity against *H. influenzae*. Children less than 2 years old are poor produces of anti-PRP antibodies, making them more susceptible to the infection.

PRP-D

Polyribosylribitol-diphtheria toxoid. Refer to Hib.

PSA (prostate-specific antigen)

A substance secreted only by the prostate epithelium that is a 34-kD glycoprotein serine protease that lyses seminal coagulum. Individuals with benign prostatic hypertrophy have a 30 to 50% elevation in PSA levels, whereas those with prostatic carcinoma have a 25 to 92% elevation. It is a more reliable indicator of prostatic carcinoma than is serum prostatic acid phosphatase (PAP). PSA levels are also valuable to signify recurrence of prostatic adenocarcinoma. Prostate cancer may occur in 22% of the individuals with PSA levels greater than 4.0 µg/l and 60% of the individuals with PSA levels greater than 10 µg/l.

pseudoalleles

Genes that are closely linked but distinct. They have functional similarity and act as alleles in complementation investigations. However, cross-over studies may separate them.

pseudoallergic reaction

A nonimmunological clinical syndrome characterized by signs and symptoms that mimic or resemble immune-based allergic or immediate hypersensitivity reactions. However, pseudoallergic reactions are not mediated by specific antibodies or immune lymphoid cells.

pseudoallergy

An anaphylaxis-like reaction in some individuals that occurs suddenly, frequently following food ingestion, and represents an anaphylactoid reaction. It may be induced by a psychogenic factor, a metabolic defect, or other nonimmunological cause. This is not an immune reaction and is classified as an anaphylactoid reaction.

pseudogene

A sequence of DNA that is similar to a sequence of a true gene, but does not encode a protein due to defects that inhibit gene expression. Thus, pseudogenes represent nonusable or junk DNA. They

may result from duplicated genes and have several defects as mutations accumulate.

pseudolymphoma

Hyperplasia of lymphoid tissue in which there is a uniform accumulation of lymphocytes. Unlike lymphomas, the cells are polyclonal. The architecture of the lymph node is well preserved with distinct cortical germinal centers with little or minimal capsular infiltration by lymphocytes. Inflammatory cells are detectable between germinal centers, but mitoses occur only within the germinal centers. The reticular framework remains intact. The lymphoid hyperplasia that characterizes pseudolymphoma may be found in various locations such as gastrointestinal tract, lung, breast, salivary gland, mediastinum, skin, soft tissue, and other areas. Pseudolymphomas may occur in individuals who later develop lymphomas.

pseudolymphomatous lymphadenitis

Hyperplasia of lymphoid organs that is similar to lymphoma except that the hyperplasia is reversible in this condition.

pseudoparaproteinemia

An elevation of transferrin to levels at least twice normal (200 to 400 mg/dl). Profound iron deficiency anemia leads to this increase. There is only one molecular form of transferrin in the serum that is concentrated into a single band that travels in the β region in serum electrophoresis, giving the appearance of a paraproteinemia. However, a real paraproteinemia is characterized by a spike on serum electrophoresis representing a monoclonal gammopathy.

psoriasis vulgaris (see figure, top of page 252)

A chronic, recurrent, papulosquamous disease. Clinical features include the appearance of a discrete, papulosquamous plaque on areas of trauma such as the elbow, knee, or scalp, although it may appear elsewhere on the skin. There is a relatively high instance of HLA-B13 and B17 antigen and decreased T suppressor cell function in psoriasis patients. It may coexist with lupus erythematosus in some individuals. Peripheral blood helper/inducer CD4[+] T lymphocytes are significantly decreased in psoriasis patients. It may be treated with psoralens and long-wave ultraviolet radiation. Psoriasis patients develop Monro microabscesses, hyperkeratosis, parakeratosis, irregular acanthosis, papillary edema, and mild chronic inflammation of the dermis. By immunofluorescence, there are focal granular or globular deposits of immunoglobulins and C3 in the stratum corneum. The finely granular deposits principally contain IgG, IgA, and C3. They are deposited in areas where stratum corneum antigens are located. C3 and properdin deposits suggest activation of the alternate complement pathway.

psychoneuroimmunology

The study of central nervous system and immune system interactions in which neuroendocrine factors modulate immune system function. For example, psychological stress as in bereavement or other causes may lead to depressed immune function through neuroendocrine immune system interaction. There are multiple bidirectional interactions among the nervous, immune, and endocrine systems. This represents an emerging field of contemporary immunological research.

PTAP

Purified diphtheria toxoid which has been adsorbed on hydrated aluminum phosphate. It has been employed to induce active immunity against diphtheria.

public antigen (supratypic antigen)

An epitope which several distinct or private antigens have in common. A public antigen such as a blood group antigen is one that is present in greater than 99.9% of a population. It is detected by the indirect antiglobulin (Coombs' test). Examples include Ve, Ge, Jr, Gyᵃ, and Okᵃ. Antigens that occur frequently, but are not public antigens include MNs, Lewis, Duffy, P, etc. In blood banking, there is a problem finding a suitable unit of blood for a transfusion to recipients who have developed antibodies against public antigens.

public idiotypic determinant (IdX or CRI)

An idiotypic determinant present on antibody molecules in numerous individuals of one species. It may occur on antibodies with a single antigen specificity, but may be present on antibodies with separate specificities as well. The terms IdX or CRI signify that these are cross-reacting idiotypic determinants. These determinants are manifestations of immunoglobulin heavy and light chain amino acid sequence similarities.

public specificity

Specificity of an epitope encoded by two or more alleles of an alloantigenic system. It is most frequently used to refer to major

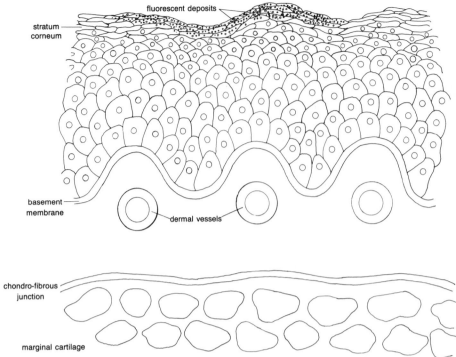

Psoriasis Vulgaris

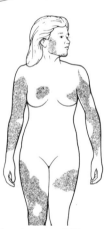

histocompatibility complex determinants such as MHC alleles where a single epitope is shared by multiple HLA molecules. In this system, antigenic products encoded by more than one allele at a single locus may carry public specificities. These specificities may also be encoded by alleles at loci that are separate but related. Thus, epitopes that represent public specificities are shared by two or more proteins in a particular group. For the specificity to be public, it must be found on at least two proteins, but may be on multiple ones.

pulsed-field gel electrophoresis

A method for separating DNA molecules that vary from a few kilobase pairs to 4000 kilobase pairs. The direction of the electric field is repeatedly altered, causing the molecules to change direction of migration and to enter new pores in the gel. Thus, both small and large DNA molecules migrate through the gel based on their size. Migration of the smaller molecules is more rapid than that of the larger molecules.

purified protein derivative (PPD)

A derivative of the growth medium in which *Mycobacterium tuberculosis* has been cultured. It is a soluble protein that is precipitated from the culture medium by trichloroacetic acid. It is used for tuberculin skin tests.

purine nucleoside phosphorylase (PNP) deficiency

A type of severe combined immunodeficiency caused by mutant types of PNP. This results in the retention of metabolites that have a toxic effect on T cells. B lymphocytes appear unaffected and their numbers are normal. All cells of mammals contain PNP, which acts as a catalyst in the phosphorolysis of guanosine, deoxyguanosine, and inosine. Insufficient PNP leads to an elevation in the concentration within cells of deoxy-guanosine, guanosine, deoxyguanosine triphosphate (dGTP), and guanosine triphosphate (GTP). dGTP blocks the enzyme ribonucleoside-diphosphate reductase which participates in DNA synthesis. T cell precursors are especially sensitive to death induced by these compounds. PNP is comprised of three 32-kD subunits. Its gene, located on chromosome 14q13.1, codes for a 289-amino acid residue polypeptide chain. Immunologic defects associated with this disorder are characterized by anergy, lymphocytopenia, and diminished T lymphocytes in the blood. By contrast, serum immunoglobulin levels and the response following deliberate immunization with various types of immunogens are within normal limits. There is an autosomal recessive mode of inheritance. Treatment is by bone marrow transplantation.

purpura

Purple areas on the skin caused by bleeding into the skin.

purpura hyperglobulinemia

Hemorrhagic areas around the ankles or legs in patients whose serum immunoglobulin levels are strikingly increased. Examples include Waldenström's macroglobulinemia, multiple myeloma, and Sjögren's syndrome.

pyogenic infection

An infection that is associated with the generation of pus. Microorganisms that are well known for their pus-inducing or pyogenic potential include *Streptococcal pyogenes, Staphylococcus aureus, Streptococcus pneumoniae,* and *Hemophilus influenzae.* Antibody-deficient patients and those having defective phagocytic cell capacity show increased susceptibility to pyogenic infections. Patients with complement deficiency such as C3 deficiency, factor I deficiency, etc. are also prone to develop pyogenic infections.

pyogenic microorganisms

Microorganisms that stimulate a large polymorphonuclear leukocyte response to their presence in tissues.

pyrogen

A substance that induces fever. It may be either endogenously produced, such as interleukin-1 released from macrophages and monocytes, or it may be an endotoxin associated with Gramnegative bacteria produced exogenously that induces fever.

pyroglobulins

Monoclonal immunoglobulins that undergo irreversible precipitation upon heating to 56°C. These monoclonal immunoglobulins are usually detected during routine inactivation of complement in serum by heating to 56°C in a water bath. Whereas most immunoglobulins are unharmed at this temperature, pyroglobulins precipitate. This may be attributable to formation of hydrophobic bonds linking immunoglobulin molecules as a consequence of diminished heavy chain polarity. Half of the pyroglobulin positive subjects have multiple myeloma, and the remaining half have a lymphoproliferative disorder such as macroglobulinemia, carcinoma, or systemic lupus erythematosus. Their relevance to disease is unknown.

pyroninophilic cells

Cells whose cytoplasm stains red with methyl green pyronin stain. This signifies large quantities of RNA in the cytoplasm, indicating active protein synthesis. For example, plasma cells or other protein-producing cells are pyroninophilic.

Q

Q fever

An acute disease caused by the rickettsia *Coxiella burnetii*. Cattle, sheep, goats, and several small marsupials serve as reservoirs. Ticks are the main vector. The microorganism is highly infectious and multiplies readily to produce clinical infection. The onset is abrupt and is accompanied by headaches, high fever, myalgia, malaise, hepatic dysfunction, interstitial pneumonitis, and fibrinous exudate. Q fever may also induce atypical pneumonia, rapidly progressive pneumonia, or be a coincidental finding to a systemic illness. The disease has a relatively low mortality. It is treated successfully with tetracycline and chloramphenicol.

Qa antigens

Class I histocompatibility antigens in mice designated Qa1, Qa2, and Qa10. They are encoded by genes in the QA region of the H-2 complex on the telomeric side. Lymphoid cells express Qa1 and Qa2 antigens, whereas hepatic cells express Q10 antigens. The Qa represents one of two regions of the Tla complex, with Tla representing the second region of this complex.

Qa-2 antigen

An antigen encoded by MHC class I genes that is produced in a GPI linked form as well as in a soluble secreted form. It shows little polymorphism and is tissue specific. Anti-Qa-2 mAbs are potent T cell activators in the presence of PMA and a cross-linking anti-Ig.

Qa locus

A subregion of the murine MHC located on the telomeric side of the H2 complex in a 1.5 centiMorgan stretch of DNA. The Qa region is a part of the Tla complex which also contains Tla. The Qa region is comprised of 220 kb that encode for class I MHC α chains that associate noncovalently with β_2 microglobulin. The Qa region is comprised of eight to ten genes designated as Q1, Q2, Q3, etc.

Qa region

Refer to Tla complex.

quantitative gel diffusion test

Estimation of the amount of antibody or antigen by a gel diffusion method such as single radial diffusion or Laurell rocket assay.

quantitative precipitin reaction

An immunochemical assay based on the formation of an antigen-antibody precipitate in serial dilutions of the reactants, permitting combination of antigen and antibody in various proportions. The ratio of antibody to antigen is graded sequentially from one tube to the next. The optimal proportion of antigen and antibody is present in the tube that shows the most rapid flocculation and yields the greatest amount of precipitate. After washing, the precipitate can be analyzed for protein content through procedures such as the micro-Kjeldahl analysis to ascertain nitrogen content, spectrophotometric assay or other techniques. Heidelberger and Kendall used the technique extensively, employing pneumococcus polysaccharide antigen and precipitating antibody in which nitrogen determinations reflected a quantitative measure of antibody content.

quaternary structure

Four components that are associated with one another.

Quarternary Structure

Two or more folded polypeptide chains packed into a configuration such as a tetramer. Quaternary antigenic determinants may be difficult to demonstrate in such structures as hemoglobulin molecules.

quaternary syphilis

The stage of syphilis which follows tertiary syphilis. It is characterized by a necrotizing encephalitis with tissues rich in spirochetes. End-stage HIV patients who have completely lost cell-mediated immunity against treponemal antigens may show this form of syphilis. Although it has been rare in the past, quaternary syphilis is being seen more and more in AIDS patients also infected with *Treponema pallidum* who show increased susceptibility to neurosyphilis.

Quellung phenomenon

Swelling of the pneumococcus capsule following exposure to antibodies against pneumococci. Refer to Quellung reaction.

Quellung reaction

Swelling of bacterial capsules when the microorganisms are incubated with species-specific antiserum. Examples of bacteria in which this can be observed include *Streptococcus pneumoniae*, *Hemophilus influenzae*, *Neisseria*, and *Klebsiella* species. The combination of a drop of antiserum with a drop of material from a patient containing an encapsulated microorganism and the addition of a small loopful of 0.3% methylene blue produces the Quellung reaction. The microorganisms are stained blue and are encircled by a clear halo that resembles swelling, but is in fact antigen-antibody complex produced at the surface of the organism. The reaction is due to an alteration of the refractory index.

quenching

When immunofluorescent cell or tissue preparations treated with fluorochrome-labeled antibodies are exposed to ultraviolet light under the microscope, the emission of fluorescent radiation from the fluorochrome label diminishes as a result of quenching. The term may also refer to diminished efficiency of assaying radioactivity in a scintillation counter by such agents as ethanol or hydrogen peroxide (H_2O_2).

Quin-2

A derivative of quinoline which combines with free Ca^{2+} to accentuate fluorescence intensity. It can be introduced into cells as an ester followed by deesterification. When T or B lymphocytes containing Quin-2 are activated, their fluorescence intensity rises, implying an elevation of free Ca^{2+} in the cytosol.

Quin-2

R

RA
Rheumatoid arthritis.

RA cell
Abbreviation for rheumatoid arthritis cell.

rabies
An infection produced by an RNA virus following a bite from an infected animal. The virus passes across the neuromuscular junction and infects the nerve from which it reaches the central nervous system. It also reaches salivary glands of lower animals. The virus infection leads to cerebral edema, congestion, round cell infiltration of the spinal cord and grey matter in the brain stem, and profound loss of Purkinje cells. Negri bodies are found prominently in the medulla oblongata, hippocampus, and cerebellum. Clinically, the fury associated with the disease is due to irritability of the central nervous system. There is fever, hyperethesia, and anoxia aggression. It may be paralytic. Human rabies is rare in the U.S. It is more common in other animals, with most of the cases appearing in skunks and raccoons. Fewer cases occur in bats and only 2% each in dogs and cats. The virus is transmitted from one person to another by inhalation or by corneal transplantation, but not by human bites.

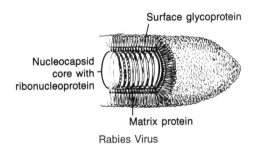

Surface glycoprotein

Nucleocapsid core with ribonucleoprotein

Matrix protein

Rabies Virus

rabies vaccination
Refer to postrabies vaccination encephalomyelitis.

rabies vaccine
In humans, significant levels of neutralizing antibody can be generated by immunization with a virus grown in tissue culture in diploid human embryo lung cells. A vaccine adapted to chick embryos, especially egg passage material, is used for prophylaxis in animals prior to exposure. The "historical" vaccine originally prepared by Pasteur made use of rabbit spinal cord preparations to which the virus had become adapted. However, they were discontinued because of the risk of inducing postrabies vaccination encephalomyelitis.

radiation chimera
Refer to irradiation chimera.

radioimmunoassay (RIA) (see below)
A technique to assay either antigen or antibody which is based on radiolabeled antigen competitively inhibiting the binding of antigen, which is not labeled, to specific antibodies. Minute quantities of enzymes, hormones, or other immunogenic substances can be assayed by RIA. Enzyme immunoassays are largely replacing RIAs because of the problems associated with radioisotope regulation and disposal.

radioimmunodiffusion
A variation of the immunodiffusion technique which uses a radioactively labeled antibody. This enhances the sensitivity of the results when read by autoradiography. Refer to single radioimmunodiffusion.

radioimmunodiffusion test
Refer to single radial immunodiffusion.

radioimmunoelectrophoresis
A type of immunoelectrophoresis that employs radiolabeled antibody or antigen to identify individual precipitin arcs by subsequent autoradiography of the arcs.

radioimmunoprecipitation assay (RIPA)
A method that demonstrates the presence of antibodies against viral constituents. Virus grown in culture in the presence of radioactive amino acid is disrupted and incubated with a test sample that might contain antibodies specific for viral antigens. This is followed by polyacrylamide gel electrophoresis of the immunoglobulins in the test sample.

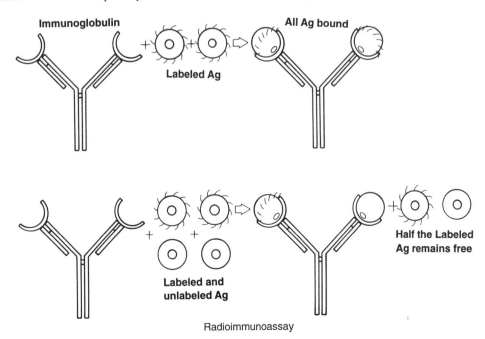

Immunoglobulin

Labeled Ag

All Ag bound

Labeled and unlabeled Ag

Half the Labeled Ag remains free

Radioimmunoassay

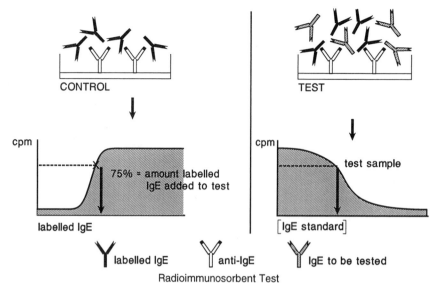

Radioimmunosorbent Test

radioimmunoscintigraphy

The use of radiolabeled antibodies to localize tumors or other lesions through use of radioactivity scanning following injection *in vivo*.

radioimmunosorbent test (RIST)

A solid-phase radioimmunoassay to determine the serum IgE concentration. A standard quantity of radiolabeled IgE is added to the serum sample to be assayed. The mixture is then combined with Sephadex® or Dextran beads coated with antibody to human IgE. Following incubation and washing, the quantity of radiolabeled IgE bound to the beads is measured. The patient's IgE competes with the radiolabeled IgE or antibody attached to the beads. Therefore, the decrease in labeled IgE attached to the beads compared to a control in which labeled IgE combines with the beads without competition represents the patient's serum concentration of IgE. The radioallergosorbent test by comparison assays IgE levels reactive with a specific allergen.

radiomimetic drug

An immunosuppressive drug such as an alkylating agent that is employed in the treatment of cancer. Its effect on DNA mimics that of ionizing radiation.

RAG-1 and RAG-2

Refer to recombination activating genes 1 and 2.

ragg

Agglutinating activity by rheumatoid factor in rheumatoid arthritis patients' sera, as revealed by a positive RA rheumatoid arthritis test.

ragocyte

Polymorphonuclear leukocytes containing IgG-rheumatoid factor-complement-fibrin conglomerates. They are found in the joints of rheumatoid arthritis patients. Also called RA cells.

ragweed

Ragweed (*Ambrosia*) is distributed throughout the warmer parts of the Western Hemisphere, but poses the greatest clinical problem in North America. Airborne ragweed pollen constitutes a troublesome respiratory allergen. Ragweed pollen appears in the air in northern states of the U.S. and adjacent eastern Canada in the latter days of July, when it reaches levels of up to thousands of grains per cubic meter by early September and then declines. Pure ragweed pollinosis has usually subsided by mid-October in the north. Short ragweed (*A. artemisiifolia*) and giant ragweed (*A. trifida*) are found widely from the Atlantic coast to the Midwest, Ozark plateau, and Gulf states. Short ragweed reaches to northern Mexico and is only sparsely found in the Pacific Northwest. Giant ragweed is abundant in

Ambrosia artemisiifolia
Ragweed

the Mississippi Delta and along the flood plains of southeastern rivers. Pollen allergens of ragweed that have been characterized include the following.

Allergen source	Allergen	Mol wt (kD) SDS-PAGE
Ambrosiae artemisiifolia (ragweed)		
	AMB α I (AGE)	38
	AMB α II (AGK)	38
	AMB α III (Ia3)	12
	AMB α IV (Ra4)	23
	AMB α V (Ra5)	5
	AMB α VI (Ra6)	8
	AMB α VII	—
Ambrosiae trifida (giant ragweed)	AMB T V (Ra6)	4.4

Raji cell assay (see facing page)

An *in vitro* assay for immune complexes in serum. The technique employs Raji cells, a lymphoblastoid B lymphocyte tumor cell line that expresses receptors for complement receptor 1, complement receptor 2, FCγ, and C1q receptors. The cell line does not express surface immunoglobulins. Following combination of Raji cells with the serum sample, the immune complex is bound and quantified using radiolabeled F(ab')₂ fragments of antibodies against IgG.

Ramon, Gaston (1886–1963)

French immunologist who perfected the flocculation assay for diphtheria toxin.

Ramon test (historical)

A rough method for assaying the activity of any given preparation of diphtheria (or tetanus) toxin. Varying quantities of antitoxin are combined with a constant quantity of toxin *in vitro*. The tubes are placed in a 44 to 46°C water bath and are observed often. The test is read by noting the tube where flocculation occurs first. This is the point of equivalence where antitoxin has neutralized the homologous toxin. However, this assay is based on antigenicity of the toxin with which the antitoxin combines, in contrast to toxicity. Therefore, it is a measure of combining power and provides an indirect idea of toxicity only insofar as toxicity and antigenicity are positively correlated. Since the two are not always closely correlated, this method is less reliable than the *in vivo* technique of Ehrlich which measures the actual toxic effect of the toxin and the ability of antitoxin to combat it. The Ramon test measures toxin in L_f (flocculating) units. The L_f unit is defined as the amount of toxin which flocculates most rapidly with one unit of antitoxin. The L_f value, in contrast to other L values described, must be calculated. To determine the L_f value for a given toxin, the following formula is used.

$$L_f / ml\ toxin = \frac{antitoxin\ units/ml \times ml\ of\ antitoxin}{ml\ of\ toxin}$$

Thus, the L_f content of a toxin may be determined if the following values are known: (1) antitoxin units per milliliter of antitoxin, (2)

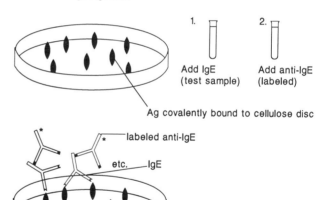

Raji Cell Assay

milliliter of antitoxin required for most rapid flocculation with toxin, and (3) milliliter of toxin employed. Although the Ramon flocculation test was classically used to determine the L_f value of toxin, it may be carried out in reverse to assay the antitoxin units in each milliliter of antitoxin which has not been previously standardized. The same formula is applicable.

$$L_f / ml \ toxin = \frac{antitoxin \ units/ml \ \times \ ml \ of \ antitoxin}{ml \ of \ toxin}$$

$$Antitoxin \ units/ml = \frac{L_f / ml \ toxin \ \times \ ml \ of \ toxin}{ml \ of \ antitoxin}$$

Varying quantities of toxin of known L_f value are combined with a constant amount of antiserum. The tube where flocculation first occurs is the point of equivalence. Therefore, the amount of toxin in a milliliter is substituted into the formula together with the known values, which include the L_f per milliliter of toxin and the number of milliliters of antitoxin held constant. By simple arithmetic, the antitoxin units per milliliter may then be calculated. In this quantitative precipitin test, antibody dilutions are varied, but antigen dilutions are kept the same. The first tube where precipitation occurs is considered the end point.

RANA (rheumatoid arthritis-associated nuclear antigen)
An antigen of Epstein-Barr virus-immortalized lymphoid cell lines that reacts with blood sera from rheumatoid arthritis patients.

random breeding
The mating of members of a population at random. The genetic diversity produced by random breeding depends upon the size of the population. If it is large, genetic diversity will be maintained. If it is small, genetic uniformity will result in spite of random breeding.

RANTES
An 8-kD protein comprised of 68 amino acid residues. It belongs to the PF4 superfamily of chemoattractant proteins. RANTES chemoattracts blood monocytes as well as CD4+/CD45RO+ T cells in vitro and is useful in research on inflammation.

rapamycin
A powerful immunosuppressive drug derived from a soil fungus on Rapa Nui on Easter Island. It resembles FK506 in structure, but it

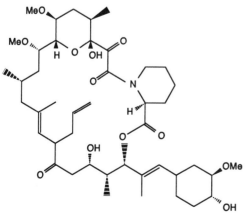

Rapamycin

has a different mechanism of action. Rapamycin suppresses B and T lymphocyte proliferation, lymphokine synthesis, and T cell responsiveness to IL-2. To achieve clinical immunosuppression, rapamycin is effective at concentrations one eighth those required for FK506 and at 1% of the levels required for cyclosporin.

RAST (radioallergosorbent test)
A technique to detect specific IgE antibodies in a patient's serum. This solid-phase method involves binding of the allergen-antigen complex to an insoluble support such as dextran particles or Sepharose®. The patient's serum is then passed over the allergen-support complex which permits specific IgE antibodies in the serum to bind with the allergen. After washing to remove nonreactive protein, radiolabeled anti-human IgE antibody is then placed in contact with the insoluble support where it reacts with the bound IgE antibody. Both the allergen and the anti-IgE antibody must be present in excess for the test to be accurate. The amount of radioactivity on the beads is proportional to the quantity of serum antibody that is allergen specific.

RAST (Radioallergosorbent Test)

Raynaud's phenomenon
Episodes of vasospasm in the fingers when the hands are exposed to cold temperatures. The ischemia of the fingers is characterized by severe pallor and is often accompanied by paresthesias and pain. It is brought on by cold, emotional stress, or anatomic abnormality. When the condition is idiopathic or primary, it is called Raynaud's disease. Raynaud's phenomenon is seen in several connective tissue diseases including systemic lupus erythematosus and systemic sclerosis. Subjects with cryoglobulinemia may also manifest the phenomenon.

RB200
Refer to lissamine rhodamine.

RCA
Regulators of complement activity.

RCA locus (regulator of complement activation)
A locus on the long arm of chromosome 1 with a 750-kb DNA segment containing genes that encode complement receptor 1, complement receptor 2, C4-binding protein, and decay-accelerating factor. These substances regulate the activation of complement

through combination with C4b, C3b, or C3dg. A separate chromosome 1 gene, not in the RCA locus, encodes factor H.

reaction of identity

Double immunodiffusion in two dimensions in gel can reveal that two antigen solutions are identical. If two antigens are deposited into separate, but adjacent wells and permitted to diffuse toward a specific antibody diffusing from a third well that forms a triangle with the other two, a continuous arc of precipitation is formed. This reveals the identity of the two antigens.

Reaction Of Identity

reaction of nonidentity

Double immunodiffusion in two dimensions in gel can show that two antigen solutions are different, i.e., nonidentical. If each antigen solution is deposited into separate, but adjacent wells and permitted to diffuse toward a combination of antibodies specific for each antigen diffusing from a third well that forms a triangle with the other two, the lines of precipitation form independently of one another and intersect, resembling crossed swords. This reaction reveals a lack of identity with no epitopes shared between the antigens detectable by these antibodies.

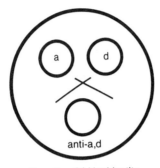

Reaction Of Nonidentity

reaction of partial identity

Double immunodiffusion in two dimensions in gel can show that two antigen solutions share epitopes, but are not identical. If each antigen is deposited into separate, but adjacent wells and permitted to diffuse toward specific antibodies diffusing from a third well that forms a triangle with the other two, a continuous arc of precipitation manifesting a spur is formed. This demonstrates that the two antigens share some epitopes, shown by the continuous arc, but not others, demonstrated by the spur.

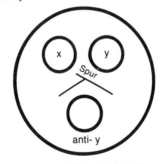

Reaction Of Partial Identity

reactive lysis

Dissolution of red blood cells not sensitized with antibody. Initiated by C5b and C6 complexes in the presence of C7, C8, and C9. The activation of complement leads to lysis as a "bystander phenomenon".

reagin (historical)

(1) Obsolete term for a complement-fixing IgM antibody reacting in the Wassermann test for syphilis. (2) A name used previously for immunoglobulin E (IgE), the anaphylactic antibody in humans that fixes to tissue mast cells leading to release of histamine and vasoactive amines following interaction with specific antigen (allergen).

Rebuck skin window

A clinical method for assessing chemotaxis used *in vivo* by making a superficial abrasion of the skin which is then covered with a glass

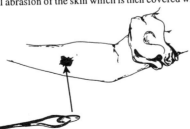

Rebuck Skin Window

slide. This is removed several hours later, air dried, and stained for leukocyte content.

receptor

A molecular configuration on a cell or macromolecule that combines with molecules that are complementary to it. Examples include enzyme-substrate reactions, the T cell receptor, and membrane-bound immunoglobulin receptors of B cells.

β-adrenergic receptor antibodies

β_2 receptor autoantibodies may have a significant role in selected diseases of man that include Chagas disease, asthma, and dilated cardiomyopathies.

recirculating pool

Refers to the continuous recirculation of T and B lymphocytes between the blood and lymph compartments.

recombinant DNA

The physical union of two or more strands of available DNA to form another DNA strand. The term describes the exchange of DNA during meiosis, mitosis, or gene conversion. It may also refer to DNA strands produced *in vitro*.

recombinant DNA technology

The technique of isolating genes from one organism and purifying and reproducing them in another organism. This is often accomplished through ligation of genomic or cDNA into a plasmid or viral vector where replication of DNA takes place.

recombinant inbred strains

Inbred strains of F_2 generation mice developed by crossing two inbred strains to yield F_1 and then F_2 generations. Progeny of the F_2 generation are inbred until they become homozygous at most loci. The progeny approach complete genetic identity and homozygosity. Recombination occurs during meiosis and consists of crossing over and recombination of parts of two chromosomes. Recombinant inbred stains help to establish genetic linkages. These genetically uniform and homozygous mice offer a means to study the consequences of reassorting various parental genes such as heavy chain genes.

recombinant vaccine

An immunogen preparation for prophylactic immunization comprised of products of recombinant DNA methodology prepared by synthesizing proteins employing cloned complementary DNA.

recombination activating genes 1 and 2 (RAG-1 and RAG-2)

Genes that activate Ig gene recombination. Pre-B cells and immature T cells contain them. It remains to be determined whether RAG-1 and RAG-2 encode the recombinases or the regulatory proteins that control recombinase function. RAG-1 and RAG-2 gene products are requisite for rearrangements of both Ig and TCR genes. In the absence of these genes neither Ig nor T cell receptor proteins are produced. This blocks the production of mature T and B cells.

recombinatorial germ line theory

An hypothesis proposed by Dreyer and Bennett which postulated that variable-region and constant-region immunoglobulin genes were separated and rejoined in DNA levels. This concept was an important step toward understanding the generation of diversity in the production of antibody molecules, a puzzle finally solved by Tonegawa.

red cell-linked antigen antiglobulin test

A passive hemagglutination test in which the red cells serve only as carriers for antigen coated on their surfaces. It can identify either agglutinating antibodies or nonagglutinating (incomplete) antibodies by the aggregation or clumping of antigen-bearing red cells. To perform the assay, the test serum is incubated with red cells treated with antigen, which are then washed, and antibody against human globulin is added.

red pulp

Areas of the spleen comprised of the cords of Billroth and sinusoids.

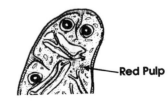

Spleen Morphology

Reed-Sternberg cells

Binucleated giant cells that contain prominent nucleoli. They are classically associated with Hodgkin's disease.

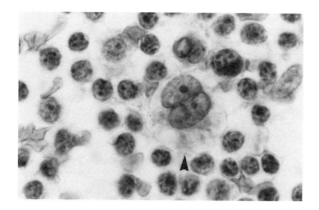

Touch Preparation
Reed-Sternberg Cell

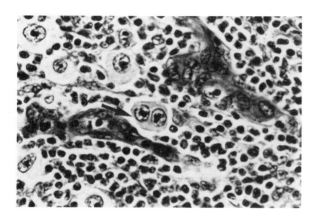

Reed-Sternberg Cell

regional enteritis

Inflammatory bowel disease characterized by chronic segmental lymphocytic and granulomatous inflammation of the gastrointestinal tract.

regulation of complement activation (RCA) cluster

A gene cluster on a 950-kb DNA segment of chromosome 1's long arm that encodes the homologous proteins CR1, CR2, C4bp, and DAF.

regulators of complement activity (RCA)

A group of proteins including C4bP, decay-accelerating factor (DAF), membrane cofactor protein (MCP), factor H, CR1 gel caps, and CR2 gel caps. They are comprised of 60- to 70-amino acid consensus repeat sequences.

Reiter complement fixation test (historical)

A diagnostic test for syphilis that used an antigen derived from a protein extract of *Treponema pallidum* (the Reiter strain). This test identified antibodies formed against *Treponema* group antigens.

rejection

An immune response to an organ allograft such as a kidney transplant. *Hyperacute rejection* is due to preformed antibodies and is apparent within minutes following transplantation. Antibodies reacting with endothelial cells cause complement to be fixed, which attracts polymorphonuclear neutrophils, resulting in denuding of the endothelial lining of the vascular walls. This causes platelets and fibrin plugs to block the blood flow to the transplanted organ, which becomes cyanotic and must be removed. Only a few drops of bloody urine are usually produced. Segmental thrombosis, necrosis, and fibrin thrombi form in the glomerular tufts. There is hemorrhage in

the interstitium, mesangial cell swelling; IgG, IgM, and C3 may be deposited in arteriole walls. *Acute rejection* occurs within days to weeks following transplantation and is characterized by extensive cellular infiltration of the interstitium. These cells are largely mononuclear cells and include plasma cells, lymphocytes, immunoblasts, and macrophages, as well as some neutrophils. Tubules become separated, and the tubular epithelium undergoes necrosis. Endothelial cells are swollen and vacuolated. There is vascular edema, bleeding with inflammation, renal tubular necrosis, and sclerosed glomeruli. *Chronic rejection* occurs after more than 60 days following transplantation and may be characterized by structural changes such as interstitial fibrosis, sclerosed glomeruli, mesangial proliferative glomerulonephritis, crescent formation, and various other changes.

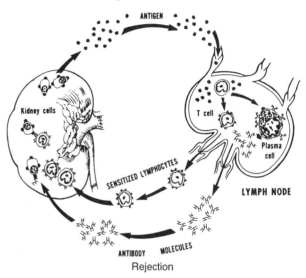

Rejection

Types of Skin Graft Rejection

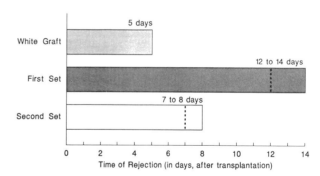

relapsing polychondritis (see figure, page 260)

Inflammation of the cartilage, especially that of the external pinnae of the ears, causing them to lose their structural integrity. It appears to have an immunological etiology, and anticollagen antibodies may be demonstrated in the serum of patients.

relative risk (RR)

Association of a particular disease with a certain HLA antigen. This represents the chance a person with the disease-associated HLA antigen has of developing the disease compared with that of a person who does not possess that antigen. Relative risk is calculated as follows:

$$RR = \frac{p^+ \times c^-}{p^- \times c^+}$$

where

p^+ = number of patients possessing a particular HLA antigen
c^- = number of controls not possessing the particular HLA antigen
p^- = number of patients not possessing the particular HLA antigen
c^+ = number of controls possessing the particular antigen

Relapsing Polychondritis

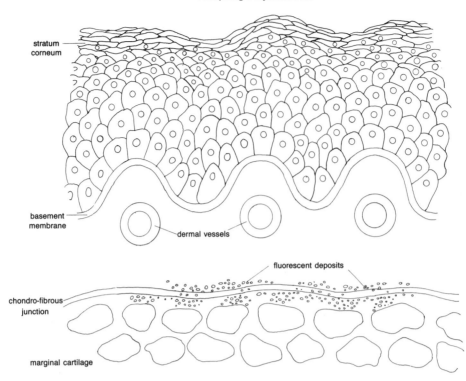

The higher the relative risk (above 1), the greater the antigen's frequency is in the patient population.

released antigen
Antigen derived from trypanosomes during an infection that may appear in the patient's serum. It corresponds to the antigenic type of the trypanosome infecting the individual.

repeating units
In immunology, this refers to antigens in which macromolecular configurations are repeated, such as the repeating units of β-1,4-glucose-β-1,3-glucuronic acid in type III pneumococcus polysaccharide. Polysaccharide antigens in the cell walls of Gram-negative bacteria also contain repeating structures. Antigens with this type of configuration are often thymus-independent antigens.

rescue graft
A replacement graft for an original graft that failed.

resident macrophage
A macrophage normally present at a tissue location without being induced to migrate there.

respiratory burst
A process used by neutrophils and monocytes to kill certain pathogenic microorganisms. It involves increased oxygen consumption with the generation of hydrogen peroxide and superoxide anions. This occurs also in macrophages that kill tumor cells. An abrupt elevation in oxygen consumption, which is followed by metabolic events in neutrophils and mononuclear cells preceding bacteriolysis. Partial reduction of oxygen by this process provides microbicidal oxidants. The initial event is a one electron reduction of oxygen by membrane-bound oxidase to form a superoxide. The hexose monophosphate shunt reaction that accompanies this reduction liberates an H^+, which unites with the oxygen to produce H_2O_2.

respiratory disease viruses
Influenza, paramyxoviruses, rhinoviruses, pneumoviruses, enteroviruses, coronaviruses, and mastadenoviruses that induce infections of the respiratory tract in man.

resting lymphocytes
By light microscopy, resting lymphocytes appear as a distinct and homogeneous population of round cells, each with a large, spherical or slightly kidney-shaped nucleus which occupies most of the cell and is surrounded by a narrow rim of basophilic cytoplasm with occasional vacuoles. The nucleus usually has a poorly visible single indentation and contains densely packed chromatin. Occasionally, nucleoli can be distinguished. The small lymphocyte variant, which is the predominant morphologic form, is slightly larger than an erythrocyte. Larger lymphocytes, ranging between 10 to 20 μm in diameter, are difficult to differentiate from monocytes. They have more cytoplasm and may show azurophilic granules. Intermediate-size forms between the two are described. By phase contrast microscopy, living lymphocytes show a feeble motility with ameboid movements that give the cells a hand-mirror shape. The mirror handle is called a uropod. In large lymphocytes, mitochondria and lysosomes are better visualized, and some cells show a spherical, birefringent, 0.5-μ diameter inclusion, called a Gall body. Lymphocytes do not spread on surfaces. The different classes of lymphocytes cannot be distinguished by light microscopy. By scanning electron microscopy, B lymphocytes sometimes show a hairy (rough) surface, but this is apparently an artifact. Electron microscopy does not provide additional information except for visualization of the cellular organelles which are not abundant. This suggests that the small, resting lymphocytes are end-stage cells. However, under appropriate stimulation, they are capable of considerable morphologic changes.

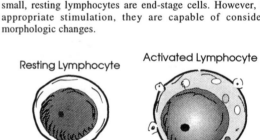

Resting Lymphocyte   Activated Lymphocyte

12 μm

restitope (see facing page)
That segment of a T cell receptor that makes contact and interacts with a class II histocompatibility antigen molecule during antigen presentation.

restriction endonuclease
Bacterial products that identify and combine with a short sequence of DNA. The enzyme acts as molecular scissors by cleaving the DNA at either the recognition site or at another location. Restriction endonucleases catalyze degradation of foreign DNA. They recognize precise base sequences of DNA and cut it into relatively few fragments termed restriction fragments. There are three major types

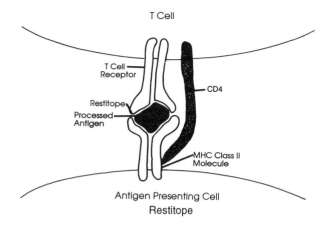

T Cell

Antigen Presenting Cell
Restitope

of restriction endonucleases. Type I enzymes identify specific base sequences, but cut the DNA elsewhere, i.e., approximately 1000 bp from the recognition site. Type II enzymes identify specific base sequences and cut the DNA either within or adjacent to these sequences. Type III endonucleases identify specific base sequences and cut the DNA approximately 25 bp from the recognition site.

restriction fragment length polymorphism (RFLP)
Genome diversity in DNA from different subjects revealed by restriction map comparisons. This is based on differences in restriction fragment lengths which are determined by sites of restriction endonuclease cleavage of the DNA molecules. This is revealed by preparing Southern blots using appropriate molecular hybridization probes. Polymorphisms may be demonstrated in exons, introns, flanking sequences, or any DNA sequence. Variations in DNA sequence show Mendelian inheritance. Results are useful in linkage studies and can help to identify defective genes associated with inherited disease.

restriction map
A diagram of either a linear or circular molecule of DNA that indicates the points where one or more restriction enzymes would cleave the DNA. DNA is first digested with restriction endonucleases, which split the DNA into fragments that can be separated by gel electrophoresis. Size determination is accomplished by comparison with DNA fragments of known size.

reticular cell
Stroma or framework cells which, together with reticular fibers, constitute the lymphoid tissue framework of lymph nodes, spleen, and bone marrow.

reticular dysgenesis
This is the most severe form of all combined immunodeficiency disorders. It is believed to be caused by a cellular defect at the level of hematopoietic stem cells. This leads to a failure in the development of B cells, T cells, and granulocytes. This condition is incompatible with life and leads to early death of affected infants. The only possibility for treatment is bone marrow transplantation. This condition has an autosomal recessive mode of inheritance.

reticulin antibodies
Antibodies of five separate types of which the R_1-RA is of greatest interest. IgA R_1-ARA show more than 98% specificity for celiac disease that has not been treated. However, sensitivity is only 20 to 25%.

reticuloendothelial blockade
The temporary paralysis of phagocytic cells of the mononuclear phagocytic system with respect to phagocytic ability by the injection of excess amounts of inert particles such as colloidal carbon, gold, or iron. Once the mononuclear phagocytes have expended their entire phagocytic ability taking up these inert particles, they can no longer phagocytize administered microorganisms or other substances that would normally be phagocytized.

reticuloendothelial system (RES)
A former term for the mononuclear phagocyte system that includes Kupffer cells lining the sinusoids of the liver as well as macrophages of the spleen and lymph nodes. Aschoff introduced the term to describe cells that could take up and retain vital dyes and particles that had been injected into the body. In addition to macrophages, less active phagocytic cells such as fibroblasts and endothelial cells were also included in the original definition. The principal function of the mononuclear phagocyte system is to remove unwelcome particles from the blood. RES activity can be measured by the

elimination rate of radiolabeled molecules or cells such as albumin or erythrocytes coated with antibody.

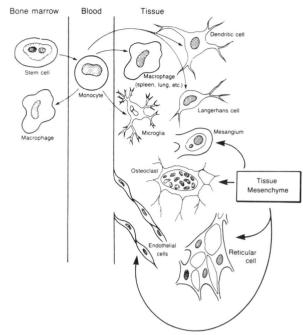

Reticuloendothelial System

reticulosis
Refer to lymphoma.

reticulum cell
Refer to reticular cell.

reticulum cell sarcoma
Obsolete term for large cell lymphoma.

retrovirus
A reverse transcriptase-containing virus such as the human immunodeficiency virus and human T cell leukemia virus. An RNA virus that can insert and efficiently express its own genetic information in the host cell genome through transcription of its RNA into DNA which is then integrated into the genome of host cells. Retroviruses are employed in research to deliberately insert foreign DNA into a cell. Thus, they have the potential for use in gene therapy when a host cell gene is either missing or defective. Retroviruses have been used to tag tumor-infiltrating lymphocytes in experimental cancer treatment.

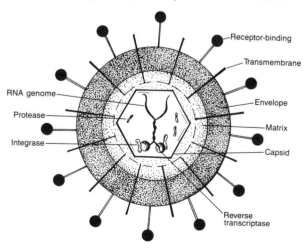

Retrovirus

reverse anaphylaxis
Anaphylaxis produced by the passive transfer of serum antibody from a sensitized animal to a normal untreated recipient only after

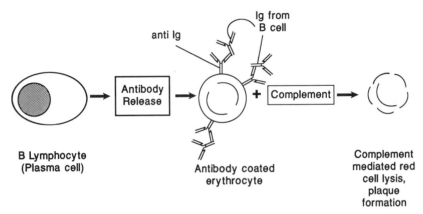

Reverse Plaque Assay

the recipient had been first injected with the antigen. Thus, the usual order of administration of antigen and antibody are reversed compared to classic anaphylaxis.

reverse Mancini technique

Refer to reverse radioimmunodiffusion.

reverse passive Arthus reaction

A reaction that differs from a classic Arthus reaction only in that precipitating antibody is injected into an animal intracutaneously, and after an interval of 30 min to 2 h, the antigen is administered intravenously. In this situation, antigen, rather than antibody, diffuses from the blood into the tissues, and antibody, rather than antigen, diffuses into the tissue, where it encounters and interacts with antigen with the consequent typical changes in the microvasculature and tissues associated with the Arthus reaction.

reverse passive cutaneous anaphylaxis (RPCA)

A passive cutaneous anaphylaxis assay in which the order of antigen and antibody administration is reversed, i.e., the antigen is first injected, followed by the antibody. In this case, the antigen must be an immunoglobulin that can fix to tissue cells.

reverse plaque assay (see above)

A method to identify antibody-secreting cells regardless of their antibody specificity. The antibody-forming cells are suspended in agarose and incubated at 37°C in petri plates with sheep red cells coated with protein A. Anti-Ig and complement are also present. Cells synthesizing and secreting immunoglobulin become encircled by Ig-anti-Ig complexes and then link to protein A on the erythrocyte surfaces. This leads to hemolytic plaques (zones of lysis). Thus, any class of immunoglobulin can be identified by this technique through the choice of the appropriate antibody.

reverse radioimmunodiffusion

A technique to quantify antibody levels that varies from the single radial diffusion test in only one detail. Samples are placed in gel containing antigen. As diffusion takes place, precipitation rings that are produced are directly proportional to the antibody concentration.

RF

Abbreviation for rheumatoid factor(s).

RFLP (restriction fragment length polymorphism)

Local DNA sequence variations of man or other animals that may be revealed by the use of restriction endonucleases. These enzymes cut double-stranded DNA at points where they recognize a very specific oligonucleotide sequence, resulting in DNA fragments of different lengths that are unique to each individual animal or person. The fragments of different sizes are separated by electrophoresis. The technique is useful for a variety of purposes, such as identifying genes associated with neurologic diseases (e.g., myotonic dystrophy) which are inherited as autosomal dominant genes or in documenting chimerism. The fragments may also be used as genetic markers to help identify the inheritance patterns of particular genes.

Rhazes (Abu Bakr Muhammad ibn Zakariya) (865–932)

Persian philosopher and alchemist who described measles and smallpox as different diseases. He also was a proponent of the theory that immunity is acquired. Rhazes is often cited as the premier physician of Islam.

rhesus antibody

An antibody reactive with rhesus antigen, especially RhD.

rhesus antigen

Erythrocyte antigen of man that shares epitopes in common with rhesus monkey red blood cells. Rhesus antigens are encoded by allelic genes. D antigen has the greatest clinical significance as it may stimulate antibodies in subjects not possessing the antigen and induce hemolytic disease of the newborn or cause transfusion incompatibility reactions.

rhesus blood group system

Rhesus monkey erythrocyte antigens such as the D antigen are found on the red cells of most humans, who are said to be Rh+. This blood group system was discovered by Landsteiner et al. in the 1940s when they injected rhesus monkey erythrocytes into rabbits and guinea pigs. Subsequent studies showed the system to be quite complex, and the rare Rh alloantigens are still not characterized biochemically. Three closely linked pairs of alleles designated Dd, Cc, and Ee are postulated to be at the Rh locus, which is located on chromosome 1. There are several alloantigenic determinants within the Rh system. Clinically, the D antigen is the one of greatest concern, since RhD− individuals who receive RhD+ erythrocytes by transfusion can develop alloantibodies that may lead to severe reactions with further transfusions of RhD+ blood. The D antigen also poses a problem in RhD− mothers who bear a child with RhD+ red cells inherited from the father. The entrance of fetal erythrocytes into the maternal circulation at parturition or trauma during the pregnancy (such as in amniocentesis) can lead to alloimmunization against the RhD antigen which may cause hemolytic disease of the newborn in subsequent pregnancies. This is now prevented by the administration of $Rh_0(D)$ immune globulin to these women within 72 h of parturition. Further confusion concerning this system has been caused by the use of separate designations by the Wiener and Fisher systems. Rh antigens are a group of 7- to 10-kD, erythrocyte membrane-bound antigens that are independent of phosphatides and proteolipids. Antibodies against Rh antigens do not occur naturally in the serum.

rhesus incompatibility

The stimulation of anti-RhD antibodies in an Rh− mother when challenged by RhD+ red cells of her baby (especially at parturition) that may lead to hemolytic disease of the newborn. The term also refers to the transfusion of RhD+ blood to an Rh− individual who may form anti-D antibodies against the donor blood, leading to subsequent incompatibility reactions if given future RhD+ blood.

rheumatic fever (RF)

An acute, nonsuppurative, inflammatory disease that is immune mediated and occurs mainly in children a few weeks following an infection of the pharynx with group A β hemolytic streptococci. M protein, a principal virulence factor associated with specific strains of the streptococci, induces antibodies that crossreact with epitopes of human cardiac muscle. These antibodies may not produce direct tissue injury, but together with other immune mechanisms evoke acute systemic disease characterized mainly by polyarthritis, skin lesions, and carditis. Whereas the arthritis and skin lesions resolve, the cardiac involvement may lead to permanent injury to the valves producing fibrocalcific deformity. Foci of necrosis of collagen with fibrin deposition surrounded by lymphocytes, macrophages, and plump modified histiocytes are termed Aschoff bodies, which are ultimately replaced years later by fibrous scars. Aschoff bodies may be found in any of the three layers of the heart. In the pericardium, they are accompanied by serofibrinous (bread and butter) pericarditis. In the myocardium, they are scattered in the interstitial connective tissue often near blood vessels. There may be dilatation of the heart and mitral valve ring. There may be inflammation of the endocardium, mainly affecting the left-sided valves. Small vegetations may form along the lines of closure. Other tissues may be affected with the production of acute nonspecific arthritis affecting the larger joints. Fewer than half of the patients develop skin lesions such as subcutaneous nodules or erythema marginatum. Subcutaneous nodules that appear at pressure points overlying extensor tendons of extremities at the wrist, elbows, ankles, and knees consist of central fibrinoid necrosis enclosed by a palisade of fibroblasts and mononuclear inflammatory cells. Rheumatic arteritis has been described in coronary, renal, mesenteric, and cerebral arteries, as well as in the aorta and pulmonary vessels. Rheumatic interstitial pneumonitis is a rare complication of the disease. Antistreptolysin O (ASO) and antistreptokinase (ASK) antibodies are found in the serum of affected individuals. Myocarditis that develops during an acute attack may induce arrhythmias such as atrial fibrillation or cardiac dilatation with potential mitral valve insufficiency. Long-term antistreptococcal therapy must be given to any patient with a history of rheumatic fever, as subsequent streptococcal infections may worsen the carditis.

rheumatoid arthritis

An autoimmune inflammatory disease of the joints which is defined according to special criteria designated 1 through 7. Criteria 1 through 4 must be present for more than 6 weeks. The "revised criteria" for rheumatoid arthritis are as follows: (1) morning stiffness in and around joints lasting at least 1 h before maximum improvement; (2) soft tissue swelling (arthritis) of three or more joints observed by a physician; (3) swelling (arthritis) of the proximal interphalangeal, metacarpal phalangeal, or

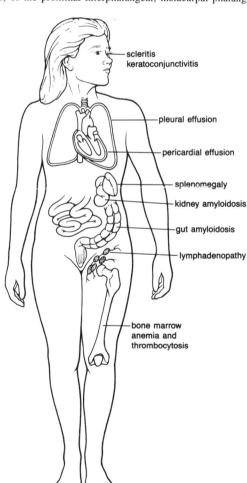

Rheumatoid Arthritis

wrist joints; (4) symmetric swelling (arthritis); (5) rheumatoid nodules; (6) presence of rheumatoid factors; and (7) roentgenographic erosions.

rheumatoid arthritis cell (RA cell)

An irregular neutrophil that contains a variable number of black-staining cytoplasmic inclusions that are 0.2 to 2.0 μ in diameter. These cells contain IgM rheumatoid factor, complement, IgG, and fibrin. They are found in synovial fluid of RA patients. Although RA cells may constitute 5 to 100% of an RA patient's neutrophils, they may also be present in patients with other connective tissue diseases.

rheumatoid factors

An autoantibody present in the serum of patients with rheumatoid arthritis, but also found with varying frequency in other diseases such as subacute bacterial endocarditis, tuberculosis, syphilis, sarcoidosis, hepatic diseases, and others, as well as in the sera of some human allograft recipients and apparently healthy persons. RF are immunoglobulins, usually of the IgM class and to a lesser degree of the IgG or IgA classes, with reactive specificity for the Fc region of IgG. This antiimmunoglobulin antibody, which may be either monoclonal or polyclonal, reacts with the Fc region epitopes of denatured IgG, including the Gm markers. Most RF are isotype specific, manifesting reactivity mainly for IgG1, IgG2, and IgG4, but only weakly reactive with IgG3. Antigenic determinants of IgG that are potentially reactive with RF include (1) subclass-specific or genetically defined determinants of native IgG (IgG1, IgG2, IgG4, and Gm determinants); (2) determinants present on complexed IgG, but absent on native IgG; and (3) determinants exposed after enzymatic cleavage of IgG. The Gm determinants are allotypic markers of the human IgG subclasses. They are located in the IgG molecule as follows: in the C_H1 domain in IgG1; in the C_H2 domain in IgG2; and in C_H2 and C_H3 domains in IgG4.

Although rheumatoid factor titers may not be clearly correlated with disease activity, they may help perpetuate chronic inflammatory synovitis. When IgM rheumatoid factors and IgG target molecules react to form immune complexes, complement is activated, leading to inflammation and immune injury. IgG rheumatoid factors may self-associate to form IgG-IgG immune complexes that help perpetuate chronic synovitis and vasculitis. IgG RF synthesized by plasma cells in the rheumatoid synovium fix complement and perpetuate inflammation. IgG RF has been shown in microbial infections, B lymphocyte proliferative disorders and malignancies, non-RA patients, and aging individuals. RF might have a physiologic role in removal of immune complexes from the circulation. Rheumatoid factors (RF) were demonstrated earlier by the Rose-Waaler test, but are now detected by the latex agglutination (or RA) test, which employs latex particles coated with IgG.

rheumatoid nodule

A granulomatous lesion characterized by central necrosis encircled by a palisade of mononuclear cells and an exterior mantle of lymphocytic infiltrate. The lesions occur as subcutaneous nodules, especially at pressure points such as the elbow, in individuals with rheumatoid arthritis or other rheumatoid diseases.

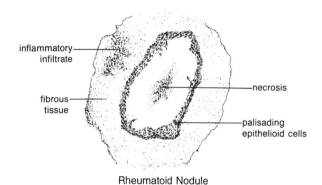

Rheumatoid Nodule

rheumatoid pneumonitis

Diffuse interstitial pulmonary fibrosis that causes varying degrees of pulmonary impairment in 2% of rheumatoid arthritis patients. It may result from the coincidental occurrence of rheumatoid arthritis and interstitial pneumonia, which is a rare situation. Gold therapy used for RA, smoking, or contact with an environmental toxin may induce interstitial pneumonia.

RhLA locus
The major histocompatibility locus in the rhesus monkey.

Rh$_{null}$
Human erythrocytes that fail to express Rh antigens due to either the homozygous inheritance of the X^0r gene which causes a regulator-type defect or the inheritance of an amorphic gene (—/—). The Rh$_{null}$ phenotype is associated with diminished erythrocyte survival.

Rh$_o$D immune globulin
Prepared from the serum of individuals hyperimmunized against Rh$_o$D antigen. It is used to prevent the immunization of Rh– mothers by Rh$_o$D+ erythrocytes of the baby, especially at parturition when the baby's red cells enter the maternal circulation in significant quantities, but also at any time during the pregnancy after trauma that might introduce fetal blood into the maternal circulation. This prevents hemolytic disease of the newborn in subsequent pregnancies. The dose used is effective in inhibiting immune reactivity against 15 ml of packed Rh$_o$(D)+ red blood cells. It should be administered within 72 h of parturition. It may be used also following inadvertent or unavoidable transfusion of RhD+ blood to RhD– recipients, especially to a woman of childbearing years.

rhodamine isothiocyanate
A reddish-orange fluorochrome used to label immunoglobulins or other proteins for use in immunofluorescence studies.

Rhodamine B Isothiocyanate

RhoGAM
Refer to Rh$_o$(D) immune globulin.

RIA
Radioimmunoassay.

ribavirin (1-β-5-D-ribofuranosyl-1,2,4-triazole-3carboxamide)
A substance that interferes with mRNA capping of certain viruses thereby restricting the synthesis of viral proteins. It is used as an aerosol to treat severe respiratory syncytial virus infection in children.

ribosome
A subcellular organelle in the cytoplasm of a cell that is a site of amino acid incorporation in the process of protein synthesis.

Richet, Charles Robert (1850–1935)
French physiologist, who with Portier, was the first to describe anaphylaxis. He received the Nobel Prize in 1913 for his work. *L'anaphylaxie*, 1911.

ricin
A toxic protein found in seeds of *Ricinus communis* (castor bean) plants. It is a heterodimer comprised of a 30-kD α chain, which

Ribavarin

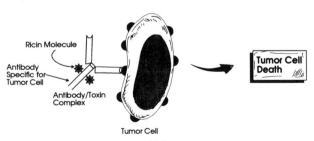

Ricin

mediates cytotoxicity, and a 30-kD β chain, which interacts with cell surface galactose residues that facilitate passage of molecules into cells in endocytic vesicles. Ricin inhibits protein synthesis by linkage of a dissociated α chain in the cytosol to ribosomes. The ricin heterodimer or its α chain conjugated to a specific antibody serves as an immunotoxin.

Ricinus communis
Refer to ricin.

Rieckenberg reaction
A trypanosome immune adherence test. Anticoagulated blood of an animal that recovered from trypanosomiasis is combined with live trypanosomes. Provided the same antigenic type of trypanosome that produced the infection is used for the test, blood platelets adhere to the trypanosomes.

rinderpest vaccines
Although several types have been used in the past, the most satisfactory contemporary vaccine contains a virus adapted to tissue culture.

ring precipitation test
Refer to ring test.

ring test
A qualitative precipitin test used for more than a century in which soluble antigen (or antibody) is layered onto an antibody (or antigen) solution in a serological tube or a capillary tube without agitating or mixing the two layers. If the antigen and antibody are specific for one another, a ring of precipitate will form at the interface. This simple technique was among the first antigen-antibody tests performed.

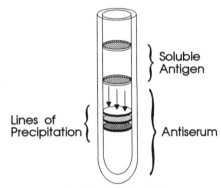

Ring Precipitation Test

RIST
Refer to radioimmunosorbent test.

Rivanol®
(2-ethoxy-6,9-diaminoacridine lactate). A commercially available substance used to isolate IgG from serum. It unites with serum proteins other than IgG to form insoluble cation complexes. When the pH is 7.0, IgG, with a high isoelectric point, fails to precipitate with Rivanol®. However, this method of IgG purification is inferior to DEAE-cellulose chromatography as a method to purify IgG.

RNA polymerase
Refer to DNA-directed RNA polymerase.

RNA splicing
The method whereby RNA sequences that are nontranslatable (known as introns) are excised from the primary transcript of a split gene. The translatable sequences (known as exons) are united to produce a functional gene product.

RNA-directed DNA polymerase (reverse transcriptase)
DNA polymerase present in retroviruses such as human immunodeficiency virus (HIV) and Rous sarcoma virus that can use an RNA template to produce DNA. The primer needed must contain a free 3′-hydroxyl group that is base paired with the template. This produces a DNA-RNA hybrid. Reverse transcriptase is critical in recombinant DNA techniques since it is employed for first strand cDNA synthesis.

rocket electrophoresis
The electrophoresis of antigen into an agar-containing specific antibody. In this electroimmunodiffusion method, lines of precipitation formed in the agar by the antigen-antibody interaction assume the shape of a rocket. The antigen concentration can be quantified since the rocket-like area is proportional to the antigen

concentration. This can be deduced by comparing with antigen standards. This technique has the advantage of speed since it can be completed within hours instead of longer periods required for single radial immunodiffusion. Also called Laurell rocket electrophoresis.

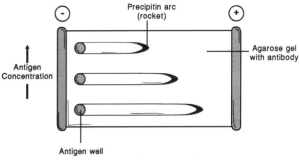

Rocket Electrophoresis

Romer reaction (historical)

Romer in 1909 described erythematous swelling following intracutaneous injection of diphtheria toxin in small quantities. The reaction was found to be neutralized by homologous antitoxin. The smallest amount of diphtheria toxin that produced a definite reaction was defined as the MRD or minimal reaction dose. In general, the MRD of a given toxin is equivalent to about 1/250 to 1/500 of the MLD (minimal lethal dose). The Lr is the smallest amount of toxin which, after mixing with one unit of antitoxin, will produce a minimal skin lesion when injected intracutaneously into a guinea pig.

Rose, Noel Richard (1927–)

American immunologist and authority on autoimmune diseases, who first discovered, with Witebsky, experimental autoimmune thyroiditis. His subsequent contributions to immunology are legion. He has authored numerous books and edited leading journals in the field.

Rose-Waaler test

Sheep red blood cells are treated with a subagglutinating quantity of rabbit anti-sheep erythrocyte antibody. These particles may be used to identify rheumatoid factor in the serum of rheumatoid arthritis patients. Agglutination of the IgG-coated red cells constitutes a positive test and is based upon immunological crossreactivity between human and rabbit IgG molecules. It may be positive in collagen vascular diseases other than rheumatoid arthritis, but it has still proven beneficial in diagnosis.

rosette

Cells of one type surrounding a single cell of another type. In immunology, it was used as an early method to enumerate T cells, i.e., in the formation of E rosettes in which CD2 markers on human T lymphocytes adhere to LFA-3 molecules on sheep red cells surrounding them to give a rosette arrangement. Another example was the use of the EAC rosette, consisting of erythrocytes coated with antibody and complement which surrounded a B cell bearing Fc receptors or complement receptors on its surface.

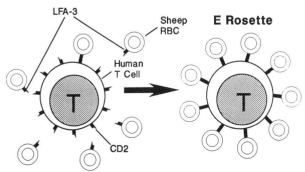

rotavirus

A double-stranded RNA virus that is encapsulated and belongs to the reovirus family. It is 70 nm in diameter and causes epidemics of gastroenteritis, which are usually relatively mild, but may be severe in children less than 2 years of age.

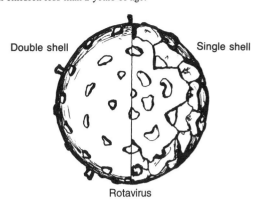

Rotavirus

rouleaux formation

Erythrocytes arranged in the form of stacked coins. This formation is observed when plasma fibrinogen and globulins are increased. It is also seen when the sedimentation rate is increased due to monoclonal immunoglobulins, as in multiple myeloma or Waldenström's disease or when dextran has been administered. Rouleaux formation may also accompany cryoglobulinemia, sarcoidosis, and cirrhosis. When rouleaux formation interferes with reading weak antigen-antibody reactions, the effect may be diminished by diluting the sample with physiological saline solution.

round cells

Term used by pathologists to describe mononuclear cells, especially lymphocytes, infiltrating tissues.

Rous sarcoma virus (RSV)

An RNA type C oncovirus that is single stranded and produces sarcomas in chickens. It is the typical acute transforming retrovirus. Within its genome are *gag, pol, env,* and v-*src* genes: *gag* encodes a core protein, *pol* encodes reverse transcriptase, and *env* encodes envelope glycoprotein. V-*scr* is an oncogene associated with the oncogenic capacity of the virus.

RPR (rapid plasma reagin) test

An agglutination test used in screening for syphilis. Antilipoidal (nontreponemal) antibodies (reagins) develop in the host usually within 4 to 6 weeks after infection with *Treponema pallidum*. Of patients with primary syphilis 93% develop a positive RPR.

RS61443 (Mycophenolate Mofetil)

An experimental immunosuppressive drug for the treatment of refractory, acute cellular allograft rejection, and possibly chronic rejection, in renal and other organ allotransplant recipients. Derived from mycophenolic acid, RS61443 interferes with guanosine synthesis, thereby blocking both T cell and B cell proliferation. It acts synergistically with cyclosporin in counteracting chronic rejection and has been reported to prevent FK506- and cyclosporin-A-induced obliterative vasculopathy.

rubella vaccine

An attenuated virus vaccine used in the MMR combination, or used alone to immunize seronegative women of childbearing age. It is not to be used during pregnancy.

runt disease

The disease process that results when neonatal mice of one strain are injected with lymph node or splenic lymphocytes from a different strain. Runt disease is accompanied by weight loss, failure to thrive, diarrhea, splenomegaly, and even death. The immune system of the neonatal animal is immature and reactivity against donor cells is weak or absent. Runt disease is an example of a graft-vs.-host reaction.

runting syndrome

Characterized by wasting, ruffled fur, diarrhea, lethargy, and debilitation. This is a consequence of thymectomy in neonatal mice that develop lymphopenia, lose lymphocytes from their lymphoid tissues, and become immunologically incompetent.

Russell body

A sphere or globule in the endoplasmic reticulum of some plasma cells. These immunoglobulin-containing structures are stained pink by eosin.

S

S

Abbreviation for sedimentation coefficient.

S-100

A heterodimeric protein comprised of α and β chains. It is present in a variety of tissues and is especially prominent in nervous system tissue including brain, neural crest, and Schwann cells. It also is positive in breast ducts, sweat and salivary glands, bronchial glands and Schwann cells, serous acini, malignant melanomas, myoepithelium, and neurofibrosarcomas.

S-100 protein

A marker, demonstrable by immunoperoxidase staining, that is extensively distributed in both central and peripheral nervous systems and tumors arising from them, including astrocytomas, melanomas, Schwannomas, etc. Most melanomas express S-100 protein. Such non-neuronal cells as chondrocytes and histiocytes are also S-100 positive.

7 S antibody

The sedimentation coefficient of immunoglobulin molecules such as IgG. 6.6 S immunoglobulins are usually referred to as 7 S immunoglobulins.

19 S antibody

The sedimentation coefficient of the IgM class of immunoglobulin.

S protein

An 83-kD serum protein in man that prevents generation of the membrane attack complex (MAC) of complement. The S protein molecule is comprised of one 478-amino acid residue polypeptide chain. Its mechanism of action is to inhibit insertion of the C5b67 complex into the membrane of a cell by first linking three of its molecules to each free C5b67 complex. It also inhibits C9 from polymerizing on C5b678 complexes. Refer also to vitronectin.

S region

The chromosomal segment of the murine MHC where genes are located that encode MHC class III molecules, including complement component C2, factor B, C4A (Slp), and C4B (Ss). The S locus within this region is the site of genes that encode a 200-kD protein termed Ss (serum substance) that corresponds to C4 in the serum of man. Also within the S region can be found the gene for Slp (sex limited protein), a protein usually found only in male mice, the gene for 21-hydroxylase, an enzyme with no known immune function, and the gene for a serum β globulin. This term likewise refers to the chromosomal segment that lies between HLA-B and HLA-D where the genes encoding the corresponding human MHC class III molecules are situated.

S value (Sverdberg unit)

The sedimentation coefficient of a protein that is ascertained by analytical ultracentrifugation.

Sabin-Feldman dye test

An *in vitro* diagnostic test for toxoplasmosis. Serial dilutions of patient's serum are combined with *Toxoplasma gondii* microorganisms, and complement is added. If specific antibodies against *Toxoplasma* organisms are present in the serum, complement interrupts the integrity of the toxoplasma membrane admitting methylene blue which has been added to the system and stains the interior of the organism. That dilution of patient's serum in which one half of the *Toxoplasma* organisms have been fatally injured is the titer.

Sabin vaccine

An attenuated live poliomyelitis virus vaccine that is administered orally to induce local immunity in the gut, which is the virus's natural route of entry, thereby stimulating local as well as systemic immunity against the causative agent of the disease.

saccharated iron oxide

Colloidal iron oxide employed to investigate the phagocytic capacity of mononuclear phagocytes.

sacculus rotundus

The lymphoid tissue-rich terminal segments of the ileum in the rabbit. It is a part of the gut-associated lymphoid tissue (GALT).

sago spleen

The replacement of lymphoid follicles by amyloid deposits that are circular and transparent in amyloidosis of the spleen.

SAIDS (simian acquired immunodeficiency syndrome)

An immunodeficiency of rhesus monkeys induced by retrovirus group D. The animals develop opportunistic infections and tumors. Their CD4$^+$ lymphocytes decrease. They suffer wasting and develop granulomatous encephalitis. The sequence homology between SIV and HIV-1 is minimal, but the sequence homology between SIV and HIV-2 is significant.

saline agglutinin

An antibody that causes the aggregation or agglutination of cells such as red blood cells or bacterial cells or other particles in 0.15 M salt solutions without additives.

Salk vaccine

An injectable poliomyelitis virus vaccine, killed by formalin, that was used for prophylactic immunization against poliomyelitis prior to development of the Sabin oral polio vaccine.

salt precipitation

An earlier method to separate serum proteins based on the principle that globulins precipitate when the concentration of sodium sulfate or ammonium sulfate is less than the concentration at which albumin precipitates. Euglobulins precipitate at concentrations that are less than those at which pseudoglobulins precipitate. This method was largely replaced by chromatographic methods using Sephadex® beads and related techniques.

salting out

Salt precipitation of serum proteins such as globulins.

SAMS

Abbreviation for substrate adhesion molecules.

Sanarelli-Shwartzman reaction

Refer to Shwartzman reaction.

Sandoglobulin®

Refer to human immune globulin.

"sandwich" methodology

Refer to sandwich technique.

sandwich technique

The identification of antibody or of antibody-synthesizing cells in tissue preparations in which antigen is placed in contact with the tissue section or smear, followed by the application of antibody labeled with a fluorochrome such as fluorescein isothiocyanate (FITC) that is specific for the antigen. This yields a product consisting of antibody layered on either side of an antigen which accounts for the name "sandwich".

saponin

A glucoside used in the past for its adjuvant properties to enhance immune reactivity to certain vaccine constituents. It was considered to slow the release of immunogen from the site of injection and to induce B cells capable of forming antibody at the site of antigen deposition.

sarcoidosis

A systemic granulomatous disease that involves lymph nodes, lungs, eyes, and skin. There is a granulomatous hypersensitivity reaction that resembles that of tuberculosis and fungus infections. Sarcoidosis has a higher incidence in African-Americans than in Caucasians and is prominent geographically in the southeastern U.S. It is of unknown etiology. Immunologically, there is a decrease in circulating T cells. There is decreased delayed-type hypersensitivity as manifested by anergy to common skin test antigens. Increased antibody formation leads to polyclonal hypergammaglobulinemia. There is a marked cellular immune response in local areas of disease activity. Tissue lesions consist of inflammatory cells and granulomas, comprised of activated mononuclear phagocytes such as epithelioid cells, multinucleated giant cells, and macrophages. Activated T cells are present at the periphery of the granuloma. CD4$^+$ T cells appear to be the immunoregulatory cells governing granuloma formation. Mediators released from T cells nonspecifically stimulate B cells, resulting in the polyclonal hypergammaglobulinemia. The granulomas are typically noncaseating, distinguishing them from those produced in tuberculosis. Patients may develop fever, polyarthritis, erythema nodosum, and iritis. They also may experience loss of weight, anorexia, weakness, fever, sweats, nonproductive cough, and increasing dyspnea on exertion. Pulmonary symptoms occur in greater than 90% of the patients. Angiotensin-converting enzyme is increased in the serum of sarcoid patients. Disease activity is monitored by measuring the level of this enzyme in the serum. The subcutaneous inoculation of sarcoidosis lymph node extracts into patients diagnosed with sarcoidosis leads to a granulomatous reaction in the skin 3 to 4 weeks after inoculation. This was used in the past as a diagnostic test, of questionable value, termed the Kveim reaction. Sarcoidosis symptoms can be treated with corticosteroids, but only in patients where disease progression occurs. It is a relatively mild disease, with 80% resolving spontaneously and only 5% dying of complications.

Sca-1

Abbreviation for stem cell antigen-1.

SCAB (single chain antigen-binding proteins)

Polypeptides that join an antibody's light chain variable sequence to the antibody heavy chain variable sequence. All monoclonal antibodies are potential sources of SCABs. They are smaller and less immunogenic than the intact heavy chains with immunogenic constant regions. Among their many possible uses are in imaging and treatment of cancer, in cardiovascular disease, as biosensors, and for chemical separations.

scarlet fever

A condition associated with production of erythrogenic toxin by group A hemolytic streptococci associated with pharyngitis. Patients develop a strawberry-red tongue, and generalized erythematous blanching areas that do not occur on the palms, the soles of the feet, or in the mouth. Patients may also develop Pastia's lines, which are petechiae in a linear pattern.

Scatchard equation

In immunology, an expression of the union of a univalent ligand with an antibody molecule. $r/c = Kn - Kr$. To obtain the average number of ligand molecules which an antibody molecule may bind at equilibrium, the bound ligand molar concentration is divided by the antibody molar concentration. This is designated as r. The free ligand molar concentration is represented by c. Antibody valence is designated by n, and the association constant is represented by K.

$$\frac{d}{[Ag]} = nk - dk$$

Scatchard Equation

Scatchard plot

A graphic representation of binding data obtained by plotting r/c against r (refer to the Scatchard equation). The purpose of this plot is to determine intrinsic association constants and to ascertain how many noninteracting binding sites each molecule contains. A straight line with a slope of $-K$ indicates that all the binding sites are the same and are independent. The plot should also intercept on the r axis of n. A nonlinear plot signifies that the binding sites are not the same and are not independent. The degree to which the sites are occupied is reflected by the slope ($-K$). An average association constant for ligand binding to heterogeneous antibodies is the reciprocal of the amount of free ligand needed for half saturation of antibody sites.

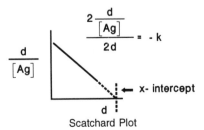

$$\frac{2\frac{d}{[Ag]}}{2d} = -k$$

Scatchard Plot

SCF (stem cell factor)

A substance that promotes growth of hematopoietic precursor cells and is encoded by the murine SI gene. It serves as a ligand for the tyrosine kinase receptor family protooncogene termed c-kit. It apparently has a role in embryogenesis in cells linked to migratory patterns of hematopoietic stem cells, melanoblasts, and germ cells.

Schick, Bela (1877–1967)

Austro-Hungarian pediatrician whose work with von Pirquet resulted in the discovery and description of serum sickness. He developed the test for diphtheria that bears his name. *Die Serumkrankheit (with Pirquet),* 1905.

Schick test

A test for susceptibility to diphtheria. Standardized diphtheria toxin is adjusted to contain 1/50 MLD in 0.1 ml, which is injected intracutaneously into the subject's forearm. Development of redness and induration within 24 to 36 h after administration constitutes a positive test if it persists for 4 days or longer. The presence of 1/500 to 1/250 or more of a unit of antitoxin per milliliter of the patient's blood will result in a negative reaction because of neutralization of the injected toxin. Neither redness nor induration appears if the test is negative. An individual with a negative test possesses sufficient antitoxin to protect against infection with *Corynebacterium diphtheriae,* whereas a positive test denotes susceptibility. A control is always carried out in the opposite forearm. For this test, toxin which has been diluted and heated to 70°C for 15 min is injected intracutaneously. Heating destroys the toxin's ability to induce local tissue injury; however, it does not affect the components of the diphtheria bacilli or of the medium which might evoke an allergic response in the individual. If the size and duration of the reaction at the injection site in the control approximates the reaction in the test arm, the result is negative. If the reaction is at least 50% larger and of longer duration on the test arm compared to the control, the individual is both allergic to the materials in the bacilli or in the medium and susceptible to the toxin. A positive Schick reaction suggests that diphtheria immunization is needed.

schistosomiasis

A schistosome infection that is characteristically followed by a granulomatous tissue reaction.

Schlepper

A name used by Landsteiner to refer to large macromolecules that serve as carriers for simple chemical molecules serving as haptens. The immunization of rabbits or other animals with a hapten-carrier complex leads to the formation of antibodies specific for the hapten as well as the carrier. T cells were later shown to be carrier specific and B cells hapten specific. Carriers are conjugated to haptens through covalent linkages such as the diazo linkage.

Diazotization

tyrosine residue
Schlepper

Schultz-Dale test (historical)

Strong contraction of the isolated uterine horn muscle of a virgin guinea pig that has been either actively or passively sensitized occurs following the addition of specific antigen to the 37°C tissue bath in which it is suspended. This reaction is the basis for an *in vitro* assay of anaphylaxis termed the Schultz-Dale test. Muscle contraction is caused by the release of histamine and other pharmacological mediators of immediate hypersensitivity following antigen interaction with antibody fixed to tissue cells.

SCID (severe combined immunodeficiency)

Refer to severe combined immunodeficiency syndrome.

SCID mouse

An autosomal recessive mutation expressed as severe combined immune deficiency in the CB-17Icr mouse strain. These mice do not have serum immunoglobulins, yet their adenosine deaminase (ADA) levels are normal. They lack T and B lymphocytes. Thus, they fail to respond to either T cell-dependent or to T cell-independent antigens when challenged. Likewise, their lymph node or spleen cells fail to proliferate following challenge by T or B lymphocyte mitogens. The lymphoid stroma in their lymph nodes and spleen is normal. Even though there is no evidence of T cell-mediated immunity, they do have natural killer cells and mononuclear phagocytes that are normal in number and function. The mutation likewise does not affect myeloid and erythroid lineage cells. B cell development is arrested at the pro-B cell stage before cytoplasmic or surface immunoglobulins are present. There are also normal numbers of macrophages in the spleen, peritoneum, and liver. The SCID mutation is associated with an intrinsic defect in lymphoid stem cells. The main characteristic of SCID mice is the failure of their lymphocytes to express antigen-specific receptors. This is due to disordered rearrangements of T cell receptors or of immunoglobulin genes. The defect in recombination of antigen-specific receptor genes may be associated with the absence of a DNA recombinase specific for lymphocytes in these mice. This mouse model may be used to investigate the effects of anti-HIV drugs as well as of immunostimulants as a substitute for human experimentation. This model is also useful for investigations of neoplasms in hosts lacking an effective immune response.

Scl-70 antibody

An antinuclear antibody found in as many as 70% of diffuse-type scleroderma (progressive systemic sclerosis) patients who experience extensive and rapid skin involvement as well as early visceral manifestations.

scleroderma

Refer to progressive systemic sclerosis.

sCR1 (soluble complement receptor type 1)

A substance prepared by recombinant DNA technology that combines with activated C3b and C4. This facilitates complement factor I's inactivation of them. sCR1 significantly diminishes myocardial injury induced by hypoxia in rats.

scratch test

Skin test for the detection of IgE antibodies against a particular allergen that are anchored to mast cells in the skin. After scratching the skin with a needle, a minute amount of aqueous allergen is applied to the scratch site, and the area is observed for the development of urticaria manifested as a wheal and flare reaction. This signifies that the IgE antibodies are specific for the applied allergen and lead to the degranulation of mast cells with release of pharmacologic mediators of hypersensitivity, such as histamine.

SDS-PAGE

Polyacrylamide gel electrophoresis in sodium dodecyl sulfate.

second messengers (IP₃ and DAG)

Upon stimulation of the T cell receptor, protein tyrosine kinase (PTK) becomes activated. PTK then phosphorylates phospholipase C (PLC), which in turn hydrolyzes phosphatidylinositol 4,5-bisphosphate (PIP_2). The products of PIP_2 hydrolysis are the intracellular second messengers inositol 1,4,5-triphosphate (IP_3) and 1,2-diacylglycerol (DAG). IP_3 leads to increased Ca^{++} release from intracellular stores and DAG leads to increased levels of protein kinase C. Protein kinase C and Ca^{++} signals, such as the interaction of CA^{++}/calmodulin to activate calcineurin, are associated with gene transcription and T cell activation. Second messenger systems are also important in B cell activation as a means of stimulating the resting B cell to enter the cell cycle. Through these and other second messenger systems, extracellular signals are received at the cell membrane and relayed to the nucleus to induce responses at the genetic level.

second-set rejection

Second-set rejection of an organ or tissue graft by a host who is already immune to the histocompatibility antigens of the graft as a consequence of rejection of a previous transplant of the same antigenic specificity as the second, or as a consequence of immunization against antigens of the donor graft. The accelerated second-set rejection compared to rejection of a first graft is reminiscent of a classic secondary or booster immune response.

second-set response

Term that describes the accelerated rejection of a second skin graft from a donor that is the same or identical with the first donor. The accelerated rejection is seen when regrafting is performed within 12 to 80 days after rejection of the first graft. It is completed in 7 to 8 days and is due to sensitization of the recipient by the first graft.

secondary allergen

An agent that induces allergic symptoms because of crossreactivity with an allergen to which the individual is hypersensitive.

secondary antibody response

IgM appears before IgG in the primary antibody response. The inoculation of an immunogen into an experimental animal which has been primed by previous immunization with the same immunogen produces antibodies following the secondary immunogenic challenge that develop more rapidly, last longer, and reach a higher titer than in the primary response. Antibodies produced in the secondary response are predominantly IgG and reach levels that are tenfold or greater than those in the primary antibody response.

secondary disease

A condition that occurs in irradiated animals whose cell population has been reconstituted with histoincompatible, immunologically competent cells derived from allogeneic donor animals. Ionizing radiation induces immunosuppression in the recipients, rendering them incapable of rejecting the foreign cells. Thus, the recipient has two cell populations, its own and the one that has been introduced, making these animals radiation chimeras. After an initial period of recovery, the animals develop a secondary runt disease, which is usually fatal within 1 month.

secondary follicle

An area in a peripheral lymphoid organ where a germinal center is located. Usually associated with a secondary immune response more often than with a primary one.

secondary granule

A structure in the cytoplasm of polymorphonuclear leukocytes that contains vitamin B_{12}-binding protein, lysozyme, and lactoferrin in neutrophils. Cationic peptides are present in eosinophil secondary granules. Histamine, platelet-activating factor, and heparin are present in the secondary granules of basophils.

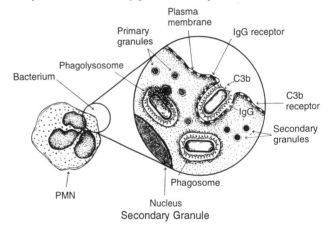

Secondary Granule

secondary immune response

A heightened antibody response following second exposure to antigen in animals that have been primed by previous contact with the same antigen. The secondary immune response depends upon immunological memory learned from the first encounter with antigen. It is characterized by a steep and rapid rise in antibody titer, usually of the IgG class, accompanied by potent cell-mediated immunity. Protein and glycoprotein immunogens stimulate this type of response. The rapid rise in antibody synthesis is followed by a gradual exponential decline in titer. This is also known as the booster response observed following administration of antigens subsequent to the secondary exposure. Refer also to anamnestic response.

secondary immunodeficiency

Immunodeficiency that is not due to a failure or intrinsic defect in the T and B lymphocytes of the immune system. It is a consequence of some other disease process and may be either transient or permanent. The transient variety may disappear following adequate treatment, whereas the more permanent type persists. Secondary immunodeficiencies are commonly produced by many effects. For example, those that appear in patients with neoplasms may result from effects of a tumor. Secondary immunodeficiencies may cause an individual to become susceptible to microorganisms that would otherwise cause no problem. They may occur following immunoglobulin or T lymphocyte loss, the administration of drugs, infections, cancer, effects of ionizing radiation on immune system cells, and other causes.

secondary lymphoid organ

The structures that include the lymph nodes, spleen, gut-associated lymphoid tissues, and tonsils where T and B lymphocytes interact with antigen-presenting accessory cells such as macrophages, resulting in the generation of an immune response.

secondary lysosome

A lysosome that has united with a phagosome.

secondary nodule

Refer to secondary follicle.

secondary response

The anamnestic or memory immune response that occurs following second exposure to an antigen in an individual who has been sensitized by a primary immunizing dose of the same antigen. The secondary response occurs more rapidly and is of much greater magnitude than is the primary response because of the memory cells that respond to the second challenge. Also called booster response.

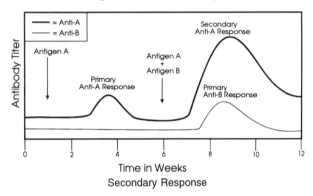

Secondary Response

secondary structure

Polypeptide chain or polynucleotide strand folding along the axis or backbone of a molecule. This is a consequence of the formation of intramolecular hydrogen bonds joining carbonyl oxygen and amide nitrogen atoms. Secondary structure is based on the local spatial organization of polypeptide chain segments or polynucleotide strands irrespective of the structure of side chains or of the relationship of the segments to one another.

secreted immunoglobulin

A product of plasma cells that is secreted as free immunoglobulin, where it may circulate as a component of blood plasma or make up part of the protein content of other body fluids.

secretor

An individual who secretes ABH blood group substances into body fluids such as saliva, gastric juice, tears, ovarian cyst fluid, etc. At least 80% of the human population are secretors. The property is genetically determined and requires that the individual be either homozygous (*Se/Se*) or heterozygous (*Se/se*) for the *Se* gene.

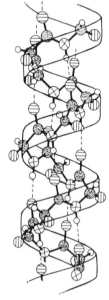

Secondary Structure

secretory component (T piece)

A 75-kD molecule synthesized by epithelial cells in the lamina propria of the gut that becomes associated with IgA molecules produced by plasma cells in the lamina propria of the gut as they move across the epithelial cell layer to reach the mucosal surface of intestine to provide local immunity. It can be found in three molecular forms: as an SIgM and SIgA stabilizing chain, as a transmembrane receptor protein, and as free secretory component in fluids.

secretory component deficiency

A lack of IgA in secretions as a consequence of gastrointestinal tract epithelial cells' inability to produce secretory component to be

linked to the IgA molecules synthesized, in the lamina propria of the gut. Secretory component normally prevents IgA destruction by the proteolytic enzymes in the gut lumen. The disorder is very infrequent, but is characterized by protracted diarrhea associated with gut infection.

secretory IgA

A dimeric molecule comprised of two IgA monomers joined by a J polypeptide chain and a glycopeptide secretory component. This is the principal molecule of mucosal immunity. IgA is the only immunoglobulin isotype that can be selectively passed across mucosal walls to reach the lumens of organs lined with mucosal cells. Specific FcαR that bind IgA molecular dimers are found on intestinal epithelial cells. The FcαR joins the antibody molecule to the epithelial cell's basal surface that is exposed to the blood. It is bound to the polyimmunoglobulin receptor on the epithelial cell's basolateral surface and facilitates vesicular transport of the anchored IgA across the cell to the surface of the mucosa. Once this complex reaches its destination, FcαR (S protein) is split in a manner that permits the dimeric IgA molecule to retain an attached secretory piece which has a strong affinity for mucous, thereby facilitating the maintenance of IgA molecules on mucosal surfaces. The secretory piece also has the important function of protecting the secreted IgA molecules from proteolytic digestion by enzymes of the gut. These latter two functions are in addition to its active role in transporting the IgA molecule through the epithelial cell. Secretory or exocrine IgA appears in the colostrum, intestinal and respiratory secretions, saliva, tears, and other secretions.

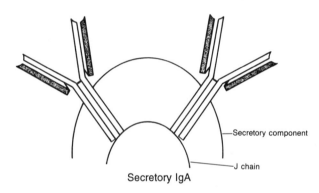

Secretory IgA

secretory immune system

A major component of the immune system that provides protection from invading microorganisms at local sites. Much of the effect is mediated by secretory IgA molecules in the secretions at the mucosal surface. Immunoglobulins may also be in clotted fluids where they protect against microorganisms.

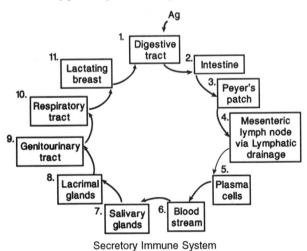

Secretory Immune System

secretory piece

A 75-kD polypeptide chain synthesized by epithelial cells of the gut for linkage to IgA dimers present in body secretions. Secretory component facilitates IgA transport across epithelial cells and protects secretory IgA released into the lumen of the gut from proteolytic digestion by enzymes in the secretions. It is not formed by plasma cells in the lamina propria of the gut that synthesize the IgA molecules with which it combines. Secretory component has a special affinity for mucous, thereby facilitating IgA's attachment to the mucous membranes. Also called secretory component.

sedimentation coefficient

The rate at which a macromolecule or particle sediments that is equivalent to the velocity per unit centrifugal field. $s = (dx/dt)/w^2x$. The sedimentation coefficient is s, the velocity is dx/dt, and the angular velocity is w. The distance from the axis of the centrifuge rotor is x. The size, shape, and weight of the macromolecule in question, as well as the concentration and temperature of the solutions, but not the centrifuge speed, determine the sedimentation coefficient. Measurement is in Svedberg units.

sedimentation pattern

The configuration of red blood cells on a test tube or plastic plate bottom at the conclusion of a hemagglutination test. The formation of a covering mat on the curved bottom of the tube or well signifies that agglutination has taken placed. The formation of a round button where the red blood cells have settled to the midpoint of the bottom of the tube or well and were not retained on the curvature constitutes a negative reaction with no agglutination.

Sedormid® purpura (historical)

In the past, a form of thrombocytopenic purpura occurring in patients, who received the drug Sedormid® (allyl-isopropyl-acetyl carbamide). The drug served as a hapten complexing with blood platelets. This platelet-drug complex was recognized by the immune system as foreign. Antibodies formed against it lysed the patient's blood platelets in the presence of complement, leading to thrombocytopenia. This was followed by bleeding, which caused the purpura manifested on the skin. This was a type II hypersensitivity reaction. Sedormid® is no longer used, but the principle of hypersensitivity it induced is useful for understanding autoimmunity induced by certain drugs.

selectins

A group of cell adhesion molecules (CAMs) that are glycoproteins and play an important role in the relationship of circulating cells to the endothelium. The members of this surface molecule family have three separate structural motifs. They have a single N-terminal (extracellular) lectin motif preceding a single epidermal growth

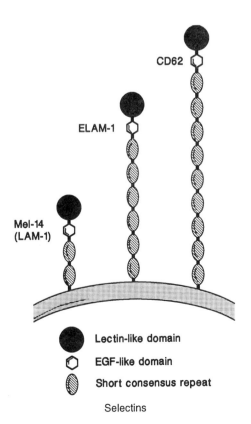

CD62

ELAM-1

Mel-14
(LAM-1)

● Lectin-like domain

⬡ EGF-like domain

◍ Short consensus repeat

Selectins

factor repeat and various short consensus repeat homology units. They are involved in lymphocyte migration.

selective IgA and IgG deficiency

This disease, which affects both males and females, is either X-linked, autosomal recessive, or acquired later in life. There may be a genetic defect in the switch mechanism for immunoglobulin-producing cells to change from IgM to IgG or IgA synthesis. Respiratory infections with pyogenic microorganisms or autoimmune states that include hemolytic anemia, thrombocytopenia, and neutropenia may occur. Numerous IgM-synthesizing plasma cells are demonstrable in both lymph nodes and spleen of affected individuals.

selective IgA and IgM deficiency

A concomitant reduction in both IgA and IgM concentrations with normal IgG levels. The IgG produced in many individuals may not be protective, and recurrent infections may result. There is an inadequate response to many immunogens in this disease which occurs in four males to every female affected.

selective IgA deficiency

The most frequent immunodeficiency disorder. It occurs in approximately 1 in 600 individuals in the population. It is characterized by nearly absent serum and secretory IgA. The IgA level is less than 5 mg/dl, whereas the remaining immunoglobulin class levels are normal or elevated. The disorder is either familial or it may be acquired in association with measles, other types of virus infection, or toxoplasmosis. The patients may appear normal and asymptomatic or they may have some form of an associated disease. IgA is the principal immunoglobulin in secretions and is an important part of the defense of mucosal surfaces. Thus, IgA-deficient individuals have an increased incidence of respiratory, gastrointestinal, and urogenital infections. They may manifest sinopulmonary infections and diarrhea. Selective IgA deficiency is diagnosed by the demonstration of less than 5 mg/dl of IgA in serum. The etiology is unknown, but is believed to be arrested B cell development. The B lymphocytes are normal with surface IgA and IgM or surface IgA and IgD. Some patients also have an IgG2 and IgG4 subclass deficiency. They are especially likely to develop infections. IgA-deficient patients have an increased incidence of respiratory allergy and autoimmune disease such as systemic lupus erythematosus and rheumatoid arthritis. The principal defect is in IgA B lymphocyte differentiation. The 12-week-old fetus contains the first IgA B lymphocytes that bear IgM and IgD as well as IgA on their surface. At birth, the formation of mature IgA B lymphocytes begins. Most IgA B cells express IgA exclusively on their surface, with only 10% expressing surface IgM and IgD in the adult. Patients with selective IgA deficiency usually express the immature phenotype, only a few of which can transform into IgA-synthesizing plasma cells. Patients have an increased incidence of HLA-A1, -B8, and -Dw3. Their IgA cells form, but do not secrete IgA. There is an increased incidence of the disorder in certain atopic individuals. Some selective IgA deficiency patients form significant titers of antibody against IgA. They may develop anaphylactic reactions upon receiving IgA-containing blood transfusions. The patients have an increased incidence of celiac disease and several autoimmune diseases as indicated above. They synthesize normal levels of IgG and IgM antibodies. Autosomal recessive and autosomal dominant patterns of inheritance have been described. It has been associated with several cancers, including thymoma, reticulum cell sarcoma, and squamous cell carcinoma of the esophagus and lungs. Certain cases may be linked to drugs such as phenytoin or other anticonvulsants. Some individuals develop antibodies against IgG, IgM, and IgA. γ Globulin should not be administered to selective IgA-deficient patients.

selective IgM deficiency

IgM is absent from the serum. Although IgM may be demonstrable on plasma cell surfaces, it is not secreted. This could be related to an alteration in secretory peptide or due to the action of suppressor T lymphocytes specifically on IgM-synthesizing and secreting cells. Gram-negative microorganisms may induce septicemia in affected individuals, since IgM's major role in protection against infection is in the intravascular compartment rather than in extravascular spaces where other immunoglobulin classes may be active.

selective immunoglobulin deficiency

An insufficient quantity of one of the three major immunoglobulins or a subclass of IgG or IgA.

Takes

1. Autograft

2. Isograft (Syngeneic graft)

A strain   A strain

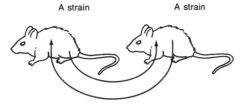

3. Semisyngeneic Graft

A strain   (AxB)F$_1$

Rejects

1. Allograft

A strain   B strain

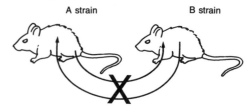

2. Semi Syngeneic Graft

A strain   (AxB)F$_1$

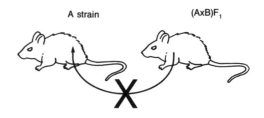

Semisyngeneic Graft

selective theory

An hypothesis that describes antibody synthesis as a process in which antigen selects cells expressing receptors specific for that antigen. The antigen-cell receptor interaction leads to proliferation and differentiation of that clone of cells which synthesizes significant quantities of antibodies of a single specificity. Selective theories included the side chain theory of Paul Ehrlich proposed in 1899, the natural selection theory proposed by Niels Jerne in 1955, and the cell selection theory proposed by Talmage and by Burnet in 1957. Burnet termed his version of the theory the clonal selection theory of acquired immunity. The basic tenets of the clonal selection theory have been substantiated by the scientific evidence. The selective theories maintained that cells are genetically programmed to react to certain antigenic specificities prior to antigen exposure. They are in sharp contrast to the instructive theories which postulated that antigen was necessary to serve as a template around which polypeptide chains were folded to yield specific antibodies. This template theory was abandoned when antibody was demonstrated in the absence of antigen.

self antigen

Autoantigen.

self marker hypothesis (historical)

A concept suggested by Burnet and Fenner in 1949 in an attempt to account for the failure of the body to react against its own antigens. They proposed that cells of the body contained a marker that identified them to the host's immunologically competent cells as self. This recognition system was supposed to prevent the immune cells of the host from rejecting its own tissue cells. This hypothesis was later abandoned by the authors and replaced by the clonal selection theory of acquired immunity which Burnet proposed in 1957.

self-recognition

The identification of autoantigens by lymphoid cells of the body's immune system. This is a consequence of the establishment of immunological tolerance of self antigens during fetal development.

self-restriction

Refer to MHC restriction.

self-tolerance

Term used to describe the body's acceptance of its own epitopes as self antigens. The body is tolerant to these autoantigens, which are exposed to the lymphoid cells of the host immune system. Tolerance to self antigens is developed during fetal life. Thus, the host is immunologically tolerant to self or autoantigens. Refer also to tolerance and to immunologic tolerance.

semisyngeneic graft (see above)

A graft that is ordinarily accepted from an individual of one strain into an F$_1$ hybrid of an individual of that strain mated with an individual of a different strain.

senescent cell antigen

A neoantigen appearing on old red blood cells that binds IgG autoantibodies. Senescent cell antigen is also found on lymphocytes, platelets, neutrophils, adult human liver cells (in culture), and human embryonic renal cells (in culture). Its appearance on aging somatic cells probably represents a physiologic process to remove senescent and injured cells that have fulfilled their function in the animal organism. Macrophages are able to identify and phagocytize dying and aging self cells that are no longer functional, without disturbing mature healthy cells.

sensitization

(1) Exposure of an animal to an antigen for the first time in order that subsequent or secondary exposure to the same antigen will lead to a greater response. The term has been used especially when the reaction induced is more of a hypersensitive or allergic nature than of an immune protective type of response. Thus, an allergic response may be induced in a host sensitized by prior exposure to the same allergen. (2) Term originally used by investigators who developed the complement fixation test to refer to the coating of cells such as red blood cells with specific antibody to "sensitize them" for subsequent lysis by complement.

sensitized

Synonymous for an animal which has been immunized or it may refer to red cells *in vitro* that have been treated with specific antibody prior to addition of complement for lysis.

sensitized cell

(1) A cell such as a lymphocyte that has been activated immunologically by interaction with a specific antigen. This type of sensitized cell is a primed lymphocyte. (2) Cells such as sheep red blood cells coated with a specific antibody. This coating renders them susceptible to lysis by complement.

sensitized lymphocyte

A primed lymphocyte that has been previously exposed to a specific antigen.

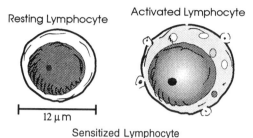

Resting Lymphocyte   Activated Lymphocyte

12 µm

Sensitized Lymphocyte

sensitizing antigen

A substance responsible for inducing hypersusceptibility or exaggerated reactivity to it.

Sephadex®

A trade name for a series of cross-linked dextrans used in chromatography.

Sepharose®

A trade name for agarose gels used in electrophoresis.

septic shock

Hypotension, with a systolic blood pressure of less than 90 mmHg or a decrease in the systolic pressure baseline of more than 40 mmHg, in individuals with sepsis.

sequential determinant

An epitope whose specificity is determined by the sequence of several residues within the antigenic determinant rather than by the antigen molecule's molecular configuration. A peptide segment of approximately six amino acid residues represents the sequential determinant structure.

sequestered antigen

An antigen that is anatomically isolated and not in contact with the immunocompetent T and B lymphoid cells of the immune system. Examples include myelin basic protein, sperm antigens, and lens protein antigens. When a sequestered antigen such as myelin basic protein is released by one or several mechanisms including viral inflammation, it can activate both immunocompetent T and B cells. An example of the sequestered antigen release mechanism of autoimmunity is found in experimental and postinfectious encephalomyelitis. Cell-mediated injury represents the principal mechanism in experimental and postviral encephalitis. In vasectomized males, antisperm antibodies are known to develop when sperm antigens become exposed to immunocompetent lymphoid cells. Likewise, lens protein of the eye that enters the circulation as a consequence of either crushing injury to an eye or exposure of lens protein to immunocompetent cells inadvertently through surgical manipulation may lead to an antilens protein immune response. Autoimmunity induced by sequestered antigens is relatively infrequent and is a relatively rare cause of autoimmune disease.

Release of sequestered antigen

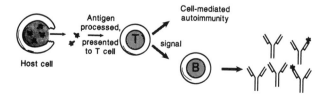

Host cell

Antigen processed, presented to T cell

T   signal

B

Cell-mediated autoimmunity

⚑ Normally sequestered antigen

Y Autoantibody to sequestered antigen

serial dilution

The successive dilution of antiserum in a row of serological tubes containing physiologic saline solution as diluent to yield the greatest concentration of antibody in the first tube and the least amount in the last tube which contains the highest dilution. For example, a double quantity of antiserum is placed in the first tube, half of which is transferred to the second tube containing an equal volume of diluent. After thorough mixing with a serological pipette, an equivalent amount is transferred to the successive tube, etc. The same volume removed from the last tube in the row is discarded. This represents a doubling dilution.

serial passage

A method to attenuate a pathogenic microorganism, but retain its immunogenicity by transfer through several animal hosts, growth media, or tissue culture cells.

seroconversion

The demonstration of specific antibody production in the serum of an individual who has been previously negative for that antibody specificity. The term frequently refers to antibodies that have been developed against such viral antigens as hepatitis B surface antigen (HBsAg) or e antigen (HBeAg) or the p24 and p41 antigens of HIV.

serological determinants

Epitopes on cells that react with specific antibody and complement, leading to fatal injury of the cells. Serological determinants are to be distinguished from lymphocyte determinants which are epitopes on the cell surface to which sensitized lymphocytes are directed, leading to cellular destruction. Although the end result is the same, antibodies and lymphocytes are directed to different epitopes on the cell surface.

serologically defined (SD) antigens

Mammalian cellular membrane epitopes that are encoded by major histocompatibility complex genes. Antibodies detect these epitopes.

serology

The study of the *in vitro* reaction of antibodies in blood serum with antigens, i.e., usually those of microorganisms inducing infectious disease.

serotherapy

A form of treatment for an infectious disease developed almost a century ago in which antiserum raised by immunizing horses or other animals against exotoxin, such as that produced by *Corynebacterium diphtheriae*, was administered to children with diphtheria. This antitoxin neutralized the injurious effects of the toxin. Thus, serotherapy was intended for prevention and treatment.

serotonin (5-hydroxytryptamine)

[5-HT]. A 176 mol wt catecholamine found in mouse and rat mast cells and in human platelets that participates in anaphylaxis in several species such as the rabbit, but not in man. It induces contraction of smooth muscle, enhances vascular permeability of small blood vessels, and induces large blood vessel vasoconstriction. 5-HT is derived from tryptophan by hydroxylation to 5-hydroxytryptophan and decarboxylation to 5-hydroxytryptamine. In man, gut enterochromaffin cells contain 90% of 5-HT, with the remainder accruing in blood platelets and the brain. 5-HT is a potent biogenic amine with wide species distribution. 5-HT may stimulate phagocytosis by leukocytes and interfere with the clearance of particles by the mononuclear phagocyte system. Immunoperoxidase staining for 5-HT, which is synthesized by various neoplasms, especially carcinoid tumors, is a valuable aid in surgical pathologic diagnosis of tumors producing it.

Serotonin

serotype

The use of specific antibodies to classify bacterial subtypes based on variations in the surface epitopes of the microorganism. Serotyping has long been used to classify *Salmonella*, streptococci, *Shigella*, and many other bacteria.

serpin

Refers to serine protease inhibitor such as C1 inhibitor.

serum

The yellow tinged fluid that forms following the coagulation of blood. Whereas plasma contains fibrinogen, serum does not. Thus, serum is the part of plasma that remains following removal of fibrinogen and clotting factors. Serum contains 35 to 55 mg/ml of serum albumin, approximately 20 mg/ml of immunoglobulins, 3 mg/ml each of transferrin and of α-1 antitrypsin, 2.5 mg/ml of α_2 macroglobulin, and 2mg/ml of haptoglobin. It is more convenient to use serum than plasma in immune reactions, as clotting might interfere with certain assays.

serum albumin

The principal protein in serum or plasma. It is soluble in water as well as in partially concentrated salt solution such as 50% saturated ammonium sulfate solution. It is coagulated by heat. It accounts for much of the plasma colloidal osmotic pressure. Serum albumin functions as a transport protein for fatty acids, bilirubin, or other large organic anions. It also carries selected hormones, including cortisol and thyroxine, and many drugs. It is formed in the liver, and levels in the serum decrease when there is protein malnutrition or significant liver and kidney disease. When the pH is neutral, albumin has a negative charge, causing its rapid movement toward the anode during electrophoresis. It is comprised of a single 585-amino acid residue chain and has a concentration of 35 to 55 mg/ml. Bovine serum albumin (BSA) and selected other serum albumins have been used as experimental immunogens in immunologic research.

serum amyloid A component (SAA)

A 12-kD protein in serum which is a precursor of the AA class of amyloid fibril protein. SAA is formed in the liver and associates with HDL3 lipoproteins in the circulation. It is a 114-amino acid residue polypeptide chain. Its conversion to AA involves splitting of peptides from both amino and carboxy terminals to yield an 8.5-kD protein that forms fibrillar amyloid deposits. There may be a 1000-fold increase in SAA levels during inflammation.

serum amyloid P component (SAP)

A protein in the serum that constitutes a minor second component in all amyloid deposits. By electron microscopy, it appears to be a doughnut-shaped pentagon with an external diameter of 9 nm and an internal diameter of 4 nm. There are five globular subunits in each pentagon. The amyloid P component in serum consists of two doughnut-shaped structures. Unlike serum amyloid A component, it does not increase during inflammation. The P component makes up 10% of amyloid deposits and is indistinguishable from normal α_1 serum glycoprotein. This 180- to 212-kD substance shows close structural homology with C-reactive protein.

serum hepatitis (hepatitis B)

An infection with a relatively long incubation period lasting between 2 and 5 months caused by a double-stranded DNA virus. Transmission may be through the administration of serum or blood products from one person to another. Additional persons at high risk include drug addicts, dialysis patients, health care workers, male homosexuals, and newborns with HIV-infected mothers. Chronic infection may occur in immunosuppressed individuals or those with lymphoid cancer. The infection may be acute or chronic, and hepatocellular carcinoma may be a complication. In acute HBV infection, the hepatocytes may undergo ballooning and eosinophilic degeneration. There may be focal necrosis of hepatocytes and lymphocytic infiltration of the portal areas and liver parenchyma. There may be central and mid-zonal necrosis of hepatocytes. Chronic type B hepatitis may progress to cirrhosis and hepatocellular carcinoma. Hepatocellular injury is induced by immunity to the virus. HBsAg, HBeAg, and IgM anti-HBc are present in the serum in acute hepatitis. In the convalescent phase of acute hepatitis, IgG anti-HBs appears in the serum. HBsAg, HBeAg, IgG anti-HBc in high titer, HBV DNA, and DNA polymerase appear in the serum in the active viral replication phase of chronic hepatitis. In the viral integration phase of chronic hepatitis, HBsAg, anti-HBc, and anti-HBe appear in the serum. Acute hepatitis B patients often recover completely. Refer to hepatitis B.

serum sickness

A systemic reaction that follows the injection of a relatively large, single dose of serum (e.g., antitoxin) into humans or other animals. It is characterized by systemic vasculitis (arteritis), glomerulo-nephritis, and arthritis. The lesions follow the deposition in tissues, such as the microvasculature, of immune complexes that form after antibody appears in the circulation between the 5th and 14th day following antigen administration. The antigen-antibody complexes fix complement and initiate a classic type III hypersensitivity reaction resulting in immune-mediated tissue injury. Patients may develop fever, lymphadenopathy, urticaria, and sometimes arthritis. The pathogenesis of serum sickness is that of a classic type III reaction. Antigen escaping into circulation from the site of injection forms immune complexes that damage the small vessels. The antibodies involved in the classic type of serum sickness are of the precipitating variety, usually IgG. They may be detected by passive hemagglutination. Pathologically, serum sickness is a systemic immune complex disease characterized by vasculitis, glomerulonephritis, and arthritis due to the intravascular formation and deposition of immune complexes that subsequently fix complement and attract polymorphonuclear neutrophils to the site through the chemotactic effects of C5a, thereby initiating inflammation. The classic reaction which occurs 7 to 15 days after the triggering injection is called the primary form of serum sickness. Similar manifestations appearing only 1 to 3 days following the injection represent the accelerated form of serum sickness and occur in subjects presumably already sensitized. A third form, called the anaphylactic form, develops immediately after injection. This latter form is apparently due to reaginic IgE antibodies and usually occurs in atopic subjects sensitized by horse dander or by previous exposure to serum treatment. The serum sickness-like syndromes seen in drug allergy have a similar clinical picture and similar pathogenesis. Refer also to type III immune complex-mediated hypersensitivity.

Experimental Serum Sickness

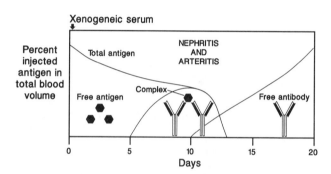

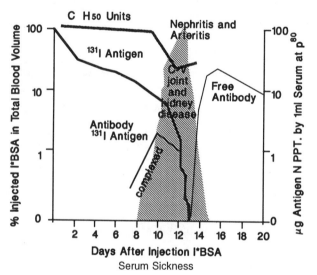

Serum Sickness

serum spreading factors

In plasma, 65- and 75-kD glycoproteins that facilitate adherence of cells as well as the ability of cells to spread and to differentiate. Refer to vitronectin.

serum thymic factor (STF)

An obsolete term for thymulin.

serum virus vaccination

A method no longer used that consisted of administering an immunizing preparation of an infectious agent such as a live vaccine virus together with an antiserum specific for the virus. It was intended to ameliorate the effects of the live infectious agent. This method was abandoned because it was considered a dangerous practice.

severe combined immunodeficiency syndrome (SCID)

Profound immunodeficiency characterized by functional impairment of both B and T lymphocyte limbs of the immune response. It is inherited as an X-linked or autosomal recessive disease. The thymus has only sparse lymphocytes and Hassal's corpuscles or is bereft of them. Several congenital immunodeficiencies are characterized as SCID. There is T and B cell lymphopenia and decreased production of IL-2. There is an absence of delayed-type hypersensitivity, cellular immunity, and of normal antibody synthesis following immunogenic challenge. SCID is a disease of infancy with failure to thrive. Affected individuals frequently die during the first 2 years of life. Clinically, they may develop a measles-like rash, show hyperpigmentation, and develop severe recurrent (especially pulmonary) infections. These subjects have heightened susceptibility to infectious disease agents such as *Pneumocystis carinii, Candida albicans,* and others. Even attenuated microorganisms, such as those used for immunization, e.g., attenuated poliomyelitis viruses, may induce infection in SCID patients. Graft-vs.-host disease is a problem in SCID patients receiving unirradiated blood transfusions. Maternal-fetal transfusions during gestation or at parturition or blood transfusions at a later date provide sufficient immunologically competent cells entering the SCID patient's circulation to induce graft-vs.-host disease. SCID may be manifested in one of several forms. SCID is classified as a defect in adenosine deaminase (ADA) and purine nucleoside phosphorylase (PNP) enzymes and in a DNA-binding protein needed for HLA gene expression. Treatment is by bone marrow transplantation or by gene therapy and enzyme reconstitution in those cases caused by a missing gene, such as adenosine deaminase deficiency.

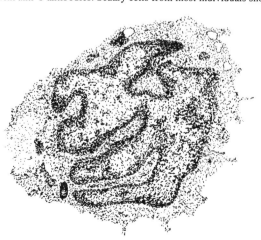

Severe Combined Immunodeficiency Syndrome (SCID)

Sezary cell

T lymphocytes that form E rosettes with sheep red cells and react with anti-T antibodies. Sezary cells from most individuals show a diminished response to plant mitogens, although those from a few demonstrate normal reactivity. Sezary cells are also poor mediators of T cell cytotoxicity, but they can produce a migration inhibitory factor (MIF)-like lymphokine. Sezary cells neither produce immunoglobulin nor do they act as suppressors; however, they have a helper effect for immunoglobulin synthesis by B cells.

Sezary syndrome

A disease that occurs in middle age, affecting males more commonly than females. It is a neoplasm, i.e., a malignant lymphoma of CD4+ T helper lymphocytes with prominent skin involvement. There is a generalized erythroderma, hyperpigmentation, and exfoliation. Fissuring and scaling of the skin on the palms of the hands and soles of the feet may occur. The peripheral blood and lymph nodes contain the typical cerebriform cells that have a nucleus that resembles the brain. There is extensive infiltration of the skin by leukocytes, with prominent clustering in the epidermis forming Pautrier's evidence. Late in the disease, T immunoblasts may appear. The so-called Sezary cells are T lymphocytes.

sheep red blood cell agglutination test

An assay in which sheep erythrocytes are either agglutinated by antibody or are used as carrier particles for an antigen adsorbed to their surface, in which case they are passively agglutinated by antibodies specific for the adsorbed antigen.

shingles (Herpes zoster)

A virus infection that occurs in a band-like pattern according to distribution in the skin of involved nerves. It is usually a reactivation of the virus that causes chickenpox.

shock organ

A particular organ involved in a specific reaction such as an anaphylactic reaction.

shocking dose

The amount of antigen required to elicit a particular clinical response or syndrome.

short-lived lymphocytes

Lymphocytes with a lifespan of 4 to 5 days, in contrast to long-lived lymphocytes which may last from months to years in the blood circulation.

Shwartzman, Gregory (1896–1965)

Russian–American microbiologist who described local and systemic reactions that follow injection of bacterial endotoxins. The Shwartzman reaction is related to disseminated intravascular coagulation. *Phenomenon of Local Tissue Reactivity and Its Immunological and Clinical Significance,* 1937.

Shwartzman (or Shwartzman-Sanarelli) reaction

A nonimmunologic phenomenon in which endotoxin (lipopolysaccharides) induces local and systemic reactions. Following the initial or preparatory injection of endotoxin into the skin, polymorphonuclear leukocytes accumulate and are then thought to release lysosomal acid hydrolases that injure the walls of small vessels, preparing them for the second provocative injection of endotoxin. The intradermal injection of endotoxin into the skin of a rabbit followed within 24 h by the intravenous injection of the same or a different endotoxin leads to hemorrhage at the local site of the initial injection. Although the local Shwartzman reaction may resemble an Arthus reaction in appearance, the Arthus reaction is immunological, whereas the Shwartzman reaction is not. In the Shwartzman

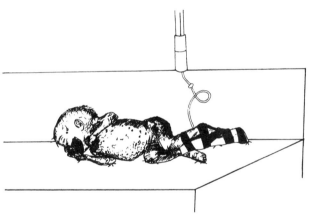

Lymphocyte from a patient with Sezary Syndrome

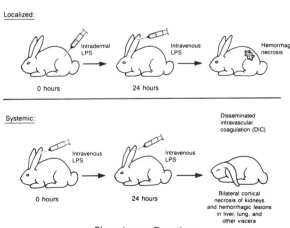

Shwartzman Reaction

reaction, there is insufficient time between the first and second injections to induce an immune reaction in a previously unsensitized host. There is also a lack of specificity since even a different endotoxin may be used for first and second injections. The generalized or systemic Shwartzman reaction again involves two injections of endotoxin. However, both are administered intravenously, one 24 h following the first. The generalized Shwartzman reaction is the experimental equivalent of disseminated intravascular coagulation that occurs in a number of human diseases. Following the first injection, sparse fibrin thrombi are formed in the vasculature of the lungs, kidney, liver, and capillaries of the spleen. There is blockade of the reticuloendothelial system as its mononuclear phagocytes proceed to clear thromboplastin and fibrin. Administration of the second dose of endotoxin, while the reticuloendothelial system is blocked, leads to profound intravascular coagulation since the mononuclear phagocytes are unable to remove the thromboplastin and fibrin. There is bilateral cortical necrosis of the kidneys and splenic hemorrhage and necrosis. Neither platelets nor leukocytes are present in the fibrin thrombi that are formed.

sia test (historical)

A former qualitative test for macroglobulinemia in which the patient's serum was placed in water in one tube and in saline in another tube. Precipitation of the serum in water, attributable to IgM's low water solubility, but not in saline constituted a positive test.

sialophorin (CD43)

A principal glycoprotein present on the surface of thymocytes, T cells, selected B lymphocytes, neutrophils, platelets, and monocytes. Monocyte and lymphocyte sialophorin is a 115-kD polypeptide chain, whereas the platelet and neutrophil sialophorin is a 135-kD polypeptide chain that only has a different content of carbohydrate from the first form. Galactose β 1-3 galactosamine in O-linked saccharides bound to threonine or serine amino acid residues represents a site of attachment for sialic acid in thymocytes in the medulla and in mature T cells. Incomplete sialylation of thymocytes in the thymic cortex accounts for their binding to peanut lectin, whereas the more thoroughly sialylated structure on T cells and thymocytes in the medulla fails to bind the lectin. In man, the sialophorin molecule is comprised of 400 amino acid residues. There is a 235-residue extracellular domain, a 23-residue transmembrane portion, and a 123-residue domain in the cytoplasm. Antibodies specific for CD43 can activate T cells. Wiskott-Aldrich syndrome patients have T cells with defective sialophorin.

sicca complex

A condition characterized by dryness of mucus membranes, especially of the eyes producing keratoconjunctivitis attributable to decreased tearing that results from lympocytic infiltration of the lacrimal glands, and by dry mouth (xerostomia), associated with decreased formation of saliva as a result of lymphocytic infiltration of the salivary glands producing obstruction of the duct. It is most frequently seen in patients with Sjögren's syndrome, but it also may be seen in cases of sarcoidosis, amyloidosis, hemochromatosis, deficiencies of vitamins A and C, scleroderma, and hyperlipoproteinemia types IV and V.

side chain theory

Concept proposed by Paul Ehrlich in 1899 which postulated that a cell possessed highly complex chemical aggregates with attached groupings, or "side chains", whose normal function was to anchor nutrient substances to the cell prior to internalization. These side chains, or receptors, were considered to permit cellular interaction with substances in the extracellular environment. Antigens were postulated to stimulate the cell by attachment to these receptors. Since antigens played no part in the normal economy of the cell, the receptors were diverted from their normal function. Stimulated by this derangement of its normal mechanism, the cell produced excessive new receptors, of the same type as those thrown out of action. The superfluous receptors were shed into the extracellular fluids and constituted specific antibodies with the capacity to bind homologous antigens. Ehrlich proposed a haptophore group that reacted with a corresponding group of an antigen, as in neutralization of toxin by antitoxin. For reactions such as agglutination or precipitation, he postulated another group termed the ergophobe group which determined the change in antigen after the antibody was anchored by its haptophore group. Ehrlich proposed receptors with two haptophore groups: one that

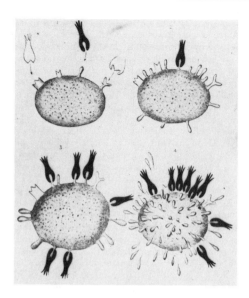

DIAGRAMMATIC REPRESENTATION OF THE SIDE-CHAIN THEORY
(PLATES I AND II)

Fig. 1   "The groups [the haptophore group of the side-chain of the cell and that of the food-stuff or the toxin] must be adapted to one another, *e.g.*, as male and female screw (PASTEUR), or as lock and key (E. FISCHER)."

Fig. 2   ". . . the first stage in the toxic action must be regarded as being the union of the toxin by means of its haptophore group to a special side-chain of the cell protoplasm."

Fig. 3   "The side-chain involved, so long as the union lasts, cannot exercise its normal, physiological, nutritive function . . ."

Fig. 4   "We are therefore now concerned with a defect which, according to the principles so ably worked out by . . . Weigert, is . . . [overcorrected] by regeneration."

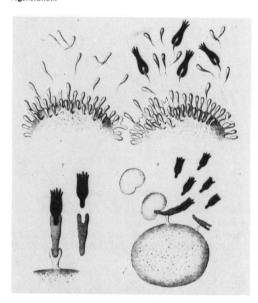

DIAGRAMMATIC REPRESENTATION OF THE SIDE-CHAIN THEORY
(cont.)

Fig. 5   ". . . the antitoxins represent nothing more than the side-chains, reproduced in excess during regeneration and therefore pushed off from the protoplasm—thus coming to exist in a free state."

Fig. 6   [The free side-chains (circulating antitoxins) unite with the toxins and thus protect the cell.]

Fig. 7   ". . . two haptophore groups must be ascribed to the 'immune-body' [haemolytic amboceptor], one having a strong affinity for a corresponding haptophore group of the red blood corpuscles, . . . and another . . . which . . . becomes united with the 'complement' . . ."

Fig. 8   "If a cell . . . has, with the assistance of an appropriate side-chain, fixed to itself a giant [protein] molecule . . . there is provided [only] one of the conditions essential for the cell nourishment. Such . . . molecules . . . are not available until . . . they have been split into smaller fragments. This will be . . . attained if . . . the 'tentacle' . . . possesses . . . a second haptophore group adapted to take to itself ferment-like material

Side Chain Theory

attached to antigen and the other to complement. The group that combined with the cell or other antigen was called the cytophilic group, and the group that combined with complement was the complementophilic group. Ehrlich named this type of receptor an amboceptor because both groups were supposed to be of the haptophore type. He considered toxins to have a haptophore group and a toxophore group. Detoxification without loss of antitoxin-binding capacity led Ehrlich to believe that a toxophore group had been altered, while the haptophore group remained intact. Ehrlich's theory assumed that each antibody-forming cell had the ability to react to every antigen in nature. The demonstration by Landsteiner that antibodies could be formed against substances manufactured in the laboratory that had never existed before in nature led to abandonment of the side chain theory. Yet its basic premise as a selective hypothesis rather than an instructive theory was ultimately proven correct.

side effect
An unwanted reaction to a drug administered for some desirable curative or other effect.

signal hypothesis
A proposed mechanism for selection of secretory proteins by and for transport through the rough endoplasmic reticulum. The free heavy and light chain leader peptide is postulated to facilitate the joining of polyribosomes forming these molecules to the endoplasmic reticulum. It also refers to the release of heavy and light polypeptide chains through the endoplasmic reticulum membrane into the cisternal space followed by immunoglobulin secretion once the immunoglobulin molecule has been assembled. This refers to the pathways whereby proteins reach their proper cellular destinations. Important to protein targeting is the signal sequence, which is a short amino acid sequence at the amino terminus of a polypeptide chain. This signal sequence directs the protein to the proper destination in the cell and is removed either during passage or when the protein arrives at its final location.

signal pathway DNA-binding proteins
Because DNA-binding proteins are made in genes distant to their sites of activity, they are termed trans-acting. These are transcription factors that interact with enhancers and promoters by binding to unique nucleotide sequences. Due to these interactions, DNA-binding proteins regulate the transcription of the genes under the influence of each promoter or enhancer. These proteins are an important means of signal transduction through which external stimuli induce responses at the nuclear level.

signal peptide
Signal sequence.

signal sequence
Refer to signal hypothesis.

"silencer" sequence
Blocks transcription of the T cell receptor α chain. This sequence is found 5' to the α chain enhancer in non-T cells and in those with γδ receptors.

silica adjuvants
In the past, silica crystals and hydrated aluminum silicate (bentonite) were occasionally used to enhance the immune response to certain antigens. They were considered to have a central action on the immune system rather than serving as depot adjuvants, yet bentonite can delay the distribution of antigen from its site of inoculation by surface adsorption.

silicosis.
The inhalation of silica particles over a prolonged period of time produces a chronic, nodular, densely fibrosing pneumoconiosis that has an insidious onset and progresses even in the absence of continued exposure to silica dust. Lymphocytes and alveolar macrophages are quickly attracted to the particles which are phagocytized by the macrophages. Some macrophages remain in the interstitial tissue or in the pulmonary lymphatic channels. Interaction of macrophages and silica particles leads over time to collagenous fibrosis and fibrosing nodules. The silica dust-macrophage interaction may cause the secretion of IL-1 which recruits T helper cells that produce IL-2 which induces proliferation of T lymphocytes. They in turn produce a variety of lymphokines. Thus, immunocompetent cells mediate the collagenous reaction. Activated T cells interact with B lymphocytes, which synthesize increased amounts of IgG, IgM, rheumatoid factor, antinuclear antibodies, and circulating immune complexes. Collagenous silicotic nodules may coalesce, producing fibrous scars. The disease may be complicated by

the development of rheumatoid arthritis, pulmonary tuberculosis, emphysema, or several other diseases, but there is no increased incidence of lung cancer in silicosis.

simian immunodeficiency virus (SIV)
Causes a disease resembling human AIDS in rhesus monkeys. The SIV sequence reveals significant homology with HIV-2, a cause of AIDS in western Africa.

Simonsen phenomenon
A graft-vs.-host reaction in chick embryos that have developed splenomegaly following inoculation of immunologically competent lymphoid cells from adult chickens. Splenic lymphocytes are increased and represent a mixture of both donor and host lymphocytes.

simple allotype
An allotype that is different from another allotype in the sequence of amino acids at a single or several positions. Alleles at one genetic locus often encode simple allotypes.

single diffusion test
A type of gel diffusion test in which antigen and antibody are placed in proximity with one another and one of them diffuses into the other, leading to the formation of a precipitate in the gel. This is in contrast to double diffusion in which both antigen and antibody diffuse toward each other.

single domain antibodies
Antibodies capable of binding epitopes with high affinity even though they do not possess light chains. They are cloned from heavy chain variable regions and can be produced in days to weeks in contrast to monoclonal antibodies which require weeks to months to develop. Their relatively small size is a further advantage. Single domain antibodies with antigen-specific VH domains are expected to find wide application in the future.

single hit theory
The hypothesis that hemolysis of red blood cells sensitized with specific antibody is induced by perforation of the membrane by complement at only one site rather than at multiple sites on the membrane.

single immunodiffusion
(1) A technique in which antibody is incorporated into agar gel and antigen is placed in a well that has been cut into the surface of the antibody-containing agar. Following diffusion of the antigen into the agar, a ring of precipitation forms at the point where antigen and antibody have reached equivalence. The diameter of the ring is used to quantify the antigen concentration by comparison with antigen standards. (2) The addition of antigen to a tube containing gel into which specific antibody has been incorporated. Lines of precipitation form at the site of interaction between equivalent quantities of antigen and antibody.

Single Immunodiffusion
(Mancini Technique)

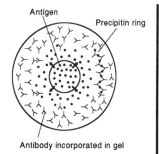
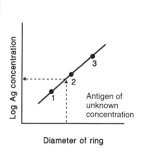

single locus probes (SLPs)
Probes which hybridize at only one locus. These probes identify a single locus of variable number of tandem repeats (VNTRs) and permit detection of a region of DNA repeats found in the genome only once and located at a unique site on a certain chromosome. Therefore, an individual can have only two alleles that SLPs will identify, as each cell of the body will have two copies of each chromosome, one from the mother and the other from the father. When the lengths of related alleles on homologous chromosomes are the same, there will be only a single band in the DNA typing pattern. Therefore, the use of an SLP may yield either a single- or

double-band result from each individual. Single locus markers such as the pYNH24 probe developed by White may detect loci that are highly polymorphic, exceeding 30 alleles and 95% heterozygosity. SLPs are used in resolving cases of disputed parentage.

single radial immunodiffusion

A technique to quantify antigens. Plates are poured in which antibody is incorporated into agar, wells are cut, and precise quantities of antigen are placed in the wells. The antigen is permitted to diffuse into the agar containing antibody and produce a ring of precipitation upon interaction with the antibody. As diffusion proceeds, an excess of antigen develops in the area of the precipitate causing it to dissolve only to form once again at a greater distance from the site of origin. At the point where antigen and antibody have reached equivalence in the agar, a precipitation ring is produced. The precipitation ring encloses an area proportional to the concentration of antigen measured 48 to 72 h following diffusion. Standard curves are employed using known antigen standards. The antigen concentration is determined from the diameter of the precipitation ring. This method can detect as little as 1 to 3 μg/ml of antigen. Known also as the Mancini technique.

Radial Immunodiffusion

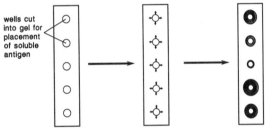

wells cut into gel for placement of soluble antigen

Agarose gel with antibody throughout

Antigen diffuses, a circular concentration gradient is produced

Precipitin rings form, antigen concentration is related to size of ring

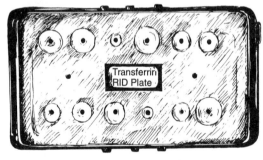

Transferrin RID Plate

Single Radial Immunodiffusion

Sips distribution

Frequency distribution of antibody association constants in a heterogeneous mixture. The Sips distribution is very similar to the Gaussian (normal) distribution. It is employed to analyze data from antigen-antibody reactions measured by equilibrium dialysis.

Sips plot

Data representation produced in assaying ligand binding to antibodies by plotting log $r/n-r$ against log c. A straight line signifies that the data are in agreement with the Sips equation. In this instance, the slope signifies heterogeneity of antibody affinity.

SIRS

Abbreviation for soluble immune response suppressor.

site-directed mutagenesis

A laboratory procedure that involves the substitution of amino acids in a protein whose function is defined for the purpose of localizing a certain activity.

SIV (simian immunodeficiency virus)

A lentivirus of primates that resembles HIV-1 and HIV-2 in morphology and attraction to cells that bear CD4 molecules such as lymphocytes and macrophages. SIV also shares with these human viruses the additional genes lacking in other retroviruses which include *vip, rev, upr, tat,* and *nef.* This virus induces the classic cytopathologic alterations of the type produced by HIV, and it can also induce chronic disease following a lengthy latency.

SIVmac239 is an SIV clone that induces a disease resembling AIDS in monkeys.

Sjögren's syndrome

A condition in which immunologic injury to the lacrimal and salivary glands leads to dry eyes (keratoconjunctivitis sicca) and dry mouth (xerostomia). It may occur alone as sicca syndrome (primary form) or together with an autoimmune disease such as rheumatoid arthritis (secondary form). The lacrimal and salivary glands show extensive lymphocytic infiltration and fibrosis. Most of the infiltrating cells are CD4+ T cells, but there are some B cells (plasma cells) that form antibody. Approximately 75% of the patients form rheumatoid factor. The LE cell test is positive in 25% of the patients. Numerous antibodies are produced, including autoantibodies against salivary duct cells, gastric parietal cells, thyroid antigens, and smooth muscle mitochondria. Antibodies against ribonucleoprotein antigens are termed SS-A (Ro) and SS-B (La). Approximately 90% of the patients have these antibodies. Anti-SS-B shows greater specificity for Sjögren's syndrome than do anti-SS-A antibodies which also occur in SLE. Sjögren's syndrome patients who also have rheumatoid arthritis may have antibodies to rheumatoid-associated nuclear antigen (RANA). There is a positive correlation between HLA-DR3 and primary Sjögren's syndrome and between HLA-DR4 and secondary Sjögren's syndrome associated with rheumatoid arthritis. Genetic predisposition, viruses, and disordered immunoregulation may play a role in the pathogenesis. About 90% of the patients are 40- to 60-year-old females. In addition to dry eyes and dry mouth, with associated visual or swallowing difficulties, 50% of the patients show parotid gland enlargement. There is drying of the nasal mucosa with bleeding, pneumonitis, and bronchitis. A lip biopsy to examine minor salivary glands is needed to diagnose Sjögren's syndrome. Inflammation of the salivary and lacrimal glands was previously called Mikulicz's disease. Mikulicz's syndrome refers to enlargement of the salivary and lacrimal glands due to any cause. Enlarged lymph nodes that reveal a pleomorphic cellular infiltrate with many mitoses are typical of Sjögren's syndrome and have been referred to as "pseudolymphoma". These patients have a 40-fold greater risk than do others of lymphoid neoplasms.

skin-fixing antibody

An antibody such as IgE that is retained in the skin following local injection, as in passive cutaneous anaphylaxis. Antibody with this property was referred to previously as reagin before IgE was described.

skin graft

Skin from the same individual (autologous graft) or donor skin that is applied to areas of the body surface that have undergone third degree burns. A patient's keratinocytes may be cultured into confluent sheets that can be applied to the affected areas, although these may not "take" because of the absence of type IV collagen 7 S basement membrane sites for binding and fibrils to anchor the graft.

skin-reactive factor (SRF)

A lymphokine that induces vasodilation and increased vascular permeability when injected into the skin.

skin-sensitizing antibody

An antibody, usually of the IgE class, that binds to the Fc receptors of mast cells in the skin, thereby conditioning this area of skin for a type I immediate hypersensitivity reaction following crosslinking of the IgE Fab regions by a specific allergen (antigen). In guinea pig skin, human IgG1 antibodies may be used to induce passive cutaneous anaphylaxis.

skin-specific histocompatibility antigen

A murine skin minor histocompatibility antigen termed Sk that can elicit rejection of skin, but not other tissues following transplantation from one parent into the other parent that has been irradiated and rendered a chimera by the previous injection of F₁ spleen cells. The two parents are from different inbred strains of mice. The rate of rejection is relatively slow. Immunologic tolerance of F₁ murine spleen cells to the skin epitope of the parent in which they are not in residence is abrogated following residence in the opposite parent.

skin test

Any of several assays in which a test substance is either injected into the skin or applied to it to determine the host response. Skin tests have long been used to determine host hypersensitivity or immunity to a particular antigen or product of a microorganism. Examples include the tuberculin test, the Schick test, the Dick test, the patch test, the scratch test, etc.

skin window

A method to observe the sequential changes in types of cells during the development of acute inflammation. Following superficial abrasion of an area of skin, sterile cover slips are applied, removed at specified intervals thereafter, stained, and observed microscopically for the types of cells present. The first cells to appear are polymorphonuclear neutrophils which comprise most of the cell population within 3 to 4 h of the induced injury. By contrast, the cover slip removed after 12 h reveals the presence of mononuclear cells such as lymphocytes, plasma cells, and monocytes. The cover slip removed after 24 h reveals predominantly monocytes and macrophages. Also termed a Rebuck window, named for the individual who perfected the method.

skin window of Rebuck

Refer to skin window.

SLE

Abbreviation for systemic lupus erythematosus.

slide agglutination test

The aggregation of particulate antigen such as red blood cells, microorganisms, or latex particles coated with antigen within 30 s following contact with specific antibody. The reactants are usually mixed by rocking the slide back and forth, and agglutination is observed both macroscopically and microscopically. The test has been widely used in the past for screening, but is unable to distinguish reactions produced by cross-reacting antibodies which can be ruled out in a tube test that allows dilution of the antiserum.

slide flocculation test

Refer to slide agglutination test.

slot blot analysis

A quick technique to detect gene amplification by determining a solution's DNA content by electrophoresis. The technique is closely similar to dot blot analysis, except that a slot instead of a punched-out hole is cut in the agar.

slow-reacting substance of anaphylaxis (SRS-A)

A 400 mol wt acidic lipoprotein derived from arachidonic acid that induces the slow contraction of bronchial smooth muscle and is produced following exposure to certain antigens. It is comprised of leukotrienes LTC_4, LTD_4, and LTE_4, which produce the effects observed in anaphylactic reactions. It is released, *in vitro*, in effluents from synthesized lung tissue of guinea pig, rabbit, and rat profused with antigen. It has also been demonstrated in human lung tissue and nasal polyps. It contracts smooth muscle of guinea pig ileum. *In vitro*, it also increases vascular permeability upon intracutaneous injection and decreases pulmonary compliance by a mechanism independent of vagal reflexes. It also enhances some of the smooth muscle effects of histamine. The source of SRS-A is mast cells and certain other cells. It is found in immediate (type I) hypersensitivity reactions. SRS-A is not stored in a preformed state and is sequentially synthesized and released. The effects have a latent period before becoming manifest. Antihistamines do not neutralize the effects of SRS-A.

slow viruses

Agents that induce infectious encephalitis following a lengthy latency. Slow viruses consist of conventional viruses and prions that are comprised of subverted cell proteins. Among the conventional group is measles, which induces subacute sclerosing panencephalitis, papovavirus, which induces progressive multifocal leukoencephalopathy, and rubella, which induces rare progressive rubella panencephalitis. The agents that cause kuru and Creutzfeldt-Jakob disease are among the nonconventional group of slow viruses.

slp

Abbreviation for sex-limited protein that is encoded by genes at the C4A complement locus in mice. It is restricted to males but may be induced in females by androgen administration.

Sm (anti-Smith) antibodies

Antibodies that occur in approximately 29% of systemic lupus erythematosus patients and are quite specific for this disease. They are specific for polypeptides of U1, U2, and U4-6 small nuclear ribonucleoproteins (snRNPs), which are significant in pre-mRNA splicing. Anti-Sm antibodies do not show the increase in titer observed with antidouble-stranded DNA antibodies during exacerbations of the disease.

small "blues"

Blue aggregates of acellular debris observed in clinical histocompatibility testing using the microlymphotoxicity test. It occurs in the wells of tissue typing trays and is due to an excess amount of trypan blue mixed with protein. This is a technical artifact.

small lymphocyte

One of the five types of leukocytes in the peripheral blood that measures 6 to 8 μm in diameter. In Wright's and Giemsa-stained blood smears, the nucleus stains dark blue and is encircled by a narrow rim of robin's egg blue cytoplasm. Even though most all of the lymphocytes look alike, they differ greatly in origin and function. They differ in other features as well. By light microscopy, T and B lymphocytes and the E rosette subpopulations look the same. However, they have different phenotypic surface markers and differ greatly in function.

smallpox

Refer to variola.

smallpox vaccination

The induction of active immunity against smallpox (variola) by immunization with a related agent, vaccinia virus, obtained from vaccinia vesicles on calf skin. Shared, and therefore crossreactive, epitopes in the vaccinia virus provide protective immunity against smallpox. This ancient disease has now been eliminated worldwide with the only laboratory stocks maintained in Atlanta (USA) and Moscow (Russia). These stocks are supposed to have been destroyed by both agencies once the virus was sequenced. Smallpox vaccination was first developed by the English physician Edward Jenner, whose method diminished the mortality from 20% to less than 1%. Following application of vaccinia virus by a multiple pressure method, vesicles occur at the site of application within 6 to 9 days. Maximum reactivity is observed by day 12. Initial vaccinations were given to 1- or 2-year-old infants with revaccination after 3 years. Children with cell-mediated immunodeficiency syndromes sometimes developed complications such as generalized vaccinia spreading from the site of inoculation. Postvaccination encephalomyelitis also occurred occasionally in adults and in babies prior to their first birthday. The procedure was contraindicated in subjects experiencing immunosuppression due to any cause.

smooth muscle antibodies

Autoantibodies belonging to the IgM or IgG class that are found in the blood sera of 60% of chronic active hepatitis patients. In biliary cirrhosis patients, 30% may also be positive for these antibodies. Low titers of smooth muscle antibodies may be found in certain viral infections of the liver.

SNagg

Agglutinating activity by rheumatoid factor in certain normal sera, as revealed by a positive RA (rheumatoid arthritis) test.

sneaking through

The successful growth of a sparse number of transplantable tumor cells that have been inoculated into a host in contrast to the induction of tumor immunity and lack of tumor growth in the same host if larger doses of the same cells are administered.

Snell-Bagg mice

A mutant strain (dw/dw) of inbred mice with pituitary dwarfism that have a diminutive thymus, deficient lymphoid tissue, and decreased cell-mediated immune responsiveness.

Snell, George Davis (1903–)

American geneticist who shared the 1980 Nobel Prize with Dausset and Benacerraf "for their work on genetically determined structures of the cell surface that regulate immunologic reactions". Snell's major contributions were in the field of mouse genetics including discovery of the H-2 locus and the development of congenic mice. *Histocompatibility (with Dausset and Nathenson)*, 1976.

SOD

Abbreviation for superoxide dismutase.

solid-phase radioimmunoassay

The attachment of antigen (or antibody) to an insoluble support, which can be used to capture antibodies (or antigens) in a specimen to be assayed. Antibodies in a serum sample are exposed to excess antigen on an insoluble support and sufficient time is allowed for antigen-antibody interaction. This is followed by washing and the application of radiolabelled anti-Fc antibodies specific for the Fc regions of the captured antibodies. After washing, quantification of the bound antibody is determined from the amount of radioactivity adhering to the insoluble support. Various materials may be used as an appropriate insoluble support. These include Sepharose® beads or tissue culture plate wells. An unrelated protein must be used to coat the insoluble support prior to application of the specific antibody to saturate areas of the insoluble support where antigen is not located.

solubilized water-in-oil adjuvant

A water-in-oil emulsion adjuvant comprised of a small volume aqueous phase compared to the volume of oil. A mixture of aqueous and oil phases results in an emulsion which is stabilized by the addition of emulsifying agents.

soluble antigen

An antigen solubilized in an aqueous medium.

soluble complex

An immune (antigen-antibody) complex formed in excess antigen and rendered soluble. Antigen excess prevents lattice formation either *in vitro* or *in vivo*. Soluble complexes may produce tissue injury *in vivo*, which is more severe if complement has been fixed. C5a attracts neutrophils, and there is increased capillary permeability. PMNs, platelets, and fibrin are deposited on the endothelium. This is followed by thrombosis and necrosis. Immune complexes induce type III hypersensitivity reactions.

somatic antigen

An antigen such as the O antigen which is part of a bacterial cell's structure.

somatic mutation

A genetic variation in a somatic cell which is heritable by its progeny. Increased somatic mutation enhances diversity of an antibody molecule's light and heavy chain variable regions. IgG and IgA antibodies reveal somatic mutations more often than do IgM antibodies.

Southern blotting

A procedure to identify DNA sequences. Following extraction of DNA from cells, it is digested with restriction endonucleases to cut DNA at precise sites into fragments. This is followed by separation of the DNA segments according to size by electrophoresis in agarose gel, denaturation with sodium hydroxide, and transfer of the single-stranded DNA to a nitrocellulose membrane by blotting. This is followed by hybridization with an ^{35}S- or ^{32}P-radiolabeled probe of complementary DNA. Alternatively, a biotinylated probe may be used. Autoradiography or substrate digestion identifies the location of the DNA fragments that have hybridized with the complementary DNA probe. Specific sequences in cloned and in genomic DNA can be identified by Southern blotting. Whereas DNA analysis is referred to as Southern blot, RNA analysis is referred to as a Northern blot, and protein analysis is referred to as a Western blot. A Northwestern blot is one in which RNA-protein hybridizations are formed.

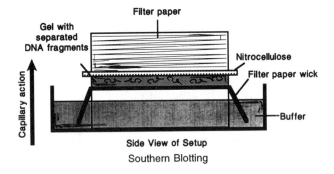

Side View of Setup
Southern Blotting

Southwestern blot

A method that combines Southern blotting that identifies DNA segments, with Western immunoblotting that characterizes proteins. A protein may be hybridized to a molecule of single-stranded DNA bound to the membrane. Southwestern blotting is helpful in delineating nuclear transcription-related proteins.

SP-40,40

A heterodimeric serum protein derived from soluble C5-9 complexes that may modulate the membrane attack complex's cell lysing action.

species specificity

In immunology, cellular or tissue antigens present in one species only and not found in other species.

specific granule

A secondary granule in the cytoplasm of polymorphonuclear leukocytes that contains lysozyme, vitamin B_{12}-binding protein, neutral proteases, and lactoferrin. It is smaller and fuses with phagosomes more quickly than does the azurophil granule.

specific immunity

An immune state in which antibody or specifically sensitized or primed lymphocytes recognize an antigen and react with it. By contrast, immunologically competent cells may interact with antigen to produce specific immunosuppression termed immunologic tolerance.

specificity

Recognition by an antibody or a lymphocyte receptor of a specific epitope in the presence of other epitopes for which the antigen-binding site of the antibody or of the lymphoid cell receptor is specific.

speckled pattern

A type of immunofluorescence produced when the serum from individuals with one of several connective tissue diseases is placed in contact with the human epithelial cell line HEp-2 and "stained" with fluorochrome-labeled goat or rabbit antisera against human immunoglobulin. The speckled pattern of fluorescence occurs in mixed connective tissue disease, lupus erythematosus, polymyositis, sicca syndrome, Sjögren's syndrome, drug-induced immune reactions, and rheumatoid arthritis. It is the most frequent and shows the greatest variations of immunofluorescent nuclear staining patterns. It is classified as (1) fine speckles associated with anticentromere antibody; (2) coarse speckles associated with antibodies against the nonhistone nuclear proteins Scl-70, nRNP, SS-B/La, and Sm; and (3) large speckles that may be limited to three to ten per nucleus that are seen in undifferentiated connective tissue disease and represent IgM antibody against class H3 histones.

spectrotype

In isoelectric focusing analysis, the arrangement of bands in a gel that is characteristic for either one protein or a category of proteins. An antibody spectrotype on an isoelectric focusing gel may signify that it is the product of a particular antibody-synthesizing clone.

sperm antibody

Antibodies specific for the head or tail of sperm. These antibodies are synthesized by 3% of infertile males and 2 to 9% of infertile females. There is a positive correlation between the titer of antibody against sperm and the couple's infertility. Treatment includes corticosteroid therapy or use of a condom to permit waning of immunologic memory in the female or washing of sperm prior to insemination.

spheroplast

A Gram-negative bacterial protoplast that contains outer membrane remains.

spherulin

An antigen derived from spherules of *Coccidioides immitis* that has been used for the delayed-type hypersensitivity skin test for coccidioidomycosis.

spleen

An encapsulated organ in the abdominal cavity which has important immunologic and nonimmunologic functions. Vessels and nerves enter the spleen at the hilum, as in lymph nodes, and travel part of their course within the fibrous trabeculae that emerge from the capsule. The splenic parenchyma has two regions that are functionally and histologically distinct. The white pulp consists of a thick layer of lymphocytes surrounding the arteries that have left the trabeculae. They form a periarterial sheath which contains mainly T cells. The sheaths then expand along their course to form well-developed lymphoid nodules called Malpighian corpuscles. The red pulp consists of a mesh of reticular fibers, continuous with the collagen fibers of the trabeculae. These fibers enclose an open system of sinusoids that drain into small veins and are lined by endothelial cells with reticular properties. The endothelium is discontinuous, leaving small slits through which cells have to pass during transit. Within the sinusoidal mesh are red blood cells, macrophages, lymphocytes, and plasma cells. The red pulp between adjacent sinusoids forms the pulp cords which are sometimes called the cords of Billroth. The marginal zone consists of a poorly defined area between the white and the red pulp where the periarterial sheath and the lymphoid nodules merge. The blood vessels branch, and at the periphery of the marginal zone, the blood empties into the pulp. Lymphocytes of the marginal zone are mainly T cells. They surround the periphery of the lymphoid nodules that comprise B cells. In this marginal region, the T and B cells contact each other. Some B cells may convert into immunoblasts. Further maturation to plasma cells occurs in the red pulp. Active follicles contain germinal centers in which lymphoblasts may be generated. They are discharged into sinusoids, and plasma cells may form. The spleen also contains dendritic cells which have long cytoplasmic extensions. Dendritic cells serve as antigen-presenting cells, interacting with lymphocytes. The spleen filters blood as the lymph nodes filter the lymph. The spleen is active in the formation of antibodies against intravenously administered particulate antigens. It has numerous additional functions, including the sequestration and destruction of senescent red blood cells, platelets, and lymphocytes.

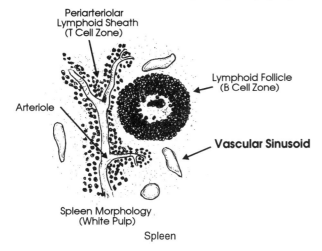

Spleen Morphology
(White Pulp)

Spleen

split thickness graft

A skin graft that is only 0.25 to 0.35 mm thick that consists of epidermis and a small layer of dermis. These grafts vascularize rapidly and last longer than do regular grafts. They are especially useful for skin burns, contaminated skin areas, and sites that are poorly vascularized. Thick split thickness grafts are further resistant to trauma, produce minimal contraction, and permit some amount of sensation, but graft survival is poor.

split tolerance

(1) Specific immunological unresponsiveness (tolerance) affecting either the B cell (antibody) limb or the T lymphocyte (cell-mediated) limb of the immune response. The unaffected limb is left intact to produce antibody or respond with cell-mediated immunity, depending on which limb has been rendered specifically unresponsive to the antigen in question. (2) The induction of immunologic tolerance to some epitopes of allogeneic cells, while leaving the remaining epitopes capable of inducing an immune response characterized by antibody production and/or cell-mediated immunity.

splits

Human leukocyte antigen (HLA) subtypes. For example, the base antigen HLA-B12 can be subdivided into the splits HLA-B44 and

Original Broad Specificities	Splits and Associated Antigens #
A2	A203#, A210#
A9	A23, A24, A2403#
A10	A25, A26, A34, A66
A19	A29, A30, A31, A32, A33, A74
A28	A68, A69
B5	B51, B52
B7	B703#
B12	B44, B45
B14	B64, B65
B15	B62, B63, B75, B76, B77
B16	B38, B39, B3901#, B3902#
B17	B57, B58
B21	B49, B50, B4005#
B22	B54, B55, B56
B40	B60, B61
B70	B71, B72
Cw3	Cw9, Cw10
DR1	DR103#
DR2	DR15, DR16
DR3	DR17, DR18
DR5	DR11, DR12
DR6	DR13, DR14, DR1403#, DR1404#
DQ1	DQ5, DQ6
DQ3	DQ7, DQ8, DQ9
Dw6	Dw18, Dw19
Dw7	Dw11, Dw17

HLA-B45. The term "split" is used to designate an HLA antigen that was first believed to be a private antigen, but later was shown to be a public antigen. The former designation can be placed in parenthesis following its new designation, i.e., HLA-B44(12).

spontaneous autoimmune thyroiditis (SAT)

Spontaneous autoimmune thyroiditis (SAT) in the obese strain (OS) of chickens, an animal model of Hashimoto's thyroiditis in man. Many mononuclear cells infiltrate the thyroid gland, leading to disruption of the follicular architecture. Immunodysregulation is critical in the etiopathogenesis of SAT. Endocrine abnormalities play a role in the pathogenesis. Disturbances in communication between the immune and endocrine systems occur in the disease. Dysregulation leads to hyperreactivity of the immune system which, combined with a primary genetic defect-induced alteration of the thyroid gland, leads to autoimmune thyroiditis. Both T and B lymphocytes are significant in the pathogenesis of SAT. T cell effector mechanisms have greater influence than do humoral factors in initiation of the disease, and most lymphocytes infiltrating the autoimmune thyroid are mature cells. Not only do autoantibodies have a minor role in the pathogenesis, but T cells rather than B cells of OS chickens are defective.

spontaneous remission

The reversal of progressive growth of the neoplasm with inadequate or no treatment. Spontaneous remission occurs only rarely.

SPOTELISA

An assay that is a variation on standard enzyme-linked immunosorbent assay (ELISA). It is used primarily for the detection of immunoglobulin secreting cells (ISC) or cytokine secreting cells (CSC), although future applications may include detection of specific hormone secreting cells. As in standard ELISA, the starting point is a plastic or nitrocellulose vessel coated with antigen or capture antibody. The ISC or CSC of interest is added then removed, following sufficient incubation time for the cell to secrete its immunoglobulin or cytokine. The secreted product binds locally to the capture protein and is subsequently detected by enzyme-linked antibody. Finally a substrate that yields an insoluble product is added and the resulting colored precipitate is quantified.

sprue

Refer to gluten-sensitive enteropathy.

spur

An extension of a precipitation line observed in a two-dimensional double-immunodiffusion assay such as the Ouchterlony test. It represents a reaction of partial identity between two antigens that crossreact with the antibody.

squamous keratin (SK)

A high mol wt keratin present in keratinizing and nonkeratinizing squamous cells and neoplasms derived from them, irrespective of site of origin or degree of differentiation.

SRBC

Abbreviation for sheep red blood cells.

SRS-A

Abbreviation for slow-reacting substance of anaphylaxis.

SRV-1

A simian AIDS virus type D that shows little similarity with HIV-1. However, they both contain genes that resemble one another. This strain was responsible for an infection among a colony of macaques in California.

SRY

The protein coded for by the sex-determining region of the Y chromosome termed the *sry* gene in man. It is equivalent to the Y chromosome's testis-determining gene. The corresponding protein in mice is termed Sry. The murine *sry* gene can cause transgenic female mice to become phenotypic males when the gene is inserted into them.

SS-A

Anti-RNA antibody that occurs in Sjögren's syndrome patients. The antibody may pass across the placenta in pregnant females and be associated with heart block in their infants.

SS-A Ro

An antigen in the cytoplasm to which 25% of lupus erythematosus patients and 40% of Sjögren's syndrome patients synthesize antibodies.

SS-A/Ro antibodies

Antibodies against SS-A/Ro antigen, which consist of 60 kD and 52 kD polypeptides associated with Ro RNAs. They may be demonstrated by immunodiffusion in the sera of 35% of systemic lupus erythematosus patients and 60% of Sjögren's syndrome patients.

SS-B

An anti-RNA antibody detectable in patients with Sjögren's syndrome as well as other connective tissue (rheumatic) diseases.

SS-B La

An antigen in the cytoplasm to which Sjögren's syndrome and lupus erythematosus patients form antibodies. Anti-SS-B antibodies may portend a better prognosis in patients with lupus erythematosus.

SS-B/La antibodies

Antibodies to SS-B/La antigen, which is a 48 kD nucleopolasmic phosphoprotein associated with selected Ro small RNA (Ro hY1-hY5). These antibodies may be demonstrated by EIA in Sjögren's syndrome that is either primary or secondary to rheumatoid arthritis or in systemic lupus erythematosus.

SSPE

Abbreviation for subacute sclerosing panencephalitis.

Ss protein

A hemolytically active substance encoded by the murine complement locus *C4B*.

SSS III

One of more than 70 types of specific soluble substances comprising the polysaccharide in capsules of *Streptococcus pneumoniae*, commonly known as the pneumococcus. It was used extensively by Michael Heidelberger and associates in perfecting the quantitative precipitation reaction.

St. Vitus dance (chorea)

Muscular twitching movements that are involuntary and may occur in acute rheumatic fever.

staphylococcal protein A

A substance derived from the cell wall of *Staphylococcus aureus* that interacts with IgG1, IgG2, and IgG4 subclasses. It stimulates human B cell activation.

status asthmaticus

A clinical syndrome characterized by diminished responsiveness of asthmatic patients to drugs to which they were formerly sensitive. Patients may not respond to adrenergic bronchodilators and are hypoxemic. Treatment is with oxygen, aminophylline, and methylprednisone.

status thymolymphaticus (historical)

A clinical condition described a half century ago as pathological enlargement of the thymus gland. Regrettably, it was treated with radiotherapy. Physicians of the time did not realize that the thymus enlarges under physiologic conditions attaining a weight of 15 to 25 g at puberty. This is followed by subsequent involution of the gland. Individuals subjected to radiation therapy were at increased risk of developing thyroid and breast cancer.

stem cells

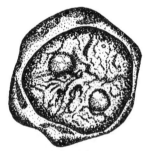

Stem Cell

Relatively large cells with a cytoplasmic rim that stains with methyl green pyronin and a nucleus that has thin chromatin strands and contains nucleoli that are pyroninophilic. They are found in hematopoietic tissues such as the bone marrow. These stem cells are a part of the colony-forming unit (CFU) pool which indicates that individual cells are able to differentiate and proliferate under favorable conditions. The stem cells, CFU-S which are pluripotent, are capable of differentiating into committed precursor cells of the granulocyte and monocyte lineage (CFU-C) of erythropoietic lineage (CFU-E and BFU-E) and of megakaryocyte lineage (CFU-Mg). Lymphocytes, like other hematopoietic cells, are generated in the bone marrow. The stem cell compartment is composed of a continuum of cells that include the most primitive with the greatest capacity for self renewal and the least evidence of cell cycle activity to the most committed with a lesser capacity for self renewal and the most evidence of cell cycle activity. Stem cells are precursor cells that are multipotential with the capacity to yield differentiated cell types with different functions and phenotypes. However, the proliferative capacity of stem cells is limited.

steric hindrance

In immunology, interference between the interaction of the paratope of an antibody molecule with the homologous epitope on antigen molecules of varying sizes based upon the shapes of the two reactants. Whereas IgM molecules potentially have an antigen-binding capacity of ten, only some of these may be able to interact with relatively large antigen molecules bearing epitopes because of the shapes of the two. By contrast, relatively small antigen molecules would be able to permit their epitopes to bind with more paratopes on the IgM molecule. Steric hindrance also refers to the blocking of ligand binding when a receptor site is already occupied by another ligand.

steroid cell antibodies

IgG antibodies that interact with antigens in the cytoplasm of cells producing steroids in the ovary, testes, placenta and adrenal cortex. Patients with Addison's disease with ovarian failure or hypoparathyroidism develop these antibodies which are rarely associated with primary ovarian failure in which organ-specific and non-organ-specific autoantibodies are prominent.

stimulated macrophage

A macrophage that has been activated *in vivo* or *in vitro*. The term activated macrophage is preferred.

Stormont test

A double intradermal tuberculin test.

street virus

A natural or genetically unmodified virus such as rabies that can be isolated from animals.

streptavidin

A protein isolated from streptomyces that binds biotin. This property makes streptavidin useful in the immunoperoxidase reaction that is employed extensively in antigen identification in histopathologic specimens, especially in surgical pathologic diagnosis.

streptococcal M protein

A cell wall protein of virulent *Streptococcus pyogenes* microorganisms which interferes with phagocytosis and also serves as a nephritogenic factor.

streptolysin O test

Refer to ASO.

striational antibodies

Antibodies demonstrable in 80 to 100% of myasthenia gravis (MG) patients with thymoma. They are not present in 82 to 100% of myasthenia gravis patients who do not have thymoma. If striational antibodies are not demonstrable in myasthenia gravis patient, the individual probably does not have a thymoma. One quarter of rheumatoid arthritis patients receiving penicillamine therapy develop IgM striational antibodies. Patients receiving immunosuppressive therapy may also be monitored for striational antibodies to signify the development of autoimmune reactions following bone marrow transplantation.

Strongyloides hyperinfection

S. stercoralis larvae may invade the tissues of immunosuppressed patients with enteric strongyloides infection to produce this condition.

STS

Abbreviation for serological test for syphilis.

subacute sclerosing panencephalitis

A slow virus disease that occurs infrequently as a complication of measles and produces progressively destructive injury to the brain through slow replication of defective viruses.

subset

Refers to a subpopulation of cells such as T lymphocytes in samples of peripheral blood. Subsets are identified by immunophenotyping through the use of monoclonal antibodies and by flow cytometry. The cells are separated based on their surface CD (cluster of differentiation) determinants, such as CD4 that identifies helper/inducer T lymphocytes and CD8 that identifies suppressor/cytotoxic T lymphocytes.

Substance P

A tachykinin which may induce joint inflammation when released at local sites. It facilitate synthesis by monocytes of Il-1, IL-6 and TNF-α and stimulates synovial cells to produce prostaglandins. Its receptor is designated NK$_1$.

substrate adhesion molecule (SAM)

Extracellular molecules that share a variety of sequence motifs with other adhesion molecules. Most prominent among these are segments similar to the type III repeats of fibronectin and immunoglobulin-like domains. In contrast to other morphoregulatory molecules, SAMS do not have to be made by the cells that bind them. SAMS can link and influence the behavior of one another. Examples include glycoproteins, collagens, and proteoglycans.

sugar cane workers lung
Refer to bagassosis and farmer's lung.

suicide, immunological
The use of an antigen deliberately labeled with high-dose radioisotope to kill a subpopulation of lymphocytes with receptors specific for that antigen following antigen binding.

Sulzberger-Chase phenomenon
The induction of immunological unresponsiveness to skin-sensitizing chemicals such as picryl chloride by feeding an animal (e.g., guinea pig) the chemical in question prior to application to the skin. Intravenous administration of the chemical may also block the development of delayed-type hypersensitivity when the same chemical is later applied to the skin. Simple chemicals such as picryl chloride may induce contact hypersensitivity when applied to the skin of guinea pigs. The unresponsiveness may be abrogated by adoptive immunization of a tolerant guinea pig with lymphocytes from one that has been sensitized by application of the chemical to the skin without prior oral feeding.

superantigen
An antigen such as a bacterial toxin that is capable of stimulating multiple T lymphocytes, especially CD4⁺ T cells, leading to the release of relatively large quantities of cytokines. Selected bacterial toxins may stimulate all T lymphocytes in the body that contain a certain family of V β T cell receptor genes. Superantigens may induce proliferation of 10% of CD4⁺ T cells by combining with the T cell receptor V β and to the MHC HLA-DR α-1 domain. Superantigens are thymus-dependent (TD) antigens that do not require phagocytic processing. Instead of fitting into the T cell receptor (TCR) internal groove where a typical processed peptide antigen fits, superantigens bind to the external region of the αβ TCR and simultaneously link to DP, DQ, or DR molecules on antigen-presenting cells. Superantigens react with multiple TCR molecules whose peripheral structure is similar. Thus, they stimulate multiple T cells that augment a protective T and B cell antibody response. This enhanced responsiveness to antigens such as toxins produced by staphylococci and streptococci is an important protective mechanism in the infected individual.

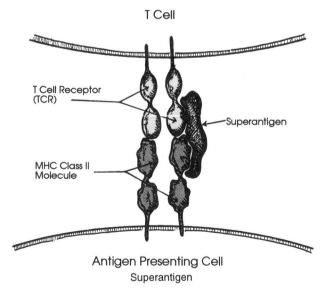

T Cell

T Cell Receptor
(TCR)

Superantigen

MHC Class II
Molecule

Antigen Presenting Cell
Superantigen

superinfection "immunity"
The inability of two related organisms, for example, plasmids, to invade a host cell at the same time.

superoxide anion
A free radical formed by the addition of an electron to an oxygen molecule, causing it to become highly reactive. This takes place in inflammation or is induced by ionizing radiation. It is formed by reduction of molecular oxygen in polymorphonuclear neutrophils (PMNs) and mononuclear phagocytes. The hexose monophosphate shunt activation pathway enhances superoxide anion generation. Superoxide anion interacts with protons, additional superoxide anions, and hydrogen peroxide. Oxidation of one superoxide anion and reduction of another may lead to the formation of oxygen and hydrogen peroxide. Superoxide dismutase, found in phagocytes,

catalyzes this reaction. Injury induced by superoxide anion is associated with age-related degeneration. It may also serve as a mutagen with implications for carcinogenesis. It plays a pivotal role in the ability of mononuclear phagocytes and neutrophils to kill microorganisms through their oxidative microbicidal function.

superoxide dismutase
An enzyme that defends an organism against oxygen-free radicals by catalyzing the interaction of superoxide anions with hydrogen ions to yield hydrogen peroxide and oxygen.

suppressin
A 63-kD, single polypeptide chain molecule with multiple disulfide linkages and a p.i. of 8.1. It is produced by the pituitary and by lymphocytes and is a negative regulator of cell growth. It inhibits lymphocyte proliferation and is more effective on T cells than on B cells. Suppressin has properties similar to TGF-β, although it is structurally different. Antisuppressin antibody leads to T or B cell proliferation.

suppressor cell
A lymphoid cell subpopulation that is able to diminish or suppress the immune reactivity of other cells. An example is the CD8⁺ suppressor T lymphocyte subpopulation detectable by monoclonal antibodies and flow cytometry in peripheral blood lymphocytes.

suppressor/inducer T lymphocyte
A subpopulation of T lymphocytes which fail to induce immunosuppression themselves, but are claimed to activate suppressor T lymphocytes.

suppressor macrophage
Macrophage activated by its response to an infection or neoplasm in the host from which it was derived. It is able to block immunologic reactivity *in vitro* through production of prostaglandins, oxygen radicals, or other inhibitors produced through arachidonic acid metabolism.

suppressor T cell factor
A soluble substance synthesized by suppressor T lymphocytes that diminishes or suppresses the function of other lymphoid cells. The suppressor factor downregulates immune reactivity.

suppressor T cells (Ts cells)
A T lymphocyte subpopulation that diminishes or suppresses antibody formation by B cells or downregulates the ability of T lymphocytes to mount a cellular immune response. Ts cells may induce suppression that is specific for antigen, idiotype, or non-specific suppression. Some CD8⁺ T lymphocytes diminish T helper CD4⁺ lymphocyte responsiveness to both endogenous and exogenous antigens. This leads to suppression of the immune response. An overall immune response may be a consequence of the balance between helper T lymphocyte and suppressor T lymphocyte stimulation. Suppressor T cells are also significant in the establishment of immunologic tolerance and are particularly active in response to unprocessed antigen. The inability to confirm the presence of receptor molecules on suppressor cells has cast a cloud over the suppressor cell; however, functional suppressor cell effects are indisputable. Some suppressor T lymphocytes are antigen specific and are important in the regulation of T helper cell function. Like cytotoxic T cells, T suppressor cells are MHC class I restricted.

supratypic antigen
Refer to public antigen.

suramin
(Antrypol, 8,8′-(carbonyl-*bis*-(imino-3,1-phenylenecarbonylimino))-*bis*-1,3,5-naphthalene trisulfonic acid). A therapeutic agent for African sleeping sickness produced by trypanosomes. Of immunologic interest is its ability to combine with C3b, thereby blocking factor H and factor I binding. The drug also blocks lysis mediated by complement by preventing attachment of the membrane attack complex of complement to the membranes of cells.

surface antigen
Epitopes on a cell surface such as the bacterial antigens Vi and O.

surface immunoglobulin
Refer to B lymphocyte receptor.

surface phagocytosis
Facilitation of phagocytosis when microorganisms become attached to the surfaces of tissues, blood clots, or leukocytes.

surrogate light chains
Invariant light chains that are structurally homologous to kappa and lambda light chains and associate with pre-B cell μ heavy chains. They are the same in all B cells. V regions are absent in surrogate

light chains. Low levels of cell surface μ chain and surrogate light chain complexes are believed to participate in stimulation of kappa or lambda light chain synthesis and maturation of B cells.

SV40 (simian virus 40)

A polyoma virus that is oncogenic. It multiplies in cultures of rhesus monkey kidney and produces cytopathic alterations in African green monkey cell cultures. Inoculation into newborn hamsters leads to the development of sarcomas. SV40 has 5243 base pairs in its genome. It may follow either of two patterns of lifecycle according to the host cell. In permissive cells, such as those from African green monkeys, the virus-infected cells are lysed, causing the escape of multiple viral particles. Lysis does not occur in nonpermissive cells infected with the virus. By contrast, they may undergo oncogenic transformation in which SV40 DNA sequences become integrated into the genome of the host cell. Cells that have become transformed have characteristic morphological features and growth properties. SV40 may serve as a cloning vector. It is a diminutive icosahedral papovavirus that contains double-stranded DNA. It may induce progressive multifocal leukoencephalopathy. It is useful for the *in vitro* transformation of cells as a type of "permissive" infection ultimately resulting in lysis of infected host cells.

SV40

Svedberg unit

A sedimentation coefficient unit that is equal to 10^{-13} s. Whereas most immunoglobulin molecules, such as IgG, sediment at 7 S, the pentameric IgM molecule sediments at 19 S.

Sweet's syndrome (acute febrile neutrophilic dermatosis)

A syndrome with neutrophilia, fever, erythematous, and painful skin plaques with pronounced dermal neutrophilic inflammation. About 10 to 15% of the cases may have an underlying malignant disease such as a myeloid proliferative disorder, acute myelogenous leukemia, or other tumor. The cutaneous lesions may become vesicular, and pustular skin lesions resemble those in bowel-bypass syndrome. Patients may develop arthritis, myalgia, conjunctivitis, and proteinuria. Immunofluorescence reveals IgG, IgM, and C3 in some lesions. Systemic steroid treatment has proven effective in improving skin lesions.

Swiss agammaglobulinemia

A type of severe combined immunodeficiency (SCID) that has an autosomal recessive mode of inheritance. Patients usually die during infancy as a consequence of severe diarrhea, villous atrophy, and malabsorption with disaccharidase deficiency. Because of severely impaired humoral and cellular immune defense mechanisms, patients have an increased susceptibility to various opportunistic infections such as those induced by *Pneumocystis carinii, Candida albicans,* measles, varicella, and cytomegalovirus. Patients are also at increased risk of developing graft-vs.-host disease following blood transfusion. A stem cell defect leads to diminished numbers of T and B lymphocytes. Patients have elevated liver enzymes, lymphopenia, anemia, and diarrhea, causing an electrolyte imbalance. They are usually treated with antibiotics, γ globulin, and an HLA 6-antigen match bone marrow transplant.

Swiss type agammaglobulinemia

Severe combined immunodeficiency disease. Refer to Swiss type of severe combined immunodeficiency.

Swiss type of severe combined immunodeficiency

A condition that results from a defect at the lymphocytic stem cell level. It results in cellular abnormalities that affect both T and B cell limbs of the immune response. This culminates in impaired cell-mediated immunity and humoral antibody responsiveness following challenge by appropriate immunogens. The mode of inheritance is autosomal recessive. Refer also to severe combined immunodeficiency syndrome.

Swiss type immunodeficiency

Refer to Swiss agammaglobulinemia.

switch

The change within an immunologically competent B lymphocyte from synthesizing one isotype of heavy polypeptide chain, such as μ, to another isotype, such as γ, during differentiation. The switch signal comes from T cells. Isotype switching does not alter the antigen-binding variable region of the chain at the N-terminus.

switch cells

A subset of T lymphocytes that governs isotype differentiation of B lymphocytes exiting the Peyer's patches to ensure that they become IgA-producing plasma cells when they home back to the lamina propria of the intestine from the systemic circulation.

switch defect disease

Refer to hyperimmunoglobulin M syndrome.

switch region

The amino acid sequence between the constant and variable portions of light and heavy polypeptide immunoglobulin chains. This amino acid segment is encoded by D and J genes. This segment of DNA controls recombination associated with immunoglobulin class switching. Specific switch region sequences are critical for switching from one immunoglobulin isotype to another, i.e., from one class to another.

switch site

Breakage points on a chromosome where gene segments unite during gene rearrangement. In immunology, it often refers to an abbreviated DNA sequence 5′ to each gene encoding a heavy chain C region. It serves as an identification site for V region gene translocation in the process of switching gene expression from one immunoglobulin heavy chain class to another. There are numerous switch sites for each gene encoding the C region.

Syk PTK

A 72-kD phosphotyrosine kinase found on B cells and myeloid cells which is homologous to the ZAP-70 PTK found on T cells and NK cells. Both Syk and ZAP-70 play roles in the functions of distinct antigen receptors.

sympathetic ophthalmia

Uveal inflammation of a healthy uninjured eye in an individual who has sustained a perforating injury to the other eye. The uveal tract reveals an infiltrate of lymphocytes and epithelioid cells, and there is granuloma formation. The mechanism has been suggested to be autoimmunity expressed as T lymphocyte-mediated immune reactivity against previously sequestered antigens released from the patient's other injured eye.

synaptophysin

A neuroendocrine differentiation marker that is detectable by the immunoperoxidase technique used in surgical pathologic diagnosis. Tumors in which it is produced include ganglioneuroblastoma, neuroblastoma, ganglioneuroma, paraganglioma, pheochromocytoma, medullary carcinoma of the thyroid, carcinoid, and tumors of the endocrine pancreas.

syngeneic

An adjective that implies genetic identity between identical twins in humans or among members of an inbred strain of mice or other species. It is used principally to refer to transplants between genetically identical members of a species.

syngeneic preference

The better growth of neoplasms when they are transplanted to histocompatible recipients than when they are transplanted in histoincompatible recipients. Refer also to allogeneic inhibition.

syngraft

A transplant from one individual to another within the same strain. Also called isograft.

synthetic antigen

An antigen derived exclusively by laboratory synthesis and not obtained from living cells. Synthetic polypeptide antigens have a backbone consisting of amino acids that usually include lysine. Side chains of different amino acids are attached directly to the backbone and then elongated with a homopolymer or, conversely, attached via the homopolymer. They have contributed much to our knowledge of epitope structure and function. They have well-defined specificities determined by the particular arrangement, number, and nature of the amino acid components of the molecule, and they may be made more complex by further coupling to haptens or derivatized with various compounds. The size of the molecule is less critical with synthetic antigens than with natural antigens. Thus, molecules as small as those of *p*-azobenzenearsonate coupled to three L-lysine residues (mol wt 750) or even of *p*-azobenzenearsonate-*N*-acetyl-L-tyrosine (mol wt 451) may be immunogenic. Specific antibodies are markedly stereospecific, and there is no crossreaction between them, e.g., poly-D-alanyl and poly-L-alanyl determinants. Studies employing synthetic antigens demonstrated the significance of aromatic, charged amino acid residues in proving the ability of synthetic polypeptides to induce an immune response.

Synthetic Polypeptide Antigen
(multichain copolymer (Phe, G)-A--L)

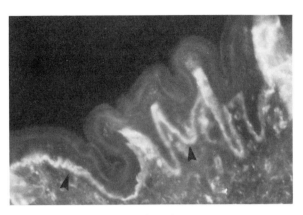

- ■ poly-L-lysine
- ▨ poly-DL-alanine
- ◉ L-phenylalanine
- ○ L-glutamic acid

to RNA, and (4) antibodies against nucleolar antigens. Indirect immunofluorescence is used to detect nuclear fluorescence patterns that are characteristic for certain antibodies. These include homogeneous or diffuse staining, which reveals antibodies to histones and deoxyribonucleoprotein; rim or peripheral staining which signifies antibodies against double-stranded DNA; speckled pattern, which indicates antibodies to non-DNA nuclear components including histones and ribonucleoproteins; and the nucleolar pattern in which fluorescent spots are observed in the nucleus and reveal antibodies to nucleolar RNA. Antinuclear antibodies most closely associated with SLE are antidouble-stranded DNA and anti-Sm (Smith) antibodies. There appears to be genetic predisposition to the disease which is associated with DR2 and DR3 genes of the major histocompatibility complex (MHC) in Caucasians of North America. Genes other than HLA genes are also important. In

immune deposits at
dermal-epidermal
junction

Systemic Lupus Erythematosus

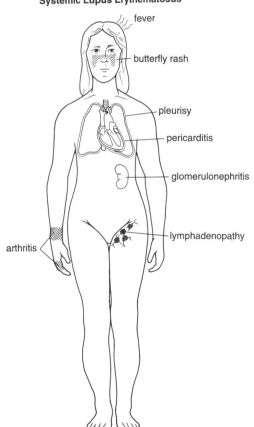

synthetic vaccines
Substances used for prophylactic immunization against infectious disease prepared by artificial techniques such as from cloned DNA or through peptide synthesis.

systemic anaphylaxis
Type I, immediate, anaphylactic type of hypersensitivity mediated by IgE antibodies anchored to mast cells that become crosslinked by homologous antigen (allergen) causing release of the pharmacological mediators of immediate hypersensitivity, producing lesions in multiple organs and tissue sites. This is in contrast to local anaphylaxis, where the effects are produced in isolated anatomical location. The intravenous administration of a serum product, antibiotic, or other substance against which the patient has anaphylactic IgE-type hypersensitivity may lead to the symptoms of systemic anaphylaxis within seconds and may prove lethal.

systemic immunoblastic proliferation
A condition characterized by gene translocation and immature lymphocyte proliferation. Clinically, patients manifest rash, dyspnea, hepatosplenomegaly, and lymphadenopathy and show an increased incidence of immunoblastic lymphoma.

systemic lupus erythematosus (SLE)
The prototype of connective tissue diseases that involves multiple systems and has an autoimmune etiology. It is a disease with an acute or insidious onset. Patients may experience fever, malaise, loss of weight, and lethargy. All organ systems may be involved. Patients form a plethora of autoantibodies, especially antinuclear autoantibodies. SLE is characterized by exacerbations and remissions. Patients often have injury to the skin, kidneys, joints, and serosal membranes. SLE occurs in 1 in 2500 people in certain populations. It has a 9:1 female to male predominance. Its cause remains unknown. Antinuclear antibodies produced in SLE fall into four categories that include (1) antibodies against DNA, (2) antibodies against histones, (3) antibodies to nonhistone proteins bound

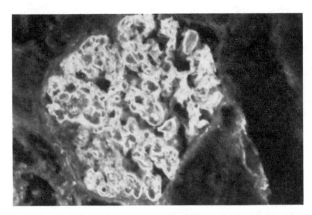

diffuse immune deposits
on peripheral capillary
loops

Systemic Lupus Erythematosus

addition to the antidouble-stranded DNA and anti-Sm antibodies, other immunologic features of the disease include depressed serum complement levels, immune deposits in glomerular basement membranes and at the dermal-epidermal junction, and the presence of multiple other autoantibodies. Of all the immunologic abnormalities, the hyperactivity of B cells is critical to the pathogenesis of SLE. B cell activation is polyclonal, leading to the formation of antibodies against self and nonself antigens. In SLE, there is a loss of tolerance to self constituents, leading to the formation of antinuclear antibodies. The polyclonal activation leads to antibodies of essentially all classes in immune deposits found in renal biopsy specimens by immunofluorescence. In addition to genetic factors, hormonal and environmental factors are important in producing the B cell activation. Nuclei of injured cells react with antinuclear antibodies, forming a homogeneous structure called an LE body or a hematoxylin body which is usually found in a neutrophil that has phagocytized the injured cell's denatured nucleus. Tissue injury in lupus is mediated mostly by an immune complex (type III hypersensitivity). There are also autoantibodies specific for erythrocytes, leukocytes, and platelets that induce injury through a type II hypersensitivity mechanism. There is an acute necrotizing vasculitis involving small arteries and arterioles present in tissues in lupus. Fibrinoid necrosis is classically produced. Most SLE patients have renal involvement which may take several forms, with diffuse proliferative glomerulonephritis being the most serious.

Subendothelial immune deposits in the kidneys of lupus patients are typical and may give a "wire loop" appearance to a thickened basement membrane. In the skin, immunofluorescence can demonstrate deposition of immune complexes and complement at the dermal-epidermal junction. Immune deposits in the skin are especially prominent in sun-exposed areas of the skin. Joints may be involved, but the synovitis is nonerosive. Typical female patients with lupus have a butterfly rash over the bridge of the nose in addition to fever and pain in the peripheral joints. However, the presenting complaints in SLE vary widely. Patients may have central nervous system involvement, pericarditis, or other serosal cavity inflammation. There may be pericarditis as well as involvement of the myocardium or of the cardiac valves to produce Libman-Sacks endocarditis. There may be splenic enlargement, pleuritis, and pleural effusion or interstitial pneumonitis, as well as other organ or system involvement. Patients may also develop antiphospholipid antibodies called lupus anticoagulants. They may be associated with a false-positive VDRL test for syphilis. A drug such as hydrazaline may induce a lupus-like syndrome. However, the antinuclear antibodies produced in drug-induced lupus are often specific for histones, a finding not commonly found in classic SLE. Lupus erythematosus induced by drugs remits when the drug is removed. Discoid lupus refers to a form of the disease limited to the skin. Corticosteroids have proven very effective in suppressing immune reactivity in SLE. In more severe cases, cytotoxic agents such as cyclophosphamide, chlorambucil, and azathioprine have been used. Refer to LE cell.

systemic sclerosis
Refer to progressive systemic sclerosis.

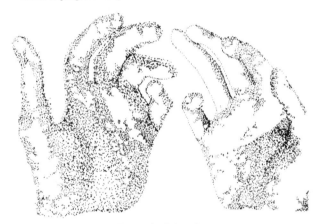

Systemic Sclerosis

T

T-200

An obsolete term for leukocyte common antigen (CD45).

T activation

The use of bacterial neuraminidase to cleave *N*-acetyl (sialic acid) residue to uncover antigenic determinants (epitopes) which have been masked or hidden. This permits the treated cells to be agglutinated by natural antibodies in the blood of most individuals. Aged blood can be used to detect T activation.

T agglutinin

An antibody, which occurs naturally in the blood serum of man, that agglutinates red blood cells expressing T antigen as a result of their exposure to bacteria or as a consequence of treatment with neuraminidase. This antibody is of interest in transfusion medicine as it may confuse blood grouping or cross-matching procedures by giving a false-positive reaction when red blood cell suspensions contaminated with microorganisms are used.

T antigen(s)

(1) An erythrocyte surface antigen that is shielded from interaction with the immune system by an *N*-acetyl-neuraminic acid residue. Thus, antibody is formed against this antigen once bacterial infection has diluted this neuraminic acid residue. Antibodies produced can cause polyagglutination of red cells bearing the newly revealed T antigen. (2) Several 90-kD nuclear proteins that combine with DNA and are critical in transcription and replication of viral DNA in the lytic cycle. T antigen participates in the change from early to late stages of transcription. (3) An epitope that shares homology at the N-terminal sequence with the SV40 virus T antigen.

T-B cell cooperation

Refers to B cell and helper T cell cooperation that leads to B cell proliferation and differentiation into plasma cells that synthesize and secrete specific antibody. B cell immunoglobulin receptors react with protein antigen. This is followed by endocytosis, antigen processing, and presentation to helper T lymphocytes. Their antigen-specific T cell receptors recognize processed antigen only in the context of MHC class II molecules on the B cell surface during antigen presentation. CD4+ helper T cells secrete lymphokines, including IL-2, that promote B cell growth and differentiation into plasma cells which secrete specific antibody. T cells are required for B cells to be able to switch from forming IgM to synthesizing IgG or IgA. B and T lymphocytes recognize different antigens. B cells may recognize peptides, native proteins, or denatured proteins. T cells are more complex in their recognition system in that a peptide antigen can be presented to them only in the context of MHC class II or class I histocompatibility molecules. Hapten-carrier complexes have been successfully used in delineating the different responses of B and T cells to each part of this complex. Immunization of a rabbit or other animal with a particular hapten-carrier complex will induce a primary immune response, and a second injection of the same hapten-carrier conjugate will induce a secondary immune response. However, linkage of the same hapten to a different carrier elicits a much weaker secondary response in an animal primed with the original hapten-carrier complex. This is termed the carrier effect. B lymphocytes recognize the hapten, and T lymphocytes the carrier.

T cell

Refer to T lymphocyte.

T cell antigen-specific suppressor factor

A soluble substance that is produced by a suppressor T cell after it has been activated. This suppressor factor has been claimed to bind antigen and cause the immune response to be suppressed in a manner that is antigen specific.

T cell chemotactic factor

Former term for IL-8.

T cell-dependent (TD) antigen

An immunogen that is much more complex than the T cell-independent (TI) antigens. They are usually proteins, protein-nuclear protein conjugates, glycoproteins, or lipoproteins. They stimulate all five classes of immunoglobulin, elicit an anamnestic or memory response, and are present in most pathogenic microorganisms. These properties ensure that an effective immune response can be generated in a host infected with these pathogens.

T cell domains

Specific areas in lymph nodes and other lymphoid organs where T lymphocytes localize preferentially.

T cell growth factor (TCGF)

Refer to interleukin-2.

T cell growth factor 1

Interleukin-2 (IL-2).

T cell growth factor 2

Interleukin-4 (IL-4).

T cell immunodeficiency syndromes (TCIS)

Decreased immune function as a consequence of complete or partial defects in the function of T lymphocytes. HUETER patients develop recurrent opportunistic infections; may manifest cutaneous anergy, wasting, diminished life expectancy, retardation in growth, and increased likelihood of developing graft-vs.-host disease; and have very serious or even fatal reactions following immunization with BCG or live virus vaccines. They also have an increased likelihood of malignancy. T cell immunodeficiencies are usually more profound than B cell immunodeficiencies. There is no effective treatment. This group of disorders includes thymic hypoplasia known as DiGeorge syndrome, cellular immunodeficiency with immunoglobulins termed Header syndrome, and defects of T lymphocytes caused by deficiency of purine nucleoside phosphorylase and lack of inosine phosphorylase.

T cell-independent (TI) antigen

An immunogen that is simple in structure, often a polysaccharide such as the polysaccharide of the pneumococcus, a dextran polyvinyl hooter, or a bacterial lipopolysaccharide. They elicit an IgM response only and fail to stimulate an anamnestic response. They are not found in most pathogenic microbes.

T cell leukemia

Adult T cell leukemia/lymphoma.

T cell leukemia viruses

Retroviruses such as HTLV-I that induce human T cell leukemia and HTLV-II which has been associated with hairy cell leukemia.

T cell lymphoma (TCL)

Neoplastic proliferation of T lymphocytes. A condition that is diagnosed by determining whether or not there has been rearrangement of the genes encoding the T lymphocyte receptor ß chain.

T cell maturation

Refer to thymus cell differentiation.

T cell migration

Cells leaving the thymus migrate to all peripheral lymphoid organs seeding in the T-dependent regions of the lymph nodes and spleen and at the periphery of the lymphoid follicles. The rate of release of thymocytes from the thymus is markedly increased following antigenic stimulation. The patterns of migration of T cells (as well as of B cells) have been studied by adoptive transfer of labeled purified cells into irradiated syngeneic mice matched for age and sex.

T cell nonantigen-specific helper factor

A substance that provides nonspecific help to T lymphocytes.

T cell receptor

Refer to T lymphocyte antigen receptor.

T cell receptor, γδ

A far less common receptor than the αβTCR. It is comprised of γ and δ chains and occurs on the surface of early thymocytes and less than 1% of peripheral blood lymphocytes. The γδTCR appears on double-negative CD4-CD8- cells. Thus, the γδ heterodimer resembles its αβ counterpart in possessing both V and C regions, but has less diversity. TCR specificity and diversity are attributable to the

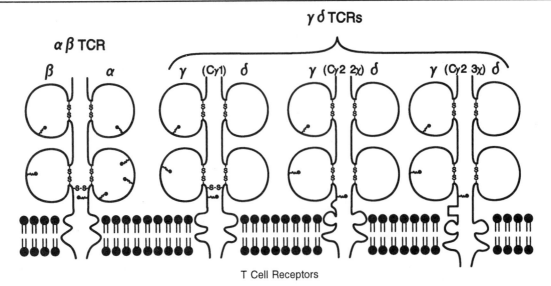

γ δ TCRs

α β TCR

β | α | γ (Cγ1) δ | γ (Cγ2 2χ) δ | γ (Cγ2 3χ) δ

T Cell Receptors

multiplicity of germ line V gene segments subjected to somatic recombination in T cell ontogeny, leading to a complete TCR gene. Cells bearing the γδ receptor often manifest target cell killing that is not MHC restricted.

Monoclonal antibodies to specific TCR V regions are being investigated for possible use in the future treatment of autoimmune diseases. γδ T cells are sometimes found associated with selected epithelial surfaces, especially in the gut.

T cell receptor genes

Four separate sets of genes encode the antigen-MHC binding region of the T cell receptor. Most (approximately 95%) peripheral T lymphocytes express α and β gene sets. Only approximately 5% of circulating peripheral blood T cells and a subset of T lymphocytes in the thymus express γ and δ genes. The αβ chains or the γδ chains, encoded by their respective genes, form an intact T cell receptor and are associated with γ, δ, ε, ζ, and η chains that comprise the CD3 molecular complex. The arrangement of TCR genes resembles that of genes that encode immunoglobulin heavy chains. The TCR δ genes are located in the center of the α genes. V, D, and J segment recombination permits TCR gene diversity. Rearrangement of a V α segment to a J α segment yields an intact variable region. There are two sets of D, J, and C genes at the β locus. During joining, marked diversity is achieved by a V-J and V-D-D-J as well as V-D-J rearrangements. Humans have eight V γ, three J γ, and the initial C γ gene. Before reaching Cγ2, there are two more J γ genes. The δ locus contains five V δ, two D δ, and six J δ genes. TCR gene recombination takes place by mechanisms that resemble those of B cells genes. B and T lymphocytes have essentially the same rearrangement enzymes. TCR genes do not undergo somatic mutation which is essential to immunoglobulin diversity.

T cell replacing factor (TRF)

An earlier term for B cell differentiation factor derived from CD4+ helper T lymphocytes that permits B lymphocytes to synthesize antibody without the presence of T lymphocytes.

T cell rosette

E rosette.

T cell specificity

Refer to MHC restriction.

T cells

Cells derived from hematopoietic precursors that migrate to the thymus where they undergo differentiation which continues thereafter to completion in the various lymphoid tissues throughout the body or during their circulation to and from these sites. The T cells are primarily involved in the control of the immune responses by providing specific cells capable of helping or suppressing these responses. They also have a number of other functions related to cell-mediated immune phenomena. Refer to T lymphocyte.

T-dependent antigen

Refers to thymus-dependent antigen.

T globulin

A serum protein found after hyperimmunization of horses that is a γ₁ 7 S globulin that appears as a prominent band of immunoglobulin when the serum is electrophoresed. It may be a subtype of IgG.

T-independent antigen

Refers to thymus-independent antigen.

T lymphocyte (T cell)

A thymus-derived lymphocyte that confers cell-mediated immunity and cooperates with B lymphocytes enabling them to synthesize antibody specific for thymus-dependent antigens, including switching from IgM to IgG and/or IgA production. T lymphocytes exiting the thymus recirculate in the blood and lymph and in the peripheral lymphoid organs. They migrate to the deep cortex of lymph nodes. Those in the blood may attach to postcapillary venule endothelial cells of lymph nodes and to the marginal sinus in the spleen. After passing across the venules into the splenic white pulp or lymph node cortex, they reside there for 12 to 24 h, exit by the efferent lymphatics, proceed to the thoracic duct, and from there proceed to the left subclavian vein where they enter the blood circulation.

Mature T cells are classified on the basis of their surface markers, such as CD4 and CD8. CD4+ T lymphocytes recognize antigens in the context of MHC class II histocompatibility molecules, whereas CD8+ T lymphocytes recognize antigen in the context of class I MHC histocompatibility molecules. The CD4+ T cells participate in the afferent limb of the immune response to exogenous antigen, which is presented to them by antigen-presenting cells. This stimulates the synthesis of IL-2, which activates CD8+ T cells, NK cells, and B cells, thereby orchestrating an immune response to the antigen. Thus, they are termed helper T lymphocytes. They also mediate delayed-type hypersensitivity reactions. CD8+ T lymphocytes include cytotoxic and suppressor cell populations. They react to endogenous antigen and often express their effector function by a cytotoxic mechanism, e.g., against a virus-infected cell. Other molecules on mature T cells in humans include the E rosette

T Cell Activation

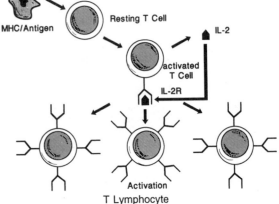

Antigen

Antigen-Presenting Cell

Resting T Cell

MHC/Antigen

IL-2

activated T Cell

IL-2R

Activation

T Lymphocyte

receptor CD2 molecule, the T cell receptor, the pan-T cell marker termed CD3, and transferrin receptors.

T lymphocyte antigen receptor (TRC)

There are two types of T cell antigen receptors: TCR1, which appears first in ontogeny, and TCR2. TCR2 is a heterodimer of two polypeptides (α and β); TCR1 consist of γ and δ polypeptides. Each of the two polypeptides comprising each receptor has a constant and a variable region (similar to immunoglobulin). Reminiscent of the diversity of antibody molecules, T cell antigen receptors can likewise identify a tremendous number of antigenic specificities (estimated to be able to recognize 10^{15} epitopes).

The TCR is a structure comprised of a minimum of seven receptor subunits whose production is encoded by six separate genes. Following transcription, these subunits are assembled precisely. Assimilation of the complete receptor complex is requisite for surface expression of TCR subunits. Numerous biochemical events are associated with activation of a cell through the TCR receptor. These events ultimately lead to receptor subunit phosphorylation.

T cells may be activated by the interaction of antigen, in the context of MHC, with the T cell receptor. This involves transmission of a signal to the interior through the CD3 protein to activate the cell.

T lymphocyte-B lymphocyte cooperation

The association of T cells and B cells through a number of receptor-ligand interactions at the surfaces of both cell types, culminating in the synthesis by B cells of antibody specific for thymus-dependent antigens.

T lymphocyte clone

Daughter cells of one T lymphocyte derived from the blood or spleen that are added to culture medium and activated by antigen. Those T cells that are stimulated by the antigen form blasts which can be separated from the remaining T cells by density gradient centrifugation. The T cells that have responded to antigen are diluted, and aliquots are dispensed into tissue culture plates to which antigen and interleukin-2 are added. Each well contains a single lymphocyte. This method provides individual T cell clones.

T lymphocyte-conditioned medium

A cell culture medium containing multiple lymphokines that have been released from T cells stimulated by antigens or lectins.

T lymphocyte hybridoma

Produced by fusing a murine splenic T cell and an AKR strain BW5147 thymoma cell using polyethylene glycol (PEG). The hybrid cell clone is immortal and releases interleukin-2 when activated by antigen provided by an antigen-presenting cell.

T lymphocyte receptor

Refer to T lymphocyte antigen receptor.

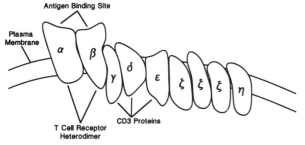

T Lymphocyte Receptor

T lymphocyte subpopulation

A subset of T cells that have a specific function and express a specific cluster of differentiation (CD) markers or other antigens on their surface. Examples include the CD4$^+$ helper T lymphocyte subset and the CD8$^+$ suppressor/cytotoxic T lymphocyte subset.

T lymphocyte-T lymphocyte cooperation

Refers to signals from one T lymphocyte subpopulation to another, such as isotype class switching in the regulation of immunologic responsiveness.

T lymphocytes

Synonym for T cells. T lymphocyte precursors are detectable in the human fetus at 7 weeks of gestation. Between 7 and 14 weeks of gestation, thymic changes begin to imprint thymic lymphocytes as T cells. The maturation (mediated by hormones such as thymosin, thymulin, and thymopoietin II) can be followed by identification of surface (cluster of differentiation [CD]) markers detectable by

immunophenotyping methods. CD3, a widespread T cell marker, serves as a signal transducer from the antigen receptor to the cell interior. Thus, the CD3 molecule is intermittently associated with the T cell receptor for antigen.

T lymphocytes in the medulla initially express both CD4 and CD8 class markers; however, these cells will later differentiate into either CD4$^+$ helper cells or CD8$^+$ suppressor cells. The CD4$^+$ cells, characterized by a 55-kD surface marker, communicate with macrophages and B cells bearing MHC class II molecules during antigen-presentation. The CD8$^+$ suppressor/cytotoxic cells interact with antigen presenting cells bearing MHC class I molecules.

T piece

Refer to secretory piece.

T1 antigen

Refer to CD5.

T3 antigen

Refer to CD3.

T4 antigen

Refer to CD4.

T6 marker

A chromosome found in CBA/H-T6 inbred mice discovered in an irradiated male. It has been used as a cell marker to trace cells transferred to other mice. It has an abbreviated length and reveals a secondary constriction adjacent to the centromere.

T8 antigen

Refer to CD8.

TAB vaccine

An immunizing preparation used to protect against enteric fever. It is comprised of *Salmonella typhi* and *S. paratyphi* A and B microorganisms that have been killed by heat and preserved with phenol. The bacteria used in the vaccine are in the smooth specific phase. They also contain both O and Vi antigens. The vaccine is administered subcutaneously. Lipopolysaccharide from the Gram-negative bacteria may induce fever in vaccine recipients. If *S. paratyphi* C is added, the vaccine is referred to as TABC. If tetanus toxoid is added, it is referred to as TABT.

Tac

A cell surface protein on T lymphocytes that binds IL-2. It is a 55-kD polypeptide (p55) that is expressed on activated T lymphocytes. Tac is an abbreviation for T activation. The p55 Tac polypeptide combines with IL-2 with a kD of about $10^{-8}\,M$. Interaction of IL-2 with p55 alone does not lead to activation. IL-2 binds to a second protein termed p70 or p75 that has a higher affinity of binding to p55. T cells expressing p70 or p75 alone are stimulated by IL-2. Yet cells that express both receptor molecules bind IL-2 more securely and can be stimulated with a relatively lower IL-2 concentration compared to stimulation when p70 or p75 alone interacts with IL-2. Anti-Tac monoclonal antibody can inhibit T cell proliferation.

Tac antigen

Refer to CD25.

TAF

Abbreviation for toxoid antitoxin floccules.

tail peptide

An immunoglobulin heavy polypeptide chain carboxy terminus that is separate from the carboxy terminal domain. This structure is present in membrane-anchored immunoglobulins. Whereas tail peptides of 20 amino acids each are present in IgM and IgA molecules that have been secreted, IgG and IgE molecules do not contain tail peptides.

Takatsy method

A technique that employs tiny spiral loops on the end of a handle which resembles those used for wire loops by bacteriologists. The loops are carefully engineered to retain a precise volume when immersed in a liquid. They are used to prepare doubling dilutions of a test liquid in microtiter wells of test plates. As the loops are passed from one well to the next, a spiral motion helps to discharge the contents into the well diluent and mix it. Several loops can be manipulated by one operator at the same time using a single plastic plate with multiple wells. This method has been applied to hemagglutination assays.

Takayasu's arteritis

Inflammation and stenosis involving large- and intermediate-sized arteries including the aortic arch. The disease occurs in the 15- to 20-year-old age group with a 9:1 female predominance. Mononuclear and giant cells infiltrate all layers of the wall of involved arteries, reflecting a true panarteritis. There may be intimal proliferation,

MHC Class II region

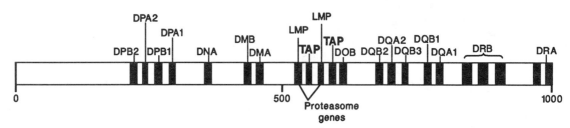

TAP1 and 2 Genes

fibrosis, elastic lamina disruption, and vascularization in the media. The disease begins with an inflammatory phase, followed within several weeks to several years (up to 8 years) by a chronic occlusive phase. In the initial inflammatory phase, patients develop fever, malaise, weakness, night sweats, arthralgias, and myalgias. Symptoms in the chronic phase are related to ischemia of involved organs. Vascular insufficiency is indicated by decreased or absent radial, ulnar, and carotid pulse. Approximately one third of the patients may have cardiac symptoms such as palpitations and congestive heart failure secondary to hypertension. IgG, IgA, and IgM may be elevated, and the erythrocyte sedimentation rate is usually increased. Corticosteroids may be helpful in controlling inflammation. Cyclophosphamide has been successfully used in those not responsive to corticosteroid therapy.

"take"

The successful grafting of skin that adheres to the recipient graft site 3 to 5 days following application. This is accompanied by neovascularization as indicated by a pink appearance. Thin grafts are more likely to "take" than thicker grafts, but the thin graft must contain some dermis to be successful. The term "take" also refers to an organ allotransplant that has survived hyperacute and chronic rejection.

Talmage, David Wilson (1919–)

American physician and investigator who in 1956 developed the cell selection theory of antibody formation. This work was a foundation for Burnet's subsequent clonal selection theory. *The Chemistry of Immunity in Health and Disease (with Cann),* 1961.

Tamm-Horsfall glycoprotein (uromodulin)

A 616-amino acid glycoprotein that produces immunosuppressive effects *in vitro*. It is formed by the kidney and contains 30% carbohydrate. It may appear in normal urine in humans.

tandem immunoelectrophoresis

A method that is a variation of crossed immunoelectrophoresis in which the material to be analyzed is placed in one well cut in the gel and the reference antigen is placed in a second well. Following electrophoresis in one direction, it is repeated at a right angle which drives the antigens that have been separated into another gel containing specific antibodies. Planes of precipitation form and are observed to determine whether or not they share identity with the reference antigen.

tanned red cell test

A passive hemagglutination assay in which red blood cells are used only as carrier particles for soluble antigens. Agglutination of the cells by specific antibody signifies a positive reaction. To render erythrocytes capable of adsorbing soluble protein antigens to their surface, the cells are treated with a weak tannic acid solution. This promotes cell surface attachment of the soluble protein antigen.

tanned red cells

The treatment of a suspension of erythrocytes with a 1:20,000 to 1:40,000 dilution of tannic acid which renders their surfaces capable of adsorbing soluble antigen. Thus, they have been widely used as passive carriers of soluble antigens in passive hemagglutination reactions. By adding toluene diisocyanate, the protein can become covalently bound to the red cell surface. However, this is not necessary for routine hemagglutination reactions.

TAP 1 and 2 genes (see above)

Refer to transporter in antigen processing 1 and 2 genes.

TAPA-1

A serpentine membrane protein that crosses the cell membrane four times. It is one of three proteins comprising the B cell co-receptor. It is also call CD81.

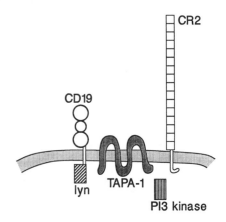

B-cell co-receptor complex
TAPA-1

Tapioca adjuvant (historical)

An immunologic adjuvant consisting of starch granules to which molecular antigen was absorbed. This permitted the adjuvant-antigen complex to form a depot in the tissues from which the antigen was slowly released to stimulate a sustained antibody response.

Taq polymerase

Thermus aquaticus polymerase. A heat-resistant DNA polymerase that greatly facilitates use of the polymerase chain reaction to amplify minute quantities of DNA from various sources into a sufficiently large quantity that can be analyzed.

target cell

A cell that is the object of an immune attack mediated either by antibodies and complement or by specifically immune lymphoid cells. The target cell must bear an antigen for which the antigen-binding regions of either antibody molecules or of the T cell receptors are specific.

tat gene

A retrovirus gene found in HIV-1. The Tat transactivating protein, which this gene encodes, gains access to the nucleus and activates viral proliferation. Additional retroviral genes become activated. Mesenchymal tumors may be induced by the *tat* genes in experimental animals.

TATA

Abbreviation for tumor-associated transplantation antigen.

TATA box

An oligonucleotide sequence comprised of thymidine-adenine-thymidine-adenine found in numerous genes that are transcribed often or rapidly.

TB

Abbreviation for tuberculosis.

Tc lymphocyte

Refer to cytotoxic T lymphocyte.

TCGF (T cell growth factor)

Interleukin-2.

TD antigen

Thymus-dependent antigen.

TdT

Abbreviation for terminal deoxynucleotidyl transferase.

T_DTH lymphocyte

A delayed-type hypersensitivity T lymphocyte.

template theory (historical)

An instructive theory of antibody formation which requires that the antigen must be present during the process of antibody synthesis. According to the refolding template theory, uncommitted and specific globulins could become refolded on the antigen, serving as a template for it. The cell thereupon releases the complementary antibodies, which thenceforth rigidly retain their shape through disulfide bonding. This theory had to be abandoned when it became clear that the specificity of antibodies in all cases is due to the particular arrangement of their primary amino acid sequence. The template theory could not explain immunological tolerance or the anamnestic (memory) immune response.

tenascin

A matrix protein produced by embryonic mesenchymal cells. It facilitates epithelial tissue differentiation and consists of six 210-kD proteins that are all alike.

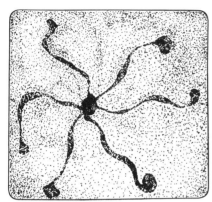

Tenascin

terminal deoxynucleotidyl transferase

Terminal deoxynucleotidyl transferase (TdT) is an enzyme catalyzing the attachment of mononucleotides to the 3′ terminus of DNA. It thus acts as a DNA polymerase. It is an enzyme present in immature B and T lymphocytes, but not demonstrable in mature lymphocytes. TdT is present both in the nuclear and soluble fractions of thymus and bone marrow. The nuclear enzyme is also able to incorporate ribonucleotides into DNA. In mice, two forms of TdT can be separated from a preparation of thymocytes. They are designated peak I and peak II. They have similar enzymatic activities and appear to be serologically related, but display significant differences in their biologic properties. Peak I appears constant in various strains of mice and at various ages. Peak II varies greatly. In some strains, peak II remains constant up to 6 to 8 months of age; in others, it declines immediately after birth. Eighty percent of bone marrow TdT is associated with a particular fraction of bone marrow cells separated on a discontinuous BSA gradient. This fraction represents 1 to 5% of the total marrow cells, but is O antigen negative. These cells become O positive after treatment with a thymic hormone, thymopoietin, suggesting that they are precursors of thymocytes. Thymectomy is associated with rapid loss of peak II and a slower loss of peak I in this bone marrow cell fraction. TdT is detectable in T cell leukemia, 90% of common acute lymphoblastic leukemia cases, and half of acute undifferentiated leukemia cells. Approximately one third of chronic myeloid leukemia cells in blast crisis and a few cases of pre-B cell acute lymphoblastic leukemia cases show cells that are positive for TdT. This marker is very infrequently seen in cases of chronic lymphocytic leukemia. In blast crisis, some cells may simultaneously express lymphoid and myeloid markers. Indirect immunofluorescence procedures can demonstrate TdT in immature B and T lymphocytes.

terminal transferase

Refer to DNA nucleotidyl exotransferase.

termination of tolerance

In several forms of tolerance, the unresponsive state can be terminated by appropriate experimental manipulation. The most common methods for breaking tolerance are the following. (1) Injection of normal T cells: Tolerance to heterologous γ globulin can be terminated by normal thymus cells. It is, however, possible only in adoptive transfer experiments with cells of tolerant animals at 81 days after the induction of tolerance and after supplementation with normal thymus cells. By this time, B cell tolerance vanishes, and only the T cells remain tolerant. Similar experiments at an earlier date do not terminate tolerance. (2) Allogeneic cells: Allogeneic cells injected at the time when B cell tolerance has vanished or has not yet been induced can also terminate or prevent tolerance. The mechanism is not specific and involves the allogeneic effect factor with activation of the unresponsive T cell population. (3) Lipopolysaccharide (LPS): This polyclonal B cell activator is also capable of terminating tolerance if the B cells are competent. It has the ability to bypass the requirements for T cells in the response to the immunogen by providing the second (mitogenic) signal required for a response. The termination of tolerance by LPS does not involve T cells at all. LPS may also circumvent tolerance to self by a similar mechanism. (4) Cross-reacting immunogens: Cross-reacting immunogens (some heterologous protein in aggregated form or a different heterologous protein) also are capable of terminating tolerance to the soluble form of the protein. Termination also occurs by a mechanism which bypasses the unresponsive T cells and is obtainable at time intervals after tolerization when the responsiveness of B cells is restored. The antibody produced to the cross-reacting antigen also reacts with the tolerogenic protein and is indistinguishable from the specificity produced by this protein in the absence of tolerance.

tertiary granule

A structure in the cytoplasm of polymorphonuclear neutrophils (PMNs) in which complement receptor 3 precursor, acid hydrolase, and gelatinase are located.

tertiary immune response

An immune response induced by a third (second booster) administration of antigen. It closely resembles the secondary (or booster) immune response.

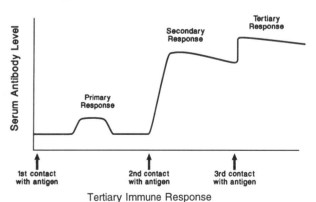

Tertiary Immune Response

tertiary structure

The folding of a polypeptide chain as a result of the interactions of its amino acid side chains which may be situated either near or

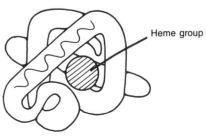

Tertiary Structure

P⌐P⌐P⌐OH + P⌐OH • • • • + P⌐OH → P⌐P⌐P⌐P⌐P⌐OH

Terminal Deoxynucleotidyl Transferase

distant along the chain. This three-dimensional folding occurs in globular proteins. Tertiary structure also refers to the spatial arrangement of protein atoms irrespective of their relationship to atoms in adjacent molecules.

test dosing

A method to determine whether or not an individual has type I anaphylactic hypersensitivity to various drugs, e.g., penicillin, or antisera prior to administration. However, the procedure is not without danger, as even a scratch test with highly diluted penicillin preparations in highly sensitized subjects has been known to produce fatal anaphylactic shock.

tetanus

A disease in which the exotoxin of *Clostridium tetani* produces tonic muscle spasm and hyperreflexia, leading to tris (lock jaw), generalized muscle spasms, spasm of the glottis, arching of the back, seizures, respiratory spasms, and paralysis. Tetanus toxin is a neurotoxin. The disease occurs 1 to 2 weeks after tetanus spores are introduced into deep wounds that provide anaerobic growth conditions.

tetanus antitoxin

Antibody raised by immunizing horses against *Clostridium tetani* exotoxin. It is a therapeutic agent to treat or prevent tetanus in individuals with contaminated lesions. Anaphylaxis or serum sickness (type III hypersensitivity) may occur in individuals receiving second injections because of sensitization to horse serum proteins following initial exposure to horse antitoxin. One solution to this has been the use of human antitetanus toxin of high titer. Treatment of the IgG fraction yields F(ab'$_2$) fragments which retain all of the toxin neutralizing capacity, but with diminished antigenicity of the antitoxin preparation.

tetanus toxin

The exotoxin synthesized by *Clostridium tetani*. It acts on the nervous system, interrupting neuromuscular transmission and preventing synaptic inhibition in the spinal cord. It binds to a nerve cell membrane glycolipid, i.e., disialosyl ganglioside. The effects of tetanus toxin are countered by specific antitoxin.

tetanus toxoid

Formaldehyde-detoxified toxins of *Clostridium tetani*. It is an immunizing preparation to protect against tetanus. Individuals with increased likelihood of developing tetanus as a result of a deep, penetrating wound with a rusty nail or other contaminated instrument are immunized by subcutaneous inoculation. The preparation is available in both fluid and adsorbed forms. It is included in a mixture with diphtheria toxoid and pertussis vaccine and is known as DTP, DPT, or triple vaccine. It is employed to routinely immunize children less than 6 years old.

tetramethylrhodamine isothiocyanate

A red fluorochrome used in immunofluorescence.

Tetramethylrhodamine Isothiocyanate

tetraparental chimera

The deliberate fusion of two-, four-, or eight-cell stage murine blastocyst ultimately yielding a mouse that is a chimera with contributions from four parents. These animals are of great value in studies on immunological tolerance.

tetraparental mouse

An allophenic mouse.

Texas red

A fluorochrome derived from sulforhodamine 101. It is often used as a second label in fluorescence antibody techniques where fluorescein, an apple-green label, is also used. This provides two-color fluorescence.

(TG)AL

Tyrosine and glutamic acid polymers fasten as side chains to a poly-L-lysine backbone through alanine residues. This substance is a synthetic antigen.

TGF (transforming growth factor[s])

Polypeptides produced by virus-transformed 3T3 cells that induce various cells to alter their phenotype.

TGF-α (transforming growth factor-α)

A polypeptide which transformed cells produce. It shares approximately one third of its 50-amino acid sequence with epidermal growth factor (EGF). TGF-α has a powerful stimulatory effect on cell growth and promotes capillary formation.

TGF-β (transforming growth factor-β)

There are five TGF-βs that are structurally similar in the C-terminal region of the protein. They are designated TGF-β1 through TGF-β5 and have similar functions with respect to their regulation of cellular growth and differentiation. After being formed as secretory precursor polypeptides, TGF-β1, TGF-β2, and TGF-β3 molecules are altered to form a 25-kD homodimeric peptide. The ability of TGF-β to regulate growth depends on the type of cell and whether or not other growth factors are also present. It also regulates deposition of extracellular matrix and cell attachment to it. It induces fibronectin, chondroitin/dermatin sulfate proteoglycans, collagen, and glycosaminoglycans. TGF-β also promotes the formation and secretion of protease inhibitors. It has been shown to increase the rate of wound healing and induce granulation tissue. It also stimulates proliferation of osteoblasts and chondrocytes. TGF-β inhibits bone marrow cell proliferation and also blocks interferon α-induced activation of natural killer (NK) cells. It diminishes IL-2 activation of lymphokine activated killer cells. TGF-β decreases cytokine-induced proliferation of thymocytes and also decreases IL-2-induced proliferation and activation of mature T lymphocytes. It inhibits T cell precursor differentiation into cytotoxic T lymphocytes. TGF-β may reverse the activation of macrophages by preventing the development of cytotoxic activity and superoxide anion formation that is needed for antimicrobial effects. In addition to suppressing macrophage activation, TGF may diminish MHC class II molecule expression. It also decreases Fcε receptor expression in allergic reactions. TGF-β has potential value as an immunosuppressant in tissue and organ transplantation. It may protect bone marrow stem cells from the injurious effects of chemotherapy. It may also have use as an antiinflammatory agent based on its ability to inhibit the growth of both T and B cells. It has potential as a possible treatment for selected autoimmune diseases. It diminishes myocardial damage associated with coronary occlusion, promotes wound healing, and may be of value in restoring collagen and promoting formation of bone in osteoporosis patients.

T$_H$0 cells

A CD4$^+$ T cell subset in both humans and mice based on cytokine production and effector functions. T$_H$0 cells synthesize multiple cytokines. They are responsible for effects intermediate between those of T$_H$1 and T$_H$2 cells, based on the cytokines synthesized and the responding cells. T$_H$0 cells may be precursors of T$_H$1 and T$_H$2 cells.

T$_H$1 cells

A CD4$^+$ T cell subset in both humans and mice based on cytokine production and effector functions. T$_H$1 cells synthesize interferon-gamma (IFN-γ), IL-2, and tumor necrosis factor (TNF)-β. They are mainly responsible for cellular immunity against intracellular microorganisms and for delayed-type hypersensitivity reactions. They affect IgG2a antibody synthesis and antibody-dependent cell-mediated cytotoxicity. T$_H$1 cells activate host defense mediated by phagocytes. Intracellular microbial infections induce T$_H$1 cell development which facilitates elimination of the microorganisms by phagocytosis. T$_H$1 cells induce synthesis of antibody that activates complement and serves as an opsonin that facilitates phagocytosis. The IFN-γ they produce enhances macrophage activation.

T$_H$2 cells

A CD4$^+$ T cell subset in both humans and mice based on cytokine production and effector functions. T$_H$2 cells synthesize IL-4, IL-5, IL-6, IL-9, IL-10, and IL-13. They greatly facilitate IgE and IgG1 antibody responses, and mucosal immunity, by synthesis of mast cell and eosinophil growth and differentiation factors and facilitation of IgA synthesis. IL-4 facilitates IgE antibody synthesis. IL-5 is an eosinophil activating substance. IL-10, IL-13, and IL-4 suppress cell-mediated immunity. T$_H$2 cells are principally responsible for host defense exclusive of phagocytes. They are crucial for the IgE and eosinophil response to helminths and for allergy attributable to activation of basophils and mast cells through IgE.

Theiler, Max (1899–1972)

South African virologist who received the Nobel Prize in 1951 "for his development of vaccines against yellow fever".

Theiler's virus myelitis

Murine spinal cord demyelination that is considered to be an immune-based consequence of a viral infection.

theliolymphocyte

Intraepithelial lymphocyte. These are small lymphocytes associated with intestinal epithelial cells.

theophylline (1,3,dimethylxanthine)

A compound used to treat acute bronchial asthma because of its powerful smooth muscle relaxing activity. Aminophylline, a salt of theophylline, is also a smooth muscle relaxant used to induce bronchodilation in the treatment of asthma.

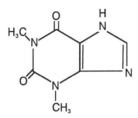

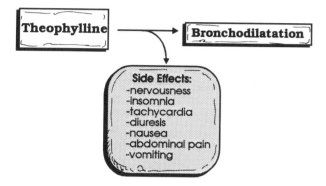

therapeutic antisera

Serum antibody preparations employed to either protect against disease or for disease therapy. They are distinct from antibody or antisera preparations used for the serological identification of microorganisms. Therapeutic antisera, such as horse antitoxin against diphtheria, were widely used earlier in the 20th century. A few specific antisera such as tetanus antitoxin are still used.

θ antigen

Refer to Thy 1 antigen.

Thomas, Edward Donnal (1920–)

American hematologist whose fundamental investigations of bone marrow transplantation led to his winning the Nobel Prize for medicine or physiology (with J. E. Murray) in 1990.

thoracic duct

A canal that leads from the cisterna chyli, a dilated segment of the thoracic duct at its site of origin in the lumbar region, to the left subclavian vein.

thoracic duct drainage

The deliberate removal of lymphocytes through drainage of lymph from the thoracic duct with a catheter.

thorotrast (thorium dioxide ^{32}THOT)

A radiocontrast medium that yields α particles. It is no longer in use since neoplasia have been attributed to the substance. It is removed by the reticuloendothelial (mononuclear phagocyte) system. It induced hepatic angiosarcoma and also cholangiocarcinoma and hepatocellular carcinoma in some patients who received it. It has been known to produce other neoplasms. In immunology, it has been used in experimental animal studies involving the blockade of the reticuloendothelial system.

threonyl-transfer RNA synthetase antibodies

An antibody against threonyl-tRNA synthetase (threonyl RS) protein that shows high specificity for myositis. These antibodies were demonstrated in 4% of polymyositis/dermatomyositis patients. Anti-threonyl RS antibodies have also been linked to the anti-synthetase syndrome in which there is fever, Raynaud's phenomenon, symmetrical arthritis, interstitial lung disease, myositis and mechanic's hands.

thrombocyte

Blood platelet.

thrombocytopenia

Diminished blood platelet numbers with values below 100,000 per cubic millimeter of blood compared to a normal value of 150,000 to 300,000 platelets per cubic millimeter of blood. This decrease in numbers of blood platelets can lead to bleeding.

thrombocytopenic purpura, idiopathic

An autoimmune disease in which antiplatelet autoantibodies destroy platelets. Splenic macrophages remove circulating platelets coated with IgG autoantibodies at an accelerated rate. Thrombocytopenia occurs even though the bone marrow increases platelet production. This can lead to purpura and bleeding. The platelet count may fall below 20,000 to 30,000 per microliter. Antiplatelet antibodies are detectable in the serum and on platelets. Platelet survival is decreased. Splenectomy is recommended in adults. Corticosteroids facilitate a temporary elevation in the platelet count. This disease is characterized by decreased blood platelets, hemorrhage, and extensive thrombotic lesions.

thrombocytosis

Elevated blood platelet numbers with values exceeding 600,000 thrombocytes per cubic millimeter of blood compared to a normal value of 150,000 to 300,000 blood platelets per cubic millimeter of blood.

thromboxanes

A group of biologically active compounds with a physiological role in homeostasis and a pathophysiological role in thromboembolic disease and anaphylactic reactions. They are cyclopentane derivatives of polyunsaturated fatty acids and are derived by isomerization from prostaglandin endoperoxide PGH_2, the immediate precursor. The isomerizing enzyme is called thromboxane synthetase. The active compound, thromboxane A_2, is unstable, being degraded to thromboxane B_2 which is stable but inactive on blood vessels; it has, however, polymorphonuclear cell chemotactic activity. The short notation is TXA_2 and TXB_2. TXA_2 and TXB_2 represent the major pathway of conversion of prostaglandin endoperoxide precursors. TXA_2, derived from prostaglandin G_2 generated from arachidonic acid by cyclooxygenase, increases following injury to vessels. It stimulates a primary hemostatic response. TXA_2 is a potent inducer of platelet aggregation, smooth muscle contraction, and vasoconstriction. TXA was previously called rabbit aorta contacting substance (RACS) and is isolated from lung perfusates during anaphylaxis. It appears to be a peptide containing less than ten amino acid residues. Thromboxane formation in platelets is associated with the dense tubular system. PMNs and spleen, brain, and inflammatory granulomas have been demonstrated to produce thromboxanes.

thy (θ)

Epitopes found on murine thymocytes and the majority of murine T lymphocytes.

Thy-1

A murine and rat thymocyte surface glycoprotein that is also found in neuron membranes of several species. Thy-1 was originally termed the θ alloantigen. Thy.1 and Thy.2 are the two allelic forms. A substitution of one amino acid, arginine or glutamine, at position 89 represents the difference between Thy1.1 and Thy1.2. Mature T cells and thymocytes in the mouse express Thy-1. The genes encoding Thy-1 are present on chromosome 9 in the mouse. Whereas few human lymphocytes express Thy-1, it is present on the surfaces of neurons and fibroblasts.

Thy 1 antigen

A murine isoantigen present on the surface of thymic lymphocytes and on thymus-derived lymphocytes found in peripheral lymphoid tissues. Central nervous system tissues may also express Thy 1 antigen.

Thy-1$^+$ dendritic cells

Cells derived from the T lymphocyte lineage that are found within the epithelium of the mouse epidermis.

thymectomy

Surgical removal of the thymus, which leads to failure to develop cell-mediated immunity and humoral immunity to thymus-dependent antigens in mice when thymectomized as neonates. T cells fail to develop, lymphoid tissues atrophy, and the failure to gain weight characterize runt or wasting disease.

thymic alymphoplasia

Severe combined immune deficiency transmitted as an X-linked recessive trait.

thymic cortex

Refer to thymus.

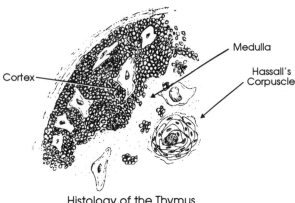

Histology of the Thymus
Thymic Cortex

thymic epithelial cell

Cells present in the cortex and, to a lesser degree, in the medulla of the thymus that are derived from the third and fourth pharyngeal pouches. They affect maturation and differentiation of thymocytes through the secretion of thymopoietin, thymosins, and serum thymic factors. Thymic epithelial cells express both MHC class I and class II molecules.

thymic hormones

Soluble substances synthesized by thymic epithelial cells that promote thymocyte differentiation. They include thymopoietin and thymosins, which are peptides that help to regulate differentiation of T lymphocytes.

thymic humoral factor(s) (THF)

Soluble substances such as thymosins, thymopoietin, serum thymic factor, etc. which are synthesized by the thymus and govern differentiation and function of lymphocytes.

thymic hypoplasia

An immunodeficiency that selectively affects the T cell limb of the immune response. Early symptoms soon after birth may stem from associated parathyroid abnormalities leading to hypocalcemia and heart defects which may lead to congestive heart failure. There is lymphopenia with diminished T cell numbers. T lymphocyte function cannot be detected in peripheral blood T cells. There is variation in antibody levels and function. The condition has been successfully treated by thymic transplantation. Some DiGeorge patients may have normal B cell immunity. All the others may have diminished immunoglobulin levels and may not form specific antibody following immunization. Clinically, DiGeorge patients may have a fish-shaped mouth, and abnormal facies with low-set ears, hypertelorism, and antimongoloid eyes, in addition to the other features mentioned above.

thymic medullary hyperplasia

The finding of germinal centers in the thymic medulla in myasthenia gravis patients. However, normal thymus glands may occasionally contain germinal centers, although the vast majority of normal thymus glands do not.

thymic nurse cell

Relatively large epithelial cells that are very near thymic lymphocytes and are believed to have a significant role in T lymphocyte maturation and differentiation.

thymic stromal-derived lymphopoietin (TSLP)

A cytokine isolated from a murine thymic stromal cell line that possesses a primary sequence distinct from other known cytokines. The cDNA encodes a 140-amino acid protein that includes a 19-amino acid signal sequence. TSLP stimulates B220+ bone marrow cells to proliferate and express surface μ. TSLP synergizes with other signals to induce thymocyte and peripheral T cell proliferation but is not mitogenic for T cells alone.

thymin

A hormone extracted from the thymus that has an activity resembling that of thymopoietin.

thymocyte

A lymphocyte in the thymus gland.

thymoma

A rare neoplasm of epithelial cells often with an associated thymic lymphoproliferation that is benign. Half of these tumors occur in patients with myasthenia gravis. They may also be associated with immunodeficiency.

thymopentin (TP-5)

A 49-amino acid polypeptide thymic hormone secreted by epithelial cells in the thymus. It affects neuromuscular transmission, induces early T lymphocyte differentiation, and affects immune regulation. Thymopentin functions biologically to normalize immune imbalances related to either hypo- or hyperresponsiveness. These could be related to thymic involution with age, thymectomy, or other factors that result in immunologic imbalance. It is a 7-kD protein that facilitates the expression of Thy-1 antigen on T lymphocytes that can interfere with neuromuscular transmission, an effect which has been implicated in myasthenia gravis patients who often develop thymoma.

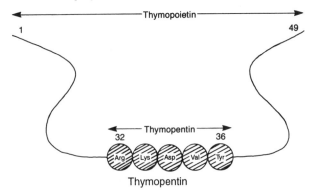

Thymopentin

thymosine (see facing page)

A 12-kD protein hormone produced by the thymus gland that can provide T lymphocyte immune function in animals that have been thymectomized.

thymosine α-1 (thymopoietin) (see facing page)

A hormone produced by the thymus that stimulates T lymphocyte helper activity. It induces production of lymphokines such as interferon and macrophage-inhibiting factor. It also enhances Thy-1.2 and Lyt-1-2-3 antigens of T lymphocytes. It may also alter thymocyte TdT concentrations.

thymotaxin

Refer to β_2 microglobulin.

thymulin

A nonapeptide (Glu-Ala-Lys-Ser-Gln-Gly-Ser-Asn) extracted from blood sera of humans, pigs, and calf thymus. Thymulin shows a strong binding affinity for the T cell receptor on the lymphocyte membrane. Its zinc-binding property is associated with biological activity. Thymulin's enhancing action is reserved exclusively for T lymphocytes. It facilitates the function of several T lymphocyte subpopulations, but mainly enhances T suppressor lymphocyte activity. Formerly called FTS.

thymus

A triangular bilobed structure enclosed in a thin fibrous capsule and located retrosternally. Each lobe is subdivided by prominent trabeculae into interconnecting lobules, and each lobule comprises two histologically and functionally distinct areas, cortex and medulla. The cortex consists of a mesh of epithelial-reticular cells enclosing densely packed large lymphocytes. It has no germinal centers. The epithelial cell component is of endodermal origin; the lymphoid cells are of mesenchymal origin. The prothymocytes, which migrate from the bone marrow to the subcapsular regions of the cortex, are influenced by this microenvironment which directs their further development. The process of education is exerted by hormonal substances produced by the thymic epithelial cells. The cortical cells proliferate extensively. Part of these cells are short lived and die. The surviving cells acquire characteristics of thymocytes. The cortical cells migrate to the medulla and from there to the peripheral lymphoid organs, sites of their main residence. The medullary areas of the thymus are even richer in epithelial cells, and the lymphocytes in the medulla are loosely packed. The lymphocytes are small cells ready to exit the thymus. Some remnants of epithelial islands, called Hassall's corpuscles, are histologically

Proposed Sites of Action of Thymosin Polypeptides
on Maturation of T-cell Subpopulations

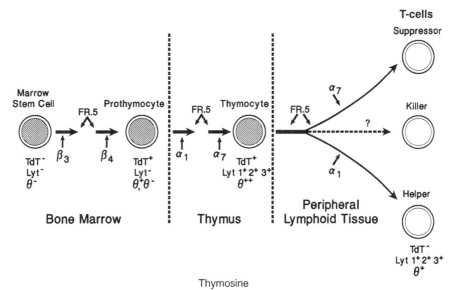

Thymosine

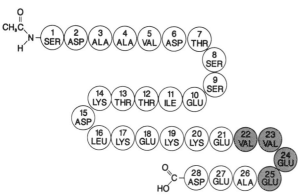

Thymosine α-1

cytes, giving a false impression of an increased number of Hassal's corpuscles. The cortex atrophies progressively. The blood-thymus barrier protects thymocytes from contact with antigen. Lymphocytes reaching the thymus are prevented from contact with antigen by a physical barrier. The first level is represented by the capillary wall with endothelial cells inside the pericytes outside of the lumen. Potential antigenic molecules which escape the first level of control are taken over by macrophages present in the pericapillary space. Further protection is provided by a third level, represented by the mesh of interconnecting epithelial cells which enclose the thymocyte population. The effects of thymus and thymic hormones on the differentiation of T cells is demonstrable in animals congenitally lacking the thymus gland (nu/nu animals), neonatally or adult thymectomized animals, and in subjects with

Blood-Thymus Barrier

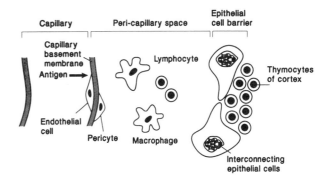

identifiable and are a marker for thymic tissue. The blood supply to the cortex comes from capillaries that form anastomosing arcades. Drainage is mainly through veins; the thymus has no lymphatic vessels. The thymus develops from the branchial pouches of the pharynx at about 6 weeks of embryonal age. In most species, it is fully developed at birth. In humans, the weight of the thymus at birth is 10 to 15 g. It continues to increase in size, reaching a maximum (30 to 40 g) at puberty. It then begins to involute with increasing age, but the adult gland is still functional. The medulla involutes first with pyknosis and beading of the nuclei of small lympho-

Structure of Thymus

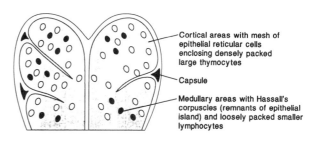

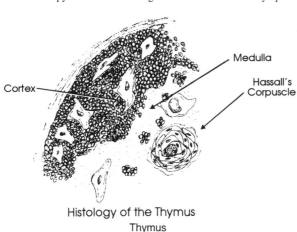

Histology of the Thymus
Thymus

immunodeficiencies involving T cell function. Differentiation is associated with surface markers whose presence or disappearance characterizes the different stages of cell differentiation. There is extensive proliferation of the subcapsular thymocytes. The largest proportion of these cells die, but the remaining cells continue to differentiate. The differentiating cells become smaller in size and move through interstices in the thymic medulla. The fully developed thymocytes pass through the walls of the postcapillary venules to reach the systemic circulation and seed in the peripheral lymphoid organs. Part of them recirculate, but do not return to the thymus.

thymus cell differentiation

Stem cell maturation and differentiation into mature T lymphocytes in the thymus is accompanied by the appearance and disappearance of specific surface CD antigens. In humans, the differentiation of CD38 positive stem cells into early thymocytes is signaled by the appearance of CD2 and CD7, followed by the transferrin receptor marker. This is followed by expression of CD1, which identifies thymocytes in the mid-stage of differentiation when T cell receptor genes γ and δ and later α and β rearrange. This is followed by the expression of CD3, CD4, and CD8 surface antigens by thymocytes, yet CD1 usually disappears at this time. Ultimately, the CD4⁺, CD8⁻, and the CD4⁻CD8⁺ subpopulations which both express the CD3 pan-T cell marker appear. An analogous maturation of T cells takes place in mice.

thymus cell education

Thymus cell differentiation.

thymus-dependent (TD) antigen

An immunogen that requires T lymphocyte cooperation for B cells to synthesize specific antibodies. Presentation of thymus-dependent antigen to T cells must be in the context of MHC class II molecules. Thymus-dependent antigens include proteins, polypeptides, hapten-carrier complexes, erythrocytes, and many other antigens that have diverse epitopes.

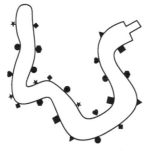

Thymus-Dependent Antigen

thymus-dependent areas

Regions of peripheral lymphoid tissues occupied by T lymphocytes. Specifically, these include the paracortical areas of lymph nodes, the zone between nodules and Peyer's patches, and the center of splenic Malpighian corpuscles. These regions contain small lymphocytes derived from the circulating cells that reach these areas by passage through high endothelial venules. Proof that these anatomical sites are thymus-dependent areas is provided by the demonstration that animals thymectomized as neonates do not have lymphocytes in these areas. Likewise, humans or animals with thymic hypoplasia or congenital aplasia of the thymus reveal no T cells in these areas.

thymus-dependent cells

Lymphoid cells that mature only under the influence of the thymus.

thymus-independent (TI) antigen

An immunogen that can stimulate B cells to synthesize antibodies without participation by T cells. These antigens are less complex than are thymus-dependent antigens. They are often polysaccharides that contain repeating epitopes or lipopolysaccharides derived

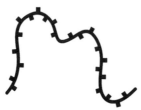

Thymus-Independent Antigen

from Gram-negative microorganisms. Thymus-independent antigens induce IgM synthesis by B lymphocytes without cooperation by T cells. They also do not stimulate immunological memory. Murine TI antigens are classified as either TI-1 or TI-2 antigens. Lipopolysaccharide (LPS), which activates murine B cells without participation by T or other cells, is a typical TI-1 antigen. Low concentrations of LPS stimulate synthesis of specific antigen, whereas high concentrations activate essentially all B cells to grow and differentiate. TI-2 antigens include polysaccharides, glycolip-

ids, and nucleic acids. When T lymphocytes and macrophages are depleted, no antibody response develops against them.

thymus-replacing factor (TRF)

Interleukin-5.

thyroglobulin

A thyroid protein demonstrable by immunoperoxidase staining that serves as a marker for papillary and/or follicular thyroid carcinomas.

thyroid antibodies

Autoantibodies present in patients with Hashimoto's thyroiditis or those with thyrotoxicosis (Graves' disease) that are organ specific for the thyroid. Antibodies against thyroglobulin and antibodies against the microsomal antigen of thyroid acinar cells may appear in patients with autoimmune thyroiditis. Antibodies against TSH receptors appear in Graves' disease patients and cause stimulatory hypersensitivity. They mimic the action of TSH. This is an IgG molecule termed long-acting thyroid stimulator (LATS). LATS levels are increased in many patients with thyrotoxicosis or Graves' disease.

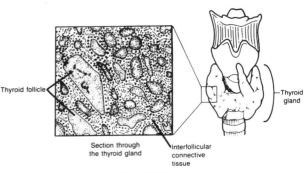

Thyroid follicle

Thyroid gland

Section through the thyroid gland

Interfollicular connective tissue

Thyroid Gland

thyroiditis, autoimmune

Hashimoto's disease (chronic thyroiditis) is an inflammatory disease of the thyroid found most frequently in middle-aged to older women. There is extensive infiltration of the thyroid by lymphocytes which completely replace the normal glandular structure of the organ. There are numerous plasma cells, macrophages, and germinal centers which give the appearance of node structure within the thyroid gland. Both B cells and CD4⁺ T lymphocytes comprise the principal infiltrating lymphocytes. Thyroid function is first increased as the inflammatory reaction injures thyroid follicles causing them to release thyroid hormones. However, this is soon replaced by hypothyroidism in the later stages of Hashimoto's thyroiditis. Patients with this disease have an enlarged thyroid gland. There are circulating autoantibodies against thyroglobulin and thy-

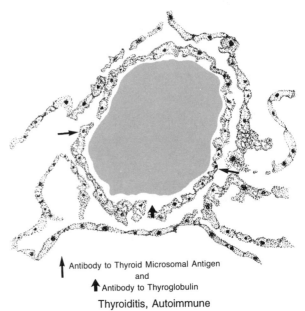

Antibody to Thyroid Microsomal Antigen
and
Antibody to Thyroglobulin

Thyroiditis, Autoimmune

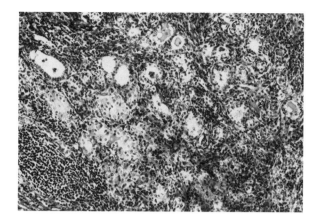

Hashimoto's thyroiditis

roid microsomal antigen (thyroid peroxidase). Cellular sensitization to thyroid antigens may also be detected. Thyroid hormone replacement therapy is given for the hypothyroidism that develops.

thyrotoxicosis

Disease of the thyroid in which there is hyperthyroidism with elevated levels of thyroid hormones in the blood and thyroid gland hyperplasia or hypertrophy. Thyrotoxicosis may be autoimmune, as in Graves' disease in which there may be diffuse goiter. Autoantibodies specific for thyroid antigens mimic-thyroid stimulating hormones (TSH) by stimulating thyroid cell function. In addition, patients with Graves' disease develop ophthalmopathy and proliferative dermopathy. The disease occurs predominantly in females (70 females to 1 male) and appears usually in the 30- to 40-year-old age group. In Caucasians, it is a disease associated with DR3. Patients may develop nervousness, tachycardia, and numerous other symptoms of hyperthyroidism. They also have increased levels of total and free T3 and T4. There is a diffuse and homogeneous uptake of radioactive iodine in these patients. Three types of antithyroid antibodies occur: (1) thyroid-stimulating immunoglobulin, (2) thyroid growth-stimulating immunoglobulin, and (3) thyroid binding-inhibitory immunoglobulin. Their presence confirms a diagnosis of Graves' disease. The thyroid gland may be infiltrated with lymphocytes. Long-acting thyroid stimulator (LATS) is classically associated with thyrotoxicosis. It is an IgG antibody specific for thyroid hormone receptors. It induces thyroid hyperactivity by combining with TSH receptors.

thyrotropin

Thyroid stimulating hormone (TSH).

TIL

Abbreviation for tumor-infiltrating lymphocytes.

tine test

A human tuberculin test that involves the intradermal inoculation of dried, old tuberculin using a four-pointed applicator that introduces the test substance 2 mm below the surface.

tingible body

Nuclear debris present in macrophages of lymph node, spleen, and tonsil germinal centers, as well as in the dome of the appendix.

tissue-fixed macrophage

Histiocyte.

tissue-specific antigen

An antigen restricted to cells of one type of tissue. Tissue-specific or organ-specific autoantibodies occur in certain types of autoimmune diseases. Organ-specific or tissue-specific antibodies are often not species specific. For example, autoantibodies against human thyroglobulin may crossreact with the corresponding molecules of other species.

tissue typing

The identification of major histocompatibility complex class I and class II antigens on lymphocytes by serological and cellular techniques. The principal serological assay is microlymphocytotoxicity using microtiter plates containing predispensed antibodies against HLA specificities to which lymphocytes of unknown specificity plus rabbit complement and vital dye are added. Following incubation, the wells are scored according to the relative proportion of cells killed. This method is employed for organ transplants such as renal allotransplants. For bone marrow transplants, mixed lymphocyte culture (MLC), also called mixed lymphocyte reaction, procedures are performed to determine the relative degree of histocompatibility or histoincompatibility between donor and recipient. Serological tests are largely being replaced by DNA typing procedures employing polymerase chain reaction (PCR) methodology and DNA or oligonucleotide probes, especia, y for MHC class II typing.

titer

An approximation of the antibody activity in each unit volume of a serum sample. The term is used in serological reactions and is determined by preparing serial dilutions of antibody to which a constant amount of antigen is added. The end point is the highest dilution of antiserum in which a visible reaction with antigen, e.g., agglutination, can be detected. The titer is expressed as the reciprocal of the serum dilution which defines the end point. If agglutination occurs in the tube containing a 1:240 dilution, the antibody titer is said to be 240. Thus, the serum would contain approximately 240 units of antibody per milliliter of antiserum. The titer only provides an estimate of antibody activity. For absolute amounts of antibody, quantitative precipitation or other methods must be employed.

TL (thymic-leukemia antigen)

An epitope on the thymocyte membrane of TL+ mice. As the T lymphocytes mature, this antigen disappears, but resurfaces if leukemia develops. TL antigens are specific and are normally present on the cell surface of thymocytes of certain mouse strains. They are encoded by a group of structural genes located at the Tla locus, in the linkage group IX, very close to the D pole of the H-2 locus on chromosome 17. There are three structural TL genes, one of which has two alleles. The TL antigens are numbered from 1 to 4 specifying four antigens: TL.1, TL.2, TL.3, and TL.4. TL.3 and TL.4 are mutually exclusive. Their expression is under the control of regulatory genes, apparently located at the same Tla locus. Normal mouse thymocytes belong to three phenotypic groups: Tl-, TL.2, and TL.1,2,3. Development of leukemia in the mouse induces a restructuring of the TL surface antigens of thymocytes with expression of TL.1 and TL.2 in TL- cells, expression of TL.1 in TL.2 cells, and expression of TL.4 in both TL- and TL.2 cells. When normal thymic cells leave the thymus, the expression of TL antigen ceases. Thus, thymocytes are TL+ (except the TL- strains) and the peripheral T cells are TL-. In transplantation experiments, TL+ tumor cells undergo antigenic modulation. Tumor cells exposed to the homologous antibody stop expressing the antigen and thus escape lysis when subsequently exposed to the same antibody plus complement.

Tla antigen

Murine MHC class I histocompatibility antigen encoded by genes that are situated near the Qa region on chromosome 17. Thymocytes may express products of up to six alleles. Leukemia cells may aberrantly express Tla antigens.

Tla complex

Genes that map to the MHC region telomeric to H2 loci on chromosome 17 in mice. These genes encode MHC class I proteins such as Qa and Tla that have no known immune function. Qa and Tla proteins that closely resemble H2 MHC class I proteins in sequence, associate noncovalently with β-2 microglobulin. Expression of Qa and Tla proteins, unlike expression of MHC H-2 class I proteins is limited to only selected mouse cells. For example, only hepatocytes express Q10 protein, only selected lymphocyte subpopulations such as activated T lymphocytes express Qa-2 proteins, and T lymphocytes express Tla proteins. Thus, Qa and Tla class I molecules differ in structure and expression from the remaining MHC class I genes and proteins in the mouse. This could account for their failure to function in antigen presentation to T lymphocytes.

T$_m$

Membrane immunoglobulin heavy chain tail (C-terminal) polypeptide.

TNF

Abbreviation for tumor necrosis factor.

TNP

Abbreviation for trinitrophenol group.

tolerance

Active state of unresponsiveness by lymphoid cells to a particular antigen (tolerogen) as a result of their interaction with that antigen. The immune response to all other immunogens is unaffected. Thus, this is an acquired nonresponsiveness to a specific antigen. When inoculated into a fetus or a newborn, an antigenic substance will be tolerated by the recipient in a manner that will prevent manifestations of immunity when the same individual is challenged with this antigen as an adult. This treatment has no suppressive effect on the response to other unrelated antigens. Immunologic tolerance is much more difficult to induce in adults whose immune systems are fully developed. However, it can be accomplished by administering repetitive minute doses of protein antigens or by administering them in large quantities. Mechanisms of tolerance induction have been the subject of numerous investigations, and clonal deletion is one of these mechanisms. Either helper T or B lymphocytes may be inactivated or suppressor T lymphocytes may be activated in the process of tolerance induction. In addition to clonal deletion, clonal anergy and clonal balance are among the complex mechanisms proposed to account for self-tolerance in which the animal body accepts its own tissue antigens as self and does not reject them. Nevertheless, certain autoantibodies form under physiologic conditions and are not pathogenic, whereas autoimmune phenomena may form under disease conditions and play a significant role in the pathogenesis of autoimmune disease.

An immunological adaptation to a specific antigen is distinct from unresponsiveness, the genetic or pathologic inability to mount a measurable immune response. Tolerance involves lymphocytes as individual cells, whereas unresponsiveness is an attribute of the whole organism. The humoral or cell-mediated response may be affected individually or at the same time. The genetic form of unresponsiveness has been demonstrated with the immune response to synthetic antigens and has led to characterization of the immune response (Ir) locus of the major histocompatibility complex. The immune response of experimental animals, which are classified as high, intermediate, or as nonresponders, is not defective, but is not reactive to the particular antigen. In some cases, suppressor cells prevent the development of an appropriate response. Unresponsiveness may also be the result of immunodeficiency states, some with clinical expression, or may be induced by immunosuppressive therapy such as that following X-irradiation, chemotherapeutic agents, or antilymphocyte sera. Tolerance, as the term is currently used, has a broader connotation and is intended to represent all instances in which an immune response to a given antigen is not demonstrable. Immunologic tolerance refers to a lack of response as a result of prior exposure to antigen. Refer to immunological tolerance.

tolerogen

An antigen that is able to induce immunologic tolerance. The production of tolerance rather than immunity in response to antigen depends on such variables as physical state of the antigen, i.e., soluble or particulate, route of administration, level of maturation of the recipient's immune system, or immunologic competence. For example, soluble antigens administered intravenously will favor tolerance in many situations, as opposed to particulate antigens injected into the skin which might favor immunity. Immunologic tolerance with cells is easier to induce in the fetus or neonate than it is in adult animals, who would be more likely to develop immunity rather than tolerance.

tolerogenic

The capacity of a substance such as an antigen to induce immunological tolerance.

tolmetin

(1-methyl-5-p-toluoylpyrrole-2-acetic acid). An antiinflammatory agent effective in the therapy of arthritis, including juvenile rheumatoid arthritis, rheumatoid arthritis, and ankylosing spondylitis.

Tonegawa, Susumu (1939–)

Japanese born immunologist working in the United States. He received the Nobel Prize in 1987 for his research on immunoglobulin genes and antibody diversity. Tonegawa and many colleagues were responsible for the discovery of immunoglobulin gene C, V, J, and D regions and their rearrangement.

tonsil

Lymphoid tissue masses at the intersection of the oral cavity and the pharynx, i.e., in the oropharynx. Tonsils contain mostly B lymphocytes and are classified as secondary lymphoid organs. There are several types of tonsils designated as palatine, flanked by the palatoglossal and palatopharyngeal arches, the pharyngeal, which are adenoids in the posterior pharynx, and the lingual, at the tongue's base.

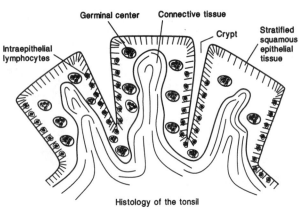

Tonsil

Histology of the tonsil

TORCH panel

A general serologic screen to identify antenatal infection. Elevated levels of IgM in a neonate reflect *in utero* infection. The panel may be further refined by determining IgM antibody specific for certain microorganisms. TORCH is an acronym for toxoplasma, other, rubella, cytomegalic inclusion virus, herpes (and syphilis). There are both false-positive and false-negative reactions in the quantitative TORCH screen. If the TORCH panel is positive, it is indicative of *in utero* infection which may have major consequences. Toxoplasmosis may result in microglial nodules, thrombosis, necrosis, and blocking of the foramina, leading to hydrocephalus. Rubella may cause hepatosplenomegaly, congenital heart disease, petechiae and purpura, decreased weight at birth, microcephaly, cataracts, and central nervous system manifestations including seizures and bulging fontanelles. Cytomegalovirus is characterized by hepatosplenomegaly, hyperbilirubinemia, microcephaly, thrombocytopenia at birth followed later by deafness, mental retardation, learning disabilities, and other manifestations. Herpes simplex can lead to premature birth. The central nervous system manifestations include seizures, chorioretinitis, paralysis that is either flaccid or spastic, and coma. Syphilis is an addendum to the TORCH designation, but congenital syphilis has increased in recent years and is not associated with specific clinical findings.

total lymphoid irradiation (TLI)

A technique to induce immunosuppression in which lymphoid organs are irradiated, whereas other organs are protected from irradiation. This method has been used in the therapy of lymphomas.

toxic complexes

Increased levels of circulating immune complexes may be harmful and trigger type III hypersensitivity reactions. The soluble complexes are pathogenic. The classic description regards such complexes as "toxic", and the term "toxic complexes" was frequently used in the literature. Such complexes are characterized by (1) formation in a zone of moderate antigen excess, (2) the antibody in the complex has no cytotropic affinity for tissues, and (3) the complex is able to activate the complement system. Complex formation is associated with conformational changes in the antibody molecule, and the activity of the complex depends on the antibody and not on the antigen. Antibodies produced in some species such as rabbit, human, and guinea pig have the above properties. Those produced in other species such as bovine, chicken, and horse are inactive in this respect. Fixation of the complexes occurs by the Fc portion of the antibody in the complex. They stick to cells and basement membranes, causing injury to the endothelium of small vessels. The injury may occur at the local site of antigen injection or may be systemic when antigen is injected intravenously. The chain of events characteristic for inflammation is set in motion with liberation of vasoactive amines and involvement of polymorphonuclear leukocytes.

toxic epidermal necrolysis

A hypersensitivity reaction to certain drugs such as allopurinol, nonsteroidal antiinflammatory drugs, barbiturates, sulfonamides such as sulfmethoxazole-trimethoprim, carbamazepine, and other agents. It may closely resemble erythema multiforme. Patients develop erythema, subepidermal bullae, and open epidermal lesions. They become dehydrated, show imbalance of electrolytes, and often develop abscesses with sepsis and shock. Toxic epidermal necrolysis may also be observed in a hyperacute type of graft-vs.-host reaction, especially in some babies receiving bone marrow transplants.

toxin neutralization (by antitoxin)

Toxicity is titrated by injection of laboratory animals, and the activity of antitoxins is evaluated by comparison with standard antitoxin of known protective ability. Antitoxin combines with toxin in varying proportions, depending on the ratio in which they are combined, to form complexes which prove nontoxic when injected into experimental animals. Mixing of antitoxin and toxin in optimal proportions may result in flocculation. If toxin is added to antitoxin in several fractions with time intervals between them instead of all at once, more antitoxin is required for neutralization to occur than would be necessary if all toxin had been added at once. This means that toxins are polyvalent. This phenomenon is explained by the ability of toxin to combine with antitoxin in multiple proportions. Neutralization does not destroy the reacting toxin. In many instances, toxin may be recovered by dilution of the toxin-antitoxin mixture. The effect of heat on a zootoxin is illustrated by the destruction of cobra venom antitoxin if cobra venom (the toxin)-antivenom (antitoxin) mixtures are subjected to boiling. The venom or toxin remains intact. Since toxins have specific affinities for certain tissues of the animal body, such as the high affinity of tetanus toxin for nervous tissue, antitoxins are believed to act by binding toxins before they have the opportunity to combine with specific tissue cell receptors.

toxins

A poison that is usually immunogenic and stimulates production of antibodies termed antitoxins with the ability to neutralize the harmful effects of the particular toxins eliciting their synthesis. The general groups of toxins include (1) bacterial toxins — those produced by microorganisms such as those causing tetanus, diphtheria, botulism, and gas gangrene, as well as toxins of staphylococci; (2) phytotoxins — plant toxins such as ricin of the castor bean, crotein, and abrin derived from the Indian licorice seed, Gerukia; and (3) zootoxins — snake, spider, scorpion, bee, and wasp venom.

toxoid

A treatment of a microbial toxin with formaldehyde to inactivate toxicity, but leave the immunogenicity (antigenicity) of the preparation intact. Toxoids are prepared from exotoxins produced in diphtheria and tetanus. These are used to induce protective immunization against adverse effects of the exotoxins in question.

toxoid-antitoxin floccules

An immunizing preparation used to induce active immunity against diphtheria in subjects who show adverse reactions to alum-precipitated toxoid. The preparation consists of diphtheria toxoid combined with diphtheria antitoxin in the presence of minimal excess antigen. It has been used in individuals who are hypersensitive to alum-precipitated toxoid alone. Horse serum in the preparation may induce hypersensitivity to horse protein in some subjects.

Tp44 (CD28)

A T lymphocyte receptor that regulates cytokine synthesis, thereby controlling responsiveness to antigen. Its significance in regulating activation of T lymphocytes is demonstrated by the ability of monoclonal antibody against CD28 receptor to block T cell stimulation by specific antigen. During antigen-specific activation of T lymphocytes, stimulation of the CD28 receptor occurs when it combines with the B7/BB1 coreceptor during the interaction between T and B lymphocytes. CD28 is a T lymphocyte differentiation antigen that four fifths of CD3/Ti positive lymphocytes express. It is a member of the immunoglobulin superfamily. CD28 is found only on T lymphocytes and on plasma cells. There are 134 extracellular amino acids with a transmembrane domain and a brief cytoplasmic tail in each CD28 monomer.

TPA

Abbreviation for tissue-plasminogen activator.

TPHA

Refer to *Treponema pallidum* hemagglutination assay.

TPI

Refer to *Treponema pallidum* immobilization test.

trace labeling

Refer to isotopic labeling.

traffic area

Thymus-dependent area.

transcobalamin II deficiency

Infants deficient in transcobalamin II, the main transport protein for vitamin B_{12}, develop megaloblastic anemia and agammaglobulinemia. B lymphocytes require vitamin B_{12} for terminal differentiation. B_{12} therapy corrects the deficiency, which has an autosomal recessive mode of inheritance.

transcobalamin II deficiency with hypogammaglobulinemia

An association of transcobalamin II deficiency with hypogammaglobulinemia, gastrointestinal, and hematologic disorders. This is inherited as an autosomal recessive trait. Patients with the condition may manifest macrocytic anemia, thrombocytopenia, leukopenia, and malabsorption resulting from small intestinal mucosal atrophy. Transcobalamin II is a protein needed for vitamin B_{12} transport in the blood. Circulating B lymphocytes are normal, but plasma cells are absent from the bone marrow. Affected subjects often fail to produce antibodies following immunogenic challenge. T cell responsiveness to PHA and skin tests are within normal limits. Replacement therapy with vitamin B_{12} given intramuscularly has improved immunoglobulin levels in the blood and rendered immunization against common antigens successful. Thus, the defect in the ability of B cells to undergo clonal expansion and to mature into antibody-producing B cells is related to a deficiency in transcobalamin II needed for the passage of vitamin B_{12} to the cell's internal environment.

transcription

RNA synthesis using a DNA template.

transcytosis

In immunology, the conveyance of immunoglobulins across epithelial cells. IgA or IgM combines with immunoglobulin receptors on epithelial cells. This is followed by internalization by receptor-mediated endocytosis. The complex is conveyed from one side of the cell to the other, where it is released by exocytosis. During this process, the receptor is split off, leaving the secretory component joined to the polymeric immunoglobulin prior to release. IgG transport by epithelial cells is by way of Fc receptors.

transduction

The use of a virus to transfer genes, such as the use of a bacteriophage to convey genes from one bacterial cell to another one. Other viruses such as retroviruses may also transfer genes from one cell type to another.

transfection

The transfer of double-stranded DNA extracted from neoplastic cells for the purpose of producing phenotypic alterations of malignancy in the recipient cells.

transfectoma

Antibody-synthesizing cells that are generated by introducing antibody genes that have been genetically engineered into myeloma cells using genomic DNA.

transfer factor (TF)

A substance in extracts of leukocytes, which was as effective as viable lymphoid cells in transferring delayed-type hypersensitivity. The active principle is not destroyed by treatment with DNase or RNase. Originally described by H. S. Lawrence, transfer factor has been the subject of numerous investigations. Transfer factor is dialyzable and is less than 10 kD. It has been separated on Sephadex®. Following demonstration of its role in humans, transfer factor was later shown to transfer delayed-type hypersensitivity in laboratory animals. It was shown to be capable of also transferring cell-mediated immunity as well as delayed-type hypersensitivity between members of numerous animal species. It also became possible to transfer delayed-type hypersensitivity across species barriers using transfer factor. Attempts to identify purified transfer factor have remained a major challenge. However, it has now been shown that transfer factor combines with specific antigen. Urea treatment of a solid-phase immunosorbent permits its recovery. Thus, specific transfer factor is generated in an animal that has been immunized with a specific antigen. T helper lymphocytes produce transfer factor. It combines with T suppressor cells as well as with Ia antigen on B lymphocytes and macrophages. It also interacts with antibody specific for V region antigenic determinants. It may be a fragment of the T cell receptor for antigen. Transfer factor has been used as an immunotherapeutic agent for many years to treat patients with immunodeficiencies of various types. It produces clinical

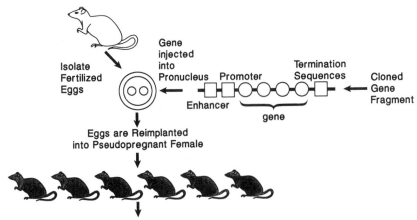

Study of hybrid gene expression occurs
with litters of mated founder animals
Transgenic Mouse

improvement in numerous infectious diseases caused by viruses and fungi. Transfer factor improves cell-mediated immunity and delayed-type hypersensitivity response, i.e., it restores decreased cellular immunity to some degree.

transferrin
A protein that combines with and competes for iron with bacteria.

transferrin receptor (T9)
The receptor on the membranes of cells for the 76-kD protein transferrin in serum that serves as a conveyer for ferric (Fe^{3+}) iron. Monoclonal antibody can detect the transferrin receptor on activated T lymphocytes, even though resting T cells are essentially bereft of this receptor. Primitive thymocytes also express it. Transferrin receptor is comprised of two 100-kD polypeptide chains that are alike and fastened together by disulfide bonds. The transferrin receptor gene is located on chromosome 3p12-ter in man.

transformation
A heritable alteration in a cell as a consequence of investigative manipulation. (1) Lymphocyte transformation. The stimulation of a resting lymphocyte with a lectin, antigen, or lymphokine to undergo blast transformation associated with the cell's division, proliferation, and differentiation. This process can be assayed quantitatively by adding ^3H (thymidine) to the cell culture which becomes incorporated into the DNA. (2) Genetic transformation by DNA. Nonvirulent living pneumococci can become virulent after taking up DNA from dead pneumococci by transformation. (3) Cells can undergo neoplastic transformation in culture and acquire the capacity for unrestricted proliferation, thereby resembling neoplastic cells.

transforming growth factor α (TGF-α)
A 5.5-kD protein comprised of 50 amino acid residues (human form). This polypeptide growth factor induces proliferation of many types of epidermal and epithelial cells. It facilitates growth of selected transformed cells.

transfusion-associated graft-vs.-host disease (TAGVHD)
Infused immunocompetent T lymphocytes react against histoincompatible immune system cells of the recipient. This is likely to occur in patients who have been either immunocompromised or receiving chemotherapy for tumors. The patient may develop a skin rash and have profound pancytopenia, as well as altered liver function tests. Three weeks following transfusion, 84% may die. To avoid graft-vs.-host reactivity induced by a blood transfusion, any blood product containing lymphocytes should be subjected to 1500 rad prior to administration.

transfusion reaction(s)
Both immune and nonimmune reactions that follow the administration of blood. Transfusion reactions with immune causes are considered serious and occur in 1 in 3000 transfusions. Patients may develop urticaria, itching, fever, chills, chest pains, cyanosis, and hemorrhage; some may even collapse. The appearance of these symptoms together with an increase in temperature by 1°C signals the need to halt the transfusion. Immune, noninfectious transfusion reactions include allergic urticaria (immediate hypersensitivity); anaphylaxis, as in the administration of blood to IgA-deficient subjects, some of whom develop anti-IgA antibodies of the IgE

class; and serum sickness, in which the serum proteins such as immunoglobulins induce the formation of precipitating antibodies that lead to immune complex formation.

transgenes
Genes that are artificially and deliberately introduced in the germ line and are foreign. Refer to transgenic mice.

transgenic animals
An animal into whose genome a foreign gene has been introduced. Introduction of the exogenous gene into a mouse can be by either microinjection into a pronucleus of an egg that has been fertilized recently or through retroviruses. The egg that has received the foreign gene is transferred to the oviduct of a pseudopregnant female. If the gene becomes integrated into a chromosome, it is passed on to the progeny through the germ line and will be expressed in all cells.

transgenic mice
Mice that carry a foreign gene that has been artificially and deliberately introduced into their germ line. The added genes are termed transgenes. Fertilized egg pronuclei receive microinjections of linearized DNA. These are placed in pseudopregnant female oviducts and development proceeds. About one fourth of the mice that develop following injection of several hundred gene copies into pronuclei are transgenic mice. Transgenic mice have been used to study genes not usually expressed *in vivo* and alterations in genes that are developmentally regulated to express normal genes and cells where they are not usually expressed. Transgenic mice are also used to delete certain populations of cells with transgenes that encode toxic proteins. They are highly significant in immunologic research.

transgenic mouse (see above)
A mouse developed from an embryo into which foreign genes were transferred. Transgenic mice have provided much valuable information related to immunological tolerance, autoimmune phenomena, oncogenesis, developmental biology, and related topics.

transgenic organisms
An animal or plant into which foreign genes that encode specific proteins have been inserted. However, controlling the site of gene insertion has not been accomplished yet. Insertion into some positions may even lead to activation of the host's own structural genes.

transgenics
The transfer of needed genes into an organism for the purpose of providing a missing protein which these genes encode.

transient hypogammaglobulinemia of infancy.
A delay in the onset of antibody synthesis by a 2 1/2- to 3-year-old child in whom maternal antibodies passed across the placenta have already disappeared. Helper T cell function is impaired, yet B cell numbers are at physiologic levels.

translation
The synthesis of a peptide chain using amino acids to produce proteins.

transplantation
The replacement of an organ or other tissue, such as bone marrow, with organs or tissues derived ordinarily from a nonself source such as an allogeneic donor. Organs include kidney, liver, heart, lung, pancreas (including pancreatic islets), intestine, or skin. In addi-

tion, bone matrix and cardiac valves have been transplanted. Bone marrow transplants are given for nonmalignant conditions such as aplastic anemia, as well as to treat certain leukemias and other malignant diseases.

transplantation antigens
Histocompatibility antigens that stimulate an immune response in the recipient that may lead to rejection.

transplantation immunology
The study of immunologic reactivity of a recipient to transplanted organs or tissues from a histoincompatible recipient. Effector mechanisms of transplantation rejection or transplantation immunity consist of cell-mediated immunity and/or humoral antibody immunity, depending upon the category of rejection. For example, hyperacute rejection of an organ such as a renal allograft is mediated by preformed antibodies and takes place soon after the vascular anastomosis is completed in transplantation. By contrast, acute allograft rejection is mediated principally by T lymphocytes and occurs during the first week after transplantation. There are instances of humoral vascular rejection mediated by antibodies as a part of the acute rejection in response. Chronic rejection is mediated by a cellular response.

transplantation rejection
The consequences of cellular and humoral immune responses to a transplanted organ or tissue that may lead to loss of function and necessitate removal of the transplanted organ or tissue. Transplantation rejection episodes occur in many transplant recipients, but are controlled by such immunosuppressive drugs as cyclosporine, rapamycin, or FK506 or by monoclonal antibodies against T lymphocytes.

transport piece
Refer to secretory piece.

transporter in antigen processing (TAP) 1 and 2 genes
Genes in the MHC class II region that must be expressed for MHC class I molecules to be assembled efficiently. TAP 1 and 2 are postulated to encode components of a heterodimeric protein pump that conveys cytosolic peptides to the endoplasmic reticulum. Here they associate with MHC class I heavy chains.

transudation
The movement of electrolytes, fluid, and proteins of low mol wt from the intravascular space to the extravascular space, as in inflammation.

Treponema pallidum hemagglutination assay
A test for antibodies specific for *T. pallidum* used formerly to diagnose syphilis. *T. pallidum* antigens were coated onto sheep red blood cells treated with tannic acid and formalin. Aggregation of the antigen-coated red cells signified that antibody was present.

Treponema pallidum immobilization test
A diagnostic test for syphilis in which living, motile *T. pallidum* microorganisms are combined with a sample of serum presumed to contain specific antibody. Complement is also present. If the serum sample contains anti-*Treponema* antibody, the motile microorganisms become immobilized. This test is much more specific for the diagnosis of syphilis than is the complement fixation test, such as the old Wassermann reaction. It has been rarely used because living *T. pallidum* microorganisms are not readily available and are hazardous to work with.

TRF
Abbreviation for T cell replacing factor.

trinitrophenyl (picryl) group
A chemical grouping that may serve as a hapten when it is linked to a protein through –NH₂ groups by reaction with picryl chloride or trinitrobenzene sulfonic acid.

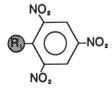

Trinitrophenyl Group

triple response of Lewis
Skin changes in immediate hypersensitivity illustrated by striking the skin with a sharp object such as the side of a ruler. The first response termed the "stroke response" is caused by the production of histamine and related mediators at the point of contact with the skin. The second response is a flare produced by vasodilation and resembles a red halo. The third response is a wheal characterized by swelling and blanching induced by histamine from mast cell degranulation. The swelling is attributable to edema between the junctions of cells that become rich in protein and fluid.

triple vaccine
An immunizing preparation comprised of three components and used to protect infants against diphtheria, pertussis (whooping cough), and tetanus. It is made up of diphtheria toxoid, pertussis vaccine, and tetanus toxoid. The first of four doses is administered between 3 and 6 months of age. The second dose is administered 1 month later, and the third dose is given 6 months after the second. The child receives a booster injection when beginning school.

Triton X-100
A quaternary ammonium salt that is surface active. Chemically, it is isooctyl phenoxy polyethoxy ethanol. It has been used as a detergent, emulsifier, wetting agent, and surfactant.

trophoblast
A layer of cells in the placenta that synthesizes immunosuppressive agents. These cells are in contact with the lining of the uterus.

tropical eosinophilia
A hypersensitivity to filarial worms manifested in the lungs. It has been reported in the Near East and in the Far East. Patients develop wheezing, productive cough, and a cellular infiltrate comprised of eosinophils, lymphocytes, and fibroblasts. Fibrosis may result.

trypan blue
A vital dye used to stain lymphoid cells, especially in the microlymphocytotoxicity test used for HLA tissue typing. Cell membranes whose integrity has been interrupted by antibody and complement permit the dye to enter and stain the cells dark blue. By contrast, the viable cells with an intact membrane exclude the dye and remain as bright circles of light in the microscope. Dead cells stain blue.

trypan blue dye exclusion test
A test for viability of cells in culture. Living cells exclude trypan blue by active transport. When membranes have been interrupted, dye enters the cells, staining them blue and indicating that the cell is dead. The method can be used to calculate the percent of cell lysis induced.

trypanosome adhesion test
Selected mammalian red blood cells or bacteria may stick to the surface of trypanosomes when specific antibody and complement are present. This has been used in the past as a test for antibodies to trypanosomes.

tryptic peptides
Peptides formed by tryptic digestion of protein.

T$_s$
(1) Suppressor T lymphocyte. (2) Secreted immunoglobulin heavy chain tail (C-terminal) polypeptide.

T$_s$1, T$_s$3 lymphocytes
Suppressor T lymphocyte subpopulations.

T$_s$F
Abbreviation for suppressor T cell factor.

TSF
T cell suppressor factor.

tube agglutination test
An agglutination assay that consists of serial dilutions of antiserum in serological tubes to which particulate antigen, such as microorganisms, is added.

tuberculid
A hypersensitivity skin reaction to mycobacteria. The lesion may be either a papulonecrotic tuberculid, with sterile papules ulcerated in the center and obliterative vasculitis, or crops of small red papules, with a sarcoid-like appearance that represent lichen scrofulosorum.

tuberculin
A sterile solution containing a group of proteins derived from culture medium where *Mycobacterium tuberculosis* microorganisms have been grown. It has been used for almost a century as a skin test preparation to detect delayed-type (type IV) hypersensitivity to infection with *M. tuberculosis*. Many tuberculin preparations have been used in the past, but only old tuberculin (OT) and purified protein derivative (PPD) are still used. Whereas OT is a heat-concentrated filtrate of the culture medium in which *M. tuberculosis* was grown, PPD of tuberculin is a trichloroacetic acid precipitate of the growth medium. Tuberculin is a mitogen for murine B lymphocytes, as well as a T lymphocyte mitogen.

tuberculin hypersensitivity
A form of bacterial allergy specific for a product in culture filtrates of *Mycobacterium tuberculosis* which, when injected into the skin, elicits a cell-mediated delayed-type hypersensitivity (type IV) response. Tuberculin-type hypersensitivity is mediated by CD4⁺ T

lymphocytes. Following the intracutaneous inoculation of tuberculin extract or purified protein derivative (PPD), an area of redness and induration develops at the site within 24 to 48 h in individuals who have present or past interaction with *M. tuberculosis.*

tuberculin reaction

A test of *in vivo* cell-mediated immunity. Robert Koch observed a localized lesion in the skin of tuberculous guinea pigs inoculated intradermally with broth from a culture of tubercle bacilli. The body's immune response to infection with the tubercle bacillus is signaled by the appearance of agglutinins, precipitins, opsonins, and complement-fixing antibodies in the serum. This humoral response is, however, not marked, and such antibodies are present in low titer. The most striking response is the development of delayed-type hypersensitivity (DTH), which has a protective role in preventing reinfection with the same organism. Subcutaneous inoculation of tubercle bacilli in a normal animal produces no immediate response, but in 10 to 14 days a nodule develops at the site of inoculation. The nodule then becomes a typical tuberculous ulcer. The regional lymph nodes become swollen and caseous. In contrast, a similar inoculation in a tuberculous animal induces an indurated area at the site of injection within 1 to 2 days. This becomes a shallow ulcer which heals promptly. No swelling of the adjacent lymphatics is noted. The tubercle bacillus antigen which is responsible for DTH is wax D, a lipopolysaccharide-protein complex of the bacterial cell wall. The active peptide comprises diaminopimelic acid, glutamic acid, and alanine. Testing for DTH to the tubercle bacillus is done with tuberculin, a heat-inactivated culture extract containing a mixture of bacterial proteins, or with PPD, a purified protein derivative of culture in nonproteinaceous media. Both these compounds are capable of sensitizing the recipient themselves. The protective role of DTH is supported by the observation that in positive reactors living cells are usually free of tubercle bacilli and the bacteria are present in necrotic areas, separated by an avascular barrier. By contrast, in infected individuals giving a negative reaction, the tubercle bacilli are found in great numbers in living tissues. The reaction is permanently or transiently negative in individuals whose cell-mediated immune responses are transiently or permanently impaired.

tuberculin test

The 24- to 48-h response to intradermal injection of tuberculin. If positive, it signifies delayed-type hypersensitivity (type IV) to tuberculin and implies cell-mediated immunity to *Mycobacterium tuberculosis.* The intradermal inoculation of tuberculin or of purified protein derivative (PPD) leads to an area of erythema and induration within 24 to 48 h in positive individuals. A positive reaction signifies the presence of cell-mediated immunity to *M. tuberculosis* as a consequence of past or current exposure to this microorganism. However, it is not a test for the diagnosis of active tuberculosis.

tuberculin-type reaction

A cell-mediated delayed-type hypersensitivity skin response to an extract such as candidin, brucellin, or histoplasmin. Individuals who have positive reactions have developed delayed-type hypersensitivity or cell-mediated immunity mediated by T lymphocytes following contact with the microorganism in question.

tuberculosis immunization

The induction of protective immunity through injection of an attenuated vaccine containing Bacille-Calmette-Guerin (BCG). This vaccine was more widely used in Europe than in the U.S. in an attempt to provide protection against development of tuberculosis. A local papule develops several weeks after injection in individuals who were previously tuberculin negative, as it is not administered to positive individuals. It is claimed to protect against development of tuberculosis, although not all authorities agree on its efficacy for this purpose. In recent years, oncologists have used BCG vaccine to reactivate the cellular immune system of patients bearing neoplasms in the hope of facilitating antitumor immunity.

tuftsin

A leukokinin globulin-derived substance that enhances phagocytosis. It is a tetrapeptide comprised of Thr-Lys-Pro-Arg. The leukokinin globulin from which it is derived represents immunoglobulin Fc receptor residues 289 through 292. Tuftsin is formed in the spleen. Its actions include neutrophil and macrophage chemotaxis, enhancing phagocyte motility, and promoting oxidative metabolism. It also facilitates antigen processing.

tuftsin deficiency

Tuftsin is a tetrapeptide that stimulates phagocytes. Tuftsin is split from an immunoglobulin by the action of one proteolytic enzyme in the spleen that cleaves the carboxy terminus between residues 292 and 293 and another enzyme that is confined to neutrophil membranes (leukokinase) that splits the molecule between positions 288 and 289. Thus, tuftsin deficiency, which is transmitted as an autosomal recessive trait, results from a lack of this splenic enzyme. Although γ globulin has been used, there is no known treatment.

tumor-associated antigens

Certain antigens designated as CA-125, CA-19-9, and CA195, among others may be linked to certain tumors such as lymphomas, carcinomas, sarcomas, and melanomas, but the immune response to these tumor-associated antigens is not sufficient to mount a successful cellular or humoral immune response against the neoplasm. Three classes of tumor-associated antigens have been described. Class 1 antigens are very specific for a certain neoplasm and are absent from normal cells. Class 2 antigens are found on related neoplasms from separate individuals. Class 3 antigens are found on malignant as well as normal cells, but show increased expression in the neoplastic cells. Assays of clinical value will probably be developed for class 2 antigens, since they are associated with multiple neoplasms and very infrequently are found in normal individuals.

tumor enhancement

The successful establishment and prolonged survival, conversely the delayed rejection, of a tumor allograft, especially in mice, as a consequence of contact with specific antibody.

tumor-infiltrating lymphocytes (TIL)

T lymphocytes isolated from the tumor they are infiltrating. They are cultured with high concentrations of IL-2, leading to expansion of these activated T lymphocytes *in vitro.* TILs are very effective in destroying tumor cells and have proven much more effective than lymphokine-activating killer (LAK) cells in experimental models. TILs have 50 to 100 times the antitumor activity produced by LAK cells. TILs have been isolated and grown from multiple resected human tumors, including those from kidney, breast, colon, and melanoma. In contrast to the non-B-non-T LAK cells, TILs nevertheless are generated from T lymphocytes and phenotypically resemble cytotoxic T lymphocytes. TILs from malignant melanoma exhibit specific cytolytic activity against cells of the tumor from which they were extracted, whereas LAK cells have a broad range of specificity. TILs appear unable to lyse cells of melanomas from patients other than those in whom the tumor originated. TILs may be tagged in order that they may be identified later.

tumor necrosis factor α (TNF-α)

A cytotoxic monokine produced by macrophages stimulated with bacterial endotoxin. TNF-α participates in inflammation, wound healing, and remodeling of tissue. TNF-α, which is also called cachectin, can induce septic shock and cachexia. It is a cytokine

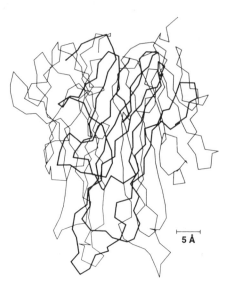

Tumor Necrosis Factor α

comprised of 157 amino acid residues. It is produced by numerous types of cells including monocytes, macrophages, T lymphocytes, B lymphocytes, NK cells, and other types of cells stimulated by endotoxin or other microbial products. The genes encoding TNF-α and TNF-β (lymphotoxin) are located on the short arm of chromosome 6 in man in the MHC region. High levels of TNF-α are detectable in the blood circulation very soon following administration of endotoxin or microorganisms. The administration of recombinant TNF-α induces shock, organ failure, and hemorrhagic necrosis of tissues in experimental animals including rodents, dogs, sheep, and rabbits, closely resembling the effects of lethal endotoxemia. TNF-α is produced during the first 3 days of wound healing. It facilitates leukocyte recruitment, induces angiogenesis, and promotes fibroblast proliferation. It can combine with receptors on selected tumor cells and induce their lysis. TNF mediates the antitumor action of murine natural cytotoxic (NC) cells, which distinguishes their function from that of natural killer (NK) and cytotoxic T cells. TNF-α was termed cachectin because of its ability to induce wasting and anemia when administered on a chronic basis to experimental animals. Thus, it mimics the action in cancer patients and those with chronic infection with human immunodeficiency virus or other pathogenic microorganisms. It can induce anorexia, which may lead to death from malnutrition.

tumor necrosis factor β (TNF-β)

A 25-kD protein synthesized by activated lymphocytes. It can kill tumor cells in culture, induce expression of genes, stimulate proliferation of fibroblasts, and mimic most of the actions of tumor necrosis factor α (cachectin). It participates in inflammation and graft rejection and was previously termed lymphotoxin. TNF-β and TNF-α have approximately equivalent affinity for TNF receptors. Both 55- and 80-kD TNF receptors bind TNF-β. TNF-β has diverse effects that include killing of some cells and causing proliferation of others. It is the mediator whereby cytolytic T cells, natural killer cells, lymphokine-activated killer cells, and "helper-killer" T cells induce fatal injury to their targets. TNF-β and TNF-α have been suggested to play a role in AIDS, possibly contributing to its pathogenesis.

tumor necrosis factor receptor

A receptor for tumor necrosis factor that is comprised of 461 amino acid residues and possesses an extracellular domain that is rich in cysteine.

tumor promoter

Refer to phorbol ester(s).

tumor rejection antigen

An antigen that is detectable when transplanted tumor cells are rejected. Also called tumor transplant antigen.

tumor-specific antigen (TSA)

An antigen present on tumor cells, but not found on normal cells. Murine tumor-specific antigens can induce transplantation rejection in mice.

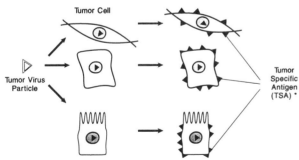

* Each tumor induced by a single virus will express the same TSA on the cell surface despite the morphology of the cell

Tumor-Specific Antigen (TSA)

tumor-specific determinants

Epitopes present on tumor cells, but identifiable also in varying quantities and forms on normal cells.

tumor-specific transplantation antigen (TSTA)

Epitopes that induce rejection of tumors transplanted among syngeneic (histocompatible) animals.

turbidimetry

Quantification of a substance in suspension based on the suspension's ability to reduce forward light transmission.

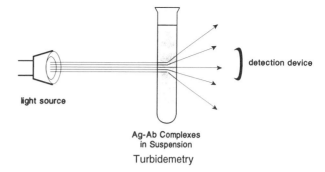

Turbidemetry

Tween

Nonionic detergent.

Tween 80®

Polyoxyethylene sorbitan monooleate. An emulsifying agent used in cultures of mycobacteria and in water-in-oil-in-water emulsion adjuvants as a stabilizing agent.

two-dimensional gel electrophoresis

A method to characterize a protein. The protein to be analyzed is subjected to isoelectric focusing by placing the soluble protein in a pH gradient and applying an electrical charge. The protein moves to the pH where it has a neutral charge. This is followed by electrophoresis in gel at a 90° angle to separate the proteins according to size. This procedure yields a pattern known as a fingerprint that is very specific.

Two-Dimensional Electrophoresis

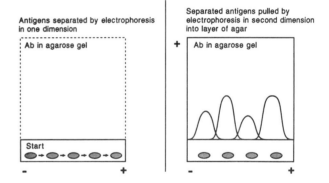

type I anaphylactic hypersensitivity

A hypersensitivity reaction mediated by IgE antibodies reactive with specific allergens (antigens that induce allergy) attached to basophil or mast cell Fc receptors. Crosslinking of the cell-bound IgE antibodies by antigen is followed by mast cell or basophil degranulation, with release of pharmacological mediators. These mediators include vasoactive amines such as histamine, which

Type I Hypersensitivity Reaction

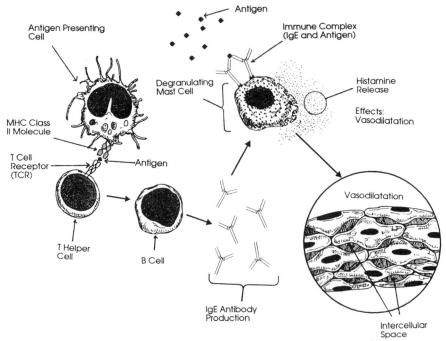

Type I Anaphylactic Hypersensitivity

causes increased vascular permeability, vasodilation, bronchial spasm, and mucous secretion. Secondary mediators of type I hypersensitivity include leukotrienes, prostaglandin D_2, platelet-activating factor, and various cytokines. Systemic anaphylaxis is a serious clinical problem and can follow the injection of protein antigens, such as antitoxin, or of drugs, such as penicillin.

type II antibody-mediated hypersensitivity

A type of hypersensitivity that is induced by antibodies and has three forms. The classic type of hypersensitivity involves the interaction of antibody with cell membrane antigens followed by complement lysis. These antibodies are directed against antigens intrinsic to specific target tissues. Antibody-coated cells also have increased susceptibility to phagocytosis. Examples of type II hypersensitivity include the antiglomerular basement membrane anti-

body that develops in Goodpasture's syndrome and antibodies that develop against erythrocytes in Rh incompatibility, leading to erythroblastosis fetalis or autoimmune hemolytic anemia.

A second variety of type II hypersensitivity is antibody-dependent cell-mediated cytotoxicity (ADCC). Killer (K) cells or NK cells, which have Fc receptors on their surfaces, may bind to the Fc region of IgG molecules. They may react with surface antigens on target cells to produce lysis of the antibody-coated cell. Complement fixation is not required and does not participate in this reaction. In addition to K and NK cells, neutrophils, eosinophils, and macrophages may participate in ADCC.

A third form of type II hypersensitivity is antibody against cell surface receptors that interfere with function, as in the case of antibodies against acetylcholine receptors in motor endplates of

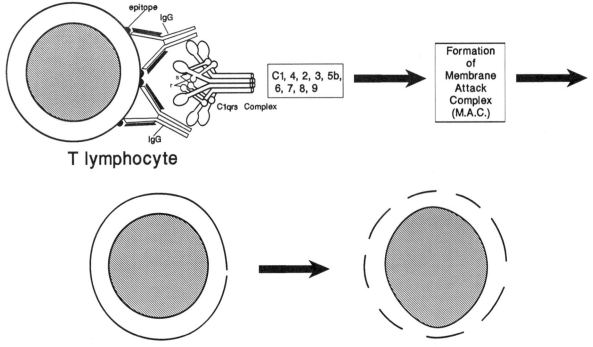

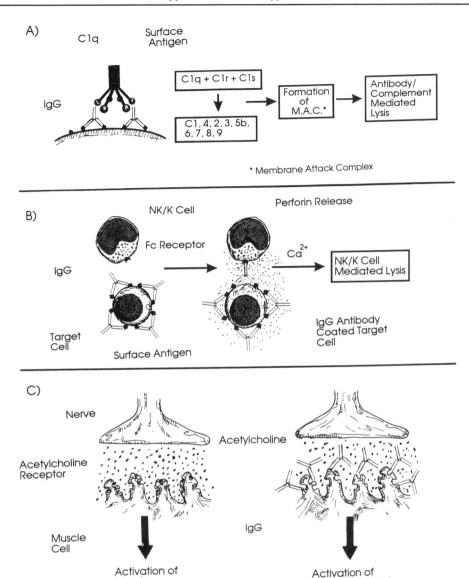

Type II Antibody-Mediated Hypersensitivity

skeletal muscle in myasthenia gravis. This interference with neuro-muscular transmission results in muscular weakness, ultimately affecting the muscles of respiration and producing death. By contrast, stimulatory antibodies develop in hyperthyroidism (Graves' disease). They react with thyroid-stimulating hormone receptors on thyroid epithelial cells to produce hyperthyroidism.

type II interferon

Synonym for interferon γ.

type III immune complex-mediated hypersensitivity

A type of hypersensitivity mediated by antigen-antibody-complement complexes. Antigen-antibody complexes can stimulate an acute inflammatory response that leads to complement activation and PMN leukocyte infiltration. The immune complexes are formed either by exogenous antigens such as those from microbes or by endogenous antigens such as DNA, a target for antibodies produced in systemic lupus erythematosus. Immune complex-mediated injury may be either systemic or localized. In the systemic variety, antigen-antibody complexes are produced in the circulation, deposited in the tissues, and initiate inflammation. Acute serum sickness occurred in children treated with diphtheria antitoxin earlier in this century as a consequence of antibody produced against the horse serum protein. When immune complexes are deposited in tissues, complement is fixed, and PMNs are attracted to the site. Their lysosomal enzymes are released, resulting in tissue injury. Localized immune complex disease, sometimes called the Arthus reaction, is characterized by an acute immune complex vasculitis with fibrinoid necrosis occurring in the walls of small vessels.

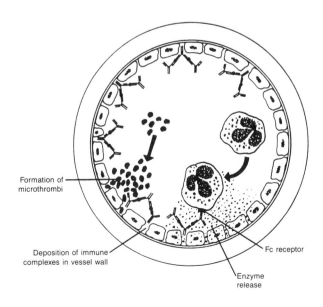

Type III Immune Complex-Mediated Hypersensitivity

type IV cell-mediated hypersensitivity

A form of hypersensitivity mediated by specifically sensitized cells. Whereas antibodies participate in type I, II, and III reactions, T lymphocytes mediate type IV hypersensitivity. Two types of reactions, mediated by separate T cell subsets, are observed. Delayed-type hypersensitivity (DTH) is mediated by CD4+ T cells, and cellular cytotoxicity is mediated principally by CD8+ T cells.

A classic delayed hypersensitivity reaction is the tuberculin or

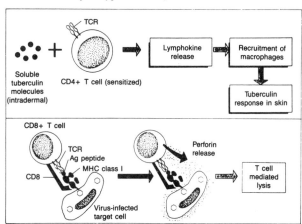

Type IV Cell-Mediated Hypersensitivity

Mantoux reaction. Following exposure to *Mycobacterium tuberculosis,* CD4+ lymphocytes recognize the microbe's antigens complexed with class II MHC molecules on the surface of antigen-presenting cells that process the mycobacterial antigens. Memory T cells develop and remain in the circulation for prolonged periods. When tuberculin antigen is injected intradermally, sensitized T cells react with the antigen on the antigen-presenting cell's surface, undergo transformation, and secrete lymphokines that lead to the manifestations of hypersensitivity. Unlike an antibody-mediated hypersensitivity, lymphokines are not antigen specific.

In T cell-mediated cytotoxicity, CD8+ T lymphocytes kill antigen-bearing target cells. The cytotoxic T lymphocytes play a significant role in resistance to viral infections. Class I MHC molecules present viral antigens to CD8+ T lymphocytes as a viral peptide-class I molecular complex, which is transported to the infected cell's surface. Cytotoxic CD8+ cells recognize this and lyse the target before the virus can replicate, thereby stopping the infection.

typhoid vaccination

Refer to TAB vaccine.

typhus vaccination

Protective immunization against typhus transmitted by lice or fleas and against Rocky mountain spotted fever by the administration of inactivated vaccines. Rickettsiae prepared in chick embryo yolk sacs or tissues are treated with formaldehyde to render them inactive. Rather than provide protective immunity that prevents the disease, these vaccines condition the host to experience a milder or less severe form of the disease than that which occurs in a nonvaccinated host.

U

U antigen

A rare MNs erythrocyte antigen present in fewer than 1% of African Americans and absent from Caucasian red blood cells. When U antigen is not present, s antigen is not expressed. Membrane sialoglycoprotein, and glycophorins A and B are requisite for U antigen expression.

U1 snRNP antibodies

Antibodies against 70 kD, A and C protein constituents of the U1 type of small nuclear ribonucleoproteins (U1 snRNPs). They are commonly found in mixed connective tissue disease or other syndromes related to SLE but infrequently in SLE itself. Formerly, these antibodies were designated as RNP (ribonucleoprotein) or nRNP (nuclear RNP) antibodies.

U2 snRNP antibodies

Antibodies against small nuclear ribonucleoparticles which are comprised of U2 snRNA and 8 other polypeptides. Anti U2 sera that react with β polypeptide are often present in overlap syndromes with myositis and may be associated with antibodies against U1 snRNP polypeptides (70 kD, A and C). U1 snRNP antibodies that interact with 70 kD polypeptide were previously designated RNP or nRNP and are a principal feature of mixed connective tissue disease.

ubiquitin

A 7-kD protein found free in the blood or bound to cytoplasmic, nuclear, or membrane proteins united through isopeptide bonds to numerous lysine residues. Ubiquitin combines with a target protein and marks it for degradation. A 76-amino acid residue polypeptide found in all eukaryotes, but not in prokaryotes. Ubiquitin is found in chromosomes covalently linked to histones, although the function is unknown. Ubiquitin is present on the lymphocyte homing receptor gp90Mel-14.

ubiquitin antibodies

Antibodies present in 79% of systemic lupus erythematosus patients that may facilitate the diagnosis of systemic lupus erythematosus when double stranded DNA (dsDNA) antibodies are also present. Ubiquitin antibodies may be inversely related to dsDNA antibodies and disease activity.

ulcerative colitis (immunologic colitis)

An ulcerative condition that may involve the entire colon, but does not significantly affect the small intestine. There is neutrophil, plasma cell, and eosinophil infiltration of the colonic mucosa. This is followed by ulceration of the surface epithelium, loss of globlet cells, and formation of crypt abscess. The etiology is unknown. An immune effector mechanism is believed to maintain chronic disease in these patients. Their serum immunoglobulins and peripheral blood lymphocytes count are usually normal. Complexes present in the blood are relatively small and contain IgG, although no antigen has been identified. The complexes could be merely aggregates of IgG. These patients have diarrhea with blood and mucus in the stool. The signs and symptoms are intermittent, and there is variation in the severity of colon lesions. The patient's lymphocytes are cytotoxic for colon epithelial cells. Antibodies against *Escherichia coli* may crossreact with colonic epithelium in these patients. However, whether or not such antibodies have a role in etiology and pathogenesis remains to be proven.

ultracentrifugation

The separation of cell components, including organelles and molecules, through high-speed centrifugation reaching 6000 rpm with a gravitational force up to 500,000g. In differential velocity centrifugation, there is a stepwise increase in gravitational force to remove selected components. Following centrifugation of a cellular homogenate at 600g for 10 min to isolate the nuclei, further spinning at 15,000g for 5 min permits isolation of mitochondria, lysozomes, and peroxisomes. Respinning at 100,000g for 1 h per-mits isolation through sedimentation of the plasma membrane, microsomal fraction, endoplasmic reticulum, and large polyribosomes. Respinning at 300,000g for 2 h permits sedimentation of ribosomal subunits and small polyribosomes. This leaves the cytosol, which is the soluble portion of the cytoplasm. Separation can also be achieved by sucrose density gradient ultracentrifugation. Cesium chloride combined with molecules to be analyzed permits the molecules to migrate to a particular density equivalent. Ultracentrifugation may be either an analytical method, which is used to identify proteins that differ in sedimentation coefficient, or as a preparative method to separate proteins based on their densities and shapes.

ultrafiltration

The passage of solutions or suspensions through membranes with minute pores of graded sizes.

umbrella effect

The masking, by relatively large amounts of IgG, of low immunoglobulin light chain concentrations in early IgM macroglobulinemia and IgA myeloma. This is observed in immunoelectrophoresis. Immunofixation electrophoresis, employing fluorochrome-labeled antibodies, can resolve this masking effect.

undifferentiated connective tissue disease

A prodromal phase of a collagen vascular or connective tissue disease. At this stage, the principal clinical manifestations have not yet become apparent.

unidentified reading frame (URF)

An open reading frame (ORF) that does not correlate with a defined protein.

unitarian hypothesis

The view that one type of antibody produced in response to an injection of antigen could induce agglutination, complement fixation, precipitation, and lysis based upon the type of ligand with which it interacted. This view is in contrast to the earlier belief that separate antibodies accounted for every type of serological reactivity described above. Usually, more than one class of immunoglobulin may manifest a particular serological reactivity such as precipitation.

univalent

A single binding site.

univalent antibody

An antibody molecule with one antigen-binding site. Although incapable of leading to precipitation or agglutination, univalent antibodies or Fab fragments resulting from papain digestion of an IgG molecule might block precipitation of antigen by a typical bivalent antibody.

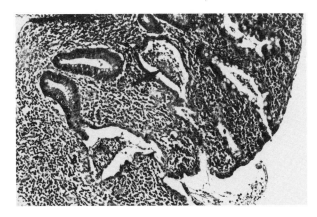

Ulcerative colitis
crypt abscess

universal donor

A blood group O RhD⁻ individual whose erythrocytes express neither A nor B surface antigens. This type of red blood cell fails to elicit a hemolytic transfusion reaction in recipients who are blood group A, B, AB, or O. However, group O individuals serving as universal donors may express other blood group antigens on their erythrocytes that will induce hemolysis. It is preferable to use type-specific blood for transfusions, except in cases of disaster or emergency.

universal recipient

An ABO blood group individual whose cells express antigens A and B, but whose serum does not contain anti-A and anti-B antibodies. Thus, red blood cells containing any of the ABO antigens may be transfused to them without inducing a hemolytic transfusion reaction, i.e., from an individual with type A, B, AB, or O. It is best if the universal recipient is Rh+, i.e., has the Rh D antigen on his erythrocytes, to avoid developing a hemolytic transfusion reaction. However, blood group systems other than ABO may induce hemolytic reactions in a universal recipient. Thus, it is best to use type-specific blood for transfusions.

unprimed

Animals or cells that have not come into contact previously with a particular antigen.

uromodulin *Tamm-Horsfall* **protein**

An 85-kD α_1 acid glycoprotein produced in Henle's ascending loop and distal convoluted tubules by epithelial cells. It is a powerful immunosuppressive protein based upon N-linked carbohydrate residues. It inhibits proliferation of T cells induced by antigen and monocyte cytotoxicity. Uromodulin is a ligand for interleukin-1 α, interleukin-1 β, and tumor necrosis factor (TNF).

uropod

Lymphocyte cytoplasm extending as an elongated tail or pseudopod in locomotion. The uropod may resemble the handle of a hand mirror. The plasma membrane covers the uropod cytoplasm.

urticaria

Pruritic skin rash identified by localized elevated, edematous, erythematous, and itching wheals with a pale center encircled by a red flare. It is due to the release of histamine and other vasoactive substances from mast cell cytoplasmic granules based upon immunologic sensitization or due to physical or chemical substances. It is a form of type I immediate hypersensitivity. It is mediated by IgE antibodies in man. The action of allergen or antigen with IgE antibodies anchored to mast cells can lead to this form of cutaneous anaphylaxis or hives. The wheal is due to leakage of plasma from venules, and the flare is caused by neurotransmitters.

urushiols

Catechols in poisoning (*Rhus toxicodendron*) plants that act as allergens to produce contact hypersensitivity, i.e., contact dermatitis at skin sites touched by the urushiol-bearing plant. The cutaneous lesion is T cell mediated and is classified as type IV hypersensitivity. There are four *Rhus* catechols that differ according to pentadecyl side chain saturation. They induce type IV delayed hypersensitivity. These substances are present in such plants as poison oak, poison sumac, and poison ivy.

uveitis

Uveal tract inflammation involving the uvea, iris, ciliary body, and choroid of the eye. It may be associated with Behcet's disease, sarcoidosis, and juvenile rheumatoid arthritis.

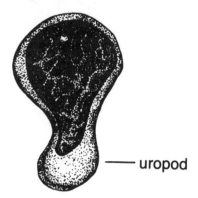

uropod

V

V gene
Gene encoding the variable region of immunoglobulin light or heavy chains. Although it is not in proximity to the C gene in germ line DNA, in lymphocyte and plasma cell DNA the V gene lies near the 5′ end of the C gene from which it is separated by a single intron.

V gene segment (see below)
DNA segment encoding the first 95 to 100 amino acid residues of immunoglobulin and T cell polypeptide chain variable regions. There are two coding regions in the V gene segment which are separated by a 100 to 400 base pair intron. The first 5′ coding region is an exon that codes for a brief untranslated mRNA region and for the first 15 to 18 signal peptide residues. The second 3′ coding region is part of an exon that codes for the terminal 4 signal peptide residues and 95 to 100 variable region residues. A J gene segment encodes the rest of the variable region. A D gene segment is involved in the encoding of immunoglobulin heavy chain and T cell receptor β and δ chains.

v-myb oncogene
A genetic component of an acute transforming retrovirus that leads to avian myeloblastosis. It represents a truncated genetic form of c-myb.

V region subgroups
Individual chain V region subdivisions based on significant homology in amino acid sequence.

vaccination
Immunization against infectious disease through the administration of vaccines for the production of active (protective) immunity in man or other animals.

vaccine
Live attenuated or killed microorganisms or parts or products from them which contain antigens that can stimulate a specific immune response consisting of protective antibodies and T cell immunity. A vaccine should stimulate a sufficient number of memory T and B lymphocytes to yield effector T cells and antibody-producing B cells from memory cells. It should also be able to stimulate high titers of neutralizing antibodies. Injection of a vaccine into a nonimmune subject induces active immunity against the modified pathogens.

vaccinia
The virus termed *Poxvirus officinale* derived from cowpox and used to induce active immunity against smallpox through vaccination. It differs from both cowpox and smallpox viruses in minor antigens.

Vaccinia gangrenosa
Chronic progressive vaccinia.

vaginal mucous agglutination test
An assay for antibodies in bovine vaginal mucous from animals infected with *Campylobacter fetus, Trichomonas fetus,* and *Brucella abortus.* The mucous can be used in the same manner as serum for a slide or tube agglutination test employing the etiologic microorganisms as antigen.

valence
The number of antigen-binding sites on an antibody molecule or the number of antibody-binding sites on an antigen molecule. The valence is equal to the maximum number of antigen molecules that can combine with a single antibody molecule. Whereas antigens are usually multivalent, most antibodies are bivalent. IgM has a valence as high as 10, and IgA has a valence that differs depending on its level of polymerization. Antibodies of the IgG, IgD, and IgE classes have a valence of 2. Due to steric hindrance, antibodies usually bind less antigen than would be expected from their valence.

van der Waals force
Weak force of attraction between atoms, ions, and molecules. It is active only at short distances, since this force varies inversely to the seventh power of the distance between ions or molecules. Thus, van der Waal's forces may be important in antigen-antibody binding.

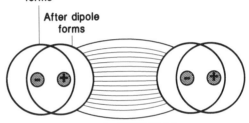

Before dipole forms

After dipole forms

van der Waals Force

variability plot
Refer to Wu-Kabat plot.

variable region
That segment of an immunoglobulin molecule or antibody formed by the variable domain of a light polypeptide chain (κ or λ) or of a heavy polypeptide chain (α, γ, μ, δ, or ε). This is sometimes referred to as the Fv region and is encoded by the V gene.

variola (smallpox)
Variola major is a *Poxvirus variolae*-induced disease that has now been eliminated from the worldwide human population. This virus-induced disease caused vesicular and pustular skin lesions, leading to disfigurement. It produced viremia and toxemia. Approximately one third of the people who were unvaccinated succumbed to the disease. *V. minor* (alastrim) is a mild form of smallpox. It was produced by a different strain that was so weak it was unable to induce the formation of pocks on the chick chorioallantoic membrane.

variolation (historical)
The intracutaneous inoculation of pus from lesions of smallpox victims into healthy, nonimmune subjects to render them immune to

Structure of heavy chain gene from IgM-producing cell

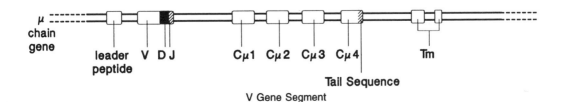

μ chain gene

leader peptide   V D J   Cμ1 Cμ2 Cμ3 Cμ4   Tm

Tail Sequence

V Gene Segment

smallpox. In China, lesional crusts were ground into a powder and inserted into the recipient's nostrils. These procedures protected some individuals, but often led to life-threatening smallpox infection in others. Edward Jenner's introduction of vaccination with cowpox to protect against smallpox rendered variolation obsolete.

vascular cell adhesion molecule 1 (VCAM-1)

A molecule that binds lymphocytes and monocytes. It is found on activated endothelial cells, dendritic cells, tissue macrophages, bone marrow fibroblasts, and myoblasts. VCAM-1 belongs to the immunoglobulin gene superfamily and is a ligand for VLA-4 (integrin α4/β1) and integrin α4/β7. It plays an important role in leukocyte recruitment to inflammatory sites and facilitates lymphocyte, eosinophil, and monocyte adhesion to activated endothelium. It participates in lymphocyte-dendritic cell interaction in the immune response.

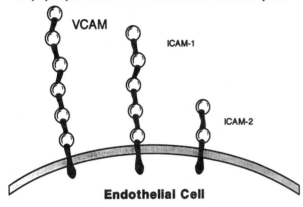

Vascular Cell Adhesion Molecule 1

vascular permeability factors

Substances such as serotonins histamines, kinins, and leukotrienes that increase the spaces between cells of capillaries and small vessels, facilitating the loss of protein and cells into the extravascular fluid.

vasculitis

Inflammation of the wall of any size blood vessel that may be accompanied by necrosis. Vasculitis with an immunologic basis is often associated with immune complex disease, in which deposition of complement-fixing microprecipitates in the vessel wall may attract polymorphonuclear neutrophils and lead to tissue injury associated with acute inflammation.

vasoactive amines

Amino group-containing substances that include histamine and serotonin that cause dilatation of the peripheral vasculature and increase the permeability of capillaries and small vessels.

vasoactive intestinal peptide (VIP)

A neuropeptide comprised of 28 residues that is a member of the secretin-glycogen group of molecules found in nerve fibers of blood vessels, in smooth muscle, and in upper respiratory tract glands. It activates adenylate cyclase and produces vasodilatation. It increases cardiac output, glycogenolysis, and bronchodilation, while preventing release of macromolecules from mucous-secreting glands. A deficiency of VIP aggravates bronchial asthma. Asthmatic patients usually do not have VIP. VIP may be increased in pancreatic islet G cell tumors.

vasoconstriction

Diminished blood flow as a consequence of contraction of precapillary arterioles.

vasodilatation (see bottom of page)

Increased blood flow through capillaries as a consequence of precapillary arteriolar dilatation.

V(D)J recombinase

An enzyme that is able to identify and splice the V (variable), J (joining), and in some cases D (diverse) gene segments that confer antibody diversity.

V(D)J recombination class switching

A mechanism to generate multiple-binding specificities by developing lymphocytes through exon recombination from a conservative number of gene segments known as variable (V), joining (J),

Class switching (from IgM to IgG production)

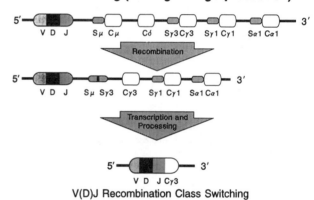

V(D)J Recombination Class Switching

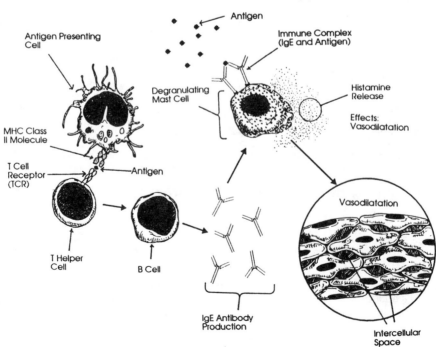

Vasodilatation (Type I Hypersensitivity)

and diversity (D) gene segments at seven different loci that include μ, κ, and λ for B cell immunoglobulin genes and α, β, γ, and δ genes for T cell receptors.

VDRL (Venereal Disease Research Laboratory) test

Reaginic screening assay for syphilis. VDRL antigen is combined with heat-inactivated serum from the patient, and the combination is observed for flocculation by light microscopy after 4 min. Reaginic assays are helpful as screening tests for early syphilis and are usually positive in secondary syphilis, although the results are more variable in tertiary syphilis. VDRL is negative in approximately half of the neurosyphilis cases. Reaginic tests for syphilis may be biologically false-positive in such conditions as malaria, lupus erythematosus, and acute infections. Biologic false-positive VDRL tests may also be seen in some cases of hepatitis, infectious mononucleosis, rheumatoid arthritis, or even pregnancy.

vector

A DNA segment employed for cloning a foreign DNA fragment. A vector should be able to reproduce autonomously in the host cell, possess one or more selectable markers, and have sites for restriction endonucleases in nonessential regions that permit DNA insertion into the vector or replacement of a segment of the vector. Plasmids and bacteriophages may serve as cloning vectors.

veiled cell

A mononuclear phagocytic cell that serves as an antigen-presenting cell. It is found in the afferent lymphatics and in the marginal sinus. It may manifest IL-2 receptors in the presence of GM-CSF.

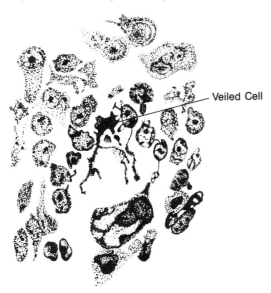

Veiled Cell

venom

A poisonous or toxic substance which selected species such as snakes, arthropods, and bees produce. The poison is transmitted to the recipient through a bite or sting.

very late activation antigens (VLA molecules)

β-1 integrins that all have the CD19 ß chain in common. They were originally described on T lymphocytes grown in long-term culture, but were subsequently found on additional types of leukocytes and on cells other than blood cells. VLA proteins facilitate leukocyte adherence to vascular endothelium and extracellular matrix. Resting T lymphocytes express VLA-4, VLA-5, and VLA-6. VLA-4 is expressed on multiple cells that include thymocytes, lymphocytes in blood, B and T cell lines, monocytes, NK cells, and eosinophils. The extracellular matrix ligand for VLA-4 and VLA-5 is fibronectin, and for VLA-6 it is laminin. The binding of these molecules to their ligands gives T lymphocytes costimulator signals. VLA-5 is present on monocytes, memory T lymphocytes, platelets, and fibroblasts. It facilitates B and T cell binding to fibronectin. VLA-6, which is found on platelets, T cells, thymocytes, and monocytes, mediates platelet adhesion to laminin. VLA-3 is a laminin receptor, binds collagen, and identifies fibronectin. It is present on B cells, the thyroid, and the renal glomerulus. Platelet VLA-2 binds to collagen only, whereas endothelial cell VLA-2 combines with collagen and laminin. Lymphocytes bind through VLA-4 to high endothelial venules and to endothelial cell surface proteins (VCAM-1) in areas of inflammation. VLA-1, which is present on activated T cells, monocytes, melanoma cells, and smooth muscle cells, binds collagen and laminin.

vesiculation

Development of minute intraepidermal fluid-filled spaces, i.e., vesicles seen in contact dermatitis.

veto cells

A proposed population of cells suggested to facilitate maintenance of self-tolerance through veto of autoimmune responses by T cells. A "veto cell" would neutralize the function of an autoreactive T lymphocyte. A T cell identifies itself as an autoreactive lymphocyte by recognizing the surface antigen on the "veto cell". No special receptors with specificity for the autoreactive T lymphocyte are required for the "veto cell" to render the T lymphocyte nonfunctional.

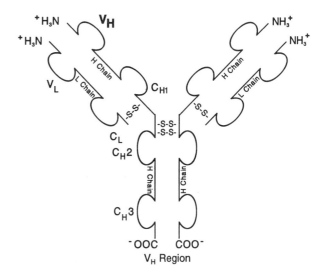

V_H region

The variable region of immunoglobulin heavy chain. That part of a variable region encoded for by the V_H gene segment.

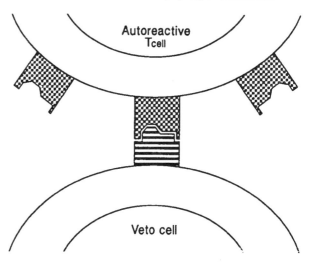

V_H Region

Vi antigen

A virulence antigen linked to *Salmonella* microorganisms. It is found in the capsule and interferes with serological typing of the O antigen, a heat-stable lipopolysaccharide of enterobacteriaceae.

vimentin

A 55-kD intermediate filament protein synthesized by mesenchyma cells such as vascular endothelial cells, smooth muscle cells, histiocytes, lymphocytes, fibroblasts, melanocytes, osteocytes, chondrocytes, astrocytes, and occasional ependymal and glomerular cells. Malignant cells may express more than one intermediate filament. For example, immunoperoxidase staining may reveal vimentin and cytokeratin in breast, lung, kidney, or endometrial adenocarcinomas.

vinblastine

A chemotherapeutic alkaloid that leads to lysis of rapidly dividing cells by disruption of the mitotic spindle microtubules. It is used for therapy of leukemia, Hodgkin's disease, lymphoproliferative disorders, and malignancies.

vincristine

A chemotherapeutic alkaloid that lyses proliferating cells through disruption of the mitotic spindle. It is used in the therapy of leukemia and other malignancies.

viral hemagglutination

Selected viruses may combine with specific receptors on surfaces of red cells from various species to produce hemagglutination. The ability of antiviral antibodies to inhibit this reaction constitutes hemagglutination inhibition, which serves as an assay to quantify the antibodies. One must be certain that inhibition is due to the antibody and not to a nonspecific agent such as mucoproteins with myxoviruses or lipoproteins with arbor viruses. Blood sera to be assayed by this technique must have the inhibitor activity removed by treatment with neuraminidase or acetone extraction, depending upon the chemical nature of the inhibitor. The attachment of virus particles to cells is termed hemadsorption. Among viruses causing hemagglutination are those which induce influenza and parainfluenza, mumps, Newcastle disease, smallpox and vaccinia, measles, St. Louis encephalitis, Western equine encephalitis, Japanese B encephalitis, Venezuelan equine encephalitis, West Nile fever, Dengue viruses, respiratory syncytial virus, and some enteroviruses. Herpes simplex virus can be absorbed to tanned sheep red cells and hemagglutinated in the presence of specific antiserum against the virus. This method is termed indirect virus hemagglutination.

viral interference

Resistance of cells infected with one virus to infection by a second virus.

viroid

A 100-kD 300-base pair subviral infectious RNA particle comprised of a circular single-stranded RNA segment. It induces disease in certain plants. Viroids may be escaped introns.

viropathic

Host tissue injury resulting from infection by a pathogenic virus.

virulence genes

Genes that govern the expression of multiple other genes with changing environmental conditions such as temperature, pH, osmolarity, etc. An example is the *toxR* gene of *Vibrio cholerae* that coordinates 14 other genes of this microorganism.

virus

An infectious agent that ranges from 10^6 D for the smallest viruses to 200×10^6 for larger viruses such as the pox viruses. Viruses contain single or double-stranded DNA or RNA that is either circular or open and linear. The nucleic acid is enclosed by a protein coat, termed a capsid, comprised of a few characteristic proteins. Most viruses are helical or icosahedral. There may be a lipid envelope which may contain viral proteins. Viruses may be incubated with cells in culture, where they produce characteristic cytopathic effects. Inclusion bodies may be produced in cells infected by viruses. Viruses infect host cells through specific receptors. Examples of this specificity include cytomegalovirus linking to β_2 microglobulin, Epstein-Barr virus linking to C3d receptor (CR2), and HIV-1 binding to CD4.

virus-associated hemophagocytic syndrome

An aggressive hemophagocytic state that occurs in both immunocompromised and nonimmunocompromised individuals, usually those with herpetic infections including CMV, EBV, and herpes virus, and may occur in infections by adenovirus and rubella, as well as in brucellosis, candidiasis, leishmaniasis, tuberculosis, and salmonellosis. There is lymphadenopathy, hepatosplenomegaly, pulmonary infiltration, skin rash, and pancytopenia. The disease is sometimes confused with malignant histiocytosis, lymphoma, sinus histiocytosis, and lymphomatoid granulomatosis.

virus neutralization test

A test based upon the ability of a specific antibody to neutralize virus infectivity. This assay can be employed to measure the titer of antiviral antibody. This test may be performed *in vivo* using susceptible animals or chick embryos or it may be done *in vitro* in tissue culture.

virus-neutralizing capacity

Serum's ability to prevent virus infectivity. Neutralizing antibody is usually of the IgM, IgG, or IgA class.

viscosity

The physical consistency of a fluid such as blood serum based on the size, shape, and conformation of its molecules. Molecular charge, sensitivity to temperature, and hydrostatic state affect viscosity.

vitamin A

Immunologically, vitamin A may serve as an adjuvant to elevate antibody responses to soluble protein antigens in mice. The adjuvant effect is produced whether vitamin A is given orally or parenterally.

vitiligo

Loss of skin pigmentation as a consequence of autoantibodies against melanocytes.

vitronectin (see below)

A cell adhesion molecule that is a 65-kD glycoprotein. It is found in the serum at a concentration of 20 mg/l. It combines with coagulation and fibrinolytic proteins and with C5b67 complex to block its insertion into lipid membranes. Vitronectin appears in the basement membrane, together with fibronectin in proliferative vitreoretinopathy. It decreases nonselective lysis of autologous cells by insertion of soluble C5b67 complexes from other cell surfaces. Also called epibolin and protein S.

V(J) recombination

Class switching.

V$_\kappa$

The variable region of an immunoglobulin κ light chain. This symbol may also be used to signify that part of a variable region encoded by the V_κ gene segment.

V$_\lambda$

The variable region of an immunoglobulin λ light chain. The symbol may designate that part of a variable region which the V_λ gene segment encodes.

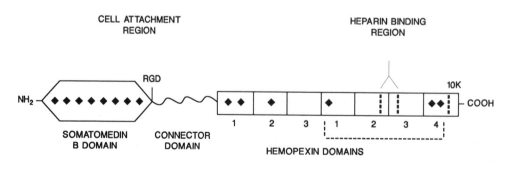

Vitronectin

V_L region

The variable region of an immunoglobulin light chain. The symbol may be used to designate the V_L gene encoded segment.

V_L Region

VLA receptors

A family of integrin receptors found on cell surfaces. They consist of α and β transmembrane chain heterodimers. There is a VLA-binding site at the arginine-glycine-aspartamine sequences of vitronectin and fibronectin. VLA receptors occur principally on T lymphocytes. They also bind laminin and collagen. They participate in cell-extracellular matrix interactions.

VLIA (virus-like infectious agent)

A mycoplasma, possibly synergistic with HIV-1, leading to profound immunodeficiency. It has been named *Mycoplasma incognitos.*

Vogt-Koyanagi-Harada (VKH) syndrome

Uveal inflammation of the eye(s) with acute iridocyclitis, choroiditis, and retinal detachment. Initial manifestations include headache, dysacusis, and sometimes vertigo. Scalp hair may show patchy loss or whitening. Vitiligo and poliosis occur often. The development of delayed-type hypersensitivity to melanin-containing tissue has been postulated. Apparently, pigmented constituents of the eye, hair, and skin are altered by some type of insult in a manner that leads to a delayed-type hypersensitivity response to them. Possible autoantigens are soluble substances from the retinal photoreceptor layer. There is a predisposition to VKH in Asians. Uveal tissue extracts have been used in delayed-hypersensitivity skin tests. Antibodies to uveal antigens may be present, but they are not specific for VKH. Corticosteroids, as well as chlorambucil, cyclophosphamide, or cyclosporin A, have been used for treatment.

Vollmer test (historical)

A tuberculin patch test employing gauze treated with tuberculin.

von Krough equation

A mathematical equation to ascertain serum hemolytic complement titer. It correlates complement with the extent of lysis of red blood cells coated with antierythrocyte antibodies under standard conditions.

VpreB and λ5

Proteins produced at the early pro-B cell stage of B cell development that are required for regulation of μ chain expression on the cell surface. VpreB and λ5 take the place of immunoglobulin light chains in pre-B cells. These proteins have an important role in B cell differentiation.

V_T region

T lymphocyte antigen receptor variable region.

W

w

Symbol for "workshop" that is used for histocompatibility (HLA) antigen and cluster of differentiation (CD) designations when new antigenic specificities have not been conclusively decided. Once the specificities have been agreed upon among authorities, the "w" is removed from the designation.

Waaler-Rose test

Refer to Rose-Waaler test.

Waldenström, Jan Gosta (1906–)

Swedish physician who described macroglobulinemia, which now bears his name. He received the Gairdner award in 1966.

Waldenström's macroglobulinemia

This paraproteinemia is second in frequency only to multiple myeloma, occurring mostly in people over 50 years of age. It may be manifested in various clinical forms. Most of the features of the disease are related to the oversynthesis of monoclonal IgM. Relatively mild cases may be characterized by anemia and weakness or pain in the abdomen resulting from enlargement of the spleen and liver. A major difference from multiple myeloma is a lack of osteolytic lesions of the skeleton, although patients may have peripheral lymphadenopathy. On bone marrow examination, many kinds of cells are found with characteristics of plasma cells and lymphocytes constituting so-called lymphocytoid plasma cells. Many are transitional or intermediate between one type or another. Patients may develop bleeding disorders of some type due to the paraproteins in their circulation. The more severe forms of the disease are characterized by features that resemble chronic lymphocytic leukemia or even lymphosarcoma with a rapidly fatal course. Many individuals may develop anemia. The large molecules of IgM with a mol wt approaching one million lead to increased viscosity of the blood. Central nervous system and visual difficulties may also be manifested.

Waldeyer's ring

A circular arrangement of lymphoid tissue comprised of tonsils and adenoids encircling the pharynx-oral cavity junction.

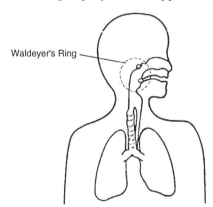

Waldeyer's Ring

warm agglutinin disease

IgG-induced autoimmune hemolysis, two fifths of which are secondary to other diseases such as neoplasia, including chronic lymphocytic leukemia or ovarian tumors. It may also be secondary to connective tissue diseases such as lupus erythematosus, rheumatoid arthritis, and progressive systemic sclerosis. Patients experience hemolysis leading to anemia with fatigue, dizziness, palpitations, exertion dyspnea, mild jaundice, and splenomegaly. Patients manifest a positive Coombs' antiglobulin test. Their erythrocytes appear as spherocytes and schistocytes, and there is evidence of erythrophagocytosis. There is erythroid hyperplasia of the bone marrow, and lymphoproliferative disease may be present.

Glucocorticoids and blood transfusions are used for treatment. Approximately 66% improve after splenomectomy, but often relapse. Three quarters of these individuals survive for a decade.

warm antibody

An antibody that reacts best at 37°C. It is usually an IgG agglutinin and shows specificity for selected erythrocyte antigens that include KELL, DUFFY, KIDD, and Rh. It may be associated with immune hemolysis.

Wassermann, August von (1866–1925)

German physician who, with Neisser and Bruck, described the first serological test for syphilis, i.e., the Wassermann reaction. *Handbook der Pathogenen Mikroorganismen (with Kolle)*, 1903.

Wassermann reaction

A complement fixation assay used extensively in the past to diagnose syphilis. Cardiolipin extracted from ox heart served as antigen which reacts with antibodies that develop in patients with syphilis. Biologic false-positive reactions using this test require the use of such confirmatory tests as the FTA-ABS test, the Reiter's complement fixation test, or the *Treponema pallidum* immobilization test. Both FTA and TPI tests use *T. pallidum* as antigen.

wasting disease

Neonatal thymectomy in mice can lead to a chronic and eventually fatal disease characterized by lymphoid atrophy and weight loss. Animals may develop ruffled fur, diarrhea, and a hunched appearance. Gnotobiotic (germ-free) animals fail to develop wasting disease following neonatal thymectomy. Thus, thymectomy of animals that are not germ free may lead to fatal infection as a consequence of greatly decreased cell-mediated immunity. Wasting disease is also called runt disease. Wasting may appear in immunodeficiency states such as AIDS as well as in graft-vs.-host (GVH) reactions.

wax D

A high mol wt (70-kD) glycolipid and peptidoglycolipid extracted from wax fractions of *Mycobacterium tuberculosis* isolated from man. It is soluble in chloroform and insoluble in boiling acetone and has the adjuvant properties which *Mycobacterium tuberculosis* organisms add to Freund's adjuvant preparations. Thus, wax D can be used to replace mycobacteria in adjuvant preparation to enhance cellular and humoral immune responsiveness to antigen.

Wegener's granulomatosis

Necrotizing sinusitis with necrosis of both the upper and lower respiratory tract. The disease is characterized by granulomas, vasculitis, granulomatous arteritis, and glomerulonephritis. The condition is believed to have an immunological etiology, although this remains to be proven. Patients develop antineutrophil cytoplasmic antibodies (ANCA). It may be treated successfully in many subjects by cyclophosphamide therapy.

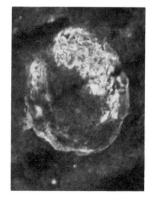

necrotic area of renal glomerulus fibrinogin - "staining"

Wegener's Granulomatosis

Weil-Felix reaction

A diagnostic agglutination test in which *Proteus* bacteria are agglutinated by the sera of patients with typhus. The reaction is based upon the crossreactivity of the carbohydrate antigen shared between *Rickettsiae* and selected *Proteus* strains. Various rickettsial diseases can be diagnosed based upon the reaction pattern of antibodies in the blood sera of rickettsial disease patients with O-agglutinable strains of *Proteus* OX19, OX2, and OX12.

Western Blot Technique

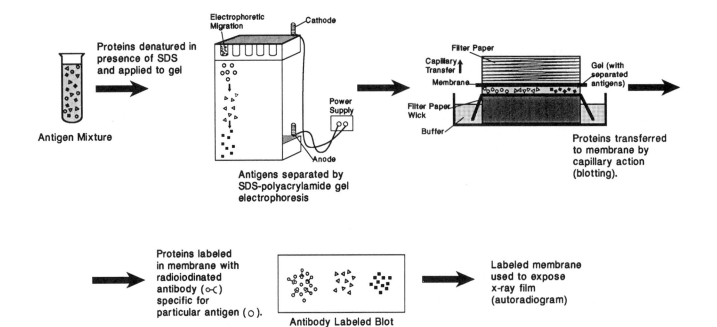

Western blot (immunoblot)

A method to identify antibodies against proteins of precise mol wt. It is widely used as a confirmatory test for HIV-1 antibody following the HIV-1 antibody screen test performed by the ELISA assay. Following separation of proteins by one- or two-dimensional electrophoresis, they are blotted or transferred to a nitrocellulose or nylon membrane followed by exposure to biotinylated or radioisotope-labeled antibody. The antigen under investigation is revealed by either a color reaction or autoradiography, respectively.

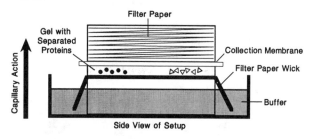

Western Blot

wheal and flare reaction

An immediate hypersensitivity, IgE-mediated (in man) reaction to an antigen. Application of antigen by a scratch test in a hypersensitive individual may be followed by erythema, which is the red flare, and edema, which is the wheal. Atopic subjects who have a hereditary component to their allergy experience the effects of histamine and other vasoactive amine released from mast cell granules following crosslinking of surface IgE molecules by antigen or allergen.

Whipple's disease

A disorder characterized immunologically by massive infiltration of the lamina propria with PAS positive macrophages. There are secondary T cell abnormalities. This is a rare infectious disease produced by one or more microorganisms which remain unidentified. Diagnosis is established by intestinal biopsy, showing microorganisms and numerous macrophages containing microbial cell wall debris in their cytoplasm. The cellular infiltration is associated with "clubbed" villi, lymphatic obstruction, malabsorption, and protein-losing enteropathy. Loss of lymphocytes into the gastrointestinal tract as a result of lymphatic obstruction may be associated with the development of lymphopenia and secondary T cell immune deficiency.

white graft rejection

Accelerated rejection of a second skin graft performed within 7 to 12 days after rejection of the first graft. It is characterized by lack of vascularization of the graft and its conversion to a white eschar. The characteristic changes are seen by day 5 after the second grafting procedure. The transplanted tissue is rendered white because of hyperacute rejection, such as a skin or kidney allograft. Preformed antibodies occlude arteries following surgical anastomosis, producing infarction of the tissue graft.

Types of Skin Graft Rejection

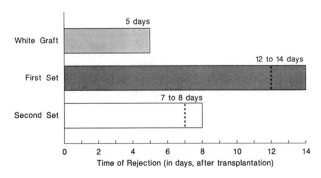

white pulp

The periarteriolar lymphatic sheaths encircled by small lymphocytes which are mainly T cells that surround germinal centers comprised of B lymphocytes and B lymphoblasts in normal splenic tissue. Following interaction of B cells in the germinal center with antigen in the blood, a primary immune response is generated within 24 h revealing immunoblastic proliferation and enlargement of germinal centers.

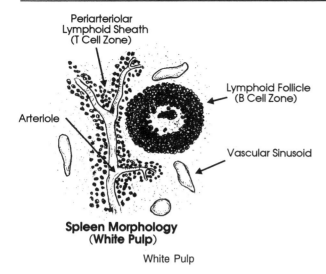

**Spleen Morphology
(White Pulp)**

White Pulp

white pulp disease

Lymphoproliferative diseases that express major anatomical changes in the splenic white pulp. Diseases producing this effect include histiocytic lymphoma, lymphocytic leukemia, and Hodgkin's disease.

whooping cough vaccine

Refer to pertussis vaccine.

Widal reaction

Bacterial agglutination test employed to diagnose enteric infections caused by *Salmonella*. Doubling dilutions of patient serum are combined with a suspension of microorganisms known to cause enteric fever such as *S. typhi*, *S. paratyphi* B, and *S. paratyphi* A and C. The microorganisms used in the test should be motile, smooth, and in the specific phase. To assay H agglutinins formalin-treated suspensions are used, and to assay O agglutinin alcohol-treated suspensions are employed. The Widal test is positive after the tenth day of the disease and may be false-positive if an individual previously received a TAB vaccine. Thus, it is important to repeat the test and observe a rising titer rather than to merely observe a single positive test. Widal originally described the test to diagnose *S. paratyphi* B infection.

wild mouse

A mouse that is free in the environment and has not been raised under laboratory conditions.

window

(1) The period between exposure to a microorganism and the appearance of serologically detectable antibody. It is observed in hepatitis B as well as in HIV-1 infections. In hepatitis B, there is a "core window" that occurs in active, but unidentified hepatitis B infection. The hepatitis B surface antigen (HBsAg) can no longer be detected, and the antibody against hepatitis B surface antigen (anti-HB$_s$) has not reached sufficiently high levels to be detected. (2) The period between the first infection with HIV-1 and synthesis of anti-p24 and anti-p41 antibodies in amounts measurable by the ELISA assay. Use of the polymerase chain reaction to demonstrate the p24

antigen can be useful to indicate infection during the window. The window period in HIV-1 infection may be between 3 and 9 months, or it may reach 36 months. Blood donated for transfusion in the U.S. is assayed for anti-HIV-1 p24 antibody. Thus, these units of blood could be in the HIV-1 infection window.

Winn assay

A method to determine the ability of lymphoid cells to inhibit the growth of transplantable tumors *in vivo*. Following incubation of the lymphoid cells and tumor cells *in vitro*, the mixture is injected into the skin of X-irradiated mice. Growth of the transplanted cells is followed. T lymphocytes that are specifically immune to the tumor cells will inhibit tumor growth and provide information related to tumor immunity.

Winn Assay

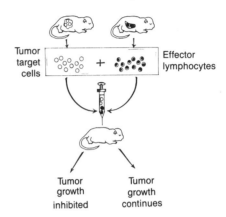

wire loop lesion

Thickening of capillary walls as a result of subendothelial immune complex deposits situated between the capillary endothelium and the glomerular basement membrane. Wire loop lesions are seen in diffuse proliferative lupus nephritis and are characteristic of class IV lupus erythematosus. They may be seen also in progressive systemic sclerosis and may appear together with crescent formation, necrosis, and scarring.

Wiskott-Aldrich syndrome

X-linked recessive immunodeficiency disease of infants characterized by thrombocytopenia, eczema, and increased IgA and IgE levels. There is decreased cell-mediated immunity (and delayed hypersensitivity), and the antibody response to polysaccharide antigens is defective, with only minute quantities of IgM appearing in the serum. There may be an inability to recognize processed antigen. Male patients may have small platelets with absent surface glycoprotein Ib. Whereas IgA and IgE are increased, IgM is diminished, although IgG serum concentrations are usually normal. By electron microscopy, T lymphocytes appear to be bereft of the markedly fimbriated surface of normal T cells. T lymphocytes have abnormal sialophorin. Patients may have an increased incidence of malignant lymphomas. Bone marrow transplantation corrects the deficiency.

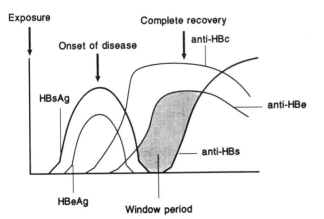

Exposure

Onset of disease

Complete recovery

anti-HBc

HBsAg

anti-HBe

anti-HBs

HBeAg

Window period

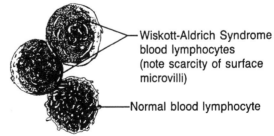

Wiskott-Aldrich Syndrome
blood lymphocytes
(note scarcity of surface
microvilli)

Normal blood lymphocyte

Wiskott-Aldrich Syndrome

Witebsky, Ernest (1901–1969)

German–American immunologist and bacteriologist who made significant contributions to transfusion medicine and to concepts of autoimmune diseases.

Witebsky's criteria

According to criteria suggested by Ernest Witebsky, an autoimmune response should be considered as the cause of a human disease if (l) it is regularly associated with that disease, (2) immunization of an experimental animal with the antigen from the appropriate tissue causes it to make an immune response (form antibodies or develop allergy), (3) associated with this response the animal develops pathological changes that are basically similar to those of the human, and (4) the experimental disease can be transferred to a nonimmunized animal by serum or by lymphoid cells.

Wright, Almroth Edward (1861–1947)

British pathologist and immunologist who made important contributions to the immunology of infectious diseases and immunization. *Pathology and Treatment of War Wounds,* 1942; *Researches in Clinical Physiology,* 1943; *Studies in Immunology, 2 vols.,* 1944.

Wu-Kabat plot

A graph that demonstrates the extent of variability at individual amino acid residue positions in immunoglobulin and T cell receptor variable regions. Division of the different amino acid number at a given position by the frequency of the amino acid which occurs most commonly at that position gives the index of variability. The index varies between 1 and 400. To show the variability graphically, a biograph is prepared where the index is plotted at each residue position. This plot indicates the extent of variability at each position and is useful in localizing immunoglobulin and T cell receptor hypervariable regions.

W,X,Y boxes (class II MHC promoter)

Three conserved sequences found in the promoter region of the HLA-DRα chain gene. The X box contains tandem regulatory sequences designated X1 and X2. Any cell that expresses MHC class II molecules will have all three boxes interacting with binding proteins, and decreased or defective production of some of these binding proteins can result in the "bare lymphocyte syndrome".

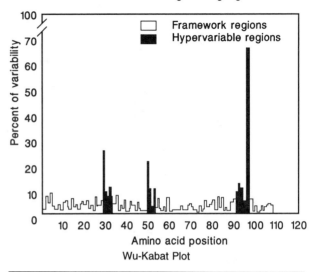

Wu-Kabat Plot of V Region of Ig Light Chain

Wu-Kabat Plot

MHC Class II

W,X,Y Boxes

X cell
See XYZ cell theory.

X-linked agammaglobulinemia
Impaired antibody-mediated immunity. Also called Bruton's X-linked agammaglobulinemia, which affects males who develop recurrent sinopulmonary or other pyogenic infections at 5 to 6 months of age after disappearance of maternal IgG. There is a defective B lymphocyte gene (chromosome Xq21.3-22). Whereas B cells and immunoglobulins are diminished, there is normal T cell function. Supportive therapy includes γ globulin injections and antibiotics. Repeated infections may lead to death in childhood. Their bone marrow contains pre-B cells with constant regions of immunoglobulin μ chains in the cytoplasm. There may be defective V_H-D-J_H gene rearrangement.

X-linked lymphoproliferative syndrome
A type of infectious mononucleosis in X-linked immunodeficiency patients who manifest EBNA positive lymphoid cells and polyclonal B lymphocyte proliferation, as well as plasma blasts in the blood. The disease may be caused by an inability to combat Epstein-Barr virus, leading to a fatal outcome.

xenoantigen
Antigen of a xenograft. Also called heteroantigen.

xenogeneic
Adjective that refers to tissues or organs transplanted from one species to a genetically different species, e.g., a baboon liver transplanted to a human.

xenograft
A tissue or organ graft from a member of one species, i.e., the donor, to a member of a different species, i.e., the recipient. It is also called a heterograft. Antibodies and cytotoxic T cells reject xenografts several days following transplantation.

Xenopus
Anuran amphibian that serves as an excellent model for investigations on the ontogeny of immunity. Comparative immunologic research employs isogeneic and inbred families of *Xenopus*. *Xenopus* is especially useful to study the role of the thymus in immune ontogeny because early thymectomy of free living larvae does not lead to runting.

xenotype
Molecular variations based on differences in structure and antigenic specificity. Examples would include membrane antigens of cells or immunoglobulins from separate species.

xid gene
An X-chromosome mutation designated *xid*. When homozygous it leads to diminished responsiveness to some thymus-independent antigens, limited decrease in responsiveness to thymic-dependent antigens, and defective terminal differentiation of B cells. Many autoimmune features are diminished when the *xid* gene is bred into autoimmune mouse strains. These include reduced anti-DNA antibody levels and diminished renal disease together with increased survival.

XYZ cell theory (historical)
An earlier concept of antibody synthesis which proposed (1) an immunocompetent "X cell" that had not participated previously in a specific immune response, (2) a "Y cell" activated immunologically by "X cell" interaction with antigen, and (3) a "Z cell" that synthesized antibodies following second contact with antigen.

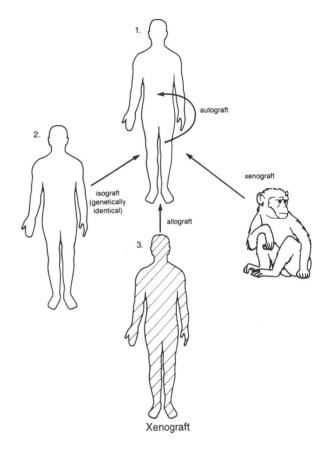

Xenograft

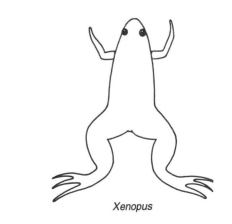

Xenopus

Y

Y cell

See XYZ cell theory.

Yalow, Rosalyn Sussman (1921–)

American investigator who shared the 1977 Nobel Prize with Guillemin and Schally for her endocrinology research and perfection of the radioimmunoassay technique. With Berson Yalow, made an important discovery of the role antibodies play in insulin-resistant diabetes.

yellow fever vaccine

A lyophilized attenuated vaccine prepared from the 17D strain of live attenuated yellow fever virus grown in chick embryos. A single injection may confer immunity that persists for a decade.

yellow jacket venom

A hymenopteran insect (bee, wasp, or ant) toxin that may induce anaphylactic shock (type I hypersensitivity), possibly leading to death, in victims of insect bites. The release of vasoactive amines may lead to urticaria, tightness in the chest, chills, fever, or even cardiovascular collapse.

Z

ZAP-70

A 70-kD tyrosine kinase present in the cytosol that is believed to participate in maintianing T lymphocyte receptor signaling. Is is similar to syk in B lymphocytes. Refer also to lck, fyn, ZAP (phosphotyrosine kinases in T cells).

Z cell

See XYZ cell theory.

zeta potential

The collective negative charge on erythrocyte surfaces that causes them to repulse one another in cationic medium. Some cations are red cell surface bound, whereas others are free in the medium. The boundary of shear is between the two cation planes, where the zeta potential may be determined as $-mV$. IgM antibodies have an optimal zeta potential of -22 to -17 mV, and IgG antibodies have an optimum of -11 to -4.5 mV. The less the absolute mV, the less the space between cells in suspension. The addition of certain proteins, such as albumin, to the medium diminishes the zeta potential.

Zeta Potential

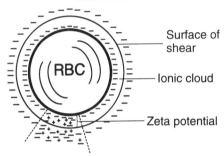

Surface of shear

Ionic cloud

Zeta potential

zidovudine (3'-azido-3'-deoxythymidine, AZT)

A reverse transcriptase inhibitor that is a thymidine analog. It is FDA approved for the treatment of acquired immune deficiency syndrome (AIDS). The mechanism of action includes phosphorylation of the drug *in vivo* to 3'-azido-3'deoxythymidine triphosphate. This combines with human immunodeficiency virus (HIV reverse transcriptase), which leads to cessation of DNA elongation.

Zidovudine

zinc

An element of great significance to the immune system, as well as to other nonantigen-specific host defenses. The interleukins of the immune system play a role in zinc distribution and metabolism in the body. As a constituent of the active site in multiple metalloenzymes, zinc is critical in chemical prothesis within lymphocytes and leukocytes. Zinc's role in the reproduction of cells is of critical significance for immunological reactions, since nucleic acid synthesis depends, in part, on zinc metalloenzymes. Zinc facilitates cell membrane modification and stabilization. Zinc deficiency is associated with reversible dysfunction of T lymphocytes in man. It causes atrophy of the thymus and other lymphoid organs and is associated with diminished numbers of lymphocytes in the T cell areas of lymphoid tissues. There is also lymphopenia. Anergy develops in zinc deficient patients. This signifies disordered cell-mediated immunity as a consequence of the zinc deficiency. There is also a decrease in the synthesis of antibodies to T cell-dependent antigens. In zinc deficiency, there is a selective decline in the number of CD4+ helper T cells and a strikingly decreased proliferative response to phytomitogens including PHA. Thymic hormonal function requires zinc. Deficiency of this element is also associated with decreased formation of monocytes and macrophages and with altered chemotaxis of granulocytes. Wound healing is impaired in these individuals who also show greatly increased susceptibility to infectious diseases, which are especially severe when they do develop.

Zinsser, Hans (1878–1940)

A leading American bacteriologist and Harvard educator whose work in immunology included hypersensitivity research and plague immunology. His famous text entitled *Microbiology (with Hiss),* 1911, is currently in its 20th edition.

zirconium granuloma

A tissue reaction in axillary regions of subjects who use solid antiperspirants containing zirconium. The granuloma develops as a consequence of sensitization to zirconium.

zonal centrifugation

The separation of molecules according to size based on molecular mass and centrifugation time.

zone electrophoresis

The separation of proteins on cellulose acetate (or on paper) based upon the charge when an electric current is passed through the gel.

zone of equivalence

That point in a precipitin antigen-antibody reaction *in vitro* where the ratio of antigen to antibody is equivalent. The supernatant contains neither free antigen nor antibody. All molecules of each have reacted to produce antigen-antibody precipitate. When a similar reaction occurs *in vivo,* immune complexes are deposited in the microvasculature, and serum sickness develops.

Precipitation Curve

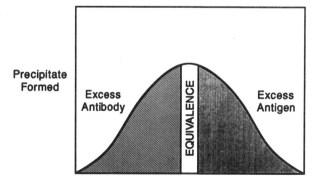

Precipitate Formed

Excess Antibody

EQUIVALENCE

Excess Antigen

Antigen Used

Zone of Equivalence

zymosan

A complex polysaccharide derived from dried cell walls of the yeast *Saccharomyces cerevisiae,* which activates the alternative complement pathway. It binds to C3b and is useful in investigations of opsonic phagocytosis.

Appendix I

	Cluster designation	Main cellular expression of antigen	Other names	Antigen molecular weight (kD)
	CD1a	Cortical Thymocytes (strong), Langerhans Cells, B-cell subset, Dendritic cells		49,12
	CD1b	Cortical Thymocytes (moderate), Langerhans Cells, B-cell subset, Dendritic cells		45,12
	CD1c	Cortical Thymocytes (weak), Langerhans Cells, B-cell subset, Dendritic cells		43,12
	CD2	T-cells, Thymocytes, NK-cells	LFA-2 (Leucocyte function antigen -2), E Rosette receptor	50
T	CD2R	Activated T-cells, NK-cells		50
	CD3	Thymocytes, mature T-cells, (Associated with T-cell receptor antigens)		16,20,25-28
	CD4	T-helper/Inducer cells, Monocytes, macrophages	Receptor for MHC class II and HIV antigens	56
	CD5	Thymocytes, T-cells, B-cell subset		67
	CD6	Thymocytes, T-cell subset, B-cell subset		105
	CD7	Majority of T-cells		40
	CD8	T-Cytotoxic/Suppressor cells	MHC Class I Receptor	32-34
P	CD9	Pre-B cells, Monocytes, Platelets		24
B	CD10	Lymphoid progenitor cells, Granulocytes	CALLA (common acute lymphoblastic leukaemia antigen)	100
	CD11a	Leucocytes	LFA-1 (Leucocyte Function Antigen) Integrin α^L subunit	180
Ad	CD11b	Granulocytes, Monocytes, NK-cells	Integrin α^M subunit, Mac-1; CR3, C3bi Receptor	170
	CD11c	Granulocytes, monocytes, NK-cells, B-cell subset, T-cell subset	Integrin α^X subunit	150
	CDw12	Granulocytes, Monocytes		90-120
M	CD13	Myeloid monocytes, Granulocytes	Aminopeptidase N	130
	CD14	Monocytes, some Granulocytes and Macrophages		55
	CD15	Granulocytes, Monocytes		multiple
Ad	CD15s	Neutrophils	Sialyl-Lewis^X (sLe^X), Ligand for CD62 structures	multiple
	CD16a	Macrophage, NK-cells	Transmembrane form, FcR IIIA/FcR IIIB	50-65
M	CD16b	Granulocyte form only	GPI-linked form, FcR IIIB	48
	CDw17	Granulocytes, Monocytes, Platelets	Lactosylceramide	Not determined
Ad	CD18	Leucocytes, Platelets negative	Integrin β2 subunit	95
	CD19	Pan B-cell		90
	CD20	Pan B-cell		33,35,37
B	CD21	Mature B-cells, Follicular Dendritic cells	C3d/EBV-receptor (CR2)	145
	CD22	Mature B-cells, Hairy cell Leukaemia cells		140
	CD23	Activated B-cells, Activated Macrophages, Eosinophils, Platelets	IgE Fc Low Affinity Receptor (RII)	50-45
	CD24	B-cells, Granulocytes		41/38
C	CD25	Activated T-cells, B-cells and Macrophages	IL-2 Receptor; Tac	55
Ac	CD26	Activated T-cells, and B-cells, Macrophages		110
T	CD27	Thymocyte subset, Mature T-cells, EBV transformed B-cells		55
	CD28	T-cell subset, activated B-cells		44
Ad	CD29	Ubiquitous (but not on erythrocytes)		130
Ac	CD30	Activated T and B-cells, Reed-Sternberg cells	Ki-1 antigen	105
P	CD31	Platelets, Monocytes, Macrophages, Granulocytes, B-cells	Platelet GPIIa' PECAM-1 (platelet endothelial cell adhesion molecule-1)	140
	CD32	Monocytes, Granulocytes, B-cells, Eosinophils	IgG FcR II, previously CDw32	40
M	CD33	Myeloid progenitor cells, Monocytes		67
	CD34	Haematopoietic precursor cells, Capillary endothelial cells		105-120
	CD35	Granulocytes (Basophils negative), Monocytes, B-cells, Erythrocytes, some NK-cells	CR1 C3b Receptor	160,190,220,250
P	CD36	Monocytes, Macrophages, Platelets, B-cells (weakly)	Platelet GPIV (IIIb)	88
B	CD37	Mature B-cells, T-cells and Myeloid cells (weakly)	gp 52-40	52-40
T	CD38	Plasma cells, Thymocytes, Activated T-cells	T10	45
	CD39	Mature B-cells, Monocytes, some Macrophages, Vascular Endothelium	gp80	100-70
B	CD40	B-cells, Monocytes (weakly), Carcinoma cells	gp50	48/44
	CD41	Platelets, Megakaryocytes	GPIIb/IIIa, alpha IIb integrin	120
	CD42a	Platelets, Megakaryocytes	GP1X	23
P	CD42b	Platelets, Megakaryocytes GP1B,α		135,23
	CD42c	Platelets, Megakaryocytes	GPIB-β	22
	CD42d	Platelets, Megakaryocytes	GPV	85
Ad	CD43	Leucocytes but not peripheral B-cells	Leukosialin, Sialophorin	90-100

T = T Cells B = B Cells M = Myeloid P = Platelets Ac = Activation
Ad = Adhesion C = Cytokine NK = NK Cells E = Endothelial BP = Blind Panel (Multi-lineage)

	Cluster designation	Main cellular expression of antigen	Other names	Antigen molecular weight (kD)
Ad	CD44	Leucocytes, Erythrocytes, Platelets (weakly), Brain cells	Pgp-1, H-Cam	80-90
	CD44R	Recognises alternatively spliced form of molecule	e.g.Exon 9, V9	
T	CD45	Pan Leucocyte	T200, LCA	180,190,205,220
	CD45RA	B-cells, T-cell subset, Monocytes	T200 Restricted	220,205
	CD45RB	B-cells, T-cell subset, Monocytes, Macrophages, Granulocytes (weakly)	T200	190,220,205
	CD45RO	T-cells, B-cell subset, Monocytes, Macrophages	T200 Restricted	180
BP	CD46	Haematopoietic and non-Haematopoietic cells, (erythrocytes negative)	MCP (membrane co-factor protein)	66/56
	CD47	All cell types	Rh group associated	47-52
	CD48	Leucocytes	Blast-1	43
	CD49a	Activated T-cells, Monocytes	VLA-1, α1 integrin chain	210
	CD49b	B-cells, Monocytes, Platelets	VLA-2 α2 integrin chain	160
	CD49c	B-cells	VLA-3, α3 integrin chain	125
Ad	CD49d	Thymocytes, B-cells	VLA-4, α4 integrin chain	150,80,70
	CD49e	Memory T-cells, Monocytes, Platelets	VLA-5, α5 integrin chain	135/25
	CD49f	Memory T-cells, Thymocytes, Monocytes	VLA-6, α6 integrin chain	120/25
	CD50	Leucocytes (Platelets and Erythrocytes negative)	ICAM-3	124
	CD51	Platelets	VNR α (Vitronectin Receptor)	125,25
	CD51/61 complex	Platelets, Endothelial cells	Vitronectin receptor, Integrin α V β 3	
BP	CD52	Leucocytes (Platelets and Erythrocytes negative)	Campath-1	21-28
B	CD53	Pan Leucocytes		32-40
Ad	CD54	Endothelial cells, many Activated cell-types	ICAM-1	90
	CD55	Many Haematopoietic and non-Haematopoietic cells	DAF (Decay Accelerating Factor)	75
NK	CD56	NK-cells	NKH1, isoform of N-CAM	220/135
	CD57	NK-cells, T-cells, B-cell subsets	HNK-1	110
Ad	CD58	Many Haematopoietic and non-Haematopoietic cells	LFA-3	65-70
	CD59	Many Haematopoietic and non-Haematopoietic cells	P18, gP18,Mac inhibitor	18-20
T	CDw60	Platelets, T-cell subset		carbohydrate
P	CD61	Platelets, Megakaryocytes	Integrin β3, GPIIIa, Vitronectin receptor β	110
Ad	CD62E	Endothelium	E-Selectin, ELAM-1, LECAM-2	140
	CD62L	B and T-cells, Monocytes, NK-cells	L-Selectin, LECAM-1, LAM-1	150
	CD62P	Platelets, Endothelial cells, Megakaryocytes	P-Selectin, GMP-140, PADGEM, LECAM-3	75-80
P	CD63	Activated Platelets, Monocytes, Macrophages	Platelet activation antigen	53
M	CD64	Monocytes	High Affinity IgGFc Receptor 1, FcγR1	75
	CDw65	Granulocytes, Monocytes	Fucoganglioside	Not confirmed
	CD66a	Neutrophil lineage cells	BGP (Biliary glycoprotein)	160-180
	CD66b	Granulocytes	CGM6, (CEA gene member 6), p100 Previously CD67	95-100
	CD66c	Neutrophils, Colon Carcinoma	NCA (Non specific cross reacting antigen)	90
	CD66d	Neutrophils	CGM1 (CEA gene member 1)	30
	CD66e	Adult Colon Epithelia, Colon Carcinoma	CEA (Carcinoembryonic antigen)	180-200
	CD67	Now CD66b		
M	CD68	Monocytes, Macrophages		110
Ac	CD69	Activated T and B-cells, Activated macrophages, NK-cells	gp 34/28, AIM (activation inducer molecule)	34/28
	CD70	Activated T and B-cells, Reed-Sternberg cells, Macrophages (weakly)	Ki-24, CD27 ligand	75,95,170
	CD71	Activated T and B-cells, Macrophages, Proliferating cells	T9, Transferrin Receptor	95
	CD72	Pan B-cell		43/39
	CD73	B and T-cell subsets	ecto-5'-NT	69
B	CD74	B-cells, Macrophages, Monocytes	gp41/35/33,li	41,35,33
	CDw75	Mature B-cells, T-cell subset (weak expression)		Not confirmed
	CDw76	Mature B-cells, T-cell subset	Previously CD76	Not confirmed
	CD77	Activated B-cells, Follicular Centre B-cells, Endothelial cells	BLA, Gb3, p^k	Not confirmed
	CDw78	B-cells	Ba	Not confirmed

T = T Cells   B = B Cells   M = Myeloid   P = Platelets   Ac = Activation

Ad = Adhesion   C = Cytokine   NK = NK Cells   E = Endothelial   BP = Blind Panel (Multi-lineage)

	Cluster designation	Main cellular expression of antigen	Other names	Antigen molecular weight (kD)
	CD79a	B-cell specific	mb-1, Igα	33, 40
	CD79b	B-cell specific	B29, Igβ	33, 40
	CD80	B-cell subset in vivo, most activated B-cells in vitro	B7, BB1	60
B	CD81	B-cells, (Broad Expression including lymphocytes)	TAPA-1 (Target of an antiproliferative antibody)	22
	CD82	Broad expression on leucocytes (weak) not erythrocytes	R2, IA4, 4F9	50-53
	CD83	Specific marker for Circulating Dendritic cells, activated B and T-cells, Germinal centre cells	HB15	43
	CDw84	Platelets and Monocytes (strong), circulating B-cells (weak)	GR6	73
	CD85	Circulating B-cells (weak), Monocytes (strong)	VMP-55, GH1/75,GR4	120, 83
	CD86	Circulating monocytes, Germinal centre cells (histol.), Activated B-cells	FUN-1, GR65, BU63	80
	CD87	Granulocytes, Monocytes, Macrophage, Activated T-cells	UPA-R (Urokinase Plasminogen Activator Receptor)	50-65
	CD88	Polymorphonuclear leucocytes, mast cells, macrophage, smooth muscle	C5a Receptor, GR10	42
	CD89	Neutrophils, monocytes, macrophage, T and B-cell subpopulation	FcαR, IgA receptor	55-70
M	CDw90	CD34+ve, subset on bone marrow, cord blood, fetal liver	Thy-1	25-35
	CD91	Monocytes and some non haemopoietic cell lines	α2M-R (α2 macroglobulin receptor)	600
	CDw92	Neutrophils, monocytes, endothelial cells, platelets	GR9	70
	CD93	Neutrophils, monocytes, endothelial cells	GR11	120
NK	CD94	NK-cells, α/β,γ/δ, T-cell subsets	KP43	43
	CD95	Variety of cell lines including myeloid and T lymphoblastoid	APO-1, FAS	42
Ac	CD96	Activated T-cells	TACTILE-(T-cell activation increased late expression)	160
	CD97	Activated cells	GR1, BL-KDD/F12	74, 80, 89
	CD98	T-cells and B-cells (weak), monocytes (strong), most human cell lines	4F2,2F3	80,40
	CD99	Peripheral blood lymphocytes, thymocytes	MIC2, E2	32
T	CD99R	B and T lymphocytes, some leukaemias	CD99 mAb restricted	32
	CD100	Broad expression on haemopoietic cells	BB18, A8, GR3	150
	CDw101	Granulocytes, macrophage	BB27, BA27, GR14	140
	CD102	Resting lymphocytes, monocytes. Vascular endothelial cells (strongest)	ICAM-2	60
Ad	CD103	Intraepithelial lymphocytes, 2-6%, PBL	HML-1, αE integrin, α6	150, 25
	CD104	Epithelia, Schwann cells, some tumor cells	β4 integrin chain, beta4	220
E	CD105	Endothelial cells, bone marrow cell subset, in vitro activated macrophage	Endoglin, TGF B1 and β3 receptor, GR7	95
	CD106	Endothelial cells	VCAM-1, INCAM-110	100, 110
P	CD107a	Activated platelets	LAMP 1 (Lysosomal associated membrane protein)	110
	CD107b	Activated platelets	LAMP 2	120
Ad	CDw108	Activated T-cells in spleen, some stromal cells	GR2	80
E	CDw109	Activated T-cells, platelets, endothelial cells	Platelet activation factor, 8A3, 7D1, GR56	170/150
BP	CD110-CD114	Nothing yet assigned to these numbers		
M	CD115	Monocytes, macrophage, placenta	M-CSFR (Macrophage colony stimulating factor receptor), CSF-1R	150
	CDw116	Monocytes, neutrophils, eosinophils, fibroblasts endothelial cells	GM-CSF R (Granulocyte, macrophage colony stimulating factor receptor), HGM-CSFR	75-85
	CD117	Bone marrow progenitor cells	Stem cell factor receptor (SCF R), cKIT	145
	CD118	Broad cellular expression	IFNα,β receptor	
	CD119	Macrophage, monocyte, B-cells, epithelial cells	IFNγR (Interferon γ receptor)	90
	CD120a	Most cell types, higher levels on epithelial cell lines	TNFR; 55kD (Tumor necrosis factor receptor)	55
	CD120b	Most cell types, higher levels on myeloid cell lines	TNF R; 75kD	75
C	CDw121a	T-cells, thymocytes, fibroblasts, endothelial cells	IL-1R (Interleukin-1 receptor), Type 1	80
	CDw121b	B-cells, macrophages, monocytes	IL-1R, Type II	68
	CD122	NK-cells, resting T-cell sub population, some B-cell lines	IL-2Rβ, IL2R; 75kD	75
	CD123	Bone marrow stem cells, granulocytes, monocytes, megakaryocytes	IL-3R	
	CDw124	Mature B and T-cells, haemopoietic precursor cells	IL-4R	140
	CD125	Eosinophils and basophils	IL-5R	
	CD126	Activated B-cells and plasma cells (strong), most leucocytes (weak)	IL-6R	80 (αsubunit)
	CDw127	Bone marrow lymphoid precursors, Pro-B-cells, mature T-cells, monocytes	IL-7R	75
BP	CDw128	Neutrophils, basophils, T-cell subset	IL-8R	58-67
	CD129	Nothing yet assigned to this number		
C	CDw130	Activated B-cells and plasma cells (strong), most leucocytes (weak), endothelial cells	IL-6R-gp 130SIG	130

T = T Cells   B = B Cells   M = Myeloid   P = Platelets   Ac = Activation

Ad = Adhesion   C = Cytokine   NK = NK Cells   E = Endothelial   BP = Blind Panel (Multi-lineage)

Appendix II

Cytokines and their receptors						
Family	Cytokine (alternative names)	Size (no. of amino acids) and form	Receptors (c denotes common subunit)	Producer cells	Actions	Effect of cytokine or receptor knock-out (where known)
Hematopoietins (four-helix bundles)	Epo (erythropoietin)	165, monomer	EpoR	Kidney	Stimulates erythroid progenitors	
	IL-2 (T-cell growth factor)	133, monomer	CD25 (α), CD122 (β),γc	T cells	T-cell proliferation	IL-2: decreased T-cell proliferation; Receptor γ chain: incomplete T-cell development
	IL-3 (multicolony CSF)	133, monomer	CD123, βc	T cells, thymic epithelial cells	Synergistic action in hematopoiesis	
	IL-4 (BCGF-1. BSF-1)	129, monomer	CD124, γc	T cells, mast cells	B-cell activation, IgE switch	Decreased IgE synthesis
	IL-5 (BCGF-2)	115, homodimer	CD125, βc	T cells, mast cells	Eosinophil growth, differentiation	
	IL-6 (IFN-β_2. BSF-2. BCDF)	184. monomer	CD126,CDw 130	T cells, macrophages	T- and B-cell growth and differentiation, acute phase reaction	Decreased acute phase reaction
	IL-7	152. monomer	CDw127, γc	Bone marrow stroma	Growth of pre-B cells and pre-T cells	
	IL-9	125. monomer	IL-9R, γc	T cells	Mast cell enhancing activity	
	IL-11	178. monomer	IL-11R, CDw130	Stromal fibroblasts	Synergistic action with IL-3 and IL-4 in hematopoiesis	
	IL-13 (P600)	132. monomer	IL-13R, γc	T cells	B-cell growth and differentiation, inhibits macrophage inflammatory cytokine production	
	IL-15 (T-cell growth factor)		IL-15R, γc	T cells	IL-2-like	
	GM-CSF (granulocyte macrophage colony stimulating family)	127. monomer	CDw116. βc	Macrophages, T cells	Stimulates growth and differentiation of myelomonocytic lineage	
	OSM (OM. oncostatin M)	196. monomer	OMR, CDw130	T cells, macrophages	Stimulates Kaposi sarcoma cells, inhibits melanoma growth	
	LIF (Leukemia inhibitory factor)	179. monomer	LIFR, CDw130	Bone marrow stroma, fibroblasts	Maintains embryonic stem cells, like IL-6, IL-11, OSM	
Interferons	IFN-γ	143. monomer	CD119	T cells, natural killer cells	Macrophage activation, increased MHC expression	Susceptibility to intracellular infection
	IFN-α	166. monomers	CD118	Leukocytes	Anti-viral, increased MHC class I expression	
	IFN-β	166. monomer	CD118	Fibroblasts	Anti-viral, increased MHC class I expression	
Immunoglobulin superfamily	B7.1 (CD80)	262 dimer	CD28. CTLA-4	Antigen-presenting cells	Co-stimulation of T-cell responses	
	B7.2 (B70)		CD28. CTLA-4	Antigen-presenting cells	Co-stimulation of T-cell responses	

Family	Cytokine (alternative names)	Size (no. of amino acids) and form	Receptors (c denotes common subunit)	Producer cells	Actions	Effect of cytokine or receptor knock-out (where known)
TNF family	TNF-α (cachectin)	157, trimers	p55, p75 CD120a, CD120b	Macrophages, natural killer cells	Local inflammation, endothelial activation	Receptor: resistance to septic shock, susceptibility to *Listeria*
	TNF-β (lymphotoxin, LT, LT-α)	171, trimers	p55, p75 CD120a, CD120b	T cells, B cells	Killing, endothelial activation	Absent lymph nodes, increased antibody
	LT-β	Transmembrane, trimerizes with TNF-β		T cells, B cells	Unknown	
	CD40 ligand (CD40-L)	Trimers	CD40	T cells, mast cells	B-cell activation, class switching	Poor antibody response, no class switch
	Fas ligand	Trimers (?)	CD95 (Fas)	T cells, stroma?	Apoptosis, Ca^{2+}-independent cytotoxicity	Lymphoproliferation
	CD27 ligand	Trimers (?)	CD27		Unknown	
	CD30 ligand	Trimers (?)	CD30		Unknown	
	4-1BBL	Trimers (?)	4-1BB		Unknown	
Chemokines	IL-8 (NAP-1)	69–79, dimers	CDw128	Macrophages, others	Chemotactic for neutrophils, T cells	
	MCP-1 (MCAF)	76, monomer(?)		Macrophages, others	Chemotactic for monocytes	
	MIP-1α	66, monomer (?)		Macrophages, others	Chemoattractant for monocytes, T cells, eosinophils	
	MIP-1β	66, monomer (?)		T cells, B cells, monocytes	Chemoattractant for monocytes, T cells	
	RANTES	66, monomer (?)		T cells, platelets	Chemoattractant for monocytes, T cells, eosinophils	
Unassigned	TGF-β	112, homo- and heterotrimers		Chondrocytes, monocytes, T cells	Inhibits cell growth, anti-inflammatory	
	IL-1α	159, monomer	CDw121a	Macrophages, epithelial cells (dead cells?)	Fever, T-cell activation, macrophage activation	
	IL-1β	153, monomer	CDw121a	Macrophages, epithelial cells (dead cells?)	Fever, T-cell activation, macrophage activation	
	IL-10 (cytokine synthesis inhibitor F)	160, homodimer		T cells, macrophages, Epstein-Barr virus	Potent suppressant of macrophage functions	
	IL-12 (natural killer cell stimulatory factor)	197 and 306, heterodimer		B cells, macrophages	Activates natural killer cells, induces CD4 T cell differentiation to T_H1-like cells	
	MIF	115, monomer		T cells, others	Inhibits macrophage migration	

Appendix III

Cytokine Family

Name	Chromosome	Amino Acid Count	Notable Features	Inflammation	Immune Response	Hematopoiesis
RANTES (SIS family)	17	60	SIS protein family, C-C group, also includes MCAF, MIP	Chemoattractant for Monocytes and memory T cells		
SCF	mouse 10	248	c-kit ligand			Stem cell factor, mast cell proliferation
LIF/HILDA	22q12.1	170	Effects on embryonic cells	Acute phase proteins Hepatocyte stimulation		Growth inhibition of some Leukemic lines
BCGF-12kDA	?	124	Presence of Alu repeats in the coding sequence		Growth of normal activated and transformed B-lymphocytes	
G-CSF	17q11-21	177				Growth factor for Granulocyte progenitors
GM-CSF	5q21-32	127		Activation of endothelial cells		Proliferation and activation of granulocytes and macrophages
TGFβ	19q3	112	Several forms, TGFβ1,β2,β3,	Stimulation and chemotaxis of inflammatory cells	Suppressor of various T & B lymphocyte functions	
M-CSF	5q31-37	522,224	Multi molecular forms (post-transcriptional modifications)			Proliferation and differentiation of macrophage progenitors
EPO	7q22	166				Erythroid lineage differentiation
TNFα TNFβ	6p21 6p21	157 171	No leader peptide Precursor 233 AA	Fever induction Activation of endothelial cells	Cytotoxic for tumour cells, T cells growth and differentiation, induction of adhesion molecules	
IFN-γ	12q24	146		Antiviral effects, Macrophage Activation	Anti-proliferative effects, NK activation MHC class II expression	
IL-1α IL-1β	2q13 2q13-q21	159 153	No leader peptide Precursor 270 AA	PGE release Fever induction Acute phase proteins Activation of endothelial cells	Co-stimulation of T and B lymphocytes	Synergy with IL-3 and GM-CSF
IL-2	4q26-q27	133			T lymphocyte proliferation and differentiation effects on B lymphocytes and NK cells	
IL-3	5q23-q31	134-140				Multipotent CSF Mast cell growth factor
IL-4	5q23-q31	120			B cell proliferation and IgG1/IgE secretion, T cell proliferation	Mast cell growth synergy with G-CSF, EPO
IL-5	5q23-q31	115	B-cell activity not proven in human		B cell proliferation in mouse	Eosinophil differentiation
IL-6	7p15-q21	189		Acute phase proteins Hepatocyte stimulation	B cell proliferation and differentiation, T cell costimulation	Plasmocyte proliferation
IL-7	8q12-13	152			Proliferation of thymocytes and mature T cells	Proliferation of lymphocyte progenitors
IL-8 (SIS family)	4q13-q21	72	SIS protein family, C-X-C group, also includes PF4 & MGCSA-gro	Neutrophil chemotaxis and activation		
IL-9	5q23-q31	126			Growth of helper CD4+ but not cytotoxic CD8+ T cells	BFU-E Stimulation
IL-10	1	160	Sequence homology with EBV		inhibits synthesis of cytokines (IFNγ and IL-2) by T (TH1) lymphocytes	
IL-11	19q13	~180				Megakaryocytopoiesis, growth of IL-6 dependant lines
IL-12 (CLMF)	?	306 + 197	Disulfide-bounded heterodimer		Synergy with IL-2 for NK (LAK) cell activation, growth factor for T cells	
IL-13	5q23-q31	132		Inhibition of cytokine synthesis by monocytes	B cell activation and proliferation	
IL-14 (HMW-BCGF)	?	483			Cell proliferation Inhibition of Ig secretion	
IL-15	?	114			T cell proliferation NK cell activation	

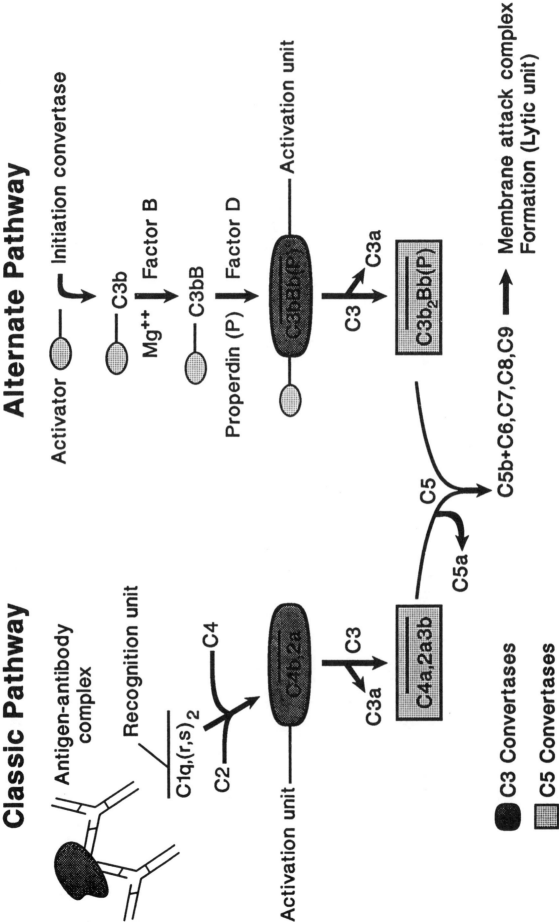

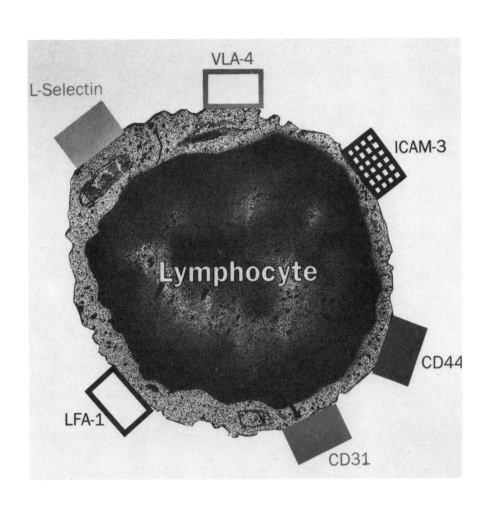

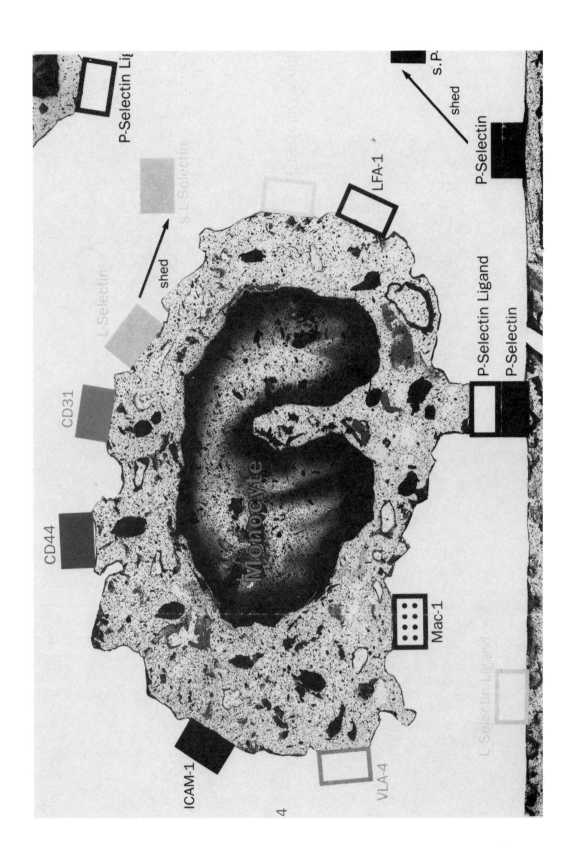